World Geomorphological Landscapes

Series Editor

Piotr Migoń, Institute of Geography and Regional Development, University of Wrocław, Wrocław, Poland

Geomorphology – 'the Science of Scenery' – is a part of Earth Sciences that focuses on the scientific study of landforms, their assemblages, and surface and subsurface processes that moulded them in the past and that change them today. Shapes of landforms and regularities of their spatial distribution, their origin, evolution, and ages are the subject of geomorphology. Geomorphology is also a science of considerable practical importance since many geomorphic processes occur so suddenly and unexpectedly, and with such a force, that they pose significant hazards to human populations. Landforms and landscapes vary enormously across the Earth, from high mountains to endless plains. At a smaller scale, Nature often surprises us creating shapes which look improbable. Many geomorphological landscapes are so immensely beautiful that they received the highest possible recognition – they hold the status of World Heritage properties. Apart from often being immensely scenic, landscapes tell stories which not uncommonly can be traced back in time for millions of years and include unique events. This international book series will be a scientific library of monographs that present and explain physical landscapes across the globe, focusing on both representative and uniquely spectacular examples. Each book contains details on geomorphology of a particular country (i.e. The Geomorphological Landscapes of France, The Geomorphological Landscapes of Italy, The Geomorphological Landscapes of India) or a geographically coherent region. The content is divided into two parts. Part one contains the necessary background about geology and tectonic framework, past and present climate, geographical regions, and long-term geomorphological history. The core of each book is however succinct presentation of key geomorphological localities (landscapes) and it is envisaged that the number of such studies will generally vary from 20 to 30. There is additional scope for discussing issues of geomorphological heritage and suggesting itineraries to visit the most important sites. The series provides a unique reference source not only for geomorphologists, but all Earth scientists, geographers, and conservationists. It complements the existing reference books in geomorphology which focus on specific themes rather than regions or localities and fills a growing gap between poorly accessible regional studies, often in national languages, and papers in international journals which put major emphasis on understanding processes rather than particular landscapes. The World Geomorphological Landscapes series is a peer-reviewed series which contains single and multi-authored books as well as edited volumes.

World Geomorphological Landscapes – now indexed in Scopus® !

More information about this series at https://link.springer.com/bookseries/10852

Christine Embleton-Hamann
Editor

Landscapes and Landforms of Austria

Springer

Editor
Christine Embleton-Hamann
Department of Geography and Regional Research
University of Vienna
Vienna, Austria

ISSN 2213-2090 ISSN 2213-2104 (electronic)
World Geomorphological Landscapes
ISBN 978-3-030-92813-1 ISBN 978-3-030-92815-5 (eBook)
https://doi.org/10.1007/978-3-030-92815-5

This Springer imprint is published by the registered company Springer Nature Switzerland AG
The registered company address is: Gewerbestrasse 11, 6330 Cham, Switzerland

Series Editor's Preface

Landforms and landscapes vary enormously across the Earth, from high mountains to endless plains. At a smaller scale, Nature often surprises us by creating shapes which look improbable. Many physical landscapes are so immensely beautiful that they have received the highest possible recognition—they hold the status of World Heritage properties. Apart from often being immensely scenic, landscapes tell stories which not uncommonly can be traced back in time for tens of million years and include unique events. In addition, many landscapes owe their appearance and harmony not solely to natural forces. For centuries, or even millennia, they have been shaped by humans who modified hillslopes, river courses and coastlines, and erected structures which often blend with the natural landforms to form inseparable entities.

These landscapes are studied by geomorphology—'the Science of Scenery'—a part of Earth Sciences that focuses on landforms, their assemblages, the surface and subsurface processes that moulded them in the past and that change them today. The shapes of landforms and the regularities of their spatial distribution, their origin, evolution and ages are the subject of research. Geomorphology is also a science of considerable practical importance since many geomorphic processes occur so suddenly and unexpectedly, and with such a force, that they pose significant hazards to human populations and not uncommonly result in considerable damage or even casualties.

To show the importance of geomorphology in understanding the landscape, and to present the beauty and diversity of the geomorphological sceneries across the world, we have launched a new book series *World Geomorphological Landscapes*. It aims to be a scientific library of monographs that present and explain physical landscapes, focusing on both representative and uniquely spectacular examples. Each book will contain details on geomorphology of a particular country or a geographically coherent region. This volume covers Austria—a country which is not particularly large in terms of area, but extremely diverse in terms of geomorphology. Its territory is dominated by high-mountain landscapes of the Alps, although even these occur in many flavours, depending on the tectonic history, rock type, the extent of Pleistocene and contemporary glaciation and last but not least, human impact. But Austria has much more to offer. The north is part of the Bohemian Massif, with granite uplands dotted by bizarre rock outcrops, deeply entrenched river canyons and gorges. The extreme east, in turn, is rather low-lying, but by no means less interesting, with fascinating stories of shifting rivers and lakes to be read from landforms and sediments.

The *World Geomorphological Landscapes* series is produced under the scientific patronage of the International Association of Geomorphologists—a society that brings together geomorphologists from all around the world. The IAG was established in 1989 and is an independent scientific association affiliated with the International Geographical Union and the International Union of Geological Sciences. Among its main aims are to promote geomorphology and to foster dissemination of geomorphological knowledge. I believe that this lavishly illustrated series, which sticks to the scientific rigour, is the most appropriate means to fulfil these aims and to serve the geoscientific community. To this end, my great thanks go to the editor of the volume, Professor Christine Embleton-Hamann, who agreed to coordinate the book and ensured that the final product exceeds all possible quality requirements. I am grateful to all individual contributors who agreed to add the task of writing chapters to their busy agendas and delivered such interesting stories to read. I also acknowledge the work of all reviewers. Their hard work meant then less work for me!

On more a personal note, I am very pleased that I was able to join the team of authors and could contribute about granite landforms in northern Austria, which are little known internationally but certainly deserve broader audience. Now is time to see other aspects of Austrian geomorphological landscapes, so wonderfully presented in this volume.

Wrocław, Poland Piotr Migoń

Preface

I always found pleasure in absorbing a scenic landscape view, for instance from the top of a mountain. As a student of German and Geography in Salzburg University, two courses became influential for my further career. The first one was a class within the German language programme that dealt with aesthetic experiences. The second one was the introductory course in geomorphology that came as a revelation: the science of geomorphology actually helps you to see more landscape details, and by increasing your knowledge about their formation, it ultimately augments your appreciation of scenic landscapes. In the past, these personal experiences resulted in a habilitation thesis exploring geomorphological techniques in assessing landscape beauty, and more recently in welcoming the challenge of organizing and editing a book on the landscapes and landforms of my home country.

This book is a joint effort of the Austrian Research Association on Geomorphology and Environmental Change, whose members compiled and wrote the majority of its chapters.

Two characteristics of Austria are (i) its reputation as a tourist's paradise and (ii) its rich cultural landscape, resulting from the presence of people since prehistoric times. The World Heritage Sites of Austria are listed as cultural properties, despite the official description of two of them stressing their spectacular landscape, which was obviously overlooked in the inscription process. Austria's tourism advertising is focused on the two tourism activities that generate the highest revenues, namely skiing and city sightseeing. We want to set this right by showcasing Austria's scenic landscapes and rich geoheritage, documented and explained in its National -, Nature- and Geoparks.

A further issue was to make regional studies in German available to the international community, according to the goals of the IAG book series. The Austrian Research Association on Geomorphology and Environmental Change has several members with deep insight into the regional development of landscapes, which they publish in national journals in German. My attempt to include these regional experts turned out to be a challenge, as they naturally delivered manuscripts that were more or less word-by-word translations. To turn their efforts into a meaningful text in English, a very time-consuming communication with the authors was needed. We finally conquered it, but this endeavour caused severe delay of the publication of the book. I therefore need to express my apologies to many colleagues who delivered their manuscripts at the initial timeline and then had to wait several years for the publication.

The chapters of this book are peer reviewed. I express my personal thanks to the following members of the international community of geomorphologists, who provided expert reviews and valuable input:

Marie-Françoise André	Isabelle Gärtner-Roer	Heinz Kollmann
Gary Brierley	Stefan Glatzel	Brian Luckman
Chris Caseldine	Andrew Goudie	John Menzies
Mike Church	Philipp Häuselmann	Jan Otto
Randy Dirszowsky	Erwin Heine	Rudolf Pavuza
Ian S. Evans	Gabor Kereszturi	Ulrike Pistotnik
Monique Fort	Jasper Knight	Peter Pisut

Jean Poesen
Tony Ramsay
Andreas Rudigier
Lothar Schrott
Marcel van der Perk
Dirk van Husen
Cees J. van Westen
Kirsten von Elverfeldt

This is a better book because of their comments. I pay tribute to the series editor Piotr Migoń, who for the Austrian volume not only provided assistance and guidance, but also an original chapter on granite landforms. I am grateful to Olav Slaymaker for fruitful discussion of scientific problems, patient help with English wording problems and continuing support. I am finally indebted to the many editorial assistants at Springer who have made this endeavour possible.

AUSTRIA
GEOMORPHOLOGIE & UMWELTWANDEL
Austrian Research Association on Geomorphology and Environmental Change
Part of the Austrian Geographical Society

Vienna, Austria
October 2021

Christine Embleton-Hamann

Contents

Editor and Contributors

About the Editor

Christine Embleton-Hamann is a retired professor at the Department of Geography and Regional Research at the University of Vienna. Her main interest is in alpine environments. Within this field, she focusses on human–environment interactions with research topics like human impact on geomorphic processes, assessment of the scenic quality of landscapes and geomorphological hazards. A second set of interest concerns the communication of geomorphological knowledge to a broader audience, in the pursuit of which objective she has written a well-received textbook on geomorphology. She is a former President of the Austrian Research Association on Geomorphology and served on the Executive Boards of the IAG and several IAG and IGU Working Groups.

Contributors

Christian Bauer Institute of Geography and Regional Science, University of Graz, Graz, Austria

Michael Doneus Department of Prehistoric and Historical Archaeology, University of Vienna, Vienna, Austria

Erich Draganits Department of Geology, University of Vienna, Vienna, Austria

Ruginia Duffy Department of Environmental Geosciences, University of Vienna, Vienna, Austria

Christine Embleton-Hamann Department of Geography and Regional Research, University of Vienna, Vienna, Austria

Ingomar Fritz Centre of Natural History, Geology and Palaeontology, Universalmuseum Joanneum, Graz, Austria

Sven Fuchs Institute of Mountain Risk Engineering, University of Natural Resources and Life Sciences (BOKU), Vienna, Austria

Thomas Geist FFG Austrian Research Promotion Agency, Vienna, Austria

Clemens Geitner Institute of Geography, University of Innsbruck, Innsbruck, Austria

Thomas Glade Department of Geography and Regional Research, University of Vienna, Vienna, Austria

Joachim Götz Fachgruppe Geomorphologie, University of Bayreuth, Bayreuth, Germany

Kati Heinrich Institute for Interdisciplinary Mountain Research, Austrian Academy of Sciences, Innsbruck, Austria

Severin Hohensinner Institute of Hydrobiology and Aquatic Ecosystem Management, University of Natural Resources and Life Sciences Vienna (BOKU), Vienna, Austria

Horst J. Ibetsberger GeoGlobe Häupl & Ibetsberger OEG, Neumarkt am Wallersee, Austria

Susan Ivy-Ochs Laboratory of Ion Beam Physics, ETH Zürich, Zürich, Switzerland

Viktor Kaufmann Institute of Geodesy, Graz University of Technology, Graz, Austria

Margreth Keiler Department of Geography, University of Innsbruck, Innsbruck, Austria;
Institute for Interdisciplinary Mountain Research, Austrian Academy of Sciences, Innsbruck, Austria

Andreas Kellerer-Pirklbauer Institute of Geography and Regional Science, University of Graz, Graz, Austria

Hanns Kerschner Institute of Geography, University of Innsbruck, Innsbruck, Austria

Gerhard Karl Lieb Institute of Geography and Regional Science, University of Graz, Graz, Austria

Martin Mergili Institute of Geography and Regional Science, University of Graz, Graz, Austria;
Institute of Applied Geology, University of Natural Resources and Life Sciences (BOKU), Vienna, Austria

Aleksandra Michniewicz Institute of Geography and Regional Development, University of Wrocław, Wrocław, Poland

Piotr Migoń Institute of Geography and Regional Development, University of Wrocław, Wrocław, Poland

Robert Peticzka Department of Geography and Regional Research, University of Vienna, Vienna, Austria

Helene Petschko Department of Geography, Friedrich Schiller University Jena, Jena, Germany

Lukas Plan Karst and Cave Group, Natural History Museum Vienna, Wien, Austria

Ronald E. Pöppl Department of Geography and Regional Research, University of Vienna, Vienna, Austria

Christoph Prager alpS GmbH, Innsbruck, Austria;
ILF Consulting Engineers Austria GmbH, Rum, Austria

Stefan Premm Department of Geography and Regional Research, University of Vienna, Vienna, Austria

Jürgen M. Reitner Geological Survey of Austria, Vienna, Austria

Doris Riedl Department of Geography and Regional Research, University of Vienna, Vienna, Austria

Reinhard Roetzel Geological Survey of Austria, Vienna, Austria

Milena Różycka Institute of Geography and Regional Development, University of Wrocław, Wrocław, Poland

Oliver Sass Fachgruppe Geomorphologie, University Bayreuth, Bayreuth, Germany

Christian Schlüchter Institut für Geologie, Universität Bern, Bern, Switzerland

Elmar Schmaltz Department of Geography and Regional Research, University of Vienna, Vienna, Austria

Ralf Schuster Geological Survey of Austria, Vienna, Austria

Wolfgang Schöner Institute of Geography and Regional Science, University of Graz, Graz, Austria

Arie Christoffel Seijmonsbergen Institute for Biodiversity and Ecosystem Dynamics, University of Amsterdam, Amsterdam, The Netherlands

Christian Semmelrock Department of Geography and Regional Research, University of Vienna, Vienna, Austria

Tobias Sprafke Institute of Geography, University of Bern, Bern, Switzerland

Stefan Steger Institute for Earth Observation, Eurac Research, Bozen/Bolzano, Italy

Hans Steyrer Department of Geography and Geology, Geology Working Group, Salzburg University, Salzburg, Austria

Erich Stocker Department of Geography and Geology, University of Salzburg, Salzburg, Austria

Kurt Stüwe Department of Earth Sciences, University of Graz, Graz, Austria

Johannes T. Weidinger ERKUDOK© Institute in the Kammerhof Museums of Gmunden (K-Hof), Gmunden, Austria

Michael Weißl Department of Geology, University of Vienna, Vienna, Austria

Martin Wenk Federal Ministry for Agriculture, Regions and Tourism, Federal Water Engineering Administration, Wien, Austria

Heinz Wiesbauer Freelance Civil Engineer for Landscape Planning and Maintenance, Vienna, Austria

András Zámolyi OMV Austria Exploration & Production GmbH, Gänserndorf, Austria

Part I

Introduction to the Physical Geography of Austria

1 Geological and Tectonic Setting of Austria

Ralf Schuster and Kurt Stüwe

Abstract

The landforms of Austria are the direct consequence of a continuous interplay between tectonic and climatic forces that have built, destroyed and reshaped the surface of the most iconic mountain belt on Earth for almost 40 Million years. As such, landforms can only be understood with a thorough geological background. This paper gives an overview of the tectonic evolution, the geological build up and the landscape evolution in the Austrian territory. The tectonic evolution of the rocks forming the major tectonic units of Austria can be traced back to some 500 Million years when they were located at different ancient continents including Gondwana, Avalonia and Laurasia. In the late Palaeozoic, the basement rocks were affected by the Variscan tectonometamorphic event during amalgamation of the supercontinent Pangaea and by a Permian extensional event. The latter is responsible for and was followed by a long-lived phase of thermal subsidence triggering the deposition of the Mesozoic sedimentary pile of the Northern Calcareous Alps. The formation and later subduction of the Neotethys and Penninic oceans began in Triassic and Jurassic times, respectively. The Alpine orogen as we know it today is largely the consequence of the head-on collision between the Adriatic and European plates once subduction had terminated around 40 Ma. The geological build up of Austria includes the Alps and its northern foreland. The foreland is composed of Variscan gneisses in the Bohemian Massif, their Mesozoic cover and Cenozoic sediments in the Molasse Basin. The Alps are made up of tectonic units derived from the European and Adriatic continents and the Neotethys and Penninic oceans that are covered by some intramontane and marginal basins that are filled with Neogene sediments. The landscape evolution evolved since the Oligocene and is highly influenced by processes in the mantle. It involved the interplay of many kilometres of rock uplift and simultaneous erosion so that few rocks at the surface today can be traced back to this time. Nevertheless, low-temperature geochronology, a series of fossil relict surfaces and enigmatic deposits like the Augenstein Formation on the plateaus of the Northern Calcareous Alps testify of a stepwise formation of the landscape over the last 25 Million years. Current research shows that up to 500 m of surface uplift may have occurred in the last 5 Million years alone.

Keywords

Austria • Geology • Geodynamics • Palaeogeography • Mantle structure • Landscape evolution

1.1 Introduction

The morphology of the Earth's surface is not only a reflection of atmospheric and hydrospheric processes, but—in fact—mostly the result of geological processes in the lithosphere and in the asthenospheric mantle. The plate tectonic environment causes compressional or extensional regimes that are responsible for the thickness of the crust and lithospheric mantle, which, in turn, are the basic controlling parameters for surface elevation on large length scales. On a smaller scale, lithologies and their sedimentary or deformation features are important parameters for the geomorphology as their physical and mechanical characteristics control the nature of weathering and ultimately erosion. Therefore, geographical and geological subdivisions of an area often go hand in hand. However, in many regions, they do not correspond (Fig. 1.1).

R. Schuster (✉)
Geological Survey of Austria, Neulinggasse 38, 1030 Vienna, Austria
e-mail: ralf.schuster@geologie.ac.at

K. Stüwe
Department of Earth Sciences, University of Graz, Universitätsplatz 2/11, 8010 Graz, Austria

C. Embleton-Hamann (ed.), *Landscapes and Landforms of Austria*, World Geomorphological Landscapes,
https://doi.org/10.1007/978-3-030-92815-5_1

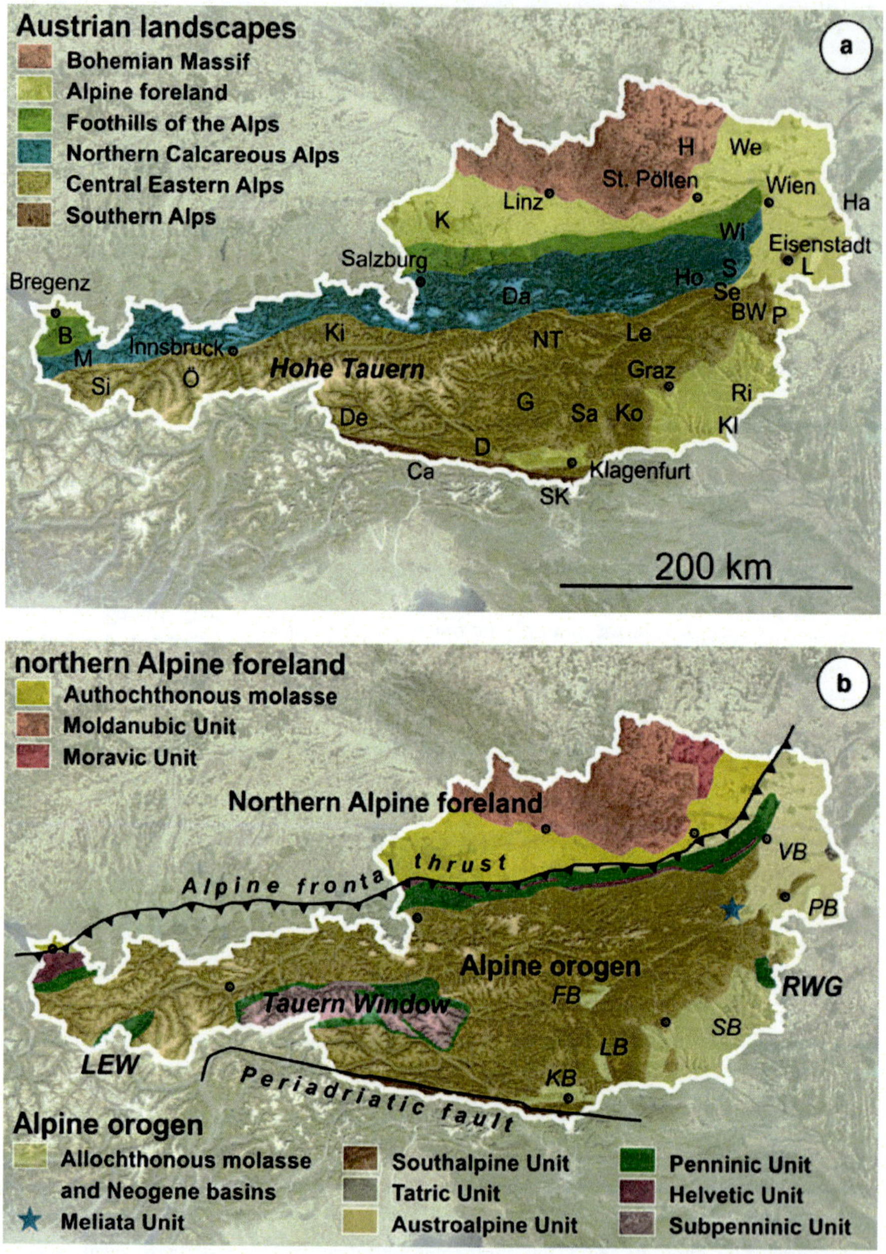

Fig. 1.1 Comparison of the distribution of landscape **a** and major tectonic **b** units in the territory of Austria. Obvious similarities of the geographical and geologic subdivisions indicate a strong influence of the geology on geomorphology. Geographic places: B = Bregenzer Wald, BW = Bucklige Welt, Ca = Carnic Alps, D = Dobratsch, Da = Dachstein, De = Deferegger Alps, G = Gurktaler Alps, H = Horn, Ha = Hainburg Mountains, Ho = Hochschwab, K = Kobernausser Wald, Ki = Kitzbühel, Kl = Klöch, Ko = Koralpe, L = Leithagebirge, Le = Leoben, M = Montafon, NT = Niedere Tauern, Ö = Ötztal, P = Pauliberg, Ri = Riegersburg, S = Schneeberg, Sa = Saualpe, Se = Semmering, Si = Silvretta, SK = South Karawanken, Wi = Wienerwald, We = Weinviertel. Geological features: LEW = Lower Engadine Window, RWG = Rechnitz Window Group, FB = Fohnsdorf Basin, KB = Klagenfurt Basin, LB = Lavanttal Basin, PB = Pannonian Basin, SB = Styrian Basin, VB = Vienna Basin

From a plate tectonic view, the territory of Austria is located at the compressional plate boundary between the Eurasian and the Adriatic plates. It covers the eastern portion of the Alpine orogenic belt and its northern foreland. This setting is the result of convergence of the African and Eurasian plates, which was more or less continuous since the Early Cretaceous. The geology of the Alpine-Mediterranean realm is complex, however, because several microplates formed and vanished between the two major plates, and the interplay between shortening processes and lateral movements makes it difficult to determine the plate tectonic arrangement through time (Froitzheim et al. 2008; Handy et al. 2010).

This complicated geology of the Austrian Alps was explained differently during three distinct periods in the geological exploration history. The first period is that of the "*Old Geology*" (as referred to by Kober 1938). It started with the first "geognostic" excursions in the late eighteenth century and was followed by the "Erste Geologische Landesaufnahme" (first geological mapping campaign), resulting in a map by Hauer (1867) on the scale of 1: 576.000. At that time, the rock series were thought to be more or less autochthonous, meaning that they were formed at the place where they still are. This implies that older rocks are overlain by younger ones. Mountain ranges were interpreted to be anticlines formed by contraction of Earth, or by the inversion of "geosynclines". Geological maps of this period show typically several zones with old "primary" rocks in the centre and younger "secondary" and "tertiary" sediments towards the sides (Sedgwick and Murchinson 1832). Terms like "Sandsteinzone", "Grauwackenzone" or "Nördliche Kalkalpen", which are still in use in a geographic sense, are remnants from this period, but they are often obsolete or problematic in the geologic nomenclature of today.

Towards the start of the twentieth century, the period of *nappe tectonics* was born when it was realized that older rocks might be thrust over younger ones, forming distinct nappe-shaped packages of rock that can travel up to hundreds of kilometres. Names like the Austrian geologists Eduard Suess or Otto Ampferer are closely connected to this development that was paralleled by equivalent discoveries by Swiss geologists. With this knowledge of nappe motion, it was possible to explain the distribution of the rocks at the surface, but the driving mechanisms were still not understood. Nevertheless, this knowledge allowed a process-driven subdivision of the Alps (e.g. Suess 1909). With little change, this general tectonic subdivision of the Alps into Helvetic, Penninic, Austroalpine and Dinaric (Southalpine) units is still in use today. However, for the internal subdivision of these units, many different suggestions exist (e.g. Kober 1938; Tollmann 1977; Neubauer et al. 2000; Schmid et al. 2004; Janák et al. 2004). For the Variscan rocks in the northern part of Austria, the major tectonic units were also established at approximately this time (Kossmat 1927). In the territory of Austria, these are the Moravian and Moldanubian units (Suess 1912), which have since been subdivided further (e.g. Fuchs 1976; Neubauer and Handler 2000; Linner 2013; Finger and Schubert 2015).

The third period is that of *modern plate tectonics*. The first interpretation of the Eastern Alps in terms of this concept was given by Frisch (1979). Today, there is a broad consensus that the Alpine orogen in central Europe developed from four major palaeogeographic realms existing in Jurassic and Cretaceous times (Fig. 1.2). These are the African and the European continental realms, and the Neotethys and the Penninic (Alpine Tethys) oceans (e.g. Froitzheim et al. 1996; Neubauer et al. 2000). While this subdivision of the Alpine orogenic cycle appears well established, the plate tectonic framework for the Variscan orogenic cycle is less clear (e.g. Matte 1986; Franke 2000; Kroner and Romer 2013).

Today, geophysical methods allow us to look at the deep structure of the Alps giving the chance to confirm some of the plate tectonic interpretations. Based on the depth of the Mohorovičić discontinuity (MOHO) and the distribution of shear wave velocities, it can be shown that the Eurasian Plate dips southward underneath the Alps, whereas the Adriatic Plate—representing a former promontory of Africa—is dipping northward below the active Alpine orogen. However, the mantle structure and its evolution in the last Million years, which is highly important for morphological changes in the past, is still a matter of debate (Lippitsch et al. 2003; Mitterbauer et al. 2011). Detailed geological maps on the scale 1:50.000 and 1:75.000 are available for the major part of the territory of Austria (http://www.geologie.ac.at/services/), and a catalogue of geological aerial photography is present (http://www.alpengeologie.org). From these, it is impressive to see how closely the morphology is related to the lithology or to lithogenetic units when field observations are compared to high-resolution digital elevation models and geological maps (Stüwe and Homberger 2012).

The following sections give an overview on the geodynamic and plate tectonic evolution as well as the lithologic content of the geologic units within the territory of Austria. Finally, the evolution of the landscape from the Eocene onward is described.

1.2 Palaeogeographic and Plate Tectonic Evolution

The following description of the geodynamic evolution of the Alpine area since the late Neoproterozoic summarizes the major tectonometamorphic events recorded in the rock series and the oceanic and continental areas from which the major tectonic units developed. It is based on Stampfli and Borel (2004), Handy et al. (2010), Faccenna et al. (2014) and others and is summarized by Schuster and Stüwe (2010) and well illustrated by Stüwe and Homberger (2012) or Schuster et al. (2014).

Following the Cadomian orogeny in the late Neoproterozoic (c. 600 Ma), all continental units present in the territory of Austria were located at the northern margin of the Gondwana continent in close proximity to the South Pole. In the late Cambrian and Ordovician (510–460 Ma; Fig. 1.3a), an abundance of magmatic rocks formed in an extensional environment (Neubauer 2002). The continental fragment of Avalonia, including the rocks of the Moravian and Brunovistulian units of the Bohemian Massif broke off from Gondwana and drifted northward. It was welded to the Laurussian continent in the frame of the Caledonian collisional event at the Ordovician/Silurian boundary (c. 420 Ma; Fig. 1.3b).

From the Late Devonian to the early Permian (380–260 Ma; Fig. 1.3c, d), the Pangaea supercontinent formed by amalgamation of all major pieces of continental crust on Earth. One important continental collision event that unified

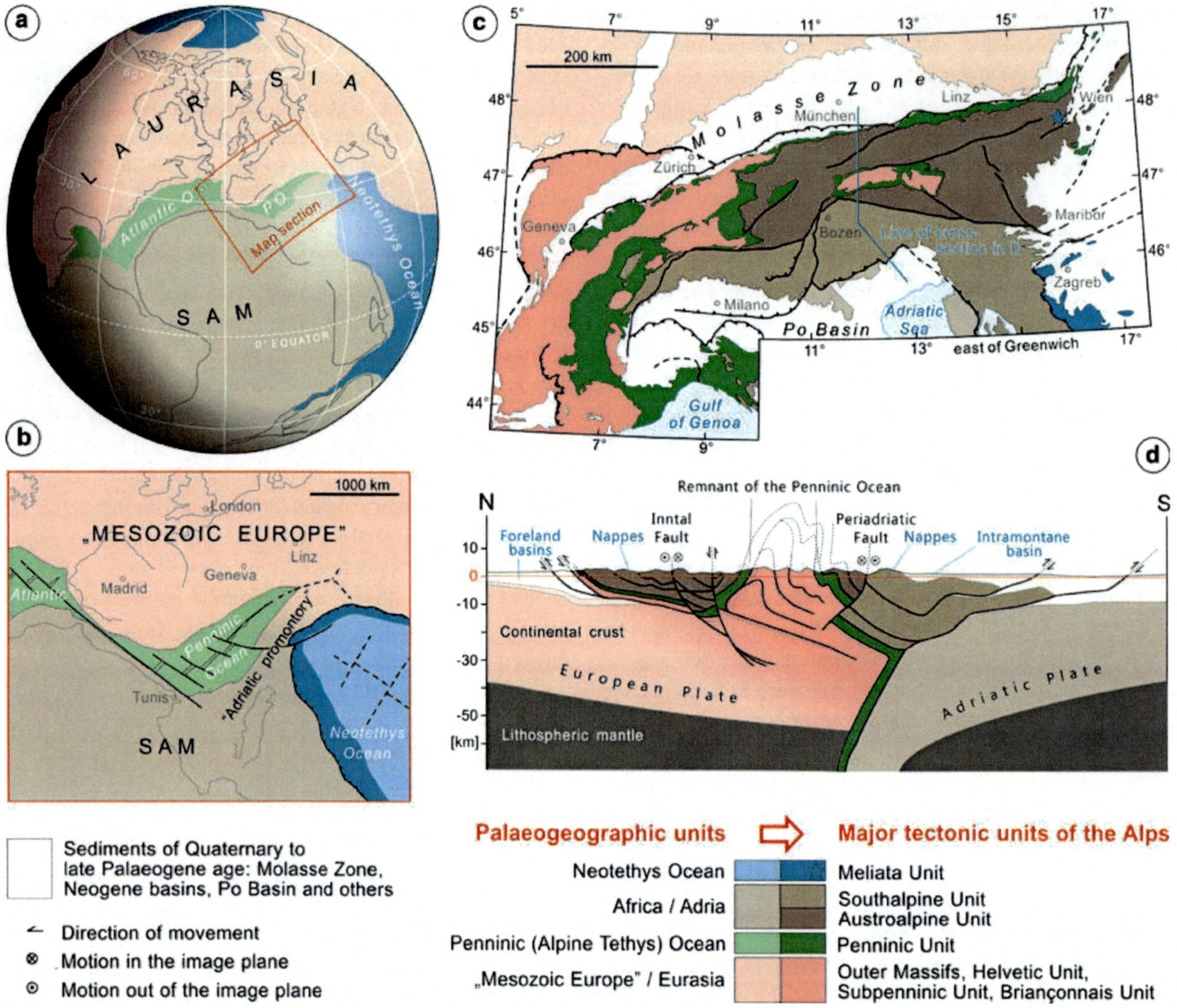

Fig. 1.2 Global **a** and regional **b** reconstructions showing the major palaeogeographic units of the Alpine realm and their transformation into the major tectonic units of the Alps, shown in map view **c** and section **d**. After Schmid et al. (2004) and Schuster et al. (2014) and references therein. SAM = southern continent assemblage that has split since that time into several continents (e.g., Africa, Adria, Australia), PO = Penninic (Alpine Tethys) Ocean

the continental masses of Laurussia and Gondwana at this time is referred to as the Variscan event. It lasted roughly from the Late Devonian to the end of the Carboniferous (380–300 Ma). In Austria, the Moldanubian and Moravian units of the Bohemian Massif represent tectonic units that were thrust during this Variscan event onto the foreland, then represented by the Brunovistulian Unit. The latter had already been consolidated during the earlier Cadomian event (Kroner and Romer 2013). Variscan metamorphism and deformation is also evident in many of the Austroalpine and Southalpine basement units.

Following the Variscan orogeny, Pangaea had a C-shape (Fig. 1.3d, e) with the Tethys Ocean forming an embayment coming in from the east (Stampfli and Borel 2004). Rock units from the Austrian territory were located at this time close to the western end of the Neotethys embayment (also called the Meliata-Hallstatt Ocean). In Permian and Triassic times (285–220 Ma), this area was affected by an extensional event. Thinning of the lithosphere caused basaltic underplating of the crust, resulting in magmatic activity and widespread high-temperature metamorphism (Schuster and Stüwe 2008). At the surface, grabens formed that accommodated new sediments and volcanoes erupted. Since the Early Triassic (c. 245 Ma) the Neotethys Ocean propagated westward. Parts of the Moldanubian and Moravian units formed an island, whereas the other units of Austria were located at a broad shelf and partly also on the slopes of the southern continental margin towards the Neotethys Ocean.

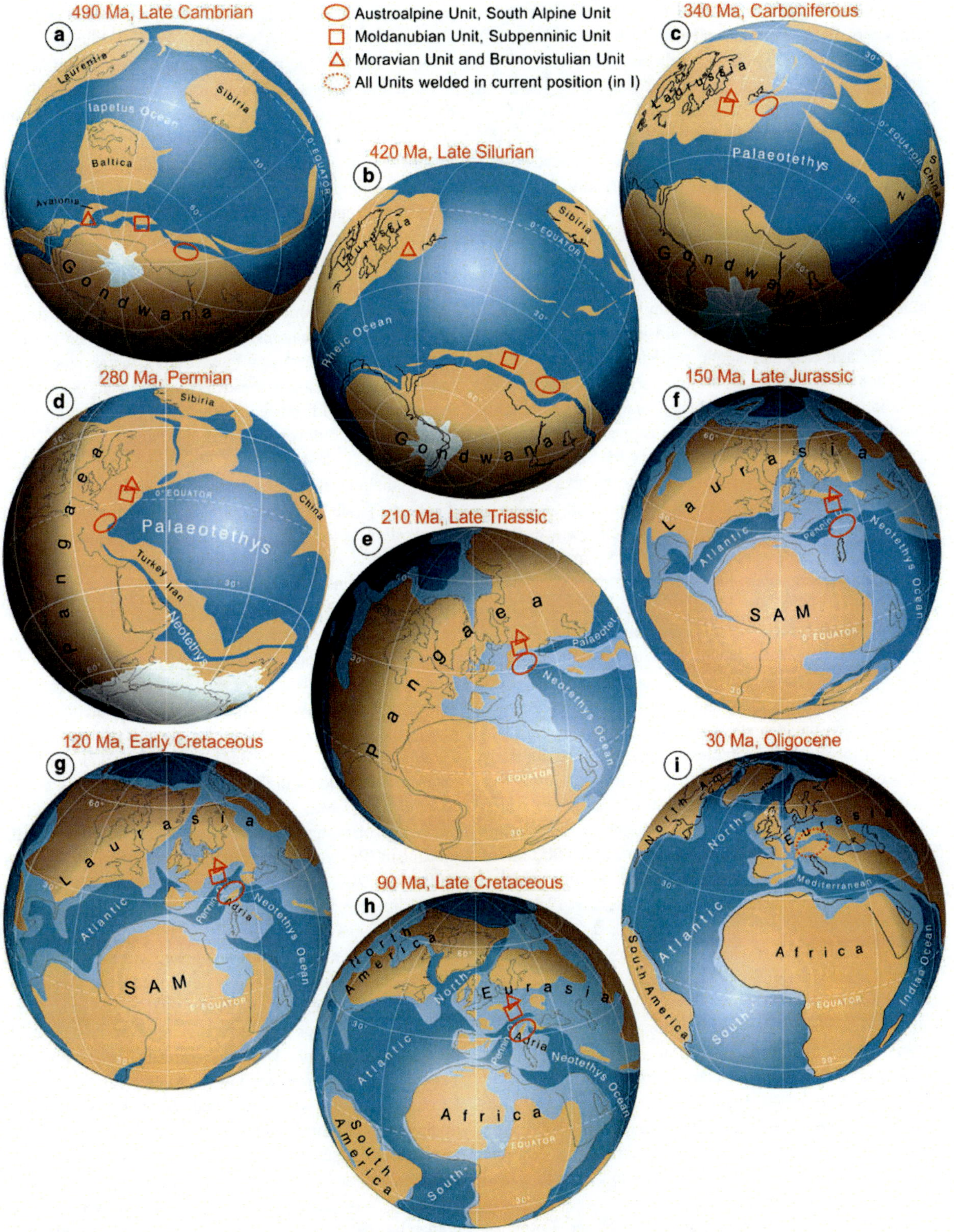

Fig. 1.3 Reconstructions of the global palaeogeography through Phanerozoic time based on Schuster et al. (2014). The positions of major continental tectonic units in the territory of Austria are shown through time. Explanation see text. SAM = southern continent assemblage that has split up since that time. Continental land masses are in brown, continental shelfs in light blue and oceanic basins in dark blue

A prolonged phase of thermal subsidence followed the termination of the Permian extensional event. During this time, the up to 3000-m-thick carbonate platforms of the Northern Calcareous Alps formed on the subsiding shelf. In the Middle Jurassic (c. 150 Ma; Fig. 1.3f), opening of the Atlantic Ocean initiated the breakup of Pangaea. The continent Laurasia was split from a southern continent assemblage including Africa, which started drifting eastwards. While the Atlantic was opening in north–south direction, a wrench corridor with sinistral offset propagated eastward into the region that forms the Alpine realm today (Frisch 1979). This early arm of the Atlantic is referred to as the Penninic Ocean, which is also known by the name "Alpine Tethys". At the same time, the closure of the Neotethys Ocean further east and south started at an intra-oceanic subduction zone. In the Late Jurassic, ophiolite nappes from this subduction zone were obducted onto the shelf margin (Missoni and Gawlick 2010). These nappes are widespread in the Dinarides. In the Eastern Alps and Western Carpathians, only tiny remnants of these obducted nappes are preserved and referred to as the Meliata Unit. In Austria, they are located in a Cretaceous out-of-sequence thrust within Permo-Mesozoic sediments of the Austroalpine Unit in the Schneeberg area in Lower Austria (Mandl and Ondrejickova 1993). In addition, eroded material from related nappes is present within Cretaceous sediments (Faupl and Wagreich 1992). Following these obduction processes (in the latest Jurassic or earliest Cretaceous around 145 Ma), a major sinistral strike-slip fault system cut the continental bridge between the Neotethys and Penninic oceans that was formed by the Adriatic promontory (Fig. 1.2b). In the Early Cretaceous (c. 130 Ma; Fig. 1.3g), this fault was activated as a south dipping intra-continental subduction zone, which ultimately became responsible for the formation of the Alps (Stüwe and Schuster 2010). Since that time, about 900 km of north–south convergence and 30° of anticlockwise rotation of the Adriatic Plate towards the Eurasian Plate was accommodated at this subduction zone (Handy et al. 2010).

In the frame of the so-called Eoalpine event in the Early to Late Cretaceous (130–80 Ma), the Austroalpine nappes and the nappes of the Central Western Carpathians formed from continental crust dragged into this subduction zone in a pro-wedge setting. Some nappes reached eclogite facies metamorphic conditions at depths of more than 80 km (Tenczer and Stüwe 2003; Janák et al. 2015) in early Late Cretaceous time (c. 95 Ma; Fig. 1.3h; Thöni 2006). As subduction proceeded, oceanic lithosphere of the Penninic Ocean was subducted into the same zone in the middle Late Cretaceous (starting at about 85 Ma). Finally, in the early Eocene (c. 45 Ma), the southern margin of the European continental lithosphere was also drawn into the same suture zone. During this process, the Penninic, the Helvetic and the Subpenninic nappes formed. Subduction related, pressure-dominated metamorphism in the Eocene and early Oligocene (at about 45–30 Ma) reached blueschist and eclogite facies conditions in parts of the Penninic and Subpenninic nappes. The final closure of the Penninic Ocean occurred in the middle to late Eocene (c. 40 Ma; Schmid et al. 2004) and a head-on collision between the Adriatic and European plates terminated the subduction process. The northern Alpine foreland basin developed on top of the down bending European continental lithosphere. In the middle Eocene to early Oligocene (40–28 Ma; Fig. 1.3i), the subducted slab, from which all the nappes were detached, broke off from the Eurasian plate at depth (von Blankenburg and Davies 1995). Around 30 Ma, this caused intense magmatism (Periadriatic magmatism) and dextral slip on the Periadriatic fault (Mancktelow et al. 2001), which since forms the boundary between the Austroalpine and Southalpine units. Surface uplift related to isostatic rebound produced the first topographic development in the range.

In the earliest Miocene (c. 21 Ma; Schmid et al. 2013), the Southalpine indenter pushed northward and caused further north–south shortening along the northern margin and in the western part of the Eastern Alps. Along the northern margin, parts of the foreland basin were sheared-off and incorporated into the orogenic wedge as the so-called Allochthonous Molasse (until the Ottnangian; c. 18 Ma). In the east (in the area of the Pannonian Basin), the last remnants of the Penninic Ocean were subducted (Faccena et al. 2014) and the retreat of the oceanic lithosphere lead to stretching in east–west direction, especially in the eastern part of the Eastern Alps (lateral escape of Ratschbacher et al. 1989, Robl et al. 2008a, Bartosch et al. 2017). East–west extension was accommodated by thinning of the lithosphere. Around 15–17 Ma, the extension in connection with the retreating subduction zone was associated with calc-alkaline volcanism (Pannonian magmatism; e.g. in the Styrian Basin; Fodor et al. 2008). In the upper part of the crust, a system of strike-slip (e.g. Mölltal, Inntal, Salzach-Enntal-Mariazell-Puchberg faults; Peresson and Decker 1996; Linzer et al. 2002; Pischinger et al. 2008) and normal faults (e.g. Brenner, Katschberg normal faults) developed (e.g. Genser and Neubauer 1989; Fügenschuh et al. 1997; Scharf et al. 2013), along which the Alps were extruded to the east. In the centre of the range tectonic windows formed (Lower Engadine, Tauern and Rechnitz windows), where Subpenninic and Penninic nappes were exhumed from below the Austroalpine. Blocks in between the faults were tilted, uplifted or subsided. A series of basins formed along the strike-slip faults and these accommodated sediments (e.g. Fohnsdorf, Tamsweg, Klagenfurt, Lavanttal, Vienna, Styrian and Pannonian basins). Since the middle Miocene, the eastern part of the Alps and surrounding areas were uplifted, most probably due to processes in the upper mantle (Wagner et al. 2010; Baran et al. 2014; Legrain et al. 2014). During further uplift, the basins in eastern Austria rose up to about 150–200 m above sea level, while the Bohemian Massif established as a new, low mountain range. Generally, these Miocene tectonic processes, in combination with the erosive overprint during the glaciation periods in the Quaternary (since 2.6 Ma), are responsible for many topographic features we can observe today.

1.3 Description of the Major Geologic Units

In this section, the major geologic units appearing within the territory of Austria are described. We begin with the units from the northern Alpine foreland. After that, those of the Alps follow. Each part is described in an order from bottom to top and from north to south (numbers in parentheses refer to Figs. 1.4 and 1.5).

1.3.1 Units in the Foreland of the Alps

The northern Alpine foreland comprises the pre-Variscan and Variscan metamorphosed basement of the Brunovistulian, Moravian and Moldanubian units. These are transgressively overlain by late Carboniferous to Cretaceous sediments and by Oligocene to Pannonian sediments of the Autochthonous Molasse.

1.3.1.1 Pre-Variscan and Variscan Metamorphosed Basement

Within the Austrian territory, the **Brunovistulian Unit** occurs in the subsurface below the Neogene sediments of the northwestern Weinviertel (Lower Austria). It received its present structural and metamorphic imprint during the Cadomian orogeny in the late Neoproterozoic (c. 600 Ma) and is the oldest unit in Austria. It consists of paragneisses, micaschists, marbles and minor quartzites as well as amphibolites that are intruded by large amounts of Cadomian granites and granodiorites. During the Variscan and Alpine orogenies, it always remained in the foreland (Finger et al. 2000).

During the Variscan event in the Late Devonian to Carboniferous (380–300 Ma), parts of the Cadomian basement were sheared-off and incorporated into the Variscan orogen at a late stage and in a marginal setting. These parts are referred to as the **Moravian Unit** (26). The Variscan metamorphic and structural overprint increases towards the tectonic higher units in the west. Within the Moravian Unit, the lower Thaya and Svratka nappes are characterized by a lower greenschist facies overprint and minor deformation. The textures of the Cadomian granites are still well-preserved (e.g. Maissau granite, Eggenburg granite) (Fritz et al. 1996). In contrast, the higher Pleißling nappe reached amphibolite facies conditions and shows a penetrative, often mylonitic schistosity (e.g. Bittesch and Weitersfeld orthogneisses).

The **Moldanubian Unit** (25) represents a deep level of the internal part of the Variscan orogen. Its eastern part is formed by the Moldanubian nappes, which are characterized by a granulite facies metamorphic imprint with metamorphic peak conditions at up to 1000 °C and 1.6 GPa at c. 340 Ma (Petrakakis 1997). The lowermost Ostrong Nappe System is formed by monotonous paragneisses ("Monotonous series"), whereas the paragneisses of the overlying Drosendorf Nappe System contain many intercalations of amphibolites, marbles and graphitic schists ("Variegated series"; Fuchs and Matura 1976). At the base of the next higher Gföhl Nappe System, an amphibolite-rich sequence (Rehberg amphibolite) is present, which is interpreted as an ophiolite remnant. Above this, there are Ordovician orthogneisses (Gföhl orthogneiss) with slices of ultramafic mantle rocks at their base. In the uppermost nappe of the Gföhl Nappe System, the orthogneisses show a mylonitic texture and are referred to as "granulites". The Moldanubian nappes were intruded by the South Bohemian Batholith at 335 to 300 Ma. The Rastenberg granodiorite, Weinsberg granite, Eisgarn granite and Mauthausen granite are the most important intrusions (Gerdes et al. 2000). In the west, the whole sequence experienced an additional high-temperature overprint at c. 330 Ma causing widespread migmatisation. This part is referred to as Bavaric Subunit.

Geographically, the Variscan basement units build up the Bohemian Massif with its elevated, but rather flat and smooth topography cut by deep river gorges like Strudengau, Wachau, Kamp Valley or Thaya Valley, indicating a young and immature landscape. The valley sides are characterized by steep cliffs of massive rocks with a widely spaced cleavage.

1.3.1.2 Carboniferous to Cretaceous Sediments and Autochthonous Molasse

The **late Carboniferous to Cretaceous sediments** (24) that overlie the Variscan basement are only visible in a few places (e.g. near to Zöbing). They were deposited during several transgressional phases in the Permian, Jurassic and Cretaceous. However, below the Autochtonous Molasse, they occur more frequently and are well-studied because of their importance for the hydrocarbon industry (Wessely 2006).

The late Oligocene to Miocene (35–18 Ma) sediments of the **Autochthonous Molasse** (23) were deposited in the foreland basin of the Alpine orogen (Rupp et al. 2006; Wessely 2006). It formed as a southward dipping basin on the Eurasian plate due to topographic loading by the northward moving Alpine nappe pile. The main part of the basin is filled by marine sediments. At the northern margin gravel,

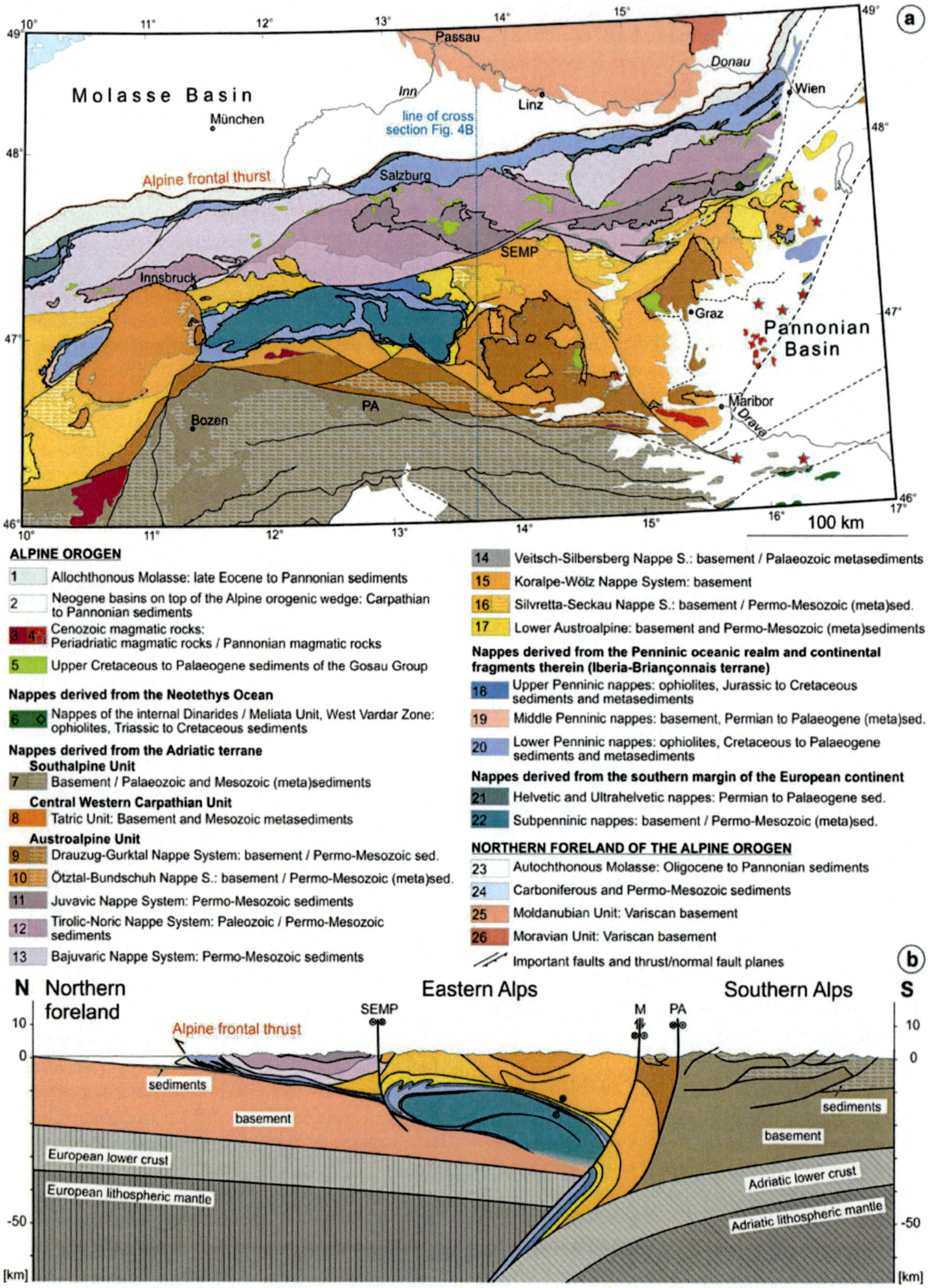

Fig. 1.4 Tectonic map **a** of the Eastern Alps and its northern foreland according to the nomenclature in Schmid et al. (2004). The numbers of the units refer to the text. Section **b** through the Eastern Alps according to Schmid et al. (2004). SEMP = Salzach-Ennstal-Mariazell-Puchberg fault, M = Mölltal fault, PA = Periadriatic fault

Fig. 1.5 Schematic diagram showing the tectonostratigraphic order of the major tectonic units of the Eastern Alps. Tectonic and lithostratigraphic units are listed in the left column. In the central column the major lithological content of the tectonic units is shown. Fields with numbers refer to the numbers of the tectonic units in the text and the colour in Fig. 1.4. Red numbers at left indicate the time of incorporation into the Alpine orogenic wedge. The metamorphic grade during the Eoalpine (Cretaceous) and Alpine (Cenozoic) events and the time of peak metamorphism is given in the right column. Note, the Meliata Unit is not part of the Austroalpine Unit, its primary emplacement was on top of the Juvavic and Tirolic-Noric nappe systems. T = Triassic, J = Jurassic, C = Cretaceous, Pa = Palaeogene

sands and clays, some fossil-rich, occur that formed in a beach or shallow water environment. The central part is dominated by sands and clays from a deeper basin, influenced by intense stream currents. In the south, coarser-grained clastic sediments derived from the Alpine nappes are present. The youngest (Pannonian, c. 7 Ma) part of the sedimentary pile consists of fluvial gravels derived from the Alps, which contain coal layers in the Kobernausser Wald (Upper Austria). Because of its composition of soft sediments, the area of the Autochthonous Molasse shows a smooth hilly landscape with larger outcrops restricted to valleys and localities affected by anthropogenic activity.

1.3.2 Units in the Alps

The tectonic lowermost element of the Alps is the Allochthonous Molasse. It is overlain by the Subpenninic and

Helvetic nappes derived from the European continental margin and the Penninic nappes that developed from the Penninic oceanic realm. The highest part of the Alpine nappe pile is made up of the Austroalpine, Tatric and Southalpine units that originate from the Adriatic terrane and also includes slices of the Meliata Unit derived from the Neotethys Ocean. Since the Oligocene, two types of magmatic rocks formed and Neogene basins developed on top of the Alpine orogenic wedge (wedge shaped Alpine nappe pile overlying the Eurasian and Adriatic plates).

1.3.2.1 Allochthonous Molasse and Units Derived from the European Continental Margin

The **Allochthonous Molasse** (1) is made up of sediments from the northern Alpine foreland basin that were deformed, sheared-off and incorporated into the Alpine orogenic wedge before the middle Miocene (until 15 Ma). It consists of conglomerates, sands and clays (Wessely 2006) forming the most external foothills along the northern margin of the Alps.

The **Subpenninic nappes** (22) developed in the Oligocene (c. 34–23 Ma) from thinned continental crust of the distal European margin (e.g. Froitzheim et al. 1996; Kurz et al. 2001) and represent ductily deformed basement and cover nappes, which lost contact with their lithospheric mantle (Schmid et al. 2004). They are characterized by a basement consisting of partly migmatitic paragneisses and amphibolites with Neoproterozoic to Carboniferous protolith ages, intruded by large masses of deformed Variscan granitoids ("Zentralgneise"). Post-Variscan transgressive sediments were deposited on top of these gneisses at different times between the late Carboniferous and the Cretaceous. Typical is a sequence with Permian and Early Triassic clastics, Middle Triassic carbonates ("Seidelwinkel Trias") and Late Triassic rauhwacke and quartzites. In other nappes, Late Jurassic marbles derived from shallow water limestones (Hochstegen Formation) appear directly on the Variscan basement. During their Alpine tectonometamorphic overprint, the Subpenninic nappes experienced greenschist to amphibolite facies metamorphic conditions in the Eocene to Miocene (45–15 Ma). The structurally lower nappes dominated by thick basement form the Venediger Nappe System, whereas nappes nearly exclusively composed of Permian to Cretaceous metasediments are referred to as Modereck Nappe System (Schmid et al. 2013). The latter also includes the so-called eclogite zone, which additionally contains material derived from the Penninic Ocean and developed in a subduction and accretion channel (Kurz and Froitzheim 2002). It experienced an eclogite facies imprint in the Eocene (c. 35 Ma). In the Hohe Tauern region, within the so-called Tauern Window, the Subpenninic nappes build up some of the highest peaks like Großvenediger (3662 m) or Hochalmspitze (3360 m). Typically, they form very steep cliffs of bright coloured, blocky orthogneisses ("Zentralgneise") (Fig. 1.6).

Fig. 1.6 Eastward view along the main crest of the Hohe Tauern in the Großglockner region (Salzburg/Carinthia). The crest is built up by the Venediger Nappe System of the Subpenninic Unit. Bright greyish orthogneisses occur at Sonnblick (3106 m). On top of Hocharn (3254 m) dark brownish paragneisses are visible. Overlying brownish-grey Jurassic and Cretaceous schists (Bündnerschiefer Group) of the Penninic and Subpenninic units are visible at Gjaidtroghöhe (2988 m) and Sandkopf (3090 m). Further up in the tectonic succession Subpenninic metasedimentary rocks including bright coloured Triassic carbonates around Rossschartenkogel (2665 m) and at Rote Wand (2855 m) follow. Photo from: www.alpengeologie.org

In the Oligocene, the **Helvetic and Ultrahelvetic nappes** (21) were detached from the distal European margin to form a thin-skinned fold and thrust belt. They are composed of Jurassic to Eocene (200–35 Ma) cover sequences, which can be subdivided with respect to their depositional depth. The Helvetic nappes sensu stricto comprise mostly carbonaceous shelf sediments up to a few hundred metres in thickness, whereas the Ultrahelvetic nappes consist of thin pelagic sediments from the deeper continental slopes (Oberhauser 1980). In the Bregenzer Wald (Vorarlberg), the Helvetic nappes form a hilly to mountainous area with an elevation between 450–2100 m, whereas the Ultrahelvetic units only appear as thin slices intercalated between Penninic nappes of the Rhenodanubian flysch zone (see below).

1.3.2.2 Nappes Derived from the Penninic Oceanic Realm

The Penninic nappes developed in the Late Cretaceous to Palaeogene (85–35 Ma) from an oceanic realm with continental fragments (terranes) therein. Based on their palaeogeographic origin and structural position, they are subdivided into the Upper, Middle and Lower Penninic nappes.

In the Palaeogene, the **Lower Penninic nappes** (20) formed from material of the northern part of the Penninic oceanic basin (Valais oceanic domain). They consist of Cretaceous ophiolites including serpentinized slices of ultramafic mantle rocks and a basaltic oceanic crust overlain by thick, Cretaceous to Eocene (135–45 Ma) flyschoid sequences (e.g. Rhenodanubian Group, Bündnerschiefer Group), which are dominated by siliciclastic or carbonaceous material. Unmetamorphosed Lower Penninic nappes build up the Rhenodanubian flysch zone along the northern margin of the Alps. Similar rocks, partly with an Eocene–Oligocene (45–35 Ma) blueschist to eclogite facies high pressure imprint and a Miocene (c. 25 Ma) greenschist facies overprint, appear in the central part of the Lower Engadine Window, the Glockner Nappe System of the Tauern Window and in the nappes of the Rechnitz Window Group.

The **Middle Penninic nappes** (19), mainly derived from the Briançonnais terrane, consist of Variscan metamorphic basement and a cover sequence with Permian fluvial to shallow marine clastics, a carbonate-rich Triassic shallow water sequence and Jurassic to Cretaceous deep-water sediments, deposited on a deep marine swell. In Austria, greenschist metamorphic Middle Penninic nappes are only present in the Lower Engadine Window in Tyrol.

Rocks that are derived from the southern part of the Penninic Ocean (Piedmont-Ligurian oceanic domain) as well as from the accretionary wedge along the southern margin of the oceanic basin towards the Adriatic terrane make up the **Upper Penninic nappes** (18). They are composed of Jurassic ophiolite slices comprising serpentinized ultramafic mantle rocks, gabbros and basalts within Late Jurassic and Cretaceous deep marine shales and flyschoid sediments (Froitzheim and Manatschal 1996). Sequences with serpentinized, exhumed mantle rocks that are directly overlain by Jurassic radiolarites and pelagic limestones are typical. Unmetamorphosed Upper Penninic slices occur in the border zone between the Rhenodanubian flysch zone and the Austroalpine nappes of the Northern Calcareous Alps (Arosa Zone, Ybbsitz klippen belt; Oberhauser 1980; Decker 1990). Further, blueschist and/or greenschist metamorphic units form the uppermost tectonic elements within the Lower Engadine and Tauern windows (e.g. Matrei Zone, Reckner Complex).

The flyschoid sequences of the Rhenodanubian flysch zone build up gentle hills along the northern margin of the Alps. Within the tectonic windows, the soft schists of the Bündnerschiefer Group cause gentle surfaces even on very steep slopes. Above the tree line, they are brownish coloured and vegetated by grass (Fig. 1.6). The stronger rocks of the ophiolite slices often protrude as cliffs forming the peaks of prominent mountains such as Großglockner (3798 m).

1.3.2.3 Nappes Derived from the Adriatic Terrane

The Austroalpine, Tatric and Southalpine are units that formed from the Mesozoic Adriatic terrane. In general, the crustal sequence of the Adriatic terrane is made up of three different types of pre-Mesozoic basement (Raumer et al. 2013) that is covered by Mesozoic sediments.

The first pre-Mesozoic basement type is dominated by biotite-plagioclase-paragneisses with intercalations of micaschists and amphibolites that developed from Neoproterozoic to Cambrian successions (1000–485 Ma). In part, they experienced a Cambro-Ordovician tectonometamorphic imprint (Neubauer 2002) and contain abundant intrusions of deformed Cambro-Ordovician (540–460 Ma) granitoids. They are affected by an amphibolite facies Variscan imprint at 380–300 Ma with minor associated granitic intrusions (Mandl et al. 2017). This basement type is typical for the Silvretta-Seckau and Ötztal-Bundschuh nappe systems. The second basement type is built up of muscovite-rich micaschists and paragneisses, typically containing intercalations of marbles, quartzites and amphibolites, that have a Cambrian to Carboniferous (540–300 Ma) depositional age. These sequences are often characterized by a first metamorphic imprint and related magmatism in the Permian (285–250 Ma), and they are frequent in the Koralpe Wölz Nappe System. Thirdly, there are weakly metamorphosed Palaeozoic sequences, for example, in the Southalpine Unit. They were deposited in a shelf environment and show a fossil record from the Ordovician to the early Carboniferous (485–340 Ma) (Schönlaub 1979). The Ordovician part is dominated by siliciclastics and felsic volcanics (e.g. Blasseneck porphyroid). In the Silurian, limestones are more common and there are some intercalations of basaltic metavolcanic rocks. Thick carbonate platform sediments and related intra-basin sediments including some basaltic extrusives are characteristic for the Devonian. From the early Carboniferous, some carbonaceous successions are preserved, but an increasing amount of clastic sediments reflects the Variscan event. Locally preserved late-orogenic Carboniferous sediments and Permian clastic sediments with salt and gypsum deposits in some units complete the depositional cycle.

The Mesozoic cover successions (Tollmann 1977; Mandl 2000) include famous units like the Northern Calcareous Alps. They start with Lower Triassic fluvial or shallow marine siliciclastic sediments and are overlain by Middle Triassic dark coloured limestones deposited in an euxinic marine environment. Above this follows the deposition of carbonate platform sediments (Wetterstein platform) in reef and lagoonal facies. After a short break with siliciclastic influx in the early Upper Triassic, limestones and dolomites of a second carbonate platform continue the sequence (Hauptdolomit-Dachsteinkalk platform). Contemporaneously, there are Middle to Upper Triassic sedimentary piles of pelagic limestones deposited on the slope towards the Neotethys Ocean (Hallstatt-facies sediments). Variegated sequences of Jurassic pelagic carbonates, olistolithes and radiolarites and some reef limestones were deposited due to increasing tectonic activity related to the opening of the Penninic Ocean and the beginning of closure of the Neotethys Ocean (Missoni and Gawlick 2010). Lower Cretaceous synorogenic clastic sediments including conglomerates and sandstones formed in foreland basins of the initial Alpine orogenic wedge (Roßfeld and Losenstein formations). Finally, Upper Cretaceous to Palaeogene conglomerates, shales, coals and reef limestones as well as turbiditic slope sediments (Gosau Group, 5) developed in piggy-back basins on top of the moving Austroalpine nappes (Faupl and Wagreich 2000).

The **Austroalpine Unit** forms a complex nappe stack that includes all three types of basement discussed above as well as its Mesozoic cover. Based on the structural position this unit is subdivided into Lower and Upper Austroalpine subunits that both are further subdivided into nappe systems (Schmid et al. 2004, Froitzheim et al. 2008). A schematic column, including the metamorphic grades according to Schuster et al. (2004), is given in Fig. 1.5.

The Lower Austroalpine Subunit (17) was derived from the continental margin towards the Penninic Ocean and was affected by extension and nappe stacking during the opening and closing of this domain, respectively. Being the structurally lowest part of the Austroalpine nappe pile, it occurs as a thin sleeve around the Penninic windows and in the Semmering area (Lower Austria/Styria). The Lower Austroalpine nappes and tectonic zones are mostly composed of crystalline basement with Neoproterozoic to Ordovician protolith ages and Permo-Mesozoic metasediments. They were sheared-off in the middle Late Cretaceous (c. 80 Ma) and experienced a lower greenschist facies metamorphic imprint during the Alpine event.

The Upper Austroalpine Subunit represents an Eoalpine nappe pile stacked in the Early to middle Late Cretaceous (135–85 Ma). Its lowermost part is the Silvretta-Seckau Nappe System (16) consisting of a basement with a dominating Variscan metamorphic imprint and remnants of Permian to Triassic cover sequences. During the Eoalpine orogenic event, it was overprinted by sub-greenschist to amphibolite facies conditions.

To the north, the Silvretta-Seckau Nappe System is overlain by the Veitsch-Silbersberg Nappe System (14), which consists of slivers of Variscan metamorphic basement rocks and Permo-Carboniferous sequences with a greenschist facies Alpine overprint. Above, the Juvavic (13), Tirolic-Noric (12) and Bajuvaric (11) nappe systems follow. These are composed of an up to 3200-m-thick pile of unmetamorphosed or lowermost greenschist facies metamorphic Permian and Mesozoic sediments and form the Northern Calcareous Alps. Additionally, the Tirolic-Noric Nappe System comprises a thick lower greenschist facies and fossiliferous Palaeozoic sequence. These Palaeozoic rocks as well as those from the Veitsch-Silbersberg Nappe System are often referred to as the Greywacke Zone. Its mostly schistose rocks form a relatively smooth landscape, e.g. in the area around Kitzbühel (Tyrol). The thick Mesozoic piles form the Northern Calcareous Alps with the typical bright coloured cliffs in between the Montafon (Vorarlberg) and the Wienerwald (Lower Austria) (Fig. 1.7a).

To the south, the Silvretta-Seckau Nappe System is overlain by three nappe systems that form the crystalline basement axis in large parts of Austria. The Koralpe-Wölz Nappe System (15) represents the Eoalpine metamorphic extrusion wedge. Its Permo-Mesozoic cover was completely stripped off during an early phase of the Eoalpine orogenic event (in the Early Cretaceous) and therefore it consists exclusively of polymetamorphic basement nappes with a Permian amphibolite high-temperature/low pressure and an Eoalpine pressure-dominated metamorphic overprint. The metamorphic grade of the latter is greenschist to eclogite facies in the individual nappes. The Ötztal-Bundschuh Nappe System (10) shows a similar lithological composition as the Silvretta-Seckau Nappe System but is positioned above the Koralpe-Wölz Nappe System. On top, there is the Drauzug-Gurktal Nappe System (9), which is made up of a Variscan metamorphic basement, anchizonal to greenschist

Fig. 1.7 Field photographs of typical landscapes in the central Eastern Alps. **a** View from above the Salzach Valley towards the town of Bischofshofen and the Tennengebirge (highest peak 2430 m, Salzburg). The area is built up by the Tirolic-Noric Nappe system of the Austroalpine unit. Palaeozoic schistose rocks (formerly known as Greywacke zone) build up the gentle hills in the foreground, whereas the bright coloured rock walls of the Tennengebirge are formed by Triassic platform carbonates (Northern Calcareous Alps). The plateau on top of the mountains represents the Oligocene Dachstein palaeosurface. **b** View from above the Gurk Valley (Carinthia) to the villages Gurk and Straßburg in the foreground and the Saualpe and Zirbitzkogel (2396 m) in the background. The hilly area in front consists of phyllite of the Drauzug-Gurktal Nappe System, whereas the mountains in the background are built up by micaschists and paragneisses of the Koralpe-Wölz Nappe system (both are part of the Upper Austroalpine subunit). As the Gurktal region formed an ice-free oasis during the Pleistocene glaciation periods (see van Husen 2011), the relatively gentle topography in the foreground preserves several Pre-Pleistocene planation surfaces that can be used to derive stages of the uplift history. Photos from: www.alpengeologie.org

facies Palaeozoic metasedimentary sequences and by unmetamorphosed Permian to Triassic sediments (Rantitsch and Russegger 2000). Within the Ötztal-Bundschuh and Drauzug-Gurktal nappe systems, the Eoalpine metamorphic grade decreases upwards from epidote–amphibolite facies at the base to diagenetic conditions at the top of the nappe pile.

The last three nappe systems form the principal basement units in the central and southern part of the Eastern Alps.

In the west, they build up high and rough mountains in the Silvretta (Vorarlberg/Tyrol) and Ötztal (Tyrol) areas, whereas further to the east, the topography is lowering via the Niedere Tauern (Salzburg/Styria), Gurktaler Alps (Carinthia/Styria; Fig. 1.7b), Saualpe and Koralpe mountains (Styria), as far as the hills of the Bucklige Welt (Styria/Lower Austria/Styria). Dependent on the elevation, the character of the landscape formed by these units differs. Typically, isolated occurrences of Permo-Mesozoic sequences appear as bright coloured, cliffy rock formations within the rather smooth topography formed by the monotonous basement rocks.

The **Tatric Unit** (8) is a tectonic element from the Central Western Carpathians, which reaches into the territory of Austria in the hills of the Hainburger Berge (Lower Austria/Burgenland). It is built up of Variscan metamorphic basement intruded by early Carboniferous granites and a Permo-Mesozoic cover series. Today, these rocks remain little deformed and are overprinted only by anchizonal or lowermost greenschist facies conditions during the Cretaceous Eoalpine event.

In contrast to all Alpine units described above, the **Southalpine Unit** (7) is considered to represent a southern external retro-arc orogenic wedge (Schmid et al. 1996). The Southalpine nappes were mobilized in the Eocene, and in the territory of Austria, they are composed of Variscan weakly metamorphosed, fossiliferous Palaeozoic rocks and Permo-Triassic sequences. The Alpine overprint reaches anchizonal conditions. They form rough mountain ranges including the Carnic Alps (Fig. 1.8a) and the South Karawanken (Carinthia).

1.3.2.4 Slices Derived from the Neotethys Oceanic Realm

The **Meliata Unit** (6) of the Eastern Alps consists of remnants from the Neotethys Ocean. It forms only small slices originally obducted onto the Adriatic continental margin in the Middle Jurassic (c. 160 Ma). Today, it is squeezed between Austroalpine nappes in the eastern part of the Northern Calcareous Alps. The Meliata Unit comprises serpentinites, basic volcanic rocks and Triassic olistolithes as well as radiolarites embedded in Jurassic turbidites. They show a sub-greenschist facies metamorphic imprint. These rocks can be correlated to those of the Meliata zone in the Western Carpathians (Mandl 2000). Material from the Neotethys Ocean is also present as detritus in Austroalpine units included in several Cretaceous formations (Faupl and Wagreich 2000), and in the "Haselgebirge", an evaporite tectonite at the base of the Juvavic and Tirolic-Noric nappe systems (Schorn et al. 2013).

1.3.2.5 Eocene to Miocene Magmatism

The **Periadriatic intrusions** (3) comprise calc-alkaline tonalites, granodiorites and granites and minor alkaline basaltic dikes present along the Periadriatic fault (e.g. Rieserferner pluton; Deferegger Alps/Carinthia). They are late Eocene to early Oligocene (40–28 Ma) in age and related to the break-off of the subducted slab from the distal European margin at depth (Davies and von Blankenburg 1995). Their intrusion is closely associated with contemporaneous strike-slip movements along the Periadriatic fault. Miocene to Quaternary volcanic rocks (18.0–1.9 Ma) related to the **Pannonian magmatism** (4) developed in the course of the extensional tectonics during the formation of the Pannonian Basin (Fodor et al. 2008; Lukács et al. 2018). In a first phase, more acidic trachytes, dacites and andesites were extruded, whereas a second phase is characterized by alkaline andesitic to basaltic rocks, e.g. at Klöch/Styria or Pauliberg/Burgenland (cf. Chap. Geomorphological Evidence of Past Volcanic Activity in the Southeast of Austria).

1.3.2.6 Oligocene and Neogene Basins Within the Alps

Remnants of Oligocene sedimentary sequences (34–23 Ma) occur on top of the plateaus of some mountains in the eastern Northern Calcareous Alps (e.g. Hochschwab, Dachstein). They are referred to as the Augenstein Formation and represent relics of locally more than 1000-m-thick sedimentary piles of quartz-rich gravels that formed in the southern continuations of the Molasse Basin (Frisch et al. 2001; Kuhlemann 2007). Also, along the Inntal Fault, a sequence dominated by conglomerates and sandstones is preserved.

Neogene basins on top of the Alpine orogenic wedge (2) occur in the eastern part of the Eastern Alps and in the neighbouring Western Carpathians and Pannonian region. They are mostly filled by clastic sediments generated during the lateral extrusion of the Eastern Alps in the Miocene (Ratschbacher et al. 1989). Most of these basins developed as half-grabens or pull-apart basins along active faults. Some, like the Vienna Basin, show a polyphase evolution with a piggy-back geometry reactivated as a pull-apart basin (Decker et al. 2005). Of special interest is the Klagenfurt Basin, which is a foreland basin of the overriding Karawanken today (Nemes et al. 1997), but originated as a pull-apart basin (cf. Chap. Klagenfurt Basin: A Large Basin in the Alps). All basins show individual sedimentary successions, generally starting in the Carpathian (17.5 Ma), with clastic detritus from the surroundings and intercalated coal deposits (e.g. Fohnsdorf Basin, Leoben Basin, Sachsenhofer et al. 2010). The Pannonian, Styrian, Vienna and Lavanttal basins contain

Fig. 1.8 Field photographs of typical landscapes. **a** View along the Gail Valley close to Villach towards the west (Carinthia). Within the Gail Valley the Periadriatic fault is located. This fault was active in the Oligocene and during the Miocene lateral extrusion of the Eastern Alps. North of it, the Austroalpine Unit with Mount Dobratsch (2166 m) is visible. The steep rock walls at its southern side consist of Triassic platform carbonates. In the south, the Southalpine Unit composed of Palaeozoic sediments of the Carnic Alps (Ca) and Triassic carbonates in the Julic Alps (Ju) in Slovenia and Italy, is visible. **b** View eastward over the Styrian Basin with the castle of Riegersburg (c. 350 m altitude) in the foreground. The castle is located on top of Miocene volcanic rocks related to the Panonnian magmatism. The gentle hills in the surrounding are formed by Neogene sediments of the Styrian Basin. Photos from: www.alpengeologie.org

marine Badenian to Sarmatian successions with more coarse-grained fan and coastal sediments at the margins and sand and clay in the central part. In many places the sediments are interlayered by volcanic rocks related to the Pannonian magmatism. Early Pannonian sediments developed in a brackish environment. Fluvial conglomerates and sands form the youngest deposits (Harzhauser and Tempfer 2004; Gross et al. 2007).

Where Quaternary sedimentation or planation progressed (e.g. Marchfeld in the Vienna Basin east of Vienna), the basins show flat topography, but more frequently they are characterized by a smooth topography due to young uplift, erosion and a superimposed drainage system. In the Styrian Basin, the pipes of the volcanic centres form steep hills representing the highest elevations of the region (e.g. Klöch or Riegersburg/Styria; Fig. 1.8b).

1.4 Mantle Structure Beneath the Eastern Alps

The processes in the Earth's mantle have an important influence on the elevation of the Earth's surface with respect to both isostatically supported and dynamic topography (e.g. Friedrich et al. 2018). Therefore, a summary of our knowledge about the mantle structure beneath the Eastern Alps is given. Much of this information was obtained from the TRANSALP project that produced a deep seismic reflection profile in a traverse across the Alps from Munich to Treviso in 1998 and 1999 (e.g. Lüschen et al. 2004). More insight can be expected in the near future as the large-scale geophysical AlpArray project (AlpArray Seismic Network 2015) is processed at present.

The Mohorovičić discontinuity (MOHO)—the boundary between the crust and the lithospheric mantle—is at a depth of about 30–35 km to the north of the Alpine arc and becomes generally deeper towards the south (Fig. 1.9a). In the west, it dips steeply and can be traced until 60 kms depth, where it disappears below another MOHO surface. The latter is bent around an east–west trending axis with a hinge at 25–30 km depth below the Po Basin. Its northern limb dips below the Alps and the southern one below the Apennines. Towards the east, from East Tyrol onward, the gap between the MOHO surfaces decreases and finally disappears in the Pannonian Basin, where the MOHO is generally at about 26 km depth (Ziegler and Dèzes 2006). The MOHO that descends below the Alpine orogen in the north belongs to the Eurasian Plate, whereas that in the south

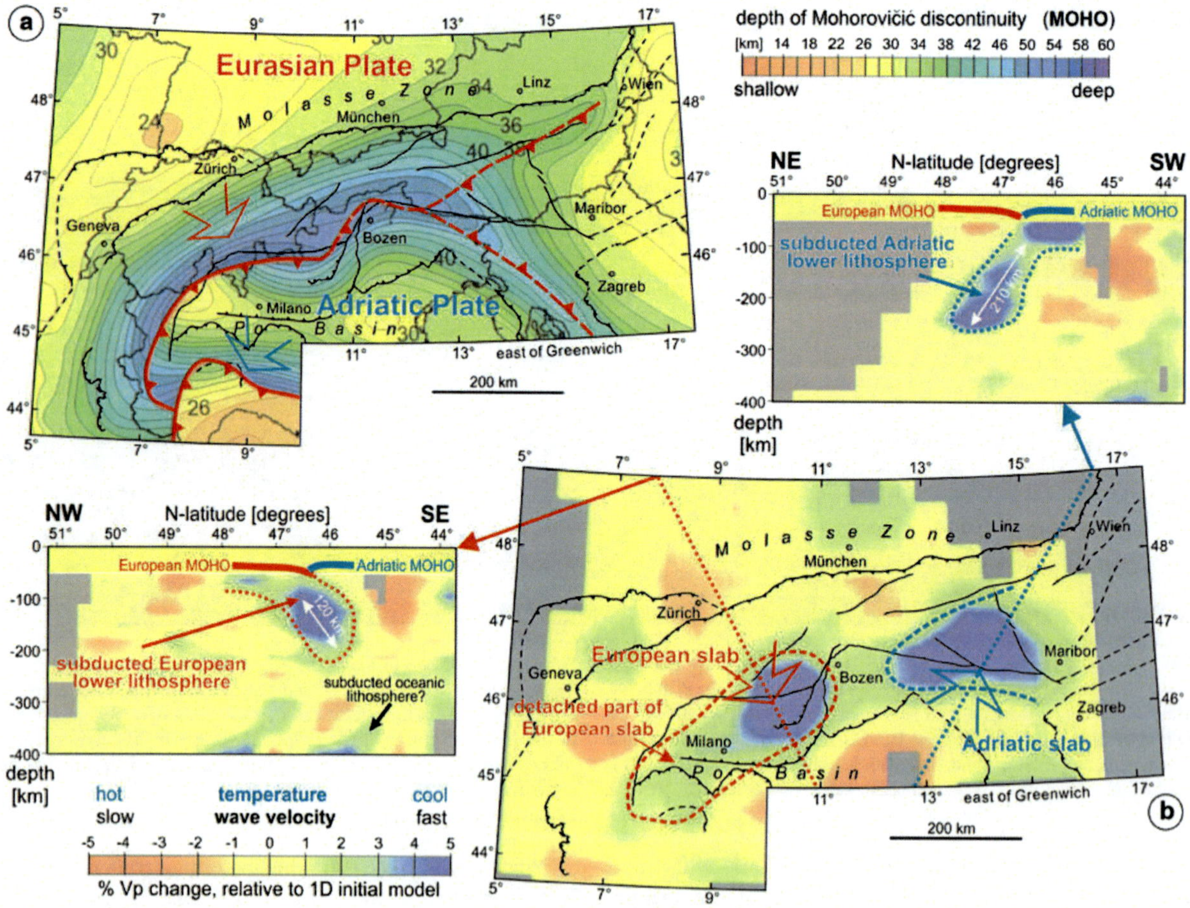

Fig. 1.9 Geophysical information on the deep structure of the Alps and surrounding areas shown for the same area as the tectonic map of Fig. 1.2c. The map **a** shows the depth of the MOHO discontinuity (Ziegler and Dèzes 2006). It may be seen that the MOHO of the Eurasian Plate dips southward and that of the Adriatic Plate dips northward below the Alpine orogenic wedge. In the west, the Adriatic Plate overrides the Eurasian Plate, whereas the situation is more diffuse in the east. Map and sections in **b** image the teleseismic mantle tomography (Lippitsch et al. 2003). Map view shows the P-wave velocity structure beneath Alps in 150 km depth. Based on the sections Schmid et al. (2004) argue for a European slab of lithospheric mantle below the western part of the Alpine arc, whreas in the eastern part an Adriatic slab is interpreted. For further explanation see text

reflects the Adriatic Plate. This indicates overriding of the Adriatic Plate onto the European Plate in the west, but a diffuse situation in the east. This may be the consequence of the Miocene east–west extension.

Teleseismic tomography data (e.g. from the ALPASS project; Brückl et al. 2007) indicate a mass of material with faster wave velocities (colder areas) below the Alpine arc (Fig. 1.9b). However, there are two cold spots, one below Lombardy and the second below East Tyrol and Carinthia. They indicate slabs of subducted lithospheric mantle hanging down deep (>200 km) into the asthenospheric mantle. While there is no doubt that the western slab is belonging to the Eurasian Plate, there is an ongoing discussion about the eastern slab, whether it belongs to the Eurasian or Adriatic Plate (Lippitsch et al. 2003; Mitterbauer et al. 2011).

In any case, these slabs are heavy and compensate the increased thickness of light continental crust accumulated in the Alpine orogenic wedge, which is indicated by the greater MOHO depth in this area. Changes in the thickness of the crust (e.g. due to erosion) as well as changes in the thickness of the lithospheric mantle (e.g. due to thermal erosion or slab break-off) have a major influence on the elevation of the area because of isostatic rebound.

1.5 Evolution of the Landscape

This section describes the evolution of surface topography and the distribution of land and sea in the Austrian territory from the Eocene to the present (Fig. 1.10). In this context, it is important to note that there is an important change in the nomenclature of geographic places in the Eocene. This is because the Mesozoic oceanic basins were completely dismembered at that time. Therefore, for the time afterwards, Paratethys is now the name for the sea to the north and on top of the Alpine orogenic wedge, whereas the Mediterranean Sea is the sea to the south.

In the middle Eocene (c. 45 Ma), the morphology, the distribution of land and sea and the spatial relationships between geographic points on the Earth's surface were very different from today. The Penninic oceanic lithosphere, formerly located to the north of the Alpine orogenic wedge, was now mostly subducted and the wedge started overriding the southern margin of the Eurasian continent with the rate of a few millimetres per year (Fig. 1.10a, 45 Ma). The flexural foreland basin (Molasse Basin) was filled by the Paratethys sea. The Paratethys got wider towards the east where it was partly floored by some remnants of Penninic oceanic lithosphere (Magura domain) in the area of the Carpathian arc and the Pannonian Basin. To the north of the Molasse Basin, flat low-lying topography characterized the southern margin of the European continent. The climate was warm and forests covered the land. Further south, in the region of the Alps, there was an island arc, but little is known about the distribution of land and sea in this area. Thermochonological data indicate that most of the Alpine metamorphic units were still at several kilometres depth and only a few isolated areas were already exhumed in the earliest Palaeogene (Hejl 1997). The preservation of up to 50 Ma old low-temperature geochronological ages in these regions suggests that erosion rates and probably topography were generally low.

During the late Eocene and Oligocene (37.8–23.0 Ma), the Alpine orogenic wedge moved continuously northward causing topographic loading at the European southern margin (Frisch et al. 1998). The result was continuing subsidence and a northward shift of the northern shoreline and of the axis of the Paratethys sea arm that paralleled the orogenic front (Fig. 1.10b; 37 Ma). In the late Eocene, the subducted lithospheric mantle slab from which all the Alpine nappes were detached, broke off from the Eurasian plate some hundreds of kilometres below the western parts of the Alps and sank into the asthenospheric mantle. This slab break-off (von Blankenburg and Davies 1995) caused important changes to many tectonic and geomorphic processes. Due to isostatic rebound from the removed heavy burden at depth, first rapid surface uplift of the range began and a moderate mountain range developed between the Western Alps and around the Inn Valley (Fig. 1.10c; 30 Ma). The Molasse Basin immediately to the north of this region also rose and marine sedimentation terminated west of Munich in the late Oligocene (Fig. 1.10d; 23 Ma). The eastern part of the Alps stayed at low altitude and formed a low-lying peninsula into the Paratethys Sea. This peninsula included the Western Carpathians, which also rose above sea level at this time. The build up of topography caused the onset of erosion and exhumation of many rocks that lie at the surface today (Kuhlemann et al. 2001). The southern part of the Eastern Alps (areas in Styria, Carinthia and Burgenland) was hilly, with rivers draining towards the north where the area of the Northern Calcareous Alps formed a coastal plain with fluvial gravel fans (Frisch et al. 2000, 2001). Remnants of these gravel plains (Augenstein Formation) are still preserved on top of the plateaus of the Northern Calcareous Alps, for example, on the Dachstein Plateau. Along the axis of the peninsula, the Periadriatic fault was active with a sinistral sense of shear until c. 30 Ma (Mancktelow et al. 2001). Granites intruded into the fault zone and especially in the western part, volcanoes erupted. Along the southern margin of the Bohemian Massif between Linz and Amstetten, the coast line was structured with sand beaches and cliffs that formed along reactivated Variscan faults. Further east, mud

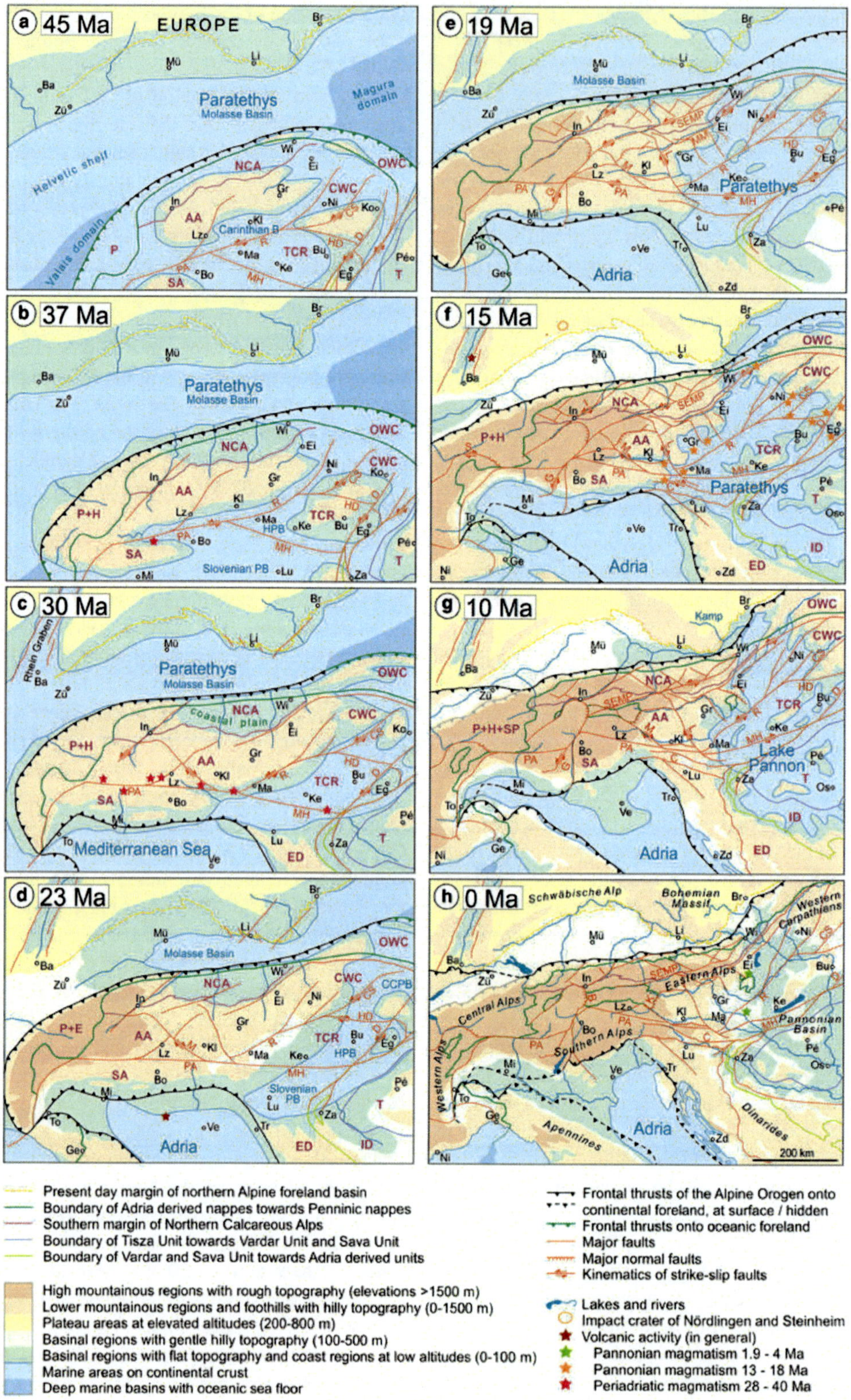
a 45 Ma
EUROPE
Paratethys
Molasse Basin
Magura domain
Helvetic shelf
Valais domain
b 37 Ma
Paratethys
Molasse Basin
Slovenian PB
c 30 Ma
Paratethys
Molasse Basin
Rhein Graben
coastal plain
Mediterranean Sea
d 23 Ma
Molasse Basin
Slovenian PB
Adria
e 19 Ma
Molasse Basin
Paratethys
Adria
f 15 Ma
Paratethys
Adria
g 10 Ma
Kamp
Lake Pannon
Adria
h 0 Ma
Schwäbische Alp
Bohemian Massif
Western Carpathians
Eastern Alps
Central Alps
Southern Alps
Western Alps
Pannonian Basin
Apennines
Adria
Dinarides
200 km
Present day margin of northern Alpine foreland basin
Boundary of Adria derived nappes towards Penninic nappes
Southern margin of Northern Calcareous Alps
Boundary of Tisza Unit towards Vardar Unit and Sava Unit
Boundary of Vardar and Sava Unit towards Adria derived units
High mountainous regions with rough topography (elevations >1500 m)
Lower mountainous regions and foothills with hilly topography (0-1500 m)
Plateau areas at elevated altitudes (200-800 m)
Basinal regions with gentle hilly topography (100-500 m)
Basinal regions with flat topography and coast regions at low altitudes (0-100 m)
Marine areas on continental crust
Deep marine basins with oceanic sea floor
Frontal thrusts of the Alpine Orogen onto continental foreland, at surface / hidden
Frontal thrusts onto oceanic foreland
Major faults
Major normal faults
Kinematics of strike-slip faults
Lakes and rivers
Impact crater of Nördlingen and Steinheim
Volcanic activity (in general)
Pannonian magmatism 1.9 - 4 Ma
Pannonian magmatism 13 - 18 Ma
Periadriatic magmatism 28 - 40 Ma

◀ **Fig. 1.10** Palaeogeographic reconstructions showing the landscape evolution of the Eastern Alps and surrounding areas since the Eocene. Changes in the spatial relationships between geographic points at the Earth's surface are indicated by geologic marker lines (see legend) and selected cities. The figure was constructed from the integrated information of Hejl (1997), Frisch et al. (2001), Rögl (2001), Wagner et al. (2010, 2011), Legrain et al. (2015), Kováč et al. (2016), Bartosch et al. (2017), Dertnig et al. (2017), Schmid et al. (2017) and several others. Explanation see text. Abbreviations: cities (Black) Ba = Basel, Bo = Bozen, Br = Brno, Bu = Budapest, Eg = Eger, Ei = Eisenstadt, Ge = Genova, Gr = Graz, In = Innsbruck, Ke = Keszethely, Kl = Klagenfurt, Ko = Košice, Li = Linz, Lu = Lubliana, Lz = Lienz, Ma = Maribor, Mi = Milano, Mü = München, Ni = Nice, Nt = Nitra, Os = Osijek, Pé = Pécs, To = Torino, Tr = Trieste, Ve = Venezia, Wi = Wien, Za = Zagreb, Zd = Zadar, Zü = Zürich. Major faults (red) B = Brenner normal fault, C = Celje fault, CS = Central Slovak fault, D = Darno fault, G = Giudicaria fault, HD = Hurbanovo-Diósjenő fault, I = Inntal fault, K = Katschberg normal fault, L = Lavanttal fault, M = Mölltal fault, MM Mur-Mürz fault, PA = Periadriatic fault, R = Rába fault, S = Simplon fault, SEMP = Salzach-Ennstal-Mariazell-Puchberg fault. Major geologic units (violett) CAA = Austroalpine nappes of the Central Eastern Alps, CWC = Central Western Carpathians, ED = External Dinarides, H = Helvetic nappes, ID = Internal Dinarides, NCA = Northern Calcareous Alps, OWC = Outer Western Carpathians, P = Penninic nappes, SA = Southalpine Unit, SP = Subpenninic nappes, T = Tisza Unit, TCR = Transdanubian Central Range. Marine basins (blue) HPB = Hungarian Palaeogene Basin, SPB = Slovenian Palaeogene Basin, CCPB = Central Carpathian Palaeogene Basin

plains with mangrove forests existed. The hinterland was drained by a precursor of the Kamp River, which entered via the Horn Basin (Steininger and Steiner 2005).

The evolution during the early Miocene (Egerian–Eggenburgian–Ottnangian, 23.0–17.5 Ma) was characterized by ongoing convergence between the Adriatic and Eurasian plates and west-directed subduction of the last remnants of Penninic oceanic lithosphere in the east. This subduction zone rolled rapidly eastward, opening up space within the Carpathian arc (Royden et al. 1983, Ren et al. 2012). Continental units coming in from the sides filled this space, for example the Alpine orogenic wedge, which experienced eastward lateral extrusion (Ratschbacher et al. 1991, Robl et al. 2008a, Wölfler et al. 2011). This lateral extrusion caused formation of large orogen-parallel strike-slips faults (Fig. 1.10e; 19 Ma) with sinistral motion in the north of the range (e.g. Inntal fault, SEMP fault, Mur-Mürz fault) and dextral motion in the south (e.g. Lavanttal fault, Mölltal fault, Periadriatic fault; Fig. 1.8a). North–south shortening in front of the Adriatic indenter caused thickening of the crust in the area around the Tauern Window and contemporaneous east–west extension led to crustal thinning in the eastern part of the Eastern Alps and the Pannonian region (Ratschbacher et al. 1989). Rapid exhumation in response to both, topography induced erosion and extension along north–south striking detachments like the Brenner and Katschberg normal faults commenced (Wölfler et al. 2011, 2012). The formerly north–south running drainage systems began to orient themselves along the newly formed east–west striking faults (Frisch et al. 1998, Robl et al. 2008a, Bartosch et al. 2017). These tectonic processes resulted in frequent changes in the distribution pattern of the sea. A global circum-equatorial seaway was established in the Eggenburgian (20.5–18.5 Ma) and caused warming of the climate (Rögl 1999) and thermophile fauna migration from the Indian Ocean. Sea level rise caused marine transgression of the Paratethys in the area around Horn in the southeastern Waldviertel. At the end of the Eggenburgian the circum-equatorial current was cut off and the climate got cooler. The Paratethys lost its contact with the Indian Ocean. However, due to further loading by the propagating Alpine orogenic wedge onto the Eurasian plate, a sea arm north of the Alps redeveloped that connected the Paratethys to the Mediterranean Sea. In the Ottnangian (at c. 18 Ma), overthrusting of the orogenic wedge onto the Eurasian Plate terminated along the northern front of the Eastern Alps, as ongoing convergence was now accommodated within the wedge. This caused further uplift of the area. Contemporaneously, the global sea level subsided due to the growth of the Antarctic ice shield. As a consequence, the western part of the Molasse Basin in Bavaria and in Upper Austria lost its marine ingression. In the east, the Paratethys became isolated and the salinity dropped dramatically, with lethal consequences for most species of the marine fauna. At that time, the Eastern Alps still had a very moderate morphology in the west and only a hilly landscape in the east.

During the middle Miocene (Carpathian–Badenian–Sarmatian, 17.5–11.6 Ma) the Tauern Window was rapidly exhumed (Wölfler et al. 2011; Fox et al. 2016) and the increasing relief of the range caused rapid erosion providing the Molasse Basins with substantial quantities of sediments (Kuhlemann et al. 2001). Cooling ages of this time are widespread in Austria, indicating final exhumation of most regions along the central axis of the Alps (Fox et al. 2016). Exhumation was accompanied by successive surface uplift that was interrupted by stagnation phases accounting for the origin of planation surfaces like the Nockberge surface (Hejl 1997). The hilly landscape in the eastern part of the Eastern Alps was now cut by distinct east–west oriented drainage systems along the newly established orogen-sized strike-slip faults (Fig. 1.10f; 15 Ma). Pull-apart basins that formed along these faults were filled by local gravel and often turned into swampy areas with alluvial forests (e.g. Tamsweg, Lavanttal, Trofaiach basins). Early forms of the Mur and Drau transported material into the Vienna and Styrian basins in the east. Due to tilting of the Koralpe (Legrain et al. 2015), large clastic fans developed along the western margin of the Styrian Basin. In the Styrian and Pannonian basins stratovolcanoes formed (e.g. Riegersburg; Gross et al. 2007). In response to the tectonic activity and sea level rise due to global temperature increase at the end of the Carpathian (c. 16.5 Ma), the Paratethys sea was re-connected to the global ocean (Rögl 1999). However, its shape was very different from before, as most of the basins that had formed on top of the orogenic wedge and the Alpine foreland basin east of the Bohemian Massif were flooded. In the Sarmatian (13.0–11.5 Ma), the Paratethys was finally isolated from the Indian Ocean and formed an inland sea reaching as far as the Caspian Sea in the east. Volcanic activity, especially in the eastern part of the Pannonian Basin, caused changes in water chemistry and faunal depletion. In the northern part of the Vienna Basin, colonies of seal populated an island at Steinberg near Zistersdorf (Lower Austria). For the last time, a shallow sea arm extended 40 km westward along the Zaya Valley into the Alpine foreland basin. Broad mudflats existed along its coast (Steininger and Steiner 2005). The western part of the Styrian Basin rose above sea level (Gross et al. 2007). Along the retreating shores, brackish environments were established. In the late Sarmatian, further uplift of the Alpine area caused the final retreat of the inland sea from the Alpine foreland basin. Large volumes of clastic material were transported into the Molasse, Vienna and Styrian basins and up to 100-m-thick gravel fan sediments were deposited. Warming climate and less fresh water input caused increased salinity and saturation in calcium carbonate so that oolithes developed along the wide coastal beaches. At the end of the Sarmatian, the climate got cooler again, the Vienna and Styrian basins fell dry and rivers cut into the sedimentary basin fill.

The time span of the late Miocene (Pannonian–Pontian, 11.6–5.3 Ma) is characterized by final retreat of the Paratethys Sea (Rögl 1999) due to inversion of the Pannonian Basin. However, in the early Pannonian (c. 11.0 Ma) a last ingression of the Paratethys still reached the Vienna and Styrian basins (Fig. 1.10g; 10 Ma). After that, the Pannonian, Vienna and Styrian basins were isolated from the Paratethys and a brackish environment was established. This stage is referred to as Lake Pannon. General consensus holds that this basin inversion was caused by the cessation of subduction underneath the Carpathian arc in connection with the onward extrusion of the Alps to the east (e.g. Bada et al. 2007). The Vienna Basin formed a large embayment as the northwestern part of Lake Pannon at this time. A proto-Danube, with a catchment area reaching to Salzburg and Upper Austria came in from the west and deposited a large alluvial fan from Krems, via Hollabrunn to Mistelbach. In the Styrian Basin, a deltaic fan formed from the Feistritz Valley, whereas there was no fan from the Mur Valley at this time (Schuster et al. 2016). In the late Pannonian (c. 8.5 Ma), Lake Pannon retreated from the Styrian and Vienna basins. In the latter, the Danube entered north of the Leiser Mountains. Large areas in the basins were vast floodplains with meandering main channels, rivulets and stagnant lakes embedded in floodplain forests. At about 7.0 Ma the floodplain forests were replaced by steppe (Harzhauser and Tempfer 2004).

Since the Miocene–Pliocene transition (c. 5.3 Ma), a remarkable new phase of rock uplift started that involved at least 500 m (e.g. Wagner et al. 2010; Legrain et al. 2014, 2015). This rock uplift caused surface uplift of the eastern end of the Alps to its present elevation and hundreds of metres of erosion in the surrounding basins, the Molasse Basin and the Styrian Basin, where the net surface uplift remained therefore small (Ebner and Sachsenhofer 1995; Genser et al. 2007). It also uplifted the Bohemian Massif by several hundreds of metres. Because of the long wavelength of this event, the causes for this rejuvenation of uplift are thought to be in the mantle (Baran et al. 2014; Legrain et al. 2014). Changes of the base level caused modelling of different landscapes dependent on the distribution of different lithologies. Rivers like the Danube (Wachau), Kamp or Thaya now began to cut gorges into the hard rocks forming the relatively flat but elevated topography of the Bohemian Massif, whereas in the area of the Molasse Basin a gentle hilly landscape developed from the weak sediments (Baumann et al. 2018). In the Alps, the stepwise uplift led to a series of planation events that are recorded as a series of successive level surfaces at different altitudes (Fig. 1.7b). The elevation of these levels is lower than that of the Miocene Nockberge surface and higher than the Pleistocene glacial terraces. They can be found in many regions at the eastern end of the Alps, for example, in the Koralpe, the Grazer Bergland, the Gurktal Alps and the Niedere Tauern (Winkler-Hermaden 1955; Wagner et al. 2011; Dertnig et al. 2017). From the Pliocene onward, important changes of the river systems can be deduced. For example, in the late Pliocene (c. 2.6 Ma), the Danube left the Vienna Basin in between the Hainburg Mountains and Leithagebirge into the Pannonian Basin, whereas the Morava entered the Pannonian Basin between the Little Carpathians and the Hainburg Mountains. Later on, in the Pleistocene, a tributary of the Morava River, captured the Danube and since that time both rivers leave the Vienna Basin through a gap between the Little Carpathians and Hainburg Mountains. Further south, another river coming in from the Pannonian Basin between Hainburg Mountains and Leithagebirge captured the Leitha River (Zámolyi et al. 2016). The river Mur that had flown across the Semmering Pass into the Vienna Basin now was captured by a small drainage near Graz and started flowing south towards the Styrian Basin.

The morphology established during the Pliocene was significantly modified during the Pleistocene glaciation periods (van Husen 2000, Robl et al. 2008b, 2015). In many parts of Austria, glaciers lowered the topography by up to several hundreds of metres. At the easten end of the Alps, however, several mountains (reaching up to 2000 m above sea level; Fig. 1.10h) were not affected by pervasive glaciation (Koralpe, parts of the Seckauer Tauern, Schneeberg, etc.). They preserved morphological features that are relicts of the surface uplift history since the Eocene (e.g. Winkler-Hermaden 1955, Frisch et al. 1998, Wagner et al. 2010, Legrain et al. 2014, 2015). Therefore, Eastern Austria forms an ideal study area for investigating the pre-Pleistocene landscape evolution of the entire Alps and for addressing modern reseach questions. Such questions are the relative importance of glacial erosion versus deep tectonic forcing on modern erosion rates and uplift in a mountain belt (e.g. Dixon et al. 2016), or the stages of mountain development from uplift to a steady state of uplift & erosion to final erosion (e.g. Hergarten et al. 2010).

References

AlpArray Seismic Network (2015) AlpArray seismic network (AASN) temporary component. AlpArray working group. Other/Seismic Network. https://doi.org/10.12686/alparray/z3_2015

Bada G, Horvath F, Dövenyi P, Szafian P, Windhoffer G, Cloetingh S (2007) Present-day stress field and tectonic inversion in the Pannonian basin. Glob Planet Change 58:165–180

Baran R, Friedrich AM, Schlunegger F (2014) The late miocene to holocene erosion pattern of the Alpine foreland basin reflects Eurasian slab unloading beneath the western Alps rather than global climate change. Lithosphere 6(2):124–131. https://doi.org/10.1130/L307.1

Bartosch T, Stüwe K, Robl J (2017) Topographic evolution of the Eastern Alps: the influence of strike-slip faulting activity. Lithosphere 9(3):384–398

Baumann S, Robl J, Prasicek G, Salcher B, Keil M (2018) The effects of lithology and base level on topography in the northern alpine foreland. Geomorphology 313:13–26

Brückl E, Bleibinhaus F, Gosar A, Grad M, Guterch A, Hrubcová P, Keller GR, Majdanski M, Sumanovac F, Tiira T, Yliniemi J, Hegedüs E, Thybo H (2007) Crustal structure due to collisional and escape tectonics in the Eastern Alps region based on profiles Alp01 and Alp02 from the ALP 2002 seismic experiment. J Geophys Res 112:B06308. https://doi.org/10.1029/2006JB004687

Davies JH, von Blanckenburg F (1995) Slab breakoff: a model of lithospheric detachment and its test in the magmatism and deformation of collisional orogens. Earth Planet Sci Lett 129:85–102

Decker K (1990) Plate tectonics and pelagic facies: late Jurassic to early cretaceous deep-sea sediments of the Ybbsitz ophiolite unit (Eastern Alps, Austria). Sed Geol 67:85–99

Decker K, Peresson H, Hinsch R (2005) Active tectonics and quaternary basin formation along the Vienna basin transform fault. Quaternary Sci Rev 24:305–320

Dertnig F, Stüwe K, Woodhead J, Stuart FM, Spötl C (2017) Constraints on the Miocene landscape evolution of the eastern Alps from the Kalkspitze region, Niedere Tauern (Austria). Geomorphology 299:24–38

Dixon JL, von Blanckenburg F, Stüwe K, Christl M (2016) Glaciation's topographic control on Holocene erosion at the eastern edge of the Alps. Earth Surf Dynam 4:895–909. https://doi.org/10.5194/esurf-4-895-2016

Ebner F, Sachsenhofer RF (1995) Paleogeography, subsidence and thermal history of the Neogene Styrian basin (Pannonian basin system, Austria). Tectonophysics 242:133–150

Faccenna C, Becker TW, Auer L, Billi A, Boschi L, Brun JP, Capitanio FA, Funiciello F, Horvàth F, Jolivet L, Piromallo C, Royden L, Rossetti F, Piromallo C (2014) Mantle dynamics in the Mediterranean. Rev Geophys 52(3):283–332

Faupl P, Wagreich M (1992) Cretaceous flysch and pelagic sequences of the Eastern Alps: correlations, heavy minerals, and paleogeographic implications. Cretaceous Res 13:387–403

Faupl P, Wagreich M (2000) Late Jurassic to Eocene palaeogeography and geodynamic evolution of the eastern Alps. Mitt Österr Geol Ges 92:79–94

Finger F, Schubert G (2015) Die Böhmische Masse in Österreich: was gibt es Neues? Abh Geol BA 64:167–179

Finger F, Hanžl P, Pin C, Von Quadt A, Steyrer HP (2000) The Brunovistulian: avalonian precambrian sequence at the eastern end of the central European variscides? Geol Soc Lon Spec Pub 179 (1):103–112

Fodor LI, Gerdes A, Dunkl I, Koroknai B, Pècskay Z, Trajanova M, Horwàth P, Vrabec M, Jelen B, Balogh K, Frisch W (2008) Miocene emplacement and rapid cooling of the Pohorje pluton at the Alpine-Pannonian-Dinaric junction, Slovenia. Swiss J Earth Sci 101:255–271

Fox M, Herman F, Willett SD, Schmid SM (2016) The exhumation history of the European Alps Inferred from linear inversion of thermochronometric data. Am J Sci 316(6):505–541. https://doi.org/10.2475/06.2016.01

Franke W (2000) The mid-European segment of the variscides: tectonostratigraphic units, terrane boundaries and plate tectonic evolution. Geol Soc Lond Spec Pub 179:35–61. https://doi.org/10.1144/GSL.SP.2000.179.01.05

Friedrich AM, Bunge HP, Rieger SM, Colli L, Ghelichkhan S, Nerlich R (2018) Stratigraphic framework for the plume mode of mantle convection and the analysis of interregional unconformities on geological maps. Gondwana Res 53:159–188. https://doi.org/10.1016/j.gr.2017.06.003

Frisch W (1979) Tectonic progradation and plate tectonic evolution of the Alps. Tectonophysics 60:121–139

Frisch W, Kuhlemann J, Dunkl I, Brügel A (1998) Palinspastic reconstruction and topographic evolution of the Eastern Alps during late tertiary tectonic extrusion. Tectonophysics 297:1–15. https://doi.org/10.1016/S0040-1951(98)00160-7

Frisch W, Székely B, Kuhlemann J, Dunkl I (2000) Geomorphological evolution of the eastern Alps in response to miocene tectonics. Z Geomorph 44(1):103–138

Frisch W, Kuhlemann J, Dunkl I (2001) The Dachstein paleosurface and the Augenstein formation in the northern Calcareous Alps—a mosaic stone in the geomorphological evolution of the eastern Alps. Int J Earth Sci 90:500–518

Fritz H, Dallmeyer RD, Neubauer F (1996) Thick-skinned versus thin-skinned thrusting: rheology controlled thrust propagation in the variscan collisional belt (The southeastern Bohemian Massif, Czech Republic—Austria). Tectonics 15(6):1389–1413

Froitzheim N, Manatschal G (1996) Kinematics of Jurassic rifting, mantle exhumation, and passive margin formation in the Austroalpine and Penninic nappes (eastern Switzerland). Geol Soc Am Bull 108:1120–1133

Froitzheim N, Schmid SM, Frey M (1996) Mesozoic paleogeography and the timing of eclogite facies metamorphism in the Alps: a working hypothesis. Eclogae Geol Helv 89:81–110

Froitzheim N, Plasienka D, Schuster R (2008) Alpine tectonics of the Alps and western Carpathians. In: McCann T (ed) The geology of

central Europe. The Geological Society of London. https://doi.org/10.1144/CEV2P.6
Fuchs G (1976) Zur Entwicklung der Böhmischen Masse. Jb Geol BA 119:45–61
Fuchs G, Matura A (1976) Zur geologie des Kristallins der Böhmischen Masse. Jb Geol BA 119:1–43
Fügenschuh B, Seward D, Mancktelow N (1997) Exhumation in a convergent orogen: the western Tauern window. Terra Nova 9:213–217
Genser J, Neubauer F (1989) Low angle normal faults at the eastern margin of the Tauern window (Eastern Alps). Mitt Österr Geol Ges 81:233–243
Genser J, Cloetingh SAPL, Neubauer F (2007) Late orogenic rebound and oblique Alpine convergence: new constraints from subsidence analysis of the Austrian Molasse basin. Glob Planet Change 58:214–223
Gerdes A, Wörner G, Henk A (2000) Post-collisional granite generation and HT-LP metamorphism by radiogenic heating: the variscan south bohemian Batholith. J Geol Soc 157(3):577–587
Gross M, Fritz I, Piller WE, Soliman A, Harzhauser M, Hubmann B, Moser B, Scholger R, Suttner TJ, Bojar BP (2007) The neogene of the styrian basin—guide to excursions. Joannea Geol Paläont 9:117–193
Handy MR, Schmid SM, Bousquet R, Kissling E, Bernoulli D (2010) Reconciling plate-tectonic reconstructions of Alpine Tethys with the geological-geophysical record of spreading and subduction in the Alps. Earth Sci Rev 102:121–158
Harzhauser M, Tempfer PM (2004) Late Pannonian wetland ecology of the Vienna basin based on molluscs and lower vertebrate assemblages (Late Miocene, MN 9, Austria). Courier-Forschungsinstitut Senkenberg 246:55–68
Hauer F (1867) Geologische Übersichtskarte der österreichisch-ungarischen Monarchie nach den Aufnahmen der K.K. Geol R.-A., M. 1:576.000. 12 Blatt. Geol RA, Wien
Hejl E (1997) Cold spots during the Cenozoic evolution of the eastern Alps: thermochronological interpretation of apatite fission-track data. Tectonophysics 272:159–173
Hergarten S, Wagner T, Stüwe K (2010) Age and prematurity of the Alps derived from topography. Earth Planet Sci Lett 297:453–460
Janák M, Froitzheim N, Yoshida K, Sasinková V, Nosko M, Kobayashi T, Hirajima T, Vrabec M (2015) Diamond in metasedimentary crustal rocks from Pohorje, eastern Alps: a window to deep continental subduction. J Metam Geol 33:495–512
Janák M, Froitzheim N, Lupták B, Vrabec M, Krogh-Ravna EJ (2004) First evidence for ultrahigh-pressure metamorphism in Pohorje, Slovenia: tracing deep continental subduction in the eastern Alps. Tectonics 23:TC5014. https://doi.org/10.1029/2004TC001641
Kober L (1938) Der geologische Aufbau Österreichs. Springer, Wien
Kossmat F (1927) Gliederung des variszischen Gebirgsbaues. Abh Sächs Geol LA 1:1–39
Kováč M, Plašienka D, Soták J, Vojtko R, Oszczypko N, Less G, Ćosović V, Fügenschuh B, Králiková S (2016) Paleogene palaeogeography and basin evolution of the western Carpathians, northern Pannonian domain and adjoining areas. Glob Planet Change 140:9–27
Kroner U, Romer RL (2013) Two plates—many subduction zones: the Variscan orogeny reconsidered. Gondwana Res 24(1):298–329
Kuhlemann J (2007) Paleogeographic and paleotopographic evolution of the Swiss and eastern Alps since the Oligocene. Glob Planet Change 58:224–236
Kuhlemann J, Frisch W, Dunkl I, Szekely B (2001) Quantifying tectonic versus erosive denudation by the sediment budget: the Miocene core complexes of the Alps. Tectonophysics 330:1–23
Kurz W, Neubauer F, Genser J, Unzog W, Dachs E (2001) Tectonic evolution of Penninic units in the Tauern window during the Paleogene: constraints from structural and metamorphic geology. In: Piller WE, Rasser MW (eds) Paleogene of the eastern Alps. Österr Akad der Wissensch, Schriftenreihe der Erdwissenschaftlichen Kommission 14:347–375
Kurz W, Froitzheim N (2002) The exhumation of eclogite-facies metamorphic rocks—a review of models confronted with examples from the Alps. Int Geol Rev 44:702–743
Legrain N, Stüwe K, Wölfler A (2014) Incised relict landscapes in a never glaciated part of the Eastern Alps. Geomorphology 221:124–138. https://doi.org/10.1016/j.geomorph.2014.06.010
Legrain N, Dixon J, Stüwe K, von Blanckenburg F, Kubik P (2015) Post-Miocene landscape rejuvenation at the eastern end of the Alps. Lithosphere 7(1):3–13. https://doi.org/10.1130/L391.1
Linner M (Sep 2013) Metamorphoseentwicklung und Deckenbau des Moldanubikums mit Fokus auf den Raum Melk—Dunkelsteinerwald. In: Gebhardt H (ed) Arbeitstagung 2013 der Geol Bundesanstalt, Melk 23.-27. Wien, pp 43–56
Linzer HG, Decker K, Peresson H, Dell'Mour R, Frisch W (2002) Balancing lateral orogenic float of the eastern Alps. Tectonophysics 54:211–237
Lippitsch R, Kissling E, Ansorge J (2003) Upper mantle structure beneath the Alpine orogen from high-resolution teleseismic tomography. J Geophys Res 108. https://doi.org/10.1029/2002JB002016
Lukács R, Harangi S, Guillong, M. Bachmann O, Fodor L, Buret Y, Dunkl I, Sliwinski J, von Quadt A, Peytcheva I, Zimmerer M (2018) Early to mid-miocene syn-extensional massive silicic volcanism in the Pannonian basin (East-Central Europe): Eruption chronology, correlation potential and geodynamic implications. Earth Sci Rev. https://doi.org/10.1016/j.earscirev.2018.02.005
Lüschen E, Lammerer B, Gebrande H, Millhan K, Nicolich R, TRANSALP working group (2004) Orogenic structure of the eastern Alps, Europe, from TRANSALP deep seismic reflection profiling. Tectonophysics 388:85–102
Mancktelow NS, Stöckli DF, Grollimund B, Müller W, Fügenschuh B, Viola G, Seward D, Villa IM (2001) The DAV and Periadriatic fault system in the eastern Alps south of the Tauern window. Int J Earth Sci 90:593–622
Mandl GW (2000) The Alpine sector of the Tethyan shelf—examples of Triassic to Jurassic sedimentation and deformation from the northern Calcareous Alps. Mitt Österr Geol Ges 92:61–78
Mandl G, Ondrejickova A (1993) Radiolarien und conodonten aus dem Meliatikum im Ostabschnitt der NKA (A). Jb Geol BA 136(4):841–871
Mandl M, Kurz W, Haunzenberger C, Fritz H, Klötzli U, Schuster R (2017) Pre-Alpine evolution of the Seckau complex (Austroalpine basement / Eastern Alps): constraints from in-situ LA-ICP-MS U-Pb zircon geochronology. Lithos 296:412–430. https://doi.org/10.1016/j.lithos.2017.11.022
Matte P (1986) Tectonics and plate tectonic model for the variscan belt in Europe. Tectonophysics 126:329–374
Missoni S, Gawlick HJ (2010) Evidence for Jurassic subduction from the northern Calcareous Alps (Berchtesgaden; Austroalpine, Germany). Int J Earth Sci 100:1605–1631
Mitterbauer U, Behm M, Brückl E, Lippitsch R, Guterch A, Keller GR, Koslovskaya E, Rumpfhuber EM, Sumanovac F (2011) Shape and origin of the East-Alp slab constrained by the ALPASS teleseismic model. Tectonophysics 510:195–206
Nemes F, Neubauer F, Cloetingh S, Genser J (1997) The klagenfurt basin in the Eastern Alps: an intra-orogenic decoupled flexural basin? Tectonophysics 282:189–203
Neubauer F (2002) Evolution of late Neoproterozoic to early Palaeozoic tectonic elements in central and Southeast European Alpine mountain belts: review and synthesis. Tectonophysics 352:87–103
Neubauer F, Handler R (2000) Variscan orogeny in the eastern Alps and Bohemian Massif: how do these units correlate? Mitt Österr Geol Ges 92:35–59

Neubauer F, Genser J, Handler R (2000) The Eastern Alps: result of a two-stage collision process. Mitt Österr Geol Ges 92:117–134
Oberhauser R (1980) Der Geologische Aufbau Österreichs. Springer, Wien-New York
Peresson H, Decker K (1996) The tertiary dynamics of the northern eastern Alps (Austria): changing paleostresses in a collisional plate boundary. Tectonophysics 272:125–157
Petrakakis K (1997) Evolution of Moldanubian rocks in Austria: review and synthesis. J Metam Geol 15:203–222
Pischinger P, Kurz W, Übleis M, Egger M, Fritz H, Brosch FJ, Stingl K (2008) Fault slip analysis in the Koralm Massif (Eastern Alps) and consequences for the final uplift of "cold spots" in Miocene times. Swiss J Geosci. https://doi.org/10.1007/s00015-008-1277-x
Rantitsch G, Russegger B (2000) Thrust-related very low grade metamorphism within the Gurktal nappe complex (Eastern Alps). Jahrb Geol BA 142(2):219–225
Ratschbacher L, Frisch W, Neubauer F, Schmid SM, Neugebauer J (1989) Extension in compressional orogenic belts: the eastern Alps. Geology 17:404–407
Ratschbacher L, Merle O, Davy P, Cobbold P (1991) Lateral extrusion in the eastern Alps: Part 1. Boundary conditions and experiments scaled for gravity. Tectonics 10:245–256. https://doi.org/10.1029/90TC02622
Ren Y, Stuart GW, Houseman GA, Dando B, Ionescu C, Hegedüs E, Radovanović S, Shen Y, South Carpathian Project Working Group (2012) Upper mantle structures beneath the Carpathian–Pannonian region: implications for the geodynamics of continental collision. Earth Planet Sci Lett 349:139–152
Robl J, Stüwe K, Hergarten S, Evans L (2008) Extension during continental convergence in the eastern Alps: the influence of orogen-scale strike-slip faults. Geology 36:603–606
Robl J, Hergarten S, Stüwe K (2008) Morphological analysis of the drainage systems in the eastern Alps. Tectonopys 460:263–277
Robl J, Prasicek G, Hergarten S, Stüwe K (2015) Alpine topography in the light of tectonic uplift and glaciation. Glob Planet Change 127:34–49. https://doi.org/10.1016/j.gloplacha.2015.01.008
Rögl F (2001) Mid-Miocene circum-Mediterranean paleogeography. Ber Inst Geol Paläont Karl-Franzens-Univ Graz 4:49–59
Rögl F (1999) Mediterranean and Parathetys. Facts and hypotheses of an Oligocene to miocene paleogeography (Short overview). Geol Carpath 50(4):339–349
Royden L, Horvath F, Rumpler J (1983) Evolution of the Pannonian basin system: 1. Tectonics Tectonics 2(1):63–90. https://doi.org/10.1029/TC002i001p00063
Rupp C, Linner M, Mandl GW (2006) Erläuterungen zur Geologischen Karte von Oberösterreich 1:200.000. Geologische Bundesanstalt, Wien
Sachsenhofer RF, Gruber W, Dunkl I (2010) Das Miozän der Becken von Leoben und Fohnsdorf. J Alp Geol 53:9–38
Scharf A, Handy MR, Favaro S, Schmid SM, Bertrand A (2013) Modes of orogen-parallel stretching and extensional exhumation in response to microplate indentation and roll-back subduction (Tauern Window, Eastern Alps). Int J Earth Sci. https://doi.org/10.1007/s00531-013-0894-4
Schmid SM, Pfiffner OA, Froitzheim N, Schönborn G, Kissling E (1996) Geophysical-geological transect and tectonic evolution of the Swiss-Italian Alps. Tectonics 15:1036–1064
Schmid SM, Fügenschuh B, Kissling E, Schuster R (2004) Tectonic map and overall architecture of the Alpine orogen. Eclogae Geol Helv 97:93–117
Schmid SM, Scharf A, Handy MR, Rosenberg CL (2013) The Tauern window (Eastern Alps, Austria)—a new tectonic map, cross-sections and tectonometamorphic synthesis. Swiss J Geosc 106:1–32
Schmid SM, Kissling E, Diehl T, van Hinsbergen DJ, Molli G (2017) Ivrea mantle wedge, arc of the western Alps, and kinematic evolution of the Alps-Apennines orogenic system. Swiss J Geosc 110(2):581–612
Schönlaub HP (1979) Das Paläozoikum in Österreich. Verbreitung, Stratigraphie, Korrelation, Entwicklung und Paläogeographie nicht metamorpher und metamorpher Abfolgen. Abh Geol BA 33:1–124
Schorn A, Neubauer F, Genser J, Bernroider M (2013) The Haselgebirge evaporitic mélange in central northern Calcareous Alps (Austria): Part of the Permian to lower Triassic rift of the Meliata ocean? Tectonophysics 583:28–48
Schuster R, Stüwe K (2008) Permian metamorphic event in the Alps. Geology 36:603–606
Schuster R, Stüwe K (2010) Die Geologie der Alpen im Zeitraffer. Mitt Nat Ver Steiermark 140:5–21
Schuster R, Koller F, Hoeck V, Hoinkes G, Bousquet R (2004) Explanatory notes to the map: metamorphic structure of the Alps—metamorphic evolution of the eastern Alps. Mitt Österr Miner Ges 149:175–199
Schuster R, Nievoll J, Rupp C, Ćorić S, Ilickovic T (2016) Neogene sedimente und landschaftsentwicklung im Umfeld der Kartenblätter GK50 Blatt 103 Kindberg und 135 Birkfeld. Arbeitstagung Der Geologischen Bundesanstalt 2015:127–143
Schuster R, Daurer A, Krenmayr HG, Linner M, Mandl GW, Pestal G, Reitner JM (2014) Rocky Austria. Geology of Austria—short and colorful. 2nd edn. Geologische Bundesanstalt, Wien
Sedgwick A, Murchison RI (1832) XVIII. A sketch of the structure of the Eastern Alps; with sections. Trans Geol Soc Lond 2–3:301–420. London
Stampfli GM, Borel GD (2004) The transmed transsects in space and time: constraints on the paleotectonic evolution of the Mediterranean domain. In: Cavazza W, Roure F, Spakman W, Stampfli GM, Ziegler PA (eds) The TRANSMED Atlas: the Mediterranean region from crust to mantle. Springer, p 141
Steininger H, Steiner E (2005) Meeresstrand am Alpenrand—Molassemeer und Wiener Becken. Verlag publication PN 1 Bibliothek der Provinz (Weitra)
Stüwe K, Homberger R (2012) High above the Alps. A bird's eye view of geology. Weishaupt, Gnas
Stüwe K, Schuster R (2010) Initiation of subduction in the Alps: continent or ocean? Geology 38(2):175–178
Suess E (1912) Die moravischen fenster und ihre Beziehung zum Grundgebirge des Hohen Gesenke. Denkschr Kk Akad Wiss, Math-Naturwiss Cl 88:541–631
Suess E (1909) Das Antlitz der Erde, vol 3. Temsky & Freytag, Prag-Wien-Leipzig
Tenczer V, Stüwe K (2003) The metamorphic field gradient in the eclogite type locality. J Metam Geol 21:377–393
Thöni M (2006) Dating eclogite-facies metamorphism in the eastern Alps—approches, results, interpretations: a review. Mineral Petrol 88:123–148
Tollmann A (1977) Geologie von Österreich, vol 1. Die Zentralalpen. Deuticke, Wien
van Husen D (2000) Geological processes during the quaternary. Mitt Österr Geol Ges 92:135–156
van Husen D (2011) Quaternary glaciations in Austria. In: Ehlers J, Gibbard PL, Hughes PD (eds) Development in quarternary sciences, vol 15. pp 15–28
von Blankenburg F, Davies HJ (1995) Slab breakoff: a model for syncollisional magmatism and tectonics in the Alps. Tectonics 14:120–131
von Raumer JF, Bussy F, Schaltegger U, Schulz B, Stampfli GM (2013) Pre-Mesozoic Alpine basements—their place in the European paleozoic framework. Geol Soc Am Bull 125(1–2):89–108

Wagner T, Fabel D, Fiebig M, Häuselmann P, Sahy D, Sheng X, Stüwe K (2010) Young uplift in the nonglaciated parts of the eastern Alps. Earth Planet Sci Lett 295:159–169

Wagner T, Fritz H, Stüwe K, Nestroy O, Rodnight H, Hellstrom J, Benischke R (2011) Correlation of cave levels, stream terraces and planation surfaces along the river Mur—timing of landscape evolution along the eastern margin of the Alps. Geomorphology 134:62–78. https://doi.org/10.1016/j.geomorph.2011.04.024

Wessely G (2006) Geologie der Österreichischen Bundesländer—Niederösterreich. Geologische Bundesanstalt, Wien

Winkler-Hermaden A (1955) Ergebnisse und Probleme der quartären Entwicklungsgeschichte am östlichen Alpensaum außerhalb der Vereisungsgebiete. Denkschr Akad Wiss, Math-Naturw Kl 110 (8):1–180

Wölfler A, Stüwe K, Danisik M, Evans NJ (2012) Low temperature thermochronology in the eastern Alps: implications for structural and topographic evolution. Tectonophysics. https://doi.org/10.1016/j.tecto.2012.03016

Wölfler A, Kurz W, Fritz H, Stüwe K (2011) Lateral extrusion in the eastern Alps revisited: refining the model by thermochronological, sedimentary and seismic data. Tectonics 30 TC4006. https://doi.org/10.1029/2010TC002782

Zámolyi A, Salcher B, Draganits E, Exner U, Wagreich M, Gier S, Fiebig M, Lomax J, Surányi G, Zámolyi F (2016) Latest Pannonian and quaternary evolution at the transition between eastern Alps and Pannonian Basin: new insights from geophysical, sedimentological and geochronological data. Int J Earth Sci 106(5):1695–1721

Ziegler PA, Dèzes P (2006) Crustal evolution of western and central Europe. Geol Soc Lond Mem 32(1):43–56

Ralf Schuster is a research scientist at the Geological Survey of Austria and private lecturer at the University of Vienna. His research focuses on geological mapping, geochronology and tectonics of the Eastern Alps and surrounding regions.

Kurt Stüwe is a professor for Earth Sciences at the University of Graz in Austria. He has worked in many branches of the Earth Sciences and has conducted research projects on mountain forming processes on all continents. In recent years he has focused on neotectonic- and geomorphological questions, mostly with a modelling approach.

Geomorphological Landscape Regions of Austria

2

Gerhard Karl Lieb and Christine Embleton-Hamann

Abstract

This chapter provides an overview of the landscapes of Austria and a general background for the individual regional chapters of Part 2 of the book. The three major landscape units of Austria are the Eastern Alps, comprising about 63% of the territory, the Alpine forelands in the north and southeast of the mountain belt (27%), and the highlands of the Bohemian Massif in the north (10%). For the designation of smaller-scale regions, the geology of these three structural units with specific attention to lithology is discussed. Based on maps of elevation, relative relief and slope inclination, four relief types are defined. The combination of geology and relief types results in ten geomorphological landscape regions. These units are finally portrayed, complemented by brief descriptions of climate and land use. The main focus of these descriptive accounts are the Quaternary and contemporary geomorphological processes and characteristic landform assemblages with references to the chapters of part 2 in which they are covered in detail.

Keywords

Austria • Tectonic units • Lithology • Relief types • Geomorphological landscape units • Landforms

2.1 Introduction: Defining the Geomorphological Landscape Units of Austria

National maps of physiographic or geological regions are abundant. They are typically based on the spatial distribution of geological units (lithology, age of rocks and tectonics or a combination of these aspects) and relief. Classic divisions of this type are frequently integrated into school curricula, thus turning into common national knowledge. Austrian school books also offer simple divisions of the national territory based on geology and relief. Naturally, the respective scientific literature presents far more complex divisions of the Austrian landscape (e.g. Fink 1986), some of them with a special purpose using only selected elements of the range of earth surface features. Regardless of the differences between them, almost all of them start with three main units, the Bohemian Massif, the Alps and the Alpine foreland, but differ in their designation of subunits.

We will start by evaluating the six structural landscape divisions, outlined in Chap. "Geological and Tectonic Setting of Austria" and Fig. 2.1a for their suitability of defining a set of geomorphological landscape units of Austria (Sect. 1.1). This will be followed by the introduction of a classification of topography and available relief, based on the spatial distribution of elevation, relative relief and slope inclination in Austria (Sect. 1.2). The combination of derived geological units and relief (Sect. 1.3) will provide a first-order system of landscape units. Further factors pertinent to geomorphological landscape units, like the Quaternary heritage of glacial landforms or the contemporary human inference with geomorphological processes, will be included in the final portraits of the derived geomorphological landscape units in Sects. 2–4, complemented by brief descriptions of climate and land use.

Lastly, the overview of Austrian geomorphological landscape regions also provides a general background for the individual regional studies of Part 2, which were selected on

G. K. Lieb (✉)
Institute of Geography and Regional Science, University of Graz, Heinrichstrasse 36, 8010 Graz, Austria
e-mail: gerhard.lieb@uni-graz.at

C. Embleton-Hamann
Department of Geography and Regional Research, University of Vienna, Universitätsstr. 7, 1010 Vienna, Austria

C. Embleton-Hamann (ed.), *Landscapes and Landforms of Austria*, World Geomorphological Landscapes,
https://doi.org/10.1007/978-3-030-92815-5_2

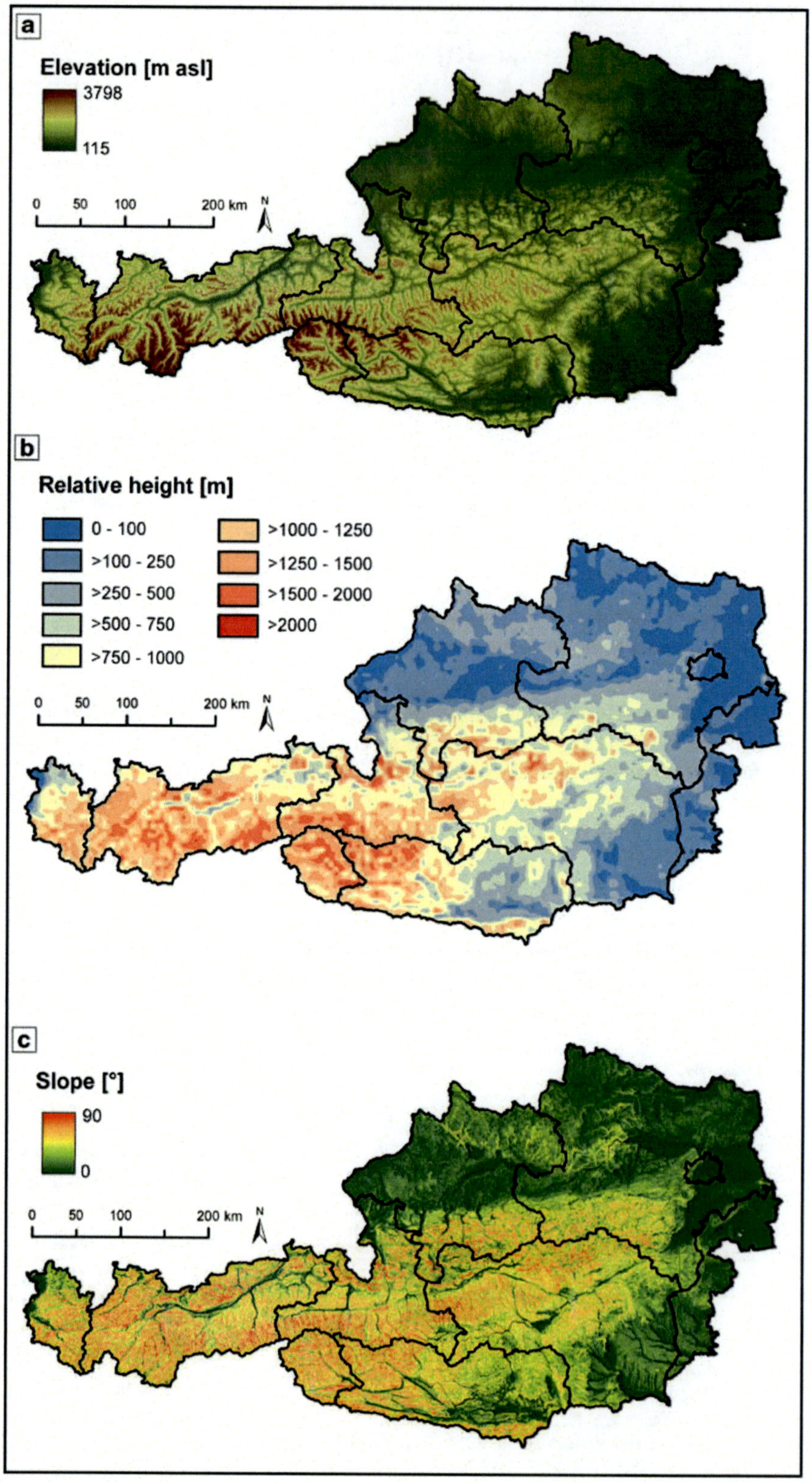

Fig. 2.1 Elevation **a**, relative height **b** and slope **c** in Austria. Elevation ranges from 3797 m asl (summit of the Großglockner) to 115 m asl (in the northeastern plains at the border to Hungary). (Geoland 2020. License CC BY 4.0 [Creative commons attribution 4.0 international]; compilation and cartography: C. Bauer)

their own criteria. These included (i) their scientific interest with respect to landforms or geomorphological process studies and (ii) the high scenic quality of their landscape making them a valuable natural heritage.

2.1.1 Geological Units of Austria

In Chap. "Geological and Tectonic Setting of Austria", Fig. 2.1a, Schuster and Stüwe present a map of Austrian landscapes that is based on the plate tectonic evolution of the Austrian territory. As the geological setting definitely influences the character of landscapes, especially the overall distribution of uplands, lowlands, basins and valley systems, we start by examining the six tectonically derived landscape units of Chap. "Geological and Tectonic Setting of Austria" with respect to their suitability for defining geomorphological landscape regions. Thereby, we will focus on their rock content, as the weathering resistance of the various rock types strongly influences the efficiency of erosional processes.

(1) *Bohemian Massif*: The northernmost geological unit within the territory of Austria is part of the pre-Variscan and Variscan metamorphosed basement of the European plate. On the surface, it belongs to two tectonic units, the Moravian Unit and the Moldanubian Unit, which have been uplifted during the Alpine orogeny. The massif consists predominantly of granites and granodiorites that intruded into the Moldanubian Unit in the Palaeozoic (South Bohemian Pluton) and are accompanied by a wide range of metamorphic rocks. Granites are regarded as typical regional rocks and play a role in cultural heritage (e.g. in traditional architecture).

(2) *Alpine foreland*: In this area, clastic sediments of Neogene age are the dominant type of rock and do not differ significantly between the Alpine forelands in the north and southeast of Austria. From a geomorphological point of view, the resistance of the diverse Neogene sediments to weathering and erosional processes is quite similar, with the exception of cemented gravels and some spatially limited occurrences of limestones, which tend to form steeper slopes. Neogene sediments are also widespread in the inner-Alpine basins.

(3) *Foothills of the Alps*: This unit lies within the Alpine orogen and is predominantly composed of rocks belonging to the Penninic Unit, but in the far west (Vorarlberg, Bregenzerwald) rocks of the Helvetic Unit also occur. A third member is the rocks of the Allochthonous Molasse. The predominant rocks in this unit are flyschoid sequences. In physiogeographic regionalizations, this area is therefore often depicted as "Flysch Zone". The Alemannic word "flysch" refers to the fact that the rocks, predominantly unmetamorphosed sandstones and marls, tend to "flow", thus pointing to their high disposition towards mass movement processes. In our division of geomorphological landscape regions, this unit will be referred to as "Northern Prealps".

(4) *Northern Calcareous Alps*: There is hardly any geological or geomorphological regionalization of Austria that does not contain this unit. The reason for this is the remarkable appearance of the relief with steep rock faces and memorable summits, many of which rise directly above the northern longitudinal valley depression (cf. Chap. "River and Valley Landscapes") and already impressed the first travellers and land surveyors of early modern times. The Northern Calcareous Alps belong to the Austroalpine Unit and consist of a thick, only partly metamorphosed Permo-Mesozoic sedimentary cover lying on top of the pre-Alpine basement. Most of the rocks are calcareous (limestones and dolomites), and a complete sedimentary sequence of the Triassic period is very typical, whereas Jurassic and Cretaceous rocks are less widespread.

(5) *Central Eastern Alps*: This very large structural unit naturally comprises terrains of different surface morphology, and for the purpose of defining landscape regions should be, if possible, divided into subunits. Along the crest of the Central Alps, weathering-resistant rocks like paragneisses, micaschists and amphibolites of the Austroalpine basement dominate. Their areas of occurrence are interrupted by three tectonic windows, in which the Penninic and Subpenninic Units appear at the surface. In general, their rocks are equally weathering resistant.
More easily erodible schistose rocks are widespread in the Palaeozoic sequences of the overlying Austroalpine nappe systems. Their largest outcrop area follows the southern margin of the Northern Calcareous Alps over a distance of 400 km between Innsbruck in the west and the Vienna Basin in the east. This zone is of different landscape appearance, has different predominant geomorphological processes and in the older literature is referred to as Greywacke Zone. On the basis of lithology, we will therefore divide the Central Alps into two subunits, named Crest of the Central Alps and Greywacke Zone (and similar units).
Finally, the mountain relief of the southernmost region of the Central Eastern Alps is dominated by carbonate rocks and shall therefore be included into the following unit.

(6) *Southern Alps*: The narrow strip south of the Periadriatic fault belongs to the Southalpine tectonic unit.

Clearly, a classification of landscape regions based on tectonic principles needs to include a dividing line along the Periadriatic fault, as the terrains north and south of it are derived from different palaeogeographical realms. However, from a lithological point of view, the sequence of rocks, consisting of basal weakly metamorphosed Palaeozoic rocks and overlying Permo-Triassic calcareous rocks, is the same on both sides of the fault, just as characteristic landforms and general landscape appearance are. Therefore, we define the southernmost geomorphological landscape region of Austria as the region south of the Drau River. Using the term "Southern Alps" for this region is the only major deviation from the delineation of landscape units in Chap. "Geological and Tectonic Setting of Austria".

2.1.2 Relief Types

In order to give a first overview, Fig. 2.1 shows three maps depicting elevation (A), relative height (B) and slope (C). For the purpose of identifying relief types, we start with the parameter of relative height, which was calculated on the basis of a 5 × 5 km grid. Typical values of relative height suggest a set of simple relief types that fit in well with the values for elevation and slope and are described in the following.

(1) *High mountains*: The relative height of this relief type is at least 1000 m, but mostly exceeds 1250 m, and maximum values in Austria are around 2500 m. High values of relative relief are recognizably associated with high values of slope inclination. High mountains rise above the regional treeline, are the most significant relief type of the Alps and in Austria occur only there. Distinctive glacial landforms are widespread and induce accentuated crests and peaks together with rock faces, wherever the bedrock is hard enough. Furthermore, the steep terrain favours current mass wasting processes.

(2) *Mountains*: The relative height of this relief type is at least 250 m, but mostly exceeds 500 m and maximum values in Austria reach c. 1000 m. Summit areas stay below the regional treeline, except where they were cleared to gain pastures. Thus, forests are the most significant land cover type. Glacial landforms may occur where the region was reached by glaciers that had their accumulation areas in the high mountains. Mass wasting processes take place on the widespread steep slopes but, in comparison with the high-mountain system, contemporary geomorphological activity is subdued.

(3) *Hills*: The relative height of this relief type is at least 50 m, predominantly around 100 m, and never exceeds 250 m. Corresponding to low relative relief, altitudes are also low, scarcely reaching 500 m asl. The moderate to low slope inclination (green colour in Fig. 2.1c) still poses a soil erosion problem to arable farming but is favourable to grassland or tree crops. Hills in most cases are composed of clastic sediments, and their current geomorphological activity is low, except for shallow mass movements under special geological conditions.

(4) *Plains*: This relief type is rare in Austria. It is only in the northeast that it stretches over larger areas; otherwise, it is restricted to the floodplains of major rivers. Slope angles are near zero, and so is relative relief. Only where Pleistocene terraces accompany the rivers, local height differences of some 50 m may appear. Austria's plains offer optimal agricultural conditions for arable land, but also for erecting all types of socioeconomic infrastructure, which frequently results in land use conflicts. Today, the only morphological process is of the fluvial type during flood events, posing a risk to all types of land use next to river channels, in spite of country-wide engineering efforts to minimize them.

Further, more complex relief types were put forward in the literature, e.g. in the map by Seger in Borsdorf (2005). One of the many suggestions, pertaining to the inner-Alpine basins of Austria, needs to be adopted. These basins occur within the larger structural unit of the Central Alps, from which they differ due to their low position in current geomorphological process domains and in Pleistocene landform heritage. Therefore, the twofold subdivision of the Central Alps will be extended by one further unit, termed "inner-Alpine basins".

A particularity of Austria's relief is the decrease of elevation and relative relief from west to east (Fig. 2.2). This applies to all relief types with the exception of plains, which in western Austria are only present in the form of valley bottoms along major rivers (e.g. Rhine and Inn). The reason for this is tectonic processes that started in the early Miocene and led to an eastward extrusion of crustal blocks (c.f. Chap. "Geological and Tectonic Setting of Austria", Sect. 5).

2.1.3 Geomorphological Landscape Regions

Figure 2 combines the geological units with the relief types. The geological units, denoted with the terms and subdivisions introduced in Sect. 1.1, are shown from left to right, while relative relief and associated relief types are indicated

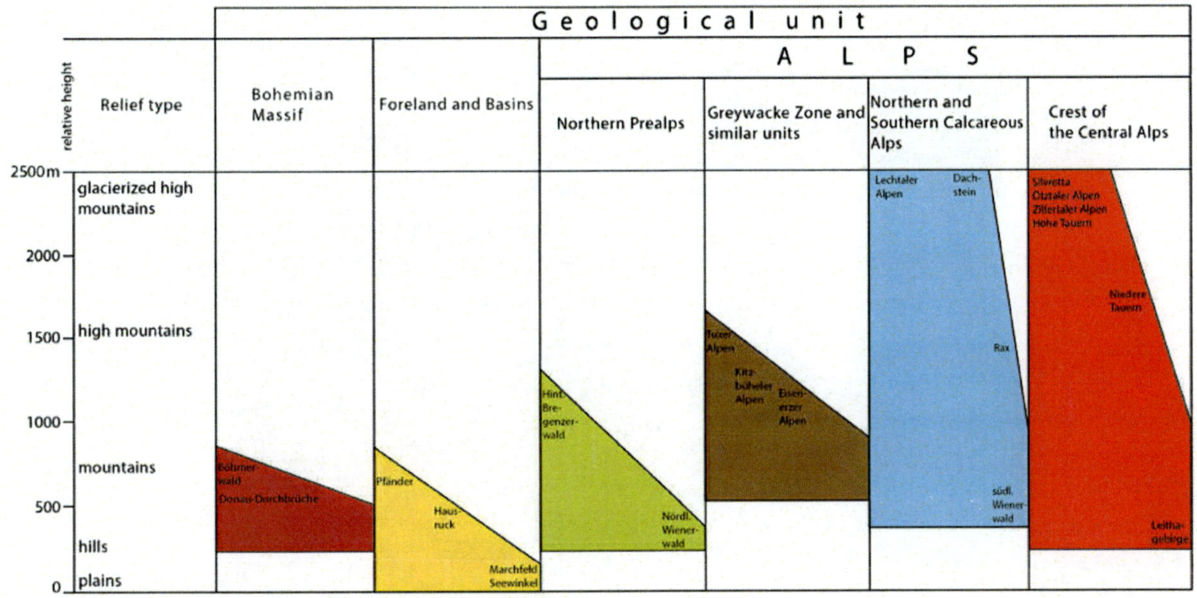

Fig. 2.2 Geomorphological landscape regions derived from geological units and relief types as explained in the text (drawing: V. Damm)

on the *y*-axis. Thus, the coloured areas within each column give combined information on geology and relief, and additionally via their right-hand slant on the particular decrease in relative relief from west (left) to east (right) within Austria. Note that the headline of the table shows the three major units, namely the Bohemian Massif, the Alpine forelands and the Alps, which are present in almost all landscape divisions suggested for the Austrian territory. Figure 2.2 is of course a schematic representation; major deviations will be mentioned in the descriptive account of each unit in Sects. 2–4.

The spatial distribution of the geomorphological landscape regions is shown in Fig. 2.3 in the same colours as in Fig. 2.2. Considering all the suggested subdivisions and finally separating the summary depiction of the Northern and the Southern Calcareous Alps on the basis of their lithology in Figs. 2.2 and 2.3, the total number of geomorphological landscape regions in Austria is ten. In the following, they will be portrayed from north to south, which implies an arrangement of units that is already indicated by the black dividing lines and the capitalized black lettering in the map (Fig. 2.3).

Section 2 deals with the Bohemian Massif as the northernmost major unit. Section 3 is devoted to the second major unit and contains individual accounts of the northern forelands, the southeastern forelands and the Vienna Basin. The Alps are covered in Sect. 4, divided into Northern, Central and Southern Alps. The Northern and the Central Alps have further subunits: the Northern Alps comprise the small belt of the Northern Prealps together with the Northern Calcareous Alps, and the Central Alps, by including the inner-Alpine basins, have a threefold subdivision. Place names mentioned in the following section are, if not otherwise indicated, shown in Fig. 2.4. The reference period for all stated temperature and precipitation means is 1971–2000.

2.2 Bohemian Massif

The Bohemian Massif exhibits the basement rocks and tectonic structures of the Variscan orogeny. It is structurally divided into the western Moldanubian Unit and the eastern Moravian Unit, overthrust by the former at a low angle. On Austrian territory, the Moldanubian Unit dominates; the Moravian Unit is only present in a c. 12 to maximal 20 km broad strip at the eastern margin. While the Moravian Unit is dominated by gneisses and other metamorphic rocks, the Moldanubian Unit mostly consists of granites, adhering to a large batholith, the so-called South Bohemian Pluton. Coarse grained granites are typical for this batholith; medium and fine grained varieties also occur but are less widespread. The granites and gneisses of the Bohemian Massif are worked in numerous quarries. Mining and processing of kaolin, a weathering product of the crystalline rocks, are of economic importance.

Two large strike-slip fault systems developed during the Variscan orogeny, a right lateral one orientated NW–SE, and a left lateral one oriented almost perpendicularly in a NE–SW direction. Both became reactivated later on, especially during the Alpine orogeny. The prominent NE–SW oriented

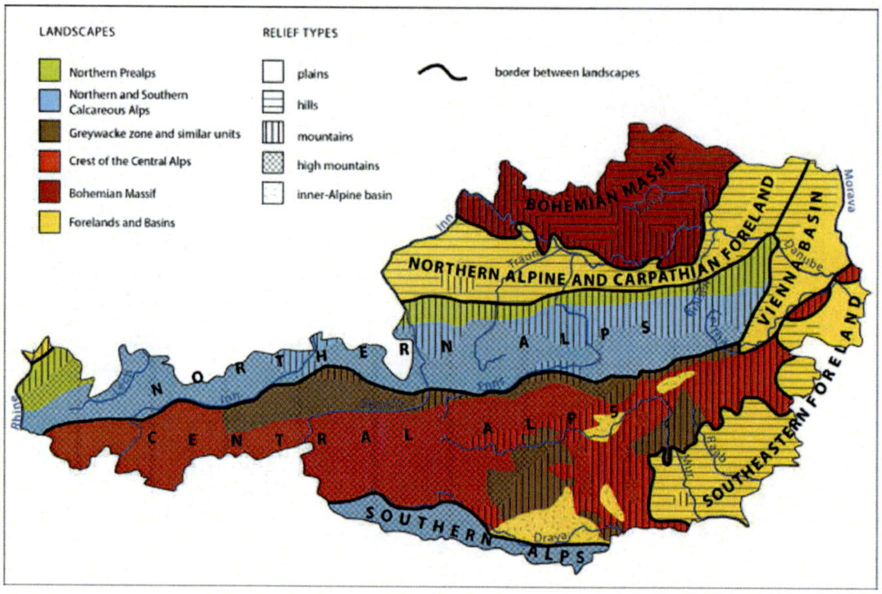

Fig. 2.3 Geomorphological landscape units and relief types of Austria. Cartography: V. Damm

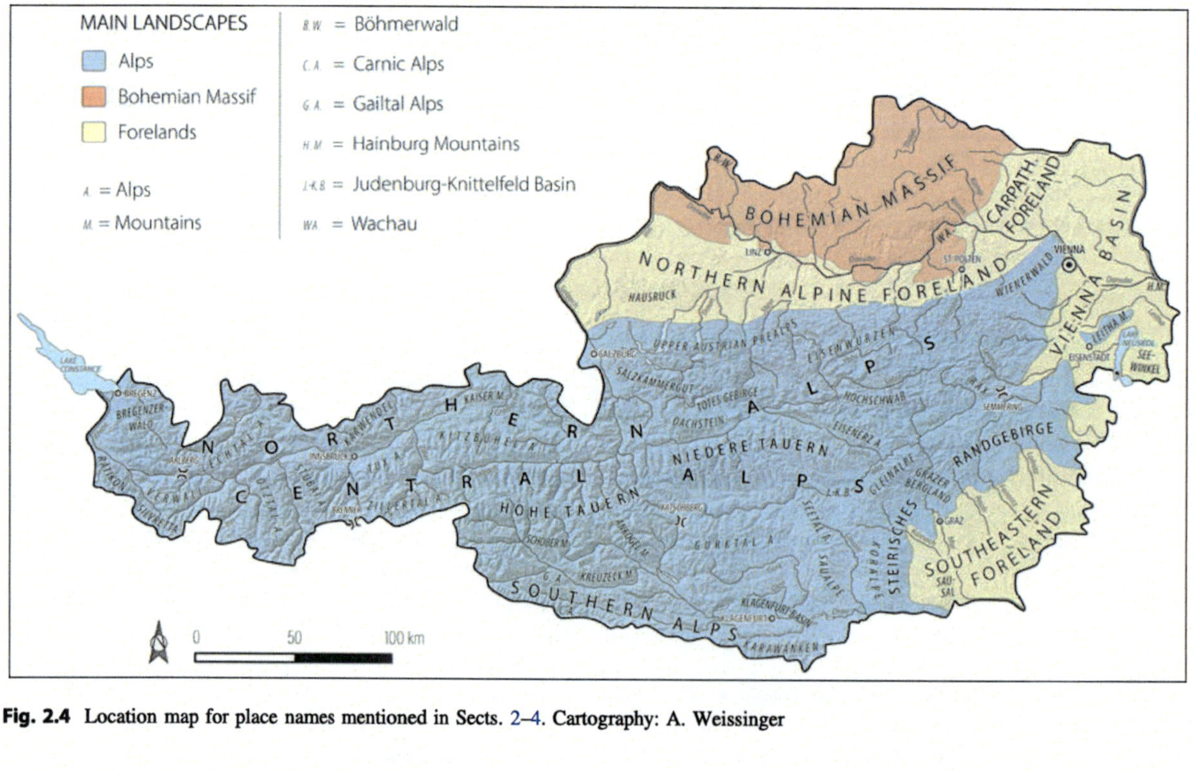

Fig. 2.4 Location map for place names mentioned in Sects. 2–4. Cartography: A. Weissinger

Diendorf fault at the eastern end of the Bohemian Massif was even active in historical times (Höck 1999).

Large parts of the Bohemian Massif appear as a hilly region, mostly extending at elevations between 400 and 600 m asl, but in the west rising stepwise to elevations of c. 1300 m asl, finally culminating at the border of the Czech Republic and Austria in 1379 m asl, the highest point of the Bohemian Massif in Austria. "Harsh" and "cool" are the key words in summary descriptions of the climatic situation. Average July temperatures are in the range of 16 to 17°.

During winter, many weather stations record 4 months of frost and something like 35 ice days. Annual precipitation on the other hand is far less uniform. As the region lies in the transition zone between maritime and continental influences, precipitation sums of 560 mm at stations in the dry east rise westwards, augmented by the higher terrain to 900 mm and finally to 1000 to 1500 mm.

At the highest elevations, forestry is the dominant land use, where it also has a long-standing tradition. At the end of the eighteenth century, a timber rafting channel was built across the continental divide in order to provide Vienna with firewood from the Bohemian forests. The landscape at lower elevations shows a mix of pastures or hay meadows and forests, whereby the forests mainly occupy the outcrops of solid bedrock (Fig. 2.5). Arable farming only becomes important in the lower eastern parts of the massif.

The geomorphic evolution of the landscape is controlled by four factors, namely by the erosion of the Variscan mountain belt to a planation surface, by the specifics of granite weathering, by strong Pleistocene downwasting in a mostly south-exposed permafrost environment with a thick active layer forming during summers and in parts by faulting.

By the middle of the Mesozoic era, the former Variscan mountain belt had been turned into a surface of low relief lying close to sea level. The warm and humid conditions of the Neogene caused an advanced decay of the granites and gneisses to great depth. During the Pliocene, the ongoing

Fig. 2.5 The typical rolling topography of the eastern part of the Bohemian highlands in Upper Austria. The area is underlain by coarse grained granites which were cut by numerous faults to a series of tilted fault blocks (Fischer 1965). Photo: L. Beckel

uplift of the Alps was accompanied by an uplift of the Bohemian Massif. Tectonic movements reactivated the old fault systems and blocks in between the faults were tilted, uplifted or subsided. Uplift also changed the formerly broad and shallow valleys of the Danube and the lower sections of its northern tributaries. Downcutting led to canyon-like valleys, whereby the Danube incision cut off four larger parts of the massif (cf. Chaps. "Wachau World Heritage Site: A Diverse Riverine Landscape" and "River and Valley Landscapes"/Sect. 4.3.3).

During the Pleistocene, a localized glaciation of the highest parts of the Bohemian Massif resulted in the formation of cirques in favourable northern positions. The generally south-facing Austrian territory was, however, mainly shaped by processes of the periglacial process domain. In many places, the old weathering mantle of the granites and gneisses was stripped down to the solid bedrock, exhuming bedrock projections (Fig. 2.5), tors and other specific granite landforms (see detailed descriptions and photos in Chap. "Geoheritage, Geotourism and Landscape Protection in Austria"/ Sect. 7.2.2). Blockfields and blockstreams were either exposed in situ or have additionally undergone downslope movement due to periglacial processes. Another important remnant of the Pleistocene is the cover of loess, of which the thickest stacks of Austria are located in the southern part of the Bohemian Massif in Lower Austria (cf. Chaps. "Wachau World Heritage Site: A Diverse Riverine Landscape" and "Sunken Roads and Palaeosols in Loess Areas in Lower Austria: Landform Development and Cultural Importance").

Today, many rivers and streams in the Bohemian Massif show a surplus of sediment, quite opposite to Alpine catchments. Again, this is caused by granite weathering. Especially, coarse grained granites decompose to grain sizes of 1 to 10 mm. This fraction is easily mobilized and causes aggradation of the riverbeds, thereby impairing freshwater ecology and aggravating the flood risk in the downstream section within the Danube floodplain (Hauer et al. 2016).

2.3 Forelands

2.3.1 Northern Alpine and Carpathian Foreland

The Northern Alpine forelands exhibit four individual landscapes, and a fifth foreland region, again with its own specific features, is that of the Carpathians. For place names in the following descriptions, see Fig. 14.1 in Chap. "Quaternary Landforms and Sediments in the Northern Alpine Foreland of Salzburg and Upper Austria" for the western part of the Alpine foreland and Fig. 11.1 in Chap. "Sunken Roads and Palaeosols in Loess Areas in Lower Austria: Landform Development and Cultural Importance" for its eastern part and the Carpathian foreland.

Hilly landscapes developed in Schlier:

The typical rock of the Molasse sediments in the Alpine foreland is the Schlier, a marly, often fine sandy siltstone. A natural monument in Upper Austria, named "Ottnangium", is devoted to this rock and its rich fossil content (cf. Sect. 7.2.3 in Chap. "Geoheritage, Geotourism and Landscape Protection in Austria"). The highest elevations of the Alpine foreland with summits at c. 800 m asl are the Hausruck and Kobernausserwald in the west, where the easily erodible Schlier is protected by a cap of Pannonian fluvial gravel. Elsewhere, the hilly Schlier landscape reaches altitudes of 300–400 m asl. A characteristic feature of the Schlier hills is their appearance as subparallel and elongated ridges between shallow valleys following the general slope. Large parts of the area of the Alpine foreland are, however, covered by Quaternary sediments.

Moraine landscape:

During the last glaciation, the Salzach Glacier in the west advanced into the foreland and left a typical till landscape, complete with drumlins, kettle holes, lakes and bogs, and bounded by a terminal moraine (for a detailed description, see Chap. "Quaternary Landforms and Sediments in the Northern Alpine Foreland of Salzburg and Upper Austria", Sect. 14.4). Further east, only isolated moraine ridges of the three previous glaciations can be found. The best preserved examples are the Mindel and Günz ramparts near Kremsmünster (Kohl 2000). The climate of the moraine landscape is influenced by moist air masses coming from the Atlantic Ocean. With an annual precipitation of 1000–1300 mm and a mean July temperature of 18.5 °C, the dominant land use is hay meadows.

Outwash terraces of the Danube tributaries coming from the Alps:

North of the Quaternary ice margin, the southern tributaries of the Danube deposited a sequence of outwash terraces (Fig. 2.6a). While along the Inn River, only the "Niederterrasse" (= lower terrace, consisting of Würm gravels) and the "Hochterrasse" (= high terrace, consisting of Riss gravels) are present, further east, starting at the Traun River, the higher and older outwash terraces of the Mindel and Günz also occur. Because of the widespread distribution of their gravels, they are named "Ältere Deckenschotter" (= older gravel sheet) and "Jüngere Deckenschotter" (= younger gravel sheet) in German. Schematic longitudinal and

Fig. 2.6 **a** Enns River (c. 3 km before its mouth) and its left bank Hochterrasse, crowned by the town of Enns. In this position, the Hochterrasse is c. 25 m high. Photo: C. Embleton-Hamann. **b**: Klippe of Staatz, situated 9 km southeast of Laa an der Thaya. The rocky outcrop is part of the Helvetic Unit of the Alpine orogeny. Photo: G. K. Lieb

transverse cross sections of the four terraces along the Traun River are shown in Figs. 3.3 and 3.5 of Chap. "The Imprint of Quaternary Processes on the Austrian Landscape".

While the lower terrace and the high terrace exhibit a flat tread, Mindel and Günz terraces at higher elevations appear as hilly landscapes, as they became entrenched by an autochthonous river network since their deposition. Especially, the older gravel sheet of the Enns River is widespread. After leaving the Alps, the palaeo-Enns turned east and only reached the Danube near the present-day mouth of the Ybbs River. The ancestral W-E stretching valley remains present as a c. 50 km long and between 4 and 12 km broad terrace, in which today the Ybbs River is entrenched, accompanied by its lower and high terrace (Fischer 1979).

Alluvial terraces in the Danube basins:

The lowest parts of the foreland are situated in the broad valley sections along the Danube, in which the altitude of the floodplain drops from 260 m asl in the west to 170 m asl in the east. In these Danube flats, the floodplain is accompanied by a c. 10 m high terrace, which breaks up into two levels downstream of Linz. The easternmost and largest Danube flat is the Tullner Feld (photo in Fig. 4.10c of Chap. "River and Valley Landscapes"). The gravel deposition of its terraces probably started in the Würm, but the surface of the upper level is younger, north of the Danube even early Holocene. This is supported by ^{14}C dates of about 9600 BP, obtained from buried tree trunks (Piffl 1971).

Carpathian foreland:

North of the Tullner Feld follows the foreland of the Carpathians. It starts with a prominent landscape feature, named the "Großer Wagram", which is a sharp, straight-lined and c. 40 m high terrace riser (for its appearance in a high resolution DTM, see Fig. 2.3a in Chap. "Sunken Roads and Palaeosols in Loess Areas in Lower Austria: Landform Development and Cultural Importance").

Situated at an altitude between 200 and 360 m asl, the Carpathian foreland exhibits a hilly landscape, except for its crowning heights between Hollabrunn and Mistelbach that consist of a flat gravel sheet of the Neogene palaeo-Danube. The underlying marine Molasse sediments are mostly concealed by a loess cover, described in more detail in Chap. "Sunken Roads and Palaeosols in Loess Areas in Lower Austria: Landform Development and Cultural Importance". With its intercalated palaeosoils, it provides valuable information about the palaeoclimate, but also hosts special landforms, originating from the interaction of natural processes and human activity (cf. Chap. "Sunken Roads and Palaeosols in Loess Areas in Lower Austria: Landform Development and Cultural Importance"). In the east, the Weinviertel Klippen Zone marks the boundary to the Vienna Basin. It consists of isolated cliff-like limestone outcrops that tower above the surrounding hills (Fig. 2.6b). The climate of the region is continental, expressed by a far lower annual precipitation of 450 to 500 mm and a slightly raised July temperature of 19.4° as compared to the Alpine foreland.

2.3.2 Southeastern Foreland

The Southeastern Foreland is bordered by the eastern margin of the Central Alps, i.e. the Steirisches Randgebirge and the two isolated small and low mountain ranges of Leithagebirge and Hainburger mountains (Sect. 4.2). As in the other forelands, Neogene sediments prevail, covered by periglacial and fluvial sediments in the valleys. In contrast to the margin of the Northern Foreland, the Alps did not overthrust the

Fig. 2.7 Hilly landscape of the Eastern Styrian basin as seen from Schoeckl (1445 m asl; Central Alps) in a southern direction. The visual impression of the undulating relief is accentuated by a shallow fog stratum in early winter. The table-shaped mountain to the right is built of Leitha limestone. The mountains at the horizon are Alpine foothills in neighbouring Slovenia (right) and Croatia (left). Photo: G. K. Lieb

sediments, but gradually descend underneath them. The Southeastern Foreland is regarded as the westernmost part of the Pannonian Basin. This has a diameter of several hundred kilometres and forms an intramontane basin with marginal bays on Austrian territory.

From a geomorphological point of view, this region is dominated by a smooth hilly landscape with decreasing elevation towards the east and relative heights of c. ± 100 m (Fig. 2.7). Occasionally, it is interrupted by plains, valley bottoms or small areas of mountainous appearance, all of them being of limited extent.

Similar to the Schlier region in the Northern Foreland, the hills have subparallel and elongated shapes. At a local scale, ridges running in a N-S direction show steeper western and less steeply inclined eastern slopes and ridges in a W-E direction have steeper northern than southern flanks. This asymmetry has been discussed for a long time resulting in different hypotheses (Morawetz 1967; Roffeis et al. 2007). Among the debated causal processes are (i) neotectonic tilting similar to that of the neighbouring mountain ranges, (ii) dipping of the Neogene strata (N-S running hill ranges) and (iii) stronger denudation processes during the glacials on south-facing slopes (W-E running hill ranges). Locally, the asymmetry may be increased by fluvial undercutting of the valley slopes.

As elsewhere in Austria, plains are restricted to small areas. In the far northeast of the Southeastern Foreland, a small part of the Pannonian Plain is situated on Austrian territory. This landscape is unique in Austria because of the non-glacial origin of its shallow lakes, including Lake Neusiedl (Chap. "Lake Neusiedl Area: A Particular Lakescape at the Boundary between Alps and Pannonian Basin"). East of it, in the so-called Seewinkel, some 40 small lakes with a high content of salt and sodium occur. Like in the Vienna Basin, summers are hot in this warmest corner of Austria, and seasonally many of the Seewinkel lakes fall dry.

Next to the low relief landscape of Lake Neusiedl and Seewinkel, terraces composed of older Pleistocene gravel are remnants of a palaeocourse of the Danube that entered the Pannonian Basin through the gate between the Leitha and Hainburger mountains (cf. Chap. "Geological and Tectonic Setting of Austria").

River catchments are of small size and rarely exhibit broader valley bottoms. The only exception is the Mur Valley that enters the Southeastern Foreland at the city of Graz. Furthermore, during the Pleistocene cold phases, the river deposited glaciofluvial gravel from its inner-Alpine catchment and thus is accompanied by well-developed outwash terraces like those in the Northern Foreland.

The small mountainous areas within the foreland are of different origin and geomorphological character. In the south, they consist of isolated outcrops of the Alps, made up of metamorphic rocks (e.g., Sausal some 20 km south of Graz). They owe their existence to regional uplift of the Alpine basement swells. Similarly, remnants of the volcanism, which occurred in two phases of the Neogene, form towering heights (Chap. "Geomorphological Evidence of Past Volcanic Activity in the Southeast of Austria").

At several locations, mountains built of the so-called Leitha limestone, which was deposited near the shoreline of the sea in the Badenian time interval of the Miocene, rise above the hills, partly uplifted to horst-like structures.

The climate is warm—with annual mean temperatures rising from 9 to 10 °C (warmest month 19–20 °C) from southwest to northeast due to decreasing elevation. Precipitation exhibits an even larger gradient in the same direction —1100 to 600 mm, because precipitation is enhanced by orographic effects along the southern section of the Steirisches Randgebirge. The climate is favourable for agricultural production, which results in the areal predominance of arable land as well as orchards and vineyards. The latter are also assets for the spa tourism that is based on the occurrence of thermal and mineral waters. These result from the extensional tectonics of the region, providing a high geothermal gradient warming up water circulating in the underground.

2.3.3 Vienna Basin

The SSW-NNE oriented and rhombohedron-shaped basin is about 200 km long and 55 km wide. It is a pull-apart basin along the Vienna Basin Transform Fault, and its subsidence started in the early Miocene. Tectonic movements continued during the Quaternary and the Holocene and are still active today, expressed in a moderate seismicity of the region. Several secondary faults bound areas with different subsidence rates, of which some subsided more than 5000 m. One deep subsidence area lies south of the Danube and is a huge groundwater aquifer (termed Mitterndorfer Basin), while three further areas are situated north of the Danube and contain economically important oil and gas fields.

The subsiding basin was invaded by the Paratethys until the area became cut off from the global sea in the middle Pannonian, and a brackish and finally freshwater environment became established. Several cycles of relative sea-level changes occurred. During low water stages, large parts of the basin fell dry and palaeorivers formed deltas in the remaining aquatic areas (Kováč et al. 2004). Periods of high water levels left distinct geomorphological features at the western rim of the basin. Their former coastlines are indicated by shallow water deposits like algal reef limestones, coarse conglomerates and sands rich in macrofossils. In two cases, these deposits are accompanied by prominent, almost level surfaces that are interpreted as shoreline platforms. One is situated at 330 m asl within the urban area of Vienna (Fig. 2.8). The second one lies 25 km further south at approximately the same elevation, is almost 2000 m broad and represents a remnant of the Pannonian lake (Fink 1973).

Geographers subdivide the Austrian part of the Vienna Basin into five landscape units.

(1) In the south lies the *Steinfeld* ("stone field") that is built by the overlapping Plio-Pleistocene fans of the rivers Schwarza and Piesting. Predominantly calcareous gravels provide a dry environment.
(2) To the north and east follows the *Feuchte Ebene* ("wet lowland") that overlies the Mitterndorfer Basin. Here, the water that sank into the gravels of the Steinfeld emerges in several springs. Recent and historical wetlands and peats characterize this area of ongoing subsidence.
(3) The third landscape unit, termed *terrace-landscape,* comprises the right bank Quaternary Danube terraces, with a staircase of terraces in the urban area of Vienna (cf. Table 12.1 and Fig. 12.5 in Chap. "The Danube

Fig. 2.8 Nussberg terrace in Vienna, which is interpreted as a shoreline platform of the Badenian Sea. **a**: topographic profile of the platform, which starts at the slope break left of the tree and ends at the band of dark green vegetation. Photo: C. Embleton-Hamann. **b**: view from above. The green areas in the middle ground are the vineyards on top of the Nussberg terrace before its break to the lower lying built-up area of Vienna in the background. Photo: G. K. Lieb

Floodplain National Park: A Fluvial Landscape with Expiration Date?"), followed towards the east by a stretch of higher ground consisting of old Pleistocene Danube gravels. These are strongly fragmented by faults, and the individual terrace segments are vertically and laterally displaced by neotectonics.

(4) The fourth landscape unit lies north of the Danube and is the monotonously flat plain of the *Marchfeld*. It is also built of Danube terraces, but in this case of three young surfaces of Late Pleistocene to Holocene age. They comprise the Gänserndorfer Flur, the Prater terrace and the current flood plain (for a description of the latter, see Chap. "The Danube Floodplain National Park: A Fluvial Landscape with Expiration Date?"). Again, neotectonic movements impede any altitudinal terrace correlation. In fact, the three surfaces are not smoothly sloping to the east in accordance with the grade of the Danube River and regionally also pass without surface break into each other. The Gänserndorfer Flur is characterized by cryoturbations up to 3 m deep and is therefore definitely older than the last glacial. Age and origin of the Prater terrace still hold unsolved scientific puzzles, as parts of the terrace consist of Holocene sediments, but on the other hand loess cover is widespread, and locally shallow cryoturbations were also mapped (Fink 1973).

(5) To the north, a c. 40 m high rise separates the Marchfeld from the fifth landscape unit, the hilly *eastern Weinviertel*. The rise is a continuation of the prominent "Großer Wagram" of the Carpathian foreland. Its eastern section is less pronounced and in places interrupted by subsidence areas, but it remains a noticeable landscape feature.

Situated in the warmest corner of Austria with a mean annual temperature of 10 °C, the Vienna Basin exhibits a continental-type climate, characterized by hot summers with a July temperature >19 °C and low annual precipitation of <600–700 mm.

Favourable climatic conditions, together with optimal soils for agricultural use, namely chernozems, make the Marchfeld Austria's most important growing area. South of the Danube, a nationally important economic area has become established, fostered by the proximity of the capital and suitable flat ground for industrial development. It also holds a story of environmental degradation associated with unrestricted economic development. In the 1970s, a third water supply system for Vienna was installed, tapping the groundwater resources of the Mitterndorfer Basin. However, after completion of the boreholes, reservoirs and water mains, it was found that the water was hopelessly polluted, mainly from landfill sites. It took 30 years to clear up the dumps at high expense and to develop a special water treatment technique for purifying the pumped-up water, which until today is only used in times of water shortage.

2.4 Alps

2.4.1 Northern Alps

2.4.1.1 Northern Prealps

The Northern Prealps follow south of the Alpine foreland as a small, c. 10–15 km broad landscape belt that only widens in the Wienerwald in the far east and the Bregenzerwald in the far west. Three tectonic units are involved in the set-up of this geomorphological region, namely the Allochthonous Molasse as the lowermost element of the Alps, overlain by Helvetic nappes and topped by the Penninic nappes. Consisting of conglomerates, sands and clays, the Allochthonous Molasse forms a marginal strip in the northeast and northwest. The main country rock of the region consists of interbedded marls and sandstones. They are typical for the Penninic Rhenodanubian Flysch, but marl/sandstone sequences are also common within the Helvetic Unit. The latter is only more widely developed within the Bregenzerwald, where it also contains limestones.

The character of the Northern Prealps is mountainous. In the east, where altitudes are low (maximum c. 550 m asl), this is caused by a dense and well incised river network providing high relative relief and a predominance of steep slopes. In the far west, both altitude and relative relief are substantial. Limestone outcrops of the Helvetic Unit culminate in peaks up to around 2000 m asl in elevation, and relative relief is in the order of 500 to 1000 m (Fig. 2.1).

This typical mountainous and mostly forest-covered landscape character is only interrupted in the area of the Salzburg and Upper Austrian Prealps. The many lakes of this region are enclosed by the hilly moraine landscape within the eastern terminal moraines of the Pleistocene Salzach Glacier and further east, in Upper Austria, by those of the Pleistocene Traun Glacier (for a detailed description of the latter, see Chap. "Quaternary Landforms and Sediments in the Northern Alpine Foreland of Salzburg and Upper Austria", Sect. 14.5.).

Mass movements are a first-order geomorphological process domain of the Rhenodanubian Flysch. This is already expressed in the term "Flysch" (cf. Sect. 1.1), but the underlying lithological disposition is also widespread in the broad Helvetic zone in the far west. Because of the importance of mass movements in these two tectonic units, a regional chapter (Chap. "The Walgau: A Landscape Shaped by Landslides") is devoted to them.

All weather stations of the region are situated in the valleys, indicating a mean annual precipitation of 1500 m in the west, of c. 1150 mm in the middle, declining to 750 mm in the east. These sums are slightly higher than in the Alpine foreland, but only mirror the situation at the bottom lowest-lying land of the Prealps. Mean July temperatures are in the order of 18 °C and of 19 °C in the Wienerwald in the east. The dominant land cover is forests and meadows, while the dominant land use is forestry and tourism. In the lake region, summer tourism is an important economic factor. The Bregenzerwald in the west hosts skiing facilities and the Wienerwald in the east constitutes a local recreation area for the citizens of Vienna.

2.4.1.2 Northern Calcareous Alps

Within Austria, the Northern Calcareous Alps form a 25–45 km broad mountain belt, stretching from the border with Switzerland in the west to Vienna in the east. Their highest summit (3036 m asl) lies in the Lechtal Alps in the west, and elevations slightly below 3000 m asl are reached as far as the Dachstein. Further east, highest elevations drop to 2000 m asl and are accompanied in the north by the lowest ridges of the Northern Calcareous Alps, staying below c. 1600 m asl. Not only height differences, but also a different appearance of the mountain landscape is typical for the western and eastern half of this landscape unit. As far as the Kaiser Mountains, ridges of the structural mountain type are present (Fig. 2.9a). Further east, the mountain character changes to karst massifs, rising with steep rock faces above the surrounding valleys and exhibiting a relatively flat topography in the summit region.

The predominant rocks are Triassic limestones and dolomites. Tributary rock types occur at the base and on top of the nappe pile. The former are Permian saltrocks and Permian to Lower Triassic shales and sandstones; the latter are the Gosau Group strata, consisting of Upper Cretaceous to Lower Palaeogene conglomerates, shales, coal and limestones. They were deposited on the already north-moving nappe pile in an interim subsidence phase. Today, they are to be found in local basins within the higher mountains built of Triassic limestone and dolomite (Wagreich and Decker 2001) (Fig. 2.9b).

The weathering behaviour of the two main country rocks, namely dolomite and limestone, is quite different. Whereas the limestone is subject to karstification (for a detailed description of the karst features of the Northern Calcareous Alps see Chap. "Karst Landscapes in Austria", Sect. 5.4.1.1), the dolomite is prone to surface erosion and characterized by high debris production. Thus, mountain ranges in dolomites are flanked by sheet screes derived from rockfall and scree cones below couloirs in the rock cliffs (Fig. 2.9a).

The impressive scenery of the Northern Calcareous Alps results from the frequent contrast of steep limestone or dolomite rock faces towering above the longitudinal courses of the rivers Inn, Salzach and Enns or, in other cases, topping the gentle relief of Gosau basins (Fig. 2.9).

The peculiar plateaus on top of the Karst massifs triggered geomorphological interest as early as the beginning of the twentieth century. One of the first researchers to propose their origin as palaeosurfaces was Lichtenecker (1924), and he started a recurrent debate of palaeosurfaces in Austria. Chapter "The Rax Karst Massif: A Typical Plateau of the Northern Calcareous Alps?", Sect. 19.2, outlines the respective history of thought. By the 1960s, discussion and mapping of palaeosurfaces was the most important national geomorphic research topic. The number of regionally mapped palaeosurfaces increased together with the speculative nature of the whole effort, and finally, the debate dried up for something like 40 years. Today, it is back, inspired by two

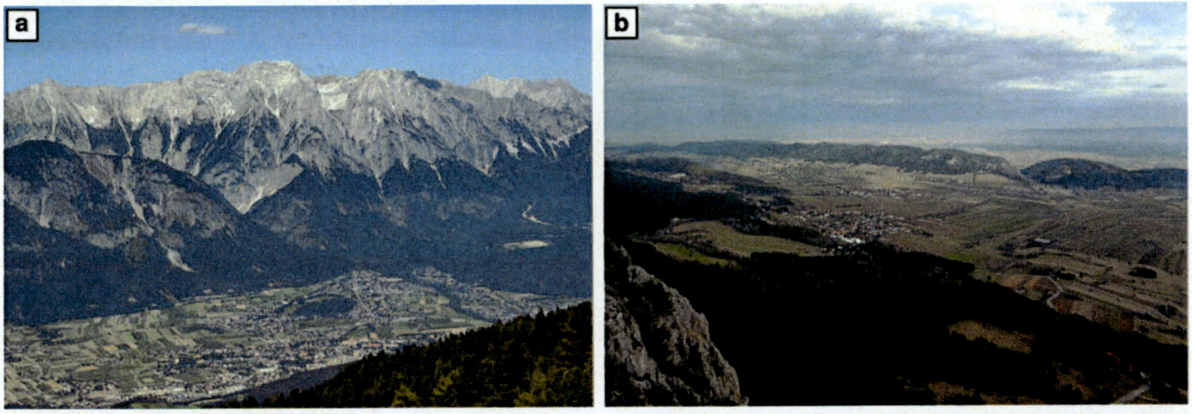

Fig. 2.9 **a**: Southernmost mountain ridge of the Karwendel range rising above the Inn Valley. The high share of dolomites in its bedrock composition favours debris production; note the debris channels and debris cones in the picture. Another characteristic feature of the Karwendel is the numerous deeply incised cirques. **b**: One of the Gosau Group basins, here embedded into the eastern end of the Northern Calcareous Alps. The forested ridge in the middle ground represents the last outcrop of the Calcareous Alps reaching c. 600 m asl; the river gap in its right half offers a glimpse of the Vienna Basin beyond. As in other Gosau Group basins, coal mining was an important economic factor in this region until the 1960s. Photos: G. K. Lieb

circumstances, namely (i) modern views of the geological evolution of the Eastern Alps and (ii) growing experience with the interpretation of the complex results of fission track dating.

Due to climate warming, the Northern Calcareous Alps have almost lost their former glaciers. Today, their largest glacier area on the Dachstein suffers from negative mass balances year after year. Otherwise, there is only one other, very small glacier left in the Lechtal Alps that still shows observable signs of ice flow.

Gravitational mass movements of the rockfall-type are widespread. Next to rockfall, the unit shows two distinct regional clusters of prehistoric and modern landslides. The first one, located in western Tyrol, encompasses huge prehistoric, 4.2–3.0 ka old rock avalanches, described in Chap. "Giant "Bergsturz" Landscapes in the Tyrol". The centre of the second cluster is the Salzkammergut, where the Mesozoic limestone and dolomite rocks rest on Permian saltrocks, and the setting of "hard on soft rocks" triggers a variety of present-day mass movements (cf. Chap. "The World Heritage Site Hallstatt-Dachstein/Salzkammergut: A Fascinating Geomorphological Field Laboratory", Sect. 17.5.2).

Some areas of the Northern Calcareous Alps belong to the wettest places in Austria. Humid Atlantic airmasses coming from the west and northwest are forced to rise at the mountain front, but in N-S oriented valleys penetrate even further south and cause annual precipitation sums of 2500–3000 mm at the summit level. Even in the dry east, the orographic effect triggers precipitation sums of 1500 mm at the highest altitudes. Mean July temperatures in the valleys depend on altitude, reach around 17.5° in areas below 600 m asl and 15–16° in the elevation range 600–1000 m asl.

The Northern Calcareous Alps hold important water and timber resources. In the east, together with the coal seams of the Gosau beds, they were the basis for the former iron-processing industry, which coined the regional place name "Eisenwurzen" (meaning source of iron products). Today, the large karst springs are used for the water supply of some of Austria's main cities. In the case of Vienna, two long pipelines, already built in 1873 and 1910, convey water from the Calcareous Alps (for details, see Chap. "Karst Landscapes in Austria", Sect. 5.5.1). Large parts of the region are under landscape/nature protection. The protected areas enclose two national parks, one geopark, nine nature parks and several water reserves.

2.4.2 Central Alps

The Central Alps are Austria's largest geomorphological landscape region. Situated between the northern and southern longitudinal valley depressions, they stretch from the western state border to the eastern one over more than 500 km. There is great variety of geology, relief and landscape appearance in this dominantly metamorphic terrain. Three subunits are distinguished:

1. The Greywacke Zone runs parallel to the northern longitudinal valley depression from near Innsbruck (west) to the southwestern edge of the Vienna Basin (east). Similar units in terms of rock type and Palaeozoic age occur in two isolated areas.
2. The crest of the Central Alps takes up the largest part of the unit. Its central ridge is an important divide and the axis of uplift of the Alpine orogen.
3. The inner-Alpine basins are concentrated in the eastern half of the Austrian Alps, because their occurrence is connected to the lateral extrusion in the Neogene. Figure 2.3 exhibits only the four largest of c. 10 inner-Alpine basins.

2.4.2.1 Greywacke Zone (and Similar Units)

Greywacke is a term taken from the mining language denoting Palaeozoic sandstones containing mostly coarse material. The Greywacke Zone is a narrow strip (maximum width c. 25 km) of predominantly schistose rocks, forming the basement of the Northern Calcareous Alps. Despite substantial relative heights, everywhere in the central and western parts exceeding 1000 m, the relief of the Greywacke Zone is relatively smooth and contrasts with that of the neighbouring landscapes. The main mountain ridges of the western section are the Kitzbuehel Alps, named after the famous skiing destination, because of its snow reliability and ideally inclined slopes for alpine skiing (Fig. 2.10a). Steeper slopes and rock faces occur where Palaeozoic calcareous sequences crop out, particularly in the Eisenerz Alps, the easternmost high-mountain section of the Greywacke Zone.

The entire Greywacke Zone is rich in ores and therefore was important for Austria's mining history. The Geopark "Ore of the Alps" in its western section (cf. Chap. "The Ore of the Alps UNESCO Global Geopark (Salzburg) Geosites and Geotourism") is centred on this topic. Even in prehistoric times, copper, silver and iron were mined and iron is still mined at the Erzberg (Chap. "The Erzberg Area: A Mining Landscape in Styria").

From a geomorphological and lithological point of view, three other mountain groups resemble the Greywacke Zone:

(1) The Tux Alps look like the Kitzbuehel Alps but are built of quartz phyllites. This is where the highest summits of the entire landscape unit are situated (reaching almost 2900 m asl).

(2) A large part of the Gurktal Alps belongs to the Drauzug-Gurktal Nappe System, with sedimentary sequences comprising not only Palaeozoic, but also Mesozoic rocks.

Fig. 2.10 a: Typical Greywacke Zone landscape of the Kitzbuehel Alps with the carbonate rocks of the Grosser Rettenstein (2366 m asl) towering above smooth relief in schistose rocks. View from the south. b: The Grazer Bergland has a similar relief but does not rise above the treeline. Hochlantsch (1710 m asl), the rocky summit in the centre, consists of carbonate rocks. View from southeast. Photos: G. K. Lieb

(3) The Grazer Bergland (Fig. 2.10b) also belongs to this nappe system. Its calcareous rocks show well-developed karst landforms, described in Chap. "Karst Landscapes in Austria", Sect. 5.2.3.

The subregions (2) and (3) have a more continental climate resembling that of the neighbouring ranges of the crest of the Central Alps, with lower annual precipitation than the Tux, Kitzbuehel and Eisenerz Alps. Due to altitudes and relative heights decreasing eastwards, subregion (3) no longer has a high-mountain character. In terms of Pleistocene glaciation, the western parts were situated within the complex of transection glaciers, whereas the Eisenerz and Gurktal Alps had local glaciers.

2.4.2.2 Crest of the Central Alps

Partly glacierized high mountains:

The western half of central ranges, stretching from the Swiss border to the Katschberg Pass (1641 m asl), is characterized by maximum elevations over 3000 m asl and current glaciation. It shows a concentration of economic activities that require high-mountain terrain and glaciers (special types of alpinism and skiing, storage power plants). The large number of storage power plants is, however, not linked to its glaciers, but to (i) high precipitation sums of 1500–2500 mm at 3000 m asl, (ii) suitable landforms (glacially shaped troughs), (iii) impermeable rocks and (iv) high relative relief. These western ranges may be further divided into two sections, separated by the deeply incised furrow of the Brenner Pass (1373 m asl).

West of the Brenner Pass, Austroalpine hard gneisses and micaschists predominate and form the highest summits. Only along the SW-NE oriented uppermost reach of the Inn River weaker schists crop out, belonging to the first of three Penninic windows along the crest of the Eastern Alps. Culmination areas are concentrated in the Silvretta range along the Swiss border (Fig. 2.11b), further east in the Ötztal Alps (Fig. 2.11a) and finally in the Stubai Alps, separated from the Ötztal Alps by the Ötz Valley, known for the huge Köfels rockslide (cf. Chap. "Giant "Bergsturz" Landscapes in the Tyrol"). The eastern fringe of the Stubai Alps has a cover of Mesozoic calcareous sediments, making some of their peaks look as sharp as those of the Calcareous Alps.

East of the Brenner Pass, the lithological setting changes to the "Tauern Window", in which the Penninic Unit outcrops for the second time, underlain by the Subpenninic Unit. In the east and west, it is bordered by the north–south striking detachments along normal faults that gave rise to the Brenner furrow and the Katschberg Pass. Its varied lithology results in differing appearances of the mountainous landscape (see examples described in Chap. "Großglockner and Pasterze Glacier: Landscape Evolution at Austria's Highest Summit and its Neighbouring Glacier System").

The Tauern Window comprises two large glacierized mountain ranges, the Zillertal Alps and the Hohe Tauern. The highest summits are built of resistant rocks, e.g. the Großvenediger (3666 m asl) of Subpenninic Variscan granitoids and the Grossglockner (3797 m asl), Austria's highest summit, of metamorphosed Penninic ophiolites. The Hohe Tauern range lies in the centre of the largest national park of the entire Alpine arc. The Hohe Tauern National Park, comprising 1856 km^2, provides excellent interpretations of geomorphological features, for instance rock glaciers, waterfalls, glacial landforms and many more on geotrails and in exhibitions.

The mountain groups between the Hohe Tauern range and the southern longitudinal valley depression have only very small glaciers (e.g. the Schober mountains) or even lack glaciers (e.g. the Kreuzeck mountains).

Fig. 2.11 **a** High-mountain landscape of the Ötztal Alps. To the left the Gepatschferner, Austria's second largest glacier is visible. Its accumulation area lies on a low inclined platform at high altitude and nourishes a glacier tongue, which in the photo appears as a light grey strip close to the left image border. The horizon line in the right half of the photo shows the summit of the Weißkugel (3738 m asl). View from the southeast. **b**: High-mountain landscape of the Silvretta. The rugged relief of the Fluchthorn (3399 m asl; Austroalpine Unit, centre) contrasts with the smoother relief in the foreground and to the right (Penninic Unit, Engadin Window). View from the south. Photos: G. K. Lieb

Fig. 2.12 **a**: High mountains without glaciation in the easternmost section of the Niedere Tauern range with summit elevations of c. 2400 m asl and slopes shaped by periglacial processes. View from the southeast. **b**: Rock tors on a flat-topped ridge of the Stubalpe (c. 1800 m asl). The cross indicates the local appreciation of such conspicuous landforms. View from the west towards the Grazer Bergland. Photos: G. K. Lieb

High mountains without current glaciation:

The typical representative of this type is the Niedere Tauern range (Fig. 2.12a), consisting of four mountain groups that differ in lithology and relief, which becomes less rugged and steep in the eastern direction. The westernmost group largely belongs to the Lower Austroalpine Unit, incorporating carbonate rocks that lead to mountainous landscapes like those in the Northern Calcareous Alps. The adjacent Schladminger Tauern group rises to 2862 m asl and has the most pronounced high-mountain relief with excellently shaped cirques, which are discussed, together with the typical valley asymmetry of steep and short west facing valley sides set against much longer east facing ones, in Chap. "The Variability and Uniqueness of Cirque Landscapes in the Schladminger Tauern", Sect. 29.4.3.

South of the Niedere Tauern follow the Gurktal Alps. Due to lithology and landscape appearance, their central part was classified as one of the external units of the Greywacke Zone (see above). Its fringes, together with the Seetal Alps and the Saualpe, are, however, built of gneisses and micaschists. Like in the Niedere Tauern range, main ridges are asymmetrical. Furthermore, they run perpendicular to the crest of the Central Alps, namely roughly N-S. This topography is related to tectonic processes linked to the lateral extrusion of the Eastern Alps in the Miocene.

Because of insufficient precipitation sums, the region has no storage power plants, but otherwise land use is comparable to that of the partly glacierized high mountains.

Mountains without Pleistocene glaciation:

The eastern end of the Pleistocene complex of transection glaciers was characterized by a few outlet glaciers occupying the low-lying areas of the Enns and Mur valleys as well as the Klagenfurt Basin, while the easternmost mountain groups of the Niedere Tauern range or the Seetal Alps only had local glaciers. The very last and very small local glaciers quarried poorly developed cirques into the highest summits of the Steirisches Randgebirge.

The Steirisches Randgebirge forms a semi-circular c. 200 km long arc, clearly visible in Fig. 2.4. It is only interrupted by the gorge of the Mur River north of Graz. As the arc rarely rises above the regional treeline, situated at c. 1700 m asl, it appears as a landscape of woodlands.

The relief of this subunit is dominated by morphological features that are relicts of the surface uplift history since the Eocene (Chap. "Geological and Tectonic Setting of Austria"). These features are flat-topped ridges at different elevations, which are a classical topic of geomorphological research. Older and younger studies (Winkler-Hermaden 1957; Lontschar and Stüwe 2020) agree in interpreting the stepwise arrangement of landscape remnants as a sequence of uplift and planation events, reflecting an uplift of c. 500 m in the last 5–10 million years.

Isolated rock tors with heights up to c. 5–20 m occur on top of the ridges (Fig. 2.12b). They are the testament of chemical weathering during the Neogene and subsequent exposure of harder rocks under Pleistocene periglacial conditions (Paschinger 1974). Hence, they were formed in the same way as the granite tors of the Bohemian Massif. In contrast to the ridges and low angle slopes, the valleys are deeply incised and partly gorge-like.

At the eastern end of the Steirisches Randgebirge, the third Penninic window ("Rechnitz Window") appears at the surface. In terms of relief, the terrain does not differ significantly from the neighbouring mountains, which drop below 500 m asl to the northeast. Isolated by a broad gap, the last spur of the crest of the Central Alps starts with the Leitha mountains, classified as hills because of their low relative relief and followed after another gap by the Hainburg mountains (480 m asl), which have a more pronounced relief due to the occurrence of carbonate rocks. They already belong to the Tatric Unit of the Carpathian mountain belt. A palaeochannel of the March River and finally the Danube detached them from the range of the Little Carpathians (cf. Chap. "Geological and Tectonic Setting of Austria", Sect. 1.5).

2.4.2.3 Inner-Alpine Basins

Limited areas within the Alpine orogen were less uplifted or experienced active subsidence. Along the large strike-slip faults of the Eastern Alps, local pull-apart basins (Fig. 2.13) formed with a spatial concentration along the Lavant, Mur and Mürz rivers. The basins were filled with local sediments,

Fig. 2.13 One of the small inner-Alpine basins, surrounded by the Niedere Tauern range in the background and by the summits of the Greywacke Zone in the middle- and foreground. In the centre of the basin the small town Trofaiach is visible. View from the east. Photo: G. K. Lieb

in which several coal seams developed. Later on, their sediment fills became eroded to a hilly relief comparable to that of the Alpine forelands. Exploiting the coal seams triggered the nineteenth-century industrialization along the Mur-Mürz valleys. The region is still one of the leading industrial areas within the entire Alps.

Stüwe and Homberger (2012) emphasized the special topographical situation of the Mur-Mürz and the Lavanttal faults, which are situated perpendicularly to each other, meeting at the spacious Judenburg-Knittelfeld Basin. Both faults border the arc of the Steirisches Randgebirge that has been moving eastwards during the last 17 Ma, thus creating the pull-apart basins along the faults. The Klagenfurt Basin (described in Chap. "Klagenfurt Basin: A Large Basin in the Alps"), Austria's largest inner-Alpine basin, owes its existence to the Periadriatic fault. In contrast to most basins along the Mur-Muerz and Lavanttal faults, it was almost entirely covered by ice during the Pleistocene cold phases and, therefore, has many glacial landforms.

There are specific climatic conditions in the basins. Next to the continental character of eastern Austria, the shielding effect of the surrounding mountains keeps annual precipitation totals low (700–1200 mm). Summer temperatures within the basins feel warm (mean July temperatures are 16–18 °C), but in winter long-lasting temperature inversions combined with fog often make weather situations uncomfortable (mean January temperatures are −4–−2 °C).

2.4.3 Southern Alps

The area south of the Drau River is referred to as "Southern Calcareous Alps", respectively as "Southern Alps". As pointed out in Sect. 1.1, the term "Southern Alps" is used in a different sense than in the structural landscape divisions of Chap. "Geological and Tectonic Setting of Austria" and Fig. 2.1a, where it is restricted to the distribution area of the Southalpine Unit. As a geomorphological landscape region, the Southern Alps encompass three mountain groups of similar structure and appearance. They consist of a heterogenous basement of slightly metamorphosed Palaeozoic rocks overlain by predominately calcareous successions of Triassic age, and with a few exceptions appear as high-mountain ridges elongated in a WNW-ENE direction. The summit region is generally built of limestones.

The landscape unit as a whole includes the southern longitudinal valley depression along the Periadriatic fault. Its western section is marked by the broad and conspicuously rectilinear Gail Valley, while further east the fault only gives rise to a series of small valleys and saddles. Opposite to the mainly metamorphic and calcareous rocks of the Southern Alps, there are outcrops of magmatic and crystalline rocks along the Periadriatic fault.

Two of the three mountain groups of the Southern Alps (numbered 1 and 2 in the following) run parallel to the Gail River west of the city of Villach and one (no. 3) borders the Klagenfurt Basin to the south.

(1) The Gailtal Alps are located between the valleys of the Drau River in the north and the Gail River in the south. They are almost entirely built of calcareous Mesozoic rocks, which also form their highest peaks. In contrast to the structural ridges in the west, the easternmost mountain group of the Gailtal Alps is an isolated karst massif, with a large plateau on its top (Dobratsch

Fig. 2.14 **a**: High-mountain relief of the crest of the Carnic Alps near Ploecken Pass with its highest summit (Hohe Warte, 2780 m asl) in the left background. The mountain landscape is built of Palaeozoic rocks, consisting of limestones in the summit region of the background and in the rockface of the right foreground. View to the west. **b**: High mountains with lower summit elevations characterizing the Karawanken in the eastern part of the Southern Alps. The valley in the centre follows the course of the Periadriatic fault that divides the Karawanken into two similar mountain ranges built of Mesozoic calcareous rocks, although they belong to different tectonic units. View to the southwest. Photos: G. K. Lieb

Massif, described in Chap. "Dobratsch—Landslides and Karst in Austria's Southernmost Nature Park".

(2) The crest of the Carnic Alps (making up a main watershed and the border with Italy) stretches south of the Gail Valley, parallel to the Periadriatic fault. On the Austrian side, Palaeozoic rocks of the Southern Alpine basement dominate, many of them famous for their fossils, a circumstance that generated a moderate level of geotourism. Despite the overall high-mountain character, landscape contrasts arise between the rocky relief in limestones (Fig. 2.14a) and the gentle lower slopes in silicate rocks.

(3) The Karawanken are the lowest mountain group. South of the Periadriatic fault they appear as a mountain ridge, which forms the border with Slovenia. In contrast, the former ridge north of the Periadriatic fault was dissected to isolated mountain massifs by retrograde erosion originating in the Klagenfurt Basin (see Chap. Klagenfurt Basin: A Large Basin in the Alps). Due to lower elevations compared to (1) and (2), the Karawanken were not entirely glaciated in the Pleistocene cold phases and partly lack high-mountain relief.

The Southern Alps show a typical vegetation succession from the submontane to the subnival zone, with an increasing areal dominance of the montane zone towards the east. Land use is therefore traditionally characterized by forestry and hay meadows or pastures, often impeded by the steep slopes. A special climatic feature of the Southern Alps is the very high amounts of annual precipitation (frequently more than 2000 mm at elevations above 1500 m) that are a consequence of humid air masses coming from the Mediterranean Sea. Because the cyclonic activity there is most intensive in autumn, the maximum mean precipitation occurs in October—in contrast to the rest of Austria where the highest precipitation values are measured in one of the summer months.

References

Borsdorf A (2005) Das neue Bild Österreichs. Strukturen und Entwicklungen im Alpenraum und in den Vorländern. Austrian Academy of Sciences, Vienna

Fink J (1973) Zur Morphogenese des Wiener Raumes. Z. Geomorph. N.F. Suppl 17:91–117

Fink MH (ed) (1986) Raumordnung und Naturgefahren. Österreichische Raumordnungskonferenz (ÖROK) vol 50. Vienna

Fischer H (1965) Geomorphologie des unteren Mühlviertels im Einzugsgebiet der Naarn. Geographischer Jahresbericht Aus Österreich 30:49–130

Fischer H (1979) Reliefgenerationen im Kristallinmassiv—Donauraum—Alpenvorland und Alpenrand im westlichen Niederösterreich. Forschungen zur Deutschen Landeskunde 213. Trier

Hauer C, Höfler S, Flödl P, Gumpinger C, Habersack H, Holzinger J, Kloibmüller A, Leitner P, Lichtneger P, Mayer T, Ottner F, Riechl D, Wagner B, Walter T, Weingraber F, Graf W (2016) Regionale Aspekte des Feststoffmanagements als Grundlage für den naturnahen Wasserbau im Mühlviertel und im Bayerischen Wald. Österr Wasser-Und Abfallwirtschaft 68:488–502

Höck V (1999) Der geologische Bau des Grundgebirges In: Steininger FF (ed) Erdgeschichte des Waldviertels (2nd ed). Waldviertler Heimatbund, Horn—Waidhofen/Thaya, pp 79–88

Kohl H (2000) Das Eiszeitalter in Oberösterreich. Schriftenreihe des OÖ Musealvereins—Gesellschaft für Landeskunde vol 17. Linz

Kováč M, Baráth I, Harzhauser M, Hlavatý I, Hudáčková N (2004) Miocene depositional systems and sequence stratigraphy of the Vienna basin Cour. Forsch-Inst Senckenberg 246:187–212

Lichtenecker N (1924) Das Bewegungsbild der Ostalpen. Naturwissenschaften 13:739–743

Lontschar K, Stüwe K (2020) Paleolandscapes of the Southern Saualpe. Mitteilungen des naturwissenschaftlichen Vereins für Steiermark vol 150. Graz, pp 129–146

Morawetz S (1967) Zur Frage der asymmetrischen Täler im Grabenland zwischen Raab und Mur. Mitteilungen des naturwissenschaftlichen Vereins für Steiermark vol 97. Graz, pp 32–38

Paschinger H (1974) Steiermark. Steirisches Randgebirge, Grazer Bergland, Steirisches Riedelland. Sammlung Geographischer Führer 10. Gebrüder Borntraeger, Berlin, Stuttgart

Piffl L (1971) Zur Gliederung des Tullner Feldes. Annalen des Naturhistorischen Museums Wien, vol 75. Wien, pp 293–310

Roffeis C, Stüwe K, Hergarten S (2007) Morphologische Analyse der Täler zwischen Raab und Mur. Mitteilungen des naturwissenschaftlichen Vereins für Steiermark, vol 136. Graz, pp 239–248

Stüwe K, Homberger R (2012) High above the Alps. A bird's eye view of geology. Weishaupt, Gnas

Wagreich M, Decker K (2001) Sedimentary tectonics and subsidence modelling of the type upper cretaceous Gosau basin (Northern Calcareous Alps, Austria). Int J Earth Sci 90(3):714–726

Winkler-Hermaden A (1957) Geologisches Kräftespiel und Landformung. Springer, Vienna

Gerhard Karl Lieb is associate professor at the University of Graz, Institute of Geography and Regional Science.. He has experience in physical geography of mountain regions with a special focus on permafrost, glaciers and geomorphic processes affected by global change as well as natural hazards. He is one of the two co-leaders of the national glacier monitoring programme of the Austrian Alpine Club. Besides those topics he works on regional geographies (Austria, Alps and Europe) and on didactics of the Austrian school subject geography and economics.

Christine Embleton-Hamann is a retired professor at the Department of Geography and Regional Research at the University of Vienna. Her main interest is in alpine environments. Within this field she focusses on human-environment interactions with research topics like human impact on geomorphic processes, assessment of the scenic quality of landscapes, and geomorphological hazards. A second set of interest concerns the communication of geomorphological knowledge to a broader audience, in the pursuit of which objective she has written a well-received textbook on geomorphology. She is a former President of the Austrian Research Association on Geomorphology and served on the Executive Boards of the IAG and several IAG and IGU Working Groups.

3 The Imprint of Quaternary Processes on the Austrian Landscape

Jürgen M. Reitner

Abstract

Austria has a diverse landscape as a result of the interplay of processes linked to tectonics and climate change. The final shaping occurred during the Quaternary (the last 2.58 Ma). This period is characterized by strong climatic variations on the global scale between glacial and interglacial conditions, which had different effects on the heterogeneous landscape, and its archives depending on the magnitude and duration of the climatic signal. The tectonic influence during the Quaternary is evident in instances of uplift as indicated by the Pleistocene terrace staircases of the Alpine foreland. In contrast, parts of the Vienna basin are characterized by subsidence linked to strike-slip faults. The oldest deposits are loess–palaeosol sequences (LPSs) which document the onset of loess accumulation with the occurrence of the Gauss/Matuyama palaeomagnetic reversal at the beginning of the Quaternary. The Early Pleistocene record consists of LPSs and gravel deposits with no indication of a glacier advance. Four major glaciations, namely Günz, Mindel, Riss and Würm (from oldest to youngest)—are known. These Ice Ages were characterized by a large complex of transection glaciers, i.e. an interconnected system of valley glaciers covering large sections of the Eastern Alps with glacier tongues terminating in the Alpine foreland. The three older glaciations are of Middle Pleistocene age, whereas the youngest happened during the Late Pleistocene. All four glaciations are recorded by Glacial Sequences genetically linking tongue basins with (subglacial) till, terminal moraine deposits and terraces consisting of proglacial outwash. Sediments of these glaciations differ in the degree of weathering and the characteristics of cover beds (e.g. LPSs). Based on geochronological data and the relation between type and magnitude of the global climate signal, the amount of reconstructed sediment production following correlation with major phases of global glaciation is used: Günz (MIS 16; 676–621 ka), Mindel (MIS 12; 478–424 ka), Riss (MIS 6; 191–130 ka) and (Late) Würm (MIS 2; 29–12 ka). Detailed knowledge of the Last Interglacial-Glacial cycle (130–12 ka) allows establishing models for climatically controlled sedimentary processes and for glacier expansion in the longitudinal valleys of the Eastern Alps. Overdeepened valleys and increased relief leading to different types of mass movements are also a legacy of glaciations. Evidence of Pleistocene permafrost (e.g. relict rock glaciers) as well as Holocene fluvial activity are further indications of the dynamic landscape development of the Austrian landscape during the Quaternary.

Keywords

Pleistocene • Holocene • Glaciation • Palaeoclimate • Geomorphology • Landforms • Processes

3.1 Introduction

The landscape of Austria bears the legacy of repeated massive climatically driven changes during the Quaternary (the last 2.58 Ma) ranging from major glaciations with glaciers covering major parts of the Eastern Alps to interglacial conditions such as those prevailing nowadays. Archives of terrestrial deposits of Quaternary age in Austria are discontinuous and do not cover the whole period. We have to deal with a patchwork of different and fragmentary records of varied genesis, ranging from glacial (e.g. tills, moraines), fluvial (e.g. terraces), lacustrine, aeolian (loess) to biological (e.g. peat) or chemical (e.g. speleothems, spring tufa). Depending on the contributing depositional, erosional or weathering processes and the topographical and tectonic setting, the covered time range and time resolution of

J. M. Reitner (✉)
Geological Survey of Austria, Neulinggasse 38,
1030 Vienna, Austria
e-mail: juergen.reitner@geologie.ac.at

C. Embleton-Hamann (ed.), *Landscapes and Landforms of Austria*, World Geomorphological Landscapes,
https://doi.org/10.1007/978-3-030-92815-5_3

archives of palaeoclimate and landscape evolution vary to a great extent.

Research on Ice Age deposits and former glacier extent in Austria started with Morlot (1847) and Simony (1847). After a phase of systematic research, Penck and Brückner (1909) produced their seminal work "Die Alpen im Eiszeitalter" ("The Alps during the Ice Age"), establishing a fourfold Alpine glacial chronology (from oldest to youngest) for the Alps and their forelands named after Bavarian rivers: Günz, Mindel, Riss (Riß in German) and Würm. Finally, major progress in understanding landscape evolution was achieved by the development of dating techniques, with their application in Austria beginning with the first radiocarbon dating efforts in the 1960s.

The current understanding of Quaternary processes and chronology has been summarized by van Husen (2000a, 2011) and that of the corresponding stratigraphy of Quaternary deposits by van Husen and Reitner (2011a). Based on these overviews, this contribution aims to show the chronology of processes and their driving factors (climate and tectonics) responsible for shaping the Austrian landscape during the Quaternary.

3.2 Quaternary—Definitions, Subdivision, Stratigraphic Frame and Principles

The onset of the Quaternary period (Fig. 3.1) and that of the Pleistocene Epoch is defined as 2.58 Ma, which is very close to the Gauss-Matuyama palaeomagnetic reversal (Gibbard et al. 2009). This definition reflects the establishment of ice sheets in the Northern Hemisphere. The further—still informal—subdivision of the Pleistocene is used according to the suggestions by Gibbard et al. (2010). The Holocene Epoch starts at 11.7 ka (Walker et al. 2009).

It is convenient to use the marine $\delta^{18}O$ record (Fig. 3.1), which documents the global climate signal and thus especially the changes of global ice volume, as a chronostratigraphic framework. Its subdivision into individual Marine Isotope Stages (MIS) goes back through geological time with odd numbers representing warm (interglacial) stages and even numbers corresponding to cold (glacial) stages (Lisiecki and Raymo 2005). Referring the last c. 120 000 years, the stratigraphic terminology derived from Greenland ice cores with numbered Greenland stadials (GS) and Greenland interstadials (GI) is also in use for correlations (Rasmussen et al. 2014) (Fig. 3.1).

The Eastern Alps and their foreland are the type area for the classical Alpine stratigraphy according to Penck and Brückner (1909), with the Günz, Mindel, Riss and Würm glaciations based on the concept of "Glaziale Serie" (Glacial Sequence), genetically linking tongue basins with till, terminal moraine deposits and terraces consisting of proglacial outwash (Fig. 3.2). Terrace bodies and moraines of the four glaciations differ in the degree of weathering (Fig. 3.3) and the characteristics of cover beds (e.g. loess/palaeosol assemblages). Cases of superposition like in the Vienna basin (Fig. 3.4 for locations) are rare. In areas affected by tectonic uplift, outwash terraces are correlated based on these differences in the context of their morphological position within the valley, with the oldest ones in the highest position.

3.3 Climatic Pacemaker

The palaeoclimatic development of the Austrian landscape reflects the global climatic development, which is characterized by a lowering of the greenhouse gas CO_2 and a transition from greenhouse to icehouse conditions in the last 40 Ma (Raymo and Ruddiman 1992). With the onset of a major Northern Hemispheric glaciation around the beginning of the Quaternary, a change of the orbitally forced cyclicity (Milanković-cycles) between warm ("interglacial") and colder periods in the marine $\delta^{18}O$ record (Raymo 1997) from the primarily 20 ka cycle to a dominant 41 ka cycle occurred (Fig. 3.1). Around 0.9 to 1 Ma, the persistent cooling, together with the variations of the earth's orbit, resulted in a change of glacial/interglacial cyclicity from 41 ka to c.100 ka (Ruddiman 2001). From this "Mid-Pleistocene transition" (c. 1.2–0.5 Ma, Head and Gibbard 2005) onwards, the large ice sheets of the Northern Hemisphere (e.g. Laurentide and Scandinavian) were big enough to withstand small warming events, whereas only intensive increase in northern summer insolation resulted in cessation of a major glaciation called termination (labelled from VII to I) (Raymo 1997). Major glaciations in the Alps started around 0.87 Ma (MIS 22), based on evidence from the Po basin (Muttoni 2003).

All major glaciations fall into the palaeomagnetic Brunhes Chron (van Husen 2000a). Based on geochronological data and the relation between type and magnitude of the global climate signal and the amount of reconstructed sediment production (van Husen and Reitner 2011a), the following correlation with the major phases of global glaciation (Raymo 1997; Lisiecki and Raymo 2005; Hughes and Gibbard 2018) is used (van Husen 2000a; van Husen and Reitner 2011a, b, Fig. 3.1): Günz (MIS 16; 676–621 ka), Mindel (MIS 12; 478–424 ka), Riss (MIS 6; 191–130 ka) and (Late) Würm (MIS 2; 29–12 ka).

All these varying global climatic conditions of the Middle to Late Pleistocene had an impact on processes shaping the landscape of the Alps in relation to the respective magnitude of the climate signal. Thus, expansion of permafrost, strong frost shattering and the vegetation cover changed simultaneously with the growth and shrinkage of valley glaciers.

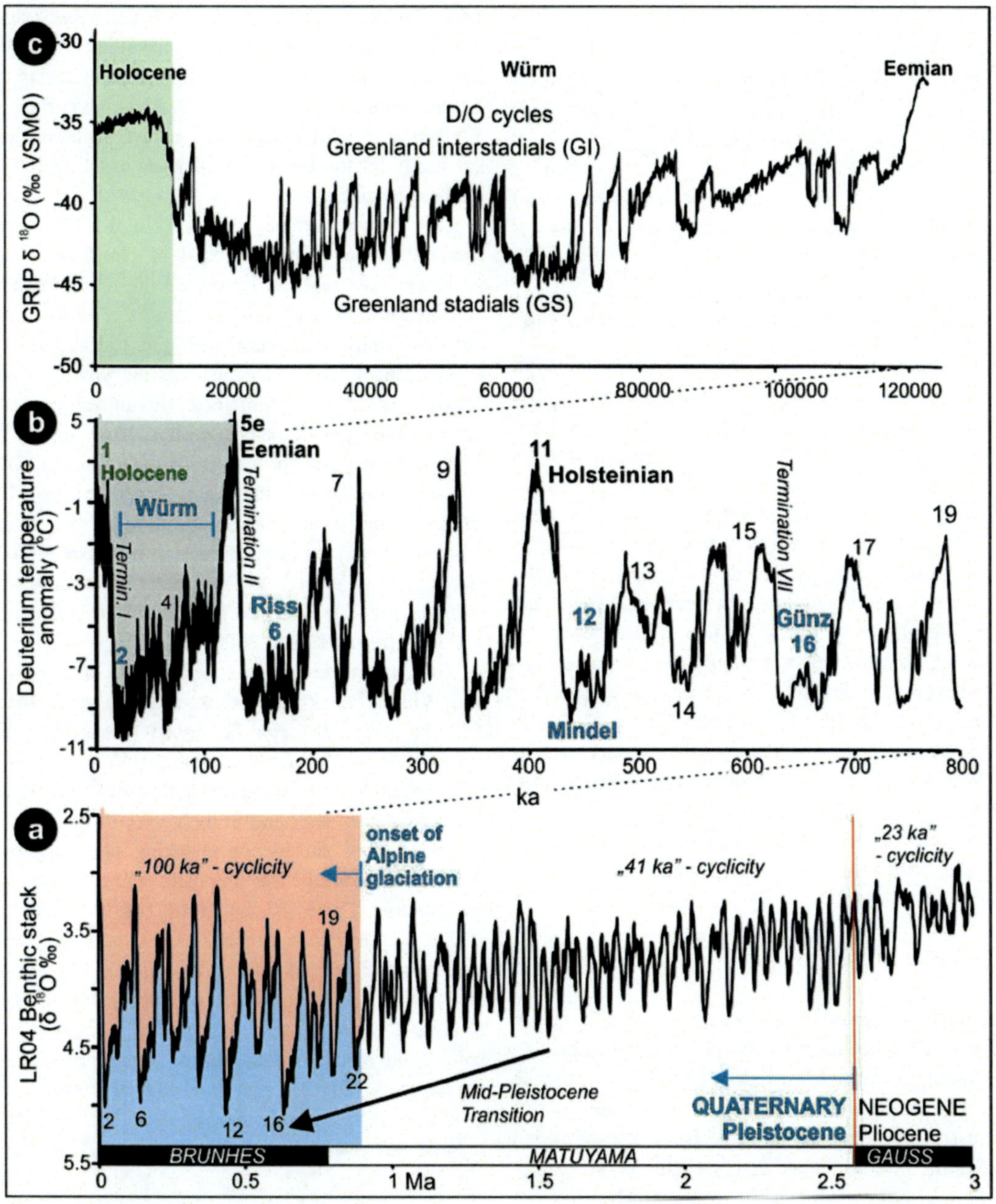

Fig. 3.1 Patterns of Quaternary climate oscillations (modified after Lee et al. 2018) on the global scale and their correlation with Alpine climato-stratigraphic units. **a** Stacked benthic $\delta^{18}O$ record (Lisiecki and Raymo 2005) for the past 3 million years showing a progressive cooling associated with a change of magnitude and frequency of climate cycles. Numbers in **a** and **b** indicate the Marine Isotope Stages (MIS). **b** Graph showing the EPICA Dome C ice-core temperature anomaly record (Jouzel et al. 2007) of the past 800,000 years. The major Alpine glaciations (in blue), important interglacials (in bold) and some of the Terminations (in italics) are indicated. **c** Greenland (NGRIP 2004) ice-core record of the past 125,000 years, highlighting the abrupt, high magnitude climatic events (Dansgaard-Oeschger cycles)

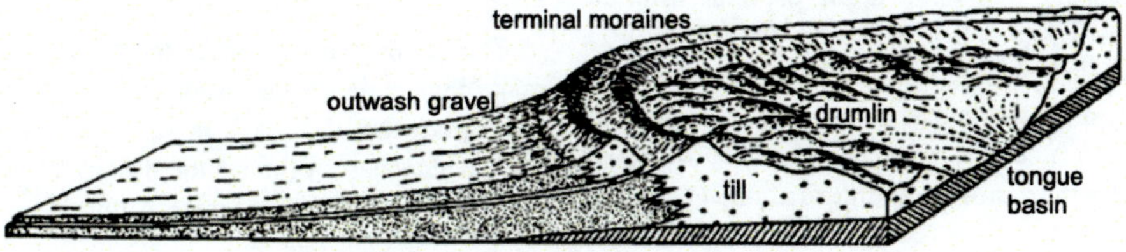

Fig. 3.2 Glacial sequence ("Glaziale Serie" in German) as defined by Penck and Brückner (1909)

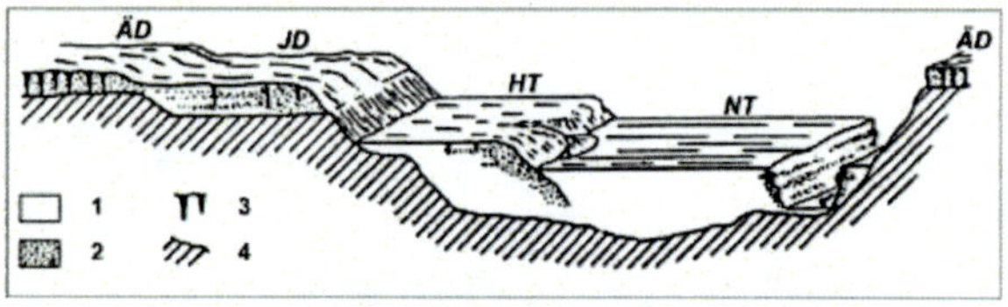

Fig. 3.3 Schematic sketch of the typical sequence of terraces in the foreland of the Eastern Alps (from van Husen 1986). 1: Unweathered gravel; 2: Conglomeratic (cemented) parts; 3: Geologische Orgel (pipe-like structure due to chemical weathering); 4: Pre-Quaternary bedrock. NT: Niederterrasse (Würm), HT: Hochterrasse (Riss), JD: Jüngere Deckenschotter (Mindel), ÄD: Ältere Deckenschotter (Günz)

These changes occurred in higher or lower parts of the Alps, or in the foreland, depending on the degree of climatic deterioration (van Husen 2000a, Fig. 5). Beside the great events (glaciations) resulting in glacier expansion into the foreland, climatic deteriorations are often documented in loess profiles only. Periglacial debris production and gravel accumulation prograded successively from the inner-Alpine areas to the foreland during climatic deterioration, finally forming extended terraces along the rivers (Danube and tributaries) probably only during the four climax periods.

In addition, high-frequency oscillations are known from Greenland ice cores for the last 120 ka with amplitudes of air temperature between 5 and 16° (Lowe and Walker 2015, with references therein). These Dansgaard-Oeschger events (Fig. 3.1) with a duration of 500–2000 years are also reflected in North Atlantic deposits, indicating changes in the ocean circulation. On a small scale, they show the same pattern as the 100 ka cycles, with a slow cooling and rapid warming.

3.4 Tectonic Influence

The plate tectonic conditions in the Austrian area have not changed since the Late Neogene (see Chap. "Geological and Tectonic Setting of Austria"). Modern GPS data display the same trend of displacement with a N-S shortening and a resulting eastward movement of large parts of central and eastern Austria (Metois et al. 2015; Reiter et al. 2018). Several of the major faults in the Eastern Alps and the Diendorf fault on the eastern edge of the Bohemian Massif are still active as indicated by the distribution of fortunately mostly weak earthquakes (Lenhardt et al. 2007; Reiter et al. 2018; Baroň et al. 2019).

Data of recent surface uplift (with Horn in the Bohemian Massif as a reference site) show that SW of the line between the cities of Salzburg and Graz uplift prevails with the highest rates in the range of 1 to 1.6 mm/a (Höggerl 2007). The other part is characterized by subsidence with a maximum around 1.4 mm/a at the eastern boarder of Austria.

If fluvial incision is considered to be in balance with rock uplift, then incision rates provide an information about uplift rates. For the area of Krems, with the palaeomagnetically correlated loess–palaeosol sequence (Fink et al. 1976), an incision of the Danube in the range of 50 m during the last c. 1.8 Ma with an average rate of 0.03 mm/a can be deduced. Wagner et al. (2010) argued for an averaged incision rate of 0.1 mm/a for the last 4 million years of the river Mur north of the city of Graz.

The famous terrace landscape in the Northern Alpine foreland is not only the result of climatically controlled processes, i.e. aggradation during the cold climatic phases and incision in the transition towards warm phases. The classical staircase pattern with the oldest terrace in the topographical highest position and the youngest one in the lowest one cannot be explained without tectonic uplift since the Late Neogene in the background. However, uplift was at least in some parts not uniform but intermittent. The fluvial terrace of the Mindel glaciation along the rivers Steyr and Enns showed a step-like interruption of the normal river gradient at major Alpine thrusts (van Husen 1971, 1981). Hence, an enhanced tectonic activity accompanied with stronger uplift is likely to have occurred between MIS 12 and MIS 6 (van Husen 2000a).

The four terraces corresponding to Günz, Mindel, Riss and Würm (Penck and Brückner 1909) along the Danube and the tributaries can be traced downstream from the classical area in Upper Austria to the Vienna basin due to a uniform and quite stable tectonic situation. East of the Leopoldsdorfer fault and at the centre of the Vienna basin, recent tectonic subsidence is taking place (Decker et al. 2005), influencing the deposits of the two youngest glaciations. Three basins north of the river Danube (Aderklaa, Obersiebenbrunn, Lassee) show morphologically evident subsidence, resulting in dissection of a fluvial surface which was the last shaped during MIS 6 (Braumann et al. 2018). In the case of the Markgrafneusiedel fault, a splay fault of the Vienna Basin Transform Fault, Hintersberger et al. (2018) could show that the offset is the result of five to six surface-breaking earthquakes, with magnitudes ranging between 6.2 ± 0.5 and 6.8 ± 0.4 during the last 120 ka.

The largest and deepest basin linked to the Vienna Basin Transform Fault is the Mitterndorf basin (indicated by the number 24 in Fig. 3.4). The Quaternary basin fill reflects distinct cyclicity, with thick sequences of coarse sediments deposited during cold periods and thin sequences of channel and flood sediments as well as palaeosols from warm periods since MIS 11 (Salcher and Wagreic 2010; Salcher et al. 2017).

The last evidence of neotectonic activity in Austria with a major impact on the landscape and on human settlement was the earthquake of Villach in AD 1348, which triggered six rock avalanches on the southern flank of the Villacher Alpe (Dobratsch) (Till 1907; Jemec Auflič et al. 2017).

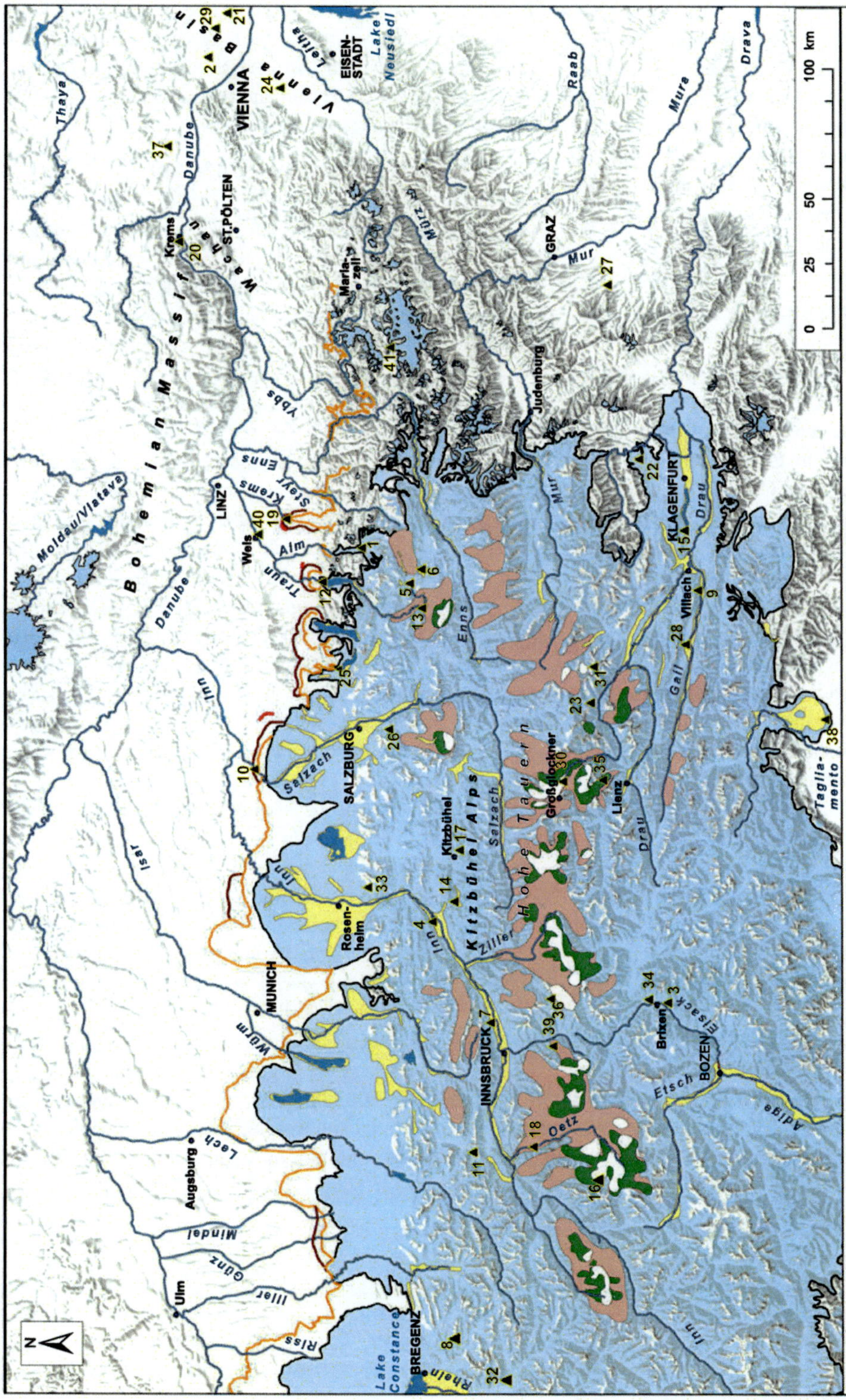

Fig. 3.4 Map indicating the ice extent during the Alpine LGM (Würm Pleniglacial) and previous glaciations (Ehlers et al. 2011). The schematic ice extent during the Alpine Lateglacial stadials Egesen and Gschnitz and that of the Little Ice Age are based on van Husen (2000a). Key locations of the Austrian Quaternary are indicated

3.5 Chronology of the Quaternary Landscape Development in Austria

At the onset of the Quaternary, the principal pattern of Eastern Alpine valleys was already established as a result of the lateral extrusion starting at c. 17 Ma. This impact of Neogene tectonics on the topographic evolution is especially evident in the presence of longitudinal valleys, which run parallel to the strike of the mountain (±E–W) and follow major strike-slip faults. These are valleys of the rivers Inn, (upper) Salzach, Enns, Mur, Mürz, Drau and Gail.

3.5.1 Early Pleistocene (2.58–0.78 Ma)

The best records of the Early Pleistocene in Austria are the loess–palaeosol sequences (LPSs) in the eastern part of the Northern Alpine foreland along the river Danube (cf. Chap. "Sunken Roads and Palaeosols in Loess Areas in Lower Austria: Landform Development and Cultural Importance"). The site of Stranzendorf is one of the rare European sites where the Quaternary/Neogene boundary has been pinned down by the discovery of the palaeomagnetic Gauss/Matuyama reversal (Fink et al. 1976). In accordance with other evidence from the Northern Hemisphere, it shows the onset of loess deposition in the Pleistocene. The site Krems-Schießstätte contains, with 40 m, the thickest LPS of Austria. According to palaeomagnetic investigation, the section started to develop at the end of the Olduvai magnetic event (c. 1.77–1.95 Ma) and continued through the Brunhes Chron, with 17 Glacial-Interglacial cycles in total (Fink and Kukla 1977). According to the fossils in the LPSs, the Early Pleistocene climatic conditions were characterized by a change between cool and dry glacials with loess sedimentation and humid and warmer interglacials with weathering (forest palaeosols), which had become more accentuated towards its younger part (Frank and Rabeder 1997).

For the Early Pleistocene, no traces of glaciations have been found. However, the succession of cool and warm periods during this time had an effect on landscape evolution, especially in the Alpine forelands in the north and southeast. Thus, gravel accumulation of braided river style along the rivers, combined with loess deposition in the surrounding area, took place more or less in the same way but under slightly warmer conditions than during the Middle Pleistocene (van Husen 2000a). The youngest deposits might correspond to the onset of major glaciations in the Alps in the course of the Mid-Pleistocene transition around 870 ka (MIS 22) (van Husen and Reitner 2011a) (Fig. 3.5).

3.5.2 Middle Pleistocene (781–130 ka)

All four "classic" Ice Ages (Günz, Mindel Riss and Würm), with glacier advances as far as the Northern Alpine foreland, are documented by the Glacial Sequence of Penck and Brückner (1909). The best areas to study these textbook examples are the Salzach Glacier tongue area north of the city of Salzburg and in the Innviertel (Weinberger 1955, Salcher et al. 2010, Fig. 3.6) and that of the Traun Glacier near the town of Gmunden (Egger 1996, 2007) (Fig. 3.5) (cf. Chap. "Quaternary Landforms and Sediments in the Northern Alpine Foreland of Salzburg and Upper Austria").

Terminal moraines of the oldest glaciation, the Günz (MIS 16), are linked to terrace bodies in the catchments of the rivers Salzach, Traun and Krems. These *Ältere Deckenschotter* (= older cover gravels) (Fig. 3.7a) are morphologically dominant between the rivers Traun and Enns, where an extensive gravel cover with a fan-shaped geometry is evident. The non-uniform gravel composition indicates a polycyclic formation of this body over several cold stages, with a final reworking and shaping during the Günz glaciation (van Husen 1981; Kohl 2000).

The further climatic development is documented in the catchments of the rivers Alm and Krems. At Kremsmünster (Fig. 3.7b), a white conglomerate (*Weiße Nagelfluh*) with angular drift boulders indicates a cold stage (MIS 14) between the Günz and Mindel glaciations (Kohl 2000, with references therein). It is separated from the Günz deposits (MIS 16) by a strongly developed interglacial soil and from the Mindel deposits by weathered soil sediments (Fig. 3.7c). The immense terminal moraine of the Mindel glaciation testifies that this glaciation was the largest in the Austrian sector.

The more than 200 000 years, lasting phase between the glaciations of Mindel and Riss led to a longer phase of river incision under the condition of regional uplift. Thus, fluvial

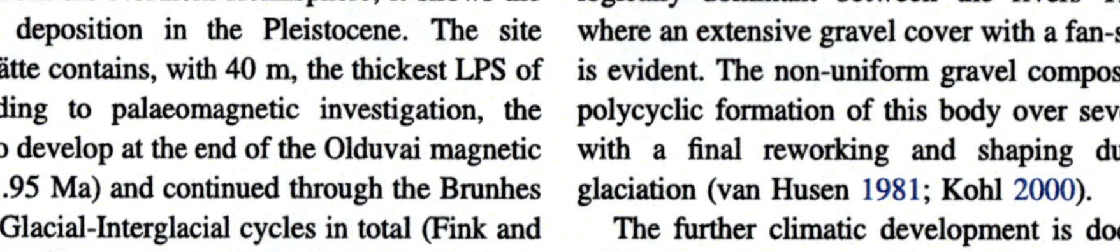

Fig. 3.5 Glacial Sequences of the Günz, Mindel, Riss and Würm glaciations along the Traun Valley in relation to the Pre-Günz terrace deposits (Kohl 1976)

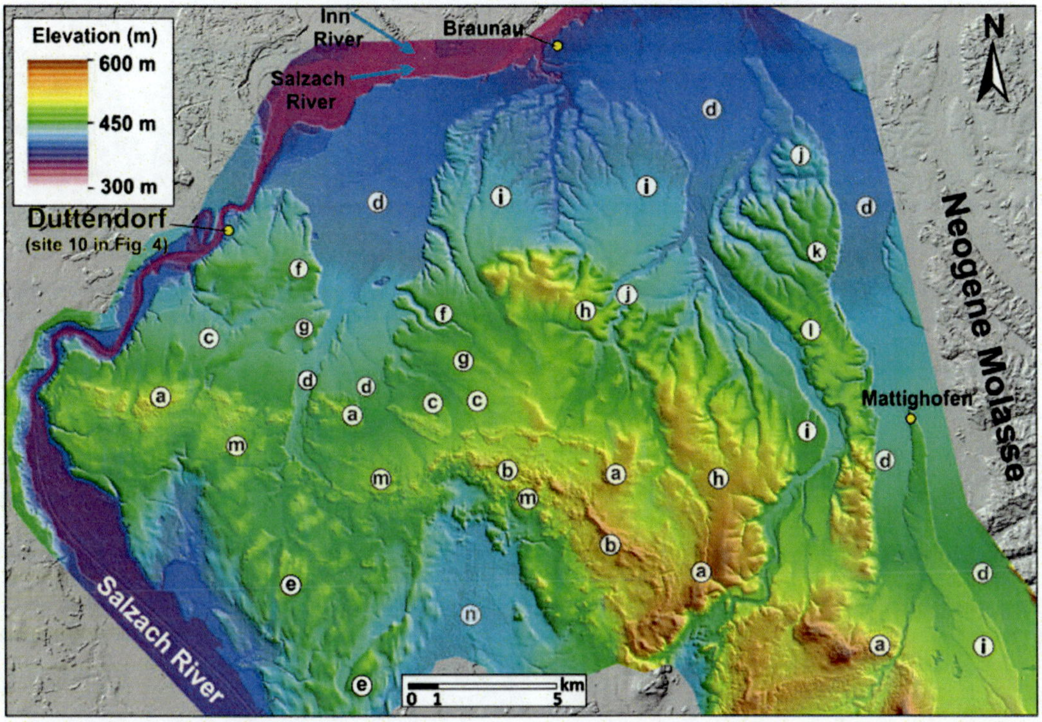

Fig. 3.6 Quaternary landforms showing the extent of the Salzach Glacier lobe north of the city of Salzburg during the glaciations of Würm (LGM), Riss, Mindel and Günz in the Northern Alpine foreland (from Salcher et al. 2018). The site Duttendorf is indicated with by the number 10 in Fig. 3.4. Note the differences in age as displayed by the morphology: **a** LGM (Würm, MIS 2) terminal moraine 1 (maximum extent); **b** LGM terminal moraine 2; **c** LGM, upper outwash *obere Niederterrasse* with sub-terrace levels; **d** LGM, lower outwash *untere Niederterrasse* with sub-terrace levels; **e** tunnel channels of the subglacial drainage system (LGM); **f** Riss terminal moraine 1 (maximum extent); **g** Riss terminal moraine 2; **h** Riss outwash, *Hochterrasse* (undifferentiated); **i** Mindel terminal moraine; **j** Mindel outwash, *Jüngere Deckenschotter* (undifferentiated); **k** Günz terminal moraine; **l** Günz outwash, *Ältere Deckenschotter* (undifferentiated); **m** Post-LGM ice-decay deposits; **n** Peat bog (Ibmer Moor)

deposits of the *Hochterrasse* (= high terrace), genetically linked to the terminal moraines of the Riss glaciation, are situated in a lower position in far narrower valleys than before. The extent of this glaciation was a little smaller than during the Mindel, but referring only to the eastern part, it was much larger than during the subsequent Würm. Such a massive difference reflects ice dynamical processes in the longitudinal valley of the river Enns (van Husen 2000a). Luminescence dating of fluvioglacial sediments in the Ybbs Valley (Bickel et al. 2015a) and in the Salzach Valley (Salcher et al. 2015), as well as of *Hochterrasse* deposits in the catchments of the rivers Inn, Traun, Krems and Steyr (Bickel et al. 2015b), provides support for correlating the penultimate glaciation with MIS 6.

There are only rare cases where deposits of the formerly glaciated area provide some hints of palaeoenvironmental development in the long phase between the Mindel and the Riss. The most prominent inner-alpine deposit of this long phase is the Hötting breccia (north of Innsbruck, Fig. 3.7d–f), a talus deposit with a complex internal stratigraphy (Penck 1920; Sanders and Ostermann 2006, Sanders and Spötl 2014). Recent radiometric dating (Spötl et al. 2013), documenting lithification and fracturing of the Hötting breccia during MIS 7 (243–191 ka) support this stratigraphic assessment. According to van Husen (2000a), this strong talus production, also found elsewhere on the southern flank of the Northern Calcareous Alps, might be linked to stronger uplift during this time span.

The gap in the glacial and fluvial records between the Mindel and the Riss regarding landscape evolution is filled by the LPSs of the Northern Alpine foreland, which are a valuable source of palaeoenvironmental and palaeoclimatic information for the Middle Pleistocene. At the reference site Wels-Aschet (van Husen and Reitner 2011b), on top of the *Ältere Deckenschotter,* results of pedological, geomagnetic weathering analysis and luminescence dating are available (Preusser and Fiebig 2009; Reitner and Ottner 2011; Terhorst et al. 2012; Scholger and Terhorst 2013). On top of the Mindel loess (MIS 12) is a prominent soil showing intense weathering during the long Holsteinian interglacial (MIS 11; 421–374 ka). Cool periods like MIS 10 and MIS 8, which were too weak to generate foreland glaciation and terrace

Fig. 3.7 Sedimentary record of the middle Pleistocene. **a** Outcrop of *Ältere Deckenschotter* (location east of Kremsmünster) with a cover of younger loess on top (outcrop height approximately 6 m). White arrows and dashed lines indicate the locations of pipe-like weathering structures ("Geologische Orgel") which occur between "pillars" of cemented gravel (conglomerate). **b** Historical quarry in Kremsmünster mining the building stone of ***Weiße Nagelfluh***. The superposition of Weiße Nagelfluh (=white conglomerate), a conglomerate predominantly made up of triassic limestone, by Graue Nagelfluh, (grey conglomerate), a conglomerate representing the proglacial sediments of the advancing Krems-Steyr Glacier during the Mindel glaciation, topped by the Mindel subglacial till is evident. **c** Weathering horizon between the white and the grey conglomerate, indicating a stable surface under supposed interglacial conditions (location: quarry south of Kremsmünster, height of the outcrop approximately 4 m). Photographs a-c: D. van Husen. **d** Nordkette, i.e. the southern flank of the Northern Calcareous Alps north of the city of Innsbruck. The terrace (with the location of (e)) is made up predominantly of Hötting breccia. **e** Historical quarry Mayr with the Hötting breccia. **f** Detail with two types of Hötting breccia (red and white). Photographs d-f: J. Reitner

deposition, are recorded either as a thin loess layer (MIS 10) or, indirectly, as the source material for the following soil formation. The latter is evident in a pedocomplex as a result of weathering during the interglacials of MIS 9 and MIS 7, which is topped by the Riss loess. Further findings of LPSs in the Danube catchment (Terhorst 2013; Sprafke et al. 2014; Sprafke 2016) as well as the stratigraphy of the Mitterndorf basin (Salcher et al. 2017) reveal the same climatic development for this time period.

3.6 Climatically Controlled Processes from the Last Interglacial-Glacial Cycle to the Present Day

The Last Interglacial-Glacial cycle together with the Holocene serves as a model for understanding the climatically driven processes, which were also active during the previous cycles (van Husen 1989, 2000a). This time period covers the Eemian (MIS 5e, c.130–115 ka) and the Würm(ian) (c.115–11.7 ka). The latter is subdivided into Early (MIS 5d-a; 115–71 ka), Middle (MIS 4–3; 71–30 ka) and Late (MIS 2; c. 30–11.7 ka) Würm, according to Chaline and Jerz (1984) (Fig. 3.8).

3.6.1 The Last Interglacial and the Early Würm

The decay of the penultimate glaciation and thus the onset of Termination II (Fig. 3.1) has been constrained by U/Th—dating of speleothems of the high-elevated Spannagel cave in the Zillertal Alps (indicated by the number 36 in Fig. 3.4) (Spötl et al. 2006). It most probably started around 136 ka, with a glacial phase similar to what is known as the Würm or Alpine Lateglacial. Full interglacial, warm and moist conditions including soil development and vegetation cover were only established at 129–130 ka and prevailed until c. 119 ka (Spötl et al. 2007). The end of the Last Interglacial appears as a time transgressive and complex sequence of events, which ended around 114 ka according to data from Entrische Kirche Cave (Meyer et al. 2008).

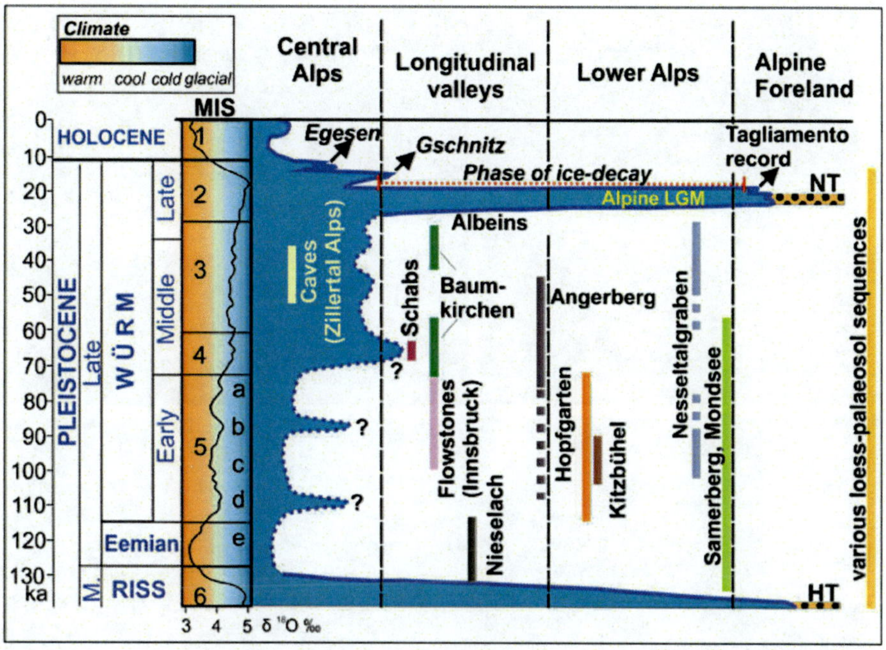

Fig. 3.8 Temporal development of the ice extent (in blue) in the Eastern Alps during the Last Interglacial/Glacial cycle (modified after Reitner 2011 and Monegato and Ravazzi 2018). The stacked $\delta\ ^{18}O$ record from benthic foraminifera and the Marine Isotope Stages (Lisiecki and Raymo 2005) is shown for comparison. NT—*Niederterrasse* (Würm), HT—*Hochterrasse* (Riss)

The section of Mondsee (outcrops, drill cores, Fig. 3.9a), located at the northern rim of the Alps, is the Austrian reference site of palaeoclimatic and palaeoenvironmental change during Termination II, the Eemian, the Early Würm and parts of the Middle Würm (~50 ka). This site was investigated in detail regarding palaeogeographic situation, sedimentology, pollen content and plant remains as well as stable isotopes (van Husen 2000b; Krenmayr 2000; Drescher-Schneider 2000; Oeggl and Unterfrauner 2000; Papesch and Rank 2000). The sedimentary record between the subglacial tills of the Riss (MIS 6) and Würm (MIS 2) consists of delta sediments showing the infill of an ancient lake, with a water surface around 50 m above the contemporary lake Mondsee.

Vegetation started with a treeless pioneer phase with herbaceous plants and grasses. The onset of the interglacial was marked by reforestation, initially consisting of shrubs and pine trees. At the climax, valleys and the foreland of the Alps were densely forested with well-developed oak-mixed forest, with a high content of fir. The temperatures at that time were 2–3 °C above the current Holocene values. The Late Interglacial was dominated by a coniferous forest with first fir, spruce and hornbeam and finally pine with increasing herbaceous vegetation. A major decrease in temperature led to the disappearance of forest and a timberline situation during the first Early Würm stadial (MIS 5d). After this cold phase, forest could recover again during the first Early Würm interstadial (MIS 5c) dominated by spruce forests with a small presence of deciduous trees and fir. Temperatures at its climax were close to current levels. The second cold period, the Early Würm stadial (MIS 5b), again shows a timberline situation with an open herbaceous communities. The second warm period (second Early Würm interstadial, MIS 5a) was also dominated by spruce forests, with rare occurrence of deciduous trees and fir.

This effect of palaeoclimate and vegetation on landscape-forming processes from MIS 5e to MIS 5a/4 is displayed at other sites with a different palaeoenvironmetal setting. For the Eemian, the site of Nieselach in the Gail Valley (van Husen 2000c) shows a meandering river system with peat formation in oxbows under interglacial conditions (Draxler 2000) (Figs. 3.9b and 3.10b). In total, the valley floor looked quite similar to its appearance during the Holocene. However, the sediments around Nieselach show a Last Interglacial valley floor of the river Gail 10–15 m above that of the Holocene. Coarse gravel on top of the peat coal with cut-and-fill structures typical for braided rivers documents the onset of the Würm (MIS 5d-MIS 2).

The sedimentary records of the Kitzbühel Alps (Tyrol) reveal the climatically controlled changes of fluvial style. Strong aggradation of braided river style during the two Early Würm stadials (MIS 5d, 5b) is regarded as the result of efficient frost shattering in the upper river catchments and thus overloading of the streams (Reitner 2005a; Reitner and Gruber 2014). However, no traces of glacier advances during these stadials have been found so far which is supported by data from the Inn Valley for MIS 5c to MIS 5a (Spötl and Mangini 2006). The valley floors in the Kitzbühel Alps during the two forested Early Würm interstadials looked like today in narrow Alpine Valleys: overbank deposits and

Fig. 3.9 **a** Interglacial (Eemian) lacustrine sediments (lake marl and banded clay) at Mondsee (from van Husen 2000b). White sheets mark different pollen zones indicating the development of vegetation. **b** Outcrop at Nieselach showing the lignite (Eemian) overlain by gravel (Würm) (from van Husen 2000c). **c** Outcrop at Hopfgarten in the Kitzbühel Alps, showing the intercalation between red sandy (distal) alluvial deposits and dark lignites (compressed peat) of Early Würm age (outcrop height: c. 40 cm). Photographs a–c: D. van Husen. **d** Panoramic view towards NW across the Inn Valley at the Angerberg Terrace, a key site for the understanding of the Würm cycle in the Lower Inn Valley (from Reitner and Gruber 2014). The subglacially shaped and till-covered Angerberg Terrace as well as the steep southern rim of the Northern Calcareous Alps with the prominent niches of two pre-LGM rockslides (Butterbichl and Mariastein) is evident

coarser fluvial deposits of the rivers were intercalated with debris flow deposits from the alluvial fans of tributary valleys which generated swampy areas with peat growth in their backwaters (Reitner and Draxler 2002) (Fig. 3.9c).

The Last Interglacial and the Early Würm climatic development is also well documented in LPSs of the Alpine foreland in the Danube catchment (Terhorst 2013). Warm interglacial climate led to intensive soil formation (e.g. Terhorst et al. 2002). The comparatively weaker soil formation during the interstadials is evident in pedocomplexes like in Stillfried in the dry eastern part of Austria (Fink 1956).

3.6.2 Middle and Late Würm

3.6.2.1 Mis 4

The time period of MIS 4 is regarded as a stadial with an increased extent of ice sheets and some mountain glaciations (Hughes et al. 2013). However, no evidence of a glacier advance exists in the records on the northern rim of the Alps, such as Mondsee (Drescher-Schneider 2000) or Samerberg (Grüger 1979). The collapse of the vegetation, with very low pollen concentrations, soil erosion and reworking seems to be a typical process, potentially with the presence of permafrost, as recorded at the Angerberg site (Fig. 3.9d) in the Lower Inn Valley (Starnberger et al. 2013).

Based on data from Schabs (Eisack Valley, Castiglioni 1964; Fliri 1978) and from Baumkirchen (Inn Valley, Barrett et al. 2017), with no subglacial till there, a limited glaciation, which may have just reached the longitudinal valleys (Fig. 3.8), seems to be plausible.

3.6.2.2 MIS 3 and Transition to the Würm Pleniglacial (LGM)

The Lower Inn Valley east of Innsbruck provides a unique sediment archive for deciphering further landscape development during MIS 3 and the transition to MIS 2. The sedimentary record of Angerberg (Fig. 3.9d) shows improved climatic conditions between c. 55 and 45 ka (MIS 3; Starnberger et al. 2013). Cold-adapted trees reappeared during interstadials, forming an open forest vegetation. MIS 3 stadials were shorter and less severe than the MIS 4.

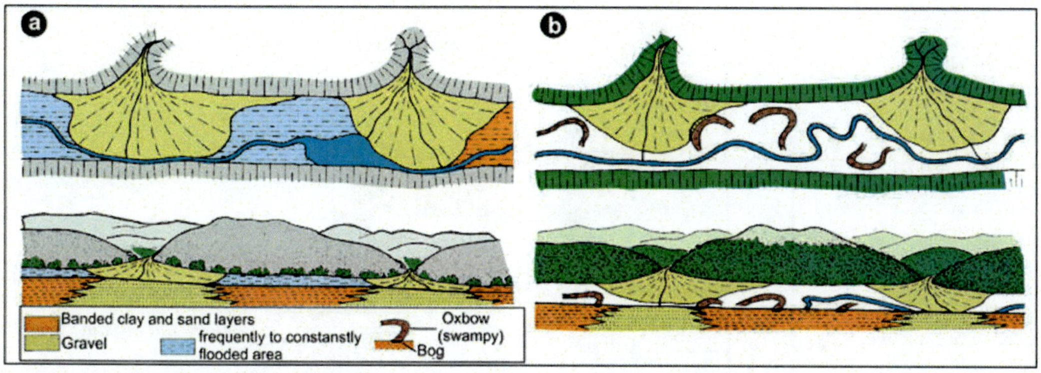

Fig. 3.10 Model showing different depositional environments during the Würm stage. Phase **a**: rapidly prograding alluvial cones block main valley and cause deposition of fine-grained flood material in inter-cone areas. Phase **b**: reduced sediment input from tributary valleys allows continuous fluvial drainage by meandering river systems on extensive alluvial floodplains (after van Husen 1983a)

The so-called Inntal-Mittelgebirge (Gnadenwald terrace) is a till-covered terrace with a top 300 m above the modern valley floor, consisting of clayey-silty lacustrine deposits, topped by a thin sand layer and proglacial gravel (*Vorstoßschotter*) (Fig. 3.11). The sequence is exposed at Baumkirchen, and this site was chosen to lithostratigraphically define the limit between the Middle and Late Würm (MIS 2) and thus the onset of the Würm Pleniglacial (= Alpine Last Glacial Maximum) with the top of the silty lacustrine deposits (Chaline and Jerz 1984). The first record of the Baumkirchen sequence with radiocarbon data of the lacustrine deposits (Fliri et al. 1970; Fliri, 1973) and pollen analysis by Bortenschlager and Bortenschlager (1978) was refined by Spötl et al. (2013) and was recently extended based on luminescence data and palynological analyses of drill cores (Barrett et al. 2017, 2018). Thus, a sequence from c. 45 ka to 33 ka is available.

The MIS 3 Baumkirchen record shows a c. 170 m thick, more or less uniform succession of laminated "banded" clayey silt. It assembles two stadials with cold and dry tundra vegetation and two very weak interstadials with well-developed open vegetation and some stands of trees. Macrofossils like branches of Scots pine, dwarf pine, green alder and sea buckthorn occur in the upper part. The sedimentation rate was high, with >3 cm/a (Bortenschlager and Bortenschlager 1978; Spötl et al. 2013; Barrett et al. 2017). The most plausible model for the infill of the Inn Valley and other (longitudinal) valleys with similar topography in the catchment is provided by van Husen (1983a; b; Fig. 3.10a). Cool periods are characterized by large quantities of sediments and high sedimentation rate. This is due to reduced vegetation on higher slopes. Such conditions, together with enhanced frost shattering, result in high debris input to the creeks of the tributary valleys, causing rapidly prograding alluvial cones to develop on the valley bottoms (Fig. 3.12b). Accordingly, the main valleys were blocked and the fluvial gradients became unbalanced. Between the cones, frequently or constantly flooded areas with fine-grained sediments were formed, and stratified clay, sand and silt were deposited, such as in Baumkirchen. The sedimentation rate depended on the growth of the cones and therefore directly on the debris input induced by climatic deterioration.

During the final climatic deterioration (Figs. 3.12c and 3.8), the valleys were filled with coarse gravels (*Vorstoßschotter*) at many places to high elevations, as a result of progressive overloading of the main river with debris (van Husen 1983a, b). The *Vorstoßschotter* extended laterally into the terraces in the foreland, which accumulated at the same time as the thick gravel bodies along the rivers, which now build up the lower parts of the *Niederterrasse* (= low terrace). In both the non-glaciated areas (e.g. Neurath, Draxler and van Husen 1991), where deposition was induced by periglacial activity (frost shattering), and in outwash zones (e.g. Duttendorf; Starnberger et al. 2011), accumulation finished at approximately the time when mountain glaciation reached its greatest extent, and climatic deterioration was at a climax.

For the Inn Valley, this final gravel accumulation under ice-free conditions occurred around 30 ka (Spötl et al. 2013) and, hence, marks the onset of the Alpine LGM. Eventually, the Inn Glacier advanced over the thick sequence as shown by the subglacial till (Fig. 3.11a, c) on top of the "Inntal-Mittelgebirge". Such a chronology is also supported by the findings from Nesseltaltalgraben (Mayr et al. 2017, 2019), a reference site for the Würm in the Salzach Glacier catchment.

Concurrently with the drop in summer insolation in the Northern Hemisphere (Fig. 3.13) and the growth of the large ice sheets, the Alpine Glaciers started to expand from the cirques to the high-elevated valleys. The glaciers eventually advanced into the main valleys. The ice began to accumulate increasingly, and ever larger areas now lay above the snow line.

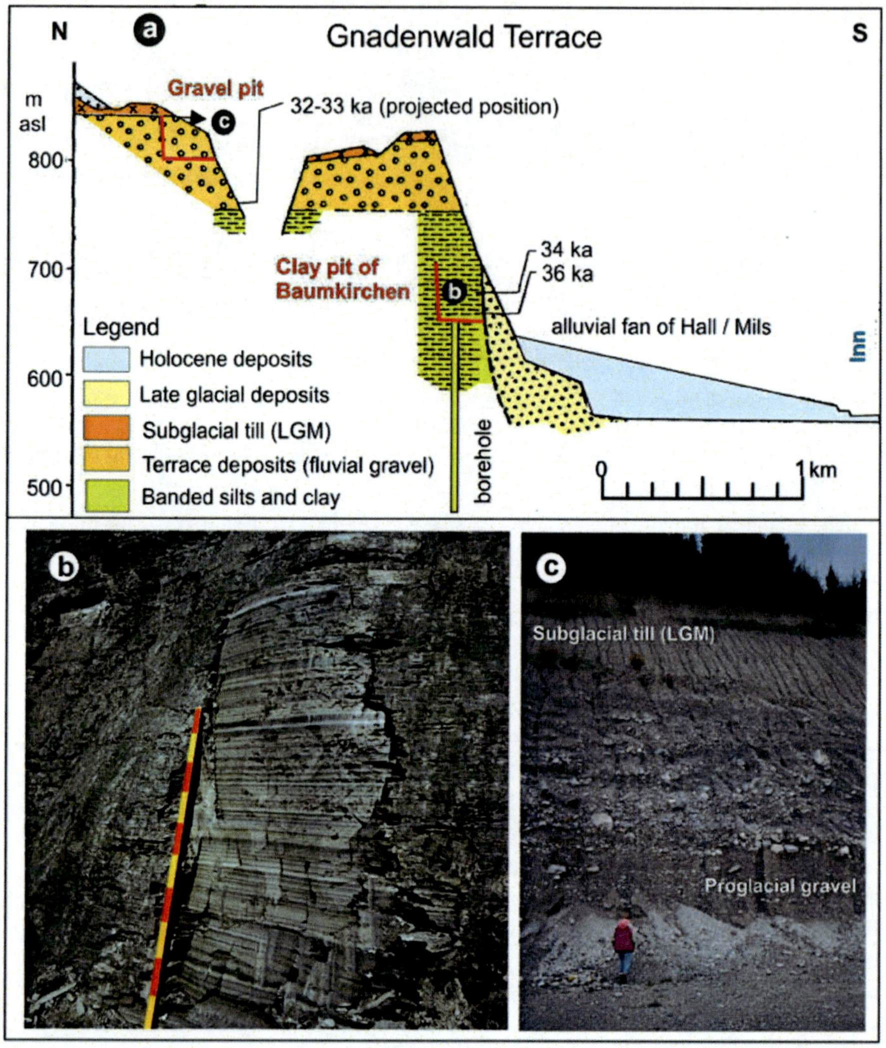

Fig. 3.11 **a** Cross-section of the Gnadenwald Terrace east of Innsbruck, at the clay pit of Baumkirchen. The sites of **b** and **c** are indicated (modified after Patzelt and Resch 1986). **b** Typical banded clayey silt of Baumkirchen. Photograph: D. van Husen. **c** Proglacial gravel (Vorstoßschotter) deposited by a braided river with a coarsening-upward trend, overlain by LGM subglacial till. Note the poor sorting and the large boulders in the gravel. Photograph: J. Reitner

The configuration of the valleys influenced the further expansion. According to van Husen (2000a), the topography of the longitudinal valleys (such as the Inn and the Salzach Valleys), especially in their relation to the tributary valleys, had a fundamental impact on the rate of ice buildup as well as on the glacier extent during the major glaciations. In the Inn Valley, the phase at the beginning of the Alpine LGM was characterized by ice congestion in the narrow valley as a result of a strong inflow of ice, especially from the southern tributaries with large, high-elevated drainage areas (e.g. Ötz Valley, Ziller Valley,). Consequently, the ice surface rose and a rapid expansion of the accumulation area occurred, favouring quick ice buildup under the condition of successive lowering of the equilibrium line. Due to this process, the Inn Glacier as well as the Salzach Glacier (Reitner 2005a) could cross watersheds to the north before these valleys were filled with local ice. In this early state of ice buildup, ice-flow directions changed dramatically with the successive growth of the glaciers and ice-dammed lakes were formed in tributary valleys (Reitner et al. 2010; Sanders et al. 2014; Menzies and Reitner 2016, 2019). Age control of this accelerated ice buildup is missing so far in the Austrian Alps. However, a larch fragment found in a thick sequence of sand and gravel at Albeins (a site in Northern Italy, indicated by the number 3 in Fig. 3.4) with a calibrated date of c. 28 ka provides a timestamp for this phase (Fliri 1988).

Finally, a large complex of transection glaciers, i.e. an interconnected system of valley glaciers (van Husen 2000a) quickly developed with outflows that traversed the present day watersheds (Fig. 3.4). At the climax of the Alpine LGM, between 26 000 and 20 000 years ago, the great Alpine Valleys such as the Inn and Salzach Valleys contained large glaciers of more than 1000 m in thickness. Similar to the modern conditions in Greenland, only the highest peaks,

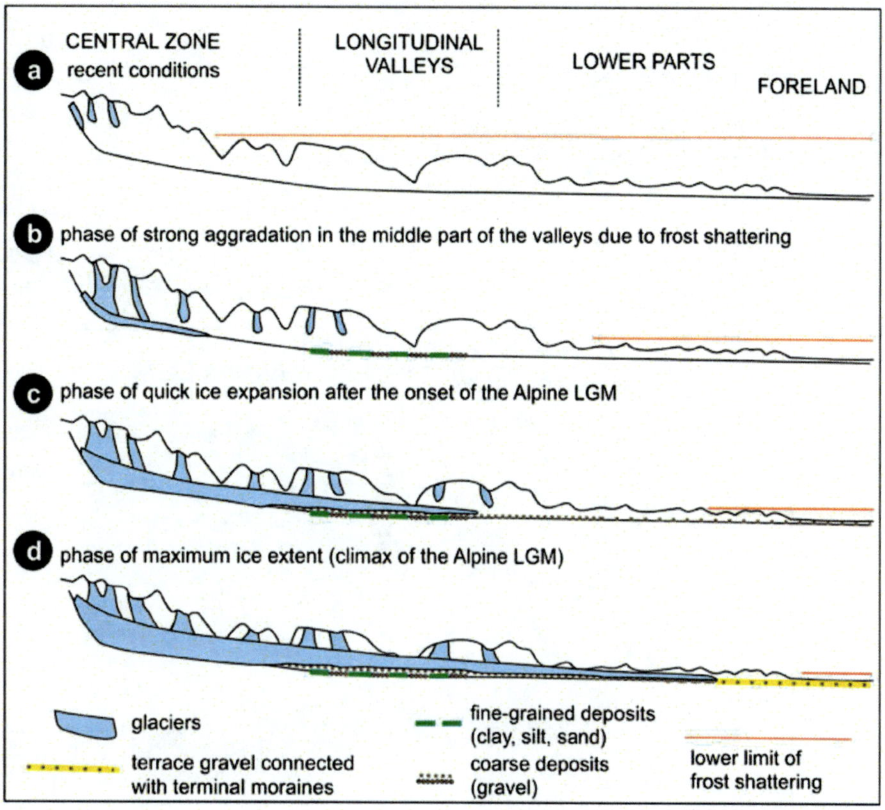

Fig. 3.12 Sketches based on van Husen (2000) showing the gradual development of climatic decay during the final phase of the Würm glaciation and associated processes in the catchments, such as lowering of the lower limit of frost shattering and resulting aggradation. Only a little lowering of the equilibrium line (c to d) is necessary to cause a very rapid and great expansion of the valley glaciers

so-called nunataks, protruded from these ice masses. The ice tongues advanced far into the Bavarian Alpine foreland (e.g. Troll 1924; Reuther et al. 2011) as well as into the South-western Innviertel region where piedmont tongues developed (Weinberger 1955). East of the river Salzach, the glacier tongues of the complex of transection glaciers just reached the edge of the Alps (e.g. lake Traunsee; van Husen 1977), whilst they ended farther to the east within the Enns, Mur and Drau Valleys as valley glaciers within the Alps (e.g. van Husen 1968). In the eastern and southeastern fringes of the Eastern Alps, only isolated glaciers existed within cirques and plateaus (van Husen 1987). The reduced glaciation of these areas is due to lower altitudes and less precipitation. This LGM glacier extent documents a depression of the equilibrium line altitude (ΔELA) with respect to the ELA of the averaged Little Ice Age in the range of 1100 m (Lichtenecker 1938), which might have reached 1500 m based on data from the Rhine Glacier (Ivy-Ochs 2015, with references therein).

In areas where the glacier tongues remained stationary for an extended period, rock material was deposited along the margin to form an end moraine. The meltwater issuing at this point became overloaded with debris during the summer floods. However, the transport capacity of the streams was not sufficient to carry away all the entrained material, and thus, it was deposited in the various braided ("wild") river channels, thereby forming huge alluvial plains. Such deposits called *Niederterrasse* (= low terrace) can be traced in the catchments of the rivers Danube, Mur and Drau from the end moraines to the Austrian border and beyond.

In the northern foreland, only a few proglacial archives recorded this enormous glacier expansion. The most important Austrian site is that of Duttendorf (cf. Chap. "Quaternary Landforms and Sediments in the Northern Alpine Foreland of Salzburg and Upper Austria"), located in the foreland of the Salzach Glacier (Traub and Jerz 1975). According to the results of luminescence dating (Starnberger et al. 2011), loess deposition started around 29 30 ka and the last outwash deposition occurred around c. 20 ka.

None of the end moraines in Austria has been dated so far. For comparison, however, an increasing number of geochronological data from ^{14}C dating of wood and exposure dating of glacially transported boulders with cosmogenic nuclides (mostly ^{10}Be) is available, from Switzerland and Italy in particular (Ivy-Ochs 2015; Monegato et al. 2017). The best age-constrained record of Alpine LGM glacier activity is found on the southern flank of the Eastern Alps, at the moraine amphitheatre of the Tagliamento Glacier (Monegato et al. 2007). The ^{14}C-based stratigraphy

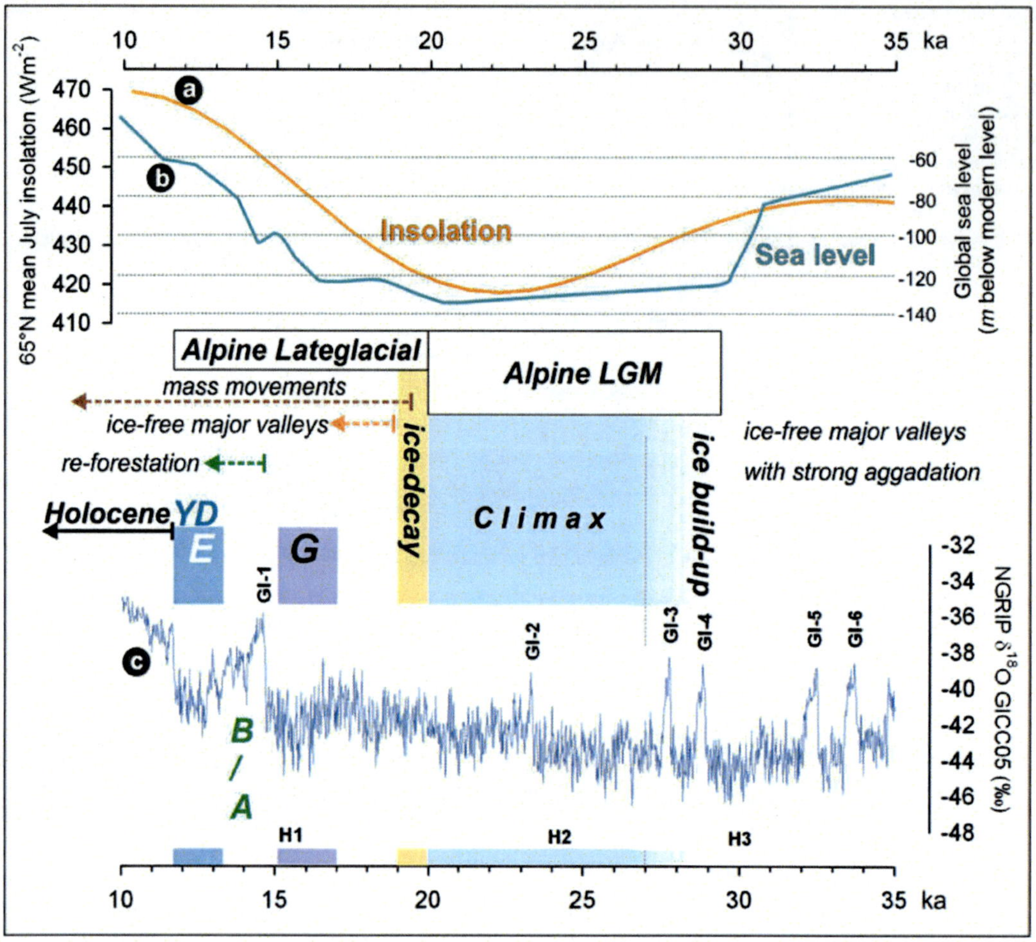

Fig. 3.13 Time span covering the Middle Würm, the Würm Pleniglacial (LGM), the Alpine Lateglacial with the phase of ice decay, the Gschnitz Stadial (G), the Bølling/Allerød Interstadial (B/A) and the Egesen Stadial (E) in the Younger Dryas (YD), and the transition to the Holocene. The main processes and events are indicated. The (a) 65°N mean July insolation (Berger and Loutre 1991) displays the major climatic forcing for the onset and decay of the LGM glaciation. Sea level changes (Lambeck et al. 2014) indicate the phases of global LGM glaciation with a minimum. (c) NGRIP isotope record with numbered Greenland Interstadials (GI) and Heinrich (H) events (Andersen 2006), showing the pace of climatic oscillations in the North Atlantic region, which had an impact on the Alpine Glacier activity

shows an early phase of ice advance between 26.5 and 23 ka, with an end moraine formation linked to a maximum extent. A second advance, during which the glaciers nearly reached the previous extent, took place between 24 and 21 ka. The last phase of glacier activity of the lobe, with the formation of recessional moraines, occurred between 21 and 19 ka.

This twofold Alpine LGM with an early climax (26–23) and a late one (<23 ka), which is also found in the Traun Glacier area (van Husen 1977, 2000a), amongst others, may reflect two different sources of precipitation for the waxing of the glaciers (Monegato et al. 2017). The first phase received an important contribution from the Mediterranean, whereas the later one had a more westerly source (Luetscher et al. 2015). According to geochronological data of the Alps (Ivy-Ochs 2015), the climax of the Alpine LGM took place during the MIS 2, in agreement with the global LGM between 26 and 19 ka. The increase in global ice volume is reflected in a sea level change with a steep drop starting around 30 ka and followed by a gradual lowering until 21 ka, leading to a sea level minimum of −134 m (Lambeck et al. 2014, Fig. 3.13). The contribution of the Alpine ice volume is in the range of 0.3 m (Sequinot et al. 2018).

During the MIS 3 stadials and the LGM, the actively aggrading braided rivers provided vegetation-free outwash plains during seasons (autumn, winter) with low or no discharge, which served as the major source area for wind-blown dust. The latter was deposited in the surrounding cold steppe areas as loess. LPSs in Upper Austria (Inn, Traun) are characterized by well-developed pedocomplexes

Fig. 3.14 Legacy of the LGM and the Alpine Lateglacial: **a** Ice wedge pseudomorphs filled with sand developed on top of fluvial gravel (Mindel), witnessing permafrost conditions in the foreland (Sierndorf, 30 km NW of Vienna), Photograph: R. Roetzel. **b** Typical subglacial till consisting of a matrix-supported diamict. Photograph: J. Reitner. **c** Remnant of a kame terrace from the phase of ice decay, consisting of delta deposits (foresets) c. 400 m above the modern valley floor (Bannberg, 8 km WSW Lienz) (from Reitner et al. 2016). **d** Drumlins of Maria Bichl and Olschitzen in the Lower Drau Valley close to Spittal/Drau were formed at the base of the Drau Glacier during the LGM. **e** This end moraine (with the Rotwand-Grundalm hut on top) was formed during the Gschnitz Stadial by a local glacier of the Kitzbühel Alps in the Upper Windau Valley (15 km SSW Hopfgarten)

from the MIS 3, most likely due to higher average precipitation compared to the sites further east (Terhorst et al. 2002). First luminescence dating efforts show phases of advanced weathering, interrupted by permafrost during MIS 3, whereas MIS 2 is represented by a series of loess and tundra gleys (Terhorst et al. 2002). In the drier area to the east, the LPSs with Palaeolithic findings (cf. Chap. "Sunken Roads and Palaeosols in Loess Areas in Lower Austria: Landform Development and Cultural Importance") e.g. Krems-Wachtberg (Lomax et al. 2014; Terhorst et al. 2014) have records that mirror the Greenland ice-core stratigraphy in more detail. Greenland stadials were phases of loess accumulation, whereas interstadials are documented by palaeosols (Terhorst et al. 2014, 2015). The thickness of loess and the intensity of soil development depend on the strength and duration of climatic phases, in agreement with central European reference sites (Rousseau et al. 2017). Cold and dry pleniglacial conditions resulted in atmospheric dust loading as much as an order of magnitude larger than today (Harrison et al. 2001) and, consequently, in high dust accumulation rates (Frechen et al. 2003). High atmospheric pressure gradients during the LGM resulted in enhanced wind speeds, which created wind-derived features, e.g. wind-polished rock surfaces (ventifacts) and wind-sculptured erosional landforms. They allow determining the dominant wind direction during their formation and show that the LGM palaeo wind system in the eastern foreland was characterized by W to NW winds (Sebe et al. 2015).

The mean annual temperatures in the non–glaciated areas, the so-called periglacial zone, were at least 10–12° below present values during the pleniglacial (Frenzel et al. 1992). This phase corresponds with the global Last Permafrost Maximum, where continuous permafrost prevailed in the periglacial areas of Austria (Vandenberghe et al. 2014). An indication of the former depth of the permafrost base has been provided by a tunnelling project in Lambach. Vertical cracks discovered 40 m below the surface are regarded as the product of ice wedge formation (van Husen 1999). Singular ice wedge pseudomorphs (Fig. 3.14a) are found in gravel pits or loess deposits. Networks of such thermal-contraction cracks are evident in the aerial photographs from archaeological prospection campaigns (e.g. Carnuntum area east of Vienna; Doneus et al. 2001), but they have not been studied so far. Permafrost degradation after the end of the LGM leading to the formation of thermokarst lakes is a plausible explanation of the frequent shallow lakes and basins in the Seewinkel area in the east (cf. Chap. "Lake Neusiedl Area: A Particular Lakescape at the Boundary between Alps and Pannonian Basin").

Multiple cycles of seasonal dynamics, with melting of the upper surface layer of rock or clastic material (active layer) during the warm season followed by refreezing, led to characteristic deposits and morphologies. Frost action on steep rock faces resulted in frost shattering, leading to an increased debris production with the resultant formation of talus cones below the cliffs. Evidence of the active layer thickness (=thaw depth) during summer is provided by the layers showing cryoturbations in e.g. gravel pits of the Hochterrasse. On slopes, mass wasting occurred mostly due to gelifluction in the meltwater-saturated active layer in summer time. In the areas affected by deep subtropical chemical weathering during the Neogene and Palaeogene, such processes led to the subsequent stripping of the loose granite grus under periglacial conditions. The legacy of these processes is the blockfields on some elevations in the Bohemian Massif and the cover of slope deposits, which is up to several metres thick (Huber 1999; Kohl 2000). All the processes related to frost action led to cryoplanation, producing more gentle and less rugged morphology. The effect can be studied by comparing end moraines of the Würm with those of older glaciations, which show a smoothed morphology due to multiple permafrost cycles (Figs. 3.5 and 3.6).

3.6.2.3 Alpine Lateglacial (MIS 2)

The end of the Alpine LGM is the starting point of the Alpine Lateglacial (Würm Lateglacial) which is the expression of Termination I (Fig. 3.1) in the Alpine context. The onset is defined by the withdrawal of the glaciers from their tongue basins (Chaline and Jerz 1984). The subdivision of the Alpine Lateglacial into glacial stadials has a long history beginning with Penck and Brückner (1909) and a stepwise development since then (Mayr and Heuberger 1968; Maisch 1982, 1987; Kerschner 1986, 2009; van Husen 1997; Ivy-Ochs et al. 2008, 2009; Ivy-Ochs 2015; Reitner et al. 2016) (cf. Chap. "The Moraine at Trins and the Alpine Lateglacial").

The onset of deglaciation (Termination I) and thus the end of the Alpine LGM were induced by a slight increase of summer insolation in the Northern Hemisphere. Although temperatures increased only slowly at the beginning of the Alpine Lateglacial around 20 ka, a rapid collapse of the complex of transection glaciers occurred. The former accumulation areas of the glaciers, where ice had been formed, were transformed abruptly into ablation zones where melting of ice predominated. Damming of glacial lakes took place at the edges of the large valley glaciers. The meltwater sought routes through and under the downwasting glaciers, so that the latter began to float, break up and disintegrate. ^{10}Be-dating of formerly ice-covered high-elevated bedrock surfaces and erratic boulders in the Zillertal Alps (Wirsig et al. 2016), together with luminescence dating of delta deposits at Hopfgarten in the Kitzbühel Alps (Klasen et al. 2007) shows that the final collapse of the Eastern Alpine ice occurred between 19 and 18 ka. ^{14}C dating of organic material from the base of peat deposits found in the former accumulation area of the Traun Glacier in Mitterndorf (Draxler 1977; van Husen 1977) as well as in the tongue area of the Drau Glacier at lake Längsee (Schmidt et al. 2002) and at lake Jeserzer See (Schultze 1984) testifies ice-free conditions already around 18.5 ka.

The most impressive geomorphic records of rapid deglaciation during this early Lateglacial phase of ice-decay (Reitner 2007) are staircases of kame terraces (Fig. 3.14c). They document the infill of ephemeral lakes formed on the rim of the first stagnant and eventually collapsing transection glacier complex. Hummocky terrains with a series of kettle holes due to decaying dead ice bodies and eskers are further characteristic elements of rapid ice decay. Impressive successions of kame terraces are present in areas of the Rhine Glacier on the flanks of the Rhätikon (Keller 1988) and in the Bregenzer Wald (de Jong et al. 1995), of the Inn Glacier in the Inn Valley (Bobek 1935) and in the Kitzbühel Alps (Reitner 2007), and of the Drau Glacier around Spittal/Drau (Reitner 2005b). Some smaller glaciers were still active and showed brief advances as documented by the presence of till on top of kame terraces (Mayr and Heuberger 1968; van Husen 1977; Reitner 2007; Reitner et al. 2016; Reitner and Menzies 2020).

With the phase of ice decay, mass wasting processes occurred under the condition of zero or very restricted vegetation. Glacially oversteepened and ice-free valley flanks in crystalline rocks were subject to deep-seated gravitational slope deformations (e.g. Reitner et al. 1993) (Fig. 3.15). In contrast, steep slopes consisting of brittle rocks (limestone, dolostone, amphibolite, etc.) reacted with rock avalanches during ice decay (e.g. Alm Valley; van Husen et al. 2007) or immediately afterwards, like in the Mallnitz Valley (Reitner et al. 2018). However, most dated large catastrophic rockslides and rock avalanches (e.g.

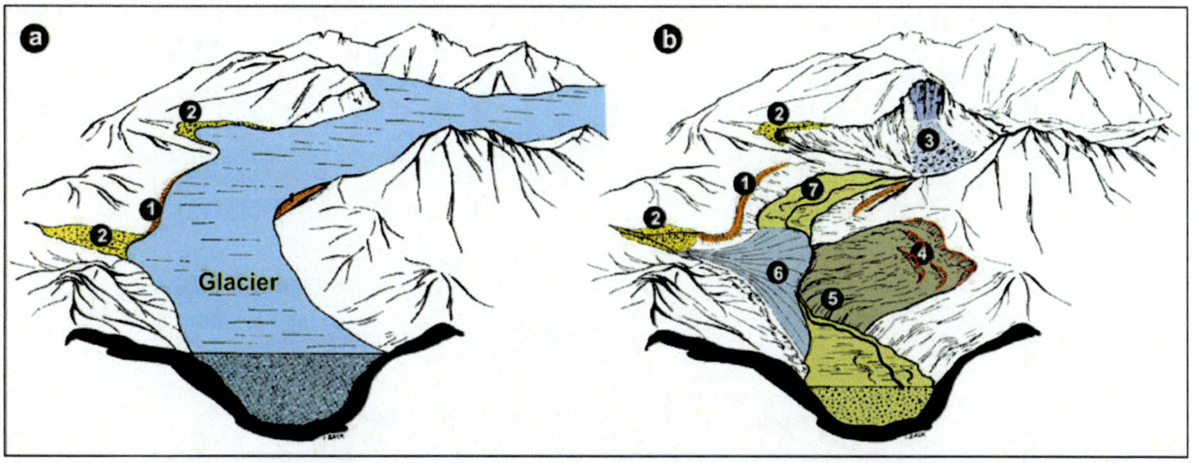

Fig. 3.15 Sketch showing an Alpine Valley filled **a** By a glacier and **b** After downwasting with the corresponding landforms (after van Husen 1987). Lateral moraines (1) and kame terraces (2) were formed at the glacier's margin. A rock avalanche (3) occurred in massive rocks (e.g. limestone) after ice decay. In contrast, a slope made up of weaker rocks like schists developed a deep-seated slope deformation with scarps (4) and a bulging toe (5) as a result of glacial oversteepening. On the opposite valley flank, massive erosion of the kame terrace led to the formation of a large alluvial fan (6). Both processes of (5) and (6) resulted in damming of the main river, with deposition and formation of mires in the backwater (7)

Köfels, Fernpass, Wildalpen) did not happen early in the Lateglacial but during the Holocene (Prager et al. 2008, with references therein).

Initially, lakes remained in the overdeepened valleys (Fig. 3.4) that were formed by subglacial erosion during the Alpine LGM (van Husen 1979; Preusser et al. 2010). However, most of them were rapidly infilled with sediment from the inflowing rivers, starting with the phase of ice decay and thereafter. Deltas of gravel and sand developed and the majority of the valleys became infilled with fine-grained suspended load (silt, clay) as early as the Alpine Lateglacial (e.g. Salzburg basin: van Husen 1979; Pomper et al. 2017). Only where catchment areas of the rivers did not provide enough material, these basins were preserved as lakes, like the lakes of the Salzkammergut area and in Carinthia.

The first re-advance of the Alpine Glaciers into a proglacial forefield free of dead ice occurred during the Gschnitz Stadial (Penck and Brückner 1909). This happened in response to the Heinrich event 1 cooling in the North Atlantic (Ivy-Ochs et al. 2006a). The corresponding end moraines at the type locality at Trins in the Gschnitz Valley have a stabilization age of 16.6 $\pm$ 1.4 ka, according to surface exposure dating with ^{10}Be (Ivy-Ochs et al. 2006a) (see Chap. "The Moraine at Trins and the Alpine Lateglacial" for more detail). At Trins, ΔELA during the Gschnitz Stadial was in the range of 700 m (Kerschner et al. 2019). Climatic conditions were considerably drier and colder than during the twentieth century, with one-third of the annual precipitation and summer temperatures lower by 8–10°. Further prominent end moraines of this stadial, partly linked to proglacial fluvial deposits, are present in many large tributary valleys of the Eastern Alps (van Husen 2000a) (Figs. 3.14e and 3.16).

After the Alpine LGM, re-vegetation started with steppe plants and alpine plants growing together as shown for the Traun Glacier area (Draxler 1987). More detailed records from the lakes Längsee and Jeserzersee in Carinthia (Schmidt et al 2002, 2012; Huber et al 2010) show an early warm period (Längsee oscillation, 18.5–18.1 ka) with the expansion of (dwarf) pine followed by the Längsee cold period (identical with the Oldest Dryas) with intercalated short warm intervals. The culmination of this cold interval between 17.6 and 16.9 ka with diatom-derived water temperatures 8–10 °C below modern values has been correlated with the Gschnitz Stadial. The main reforestation occurred with the onset of the Bølling/Allerød Interstadial (14.7–12.9 ka) during warmer and wetter climatic conditions (see compilation in Heiri et al. 2014). At lake Mondsee, this is reflected by a pine forest, before the climatic deterioration of the Younger Dryas, again linked to a major cooling in the North Atlantic due to ocean circulation change, led to an expansion of shrub, grassland and herb communities and thus to opening of forests (Lauterbach et al. 2011).

With the temperature drop at the transition to the Younger Dryas (c. 12.9–11.7 ka), Alpine Glaciers showed a massive re-advance which was termed Daun Stadial by Penck and Brückner (1909) and is now known as Egesen Stadial (see discussion in Reitner et al. 2016). Overridden landslide deposits of Allerød Age (Bichler et al. 2016) indicate a starting position in the high cirques. Glaciers reached their maximum early in the Younger Dryas as documented by dated moraines followed by sequences of end moraines

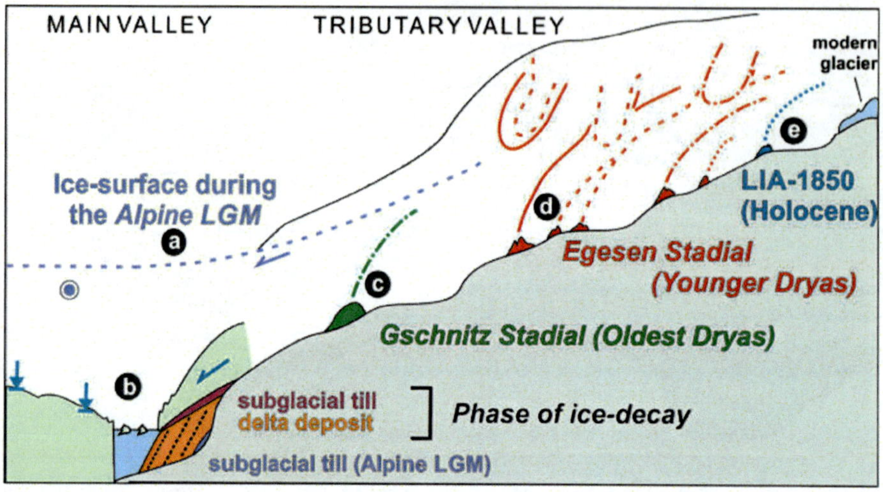

Fig. 3.16 Schematic sketch of the characteristic Lateglacial development of a tributary valley from the Alpine LGM to the Holocene in different scenes (**a**–**e**), from Reitner et al. (2016). **a** Alpine LGM: ice surface during the climax of the Alpine LGM (c. 27–20 ka). The ice flow of the tributary valley was perpendicular to that of the main valley. Deposition of subglacial till on bedrock occurred. **b** Phase of ice decay (c. 19–20 ka): the glacier in the main valley became stagnant and eventually transformed to dead ice resulting in massive downwasting. Short-lived ice-dammed lakes developed which were filled by delta deposits from the tributary valley. The still active tributary glacier advanced over the delta sediments towards the lake and deposited its subglacial till, whilst it reached its Lateglacial maximum position. **c** Gschnitz Stadial (c. 16–17 ka): after a major re-advance, the glacier of the tributary valley stabilized for a considerable time forming large latero-frontal moraines. **d** Egesen Stadial (Younger Dryas; c. 12.9–11.7 ka): the glacier of the tributary valley reached its maximum extent, documented by latero-frontal moraines. Multiple moraines document various halts during recession. **e** Holocene (11.7 ka until now): the Little Ice Age—AD 1850 moraines exemplify similar maximum glacier extents during the Holocene compared to modern conditions

documenting recessing glaciers with Stillstands and small re-advances, with the last glacier stabilization around the Younger Dryas/Holocene transition (Ivy-Ochs et al. 2006b, 2009; Bichler et al. 2016; Moran et al. 2016a; Reitner et al. 2016). Hence, successions of multiple sharp-crested moraines, like in the Schober Gruppe (Buchenauer 1990, Fig. 15), are a common feature in the high-elevated parts of the Alps. ΔELA depression of the maximum extent ranged between 200 m in the dry inner-Alpine parts and 400 m on the wetter northern fringe (Kerschner 2009). Summer temperature was 3.5–4° lower than modern values, and precipitation varied from +10% on the northern fringe to -30% in the dry areas compared to modern conditions (Kerschner and Ivy-Ochs 2008). The successive reduction in glacier size in the course of the Younger Dryas reflects a change to drier conditions.

The stadials of the Alpine Lateglacial were also characterized by permafrost in the Alps, which is evident due to the presence of relict rock glaciers in areas some hundred metres below the modern ones. In principal, rock glaciers (cf. Chap. "Rock Glaciers in the Austrian Alps: A general Overview with a Special Focus on Dösen Rock Glacier, Hohe Tauern Range") consisting of an ice-debris mixture are a common morphological expression of discontinuous permafrost in the periglacial environment of the Eastern Alps (Lieb 1996, Keller-Pirklbauer et al. 2012, Krainer and Ribis 2012). Stabilization ages derived from exposure age dating with cosmogenic nuclides range from the mid-Younger Dryas to the Younger Dryas/Holocene transition (Ivy-Ochs et al. 2009; Moran et al. 2016b). However, relict rock glaciers 1200 m below the present permafrost limit in the Reißeckgruppe pinpoint together with first ^{10}Be-ages to rock glacier stabilization before the Bølling/Allerød Interstadial (Steinemann et al. 2020).

3.6.3 Holocene

The rapid warming at the Holocene/Lateglacial transition (Fig. 3.13) led to an early recovery of the pine forests at the northern fringe of the Alps (Lauterbach et al. 2011) and eventually to a rapid rise of the inner-Alpine timberline. Already by 10.2 ka, the largest glacier in the Austrian Alps, the Pasterze (cf. Chap. "Großglockner and Pasterze Glacier: Landscape Evolution at Austria's Highest Summit and its Neighbouring Glacier System"), was smaller than it is today, and trees like larch and Swiss pine were growing in the area presently still covered by ice (Nicolussi and Patzelt 2001). In the Kauner Valley in western Tyrol, dendrochronologically investigated logs prove that already around 9 ka, the timberline was even higher than around the year AD 2000 (Nicolussi et al. 2005, Fig. 3.17). An elevated timberline, 100 m and more above the 1980s level, remained until around 4 ka. In general, the time period from c. 10 to 5 ka was characterized by an orbitallly driven higher summer insolation which together with additional phases of higher solar activity led to glacier-hostile conditions, with long glacier recession phases and only short periods of advances (Joerin et al. 2006; Nicolussi et al. 2014, Fig. 3.17). These warm conditions allowed humans like the c. 5.3 ka old "Alpine Iceman" (Bortenschlager and Oeggl 2000) to use high-elevated passes.

The trend to a cooler climate in the Late Holocene (last 4.2 ka) led to the opposite pattern in the glacial records, with relatively short recessions and prolonged advances (Joerin et al. 2006), along with the lowering of tree lines (Nicolussi et al. 2005, Fig. 3.17). Significant end moraine ramparts from the Little Ice Age, a cool period lasting from the late thirteenth until the middle of the nineteenth century, are the product of several glacier advances of similar size to the maximum extent of the last 10 000 years. In addition, they testify to a glaciated area during the Little Ice Age that was more than twice as great as at present.

In general, the long-term Holocene climatic values have not exceeded a fluctuation range of the temperature of 2 °C (Auer et al. 2014). However, the impact of temperature and precipitation variation did have an influence on processes like mass wasting, erosion and sedimentation in Alpine Valleys. Similarly to the Würm, but on a much smaller scale, cool and wet climatic phases with a lowered timberline led to higher sediment load of creeks, and thus to prograding alluvial fans of tributary valleys as shown for the Inn Valley (Patzelt 1987, 1994) (Fig. 3.17). Reconstructed palaeo-denudation from a lake record in a high Alpine catchment in the Hohen Tauern using ^{10}Be (Grischott et al. 2017) shows the coupling of climate and hillslope erosion, with a doubling of the denudation rate in the last 4 000 years (0.8 mm/a) compared to that of the 7–5 ka interval (0.4 mm/a). However, in addition to the climatic steering, a limited anthropogenic influence on riverine sediment budgets due to forest clearance in connection with the extension of Alpine pasture (Gilck and Poschlod 2019) cannot be ruled out, at least since the Bronze Age (Patzelt 1987).

All these climatic changes had a severe impact on river activity in the Alpine foreland, as documented for the river Danube and its tributaries (van Husen 2000a). The collapse of the glaciers at the onset of the Lateglacial resulted in the rapid formation of lakes in the former tongue basins, which then acted as a "clarifying basin" for the inflowing meltwater streams. This effect, together with the climatically induced decrease in delivery of frost-shattered debris, led to virtually unloaded streams at the outlet, which immediately dissected the fluvial deposits of the *Niederterrasse*. Erosion and the following transport by all the tributaries resulted in an overloading of the receiving stream, the river Danube, during the Lateglacial and the inception of the Holocene. Due to this bedload (gravel, sand), the river was not able to incise its bed. For a considerable time, the Danube remained at the same level as the glacial terrace but changed its style from braided to meandering river. Accumulations of drift boulders (Eppensteiner et al. 1973) and subfossil logs (Becker 1982) at the stream bed document the reworking of the gravel body whilst laterally shifting its course. This occurred west of Vienna until the Early Holocene (Piffl 1971) and east of it until the Middle Holocene (Fink and Kukla 1977). Incision did not start before the transition towards the Late Holocene.

3.7 Legacy of Quaternary Processes

The Austrian landscape shows the imprint of the various Quaternary processes on different scales:

In the Alpine foreland, staircases of fluvial terraces witness the combined action of climatically controlled aggradation and erosion, together with tectonic uplift. Aeolian action is evident with the LPSs along the Danube and its tributaries but also documented from inside the Alps (Gild et al. 2018). In addition, aeolian erosion might be manifested between the Wachau and the northern fringe of the Alps by unidirectionally shaped hills resembling features of yardangs, which are already known from Hungary (Sebe et al. 2011).

Whilst considering the formerly glaciated part of the Eastern Alps, manifold small to medium-scale morphological and sedimentary remnants are obvious: subglacially moulded valley flanks separated by a trimline from the areas above, affected by frost shattering. Similarly, the presence of narrow ridges (arêtes) and pyramidal peaks (horns) in the high mountain topography bear witness to the erosional work of glaciers. Drumlins (Fig. 3.14d) made up of subglacial till (Fig. 3.14b) and roche moutonnées indicate former ice-flow directions which are also evident from the distribution of erratic boulders. Assemblages of terminal moraines like those north of the city of Salzburg and in Southwestern Upper Austria (cf. Chap. "Quaternary Landforms and Sediments in the Northern Alpine Foreland of Salzburg and Upper Austria") witness the extent of the major glaciations. In contrast, cirques (cf. Chap. "The variability and uniqueness of cirque landscapes in the Schladminger Tauern") are in most cases the product of glacial erosion in times of much smaller glacier extents, like during the Alpine Lateglacial and similar phases, which quite often happened during the Pleistocene (cf Porter 1989).

The huge scouring effect of the debris-laden ice and meltwater at the glacier base formed trough valleys with a U-shaped cross-section. The same processes shaped the overdeepened valleys and basins, with bedrock surfaces as far as several hundred metres below the present surface. In the Alps, overdeepened features are mainly associated with tectonic structures, weak lithologies and/or Pleistocene ice confluence or diffluence situations (Preusser et al. 2010). For the Eastern Alps, two settings are characteristic (for

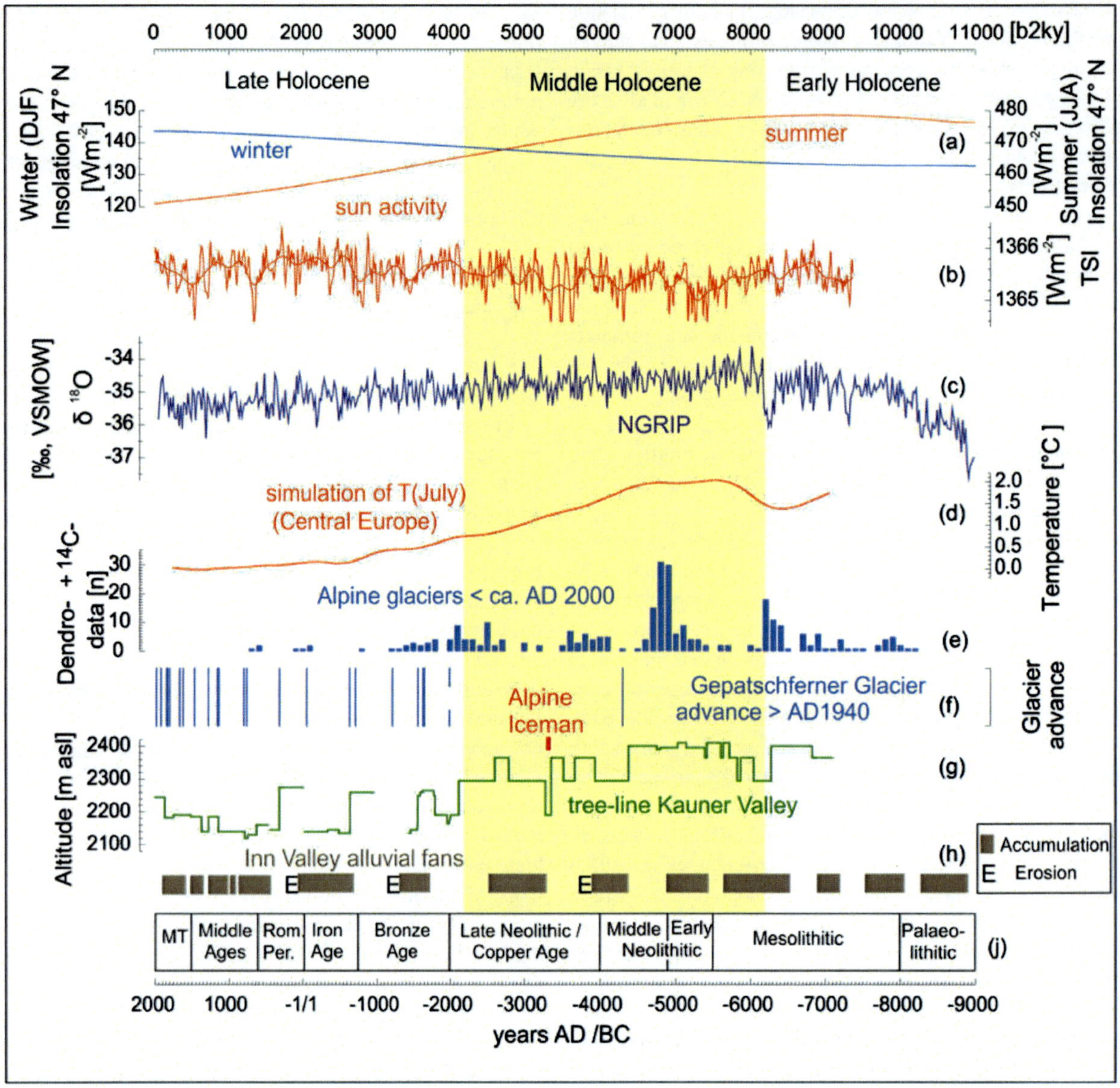

Fig. 3.17 Holocene environmental records and proxy-based climate reconstructions from Austria, the Alps and Greenland in comparison with selected climate forcings (a-g based on Auer et al 2014). **a** Evolution of insolation during summer (June–July–August) and winter (December–January–February) at 47°N; **b** Reconstruction of solar variability for the last 9, 000 years (Steinhilber et al. 2009); **c** Oxygen isotope record of the NGRIP ice core, Central Greenland (Vinther et al. 2009); **d** Simulation of the temperature evolution in July in Central Europe over the last 9,000 years until the pre-industrial period (Renssen et al. 2009); **e** Dendrochronologically, i.e. calendar-dated and ^{14}C-dated evidence for shorter glaciers than today (~AD 1990/2010) (Hormes et al. 2001; Nicolussi and Patzelt 2001; Joerin et al 2006, 2008; Drescher-Schneider and Kellerer-Pirklbauer 2008; Nicolussi 2009, 2011; Nicolussi and Schlüchter 2012); **f** Established advances of the Gepatschferner Glacier (Kauner Valley) beyond the glacier extent in AD 1940 (Nicolussi and Patzelt 2001); **g** Tree-line record in the Kauner Valley based on finds of wood remains (Nicolussi et al. 2005); **h** Phases of sedimentation/accumulation and erosion in the Inn Valley (Patzelt 1995). The cultural stages are simplified based on Gassner et al. (2002) and Urban (2000)

distribution see Fig. 3.4): (1) glacially scoured basins in the ablation area of glaciers, the classical tongue basins of Penck and Brückner (1909) like the Salzburg basin (van Husen 1979, Pomper et al. 2017) filled up with sediment, or the lakes Traunsee (at Gmunden) and Hallstatt (van Husen 1979;). (2) Buried elongated valleys in the longitudinal valleys of the Eastern Alps (e.g. Inn Valley, Preusser et al. 2010; Drau Valley, Burschil et al. 2019). The infill of some of these basins assembles the evidence of multiple glaciations (e.g. basin of Hopfgarten, Reitner et al. 2010). However, the most extraordinary example of overdeepening with respect to its geometry and lithology is reported from Bad Aussee

(Styria). In a narrow basin, a minimum depth of the bedrock surface of 880 m has been revealed by drilling in an area made up by evaporitic bedrock (van Husen and Mayer 2007).

Alpine glaciations modified the shapes of the originally fluvial and thus V-shaped valleys. On a local scale, the frequent glacial events enabled river captures due to the successive lowering of drainage divides (Robl et al. 2008). On a regional scale, it is evident that glacial erosion resulted in increased relief (e.g. Sternai et al. 2012). On the orographic scale, it is still a matter of considerable scientific debate to what extent multiple glaciations changed the hypsometric distribution and whether glacial erosion enhances isostatic uplift (Molnar and England 1990; Champagnac et al. 2007) or at least keeps pace with rock uplift. There is still an ongoing debate to which extent the observed uplift rate of the Eastern Alps can be explained by isostatic adjustment to deglaciation and erosion and by crustal shortening (Sternai et al. 2019).

After more than 150 years of research, such debates as well as the increase of knowledge gained over the last decade show that the Austrian landscape, as the result of tectonics and climate change, is still a challenging field of geological and geomorphological research.

References

Andersen KK, Svensson A, Johnsen SJ et al (2006) The greenland ice core chronology 2005, 15–42 ka. Part 1: constructing the time scale. Quatern Sci Rev 25:3246–3257

Auer I, Foelsche U, Böhm R et al (2014) Kapitel 3: Vergangene Klimaänderung in Österreich. In: Kromp-Kolb H et al (eds) Österreichischer Sachstandsbericht Klimawandel 2014. Verlag der Österreichischen Akademie der Wissenschaften, Vienna, pp 228–299

Baroň I, Plan L, Sokol L, Grasemann B et al (2019) Present-day kinematic behaviour of active faults in the Eastern Alps. Tectonophysics 752:1–23

Barrett S, Starnberger R, Tjallingii R et al (2017) The sedimentary history of the inner-alpine Inn Valley, Austria: extending the Baumkirchen type section further back in time with new drilling. J Quat Sci 32(1):63–79

Barrett SJ, Drescher-Schneider R, Starnberger R et al (2018) Evaluation of the regional vegetation and climate in the Eastern Alps (Austria) during MIS 3–4 based on pollen analysis of the classical Baumkirchen paleolake sequence. Quat Res 90:153–163

Becker B (1982) Dendrochronologie und Paläoökologie subfossiler Baumstämme aus Flussablagerungen. Ein Beitrag zur nacheiszeitlichen Auenentwicklung im südlichen Mitteleuropa. Mitt. Komm. Quartärforschg. Österr Akad Wiss 5:1–120, Wien

Berger A, Loutre MF (1991) Insolation values for the climate of the last 10 million years. Quat Sci Rev 10:297–317

Bichler MG, Reindl M, Reitner JM et al (2016) Landslide deposits as stratigraphical markers for a sequence-based glacial stratigraphy: a case study of a Younger Dryas system in the Eastern Alps. Boreas 45(3):537–551

Bickel L, Lüthgens C, Lomax J et al (2015a) Luminescence dating of glaciofluvial deposits linked to the penultimate glaciation in the Eastern Alps. Quat Int 357:110–124

Bickel L, Lüthgens C, Lomax J et al (2015b) The timing of the penultimate glaciation in the northern Alpine Foreland: new insights from luminescence dating. Proc Geol Assoc 126(4–5):536–550

Bobek H (1935) Die jüngere Geschichte der Inntalterrasse und der Rückzug der letzten Vergletscherung im Inntal. Jb Geol BA 85:135–189

Bortenschlager I, Bortenschlager S (1978) Pollenanlytische Untersuchungen am Bänderton von Baumkirchen. Z Gletscherk Glazialgeol 14:95–103

Bortenschlager S, Oeggl K (eds) (2000) The Iceman and his natural environment: palaeobotanical results, vol 4. Springer, Wien, New York

Braumann SM, Neuhuber S, Fiebig M et al (2018) Challenges in constraining ages of fluvial terraces in the Vienna Basin (Austria) using combined isochron burial and pIRIR225 luminescence dating. Quat Int. https://doi.org/10.1016/j.quaint.2018.01.009

Buchenauer HW (1990) Gletscher- und Blockgletschergeschichte der westlichen Schobergruppe (Osttirol). Marburger Geographische Schriften 117:1–276

Burschil T, Tanner DC, Reitner JM et al (2019) Unravelling the shape and stratigraphy of a glacially-overdeepened valley with reflection seismic: the Lienz Basin (Austria). Swiss J Geosci 112(2–3):341–355. https://doi.org/10.1007/s00015-019-00339-0

Castiglioni GB (1964) Osservazioni sui depositi quaternari di Scives (Bressanone, Alto Adige) e su alcuni fenomeni di defor mazione in strati argilosi. Studi Trentini Di Scienze Naturali 1:3–24

Chaline J, Jerz H (1984) Arbeitsergebnisse der Subkommission für Europäische Quartärstratigraphie. Stratotypen Des Würm-Glazials. Eiszeitalt Ggw 35:185–206

Champagnac JD, Molnar P, Anderson RS et al (2007) Quaternary erosion-induced isostatic rebound in the western Alps. Geology 35 (3):195–198

de Jong MGG, de Graaff LWS, Rupke J (1995) Der Eisabbau im Vorderen Bregenzerwald und in den Nachbargebieten (Vorarlberg, Österreich; Bayern, Deutschland) nach dem letzteiszeitlichen Eishochstand. J Geol B-A 138(1):27–54

Decker K, Peresson H, Hinsch R (2005) Active tectonics and quaternary basin formation along the Vienna Basin transform fault. Quat Sci Rev 24(3–4):305–320

Doneus M, Scharrer G, Neubauer W (2001) Archäologische Prospektion der Landschaft Carnuntum. Möglichkeiten Der Luftbildarchäologie Carnuntum Jahrbuch 2000:53–72

Draxler I (1977) Pollenanalytische Untersuchungen von Mooren zur spät-und postglazialen Vegetationsgeschichte im Einzugsgebiet der Traun. JbGeol BA 120(1):131–163

Draxler I (1987) Zur Vegetationsgeschichte und Stratigraphie des Würmspätglazials des Traungletschergebietes. Mitt Komm Quartärforsch Österr Akad Wiss 7:37–49

Draxler I, van Husen D (1991) Ein ^{14}C-datiertes Profil in der Niederterrasse bei Neurath (Stainz, Stmk.) Z Gletscherk Glazialgeol 25:123–130

Draxler I (2000) Pollenanalytische Untersuchung der schieferkohleführenden Sedimentfolge von Nieselach bei St. Stefan im unteren Gailtal, Kärnten. Mitt. Komm Quartärforsch Österr Akad Wiss 12:155–179, Wien

Drescher-Schneider R (2000) Die Vegetations- und Klimaentwicklung im Riß/Würm-Interglazial und im Früh- und Mittelwürm in der Umgebung von Mondsee: Ergebnisse der pollenanalytischen Untersuchungen. Mitt. Komm Quartärforsch Österr Akad Wiss 12:39–92, Wien

Drescher-Schneider R, Kellerer-Pirklbauer A (2008) Gletscherschwund einst und heute—Neue Ergebnisse zur holozänen Vegetations- und Gletschergeschichte der Pasterze (Hohe Tauern, Österreich). Abh Geol BA 62:45–51

Egger H (1996) Geologische Karte der Republik Österreich 1:50,000, Blatt 66 Gmunden Geologische Bundesanstalt, Wien

Egger H (2007) Erläuterungen zu Blatt 66 Gmunden. Geologische Bundesanstalt, Wien
Ehlers J, Gibbard PL, Hughes, PD (eds) (2011) Quaternary glaciations-extent and chronology: a closer look. Developments in Quaternary Science 15). Elsevier
Eppensteiner W, van Husen D, Krzemien R (1973) Beobachtungen an pleistozänen Driftblöcken des Marchfeldes. Verh. Geol BA 1973:331–336
Fink J (1956) Zur Korrelation der Terrassen und Lösse in Österreich. Eiszeitalt Ggw 7:49–77
Fink J, Kukla GJ (1977) Pleistocene climates in Central Europe: at least 17 interglacials after the Olduvai event. Quat Res 7(3):363–371
Fink J, Fischer H, Klaus W et al (1976) Exkursion durch den österreichischen Teil des nördlichen Alpenvorlandes und den Donauraum zwischen Krems und Wiener Pforte. Mitt Komm Quartärforsch Österr Akad Wiss 1:1–113
Fliri F (1973) Beiträge zur Geschichte der alpinen Würmvereisung: Forschungen am Bänderton von Baumkirchen (Inntal, Nordtirol). Z Geomorph Suppl 16:1–14
Fliri F (1978) Die Stellung des Bändertonvorkommens von Schabs (Südtirol) in der alpinen Würm-Chronologie. Z Gletscherk Glazialgeol 14:115–118
Fliri F (1988) Die Schottergrube von Albeins bei Brixen, eine neue Schlüsselstelle zur Kenntnis der Chronologie der Würmvereisung in Südtirol. Z Gletscherk Glazialgeol 24:137–142
Fliri F, Bortenschlager S, Felber H et al (1970) Der Bänderton von Baumkirchen (Inntal, Tirol). Eine neue Schlüsselstelle zur Kenntnis de Würmvereisung der Alpen. Z Gletscherk Glazialgeol 6:5–35
Frank C, Rabeder G (1997) Stranzendorf. In: Döppes D, Rabeder G (eds) Pliozäne und Pleistozäne Faunen Österreichs. Mitt Komm Quartärforsch Österr Akad Wiss 10:130–139
Frechen M, Oches EA, Kohfeld KE (2003) Loess in Europe—mass accumulation rates during the last glacial period. Quat Sci Rev 22 (18–19):1835–1857
Frenzel B, Pesci M, Velichko AA (1992) Atlas of paleoclimates of the Northern Hemisphere: late pleistocene-holocene. Fischer, Stuttgart
Gassner V, Jilek S, Ladstätter S (2002) Am Rande des Reiches. Die Römer in Österreich. In: Wolfram H (ed) Österreichische Geschichte 15 v. Chr. – 378 n. Chr., Carl Ueberreuter, Vienna, p 488
Gibbard PL, Head MJ, Walker MJC et al (2010) Formal ratification of the quaternary system/period and the pleistocene series/Epoch with a base at 2.58 Ma. J Quat Sci 25:96–102
Gilck F, Poschlod P (2019) The origin of alpine farming: a review of archaeological, linguistic and archaeobotanical studies in the Alps. The Holocene 29(9):1503–1511
Gild C, Geitner C, Sanders D (2018) Discovery of a landscape-wide drape of late-glacial aeolian silt in the western Northern Calcareous Alps (Austria): first results and implications. Geomorphology 301:39–52
Grischott R, Kober F, Lupker M et al (2017) Millennial scale variability of denudation rates for the last 15 kyr inferred from the detrital ^{10}Be record of Lake Stappitz in the Hohe Tauern massif Austrian Alps. The Holocene 27(12):1914–1927
Grüger E (1979) Spätriß, Riß/Würm und Frühwürm am Samerberg in Oberbayern–ein vegetationsgeschichtlicher Beitrag zur Gliederung des Jungpleistozäns. Geologica Bavarica 80:5–64
Harrison SP, Kohfeld KE, Roelandt C et al (2001) The role of dust in climate changes today, at the last glacial maximum and in the future. Earth-Sci Rev 54(1–3):43–80
Head MJ, Gibbard PL (2005) Early-Middle Pleistocene transitions: an overview and recommendation for the defining boundary. Geol Soc Spec Publ 247(1):1–18
Heiri O, Koinig KA, Spötl C et al (2014) Palaeoclimate records 60–8 ka in the Austrian and Swiss Alps and their forelands. Quat Sci Rev 106:186–205
Hintersberger E, Decker K, Lomax J et al (2018) Implications from palaeoseismological investigations at the Markgrafneusiedl Fault (Vienna Basin, Austria) for seismic hazard assessment. Nat Hazards Earth Syst Sci 18(2):531–553
Höggerl N (2007) Höhenänderungen in Österreich. In: Hofmann T, Schönlaub HP (eds) Geo-Atlas Österreich: Die Vielfalt des geologischen Untergrundes. Böhlau, Wien, pp 56–57
Hormes A, Müller BU, Schlüchter C (2001) The Alps with little ice: evidence for eight Holocene phases of reduced glacier extent in the Central Swiss Alps. The Holocene 11:255–265
Huber K, Weckström K, Drescher-Schneider R et al (2010) Climate changes during the last glacial termination inferred from diatom-based temperatures and pollen in a sediment core from Längsee (Austria). J Paleolimnol 43(1):131
Huber KH (1999) Känozoikum (Erdneuzeit): Zum Formenschatz der Granitverwitterung und -abtragung im nordwestlichen Waldviertel. In: Steininger FF (ed) Erdgeschichte des Waldviertels, 2nd edn, Waldviertler Heimatbund, Horn, p 113–132
Hughes PD, Gibbard PL (2018) Global glacier dynamics during 100 ka Pleistocene glacial cycles. Quat Res 90:1–22
Hughes PD, Gibbard PL, Ehlers J (2013) Timing of glaciation during the last glacial cycle: evaluating the concept of a global 'Last Glacial Maximum'(LGM). Earth-Sci Rev 125:171–198
Ivy-Ochs S (2015) Glacier variations in the European Alps at the end of the last glaciation. Cuadernos De Investigación Geográfica 41:295–315
Ivy-Ochs S, Kerschner H, Kubik PW et al (2006a) Glacier response in the European Alps to Heinrich event 1 cooling: the Gschnitz stadial. J Quat Sci 21(2):115–130
Ivy-Ochs S, Kerschner H, Reuther A et al (2006b) The timing of glacier advances in the northern European Alps based on surface exposure dating with cosmogenic ^{10}Be, ^{26}Al, ^{36}Cl, and ^{21}Ne. Geol Soc Am Spec Pap 415:43–60
Ivy-Ochs S, Kerschner H, Reuther A et al (2008) Chronology of the last glacial cycle in the European Alps. J Quat Sci 23:559–573
Ivy-Ochs S, Kerschner H, Maisch M et al (2009) Latest Pleistocene and Holocene glacier variations in the European Alps. Quat Sci Rev 28:2137–2149
Jemec Auflič M, Jež J, Popit T et al (2017) The variety of landslide forms in Slovenia and its immediate NW surroundings. Landslides 14(4):1537–1546
Joerin UE, Stocker TF, Schlüchter C (2006) Multicentury glacier fluctuations in the Swiss Alps during the Holocene. Holocene 16 (5):697–704
Joerin UE, Nicolussi K, Fischer A, Stocker TF, Schlüchter C (2008) Holocene optimum events inferred from subglacial sediments at Tschierva Glacier, Eastern Swiss Alps. Quat Sci Rev 27:337–350
Jouzel J, Masson-Delmotte V, Cattani O et al (2007) Orbital and millennial Antarctic climate variability over the past 800,000 years. Science 317:793–796
Keller O (1988) Ältere spätwürmzeitliche Gletschervorstöße und Zerfall des Eisstromnetzes in den nördlichen Rhein-Alpen (Weißbad-Stadium/Bühl-Stadium). Physische Geographie 27:1–241
Kellerer-Pirklbauer A, Lieb GK, Kleinferchner H (2012) A new rock glacier inventory of the eastern European Alps. Austrian J Earth Sci 105(2):78–93
Kerschner H, Ivy-Ochs S (2008) Palaeoclimate from glaciers: examples from the Eastern Alps during the Alpine Lateglacial and early Holocene. Glob Planet Change 60(1–2):58–71
Kerschner H (1986) Zum Sendersstadium im Spätglazial der nördlichen Stubaier Alpen, Tirol. – Z Geomorphol Suppl 61:65–76
Kerschner H (2009) Gletscher und Klima im Alpinen Spätglazial und frühen Holozän. Alp Space—Man Environ 6:5–26
Klasen N, Fiebig M, Preusser F et al (2007) Luminescence dating of proglacial sediments from the Eastern Alps. Quat Int 164:21–32

Kohl H (2000) Das Eiszeitalter in Oberösterreich: Abriss einer Quartärgeologie von Oberösterreich. Schriftenreihe Des Oberösterreichischen Musealvereins 17:1–487

Krainer K, Ribis M (2012) A rock glacier inventory of the Tyrolean Alps (Austria). Austrian J Earth Sci 105(2):32–47

Krenmayr HG (2000) Sedimentologie der letzt-interglazialen bis Mittelwürm-zeitlichen Seesedimente bei Mondsee. Mitt. Komm Quartärforsch Österr Akad Wiss 12:13–38, Wien

Lambeck K, Rouby H, Purcell A et al (2014) Sea level and global ice volumes from the last glacial maximum to the Holocene. Proc Natl Acad Sci 111(43):15296–15303

Lauterbach S, Brauer A, Andersen N et al (2011) Environmental responses to Lateglacial climatic fluctuations recorded in the sediments of pre-alpine lake Mondsee (northeastern Alps). J Quat Sci 26(3):253–267

Lee JR, Candy I, Haslam R (2018) The Neogene and Quaternary of England: landscape evolution, tectonics, climate change and their expression in the geological record. Proc Geol Assoc 129(3):452–481

Lenhardt WA, Švancara J, Melichar P et al (2007) Seismic activity of the Alpine–Carpathian–Bohemian Massif region with regard to geological and potential field data. Geol Carpath 58(4):397–412

Lichtenecker N (1938) Die gegenwärtige und die eiszeitliche Schneegrenze in den Ostalpen. In: Götzinger G (ed) Verhandlungen der III Internationalen Quartärkonferenz (INQUA), September 1936. Geologische Landesanstalt, Wien, pp 141–147

Lieb GK (1996) Permafrost und Blockgletscher in den östlichen österreichischen Alpen. Arbeiten Aus Dem Institut Für Geographie in Graz 33:9–125

Lisiecki LE, Raymo ME (2005) A Pliocene-Pleistocene stack of 57 globally distributed benthic δ18O records. Paleoceanography 20(1). https://doi.org/10.1029/2004PA001071

Lomax J, Fuchs M, Preusser F et al (2014) Luminescence based loess chronostratigraphy of the Upper Palaeolithic site Krems-Wachtberg, Austria. Quat Int 351:88–97

Lowe JJ, WalkerMJ (2015) Reconstructing quaternary environments. Routledge, London & New York

Luetscher M, Boch R, Sodemann H et al (2015) North Atlantic storm track changes during the last glacial maximum recorded by Alpine speleothems. Nat Comm 6:6344. https://doi.org/10.1038/ncomms7344

Maisch M (1982) Zur Gletscher- und Klimageschichte des alpinen Spätglazials. Geogr Helv 37:93–104

Maisch M (1987) Zur Gletschergeschichte des alpinen Spätglazials: analyse und Interpretation von Schneegrenzdaten. Geog Hel 42: 63–71

Mayr C, Brandlmeier B, Diersche V et al (2017) Nesseltalgraben, a new reference section of the last glacial period in southern Germany. J Paleolimnol 58(2):213–229

Mayr C, Stojakowits P, Lempe B et al (2019) High-resolution geochemical record of environmental changes during MIS 3 from the northern Alps (Nesseltalgraben, Germany). Quat Sci Rev 218:122–136

Mayr F, Heuberger H (1968) Type areas of Lateglacial and Postglacial deposits in Tyrol, Eastern Alps. In: Richmond GM (ed) Glaciations of the Alps. Series in earth sciences 7. University of Colorado, Boulder, pp 143–165

Menzies J, Reitner JM (2016) Microsedimentology of ice stream tills from the Eastern Alps, Austria–a new perspective on till microstructures. Boreas 45(4):804–827

Menzies J, Reitner JM (2019) Microstructures, subglacial till deposition, and shear band development revealing up-section changes in shear—a study from Weissbach, Austria. PGA Proc Geol Assoc 130(2):196–209

Métois M, D'Agostino N, Avallone A et al (2015) Insights on continental collisional processes from GPS data: dynamics of the peri-Adriatic belts. J Geophys Res Solid Earth 120:8701–8719. https://doi.org/10.1002/2015JB012023

Meyer MC, Spötl C, Mangini A (2008) The demise of the Last Interglacial recorded in isotopically dated speleothems from the Alps. Quat Sci Rev 27(5–6):476–496

Molnar P, England P (1990) Late Cenozoic uplift of mountain ranges and global climate change: chicken or egg? Nature 346(6279):29

Monegato G, Ravizzi C (2018) The Late Pleistocene multifold glaciation in the Alps: update and open questions. Alp Mediterr Quat 31:225–229

Monegato G, Ravazzi C, Donegana M et al (2007) Evidence of a two-fold glacial advance during the last glacial maximum in the Tagliamento end moraine system (eastern Alps). Quat Res 68:284–302

Monegato G, Scardia G, Hajdas I et al (2017) The Alpine LGM in the boreal ice-sheets game. Sci Rep 7(1):2078. https://doi.org/10.1038/s41598-017-02148-7

Moran AP, Ivy-Ochs S, Schuh M et al (2016a) Evidence of central Alpine glacier advances during the younger dryas—early Holocene transition period. Boreas 45(3):398–410

Moran AP, Ivy-Ochs S, Vockenhuber C et al (2016b) Rock glacier development in the Northern Calcareous Alps at the Pleistocene-Holocene boundary. Geomorphology 273:178–188

Morlot AV (1847) Erläuterungen zur geologischen Übersichtskarte der nordöstlichen Alpen. Ein Entwurf zur vorzunehmenden Bearbeitung der physikalischen Geographie und Geologie ihres Gebietes. Braumüller & Seidel, Wien

Muttoni G, Carcano C, Garzanti E et al (2003) Onset of major Pleistocene glaciations in the Alps. Geology 31(11):989–992

NGRIP project members (2004) High-resolution record of Horthern Hemisphere climate extending into the last interglacial period. Nature 431:147–151

Nicolussi K (2011) Gletschergeschichte der Pasterze—Spurensuche in die nacheiszeitliche Vergangenheit. In: Lieb GK, Slupetzky H (eds) Die Pasterze. Verlag Anton Pustet, Der Gletscher am Großglockner, pp 24–27

Nicolussi K, Patzelt G (2001) Untersuchungen zur holozänen Gletscherentwicklung von Pasterze und Gepatschferner (Ostalpen). Z Gletscherk Glazialgeol 36:1–87

Nicolussi K, Schlüchter C (2012) The 8.2 ka event—calendar-dated glacier response in the Alps. Geology 40:819–822

Nicolussi K, Kaufmann M, Patzelt G et al (2005) Holocene tree-line variability in the Kauner Valley, Central Eastern Alps, indicated by dendrochronological analysis of living trees and subfossil logs. Veg Hist Archaeobot 14(3):221–234

Nicolussi K, Drescher-Schneider R, Le Roy M et al (2014) Alpine Gletscherschwankungen während des Holozäns. Geogr Rundsch 2014(7–8):16–20

Nicolussi K (2009) Alpine Dendrochronologie—Untersuchungen zur Kenntnis der holozänen Umwelt- und Klimaentwicklung. In: Schmidt R, Matulla C, Psenner R (eds) Klimawandel in Österreich: Die Letzten 20.000 Jahre – und ein Blick voraus, Alpine space—man & environment. Innsbruck University Press, Innsbruck, pp 41–54

Oeggl K, Unterfrauner H (2000) Die Pflanzenreste des Riß/Würm-Interglazials und des Würmglazials von Mondsee. Mitt. Komm Quartärforsch Österr Akad Wiss 12:93–122. Wien

Papesch W, Rank D (2000) Isotopenuntersuchungen an den Seekreiden aus den Bohrkernen Mondsee. Mitt. Komm Quartärforsch Österr Akad Wiss 12:123–127. Wien

Patzelt G (1987) Untersuchungen zur nacheiszeitlichen Schwemmkegel-und Talentwicklung in Tirol. Veröffentlichungen Des Tiroler Landesmuseum Ferdinandeum 67:93–123

Patzelt G (1994) Holocene development of alluvial fans and the floor of the Inn valley, Tyrol. Mt Res Dev 14:283–284

Patzelt G, Resch W (1986) Quartärgeologie des mittleren Tiroler Inntales zwischen Innsbruck und Baumkirchen (Exkursion C am 3. April 1986). Jahresberichte und Mitteilungen des Oberrheinischen Geologischen Vereins, pp 43–66

Patzelt G (1995) Eastern Alps Traverse. In: Schirmer W (ed) Quaternary field trips in Central Europe. International union for quaternary research, XIV international congress, vol 1. Berlin, p 385

Penck A (1920) Die Höttinger Breccie und die Inntalterrasse nördlich Innsbruck. Abhandlungen Der Preussischen Akademie Der Wissenschaften Physikalisch-Mathematische Klasse 1920:1–136

Penck A, Brückner E (1909) Die Alpen im Eiszeitalter. Tauchnitz, Leipzig

Piffl L (1971) Zur Gliederung des Tullner Feldes. Annalen des Naturhistorischen Museums Wien 75:293–310. Wien

Pomper J, Salcher BC, Eichkitz C et al (2017) The glacially overdeepened trough of the Salzach Valley, Austria: Bedrock geometry and sedimentary fill of a major Alpine subglacial basin. Geomorphology 295:147–158

Porter SC (1989) Some geological implications of average quaternary glacial conditions. Quat Res 32(3):245–261

Prager C, Zangerl C, Patzelt G et al (2008) Age distribution of fossil landslides in the Tyrol (Austria) and its surrounding areas. Nat Hazard 8(2):377–407

Preusser F, Fiebig M (2009) European Middle Pleistocene loess chronostratigraphy: some considerations based on evidence from the Wels site, Austria. Quat Int 198(1–2):37–45

Preusser F, Reitner JM, Schlüchter C (2010) Distribution, geometry, age and origin of overdeepened valleys and basins in the Alps and their foreland. Swiss J Earth Sci 103(3):407–426

Rasmussen SO, Bigler M, Blockle SP et al (2014) A stratigraphic framework for abrupt climatic changes during the Last Glacial period based on three synchronized Greenland ice-core records: refining and extending the INTIMATE event stratigraphy. Quat Sci Rev 106:14–28

Raymo ME (1997) The timing of major climate terminations. Paleoceanography 12(4):577–585

Raymo ME, Ruddiman WF (1992) Tectonic forcing of late Cenozoic climate. Nature 359:117–122

Reiter F, Freudenthaler C, Hausmann H et al (2018) Active seismotectonic deformation in front of the Dolomites indenter, Eastern Alps. Tectonics. https://doi.org/10.1029/2017TC004867

Reitner JM (2005a) Quartärgeologie und Landschaftsentwicklung im Raum Kitzbühel-St. Johann i.T-Hopfgarten (Nordtirol) vom Riss bis in das Würm-Spätglazial (MIS 6–2). Dissertation, University of Vienna

Reitner JM (2005b) Landschaftsentwicklung im Quartär. In: Schuster R (ed) Arbeitstagung 2005 der Geologischen Bundesanstalt Blatt 182 Spittal an der Drau, Gmünd/Kärnten. Geologische Bundesanstalt, Vienna, pp 63–81

Reitner JM (2007) Glacial dynamics at the beginning of termination I in the Eastern Alps and their stratigraphic implications. Quat Int 164:64–84

Reitner JM (2011) Das Inngletschersystem während des Würm-Glazial. In: Gruber A (ed) Arbeitstagung 2011 der Geologischen Bundesanstalt Blatt 88 Achenkirch. Geologische Bundesanstalt, Vienna, pp 79–88

Reitner JM, Draxler I (2002) Die klimatisch-fazielle Entwicklung vor dem Würm-Maximum im Raum Kitzbühel-St. Johann—Hopfgarten (Nordtirol/Osterreich). Terra Nostra 6:298–304

Reitner JM, Gruber A (2014) Glacial dynamics and large pre-LGM rock-slides in the lower Inn Valley and in the Brixen Valley. In: Kerschner H, Krainer K, Spötl C (eds) From the foreland to the central Alps. Field trips to selected sites of quaternary research in the Tyrolean and Bavarian Alps. Geozon Science Media, Berlin, pp 46–67

Reitner, JM, Menzies J (2020) Microsedimentology of tills near Ainet, Austria-were palaeo-ice streams in the European Alps underlain by soft deforming bed zones? Austrian J Earth Sci 113(1):71–86

Reitner JM, Ottner F (2011) Geochemische Charakterisierung der Verwitterungsintensität der Löss–Paläoboden-Sequenz von Wels/Aschet. Mitt Komm Quartärforsch Österr Akad Wiss 19:37–45

Reitner J, Lang M, van Husen D (1993) Deformation of high slopes in different rocks after Würmian deglaciation in the Gailtal (Austria). Quat Intern 18:43–51

Reitner JM, Ivy-Ochs S, Drescher-Schneider R et al (2016) Reconsidering the current stratigraphy of the Alpine lateglacial: implications of the sedimentary and morphological record of the Lienz area (Tyrol/Austria). E&G Quat Sci J 65(2):113–144

Reitner JM, Ostermann M, Schuster R et al (2018) Der Bergsturz vom Auernig (Mallnitz/Kärnten), seine Altersstellung und Folgen. Carinthia II 207(127):503–548

Reitner JM, Gruber W, Römer A et al (2010) Alpine overdeepenings and paleo-ice flow changes: an integrated geophysical-sedimentological case study from Tyrol (Austria). Swiss J Geosci 103(3):385–405

Renssen H, Seppä H, Heiri O et al (2009) The spatial and temporal complexity of the Holocene thermal maximum. Nat Geosci 2:411–414

Reuther AU, Fiebig M, Ivy-Ochs S et al (2011) Deglaciation of a large piedmont lobe glacier in comparison with a small mountain glacier–new insight from surface exposure dating. Two studies from SE Germany. E&G Quat Sci J 60(2–3):248–269

Robl J, Hergarten S, Stüwe K (2008) Morphological analysis of the drainage system in the Eastern Alps. Tectonophysics 460(1–4):263–277

Rousseau DD, Boers N, Sima A et al (2017) (MIS3 & 2) millennial oscillations in Greenland dust and Eurasian aeolian records—a paleosol perspective. Quat Sci Rev 169:99–113

Ruddiman WF (2001) Earth's climate: past and future. Macmillan

Salcher BC, Wagreich M (2010) Climate and tectonic controls on Pleistocene sequence development and river evolution in the Southern Vienna Basin (Austria). Quat Int 222(1–2):154–167

Salcher BC, Hinsch R, Wagreich M (2010) High-resolution mapping of glacial landforms in the North Alpine Foreland, Austria. Geomorphology 122(3–4):283–293

Salcher BC, Starnberger R, Götz J (2015) The last and penultimate glaciation in the North Alpine Foreland: new stratigraphical and chronological data from the Salzach glacier. Quat Intern 388:218–231

Salcher BC, Frank-Fellner C, Lomax J et al (2017) Middle to late Pleistocene multi-proxy record of environmental response to climate change from the Vienna Basin, Central Europe (Austria). Quat Sci Rev 173:193–210

Salcher B, Starnberger R, Götz J (2018) Field trip post-EX-3: sediment-landform associations of major glaciations in the North Alpine Foreland In: Neubauer (ed) XXI international congress of the Carpathian Balkan geological association: guidebook to pre- and post-conference excursions: Salzburg. Berichte der Geologischen Bundesanstalt 126(2018):289–303

Sanders D, Ostermann M (2006) Depositional setting of the sedimentary rocks containing the "warm-interglacial" fossil flora of the Höttinger Brekzie (Pleistocene, Northern Calcareous Alps, Austria): a reconstruction. Veröffentlichungen Tiroler Landesmuseums Ferdinandeum 86:91–118

Sanders S, Spötl C (2014) The Hötting Breccia—a Pleistocene key site near Innsbruck, Tyrol. In: Kerschner H, Krainer K, Spötl C (eds) From the foreland to the central Alps. Field trips to selected sites of Quaternary research in the Tyrolean and Bavarian Alps. Geozon Science Media, Berlin, pp 82–94

Sanders D, Wischounig L, Gruber A et al (2014) Inner gorge–slot canyon system produced by repeated stream incision (eastern Alps):

significance for development of bedrock canyons. Geomorphology 214:465–484
Schmidt R, van den Bogaard C, Merkt J et al (2002) A new Lateglacial chronostratigraphic tephra marker for the south-eastern Alps: the Neapolitan Yellow Tuff (NYT) in Längsee (Austria) in the context of a regional biostratigraphy and palaeoclimate. Quatern Int 88 (1):45–56
Schmidt R, Weckström K, Lauterbach S et al (2012) North Atlantic climate impact on early late-glacial climate oscillations in the south-eastern Alps inferred from a multi-proxy lake sediment record. J Quat Sci 27(1):40–50
Scholger R, Terhorst B (2013) Magnetic excursions recorded in the middle to upper Pleistocene loess/palaeosol sequence Wels-Aschet (Austria). E&G Quaternary Sci J 62(1):14–21
Schultze E (1984) Neue Erkenntnisse zur spät- und frühpostglazialen Vegetations- und Klimageschichte im Klagenfurter Becken. Carinthia II 174(94):261–266
Sebe K, Csillag G, Ruszkiczay-Rüdiger Z et al (2011) Wind erosion under cold climate: a Pleistocene periglacial mega-yardang system in Central Europe (Western Pannonian Basin, Hungary). Geomorphology 134(3–4):470–482
Sebe K, Roetzel R, Fiebig M et al (2015) Pleistocene wind system in eastern Austria and its impact on landscape evolution. CATENA 134:59–74
Seguinot J, Ivy-Ochs S, Jouvet G et al (2018) Modelling last glacial cycle ice dynamics in the Alps. Cryosphere 12(10):3265–3285
Simony F (1847) Über die Spuren der vorgeschichtlichen Eiszeit im Salzkammergute. Berichte über Mittheilungen von Freunden der Naturwissenschaften in Wien I:215–248. Wien
Spötl C, Mangini A (2006) U/Th age constraints on the absence of ice in the central Inn Valley (eastern Alps, Austria) during Marine Isotope Stages 5c to 5a. Quat Res 66(1):167–175
Spötl C, Holzkämper S, Mangini A (2007) The last and the penultimate interglacial as recorded by speleothems from a climatically sensitive high-elevation cave site in the alps. In: Sirocko F, Claussen M, Sánchez Goñi MF et al (eds) The climate of past interglacials. Developments in Quaternary Sciences, vol 7. Elsevier, pp 471–491
Spötl C, Reimer PJ, Starnberger R et al (2013) A new radiocarbon chronology of Baumkirchen, stratotype for the onset of the Upper Würmian in the Alps. J Quat Sci 28(6):552–558
Sprafke T (2016) Löss in Niederösterreich: Archiv quartärer Klima- und Landschaftsveränderungen. Würzburg University Press
Sprafke T, Thiel C, Terhorst B (2014) From micromorphology to palaeoenvironment: the MIS 10 to MIS 5 record in Paudorf (Lower Austria). CATENA 117:60–72
Starnberger R, Rodnight H, Spötl C (2011) Chronology of the last glacial maximum in the Salzach Palaeoglacier area (Eastern Alps). J Quat Sci 26(5):502–510
Starnberger R, Drescher-Schneider R, Reitner JM et al (2013) Late Pleistocene climate change and landscape dynamics in the Eastern Alps: the inner-alpine Unterangerberg record (Austria). Quat Sci Rev 68:17–42
Steinemann O, Reitner JM, Ivy-Ochs S et al (2020) Tracking rockglacier evolution in the Eastern Alps from the Lateglacial to the early Holocene. Quatern Sci Rev 241:106424. https://doi.org/10.1016/j.quascirev.2020.106424
Steinhilber F, Beer J, Fröhlich C (2009) Total solar irradiance during the Holocene. Geophys Res Lett 36. https://doi.org/10.1029/2009GL040142
Sternai P, Herman F, Champagnac JD et al (2012) Pre-glacial topography of the European Alps. Geology 40(12):1067–1070
Sternai P, Sue C, Husson L et al (2019) Present-day uplift of the European Alps: evaluating mechanisms and models of their relative contributions. Earth-Sci Rev 190:589–604
Terhorst B (2013) A stratigraphic concept for middle Pleistocene quaternary sequences in upper Austria. E&G Quat Sci J 62(1):4–13
Terhorst B, Frechen M, Reitner J (2002) Chronostratigraphische Ergebnisse aus Lossprofilen der Inn-und Traun-Hochterrassen in Oberosterreich. Z Geomorphol Suppl 127:213–232
Terhorst B, Ottner F, Wriessnig K (2012) Weathering intensity and pedostratigraphy of the middle to upper Pleistocene loess/palaeosol sequence of Wels-Aschet (Upper Austria). Quat Int 26:142–154
Terhorst B, Kühn P, Damm B et al (2014) Paleoenvironmental fluctuations as recorded in the loess-paleosol sequence of the upper Paleolithic site Krems-Wachtberg. Quat Int 351:67–82
Terhorst B, Sedov S, Sprafke T et al (2015) Austrian MIS 3/2 loess–palaeosol records—key sites along a west–east transect. Palaeogeogr Palaeoclimatol Palaeoecol 418:43–56
Till A (1907) Das große Naturereignis von 1348 und die Bergstürze des Dobratsch. Mitteilungen der k.k. Geographischen Gesellschaft in Wien 10–11; pp 534–645. Wien
Traub F, Jerz H (1975) Ein Lößprofil von Duttendorf (Oberösterreich) gegenüber Burghausen an der Salzach. Z Gletscherk Glazialgeol 11:175–193
Troll C (1924) Der diluviale Inn-Chiemsee-Gletscher: das geographische Bild eines typischen Alpenvorlandgletschers. Forschungen Zur Deutschen Landeskunde 23:1–121
Urban OH (2000) Der lange Weg zur Geschichte. Die Urgeschichte Österreichs. In: Wolfram H (ed) Österreichische Geschichte bis 15 v. Chr. Carl Ueberreuter, Vienna, p 511
van Husen D (1968) Ein Beitrag zur Talgeschichte des Ennstales im Quartär. Mitt Ges Geol Bergbaustud 18:249–286
van Husen D (1971) Zum Quartär des unteren Ennstales von Großraming bis zur Donau. Verh Geol BA 1971:511–521
van Husen D (1977) Zur Fazies und Stratigraphie der jungpleistozänen Ablagerungen im Trauntal. Jb Geol BA 120:1–130
van Husen D (1979) Verbreitung, Ursachen und Füllung glazial übertiefter Talabschnitte an Beispielen in den Ostalpen. Eiszeitalt Ggw 29:9–22
van Husen D (1981) Geologisch-sedimentologische Aspekte im Quartär von Österreich. Mitt Österr Geol Ges 74(75):197–230
van Husen D (1983a) A model of valley bottom sedimentation during climatic changes in a humid alpine environment. In: Evenson EB, Schlüchter C, Rabassa J (eds) Tills and related deposits. Balkema, Rotterdam, pp 341–344
van Husen D (1983b) General sediment development in relation to the climatic changes during Würm in the eastern Alps. In: Evenson EB, Schlüchter C, Rabassa J (eds) Tills and related deposits. Balkema, Rotterdam, pp 345–349
van Husen D (1986) Bau- Und Hydrogeologische Bedeutung Eiszeitlicher Vorgänge Mitteilungen Der Gesellschaft Der Geologie- Und Bergbaustudenten in Österreich 33(1986):23–45
van Husen D (1987) Die Ostalpen in den Eiszeiten. Veröff Geol Bundesanst 2:1–24
van Husen D (1989) The last interglacial-glacial cycle in the Eastern Alps. Quat Int 3:115–121
van Husen D (1997) LGM and late-glacial fluctuations in the Eastern Alps. Quat Int 38:109–118
van Husen D (1999) Geologisch-baugeologische Erfahrungen beim Bau des Eisenbahntunnels Lambach, OÖ. Mitt Österr Geol Ges 90:137–154
van Husen D (2000a) Geological processes during the Quaternary. Mitt Österr Geol Ges 92:135–156
van Husen D (2000b) Die paläogeographische Situation des Mondsees im Riß/Würm Inter-glazial und Frühwürm. Mitt. Komm Quartärforsch Österr Akad Wiss 12:9–12. Wien
van Husen D (2000c) Die Schieferkohle von Nieselach (Gailtal Kärnten): Geologischer Rahmen, Sedimentationsbedingungen im

Talboden, zeitliche Stellung. Mitt. Komm Quartärforsch Österr Akad Wiss 12:131–139. Wien

van Husen D (2011) Quaternary glaciations in Austria. In: Ehlers J, Gibbard PL, Hughes PD (eds) Quaternary Glaciations—extent and chronology a closer look. Develop Quat Sci 15:15–28. Elsevier

van Husen D, Mayer M (2007) The hole of Bad Aussee, an unexpected overdeepened area in NW Steiermark, Austria. Austrian J Earth Sci 100:128–136

van Husen D, Reitner JM (2011a) An outline of the quaternary stratigraphy of Austria. E&G Quat Sci J 60:366–387

van Husen D, Reitner, JM (2011b) Klimagesteuerte Terrassen-und Lössbildung auf der Traun-Enns-Platte und ihre zeitliche Stellung (Das Profil Wels/Aschet). In: van Husen D, Reitner JM (eds) Die Löss-Sequenz Wels/Aschet (ehemalige Lehmgrube Würzburger). Mitt Komm Quartärforsch Österr Akad Wiss, vol 19, Verlag der Österreichischen Akademie der Wissenschaften, pp 1–11

van Husen D, Ivy-Ochs S, Alfimov V (2007) Mechanism and age of late glacial landslides in the Cal-careous Alps; The Almtal, Upper Austria. Austrian J Earth Sci 100:114–126

Vandenberghe J, French HM, Gorbunov A et al (2014) The Last Permafrost Maximum (LPM) map of the Northern Hemisphere: permafrost extent and mean annual air temperatures, 25–17 ka BP. Boreas 43(3):652–666

Vinther BM, Buchardt SL, Clausen HB et al (2009) Holocene thinning of the Greenland ice sheet. Nature 461:385–388

Wagner T, Fabel D, Fiebig M et al (2010) Young uplift in the non-glaciated parts of the Eastern Alps. Earth Planet Sci Lett 295(1–2):159–169

Walker M, Johnsen S, Rasmussen SO et al (2009) Formal definition and dating of the GSSP (Global Stratotype Section and Point) for the base of the Holocene using the Greenland NGRIP ice core, and selected auxiliary records. J Quat Sci 24:3–17

Weinberger L (1955) Exkursion durch das österreichische Salzachgletschergebiet und die Moränengürtel der Irrsee- und Attersee-Zweige des Traungletschers. Verh Geol BA 1955:7–34

Wirsig C, Zasadni J, Christl M et al (2016) Dating the onset of LGM ice surface lowering in the High Alps. Quat Sci Rev 143:37–50

Jürgen M. Reitner is a senior scientist in charge of Quaternary geology with the Geological Survey of Austria and external lecturer for Quaternary sciences at the University of Vienna. His research focuses on geological mapping and sedimentology, morphology and stratigraphy of glacial and periglacial deposits as well as on the chronology and dynamics of deep-seated slope deformations and rockslides.

4 River and Valley Landscapes

Gerhard Karl Lieb, Wolfgang Schöner,
and Christine Embleton-Hamann

Abstract

Rivers and valleys contribute to Austria's proverbial beauty. Driven by climatological characteristics both elements are closely linked to each other. The section first gives a hydro-climatological characterization of Austria with special emphasis on rivers. Different runoff regimes are distinguished and can be grouped according to the influence of precipitation, evapotranspiration and storage. Snow cover and glaciers play a crucial role in hydrology due to the predominately mountainous character of Austria. Climate change will have significant consequences for hydrological processes and will be challenging for flood management in the future. Rivers have formed a complicated valley network which mirrors past tectonic processes, especially in mountainous areas. Younger geomorphological processes, in particular the Pleistocene glaciation and the glaciofluvial sedimentation combined with it, shaped the recent character of the valleys. Based on morphology and sedimentology, we provide a classification of valleys into 5 types. The sediment fills of the valleys are important for drinking water supply.

Keywords

Hydroclimatology • Runoff regimes • Floods • River network • Valley types

G. K. Lieb (✉) · W. Schöner
Institute of Geography and Regional Science,
University of Graz, Heinrichstrasse 36, 8010 Graz, Austria
e-mail: gerhard.lieb@uni-graz.at

C. Embleton-Hamann
Department of Geography and Regional Research,
University of Vienna, Universitätsstr. 7, 1010 Vienna, Austria

4.1 Introduction

Rivers and valleys are characteristic elements of Austria's landscape. Besides the Alps, the Danube is another dominant natural feature which is even part of the national anthem. As rivers form valleys, both are causally related to each other and therefore are presented together in this chapter. Considering rivers, runoff is the most obvious quantitative characteristic. The runoff of Austrian rivers is governed by the components of the water balance, which in turn are strongly influenced by the orography of the Alps. The main drivers of the runoff and its changes over time and space are, on the one hand, precipitation and evapotranspiration and, on the other hand, the storage of water by the snow cover (and glaciers) as well as subterranean runoff. Seasonal fluctuations in the quantity of these components also lead to a similar inter-seasonal discharge fluctuation, which is particularly large in the high mountains. As a special feature of runoff in Austria, the solubility of the rock in karst areas leads to a special form of subterranean runoff, the share of which is particularly large in Austria and is of great importance for water supply.

Rivers form valleys through fluvial processes. These lead to erosion of the underlying substrate and to the incision of valleys in the upper parts of the river course, while downstream deposition of sediments dominates, either in the form of distinct fans or spread out wider in valley fills and alluvial plains. In the context of the Alps, the first rivers and valleys developed as soon as parts of the Alpine orogen emerged from the Tethys Ocean. Remnants of a fluvial network date back to the Cretaceous period when the Gosau Sea filled the depressions between the hills and ridges of the precursors of the Alps. However, no valleys in a morphological sense have been preserved from that time and only conglomerates of fluvial origin bear witness to the existence of rivers. The starting point of the development of the current valley pattern was the Oligocene uplift and simultaneous brittle deformation of the Alpine orogen.

C. Embleton-Hamann (ed.), *Landscapes and Landforms of Austria*, World Geomorphological Landscapes,
https://doi.org/10.1007/978-3-030-92815-5_4

4.2 Hydrology of Rivers

In short, hydrology is the science that deals with water on and below the Earth's surface. Considered in more detail, hydrology covers the water in its various forms, how it circulates, its spatial and temporal distribution, its biological, chemical and physical characteristics and its interaction with the environment (Spreafico and Weingartner 2005). Hydrology of rivers, in particular, describes the surface water flows with respect to the quantity and quality of the water. Consequently, it aims at quantifying the volume of the surface water flow and the related solute transport, sediment transport and erosion. Surface water flow includes both diffuse overland flow and flow in recognizable river channels. In specific cases, the flow in large rivers is considered as a topic of hydraulics or hydrodynamics, which focuses more on the hydro-dynamic properties of liquids. The water quantity of a river is commonly described by water level, discharge (runoff), runoff coefficient or similar measures. Quantity also includes the sediment transport. Quality refers to the chemical, physical, biological and radiological characteristics of water. Generally, quality is defined as reference to a set of standards, e.g., drinking water or health of ecosystems.

The hydrology of rivers is dependent on various properties within the catchment area, including meteorology (precipitation and available energy, e.g., for evaporation), orography, land use, soil and geology. A key characteristic of rivers, which directly links to valley landscapes, is the catchment area which is governed by the orography as well as the subsurface geology (particularly in karst areas).

The interaction between rivers and the water flowing underground (groundwater in aquifers) can be rather complex and the direction of net water exchange (either into the surface water or into the aquifer) may vary spatially along a stream. Additionally, it may vary over time at any location of the stream, depending on the stream and groundwater levels, respectively (or more precisely the hydraulic gradient). The interaction between surface water and groundwater becomes particularly complex in karst areas, since rocks such as limestone or dolomite can be corroded through the dissolving effect of the carbonic acid contained in the water (cf. Chapter 5). Since 20% of the Austrian area is covered by karst regions, the importance of karst hydrology for the general hydrology of rivers is obvious. This statement becomes even more important as the water supplies of a large number of villages and towns in Austria come from karst areas. This is particularly relevant for Vienna, where almost all the drinking water originates from the limestone/dolomite areas of the Northern Calcareous Alps.

4.2.1 Brief Hydro-Climatological Characterization of Austria

The hydro-climatological structuring of Austria is primarily caused by the spatial variability of precipitation amounts and the transition from liquid to solid precipitation with increasing elevation. Additionally, evapotranspiration, land use, geology, soil properties as well as various feedbacks are important, depending on the spatial scale considered. Generally, altitude affects not only the amount of precipitation (which increases with altitude), but also air temperature and evapotranspiration (which decrease with altitude) as well as the amount of snow (which increases with altitude because of the previous two effects).

The reason for the usually observed increase of precipitation with elevation is complex and is caused by the influence of mountain orography on precipitation-forming processes (Rotunno and Houze 2007). Consequently, weather conditions and the direction of inflow of the precipitation-forming air masses and interaction with the orography of the Alps play an important role for the amount and spatial variability of precipitation. However, there are also precipitation events in which this influence has little or no effect. This is particularly evident for convective precipitation events in summer (e.g., thunderstorms), when weather patterns with weak large-scale surface pressure-gradients occur. In general, a high correlation between the annual precipitation totals and the number of precipitation days can be observed for a large part of Austria. However, for the south of Austria annual precipitation totals are significantly related to the intensity of precipitation events (Auer et al. 2001).

Figure 4.1 shows the elevation dependency of selected components of the water cycle for Austria. Obviously, the annual precipitation totals are spatially significantly more variable and less elevation-dependent compared to the other variables shown. While precipitation, accumulated new snow and potential evapotranspiration are components of the water balance, air temperature, as a useful proxy of surface energy balance, can explain the elevation dependency of evaporation and snow melt as well as the type of precipitation (solid or liquid) rather well. As shown by several studies, the elevation dependency of annual potential evapotranspiration in the Alps is strongly linked to the length of the vegetation period, which in turn is governed by the number of snow-free days (e.g., Menzel et al. 1998; Körner 2004).

4.2.2 Hydrology and Runoff Regimes in Austria

Runoff regimes are used to classify the runoff patterns of rivers, which describe the changes of discharge in time

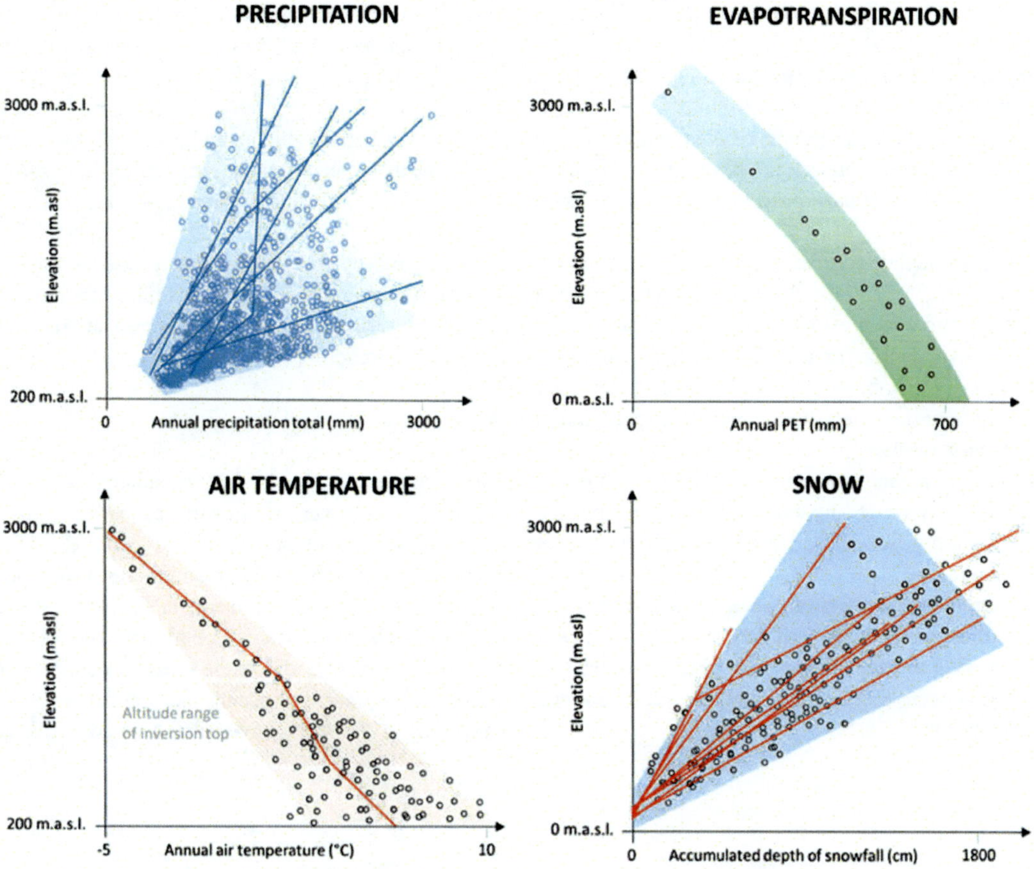

Fig. 4.1 Elevation dependency of hydrological key variables (30-year annual averages). From upper right to lower left: precipitation totals, potential evapotranspiration, mean air temperature and accumulated new snow. Dots symbolize measurements from a station, lines indicate linear regression for sub-regions within Austria and shaded areas symbolize the variability distribution with elevation

(usually over the year), resulting from the interplay of climate, vegetation, geological and geomorphological setting together with human impact in the respective catchment area. In Austria, climate, namely annual course of precipitation, evapotranspiration and snow melt/retention, is the main driver of runoff and, consequently, runoff regimes.

Generally speaking, runoff of rivers in Austria is strongly determined by precipitation, which peaks during summer for a large part of Austria. Only a small region in the south is characterized by a second precipitation peak in autumn, generated by cyclones approaching the Alps from the south. For mountain areas of the Alps, a typical feature of the runoff is that a significant part of precipitation is stored as snow or glacier ice, with time scales of one month to decades or even significantly longer. According to the storage time of the precipitation in the snow cover (which can be transferred further to firn and glacier ice) and glacier ice, the precipitation-runoff relationship is time-delayed. In this context, it is important to know that snow cover and glaciers can generate runoff by melt even in periods without precipitation, provided that enough energy is available from the atmosphere (and snow or glaciers still exist). This compensatory effect is particularly important during hot drought conditions in the Alps.

Following Mader et al. (1996), the following runoff regimes characterize Austria's rivers and their catchments:

(1) Simple runoff regimes (with a single runoff peak). They are characterized by two periods, namely a high-flow versus a low-flow period:
- Glacial regime (with runoff maximum in either July or August)
- Nivo-glacial regime (with runoff maximum in either June or July)
- Nival regime (with runoff maximum in June)
- Moderate nival regime (with runoff maximum in either May or June)
- Nival transition regime (with runoff maximum in either April or May)

- Pluvial transition regime (with runoff maximum in March)

(2) Complex runoff regimes (with two or more runoff peaks over the year):
- Summer pluvial regime (runoff peak in summer due to precipitation without significant contribution of glacier ice melt, weak secondary peak in spring due to snow melt)
- Winter nival regime (primary runoff peak in spring due to snow melt, weak secondary peak in winter due to temporary snow melt events and precipitation)
- Autumn nival regime (rather similar to the nival regime with primary runoff peak in spring caused by snow melt, but with secondary runoff peak in autumn due to precipitation events)
- Nivo-pluvial regime (characterized by runoff peak in April and May caused by snow melt, at least one secondary runoff peak either in summer or winter, caused by precipitation)
- Pluvio-nival regime (runoff peak in March/April or even earlier caused by precipitation and additional snow melt as well as secondary runoff peak dependent on runoff generating mechanisms, e.g., caused by summer precipitation)
- Winter pluvial regime (runoff peak in winter caused by precipitation and low evapotranspiration, secondary peak mainly in autumn, again due to precipitation)

Figure 4.2 shows the annual course of the water balance components—precipitation, evaporation and snow storage (SWE)—for the nival and pluvial regimes, which are typical runoff regimes in Austria. The nival regime is strongly characterized by the storage and release of water from the snow cover over the year, whereas runoff for pluvial regimes is generally forced by the annual course of precipitation and evapotranspiration.

Alps

The Alps are characterized by runoff regimes in which the seasonal snow cover and, in some catchments, glaciers store water from precipitation in frozen state. The nival runoff regime shown in Fig. 4.2 can be used as a suitable example to describe the runoff of rivers in the Alps. However, the varying percentage of snow or glacier cover as part of the total catchment area affects the seasonal concentration of the runoff. Temporarily varying accumulation and ablation processes of the snow cover play a decisive role for runoff

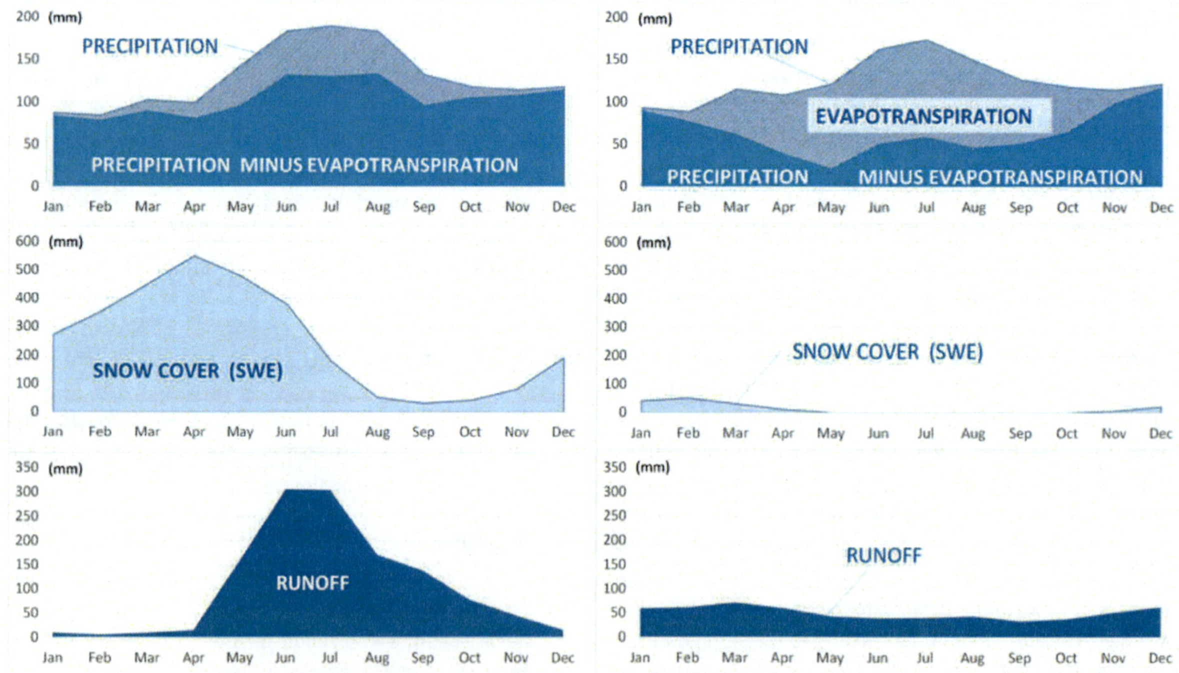

Fig. 4.2 Comparison of two different runoff regimes in Austria. Left: Nival runoff regime (typical high mountain catchment in the Alps) with significant influence of the snow cover. Right: Pluvial runoff regime (typical non-Alpine catchment) with relatively constant discharge over the year. SWE = Snow water equivalent

generation and thus the annual course of the runoff. The strong seasonality of the runoff (high variation coefficient) leads to low-flow conditions during winter, whereas runoff during snow melt in spring is particularly large.

On an annual average, the Alps are characterized by large water resources, which are used for hydroelectric power generation and water supply. Many of Austria's largest cities (including Vienna) receive their drinking water from the Alps, especially from karst areas (cf. Chapter "Karst Landscapes in Austria", Sect. 5.5.1). Next to karst areas, pore groundwater in the Pleistocene valley fills of the rivers of the Alps and the Alpine forelands is an important water source. Recharge of these groundwater systems is best during snow melt.

Non-Alpine areas

The non-Alpine areas of Austria benefit considerably from the water wealth of the Alps. For these areas, however, the storage of water in the snow cover plays no or only a minor role compared to the Alpine region. The runoff regime is usually governed by the annual cycle of precipitation and evaporation. Larger rivers such as the Inn, Salzach, Enns, Drau, Mur or Danube still have a strong footprint from the Alpine upstream catchments in the form of snow melt during spring. This footprint levels out downstream of the larger rivers. As shown in Fig. 4.2, evapotranspiration becomes more important with decreasing elevation and, consequently, for the non-Alpine areas. This is clearly related to the increasing amount of available energy from the atmosphere and the minor role of energy loss due to snow melt. Compared to the rivers in the Alps, runoff in non-Alpine areas is far more stable over the whole cycle of a year. Precipitation, preferably liquid, plays a much larger role, which is why the related runoff regime is called pluvial.

4.2.3 Floods and River Engineering Measures

Floods are extreme events of water level or runoff in a river. Generally, these extreme values of runoff or water level are described by their statistical return period, e.g., HQ100 means a rare runoff event recurring once in a hundred years. HQ30, HQ100 and HQ150 are used as design floods for the dimensioning of infrastructure and for the designation of flood protection zones in Austria. Floods in Austria are caused by different meteorological and hydrological mechanisms which show typical seasonality in their relevance and result in the following classification of main types of floods in Austria (Merz and Blöschl 2003):

- Floods forced by large-scale precipitation (associated with cyclones interacting with Alpine orography)
- Floods generated by small-scale convective precipitation (associated with moving cells and thunderstorms)
- Rain-on-snow triggered floods
- Snow melt triggered floods.

Floods caused by large-scale precipitation are the most relevant in Austria (43% of all floods, according to Merz and Blöschl 2003). Because of the linkage to climate causes, floods are also related to the runoff regimes (see Sect. 2.2). However, while runoff regimes are entirely characterized by atmospheric variables (precipitation, snow melt, evapotranspiration), flood regimes are governed by additional characteristics, such as land use and retention volume (e.g., hydraulic structures) in the catchment. Based on a number of catchments in Upper Austria Viglione et al. (2016) showed that the relative contribution of flood trend drivers in Austria varies with the size of catchments (Fig. 4.3). For small catchments, flood trends are dominated by land-use changes and climate (precipitation, snow melt, evapotranspiration), flood trends in medium-sized catchments (100–1000 km^2) are dominated by changing climate, whereas in large catchments (10 000–100 000 km^2) flood trends are mainly forced by changing climate and retention volume.

The range of the runoff for hundred-year floods in Austria is shown in Fig. 4.4. As runoff generally increases with catchment size, the volume of a flood follows the same relationship (increase with catchment size). Consequently, the largest floods are observed in the largest rivers of Austria, e.g., the Danube, the Inn, the Salzach, the Drau or the Mur (see Fig. 4.4). Given the flood drivers in Austria

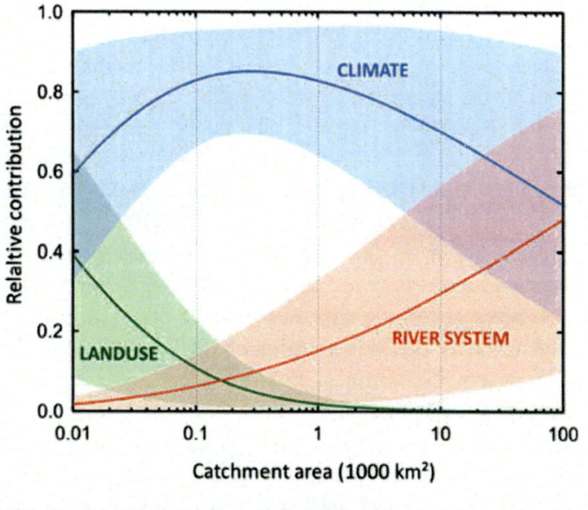

Fig. 4.3 Modelled flood trend attribution for c. 100 catchments in Upper Austria where distinction is made between atmospheric drivers (blue, e.g., precipitation), catchment drivers (green, e.g., land use) and river system drivers (red, e.g., loss of retention volume). Mean (dark line) and 90% credible bounds (light transparent areas). (modified from Viglione et al. 2016)

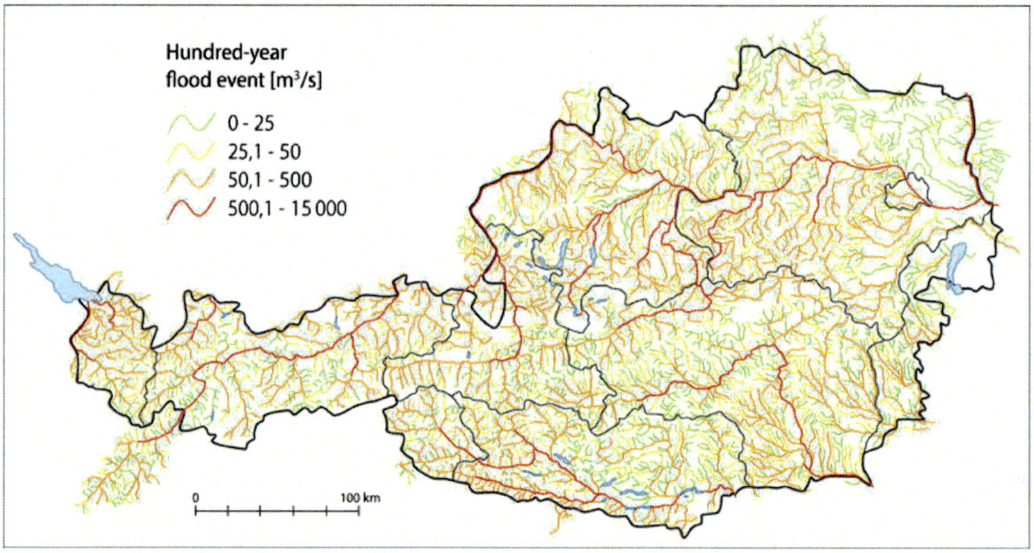

Fig. 4.4 Hundred-year flood (m^3/s) for 26 000 km of streams in Austria as a result of a modelling approach based on observed runoff data. Modified from Merz et al. (2008). (Cartography: A. Weissinger)

described above, climate change will impact floods at all catchment sizes. On the one hand, changing precipitation in terms of both amount and seasonal timing (and also precipitation type, e.g., large-scale frontal versus small-scale highly convective) is relevant. On the other hand, a raised snowfall line and increased snow melt due to higher air temperatures are significantly changing key drivers of floods in Austria. Quantifying these mechanisms, BMLFUW (2017) concluded that under future climate change HQ100 will slightly increase in eastern and southern Austria, will be rather unchanged in the western part of Austria and will increase by about 10% in the rest of the country.

Due to the high level of Austria's socioeconomic development, the status of protection against floods—and other natural hazards—is largely satisfactory. However, river engineering measures in the past did not sufficiently consider the entire hydrological system. For instance, feeling safe due to the protection measures designed for HQ100, settlements expanded close to the rivers, which resulted in reduced retention areas. However, these retention areas would have been necessary for the large floods which occurred in recent years (such as the Danube floods of 2002 and 2013). Flood management in Austria is also dependent on hydropower plants and their potential to reduce flood waves and related water levels. Today, the rivers of Austria are extensively used for power generation by a large number of power plants of different sizes and technical type (run-of-river power plant, storage power plant) (Fig. 4.5).

4.3 The River Network

Slope and related potential energy form the key force for initiating drainage systems, directing the flow of the uppermost reaches of mountain streams perpendicular to the direction of the mountain ridges. This simplest type of parallel drainage following the general slope of a region becomes rare in higher-order channels (an example within Austria is the parallel rivers draining the southern side of the Niedere Tauern range). Since the development of a drainage network depends on several other factors, river patterns are usually far more complex (Fig. 4.6).

4.3.1 Catchment Areas

Position, shape and size of catchment areas have already played an important role for interpreting flood discharges of rivers (see Sect. 2.3). Over time, processes like headward erosion and river capture change the size and shape of a drainage basin. For instance, the upper course of the Moldau/Vltava River may have drained to the Danube until the Neogene period. This is suggested by a string of old gravel, starting south of the prominent right angle turn of the Moldau/Vltava River to the north. From there, the old fluvial deposits can be followed across the Bohemian highlands of Upper Austria towards the Danube (Chábera and Huber 2000).

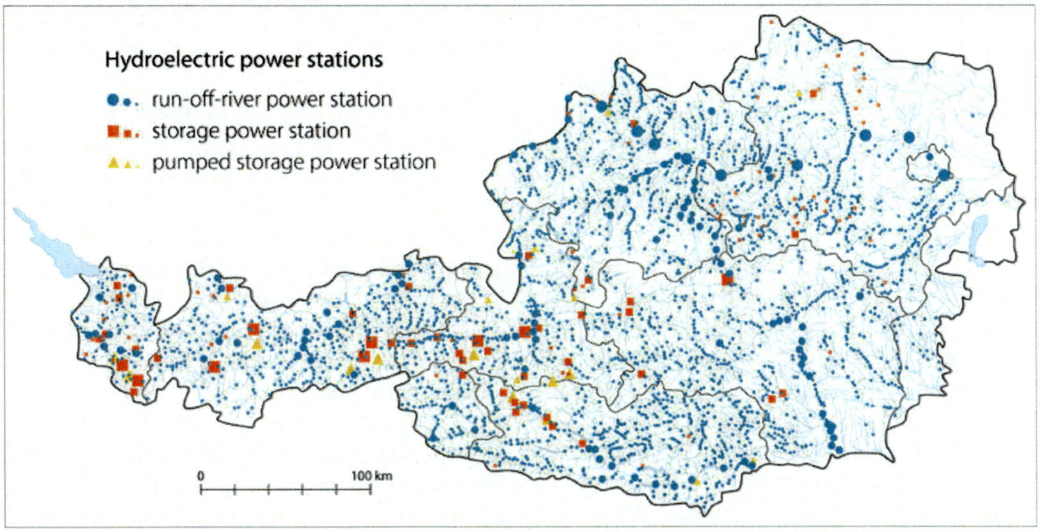

Fig. 4.5 Location of hydropower plants in Austria in 2011. Power plant capacity is roughly indicated by symbol size. Modified from Habersack et al. (2012). (Cartography: A. Weissinger)

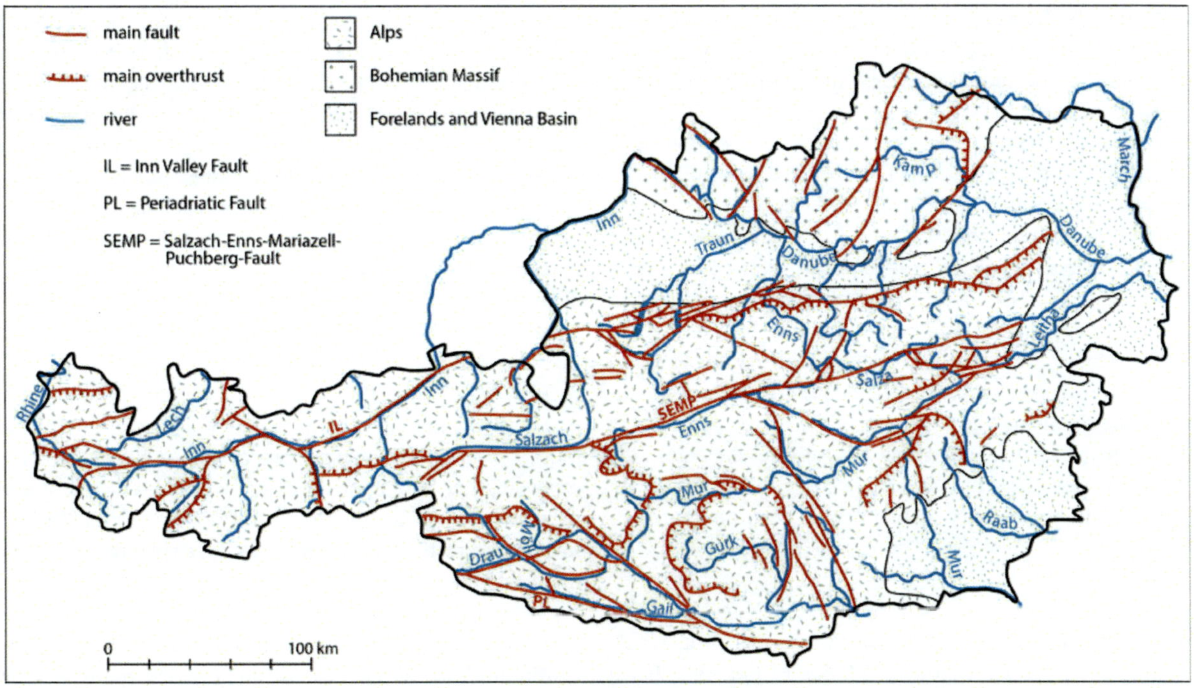

Fig. 4.6 River network, main landscape regions (cf. Chapter "Geomorphological Landscape Regions of Austria") and major tectonic structures (faults and nappe boundaries after Geologische Bundesanstalt 2013) of Austria (Cartography: V. Damm)

Examples of more recent changes of river basins can be found at many locations in Austria. The subsidence of the Klagenfurt Basin, for instance, triggered a northward shift of the watershed between the catchments of the rivers Drau and Mur, thus increasing the size of the Drau River basin at the expense of the Mur River basin. Eicher (1982) suggested that the watershed was additionally lowered by glacial abrasion, caused by an overspill of ice from the north. The current shape of many river basins largely dates back to the period after the Last Glacial Maximum (LGM). In the geological future, erosion will further change river basins. A potential candidate of river capture is the low watershed between the rivers Enns and Salzach near Eben im Pongau (province of Salzburg). A minor first-order stream development towards the Salzach River basin would deflect the uppermost reach of the Enns River towards the Salzach River!

4.3.2 Causal Processes for the Development of Austria's River Pattern

The rivers of Austria drain to three major river basins. 96.1% of the Austrian territory drains to the Danube River basin (and finally to the Black Sea), 2.8% drain to the Rhine River basin and a further 1.1% to the Elbe/Labe system (both oriented to the North Sea) (BMLFUW 2011).

Hence, the most important receiving river is the Danube, which flows through the northern part of the country in a roughly west–east direction parallel to the main Alpine crest (which also acts as the main watershed of the Alps). In terms of runoff, the largest tributaries from the northern side of the Alps are the rivers Inn, Traun and Enns. After leaving Austria and turning south in the Hungarian Dunakanyar ("knee of the Danube"), the Danube finally also collects the drainage from the south-eastern and eastern slopes of the Alps, where the most important tributaries are the rivers Drau and Raab).

Likewise, south of the main divide of the Bohemian Massif the rivers drain to the Danube. The main tributaries are the rivers Kamp and March. Because of low regional precipitation, their contribution to the discharge of the Danube is much smaller than that of the rivers coming from the Alps. Only a small area of the Bohemian Massif drains to the Elbe/Labe system (and consequently to the North Sea), and a small part of the Alps situated in the westernmost province of Austria is part of the Rhine River basin.

In the following, the evolution of the drainage pattern in three major landscape regions of Austria, the Alps, the forelands (including the Vienna Basin) and the Bohemian Massif, will be presented in more detail (Fig. 4.6).

Alps

The Alps show quite a complicated river network, which above all reflects the tectonic evolution of the Alps. For a first characterization, it is useful to distinguish between longitudinal and transversal valleys. Longitudinal valleys are oriented parallel to the direction of the Alpine mountain chains, i.e., west–east. Frisch et al. (2000) addressed them as one of five principal geomorphological characteristics and thus as a dominant feature of the Eastern Alps. Consequently, they are used as important demarcations in all physical geographical/geomorphological regionalizations of Austria. There are two main valley depressions, of which each consists of a series of valleys, generally more connected with than divided from each other by low watersheds:

- The *northern longitudinal valley depression* stretches across almost the entire Austrian territory (Fig. 4.9). The rivers, which partly flow within this valley depression, are from west to east: Ill, Alfenz, Rosanna, Inn, Salzach, Enns, Palten, Liesing, Mur and Mürz. It starts in Vorarlberg and finally reaches the southeastern edge of the Vienna Basin across the Semmering Pass.
- The *southern longitudinal valley depression* starts in Southern Tyrol. Its Austrian section is drained by the rivers Drau and Gail.

Both longitudinal valley depressions are characterized by wide valley floors, allowing the accommodation of settlements (for instance, the urban agglomeration of Innsbruck/Tyrol), traffic lines and industrial plants. Besides the Rhine Valley, they are therefore the most densely populated areas within the Alps.

There is a strong structural control of the river network in the Austrian Alps. In their western part, the Inn River follows a fault line, the Inn Valley Fault (Fig. 4.6). It is considered as one of the oldest Alpine faults and dates back at least to the Oligocene (Frisch et al. 2000). At the turn from the Oligocene to the Miocene, the eastern part of the Austrian Alps was not yet a high mountain area but a hilly landscape with a drainage pattern that had nothing in common with the current one. Neither the Northern Calcareous Alps nor the longitudinal Enns Valley existed. Thus, the rivers coming from the (already slightly uplifted) Central Alps flowed in a south-north direction and deposited the Augenstein sediments on the calcareous country rock.

However, the situation completely changed when large blocks of the Austroalpine nappe system began to move eastwards due to the lateral extrusion caused by the northward motion of the Adriatic plate (Frisch et al. 2000). The main consequences of this process on the river network were the following:

- Two large fault systems evolved, the Salzach-Ennstal-Mariazell-Puchberg fault (SEMP) in the north and the Periadriatic fault (PL) in the south (Fig. 4.6). Between them, the crustal blocks moved towards the Pannonian Basin, and the brittle rock of the shear zones became the guideline for a new valley incision in a west–east direction. Vertical movements also occurred—especially the uplift of the Calcareous Alps—providing further incentive for erosional downcutting along the master faults. Thus, along the SEMP, the Salzach and Enns valley sections of the northern longitudinal valley depression developed, as did the Gail Valley section of the southern longitudinal valley depression along the PL.

- Between the SEMP and the PL, many secondary fault systems developed, of which the main ones are depicted in Fig. 4.6. One important branch starts in the longitudinal section of the Enns River and stretches to the southeast where it finally merges with another secondary fault along the Mur River. As the country rock (greywacke) along this branch is more prone to the development of broad valleys than the limestone along the final eastern section of the SEMP, the northern longitudinal valley depression at this point turns southeast and follows this branch.
- The simultaneous uplift of the Alps intensified erosion and deeper lithological units were exhumed. As the boundaries between different rocks and nappes interacted with fluvial processes, they shaped the river network on a local scale.

All rivers that flow along the longitudinal valley depression sooner or later must cross the peripheral mountain ranges to reach the forelands and finally the Danube River. To achieve this, the rivers had to cut narrow transversal valleys through the mountains. In Fig. 4.6, this is apparent by the almost right angle turns that the rivers Inn, Salzach, Enns and Mur take when leaving their longitudinal valley sections. The transversal valleys are much narrower than the longitudinal ones and, in the case of the rivers Enns and Mur, have a winding course due to adaption to local tectonic structures.

Forelands

The river network of the forelands is in most cases much simpler than that of the Alps. After leaving the Alps, the rivers follow the general slope towards the Danube River. Only in some places has younger tectonic uplift initiated different river courses, e.g., where the Mur River turns to the east at the Austrian–Slovenian border. A peculiarity of the highest elevations of the northern Alpine foreland is their local radial drainage (Hausruck and Kobernaußerwald, cf. map in Chapter "Quaternary Landforms and Sediments in the Northern Alpine Foreland of Salzburg and Upper Austria ", Fig. 14.1). They have a Pannonian cap of fluvial gravel, in which precipitation seeps away and resurfaces in numerous springs at its base, feeding streams in all directions.

Bohemian Massif

Similar to the Alps, the drainage pattern of the Bohemian Massif largely mirrors the fault systems. In the Austrian part of the massif two fault directions prevail: (i) northwest-southeast and (ii) northeast-southwest. Both also influence the river course of the Danube. A special feature is the meandering course of many rivers in the Bohemian Massif (cf. Chapter "Deeply Incised Valley Meanders of the Bohemian Massif").

4.3.3 The Course of the Danube River

The Danube River flows over 350 km through the territory of Austria. The geomorphological character of its valley is very diverse as there is a steady alternation between narrow, partly gorge-like and wide, partly basin-like sections with large alluvial plains. Interpreting the Danube's course in Austria has been a "classical" geomorphological research topic in the twentieth century (e.g., Blühberger 1996). In the following, the Danube sections upstream and downstream of Vienna will be briefly addressed.

- In the western two-thirds of its course through Austria, the Danube has incised a gorge-like valley at four sections into the solid granites and gneisses of the Bohemian Massif (Fig. 4.7), although only a few kilometres to the south the soft Neogene sediments would have allowed a far easier valley formation. Obviously, the sediment yield of the rivers coming from the Alps pushed the Danube to the north, where the river established a new course, mainly following faults. Probably each of the four canyon sections has its individual morphogenesis (see Chapter "Wachau World Heritage Site: A Diverse Riverine Landscape" for the evolution of the Wachau Gorge).
- The lowest, c. 50 km long reach of the Danube in Austria crosses the Vienna Basin after entering it at the so-called Wiener Pforte (Vienna Gate, Fig. 4.8). The latter is a narrow and geologically young incision into Flysch rocks, again most probably facilitated by fault lines. The Miocene palaeochannel of the Danube can be traced c. 25 km north of the Vienna Gate. Until the beginning of the Quaternary, it gradually shifted southwards to its current position. Simultaneously, it deposited its sediments into the subsiding Vienna Basin (Blühberger 1996), initially as delta deposits into the Parathetys, followed by delta deposits into the Pannonian Lake and finally as alluvial fans on dry land.

4.4 Valley Shapes and Valley Sediment Fills

The characteristics of a valley landscape evolve from the interplay of erosion and deposition of rivers (or glaciers) over time. The legacy of tectonic history and former glaciation helps to explain large-scale valley features,

Fig. 4.7 River bend of Schlögen is the most famous part of the gorge-like uppermost section of the Danube Valley (Passauer Tal) in Austria. The double-bend is caused by the river's adaptation to the direction of northwest-southeast trending faults. View towards northwest, flow direction is from left to right. (Photo: G. K. Lieb)

whereas small-scale and regional erosion versus accumulation processes are nowadays driven by human-induced channel changes, which are widespread in Austria (see Fig. 4.8, depicting the complete change of the Danube course for flood prevention). Special emphasis will be given to valley sediment fills, because they play a key role for the water supply in Austria.

4.4.1 Valley Types and Their Spatial Distribution

In order to provide a country-wide overview of the character of Austria's main valleys, Fig. 4.9 differentiates between four types of valleys with wide valley bottoms, and one type of deeply incised valleys and gorges with mainly bedrock channels. The latter can therefore be considered as an expression of fluvial downcutting that has been taking place until recent times. Additionally, low watersheds within the valley systems (like the example discussed in Sect. 3.1) are indicated. The three main landscape units of Austria are marked in the same way as in Fig. 4.6.

- *Glacially-shaped valleys with wide valley floors*: This valley type occurs in the western and central parts of the Alps only and its distribution mirrors the extent of the complex of transection glaciers during the LGM. These valleys are extremely overdeepened (e.g., the depth to bedrock of the Inn Valley in the section shown in Fig. 4.10a is c. 900 m), and the fluvial dynamics of the main rivers are strongly influenced by the sediment input of the tributaries. In most cases, these are mountain creeks draining small and steep basins from which large amounts of debris can be mobilized by debris flows during heavy rainfall events. Before river regulation, which already started in the nineteenth century, the valley floors were wetlands prohibiting agricultural use and settlements.
- *Valleys with glaciofluvial terraces*: The large rivers adjacent to the complex of transection glaciers in the LGM received the meltwater from the glaciers. These rivers transported a high sediment load which was provided by glacial as well as by periglacial processes. The combination of high sediment load and only seasonal discharge of the rivers during summer resulted in thick valley fills. However, during Interglacial conditions,

Fig. 4.8 Wiener Pforte (Vienna Gate) as seen from the observation tower "Donauturm" (Vienna) towards the northwest. After regulation work, the current main channel of the Danube is the left one of the three visible water courses; "Alte Donau" (Old Danube) to the right represents a former meander. (Photo: G. K. Lieb)

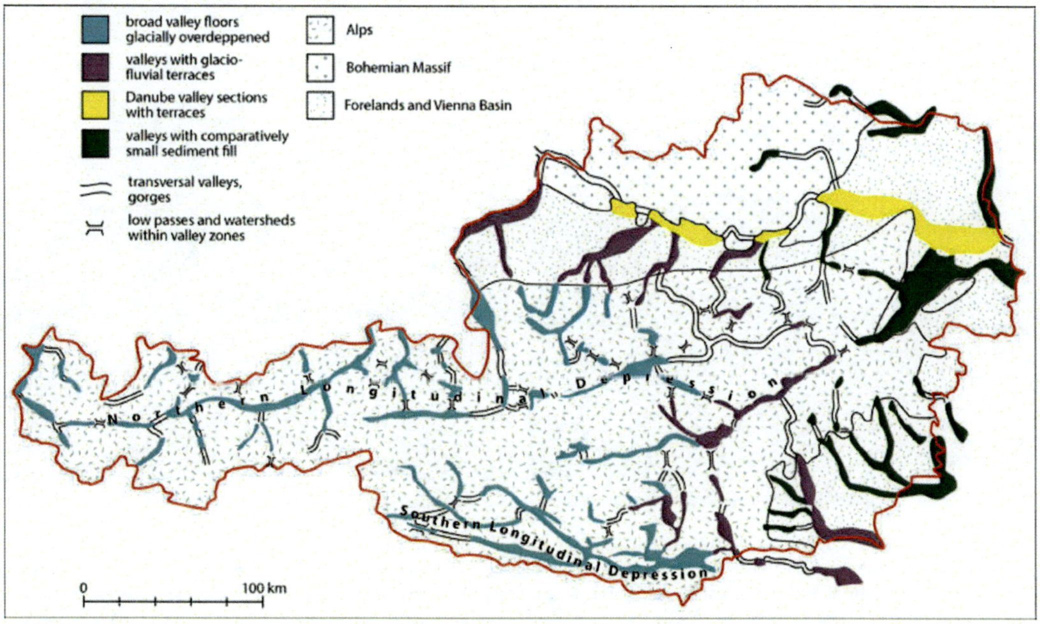

Fig. 4.9 Valley types and main landscape regions of Austria (cf. Chapter "Geomorphological Landscape Regions of Austria") (Cartography: V. Damm)

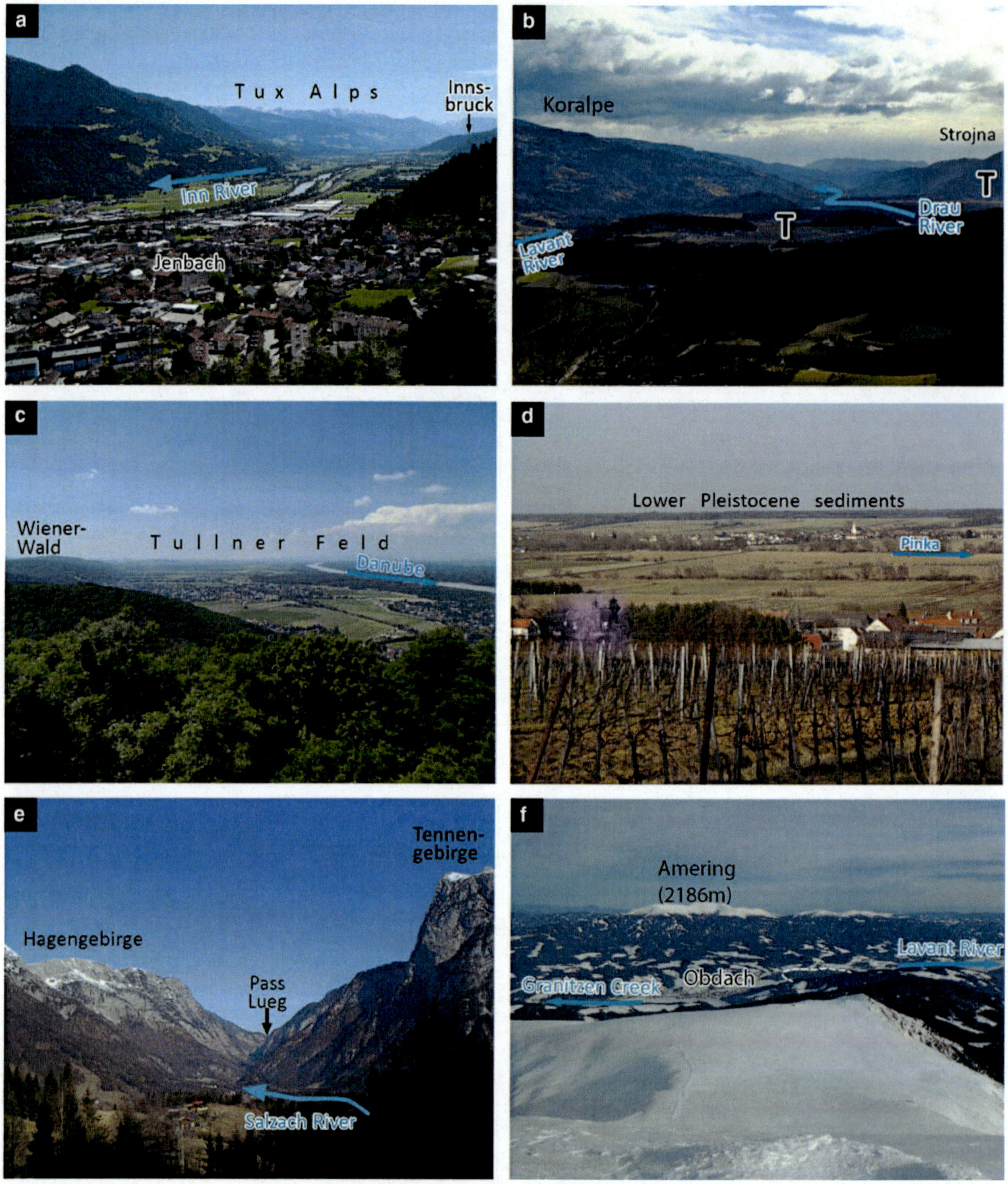

Fig. 4.10 Valley types shown in Fig. 4.9 and explained in the text. **a**: the central part of the Inn Valley east of Innsbruck is an example of a glacially-shaped valley (view from Jenbach towards southeast). **b**: the Drau Valley downstream of the Klagenfurt Basin is a valley characterized by glaciofluvial terraces, in the photo marked with the letter "T" (view in a south-eastern direction towards Slovenia). **c**: example of a broad plain valley reach of the Danube; the lowest part of the plain is covered by riparian forest (view to the west). **d**: the Pinka Valley is filled with fluvial sediments forming a wide valley floor (view in a south-eastern direction from Burgenland towards Hungary in the background). **e**: the deep gorge of Pass Lueg was incised into the Northern Calcareous Alps by the transversal reach of the Salzach River (view to the north). **f**: example of a low divide (dotted line) between two similarly aligned river courses (Styria, view in a south-eastern direction). Photos: G. K. Lieb

when the climate became warmer and higher amounts of precipitation increased runoff, rivers were able to partly erode their former deposits. This resulted in the formation of prominent terraces, especially in areas with calcareous substrate, where the sediments became cemented by calcium carbonate (Fig. 4.10b). In most valleys, this sequence of processes took place in each of the Interglacial-Glacial cycles resulting in terrace sequences consisting of several terraces, with the oldest at the top.

- *Valley reaches of the Danube with terraces*: The overall character of the Danube Valley has been described in Sect. 3.3 as a sequence of gorges and broad valley sections. In the latter, a pronounced terrace system developed during the Pleistocene (e.g., the staircase of terraces in Vienna, see Table 12.1 and Fig. 12.5 in Chapter "The Danube Floodplain National Park: A Fluvial Landscape with Expiration Date?"). Due to neotectonics and a greater distance from the Alpine ice centre, terrace formation is more complex than that of the glaciofluvial terraces of the previous valley type. A further difference is that the Tullner Feld (Fig. 4.10c) upstream of Vienna and the Marchfeld downstream of Vienna are the only areas in Austria where valley floors extend over several kilometres, so that they appear as alluvial plains.
- *Valleys with comparatively small sediment fill*: Periglacial slope and fluvial processes dominated the Pleistocene history of river catchments without or with insignificant glaciation. Compared to the previous valley types, they received a much smaller sediment input. Therefore, valley fills are shallow and terrace systems only developed to a limited extent. This valley type occurs in the easternmost part of the Alps, its adjacent foreland areas (Fig. 4.10d) and in the Bohemian Massif.
- *Transversal valleys without valley bottoms, gorges*: This type represents valleys in which fluvial incision has been the predominant geomorphological process until recent times. It occurs in three settings, namely (i) where longitudinal river channels break through adjacent mountain chains (e.g., the Salzach River (Fig. 4.10e) or the Drau River just after leaving the Austrian territory (Fig. 4.9)), (ii) in valleys developed in highly resistant rocks (e.g., the Kamp River in the granites and gneisses of the Bohemian Massif, or the rivers Salza and Erlauf in the carbonatic rocks of the Northern Calcareous Alps), and (iii) where valleys became blocked by major mass movements (e.g., the Taxenbach Gorge in the longitudinal section of the Salzach Valley). Many of these valley sections are famous for their scenic landscape (e.g., the Gesäuse Gorge of the Enns River, described in Chapter "Gesäuse—River Gorge, Limestone Massifs and Sediment Cascades") and all of them are known as notorious challenges for the construction and maintenance of road or railway lines.

4.4.2 Valley Sediment Fills and Their Relevance for Water Supply

Van Husen (1979) studied the processes involved in the formation of the sediment fill of the first valley type in detail. In most cases, a complex sequence of different processes was involved, including ice decay and moraine deposition, fluvial deposition of the main river, torrential deposition from the tributaries and silting up of lakes. These various materials filled the glacially overdeepened depressions. Thus, thick, but inhomogeneous sediment bodies formed. Despite their thickness, they do not have useful hydrological properties as a water resource because of the occurrence of fine-grained material like silts forming aquicludes in between gravel aquifers.

All other valley types with considerable sediment fills show a more homogenous structure of their deposits, consisting predominantly of sands and pebbles. These substrates, especially the sediments of the valley type with glaciofluvial terraces, have perfect hydrological properties as aquifers, bearing large amounts of groundwater that may easily be extracted for water supply. In fact, these pore water aquifers provide drinking water for about one-half of the Austrian population, whereas the other half is supplied with water from karst aquifers. Karst aquifers are situated predominantly in remote areas of the Calcareous Alps and are thus not very prone to contamination. Pore water aquifers, however, are located in the valley floors where socioeconomic activities (agriculture, industry, trade, traffic and others) as well as the settlements are concentrated. Furthermore, the gravel is also mined as raw material for construction works. Thus, the groundwater vulnerability is high and affords sophisticated tools of spatial planning and protection measures. As in the case of karst aquifers, they became implemented in all imperiled areas of groundwater reservoirs.

References

Auer I, Böhm R, Mohnl H, Potzmann R, Schöner W, Skomorowski P (2001) ÖKLIM—Der digitale Klimaatlas Österreichs. Proceedings DACH 2001, Wien

Blühberger G (1996) Wie die Donau nach Wien kam. Böhlau Verlag, Wien, Köln, Weimar, p 283

BMLFUW (Bundesministerium für Land- und Fortstwirtschaft, Umwelt und Wasserwirtschaft) (2011) Flusseinzugsgebiet und Flussgebietseinheit. https://www.bmlfuw.gv.at/wasser/wasser-oesterreich/plan_gewaesser_ngp/umsetzung_wasserrahmenrichtlinie/feg_fge.html. (Accessed Jan 2018)

BMLFUW (Bundesministerium für Land- und Fortstwirtschaft, Umwelt und Wasserwirtschaft) (2017) Klimawandel in der Wasserwirtschaft—Follow up zur ZAMG/TU-Wien Studie (2011) Anpassungsstrategien an den Klimawandel für Österreichs Wasserwirtschaft im Auftrag von Bund und Ländern. https://www.bmlrt.gv.at/dam/jcr:

e137df9c-4431-458b-a84e-e97c31385648/Klimawandel_2017_final29112017fv4.pdf. (Accessed Jan 2018)

Chábera S, Huber KH (2000) Ein Beitrag zur Frage der oberen Moldau im Tertiär. Jb Oö Mus-Ver 145/1, Linz, 339–367

Eicher H (1982) Zur Flussnetzentwicklung im Klagenfurter Becken und in der Norischen Senke. Carinthia II, 171/92, Klagenfurt 153–172

Frisch W, Székely B, Kuhlemann J, Dunkl I (2000) Geomorphological evolution of the Eastern Alps in response to Miocene tectonics. Z Geomorph N.F. 44:103–138

Geologische Bundesanstalt (ed) (2013) Geologische Übersichtskarte der Republik Österreich 1: 1,500,000 (ohne Quartär): Wien

Habersack H, Wagner B, Hauer C, Jäger E (2012) Wasserkraft in Österreich—aktueller Bestand und decision support system (DSS WASSERKRAFT). Österreichische Wasser- und Abfallwirtschaft 64(5–6):336–343

Körner C (2004) Alpine plant life, 2nd edn. Springer, Berlin, p 344

Mader H, Steindl T, Wimmer R (1996) Abflussregime österreichischer Fließgewässer—Beitrag zu einer bundesweiten Fließgewässertypologie. Umweltbundesamt, Wien, p 192

Menzel L, Lang H, Rohmann M (1998) Mean annual evapotranspiration 1973–1992. Hydrological atlas of Switzerland, plate 4.1, federal office of topography, Bern

Merz R, Blöschl G (2003) Hochwasserabflüsse in Österreich—das HORA Projekt. Oesterr Wasser Abfallwirtsch 60(9/10):129–138

Merz R, Blöschl G, Humer G (2008) National flood discharge mapping in Austria. Nat Hazards 46:53–72. https://doi.org/10.1007/s11069-007-9181-7

Rotunno R, Houze RA (2007) Lessons on orographic precipitation from the Mesoscale Alpine Programme. Q.J.R. Meteorol Soc 133:811–830. https://doi.org/10.1002/qj.67

Spreafico M, Weingartner R (2005) Hydrologie der Schweiz—Ausgewählte Aspekte und Resultate. Berichte des BWG, Serie Wasser Nr. 7, Bern

van Husen D (1979) Verbreitung, Ursachen und Füllung glazial übertiefter Talabschnitte an Beispielen in den Ostalpen. Eiszeitalter u. Gegenwart 22, Hannover, 9–22

Viglione A, Merz B, Viet Dung N, Parajka J, Nester T, Blöschl G (2016) Attribution of regional flood changes based on scaling fingerprints. Water Resour Res 52:5322–5340. https://doi.org/10.1002/2016WR019036

Gerhard Karl Lieb is associate professor at the University of Graz, Institute of Geography and Regional Science. He has experience in physical geography of mountain regions with a special focus on permafrost, glaciers and geomorphic processes affected by global change as well as natural hazards. He is one of the two co-leaders of the national glacier monitoring programme of the Austrian Alpine Club. Besides those topics he works on regional geographies (Austria, Alps and Europe) and on didactics of the Austrian school subject geography and economics.

Wolfgang Schöner is professor at the University of Graz, Institute of Geography and Regional Science. His research interest is on climate change in mountain and polar regions and related impacts, in particular on the cryosphere and water cycle. He is currently an Austrian delegate for the International Arctic Science Committee and co-lead of the Observations working group of the Global Cryosphere Watch programme of the WMO.

Christine Embleton-Hamann is a retired professor at the Department of Geography and Regional Research at the University of Vienna. Her main interest is in alpine environments. Within this field she focusses on human-environment interactions with research topics like human impact on geomorphic processes, assessment of the scenic quality of landscapes, and geomorphological hazards. A second set of interest concerns the communication of geomorphological knowledge to a broader audience, in the pursuit of which objective she has written a well-received textbook on geomorphology. She is a former President of the Austrian Research Association on Geomorphology and served on the Executive Boards of the IAG and several IAG and IGU Working Groups.

5 Karst Landscapes in Austria

Christian Bauer and Lukas Plan

Abstract

The term karst refers to specific geomorphological processes and thereof resulting in characteristic landscapes. About one-fifth of Austria consists of lithologies—mainly limestone, dolomite, and marble—susceptible to dissolution (i.e. karstification). Those lithologies occur in every landscape including the complete range of altitude: Northern Calcareous Alps, Southern Calcareous Alps, Helvetic units, Central Eastern Alps, Bohemian Massif, Alpine forelands, and Neogene basins. Consequently, Austrian karst landscapes are also subjected to a suite of non-karstic geomorphological processes, resulting in a great variety of endo- and exokarst features with distinct modifications. Small-scale solution features (karren) and dolines are very common. At high altitudes, these features have been exposed to processes related to Pleistocene glaciations. Moreover, karst springs with high discharge variabilities are well-known hydrologic features. Austria hosts 18,100 caves, and some of them are amongst the longest and deepest in the world. Although the majority of the caves are of epigenetic origin, some caves are also related to hypogene speleogenesis. Imprints from human activities on the karst environment can be traced back to the Palaeolithic period. Today, human-karst interactions are of particular importance: karst aquifers provide the catchment areas for drinking water supply for several municipals and karst landscapes represent resources for tourism, recreation, and furthermore. Not least, research in karst and caves makes an important contribution to science (e.g. palaeoclimatology).

C. Bauer (✉)
Institute of Geography and Regional Science, University of Graz, Heinrichstrasse 36, 8010 Graz, Austria
e-mail: christian.bauer@uni-graz.at

L. Plan
Karst and Cave Group, Natural History Museum Vienna, Burgring 7, 1010 Wien, Austria

Keywords

Karst • Karst landscape types • Distribution of karst areas • Surface karst features • Caves • Human-karst interaction

5.1 Introduction

The term *karst* is derived from the *Kraš* region near Trieste—a rural region that lacks surface drainage but shows unusual landforms. It was not, however, until the dissertation of Jovan Cvijić at the University of Vienna (Cvijić 1893) that karst became an international term for landscapes containing distinct surface and subsurface features and underground drainage due to the solubility of rocks. Therefore, many terms in karstology such as doline, ponor, or polje have their origins in Slovenian or Serbo-Croatian languages (Ford and Williams 2007).

About 20% of the Austrian territory is built up by lithologies susceptible to karstification: mainly limestone, dolomite, marble, and a few relatively small gypsum occurrences (Spötl and Plan 2016). Due to the humid climate, rock salt can only persist in the subsurface where it is protected from dissolution by impermeable sediments. Besides other geological factors as bedding of sediments and tectonic faulting, karst development depends on interrelated environmental conditions, such as the epikarst development (e.g. bare rock versus closed vegetation cover), direct climate factors (e.g. availability of water, temperature variations), and relief (hydraulic gradient towards the base level). Austria's great landscape diversity and the widespread occurrence of carbonates result in a great variety of karst settings (Fig. 5.1). These karst landscapes can be found in every geographical unit although their distribution and development are quite heterogeneous. The most prominent Austrian karst landscapes—mainly in the form of high alpine karst plateaus—are located in the Northern Calcareous Alps (NCA). Other karst areas are found in the Southern

C. Embleton-Hamann (ed.), *Landscapes and Landforms of Austria*, World Geomorphological Landscapes,
https://doi.org/10.1007/978-3-030-92815-5_5

Calcareous Alps (SCA), the Helvetic units, and many smaller and bigger (Central Styrian Karst) occurrences of low- to medium-grade metamorphic carbonates in the Central Eastern Alps (CEA). Much less important are small marble outcrops in the Bohemian Massif as well as limestones of the Alpine foreland and basins including the Klippen Belt.

This chapter is structured according to these geographical units and gives examples of various karst landscape types, their morphology, and hydrology. Furthermore, some considerations on human-karst interactions, including the value of karst research, are given.

5.2 Surface Karst Features

In Austria, even though most surface karst features are not as big and impressive as for example in the Dinaric Karst, they show the same diversity (Plan 2016). Closed depressions are the most striking surface features in karst landscapes, and dolines in particular are common in Austria. Especially in the formerly non-glaciated summit areas of the NCA (Dachstein-palaeosurfaces sensu Frisch et al. 2001; cf. Chapter "The Rax Karst Massif: A Typical Plateau of the Northern Calcareous Alps?"), dolines reach up to several hundred metres in diameter. In contrast, in the formerly glaciated areas of the NCA, some kilometre-sized elongated depressions were formed by glacial erosion and significantly reshaped by karstification. Poljes of similar size (which is smaller than in the Dinaric Karst) are also of polygenetic origin (often including glacial erosion), and therefore, it is a matter of debate whether the term polje is appropriate or not. Karren features are very common, especially on pure, thick-bedded limestones such as the Dachsteinkalk. Therefore, also international studies on karren development were conducted on Austrian karst plateaus (e.g. Veress et al. 2013).

The distinct subterranean drainage results in karst springs with discharge variabilities of a few orders of magnitude, reflecting the aquifer response to storm events or snowmelt. Some karst massifs are mainly drained by a single large spring at the valley bottom. For example the catchment area of Pießling-Ursprung (Fig. 5.2a, b) comprises several tens of square kilometres of the Warscheneck karst plateau (east of Totes Gebirge), and under peak flood condition, a maximum discharge of 41 m^3/s was measured. The output of the spring

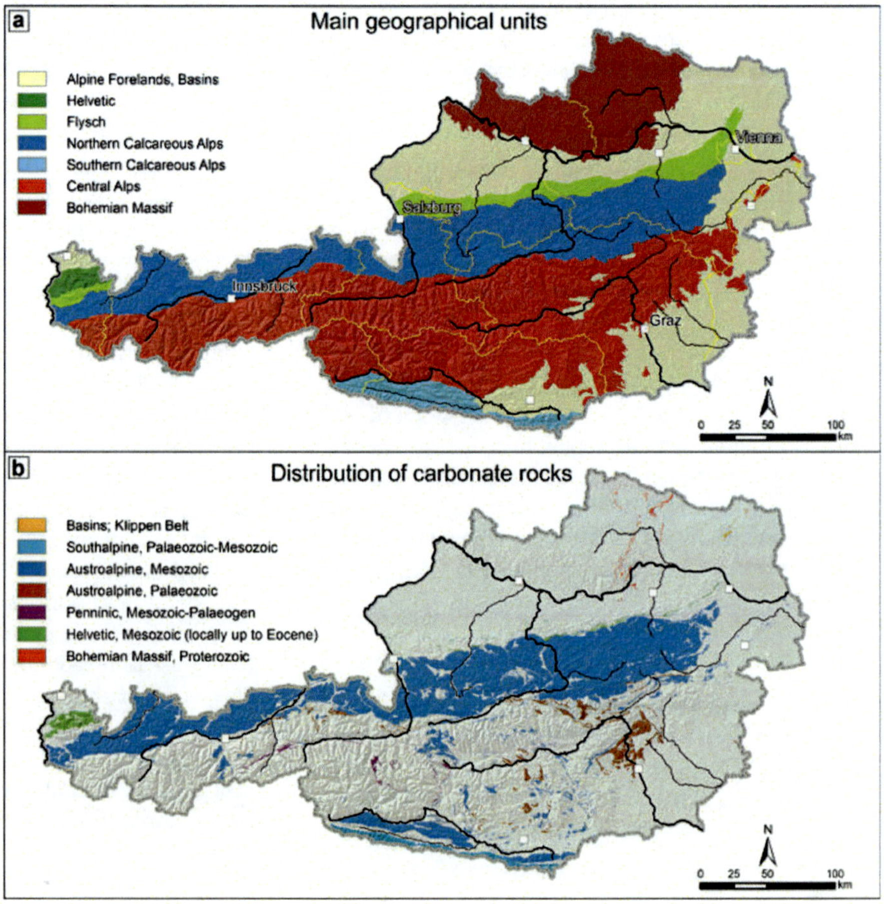

Fig. 5.1 **a** Main geographical units in Austria. **b** Distribution of carbonate outcrops and their tectonic affiliation based on Weber (1997)

Fig. 5.2 **a**, **b** Karst springs at low and high discharge. Pießling-Ursprung, (Upper Austria) Photographs: **a** G.K. Lieb, **b** H. Steinmaßl, **b**, **c** Free draining karst spring Totes Weib (Styria). Photograph **c** C. Bauer, Photograph **d** L. Plan

is controlled by a siphoning reservoir (referred to as vauclusian spring), reaching 87 m below the overflow (Benischke et al. 2016b). Where incision due to river erosion or glacial scour was faster than karstification, springs emerge from caves lying above the valley bottoms. Examples are Gollinger Wasserfall east of the Göll massif or Totes Weib south of Hinteralm (Fig. 5.2c, d).

In contrast to springs, at ponors, surface water infiltrates into the karst aquifer. They are not as common in Austria as for example in the Dinaric Karst where entire rivers disappear underground. The biggest blind valley in Austria is Lurgrotte near Semriach (see below).

5.3 Caves

Probably, one of the most prominent features in karst terrains are caves, and karst researchers have always paid special attention to them. Modern process-based definitions of karst consider the self-organizing groundwater flow system due to speleogenesis as the primary mechanism of karst formation (Klimchouk 2015). Caves rank amongst the best-documented karst features, and the same holds true for Austria: towards the end of the nineteenth century, the first speleological societies were founded and caves were studied in a scientific way (e.g. Kraus 1894). Since 1949, individual caving clubs have been united in the Austrian Speleological Association (Verband Österreichischer Höhlenforscher; e.g. Stummer et al. 2016). Today, structured and standardized cave documentation—archived in the Austrian Cave Register (ACR)—facilitates an area-wide quantification of caves. As usual, the definition of a cave is based on morphology (and not on genetic factors); hence, caves of non-karstic origin are included in the ACR as well. However, the majority of caves in Austria are related to karst processes (i.e. dissolution); even though in some carbonate areas, non-karstic caves (crevice caves that formed due to gravitational mass movements, shelter caves that mainly formed due to frost weathering, etc.) are dominant (Oberender and Plan 2018).

By the end of 2021, 18,100 caves with a minimum length of 5 m have been registered in Austria, and every year, 300–400 caves are newly documented. The distribution of caves based on the ACR is illustrated in Fig. 5.3. The karst regions of the NCA host 79% of all caves (Plan and Oberender 2016). These areas include prominent karst plateaus with thick limestone sequences (e.g. Hagengebirge, Tennengebirge, Dachstein, Totes Gebirge, and Hoch-schwab). Further, the Central Styrian Karst and the Helvetic unit host lots of caves, whereas they are rather rare in the SCA (except for the Dobratsch and Obir massif). The heterogeneous occurrence of carbonates within the CEA is reflected by the distribution of caves. Apart from the high cave density in Palaeozoic meta-limestones in the Central Styrian Karst, there are many occurrences in the Semmering-Wechsel area, where Mesozoic marbles are present, whereas the high alpine karst regions of the Penninic units host only a few, but sometimes extensive caves. Especially in the CEA and SCA, the cave density is partly an artefact, as many areas are speleologically not investigated yet. Compared to the marginal outcrops of marbles in the Bohemian Massif, the percentage of caves in total is remarkable. The reason for this is that the genesis of most caves is related to weathering and mechanical erosion.

In addition to the high number of caves in Austria, their dimension ranks amongst the top of the world. With a total of 153 km of surveyed passages, the Schönberg-Höhlensystem in Totes Gebirge (Styria/Upper Austria) is the longest cave in the European Union and fourteenth longest in the world (Gulden 2021; Table 5.1). The deepest cave in Austria is the Lamprechtsofen in Leoganger Steinberge (Salzburg), reaching a vertical difference of 1727 m (fifth deepest in the world) (Table 5.2). Thirty-six caves are longer than 10 km, and 123 exceed 2 km in length. Concerning depth, 18 caves reach more than 1 km and 271 caves are more than 200 m deep (Pfarr et al. 2021). In Austria, the cumulative length of all mapped cave passages is 2450 km. A comprehensive review of various aspects and regions concerning caves and karst in Austria is given by Spötl et al. (2016).

Most karst caves, especially all the long and deep ones mentioned above, are epigean: it means that precipitation water becomes enriched with CO_2 in the soil and dissolves carbonate rocks on its subsurface path within a local flow regime. In contrast, hypogene caves are formed by water that rise from depth, acquiring the dissolutional potential either from (a) cooling of hydrothermal water, (b) carbon dioxide from deep sources, or (c) from sulphuric acid. As hypogene karst is not related to surface hydrology, most of the hypogene caves do not even have natural entrances but were opened in quarries or mines.

5.4 Karst Landscapes

5.4.1 High Alpine Karst Regions in the Austro-Alpine, South Alpine, and Helvetic Units

This section focuses on high-elevated alpine regions of rather pure and relatively thick carbonate sequences and their setting of interrelated environmental variables: contemporary cover (e.g. bare or poorly developed soils); specific climate modifications (e.g. temporary cover of snow or frost weathering); and geomorphological history (e.g. landscape evolution and former glaciation). During the Pleistocene, most areas of the high alpine regions in Austria were affected by glacial and periglacial processes. Amongst

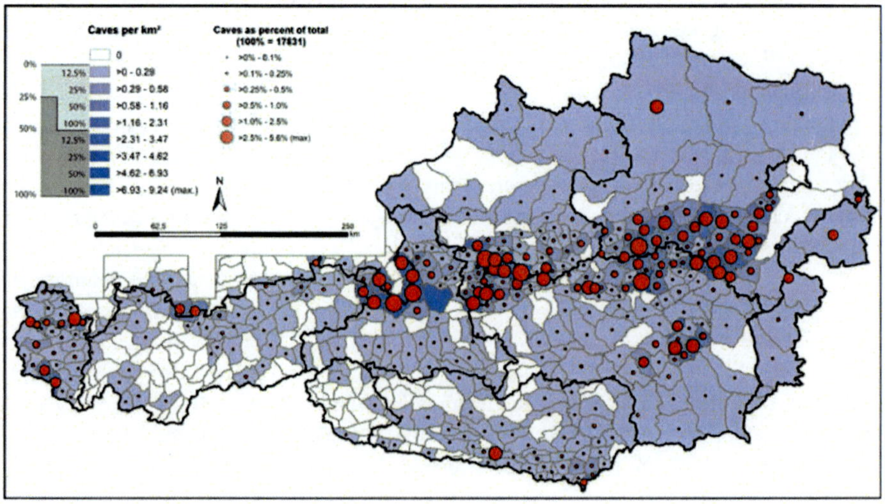

Fig. 5.3 Distribution of caves in Austria. Cave density and caves as percentage of total listed in the Austrian Cave Register. The cave density (caves/km²) is related to the mountain ranges of the ACR

Table 5.1 Five longest caves in Austria by 1/12/2021 (Pfarr et al. 2021)

	Cave name	Area	Province	ACR no	Length (km)	Depth (m)
1	Schönberg-Höhlensystem	Totes Gebirge	St/UA	1626/300	153.1	1061
2	Schwarzmooskogel-Höhlensystem	Totes Gebirge	St	1623/40	135.7	1125
3	Hirlatzhöhle	Dachstein	UA	1546/7	113.6	1560
4	Dachstein-Mammuthöhle	Dachstein	UA	1547/9	67.7	1207
5	Lamprechtsofen	Leoganger Steinberge	Sbg	1324/1	60.0	1727

St Styria, *UA* Upper Austria, *Sbg* Salzburg

Table 5.2 Five deepest caves in Austria by 1/12/2021 (Pfarr et al. 2021)

	Cave name	Area	Province	ACR no	Length [km]	Depth [m]
1	Lamprechtsofen	Leoganger Steinberge	Sbg	1324/1	60.0	1727
2	Hirlatzhöhle	Dachstein	UA	1546/7	113.6	1560
3	Hochscharten-Höhlensystem	Hoher Göll	Sbg	1336/153	14.7	1394
4	Berger-Platteneck-Höhlensystem	Tennengebirge	Sbg	1511/162	30.4	1291
5	Schwer-Höhlensystem	Tennengebirge	Sbg	1511/268	6.3	1219

St Styria, *UA* Upper Austria, *Sbg* Salzburg

others, the resulting impacts on karst features are: (1) destruction of features due to glacial erosion; (2) infilling of features with glacial detritus (e.g. suffosion dolines); (3) modification of karst drainage (e.g. sealing of the epikarst). In high alpine karst areas of the NCA, Holocene karst features are abundant (Fig. 5.4a, b). The resulting specific modifications are evident in microscale karst features such as *karren* (Fig. 5.4c–e), which cover wide areas of high alpine karst regions. The hydrochemical control of bare rock karren differs from dissolution under closed soil cover (e.g. Ford and Williams 2007). In comparison to covered karst areas, where carbonic acid becomes much stronger, the lack of biogenic CO_2 is compensated by snow accumulation and higher amounts of rainfall in more elevated regions. In general, bare rock karren are distinctly sharper than in covered karst areas where they are round and smooth. Bauer (1962) introduced an altitudinal zonation of karst features in formerly glaciated terrains in the NCA. According to this, the main karren zone is situated between 1600 and 2000 m asl, and above 2000 m, frost shattering limits the development of karren.

5.4.1.1 Northern Calcareous Alps

In Austria, the largest and most prominent karst landscapes have developed in the Northern Calcareous Alps (NCA). They mainly consist of Upper Austroalpine carbonate sequences dominated by Triassic limestone and dolomite that can reach thicknesses of 2 km and more. East of the Saalach River, the NCA typically host plateau-like high alpine karst massifs with several levels of kilometre long caves. The elevations of these plateau systems mostly range between 1500 and 2500 m asl, and only in a few parts, the summits reach almost 3000 m. Typically, the valley bottoms are incised down to around 500 m asl, which result in a potential cave depth of up to 2.5 km at Dachstein for example. The most important karst massifs are Steinernes Meer, Hagengebirge, Untersberg, Tennengebirge (Fig. 5.5a, b), Dachstein, Totes Gebirge, Hochschwab (Fig. 5.5c), and Schneeberger Alpen. In addition, the pre-alpine landscapes of the NCA in the north of the high alpine plateaus show abundant karst features and partly extensive cave systems as well. The highest summit of the NCA (Parseierspitze, 3036 m asl) is located in the western part of Austira, where as a result of a different tectonic history, extensive karst plateaus are missing, and the cave density is much lower.

Due to the importance of the NCA, a chapter in the regional part of this book deals with the landscape evolution of these plateaus since the Palaeogene, focussing on the surface karst morphology of the Rax plateau at the eastern end of the NCA (cf. Chapter "The Rax Karst Massif—A Typical Plateau of the Northern Calcareous Alps?"). Subsurface morphology of the NCA is not only remarkable because of the dimension of the caves: Austria's 27 longest and seven deepest caves are located there. In thick carbonate sequences, (epi)phreatic caves tend to adjust to the fluvial base level, and therefore, most of the passages are arranged in distinct levels. In the NCA, three major cave levels can be distinguished of which the so-called "giant cave level" is the most prominent one (Haseke-Knapczyk 1989). It has developed at 1500 ± 300 m asl, and many extensive caves like Dachstein-Mammuthöhle, Eisriesenwelt, Schönberg-Höhlensystem, Schwarzmooskogelhöhlensystem were entirely or predominantly developed

Fig. 5.4 Surface karst features of high alpine karst areas. All examples are from the area In den Karen (Southern Totes Gebirge; bedded Dachstein Limestone). **a** Karrenfeld. Glacial erosion exposed this bedding plane and Rinnenkarren were formed by dissolution. **b** Karren tables. These erratic boulders were left after glacial retreat on a planar limestone surface. The boulder protects the rock underneath from ongoing solutional lowering and a pedestal forms. Its height shows Holocene surface dissolution. **c** Firstrillenkarren (width of picture 30 cm). **d** Rinnenkarren. **e** Meandering Hohlkarre. Accumulation of soil and subsequent vegetation growth sculptured a channel with overhanging walls that were later exposed. Photographs: C. Bauer

within that level. Morphological and tectonic correlations (Frisch et al. 2002), as well as burial ages of quartz pebbles of the Augenstein Formation (Häuselmann personal communication), suggest that the giant cave level has formed about 5 million years ago.

5.4.1.2 Helvetic Units

In contrast to the plateau-like alpine karst, which dominates the NCA, the karst landscapes of the Helvetic nappes in the westernmost part of Austria differ significantly in their subsurface drainage pattern and were classified as folded alpine karst by Hötzl (1992). Karst features and subsurface drainage networks are dominantly developed in the Lower Cretaceous Schrattenkalk, a thin-bedded limestone with a purity of 97% $CaCO_3$ and thickness of up to 125 m. This karst rock is underlain by impermeable marls of the Drusberg Formation. The area is typified by E-W trending folds with a wavelength of 1 to 2 km which mainly control the subsurface drainage pattern (Goldscheider and Hötzl 1999). In Austria, at the border with Germany, the most important karst area is the Hoher Ifen massif (Fig. 5.6a–c) and parts of the Bregenzerwald west of it. Especially, the elevated parts of Hoher Ifen (summit at 2230 m asl) are characterized by extensive karren fields, and this part of the plateau is called Gottesackerplateau (Fig. 5.6c). As this region is tributary to the catchment of the Danube (through Schwarzwasser Valley) and to the Rhine (through Subersach Valley), tracer tests were carried out to locate the European watershed (Goldscheider 2005).

According to the ACR, 185 caves are registered in the Hoher Ifen massif including both the Austrian (113) and the German (72) sides. Eight caves have passage lengths of more than 500 m (Fig. 5.6a), of which Hölloch im Mahdtal, with the upper entrance situated in Germany, is the longest (12.9 km) and deepest (−452 m). Within the Austrian territory, the longest cave is the Schneckenloch. It is 3.6 km long and ascends +148 m from the entrance. The speleogenesis of partly huge galleries is controlled by the contact of the well-karstified Schrattenkalk with the impermeable Drusberg Formation below (Klampfer et al. 2017). A rare karst hydrological phenomenon can be observed at Schwarz-wasserhöhle. It is a so-called estavelle that acts as a ponor (the river sinks into the cave entrance) during low flow conditions and as a spring that discharges up to 4 m^3/s during floods (Goldscheider 2005).

5.4.1.3 Southern Calcareous Alps

The Southern Calcareous Alps (SCA) represents a 190 km long, W-E trending mountain range along the southern border of Austria towards Slovenia and Italy. In contrast to the NCA, they consist of two different tectonic units, divided by the large strike-slip fault system of the Periadriatic Line: (1) the mountains north of the Periadriatic Line (Lienzer Dolomiten, Gailtaler Alps, and the Northern Karawanken)

Fig. 5.5 Karst landscapes of the central Northern Calcareous Alps. **a** Typical high alpine karst plateau, Tennengebirge (Salzburg). View from Mittleres Streitmandl (2360 m asl) towards the east. Photograph C. Bauer. **b** Different sorts of bare rock karren chiselling limestone surfaces in the Dachstein region (Upper Austria). Photograph: L. Plan. **c** Hochschwab plateau (Styria). The surface is characterized by large dolines. Photograph C. Bauer

are part of the Austroalpine units; (2) the mountains to the south (Carnic Alps and Southern Karawanken) belong to the South Alpine units (Fig. 5.7a). In the Northern Karawanken, the dominant carbonates are Mesozoic limestone and dolomite of the Wetterstein Formation and in the Southern Karawanken, the Mesozoic Schlerndolomite and Dachsteinkalk. In the eastern part of the Southern Karawanken, near Bad Vellach, Palaeozoic carbonates are exposed. Even though the carbonate-dominated regions are characterized by distinct subsurface drainage (Brenčič and Poltnig 2008), exo- as well as endokarst features are not as abundant as in similar lithologies of the NCA and heterogeneously distributed (Spötl et al. 2016). This spatial heterogeneity is clearly pronounced in the eastern part of the Karawanken. In contrast to the Mesozoic carbonates of the South Alpine units, the weakly metamorphosed carbonates of Devonian and Silurian age host a remarkable number of caves (Fig. 5.7a). Plateaus with distinct karst depressions are only rudimentarily developed (e.g. on Mt. Petzen), and large epigean cave systems, which are widespread in analogue lithologies of the NCA, are unknown for the Karawanken (Spötl 2016). The morphology of many bigger caves is related to a hypogene origin (Fig. 5.7b, c): the Kotzakhöhle (South Alpine) or the Obir-Tropfsteinhöhlen (Austroalpine). The latter comprises a couple of natural caves with a total length of some kilometres that were encountered due to Pb–Zn mining in the nineteenth century. They are rich in speleothems, and parts of these caves and mining galleries are accessible to visitors. Distinct cave morphologies and alterations of stable isotopes in the host rock as well as the evidence of ascending water enriched in CO_2 in the area of Bad Vellach point towards hypogene speleogenesis (Spötl 2016). Another area in the SCA with plenty of caves is the Dobratsch (cf. Chapter "Dobratsch—Landslides and Karst in Austria's Southernmost Nature Park").

5.4.2 Central Eastern Alps

The Central Eastern Alps (CEA) extends from high alpine regions in the west (e.g. Silvretta, Hohe Tauern) to the low altitude margin of the Vienna basin in the east (Fig. 5.1a), forming an altitude-range between c. 150 m asl at the lowest part of the basin to 3798 m asl at Großglockner. This part of the Eastern Alps is mainly built by polyphase deformed mid- to low-grade metamorphic rocks of Palaeozoic to Mesozoic age

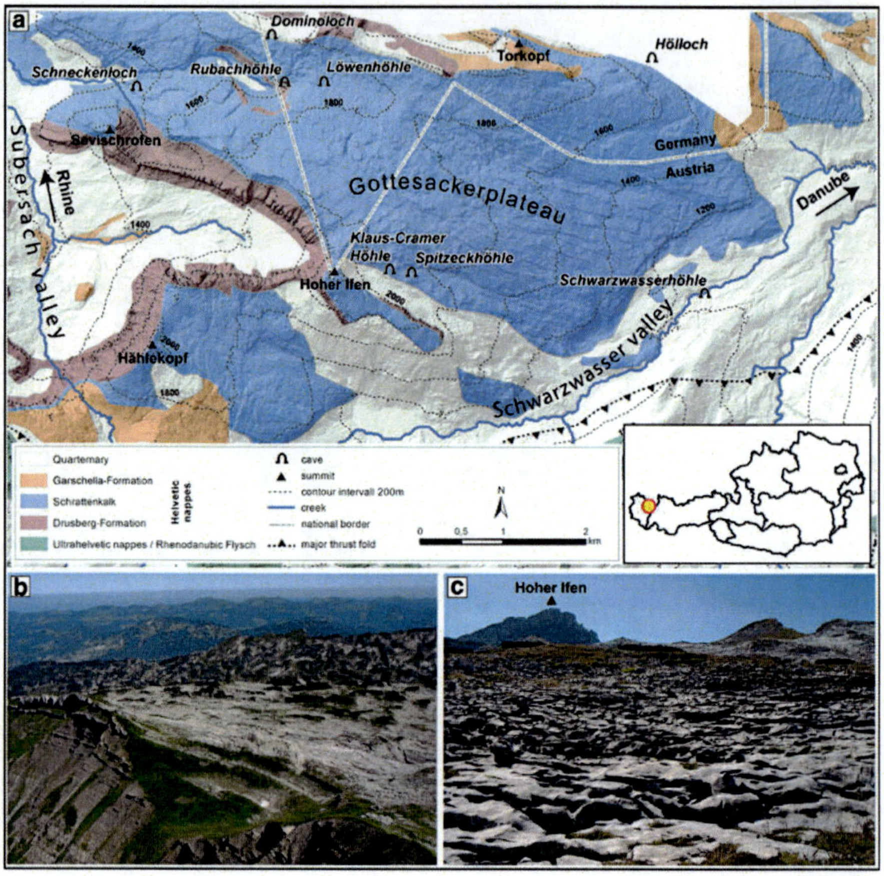

Fig. 5.6 Karst of the Helvetic unit: Hoher Ifen (2230 m). **a** Geological situation (Zacher 1990). Hillshade visualization. Only caves with more than 500 m length are marked. **b** View from Hoher Ifen towards NNW. Alternating marl and limestone sequences of the Drusberg Formation are clearly visible. **c** Gottesackplateau; Photographs: E. Büchel

(Pfiffner 2015). The CEA are characterized by the dominance of silicate lithologies (e.g. gneisses, schists, and amphibolites). Karst areas are developed where carbonates are intercalated in these silicate rocks (e.g. Permo-Mesozoic carbonates in the Hohe Tauern) or largely in Palaeozoic sediments (e.g. Greywacke Zone, Palaeozoic of Graz). Often, at the border of silicate and carbonate rocks, relatively big quantities of water from non-karstic catchments, which are undersaturated with respect to calcite, aliment the soluble rocks and lead to localized and enhanced karstification. This phenomenon is widespread in the CEA and is termed contact karst.

5.4.2.1 The Local Occurrence of Karst in High Alpine Penninic Units (Seidlwinkeltrias)

The high alpine karst regions in the CEA (e.g. Hohe Tauern) are strongly affected by competing geomorphological processes that are related to glacial (e.g. Pleistocene glaciation) and/or periglacial environments (referred to as glacio-karst). The effects of these modifications are evident in the karst area of the Permo-Mesozoic Seidlwinkeltrias. The type locality is situated above 2000 m asl, at the border of Salzburg and Carinthia close to the Hochtor mountain pass (2576 m asl), and the karst area covers 2.5 km^2 (Fink 1984). Karst rocks comprise low-metamorphic limestones, dolomites, calcareous schists, and gypsum (Krainer 2005). The polygenetic origin of the karst landscape in the Seidlwinkeltrias comprises: (a) terraced limestone pavements due to glacial erosion/scour (so-called Schichttreppenkarst, Fig. 5.8a); (b) covered karst surface due to periglacial processes and partial burying by rock glacier detritus (Fig. 5.8 b); (c) debris cover produced by frost shattering (so-called Scherbenkarst); (d) caves on the plateau are often unroofed due to glacial erosion. For the moment, 23 small caves (<100 m) with predominantly vertical development are registered.

5.4.2.2 Marble Stripe Karst at the Mur Spring

The Mur River is the second longest river in Austria and emerges from a karst spring at 1900 m asl in the Federal Province of Salzburg. The associated karst area has developed in a 10 to 50 m-thick marble layer of the Mesozoic Silbereck Group sandwiched between gneisses. This special type of contact karst, where karstification is even more localized, is termed marble stripe karst (Lauritzen 2001) (Fig. 5.9a, b). Within a narrow, but 1.7 km long-band seven caves with a cumulative length of 640 m, 150 dolines, some

Fig. 5.7 Karst in the Southern Calcareous Alps, the Karawanken. **a** Simplified geological map based on Brenčič and Poltnig (2008). Hillshade visualization. **b** Hypogene morphology with inclined walls and watertable marks in the Kotzakhöhle. **c** Abundant speleothem decoration below a flat ceiling due to hypogene speleogenesis (Obir-Tropfsteinhöhlen). Photographs: L. Plan

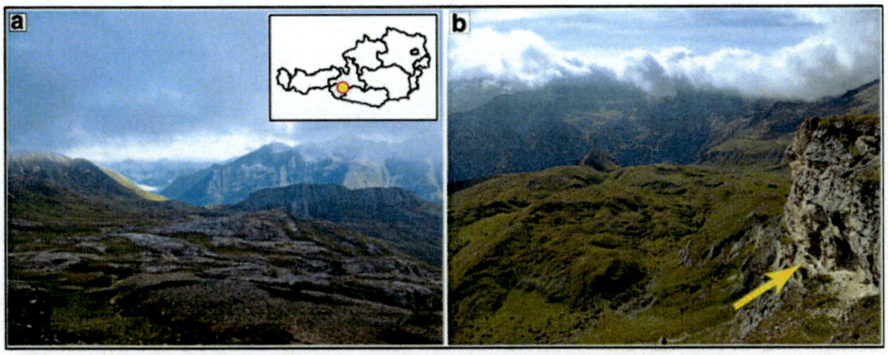

Fig. 5.8 Karst landscape of the Central Eastern Alps/Hochtor. **a** Limestone pavement as effect of glacier action on the karst surface. Baumgartlkar view from NE. **b** Carbonate outcrop (yellow arrow) of the Seidlwinkeltrias that further left in the photograph is covered by a rock glacier. Photographs: G. K. Lieb

ponors, and numerous karren assemblages were documented (Steinwender and Plan 2011), amongst the latter, so-called Strukturkarren prevail, where dissolution retraces the mylonitic foliation of the metamorphic limestone.

5.4.2.3 Central Styrian Karst

The Central Styrian Karst (as defined by Bock 1913) represents the major karst region in the CEA. The area, also known as Palaeozoic of Graz, is located in the central part of the Federal Province of Styria, ranging approximately 40 km from north to south and 50 km from west to east, altogether comprising about 2000 km^2. The carbonate lithology consists of low-grade metamorphic Upper Devonian limestones (e.g. Schöcklkalk) and dolomites overlying sandstone and slates (Gasser et al. 2010). Due to the relatively low elevations, large parts are covered by vegetation, which results in

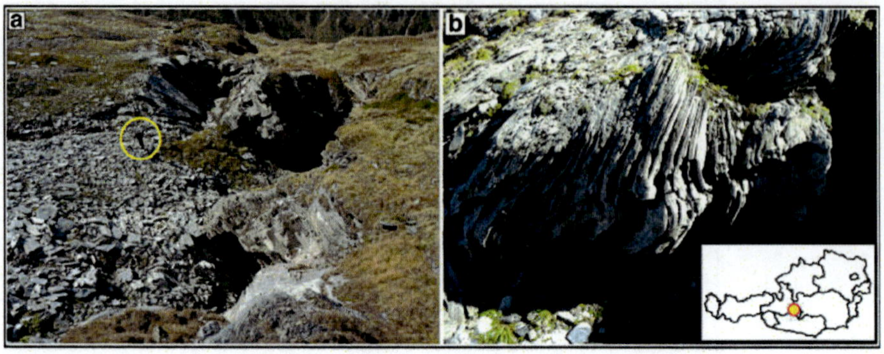

Fig. 5.9 Marble stripe karst above the Mur spring. **a** Dolines and ponors. Person for scale in the circle. **b** Strukturkarren (width of the picture 0.8 m). Photographs: L. Plan

an abundant speleothem decoration of the caves. Again, the contact of karstic and non-karstic rocks leads to intense karstification. More than 900 caves are registered, and about one-third of them is located in the 14.5 km^2 Tanneben massif (15 km north of the city Graz) and its surroundings (Benischke et al. 2016a). Five caves of the Tanneben massif (Fig. 5.10), which is the most thoroughly investigated area, have passage lengths of more than 500 m. Speleogenesis of these caves, and especially of the 6 km long Lurgrotte, is strongly related to the Mur River. The incision due to efficient erosion has created a multi-level cave system (Wagner et al. 2011). The area is drained by two main karst springs towards the west: (i) the Schmelzbach-Ursprung originates in the Lurgrotte cave and drains via the Peggauer cave entrance and the (ii) Hammerbach-Ursprung situated 800 m to the south. The related karst aquifers are a mixture of 8.5 km^2 autochthonous (i.e. carbonates of the Tanneben massif) and 14.5 km^2 allochthonous input via the Lurbach creek. The Lurbach drains a basin of impermeable schists, and at the contact with the Schöcklkalk, the creek loses water to the different subsurface drainage systems in the Tanneben massif. The main drainage of the karst system is characterized by three subterranean flow paths (Behrens et al. 1992):

(1) During low flow conditions, the Hammerbach drains parts of the autochthonous recharge of the Tanneben massif and the total amount of allochthonous input from Lurbach. The latter infiltrates into several ponors in front of the Semriach entrance of Lurgrotte or at higher discharge 200 m inside Lurgrotte. The Schmelzbach is only supplied by autochthonous recharge of parts of the Tanneben massif. At this stage, both aquifer systems are separated (Fig. 5.10, subterranean flow path ①).
(2) Increasing discharge at the Hammerbach comes along with the alimentation of the Hammerbach aquifer into the Schmelzbach aquifer (Fig. 5.10, subterranean flow path ②).
(3) During flood conditions, the allochthonous input of the Lurbach exceeds the capacity of the ponors resulting in a rapid flow through cave system towards the Peggau entrance (Fig. 5.10, subterranen flow path ③).

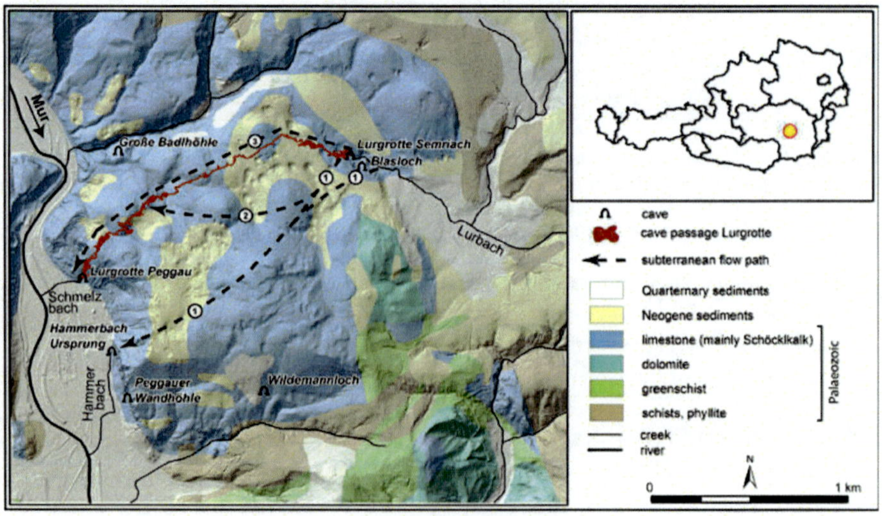

Fig. 5.10 Tanneben massif (CSK). Simplified geological map based on Flügel et al. (2011), 1 m Hillshade visualization. Most important caves are displayed. Cave passage Lurgrotte based on Bock and Dolischka 1953. The meaning of subterranean flow paths ①, ②, and ③ is explained in the text

Defining discharge thresholds for these three flow paths is difficult. Recent studies revealed the complex hydrologic pattern (e.g. varying threshold discharge values) of the Lurgrotte aquifer due to the randomly blocked stream sink and to the sealed conduit networks in the Hammerbach aquifer (Mayaud et al. 2014).

5.4.3 Alpine Forelands and Neogene Basins

Approximately, 14,000 km^2 (17%) of Austria are represented by Cenozoic basins. The Molasse Zone in the north of the Alps dates back to the Palaeogene, whereas small intermountain basins, the large Vienna Basin in the east and the Styrian Basin in the southeast of the Alps, formed in the Neogene. In the Molasse Zone, karst occurs almost exclusively in the so-called Folded Molasse in the north of Bregenzerwald. There, carbonatic conglomerates host a suite of karst features including karren, dolines, ponors, and small caves (Göppert 2011).

The development of the Vienna and the Styrian Basins started in the Miocene and was initiated by the eastward lateral extrusions of crustal wedges of the Austroalpine. The process caused subsidence, accompanied by a sequence of regressions and transgressions (e.g. Salcher et al. 2012; Hohenegger et al. 2009). Despite different tectonic settings, the facies of the two basins are similar and consist of sandstones and mudstones. The carbonate facies—the so-called Leitha Limestone—was sedimented under shallow marine conditions during the Middle Miocene (Badenian stage) and represents isolated carbonate platforms with thicknesses up to 200 m in the Vienna Basin and 60 m in the Styrian Basin (Wiedl et al. 2012).

5.4.3.1 Vienna Basin

In this 200 km long and up to 50 km wide pull-apart basin, Neogene sediments reach a thickness of 6 km. Carbonates only crop out at the margins, and some karstic caves are known. Only a few of them are attributed to a "normal" epigean speleogenesis. An example is Weingartenbachschwinde, a ponor with a small cave that developed in Leitha Limestones (Pavuza 1998). Some 70 karst caves in the surroundings of the Vienna Basin are attributed to a hypogene speleogenesis either by ascending thermal or by H_2S-bearing water, where H_2S is subsequently oxidized into sulphuric acid (Plan and Spötl 2016). They have developed in different carbonates of the basin sediments and the adjacent Austroalpine. The longest one is Eisensteinhöhle that has formed in a Miocene carbonate breccia (Gainfarner Brekzie). It is 2.4 km long and leads down to a lukewarm spring 87 m below the entrance. The host rock of the 333 m long Fraisloch for example is the Wetterstein Limestone and for Nasser Schacht it is both the Miocene Leitha Limestone for the upper parts and Triassic Dolomite for the lower ones. Surface features are rare or absent in the Vienna Basin.

5.4.3.2 Styrian Basin

The few isolated areas in the Leitha Limestone in the Styrian Basin are characterized by intense surface karstification. The Wildoner Buchkogel, 30 km south of Graz, is coined by numerous dolines, and their density is even comparable to those of the high alpine karst areas in the NCA (Fig. 5.11a, Bauer 2015). Northeast of Wildon, a small, but well-developed contact karst area with distinct hydrology is remarkable (Fig. 5.11a–c). Karstification is associated with allochthonous input from the surrounding impermeable Neogene sediments into ponors. One of these ponors drains into the 54 m long Frauenhöhle; the only known dissolution cave in this area (Fig. 5.11a–c).

5.4.4 Bohemian Massif

The Bohemian Massif represents a deeply eroded, crystalline basement of the Variscan orogeny. Crystalline rocks dominate, and thus, karstification is limited to marble layers interbedded in impermeable rocks (Fig. 5.12a). They are rather thin, but often have great lateral extent (Tollmann 1985). Well-developed surface karst features are almost missing, but epikarst features can be observed at some artificial marble outcrops (Fig. 5.12b). The speleogenesis of the marble caves is unclear. For some of them, dissolution seems to be the main factor, but for several well-known caves near the Krems River (Mayer et al. 1993), the influence of mechanical erosion of the river may play an important role (Fig. 5.12c).

5.5 Human-Karst Interaction

Human activities have affected the karst environment in Austria since the Palaeolithic period. The earliest traces are remains of hunter-gatherer societies in caves. Excavated fragments from the Repolusthöhle (Acheulian culture) in the Central Styrian Karst and the Gudenushöhle (Mousterrian culture) in the Bohemian Massif (cf. Figure 5.12c) indicate human presence in Austria since more than 100,000 years ago (e.g. Modl et al. 2014; Neugebauer-Maresch 1993). The widespread occurrence of karst rocks, even in densely populated areas, induces various direct effects on the karst environment. Four examples are given, and the value of cave and karst science is briefly discussed.

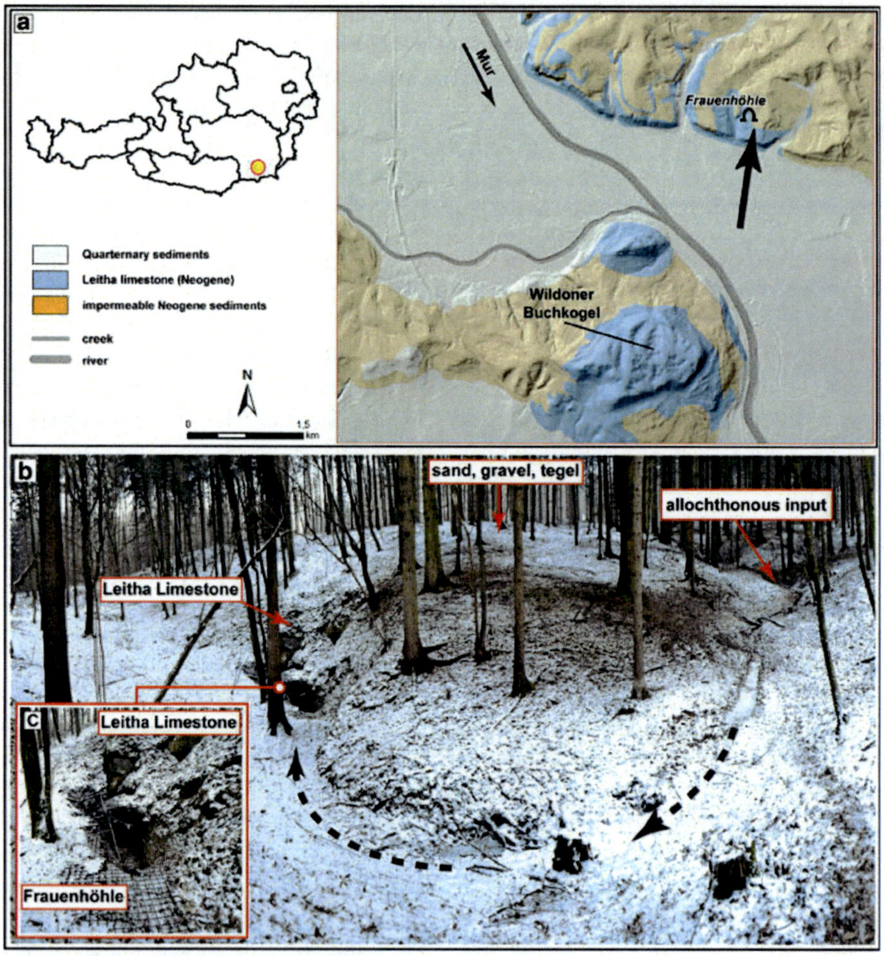

Fig. 5.11 Karst in the Styrian basin. **a** Generalized geological map. It is clearly visible that the Wildoner Buchkogel is dissected by dolines. The black arrow indicates the position of Fig. 5.11c; 10 m hillshade visualisation. **b** Contact karst area in the Styrian basin. The catchment area of the small stream (black arrows with dashed line indicate flow direction) is situated in impermeable Neogene sediments. At the contact with the Leitha Limestone, the stream is drained subsurface through the cave Frauenhöhle. **c** Detail insert, illustrating the ponor at Frauenhöhle. Photographs: C. Bauer

Fig. 5.12 Karst landscapes in the Bohemian Massif. **a** Subjacent bed of banded marble, with overlying impermeable rocks, located at the Krems River. **b** Marble quarry face near Arzwiesen/Krems, illustrating the epikarst development. It is clearly visible that some joints are solutionally enlarged. **c** The 30 m long Gudenushöhle is situated in banded marble approximately 7 m above the current river bed. Photographs: C. Bauer

5.5.1 Water Supply

Especially, the high alpine karst massifs of the NCA and the SCA provide catchment areas for large karst aquifers, and about 50% of Austria's inhabitants are supplied by karst water. Beside numerous local utilizations, the cities of Innsbruck, Salzburg, Villach, and Graz are significantly supplied by drinking water from karst aquifers. For the city of Vienna, epidemics caused by drinking water from contaminated wells led to the idea of a transregional water supply. In 1869, the geologist and politician Eduard Suess drew up the plan to tap water from high alpine karst catchments of Schneeberg and Rax. Already in 1873, the *1. Hochquellenleitung,* a 90 km long pipeline, was put into operation (Fig. 5.13, Drenning 1973). Water shortage led to the construction of a second 200 km long water main (*2. Hochquellenleitung*) to provide karst water from the north of the Hochschwab massif, which was finished in 1910 (Drenning 1988). To ensure the growing demands, this water supply system was enlarged successively by capturing further springs. Nowadays, 1.9 million inhabitants in Austria's capital at average obtain 94% of the total amount of drinking water (144 million m^3/a) from karst aquifers.

The catchment areas at Schneeberg, Rax, Schneealpe, and Hochschwab are karst plateaus, reaching altitudes of 2277 m asl at the latter massif. As the soil cover is scarce or absent in many parts and subsurface flow velocities are high, the aquifers are highly vulnerable to pollution (Plan et al. 2008). Even though these karst massifs are not inhabited, strong efforts are required to ensure high water quality. For more than three decades, Vienna Water has initiated research projects on karst hydrology, geology, karst morphology, meteorology, and forestry in order to assure quality and reliance of long-term water supply. The results support planning strategies, including measures for the catchment protection as well as appropriate land-use management (e.g. tourism, silviculture, pasture) and a permanent spring monitoring system, especially concerning quality parameters like dissolved organic carbon, total organic carbon, spectral absorption coefficient, turbidity.

5.5.2 Destruction of Karst Landscapes

Human activities affect all types of landscapes, but especially, the rugged morphology of karst areas is often destroyed in order to be adopted for the needs of modern

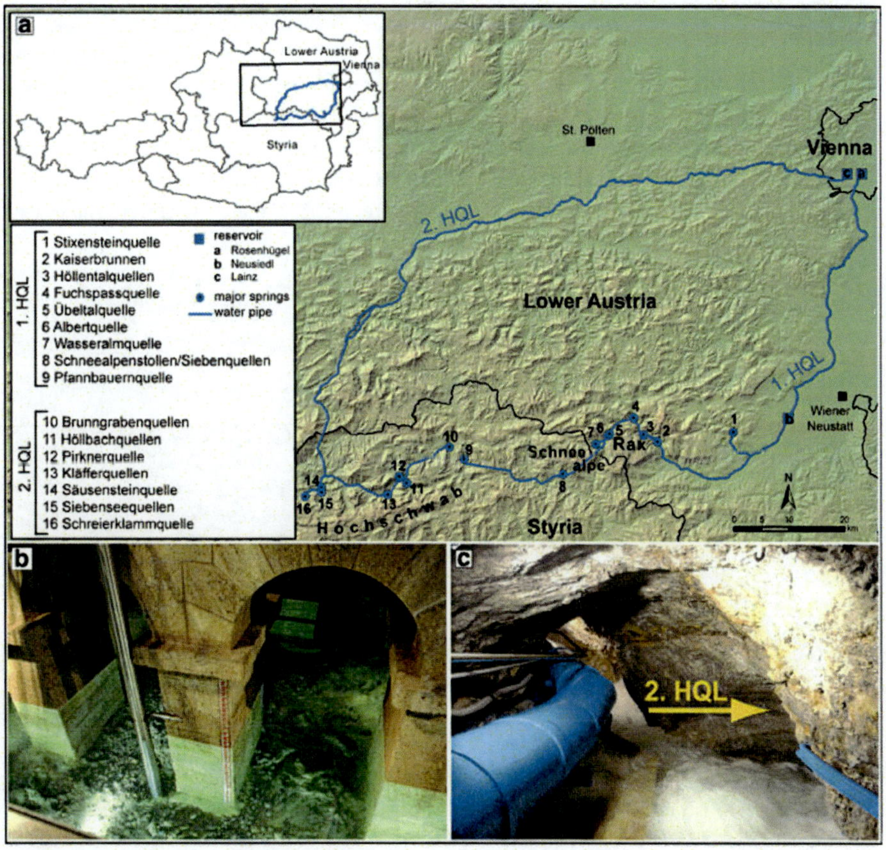

Fig. 5.13 Vienna water supply. **a** Map and main developed springs (HQL = Hochquellenleitung). **b** Spring tapping of the Kaiserbrunnen. **c** Spring tapping of the Kläfferquellen; Photographs: C. Bauer

society. Karst relief has to be bulldozed for ski-slopes or to enable the mechanical mowing of meadows. Dolines in particular have to be filled and not seldom, waste was or is used which can have severe negative effects on these highly vulnerable sites concerning the quality of the karst groundwater.

Further, carbonates are the most quarried rocks in Austria, and 29 million tonnes of limestone, dolomite, and marble were produced in 2016 (BMWFW 2017). Assuming a density of 2.7 g/cm^3, this equates a cube with an edge length of 0.2 km, which is an order of magnitude more than the natural bulk dissolution in Austrian karst areas ($\sim$17,000 km^2), assuming an average denudation rate of 0.7 mm/a (Plan 2016). Therefore, it is obvious that the quarries destroy significant portions of karst landscapes including surface karst features and caves. Examples are the quarry in Peggau (Styria) that has eaten away the landscape around the entrance of Lurgrotte (Fig. 5.14a, Bauer and Kellerer-Pirklbauer 2010) or the one near Bad Deutsch Altenburg (Lower Austria) where most of the Pfaffenberg Hill, including the summit, has been removed. On the other hand, it has to be mentioned that many interesting caves—especially the hypogene ones around the Vienna basin—would have remained unknown without quarrying. A review on protection strategies of karst landscapes is given by Trimmel (1998).

5.5.3 Gypsum Karst

In Austria, gypsum (partly in the form of anhydrite) only locally forms the bedrock in rather small patches. The most important evaporite-bearing sediments occur in the NCA, and these are the Upper Permian Haselgebirge Formation, the Lower Triassic Werfen Formation, and the Upper Triassic (Carnian) Raibl and Opponitz Formations. The low-grade metamorphic "Bunter Keuper" (Upper Triassic) of the Lower Austroalpine also contains evaporites. Due to its higher solubility and dissolution kinetics, karst processes in gypsum are almost 10 times faster than in limestone. The most striking karst landforms are dolines that are often funnel-shaped and can reach diameters of several tens of metres even on steep slopes. Due to vegetation and sediments associated with the gypsum-bearing formations, karren are rare. The same accounts for caves, which are relatively small, unstable, and change their shape quite rapidly.

Despite its high importance as a natural resource (0.7 million tonnes were quarried in 2016; BMWFW 2017), gypsum can cause serious hazards to the infrastructure. The rapid dissolution by natural or human-induced water—ranging from mining water to the emptying of swimming pools—can cause fast-developing cavities and subsequently collapse features. In some regions, reports of freshly collapsed sinkholes are frequent and sometimes infrastructure is damaged. Major problems occur e.g. in the towns of Reutte (Tirol) and Hinterbrühl (Lower Austria, Fig. 5.14b), and special protective measures are necessary for construction sites (e.g. Posch-Trotzmüller et al. 2017).

5.5.4 Show Caves

Caves and their interior deposits have attracted visitors for a long time, not only in Austria. For example due to stone inscriptions in the Drachenhöhle at Mixnitz, the evidence of visitations can be traced back to AD 1387 (Klebel 1931). These inscriptions allow to infer that until AD 1530, most of the ancient tourists were members of the nobility. However, these visits were rather expeditions than touristic visits to the cave.

If show caves are defined as caves where (a) anthropogenic modifications (e.g. pathways) are made to provide easier access into the cave and (b) an entrance fee has to be

Fig. 5.14 Human-karst interactions. **a** Destruction of karst landscapes. The cave Lurgrotte is wedged between limestone quarries. The white dashed lines indicate the protection area of the cave system. Photograph C. Bauer. **b** Infrastructural damage related to the subsurface dissolution of gypsum (Hinterbrühl, Lower Austria). Photograph G. Posch-Trötzmüller

paid, then the Hohlensteinhöhle east of Mariazell is probably the first Austrian show cave. According to Kraus (1887), the development of this cave included the installation of ladders and passage enlargement using explosives. The first guided tours into Hohlensteinhöhle are reported for 1832. In 1882, the Kraushöhle near Gams/Hieflau became the first show cave in the world with electric lighting—two years earlier than the famous Postojna Cave in Slovenia (Trimmel 1968).

Presently, 31 show caves are listed on the webpage of the Austrian Speleological Association (VÖH 2019) (Table 5.3). Due to the diversity of karst regions and the large altitudinal range, the Austrian show caves present a great diversity. Austria is well known for its spectacular ice caves, but some are also abundantly decorated with dripstones. Others show gypsum or aragonite crystals, archaeological, or palaeontological remains. In addition, active subsurface brooks that eventually emerge from the entrance or vast underground passages or chambers attract visitors (Fig. 5.15a–d). In some caves, little or no installations or artificial light is present, and moving through the cave partly via ladders or on ropes is exciting.

There are various impacts on the fragile cave eco- and geosystem due to installations or the visitors. Modifications or damages to the morphology are often necessary when the cave is developed as a show cave. The ongoing impacts range from the so-called Lampenflora—i.e. plants that grow due to the artificial light—via the input of nutrition from decomposing wooden installations to heat emissions by the

Table 5.3 Show caves in Austria

Name	Province	Entrance elevation (m asl)	Features/interior deposits
Griffener Tropfsteinhöhle	Carinthia	485	Colourful dripstones, archaeologic site
Obir-Tropfsteinhöhlen	Carinthia	1100	Abundant dripstones, show mine
Allander-Tropfsteinhöhle	Lower AUT	400	Dripstone, moonmilk, brown bear skeleton
Einhornhöhle	Lower AUT	580	Dripstones, subfossile bones
Eisensteinhöhle	Lower AUT	380	Cave popcorn, aragonite crystals, thermal pond
Hermannshöhle	Lower AUT	627	Abundant dripstones, bats
Hochkarschacht	Lower AUT	1620	Giant chamber, dripstones
Nixhöhle	Lower AUT	556	Moonmilk, dripstones, cave bear skeleton
Ötschertropfsteinhöhle	Lower AUT	710	Dripstones, lake
Eiskogelhöhle	Salzburg	2100	Massive ice, giant passages, adventure trip
Eisriesenwelt	Salzburg	1641	Massive ice
Entrische Kirche	Salzburg	1040	Dripstones, brook
Feuchter Keller	Salzburg	1400	Adventure trip
Lamprechtsofen	Salzburg	660	Active brook, dripstones
Prax Eishöhle	Salzburg	1600	Ice
Arzberghöhle	Styria	730	Cave bear remains, moonmilk, adventure trip
Frauenmauerhöhle	Styria	1467	Giant cave passages, ice
Grasslhöhle	Styria	740	Abundant dripstones
Hohlensteinhöhle	Styria	1031	Dripstones, moonmilk, small brook
Katerloch	Styria	900	Abundant dripstones, cave lake
Kraushöhle	Styria	600	Gypsum, dripstones, moonmilk
Lurgrotte Peggau	Styria	400	Dripstones, brook
Lurgrotte Semriach	Styria	640	Abundant dripstones, huge chamber, episodic brook
Odelsteinhöhle	Styria	1084	Aragonite helictites, dripstones
Rettenwandhöhle	Styria	630	Dripstones, archaeological site
Hundsalm Eis- und Tropfsteinhöhle	Tirol	1520	Ice, moonmilk, marble structures
Spannagelhöhle	Tirol	2521	Dripstones, potholes
Dachstein-Mammuthöhle	Upper AUT	1368	Giant passages
Dachstein-Rieseneishöhle	Upper AUT	1455	Massive ice, cave bear dummy
Gassel-Tropfsteinhöhle	Upper AUT	1229	Abundant dripstones
Koppenbrüllerhöhle	Upper AUT	580	Active brooks with waterfalls, dripstones

Fig. 5.15 Show caves in Austria. **a** Cave ice formation in the Parsivaldom/Dachstein-Rieseneishöhle; Photograph: C. Bauer. **b** Abundance of calcite-speleothems in the Fantasiehalle/Katerloch: Soda straws and cone stalactites, columns, palm-tree trunk, and (very high) thin candle-style stalagmites. Photograph: C. Bauer. **c** So-called Paläotraun, a gallery of paragenetic origin in the Mammuthöhle (Dachstein). Photograph: A. Neumann. **d** View from the tourist trail down to Klingfall in Koppenbrüllerhöhle. Photograph: H. Thaler

visitors that affect the cave climate, including ice formations. In general, show caves are protected by the cave or nature protection laws, and measures have to be approved to minimize these impacts.

5.5.5 Value of Karst and Cave Research

Karst and especially cave science are interdisciplinary fields making wide-ranging contributions in geo- and biosciences, climatology, archaeology, and even medicine (speleotherapy) or cosmonautics (training of astronauts in caves).

With the background of climate change, palaeoclimatology has experienced rising interest in the past decades, and caves have proven to be suitable archives. Studies include stable isotope analysis and dating of speleothems, monitoring and studying of cave ice, or the occurrences and age of cryogenic cave carbonates (e.g. Spötl and Boch 2016). Caves and their deposits are also archives for active tectonics and palaeo-earthquakes (e.g. Plan et al. 2010), and due to stable climatic conditions, they have proven to be ideal to measure active fault displacements (e.g. Baron et al. 2019).

For geomorphologists, caves and karst landscapes offer the opportunity to measure valley incision or mountain uplift rates, if the deposition of cave sediments can be associated to the base level, which is normally the valley bottom. At the Hainburger Berge at the margin of the Vienna basin, a cave filled with quartz pebbles that are very similar to the ones of the nearby modern Danube gave incision rates of ca. 40 m/Ma for the past 4.3 Ma (Neuhuber et al. 2020). Values derived from burial age dating of cave sediments in the Mur Valley gave an average rate of 125 m/Ma for the past 4 Ma (Wagner et al., 2012). At Kirchberg am Wechsel (Lower Austria), also a formerly non-glaciated area, the onset of speleothem growth indicates similar values of around 100 m/Ma for the past 0.5 Ma (Plan et al. 2015). For the glaciated Alpine valleys at Dachstein or Hagengebirge, burial age dating of Augenstein sediments gave rates of 120 to 210 m/Ma for the past 4–6 Ma (Häuselmann et al. 2020). At Kraushöhle near Gams (Styria), the dating of speleothems and the mineral alunite, that formed during the sulphuric acid speleogenesis, suggests that the narrow Notklamm Gorge incised by 1000 mm/Ma within the past 80 ka (De Waele et al. 2009).

For comparison, Eastern Alpine karst denudation rates measured by different methods range from less than 10 µm/a (= m/Ma) for single bare limestone surfaces up to 100 µm/a for whole catchment areas (Pavuza and Oberender 2013).

References

Bauer C (2015) Analysis of dolines using multiple methods applied to airborne laser scanning data. Geomorphology 250:78–88. https://doi.org/10.1016/j.geomorph.2015.08.015

Bauer C, Kellerer-Pirklbauer A (2010) Human impacts on karst environment: a case study from Central Styria. Z Geomorphol 54:1–26

Bauer F (1962) Nacheiszeitliche Karstformen in den österreichischen Kalkalpen. In: Proceedings of the 2nd International Congress of speleology, pp 299–328

Baroň I, Plan L, Sokol L, Grasemann B, Melichar R, Mitrovic I, Stemberk J (2019) Present-day kinematic behaviour of active faults in the Eastern Alps. Tectonophysics 752:1–23. https://doi.org/10.1016/j.tecto.2018.12.024

Behrens H, Benischke R, Bricelj M, Harum T, Käss W, Kosi G. Leditzky HP, Leibundgut C, Małoszewski P, Maurin N, Rajner V, Rank D, Reichart B, Stadler H, Stichler W, Trimborn P, Zojer H, Zupan M (1992) Investigations with Natural and Artificial Tracers in the Karst Aquifer of the Lurbach System (Peggau-Tanneben-Semriach, Austria). Steir. Beitr. Hydrogeologie 43 Graz

Benischke R, Kusch H, Wagner T (2016a) Mittelsteirischer Karst. In: Spötl C, Plan L, Christian E (eds) Höhlen und Karst in Österreich. Oberösterreichisches Landesmuseum, Linz, pp 701–718

Benischke R, Stadler H, Völkl G (2016b) Karstquellen. In: Spötl C, Plan L, Christian E (eds) Höhlen und Karst in Österreich. Oberösterreichisches Landesmuseum, Linz, pp 73–96

BMWFW (2017) Österreichisches Montan-Handbuch 2017. Bundesministerium für Wissenschaft, Forschung und Wirtschaft. www.bmnt.gv.at/energie-bergbau/bergbau/Montanhandbuch.html. abgerufen 21 Jan 2019

Bock H (1913) Charakter des Mittelsteirischen Karstes. Mitteilungen Für Höhlenkunde 6:5–19

Bock H, Dolischka A (1953) Plan der Lurgrotte Peggau – Semriach, M = 1:2.500, Graz

Brenčič M, Poltnig W (2008) Grundwasser der Karwanken. Versteckter Schatz. Vehling, Ljubliana

Cvijić J (1893) Das Karstphänomen. Versuch Einer Morphologischen Monographie Geograph Abh 5:218–329

De Waele J, Plan L, Audra P, Rossi A, Spötl C, Polyak V, Mcintosh B (2009) Kraushöhle (Austria): morphology and mineralogy of an alpine sulfuric acid cave. In: 15th International Congress of speleology, Kerville, Proceedings Part 2, pp 831–837

Drenning A (1973) Die I. Wiener Hochquellenleitung. Magistrat der Stadt Wien, Abt. 31, Wasserwerk, Vienna

Drenning A (1988) Die II. Wiener Hochquellenleitung. Magistrat der Stadt Wien, Abt. 31, Wasserwerk, Vienna

Fink MH (1984) Das Karstgebiet beim Hochtor, Hohe Tauern (Salzburg-Kärnten). Die Höhle 35:127–134

Flügel H, Nowotny A, Gross M (2011) Geologische Karte der Republik Österreich 1:50.000, Blatt 164 Graz. Austrian Geological Service, Vienna

Ford D, Williams P (2007) Karst hydrogeology and geomorphology. Wiley, Chichester

Frisch W, Kuhlemann J, Dunkl I, Székely B (2001) The Dachstein paleosurface and the Augenstein Formation in the Northern Calcareous Alps—a mosaic stone in the geomorphological evolution of the Eastern Alps. Int J Earth Sci 90:500–518

Frisch W, Kuhlemann J, Dunkl I, Székely B, Vennemann T, Rettenbacher A (2002) Dachstein-Altfläche, Augenstein-Formation und Höhlenentwicklung – die Geschichte der letzten 35 Millionen Jahre in den zentralen Nördlichen Kalkalpen. Die Höhle 53:1–36

Gasser D, Stüwe K, Fritz H (2010) Internal structural geometry of the Paleozoic of Graz. Int J Earth Sci 99:1067–1081. https://doi.org/10.1007/s00531-009-0446-0

Goldscheider N (2005) Fold structure and underground drainage pattern in the alpine karst system Hochifen-Gottesacker. Eclogae Geol Helv 98:1–17

Goldscheider N, Hötzl H (1999) Hydrological characteristics of folded alpine karst systems exemplified by the Gottesacker plateau (German–Austrian Alps). Acta Carsologica 28(1):87–103

Göppert N (2011) Karst geomorphology of carbonatic conglomerates in the Folded Molasse zone of the Northern Alps (Austria/Germany). Geomorphology 130:289–298

Gulden B (2021) World's longest and deepest caves. www.caverbob.com. Accessed 22 Dec 2021

Haseke-Knapczyk H (1989) Der Untersberg bei Salzburg. Veröffentl Des Österr Mab-Programms 15:1–223

Häuselmann P, Plan L, Pointner P, Fiebig M (2020) Cosmogenic nuclide dating of cave sediments in the Eastern Alps and implications for erosion rates. Int J Speleol 49:107–118. https://doi.org/10.5038/1827-806X.49.2.2303

Hohenegger J, Rögl F, Ćoric S, Pervesler P, Lirer F, Roetzel R, Scholger R, Stingl K (2009) The Styrian Basin: a key to the Middle Miocene (Badenian/Langhian) Central Paratethys transgressions. Austrian J Earth Sci 122:102–132

Hötzl H (1992) Karstgrundwasser. In: Käss W (ed) Geohydrologische Markierungstechnik, Lehrb. d. Hydrogeol. 9, Gebrüder Bornträger, Berlin/Stuttgart, pp 374–406

Klampfer A, Plan L, Büchel E, Spötl C (2017) Neubearbeitung und Forschung im Schneckenloch, der längsten Höhle in Vorarlberg. Die Höhle 68:14–30

Klebel E (1931) Alte Inschriften und Wappen. – Speläologische Monographien VII/IX: 98–105

Klimchouk A (2015) The karst paradigm: changes, trends and perspectives. Acta Carsologica 44:289–313. https://doi.org/10.3986/ac.v44i3.2996

Krainer K (2005) Nationalpark Hohetauern, Geologie. Carinthia, Inssbruck

Kraus F (1887) Der Hohlenstein bei Mariazell. Österreichische Touristen Zeitung 7(4)

Kraus F (1894) Höhlenkunde. Facsimile reprint 2009, Die Höhle, Supplement 56

Lauritzen SE (2001) Marble stripe karst of the Scandinavian Caledonides: an end-member in the contact karst spectrum. Acta Carsologica 30(2):47–79

Mayaud C, Wagner T, Benischke R, Birk S (2014) Single event time series analysis in a binary karst catchment evaluated using a groundwater model (Lurbach system, Austria). J Hydrol 511:628–639

Mayer A, Raschko H, Wirth, J (1993) Die Höhlen des Kremstales. Die Höhle, supplement 33

Modl D, Brandl M, Pacher M, Drescher-Schneider R (2014) Abriss der Erforschungsgeschichte der Repolusthöhle (Steiermark, Österreich) mit einem Bericht zur Feststellungsgrabung im Jahr 2010. Schild Von Steier 26:97–107

Neugebauer-Maresch C (1993) Altsteinzeit im Osten Österreichs. Wiss Schriftenreihe Niederösterreichs 95(96/97):1–96

Neuhuber S, Plan L, Gier S, Hintersberger E, Lachner J, Scholz D, Lüthgens C, Braumann S, Bodenlenz F, Voit K, Fiebig M (2022) Numerical age dating of cave sediments quantify vertical movement at the Alpine-Carpathian transition in the Plio- and Pleistocene. Geol Carpathica 71:539–557. https://doi.org/10.31577/GeolCarp.71.6.5

Oberender P, Plan L (2018) A genetic classification of caves and its application in eastern Austria. In: Parise M, Gabrovsek F, Kaufmann G, Ravbar N (eds) Advances in karst research. Theory, fieldwork and applications. Geological Society, London, Special Publications Geological Society of London, vol 466, pp 121–136

Pavuza R (1998) Höhlen und Verkarstung. Die Höhle, Supplement 51:14–19

Pavuza R, Oberneder P (2013) Karst denudation data from the Northern Calcareous Alps (Austria). Geomorfologický Sborník 11:61–62

Pfarr T, Seebacher R, Plan L (2021) Die längsten und tiefsten Höhlen Österreichs. http://hoehle.org/laengste-tiefste. Accessed 20 Dec 2021

Pfiffner A (2015) Geologie der Alpen, 3. Auflage. UTB, Bern

Plan L (2016) Oberflächenkarstformen. In: Spötl C, Plan L, Christian E (eds) Höhlen und Karst in Österreich. Oberösterreichisches Landesmuseum, Linz, pp 23–34

Plan L, Decker K, Faber R, Wagreich M (2008) Karst morphology and groundwater vulnerability of high alpine karst plateaus. Environ Geol 58:285–297. https://doi.org/10.1007/s00254-008-1605-5

Plan L, Grasemann B, Spötl C, Decker K, Boch R, Kramers J (2010) Neotectonic extrusion of the Eastern Alps: Constraints from U/Th dating of tectonically damaged speleothems. Geology 38(6):483–486. https://doi.org/10.1130/G30854.1

Plan L, Oberender P (2016) Höhlen in Österreich. In: Spötl C, Plan L, Christian E (eds) Höhlen und Karst in Österreich. Oberösterreichisches Landesmuseum, Linz, pp 11–22

Plan L, Schober A, Scholz D, Spötl C, Pruner P, Bosák P (2015) Speleogenesis of the Hermannshöhle cave system (Austria): Constraints from 230Th/ U-dating and palaeomagnetic analysis. Int J Speleol 44:315–326. https://doi.org/10.5038/1827-806X.44.3.8

Plan L, Spötl C (2016) Hypogene Karsthöhlen. In: Spötl C, Plan L, Christian E (eds) Höhlen und Karst in Österreich. Oberösterreichisches Landesmuseum, Linz, pp 49–60

Posch-Trotzmüller G, Atzenhofer B, Hobiger G (2017) Gipsvorkommen in den Kalkalpen: Erdfallprävention mittels Geologie und Hydrochemie. In: Wimmer-Frey I, Römer A, Janda C (eds) Arbeitstagung 2017, Geol. Bundesanstalt, Wien, pp 42–45

Salcher BC, Meurers B, Smit J, Decker K, Hölzel M, Wagreich M (2012) Strike-slip tectonics and Quaternary basin formation along the Vienna Basin fault system inferred from Bouguer gravity derivatives. Tectonics 31:1–20. https://doi.org/10.1029/2011TC002979

Spötl C (2016) Südliche Kalkalpen. In: Spötl C, Plan L, Christian E (eds) Höhlen und Karst in Österreich. Oberösterreichisches Landesmuseum, Linz, pp 719–730

Spötl C, Boch R (2016) Höhlen in der Paläoklimaforschung. In: Spötl C, Plan L, Christian E (eds) Höhlen und Karst in Österreich. Oberösterreichisches Landesmuseum, Linz, pp 155–170

Spötl C, Plan L (2016) Karst und Höhlen. In: Spötl C, Plan L, Christian E (eds) Höhlen und Karst in Österreich. Oberösterreichisches Landesmuseum, Linz, pp 1–10

Spötl C, Plan L, Christian E (2016) Höhlen und Karst in Österreich. Oberösterreichisches Landesmuseum, Linz

Steinwender C, Plan L (2011) Kontaktkarst im Bereich Murursprung-Rosskar (Lungau, Salzburg). Die Höhle 62:15–26

Stummer G, Plan L, Mattes J (2016) Höhlenkundliche Organsiationen. In: Spötl C, Plan L, Christian E (eds) Höhlen und Karst in Österreich. Oberösterreichisches Landesmuseum, Linz, pp 391–498

Tollmann A (1985) Geologie von Österreich, vol 2. Franz Deuticke, Vienna

Trimmel H (1968) Höhlenkunde. Friedrich Vieweg & Sohn, Braunschweig

Trimmel H (1998) Karstlandschaftsschutz. CIPRA-Österreich, Wien

Veress M, Zentai Z, Péntek K, Mitre Z et al (2013) Flow dynamics and shape of rinnenkarren systems. Geomorphology 198:115–127

VÖH (2019) Schau! – Höhlen in Österreich. www.schauhoehlen.at. Accessed 20 Jan 2019

Wagner T, Fritz H, Stüwe K, Nestroy O, Rodnight H, Hellstrom J, Benischke R (2011) Correlations of cave levels, stream terraces and planation surfaces along the River Mur—timing of landscape evolution along the eastern margin of the Alps. Geomorphology 134:62–78. https://doi.org/10.1016/j.geomorph.2011.04.024

Weber L (1997) Basiskarte Geologie der Metallogenetischen Karte von Österreich 1:500000. Austrian Geological Service, Vienna

Wiedl T, Harzhauser M, Piller WE (2012) Facies and synsedimentary tectonics on a Badenian carbonate platform in the southern Vienna Basin (Austria, Central Paratethys). Facies 58(4). https://doi.org/10.1007/s10347-011-0290-0

Zacher W (1990) Geologische Karte der Republik Österreich 1:50.000, Blatt 113 Mittelberg. Austrian Geological Service, Vienna

Christian Bauer is senior lecturer for physical geography at the University of Graz, Institute of Geography and Regional Science. His research focuses on geomorphology and geomorphometry in karst terrains, applied geomorphological topics using GIS (natural hazards) as well as human–environment interactions. His interdisciplinary research experience involves different disciplines, including Geosciences, System Science and History.

Lukas Plan is a research scientist at the Karst and Cave Group at the Natural History Museum Vienna. As geologist, he mainly deals with speleogenesis, evolution of karst features and landscapes, karst hydrology, and tectonics. He has explored and surveyed many kilometres of caves predominantly in the Eastern Alps and was co-editor of a book on caves and karst in Austria.

Geomorphic Hazards in Austria

6

Sven Fuchs, Martin Wenk, and Margreth Keiler

Abstract

Endogenic and exogenic geomorphic processes of different types and spatiotemporal dynamics can be observed within the territory of Austria. If these processes affect assets such as exposed buildings or infrastructure lines, they turn into hazards. In particular in the mountainous parts and in the Alpine foreland geomorphic hazards of different magnitude and frequency have repeatedly led to economic losses and fatalities. Together, the mountains and the Alpine foreland account for approximately 70% of the Austrian territory. Consequently, geomorphic hazards are an important issue in Austria. In the following, a brief overview of the major types of these hazards and their characteristics is given, including river and torrential flooding, gravitational mass movements, snow avalanches, hazards associated with glaciers and permafrost, as well as seismic hazards. Furthermore, information on the temporal and spatial occurrence of major event types and associated losses is provided.

S. Fuchs (✉)
Institute of Mountain Risk Engineering, University of Natural Resources and Life Sciences (BOKU), Peter-Jordan-Str. 82, 1190 Vienna, Austria
e-mail: sven.fuchs@boku.ac.at

M. Wenk
Federal Ministry for Agriculture, Regions and Tourism, Federal Water Engineering Administration, Marxergasse 2, 1030 Vienna, Austria

M. Keiler
Department of Geography, University of Innsbruck, Innrain 52f, 6020 Innsbruck, Austria

M. Keiler
Institute for Interdisciplinary Mountain Research, Austrian Academy of Sciences, Innrain 25, 6020 Innsbruck, Austria

Keywords

Flooding • Gravitational mass movements • Snow avalanches • Soil erosion • Glacier hazards • Permafrost hazards • Seismic hazards

6.1 Introduction

Mountain areas are typically characterized by steep slopes, which can result in highly dynamic geomorphic processes such as landslides and other gravitational mass movements, sometimes also triggering further hazards in combination with cascading processes, leading to multi-hazard threats (Kappes et al. 2012). At the same time, significant proportions of mountain regions are used for human settlements with associated economic and transport infrastructure, which may be at risk from geomorphic processes. The latter is particularly true for Austria, dominated by high mountains on one hand, and relatively densely populated on another one.

Following Varnes (1984) and Fell et al. (2008), a natural hazard in the geomorphic context is rooted in either endogenous or exogenous processes, where the first generally result in an increase and the latter in a decrease in relief, both endangering any exposed element at risk. A geomorphic hazard, therefore, represents the potential interaction between the landscape processes and their impact on the human environment (Keiler and Fuchs 2016).

With respect to geomorphic processes, the description of hazard should include the locality, volume (or area), classification, and velocity (or pressure). Hence, information is required on its probability of occurrence within a given period of time for a specific location, referred to as frequency, and on magnitude, which refers to scientifically based measures of the strength of physical processes. If measures of magnitude concern impacts of an event on the anthroposphere (such as elements at risk exposed to natural hazards), intensity is used instead (Giles 2013). Assessments

C. Embleton-Hamann (ed.), *Landscapes and Landforms of Austria*, World Geomorphological Landscapes,
https://doi.org/10.1007/978-3-030-92815-5_6

are repeatedly based on intensity estimates that incorporate human variables as indices of destruction, since direct measurements of process magnitude are often not available with respect to those geomorphic hazards which occur in mountain regions (van Westen et al. 2006).

In Austria, hazards of flooding, gravitational mass movements, snow avalanches and soil erosion can be found, as well as hazards associated with glaciers and permafrost and finally those connected with seismic activity (Embleton-Hamann 2007). In the following, a brief overview of the characteristics of these hazards is given, reflecting on some recent insights into the distribution and the influence of environmental change with a focus on the Austrian Alps and the Alpine foreland. Afterwards, a brief overview illustrates the consequences of these geomorphic hazards in relation to the human environment.

6.2 Characteristics of Hazards and Their Distribution in Austria

The spatial distribution of geomorphic processes inducing damage for hydrological hazards, landslides, rockfall and snow avalanches is shown in Fig. 6.1. Landslides are prominent along the Alpine margins within the Flysch zone composed from sandstones and shale/mudstones. Rockfall and snow avalanches are typical geomorphologic processes in the mountainous parts of Austria, while hydrological hazards (river flooding and torrential flooding) are relatively evenly distributed, with an increasing density in the mountainous areas.

6.2.1 River Flooding

A flood is a relatively high flow which overtaxes the natural channel provided for the runoff (Chow 1956) and is usually described in terms of its magnitude and frequency. The importance of a flood relative to smaller flows in shaping channel and valley morphology is dependent on the magnitude and duration of the hydraulic forces generated during the high discharge in comparison with the erosional resistance of the channel boundaries (Wohl 2004). Floods are related to or triggered by heavy or prolonged rainfall and rapid snowmelt, ice jams or ice break-up, damming of river valleys by landslides or avalanches, and the failure of natural or man-made dams (Arnell 2002). In hazard management in the European Alps two types of river flooding are distinguished: static and dynamic. Static flooding occurs in areas with relatively flat topography (Fig. 6.2). Water levels rise slowly and flow velocity is very low, if the water is moving at all. The damage caused by static flooding is regularly due either to the influence of the water on existing structures or on agricultural production. In dynamic floods, the water movement is much more rapid and affects the elements at risk due to erosion or direct impact of sediment load (Parker

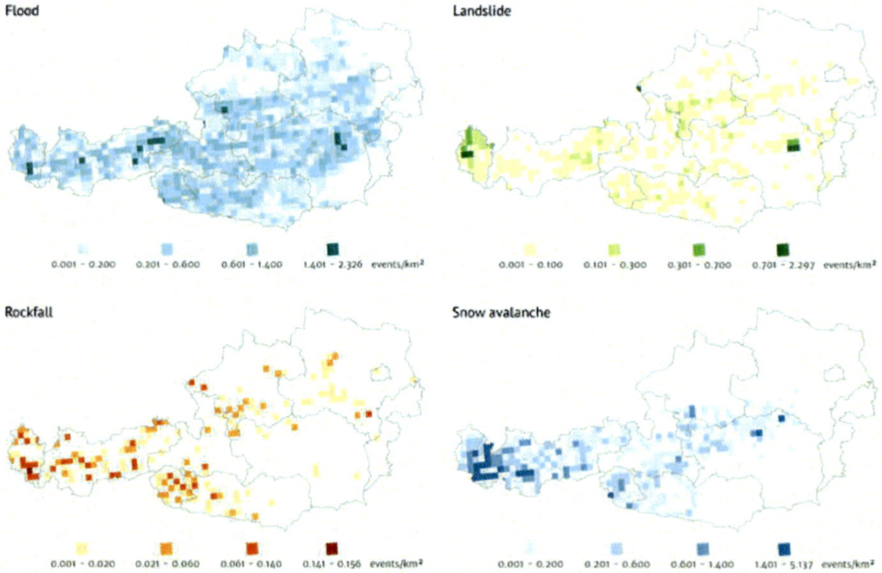

Fig. 6.1 Spatial distribution of geomorphological hazards causing damage between 1850 and 2014 in Austria, upper left: hydrological hazards, upper right: landslides, lower left: rockfall, lower right: snow avalanches. The grid cell size is 8000 × 8000 m. This material was originally published in 'Challenges for natural hazard and risk management in mountain regions of Europe' by M. Keiler and S. Fuchs, and has been reproduced by permission of Oxford University Press [https://oxfordre.com/naturalhazardscience]

Fig. 6.2 Static inundation along the Danube, Austria: flood event of August 2002. *Picture credits* Austrian Armed Forces, with permission

2000). According to the United Nations (2002), floods are among the most common, most costly and most deadly hazards world-wide. Barredo (2009) estimated an average annual loss of approximately 3.4 billion € due to major flood events over the period 1970–2006 in Europe. Flood events in Austria are observed as a result of prolonged heavy rainfall, sometimes in combination with (seasonally unusual) snow-melt in high-mountain regions, and can be recorded along all rivers of the country. As a result, the exposure of elements at risk is significantly high. According to Fuchs et al. (2015), almost 220,000 buildings are exposed to river flooding, which equates to 9% of the entire building stock in the country. The majority of these buildings are commercial buildings located in the floodplains along larger rivers.

6.2.2 Torrential Flooding

Mountain torrents repeatedly lead to debris flows and hyperconcentrated flows due to the considerable amount of sediment being transported either from the catchment downstream or eroded from the stream bed (Wohl 2004, see Fig. 6.3). Sediment load ranges from clay-sized particles to boulders measuring several metres in diameter. The destructive nature of mountain torrents is a result of the high density, combined with flow velocity and discharge (Fuchs et al. 2008). A general classification can be made depending on the relative concentration of water, fine and coarse sediment, as first suggested by Phillips and Davies (1991).

As defined by the Austrian Standards Organization in document No. 24800 (Austrian Standards 2009), debris flows are highly concentrated mixtures of water, fine and coarse sediment, and frequently woody debris (Mazzorana and Fuchs 2010). The coarse sediment is usually concentrated in the upper layers and at the front of the flow. The sediment concentration reaches values up to the plastic limit, but is often between 40 and 70% by volume. The specific bulk density of the mixture amounts to 1.7–2.4 kg/m^3. The flow is characterized as unsteady and non-uniform, and debris flows typically occur as surges (e.g. Pierson 1986). The flow behaviour is generally termed 'non-Newtonian', indicating that standard hydraulic models are not capable of describing the flow satisfactorily. The event volumes of debris flows vary considerably between several thousand to some hundred thousand cubic metres. Debris flows can be roughly classified due to the relative concentration of fine and coarse sediment by the prefix 'viscous/muddy' or 'granular/stony' to describe the main flow behaviour (e.g. Takahashi 1991).

The term hyperconcentrated flow was originally used for streamflow with sediment concentrations between 20 and 60% by volume (Neall 2004), and rheologically the fluid appears to be slightly plastic but flows like water (Pierson and Costa 1987). Consequently, hyperconcentrated flows possess fluvial characteristics, yet are nonetheless capable of carrying very high sediment loads.

Alpine settlements are often located in the run-out area on torrential fans, and as such an estimated 112,000 buildings are exposed to torrential flooding, most of them belonging to the category of residential buildings, hotels and guest houses as well as agricultural buildings (Fuchs et al. 2015).

6.2.3 Gravitational Mass Movements

Gravitational mass movements are defined as the downward and outward movement of slope-forming material under the

Fig. 6.3 Dynamic flooding of Trisanna river in Kappl, Austria, after the flood events of August 2005. *Picture credits* Austrian Armed Forces, with permission

influence of gravity and not requiring other moving agents such as water, air or ice. The term landslide is often used synonymously for any mass movement phenomena. However, in a pure sense, the term Iandslide is understood as a generic term describing downward movements of slope-forming material as a result of shear failure occurring along a well-defined shear plane (Dikau 2004). Gravitational mass movements are common natural hazards in mountain regions (Crozier 1999). Landslides are often connected to triggers such as heavy rainfall, rapid snow melting, earth tremors, slope undercutting, etc. Their impact on society ranges from low (small and shallow landslides in remote regions and less used areas) to high (collapse or burial of buildings and infrastructure, loss of life and loss of agricultural land). Although large magnitude landslides have a low probability, these events tend to result in significant loss of human life throughout Europe's mountain regions (Kilburn and Pasuto 2003). Society contributes to the occurrence of landslides (e.g. through land-use change, road construction involving slope undercutting or loading) as well as exacerbating the impact of the events by the construction of properties on steep slopes. Several classifications of mass movements have been published (e.g. Varnes 1978; Hutchinson 1988; Cruden and Varnes 1996), and most of them are based on morphology, type of material involved, mechanism of movement and rate of motion (Fig. 6.4). Falls, topples and rotational as well as translational slides can be found throughout the country, apart from flows and—quite rarely—large rock slides (Bergstürze). Whereas the first types are relatively frequent, reflecting the geology of Austria, rock slides are rather non-contemporary phenomena (Embleton-Hamann 2007).

Deep-seated and shallow landslides are common in many regions of Austria, in particular in those areas situated in the Flysch unit (Vorarlberg, Lower Austria and Styria) and areas of the Central Alps comprised of metamorphic rock types (phyllites and mica schist) (Fig. 6.5). Both deep-seated and shallow landslides are discussed in more detail in Chapter "The Walgau: A Landscape Shaped by Landslides". Moreover, shallow, often translational landslides, may also be associated with Quaternary deposits and weathered nappes.

As reported by Abele (1974), the total number of processes of the Bergsturz type is higher in the Northern and Southern Limestone Alps than in the crystalline Central Alps, which may be a result of the different weathering behaviour (for details, see Chapter "Giant "Bergsturz" Landscapes in the Tyrol"). In terms of volume, the Bergstürze of Köfels, Fernpass and Dobratsch are among the ten largest within the European Alps (Embleton-Hamann 1997); the latter is discussed in Chapter "Dobratsch—Landslides and Karst in Austria's Southernmost Nature Park". The Bergsturz of Köfels is the largest to have occurred in the crystalline Alps. Generally, Late Pleistocene glacier retreat resulting in an unbalanced relief due to oversteepened slopes was assumed to be the dominant Bergsturz trigger. More recent studies, however, suggest a rather continuous temporal distribution of landslide activities, with (i) some peaks of activity in the early Holocene at about 10 500–9 400 cal BP (calibrated years before present, present = 1950) and (ii) in the federal state of Tyrol a significant increase of deep-seated rockslides in the Subboreal at about 4200–3000 cal BP (Prager et al. 2008). Accordingly, Heuberger (1966) dated the Köfels Bergsturz as an early Holocene

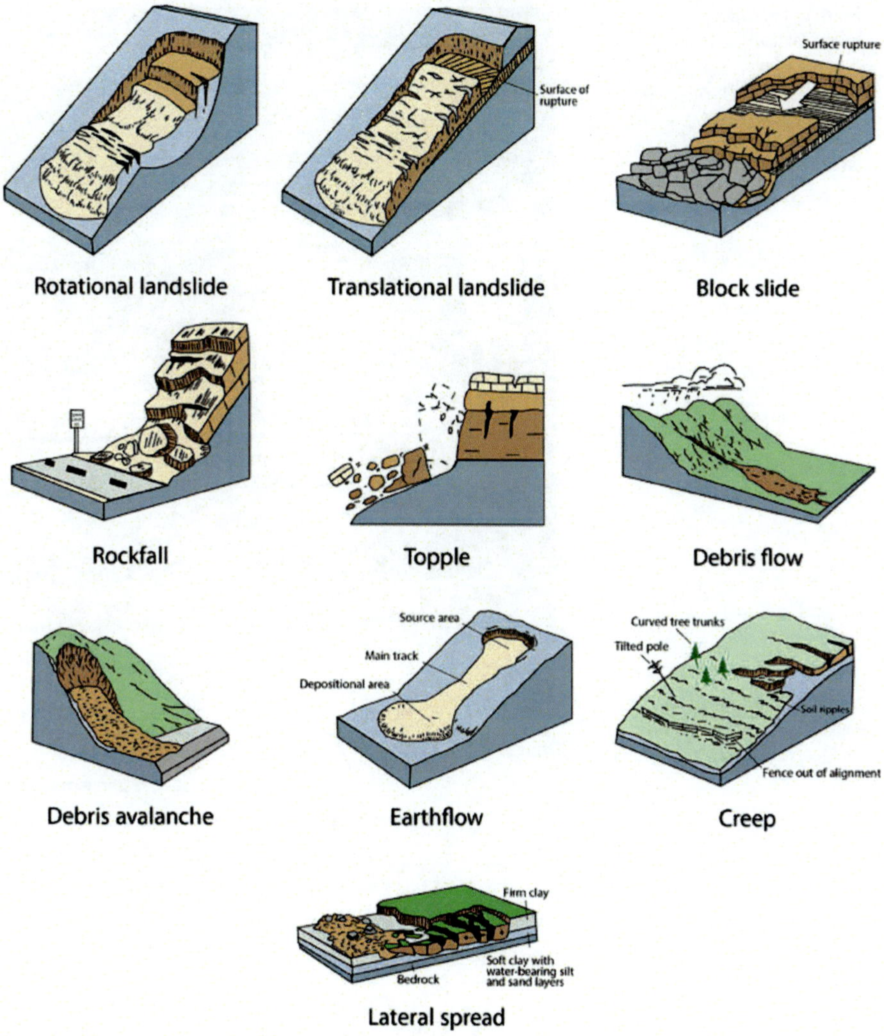

Fig. 6.4 Schematics illustrating the major types of landslide movement (Department of the Interior/USGS, with permission)

event at around 8710 ^{14}C years BP, whereas more recent dating suggests an age of 9527–9498 cal BP (Nicolussi et al. 2015). The new dating of the Köfels Bergsturz is close to the less well constrained age of the Flims landslide in the east of Switzerland. Flims is the largest Bergsturz in the Alps and is located 130 km west of Köfels. Thus, this near-synchronicity of these Bergstürze raises the question of a possible common trigger such as a strong earthquake (Nicolussi et al. 2015).

6.2.4 Snow Avalanches

Snow avalanches are a well-known hazard type related to snow. They are defined as the sudden release of snow masses and ice on slopes and may contain a certain proportion of rocks, soil and vegetation; the dislocation on the trajectory is thereby more than 50 m downhill (Wilhelm 1975). Due to the speed of the moving mass, snow avalanches can be distinguished from creeping and gliding movements of snow. A number of classifications of snow avalanches exist, developed in different countries and based on different classification principles (e.g. Kuroda 1967; de Quervain et al. 1981; Dzyuba and Laptev 1984). De Quervain et al. (1981) suggested a scheme to classify avalanches according to their release type, the shape of the trajectory and the type of movement, which is still used by the majority of scientists and practitioners in the field (Table 6.1). The snowpack evolution, from the beginning of solid precipitation accumulation until the snow cover melt, is crucial in relation to the release of snow avalanches. The conditions that lead to the release of avalanches and also a possible increase in avalanche hazard are often pervasive, but the prediction of individual avalanche events is extremely

Fig. 6.5 Landslide in Gasen/Haslau, Central Alps, c. 35 km northeast of Graz, in August 2005. *Picture credits* Arben Koçiu, Austrian Geological Survey, with permission

Table 6.1 Morphological avalanche classification system (de Quervain et al. 1973; Fuchs et al. 2019)

Zone	Criterion	Characteristic and denomination	
Origin	Manner of starting	From a point	From a line
		Loose snow avalanche	*Slab avalanche*
	Position of failure layer	Within the snowpack	On the ground
		Surface-layer avalanche	*Full-depth avalanche*
	Liquid water in snow	Absent	Present
		Dry-snow avalanche	*Wet-snow avalanche*
Transition	Form of path	Open slope	Gully or channel
		Unconfined avalanche	*Channelled avalanche*
	Form of movement	Snow dust cloud	Flowing along ground
		Powder snow avalanche	*Flowing snow avalanche*
Deposition	Surface roughness of deposit	Coarse	Fine
		Coarse deposit	*Fine deposit*
	Liquid water in deposit	Absent	Present
		Dry deposit	*Wet deposit*
	Contamination of deposit	No apparent contamination	Rock debris, soil, branches, trees
		Clean deposit	*Contaminated deposit*

difficult due to the high spatial variability and transient/dynamic nature of the snowpack (Schweizer et al. 2003). As a result, however, whole valleys may be endangered by snow avalanches during a winter season (Fig. 6.6). Different mechanisms of snow avalanche formation correspond to different volumes, repeatability and dynamic characteristics of the events (McClung and Schaerer 2006). In general, snow avalanches start from terrain that favours snow accumulation and is steeper than about 28°–60° (McClung and Schaerer 2006). On terrain inclined less than about 15°, snow avalanches start to decelerate and finally come to a stop.

Regarding the release type, snow avalanches are classified into two main groups: loose snow avalanches and slab

Fig. 6.6 Search and rescue team of the Austrian Armed Forces after the avalanche event of February 1999 in Galtür, Austria. *Picture credits* Austrian Armed Forces, with permission

avalanches. Loose snow avalanches are released from a more or less definable point in a relatively cohesionless surface layer of either dry or wet snow (Fuchs et al. 2015). The elements at risk are affected by the air pressure plume in front of the avalanche and/or by the high impact pressure of the snow in motion. Slab avalanches, in contrast, involve the release of a cohesive slab over an extended plane of weakness. Typically, natural slab avalanche activity is at the highest soon after snowstorms because of the additional load of the deposited snow (Schweizer et al. 2003). The existence of a weak layer below a cohesive slab layer is a prerequisite for the development of dry snow slab avalanches. This weak layer is either buried surface hoar or the result of metamorphism in the snowpack; during this metamorphism the properties of the snowpack change. Crystals formed by kinetic grain growth such as surface hoar or depth hoar (Fierz et al. 2009), together with changes in response to temperature and variability in water vapour gradients, can also be accompanied by formation of solid and icy layers on top of the snowpack. Such surfaces restrict the connection of new-fallen snow with the older snow below the solid layer and often form the horizon at which the snow masses start to move downhill. Slab thickness is usually less than 1 m, typically about 0.5 m, but can reach several metres in the case of large, disastrous avalanches (Bründl et al. 2010).

Exposure to snow avalanches is mainly a challenge in the western part of Austria and almost 10 000 buildings are at risk (Fuchs et al. 2015). Additionally, snow avalanches result in temporary road closures and in the interruption of train connections.

6.2.5 Soil Erosion

Water and wind can erode, transport and eventually redeposit soil. The initial impact of raindrops can break soil aggregates into primary particles by the translation of kinetic energy from the drops to the soil aggregates. Due to the influence of gravity, more soil particles are splashed downslope than upslope, and detached particles are splashed further downslope. The cumulative effect is a net downslope transfer of soil particles, known as splash erosion (Torri and Borselli 2011). When rainfall intensity exceeds soil infiltration capacity, runoff occurs. If flowing water concentrates in surface depressions, it incises into the soil and where the flowing water concentrates in shallow channels, rill erosion starts and may continue to the development of gullies (Fullen and Catt 2004). There is consensus among scholars that the distinction between these two forms of erosion is scale: while rills are incised into the topsoil (the A horizon), gullies are deeper and incise into the subsoil or parent material (the B or C horizons). As reported by Embleton-Hamann (2007), soil erosion has increased in Austria during the last decades as a result of intensification of agriculture and around 12% of the agricultural area in Austria has been classified as 'erodible land' (Strauss and Klaghofer 2006).

Focusing on geomorphology, the most prominent examples of soil erosion include the formation of large gullies and sunken roads ('Hohlwege') on loess soils of Lower Austria, which can be found especially in the wine-growing region of the Wagram, a distinct terrace landscape stretching east of the city of Krems along the Danube river (see Chapter

"Sunken Roads and Palaeosols in Loess Areas in Lower Austria: Landform Development and Cultural Importance" for details).

Less prominent, but still important, are shallow eroded areas, also known as 'Blaiken'. These usually have a size between 2 m^2 and 200 m^2 and a depth between a few decimetres and 2 m (Laatsch and Grottenthaler 1972; Schauer 1975) and are defined as phenomena where a loss of vegetation cover or topsoil has exposed the underlying, often unconsolidated material. Even if these erosion forms are relatively small in dimension, they nevertheless affect larger areas in the Austrian Alps and therefore lead to substantial material transfer. Moreover, a general destabilization of slopes results from 'Blaiken', leading to second-order processes such as an increase in snow gliding and a general increase in erosion activity, which in turn deepens and widens the initially limited dimensions of these shallow eroded areas (Wiegand and Geitner 2013).

Since the interactions between highland and lowland have a high relevance in mountainous areas and very often the highland is seen as the main cause of intensified hazardous conditions in the lowland, soil erosion is one of the main sources for sediment being further transported during flood events in mountain and foreland rivers. Therefore, it is crucial to avoid soil erosion, in particular on agricultural land. Practices such as mulch and direct seeding are promising, and the aim of these measures is to maintain a complete soil cover throughout the whole year. Soil management systems with reduced tillage intensity in combination with cover crops during winter are further effective solutions for reducing soil erosion brought about by water (Klik and Eitzinger 2010).

6.2.6 Glacier Hazards

Glaciers are accumulations of snow and ice on the surface of the Earth as a result of temperature and precipitation, covering high elevations in the Alps where the annual amount of snowfall (predominantly during the cold season) outweighs the annual amount of snow melt (predominantly during the warm season). Spatial and temporal variability of snow cover and snow depth are strongly related to regional and local precipitation patterns and temperature regimes, both parameters interacting with the terrain (Fuchs et al. 2015, 2019). Under these conditions, consecutive annual snow layers develop. The pressure of these layers forces snow at depth to change structure and density (recrystallization), first producing firn (density of 0.400–0.830 kg/m^3) and, when pores are sealed off to create air bubbles, ice (density of 0.830–0.917 kg/m^3, Stroeven 2004). When ice is thick enough to deform plastically under its own weight, glacier flow occurs, which is an effective erosive agent and has been responsible for shaping alpine valleys and producing distinct surface forms such as cirques, U-shaped valleys and moraines (Benn and Evans 2010).

From a geomorphic point of view, two major glacier hazards can be distinguished: (i) ice avalanches resulting from collapses and falls from glacier tongues and (ii) hazards related to glacier meltwater, such as a high sediment concentration of different grain size distribution in mountain torrents and glacial lake outburst floods (GLOFs). Ice avalanches are relatively scarce in the Austrian Alps but there are many records for the latter hazard type (Embleton-Hamann 2007). Hazards related to glacier meltwater are primarily determined by snowmelt and thus by spring temperature (Stewart 2009) and, during summer, also by ice melt of the glaciated areas. Glacifluvial processes are omnipresent in many catchments of the Austrian Alps and include erosion, transport and deposition, leading to the origin of new landforms or remodelling of existing ones. The type, rate and effectiveness of meltwater erosion are influenced by the nature of the basal substrate (sediment and bedrock), meltwater supply and pathway, and sediment supply (Brennand 2004). Recent studies have analysed the evolution of glacial lakes in the Austrian Alps since the Little Ice Age (Buckel et al. 2018). The latter study indicates that the formation of new glacial lakes is strongly related to glacier retreat and increasing temperatures, especially in the last 35 years, but also the local topography of the deglaciated area supports or counteracts the lake formation. Recently, GLOF hazards have received more attention since there may be increasing evidence of new proglacial lakes being formed as a result of glacier retreat, such as in the case of the Grindelwald Glacier in the Swiss Alps (Huggel et al. 2012) or in some of the high-mountain regions of Tyrol and Salzburg in Western Austria (Emmer et al. 2015). With respect to observed GLOFs, Aulitzky et al. (1994) and Embleton-Hamann (2007) reported two periods within historical time when relatively frequent lake outbursts were recorded, towards the end of the seventeenth century and during the nineteenth century, with a maximum in the 1860s. The outburst of some of these floods was related to supraglacial lakes and others related to marginal lakes dammed by glacier ice, such as the Rofener Ice Lake and the Gurgler Ice Lake in the Ötztaler Alps (see Chapter "The Upper Ötz Valley: High Mountain Landscape Diversity and Long Research Tradition" for details).

Major landforms giving the appearance of the Austrian Alps as a whole originate from the last (Würmian) glaciation, such as the glacial basins of the former outlet glaciers along the northern Alpine margin covered by lakes today and the associated large (terminal) moraines further northwards, as well as the associated fluvioglacial sandars (outwash plains), such as the 'Münchener Schotterebene' along parts of the northern Alpine margin (see Fig. 6.7). These

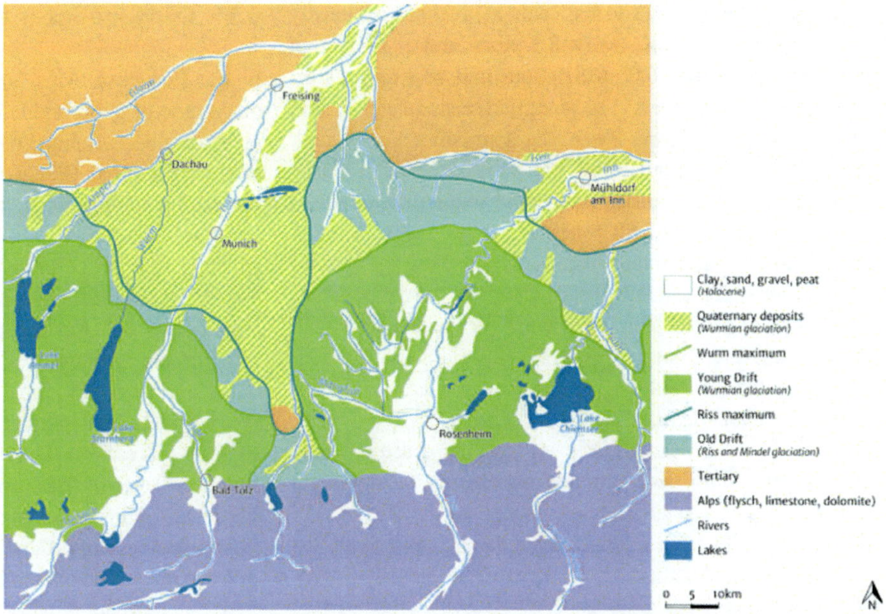

Fig. 6.7 Geological overview of the Munich region with Alpine margin, Young and Old Drift and the 'Münchener Schotterebene' (based on GK500 Bavaria)

landforms are valuable sources for palaeoenvironmental studies and may be a source for hazard types associated with sediment erosion and transport, such as dynamic flooding.

6.2.7 Permafrost

Permafrost is defined as ground (soil or rock) that remains below 0 °C for at least two years, and the term is defined purely in terms of temperature rather than the presence of frozen water (Harris 2004). As a result, apart from modelling its spatial distribution, permafrost is only detectable indirectly by assessing the distribution of distinct surface forms, such as rock glaciers, perennial snow patches and the occurrence of protalus ramparts. In the Austrian Alps, permafrost may occur at elevations higher than 2500 m asl, amounting to around 2% of the area of the territory or approximately 1600 km^2. According to a modelling exercise by Ebohon and Schrott (2008), most of the permafrost is located in the mountains of Tyrol (taking up 9.82% of the area of this province), followed by Salzburg (2.76%), Vorarlberg (1.90%) and Carinthia (1.65%). It has repeatedly been reported that mountain regions affected by permafrost could turn into hazardous areas due to warming associated with climate change (Harris et al. 2003, 2009; Haeberli 2013), resulting in melting of sub-surface ice and a subsequent destabilization of material. Hazard types may include any size of rockfall, such as those events observed during the summer of 2003 throughout the European Alps (Huggel et al. 2012), or changes in debris flow activity in mountain catchments (Sattler et al. 2011). Moreover, constructions in the Alps may suffer from melting permafrost, such as pillars of cable cars, snow rakes in avalanche starting zones and buildings. To give an example, the alpine hut 'Hochwildehaus', located at 2883 m asl in the Inner Ötz Valley east of the Gurgler Glacier, had to be closed in 2016 due to structural damage resulting from movements due to permafrost melt.

6.2.8 Seismic Hazards

Seismic activity in Austria is predominantly linked to alpine tectonics. Present-day tectonics of the Eastern Alps is characterized by strike-slip faulting regimes in a complex transition zone between the European, the Pannonian and the Adriatic stress provinces. Manifestations of vertical as well as horizontal stress decoupling within the orogen are due to a thermally and mechanically weakened crust. Differential vertical uplift derived from repeated precise levellings relative to the reference point in the Bohemian massif can be observed in western Austria including the Tauern Window, and subsidence, on the other hand, in the Vienna Basin and in the Styrian basin. This behaviour of vertical motions is related to the effects of isostatic response to active plate convergence and strain partitioning as well as to rebound occurring in response to Quaternary deglaciation and ice-induced erosion (Székely et al. 2002).

The list of major historical Austrian earthquakes includes 73 events for the period 1201–1978 (Drimmel 1980). According to the underlying statistics, a strong earthquake with a maximum intensity higher than 8.0° MSK (the Medvedev-Sponheuer-Karnik scale, also known as the MSK, is a macroseismic intensity scale used to evaluate the severity of ground shaking on the basis of observed effects in the fault area) occurs every 46.3 years, with a maximum intensity higher than 7.0° MSK every 8.5 years and a maximum intensity higher than 6.0° MSK (the limit of strong earthquakes) every 1.56 years. The identified seismotectonic fault lines include the Vienna Basin, the Mur-Mürz Valley fault contributing to the origin of the Vienna Basin, the Inn Valley (in particular at the intersection with the Wipptal disturbance) and the Lavant Valley fault zones (Aric et al. 1980; Grünthal et al. 1998). In parts of the Vienna Basin striking southwards to the Mur-Mürz tectonic disturbance, the so-called Thermenlinie is of considerable importance for seismic activity in the country, and around 50% of the seismic activity is registered in this area that is 250 km long and around 30 km wide (Embleton-Hamann 2007).

Even if strong earthquakes are rare in Austria, some major events have been reported starting in the thirteenth century. The so-called Neulengbacher earthquake of 15 September 1590 caused considerable damage and some lost lives in Vienna. Another earthquake with an intensity of 8.0° MSK occurred on 8 October 1927 in Schwadorf east of Vienna (the Vienna international airport 11/29 runway terminates there). On 7 October 1930 a 8.0° MSK earthquake resulted in heavy destruction in the village of Namlos (Tyrol), and on 3 October 1936 a 7.0°–8.0° MSK event was recorded at Obdacher Sattel (Styria). Finally, on 16 April 1972 a strong earthquake in Seebenstein (around 70 km south of Vienna) resulted in considerable losses in the city of Vienna. Further strong earthquakes have not been recorded so far in Austria (Grünthal et al. 1998). Nevertheless, some of the ground movements led to second-order consequences and can be interpreted in terms of multi-hazards (Kappes et al. 2012). Among them, the Puchberg earthquake of 1939 can be connected to the Losenheim rock avalanche, and the well-known Friuli earthquake of 1348 resulted in the Dobratsch Bergsturz (see Chapter "Dobratsch: Landslides and Karst in Austria's Southernmost Nature Park"). As a result of the 1976 Friuli earthquake, numerous smaller mass movements (mainly falls and topples) were reported to occur in Carinthia (Grünthal et al. 1998).

6.3 Consequences of Geomorphic Hazards

In the European Alps, an increase in the frequency and magnitude of geomorphic hazards and associated losses has repeatedly been claimed (i) as a result of increasing exposure of elements at risk (Fuchs et al. 2015, 2017), (ii) to be due to natural fluctuations in hazard occurrence (Schmocker-Fackel and Naef 2010) and (iii) due to the effects of climate change affecting triggers of these hazards (Huggel et al. 2012). In Fig. 6.8, the annual number of geomorphic hazardous events causing losses in Austria is shown for the period 1900–2014 with respect to snow avalanches, torrential flooding, landslides and river flooding, as well as the 10 years moving average of the total number per year. While between 1900 and 1959 an increase in the annual number of hazard events of around a factor of four can be concluded—presumably also due to an improved event observation—between 1960 and 1964 a decrease of around 50% is traceable, followed by an increase due to the excessive events in 1965 and 1966. Since then, the 10 years moving average steadily decreased again, which is in line with the increasing efforts invested in technical mitigation measures since the mid-1960s (Fuchs 2009; Holub and Fuchs 2009). Due to the high number of hazard events in 1999, 2002, 2005 and 2009, however, the curve is once more increasing to around 440 events per year. During the period of investigation, specific years with an above-average occurrence of individual hazard types can be traced, for example with snow avalanches in 1951, 1954, 1999 and 2009, torrential flooding in 1965, 1966, 2005 and 2013, and river flooding in 1904, 1959, 1966 and 2002. The trend reported in Fig. 6.8 is in clear contrast to the trends repeatedly presented for world-wide data and indicating an exponential increase in the number of events since the 1950s (Keiler 2013). Apart from hazard dynamics (the natural frequency and magnitude of events), decreasing dynamics in mountain hazard losses may result from (i) increased investments into technical mitigation (Holub and Fuchs 2009), (ii) an increased awareness of threats being consequently considered in land-use planning (Wöhrer-Alge 2013), both

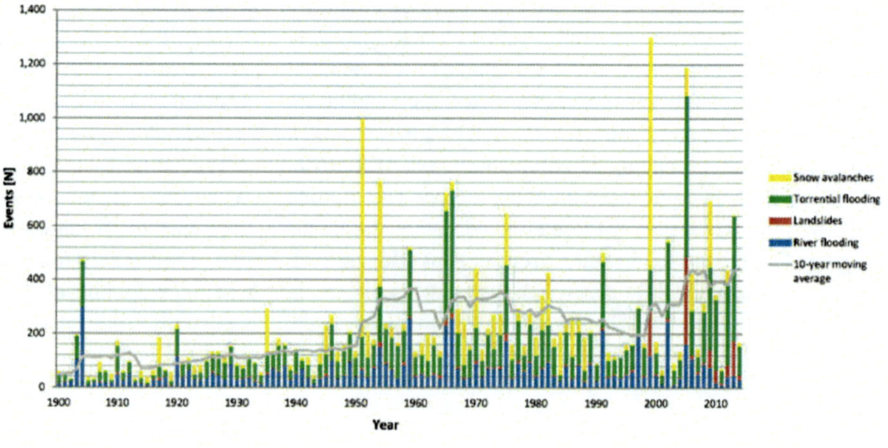

Fig. 6.8 Annual number of documented natural hazards causing losses in Austria (Fuchs et al. 2015, with permission)

leading to less exposure and (iii) decline in vulnerability (Jongman et al. 2015). Apart from the ongoing discussion of the effects of climate change influencing the hazard triggers (e.g. Auer et al. 2007; Keiler et al. 2010; Lung et al. 2013), the effects of dynamics in exposure have so far not been sufficiently studied in the context of a possible influence on dynamics of geomorphic hazards (Keiler and Fuchs 2016).

In general, the main challenge of risk reduction is rooted in the inherently connected dynamic systems driven by both geophysical and social forces, leading to the call for an integrative management approach based on multi-disciplinary concepts that take into account different theories, methods and conceptualizations (Fuchs and Keiler 2013).

References

Abele G (1974) Bergstürze in den Alpen. DAV und ÖAV, München
Aric K, Duma G, Gutdeutsch R (1980) Untersuchung der Bebentätigkeit in Kärnten, Friaul und im weiteren ostalpinen Raum. Mitteilungen Der Österreichischen Geologischen Gesellschaft 71 (72):261–268
Arnell N (2002) Hydrology and global environmental change. Pearson, Essex
Auer I, Böhm R, Jurkovic A, Lipa W, Orlik A, Potzmann R et al (2007) HISTALP—historical instrumental climatological surface time series of the Greater Alpine Region. Int J Climatol 27(1):17–46
Aulitzky H, Heuberger H, Patzelt G (1994) Mountain hazard geomorphology of Tyrol and Vorarlberg, Austria. Mt Res Dev 14(4):273–305
Austrian Standards (ed) (2009) Schutzbauwerke der Wildbachverbauung – Begriffe und ihre Definitionen sowie Klassifizierung. Technical paper ONR 24800, 75 p
Barredo J (2009) Normalised flood losses in Europe: 1970-2006. Nat Hazards Earth Syst Sci 9(1):91–104
Benn DI, Evans DJA (2010) Glaciers and glaciation. Hodder Education, London
Brennand TA (2004) Glacifluvial. In: Goudie A (ed) Encyclopedia of geomorphology. Routledge, London, pp 459–465
Bründl M, Bartelt P, Schweizer J, Keiler M, Glade T (2010) Review and future challenges in snow avalanche risk analysis. In: Alcántara-Ayala I, Goudie A (eds) Geomorphological hazards and disaster prevention. Cambridge University Press, Cambridge, pp 49–61
Buckel J, Otto JC, Prasicek G, Keuschnig M (2018) Glacial lakes in Austria—distribution and formation since the Little Ice Age. Global Planet Change 164:39–51
Chow VT (1956) Hydrologic studies of floods in the United States. Int Assoc Sci Hydrol Publ 42:134–170
Crozier M (1999) The frequency and magnitude of geomorphic processes and landform behaviour. Zeitschrift Für Geomorphologie n.f Suppl-Bd 115:35–50
Cruden D, Varnes D (1996) Landslide types and processes. In: Schuster R, Turner R (eds) Landslides. Investigation and mitigation). National Academy Press, Washington, pp 36–75
de Quervain MR, de Crécy L, LaChapelle ER, Lossev K, Shoda M, Nakamura T (1981) Avalanche atlas. Illustrated international avalanche classification. UNESCO, Paris
Dikau R (2004) Mass movement. In: Goudie A (ed) Encyclopedia of geomorphology. Routledge, London, pp 644–653
Drimmel J (1980) Die zeitliche und räumliche Verteilung der wichtigsten Ostalpenbeben. In: Österreichs DGA (ed) Geologische Bundesanstalt. Springer, Vienna, pp 513–524
Dzyuba VV, Laptev MN (1984) Geneticheskaya klassifikatsiya i diagnosticheskie priznaki snezhnykh lavin [Genetic classification and diagnostic features of snow avalanches]. Materialy Glyatsiologicheskikh Issledovanii [data of Glaciological Studies] 50:97–104
Ebohon B, Schrott L (2008) Modelling mountain permafrost distribution: a new permafrost map of Austria. In: Kane DL, Hinkel KM (eds) Proceedings of the 9th international conference on permafrost. University of Alaska, Fairbanks, pp 397–402
Embleton-Hamann C (1997) Austria. In: Embleton C, Embleton-Hamann C (eds) Geomorphological hazards of Europe. Elsevier, Amsterdam, pp 1–30
Embleton-Hamann C (2007) Geomorphological hazards in Austria. In: Kellerer-Pirklbauer A, Keiler M, Embleton-Hamann C, Stötter J (eds) Geomorphology for the future. Innsbruck University Press, Innsbruck, pp 33–56
Emmer A, Merkl S, Mergili M (2015) Spatiotemporal patterns of high-mountain lakes and related hazards in western Austria. Geomorphology 246:602–616
Fell R, Corominas J, Bonnard C, Cascini L, Leroi E, Savage W (2008) Commentary on Guidelines for landslide susceptibility, hazard and risk zoning for land-use planning. Eng Geol 102(3–4):99–111
Fierz C, Armstrong R, Durand Y, Etchevers P, Greene E, McClung DM et al (2009) The international classification for seasonal snow on the ground. UNESCO, Paris
Fuchs S (2009) Susceptibility versus resilience to mountain hazards in Austria—paradigms of vulnerability revisited. Nat Hazard 9 (2):337–352
Fuchs S, Kaitna R, Scheidl C, Hübl J (2008) The application of the risk concept to debris flow hazards. Geomech Tunnell 1(2):120–129
Fuchs S, Keiler M (2013) Space and time: coupling dimensions in natural hazard risk management? In: Müller-Mahn D (ed) The spatial dimension of risk—how geography shapes the emergence of riskscapes. Earthscan, London, pp 189–201
Fuchs S, Keiler M, Sokratov S (2015a) Snow and avalanches. In: Huggel C, Carey M, Clague JJ, Kääb A (eds) The high-mountain cryosphere: environmental changes and human risks. Cambridge University Press, Cambridge, pp 50–70
Fuchs S, Keiler M, Zischg A (2015b) A spatiotemporal multi-hazard exposure assessment based on property data. Nat Hazard 15 (9):2127–2142
Fuchs S, Thaler T, Röthlisberger V, Zischg A, Keiler M (2017) Natural hazard management from a co-evolutionary perspective: the cycle of exposure and policy response in the European Alps. Ann Assoc Am Geogr 107(2):382–392
Fuchs S, Keiler M, Sokratov S (2019) Snow avalanches. In: Maggioni V, Massari C (eds) Extreme hydroclimatic events and multivariate hazards in a changing climate. Elsevier, Amsterdam, pp 369–389
Fullen MA, Catt JA (2004) Soil erosion. In: Goudie A (ed) Encyclopedia of geomorphology. Routledge, London, pp 977–981
Giles D (2013) Intensity scales. In: Bobrowski P (ed) Encyclopedia of natural hazards. Springer, Dordrecht, pp 544–552
Grünthal G, Mayer-Rosa D, Lenhardt WA (1998) Abschätzung der Erdbebengefährdung für die D-A-CH-Staaten - Deutschland, Österreich, Schweiz. Bautechnik 75(10):753–767
Haeberli W (2013) Mountain permafrost—research frontiers and a special long-term challenge. Cold Reg Sci Technol 96:71–76
Harris C (2004) Permafrost. In: Goudie A (ed) Encyclopedia of geomorphology. Routledge, London, pp 777–779
Harris C, Arenson LU, Christiansen HH, Etzemüller B, Frauenfelder R, Gruber S et al (2009) Permafrost and climate in Europe: monitoring and modelling thermal, geomorphological and geotechnical responses. Earth Sci Rev 92(3–4):117–171
Harris C, Vonder Mühll D, Isaksen K, Haeberli W, Sollid JL, King L et al (2003) Warming permafrost in European mountains. Global Planet Change 39(3–4):215–225

Heuberger H (1966) Gletschergeschichtliche Untersuchungen in den Zentralalpen zwischen Sellrain und Ötztal. Wiss Alpenvereinshefte 22:1–126
Holub M, Fuchs S (2009) Mitigating mountain hazards in Austria—legislation, risk transfer, and awareness building. Nat Hazard 9 (2):523–537
Huggel C, Clague J, Korup O (2012) Is climate change responsible for changing landslide activity in high mountains? Earth Surf Proc Land 37(1):77–91
Hutchinson J (1988) General report: morphological and geotechnical parameters of landslides in relation to geology and hydrogeology. Paper presented at the 5th International Symposion on Landslides, Lausanne
Jongman B, Winsemius HC, Aerts JCJH, de Perez EC, van Aalst MK, Kron W et al (2015) Declining vulnerability to river floods and the global benefits of adaptation. Proc Natl Acad Sci USA 112(18): E2271–E2280
Kappes M, Keiler M, von Elverfeldt K, Glade T (2012) Challenges of analyzing multi-hazard risk: a review. Nat Hazards 64(2):1925–1958
Keiler M (2013) World-wide trends in natural disasters. In: Bobrowski P (ed) Encyclopedia of natural hazards. Springer, Dordrecht, pp 1111–1114
Keiler M, Fuchs S (2016) Vulnerability and exposure to geomorphic hazards—some insights from mountain regions. In: Meadows M, Lin J-C (eds) Geomorphology and society. Springer, Tokyo, pp 165–180
Keiler M, Knight J, Harrison S (2010) Climate change and geomorphological hazards in the eastern European Alps. Philos Trans R Soc Lond Ser a: Math Phys Eng Sci 368:2461–2479
Kilburn CRJ, Pasuto A (2003) Major risk from rapid, large-volume landslides in Europe (EU Project RUNOUT). Geomorphology 54:3–9
Klik A, Eitzinger J (2010) Impact of climate change on soil erosion and the efficiency of soil conservation practices in Austria. J Agric Sci 148(5):529–541
Kuroda M (1967) Classification of snow avalanches Physics of snow and ice. In: Proceedings (International conference on low temperature science, Sapporo, Aug. 14–19, 1966), vol 1, pp 1277–1290. Institute of Low Temperature Science, Hokkaido University, Sapporo
Laatsch W, Grottenthaler W (1972) Typen der Massenverlagerung in den Alpen und ihre Klassifikation. Forstwissenschaftliches Centralblatt 91(1):309–339
Lung T, Lavalle C, Hiederer R, Dosio A, Bouwer LM (2013) A multi-hazard regional level impact assessment for Europe combining indicators of climatic and non-climatic change. Glob Environ Chang 23(2):522–536
Mazzorana B, Fuchs S (2010) Fuzzy Formative scenario analysis for woody material transport related risks in mountain torrents. Environ Model Softw 25(10):1208–1224
McClung D, Schaerer P (2006) The avalanche handbook. The Mountaineers, Seattle
Neall VE (2004) Hyperconcentrated flow. In: Goudie A (ed) Encyclopedia of geomorphology. Routledge, London, p 542
Nicolussi K, Spötl C, Thurner A, Reimer PJ (2015) Precise radiocarbon dating of the giant Köfels landslide (Eastern Alps, Austria). Geomorphology 243:87–91
Parker DJ (ed) (2000) Floods. Routledge, London
Phillips C, Davies T (1991) Determining rheological parameters of debris flow material. Geomorphology 4(2):101–110
Pierson TC (1986) Flow behavior of channelized debris flows, Mount St. Helens, Washington. In: Abrahams A (ed) Hillslope processes. Allen and Unwin, Boston, pp 269–296
Pierson TC, Costa JE (1987) A rheologic classification of subaerial sediment-water flows. Geol Soc America Rev Eng Geol 7:1–12
Prager C, Zangerl C, Patzelt G, Brandner R (2008) Age distribution of fossil landslides in the Tyrol (Austria) and its surrounding areas. Nat Hazard 8(2):377–407
Sattler K, Keiler M, Zischg A, Schrott L (2011) On the connection between debris flow activity and permafrost degradation: A case study from the Schnalstal, South Tyrolean Alps, Italy. Permafrost Periglac Process 22:254–265
Schauer T (1975) Die Blaikenbildung in den Alpen. Bayerisches Landesamt für Wasserwirtschaft, München
Schmocker-Fackel P, Naef F (2010) Changes in flood frequencies in Switzerland since 1500. Hydrol Earth Syst Sci 14(8):1581–1594
Schweizer J, Jamieson B, Schneebeli M (2003) Snow avalanche formation. Rev Geophys 41(4):1016
Stewart IT (2009) Changes in snowpack and snowmelt runoff for key mountain regions. Hydrol Process 23(1):78–94
Strauss P, Klaghofer E (2006) Austria. In: Boardman J, Poesen J (eds) Soil erosion in Europe. Wiley, Chichister, pp 205–212
Stroeven AP (2004) Glaciers. In: Goudie A (ed) Encyclopedia of geomorphology. Routledge, London, pp 454–459
Székely B, Reinecker J, Dunkl I, Frisch W, Kuhlemann J (2002) Neotectonic movements and their geomorphic response as reflected in surface parameters and stress patterns in the Eastern Alps. Stephan Mueller Special Publ Ser 3:149–166
Takahashi T (1991) Debris flow. Balkema, Rotterdam
Torri D, Borselli I (2011) Water erosion. In: Huang PM, Li Y, Sumner ME (eds) Handbook of soil sciences: Resourcemanagement and environmental impacts. CRC Publications, Boca Raton, pp 22-21–22-19
UN (ed) (2002) Guidelines for reducing flood losses. United Nations, Geneva
van Westen C, van Asch TWJ, Soeters R (2006) Landslide hazard and risk zonation—why is it still so difficult? Bull Eng Geol Env 65 (2):167–184
Varnes D (1978) Slope movement. Types and processes. In: Schuster R, Krizek R (eds) Landslides: analysis and control. National Academy of Sciences, Washington, pp. 11–33
Varnes D (1984) Landslide hazard zonation: a review of principles and practice, vol 3. UNESCO, Paris
Wiegand C, Geitner C (2013) Investigations into the distribution and diversity of shallow eroded areas on steep grasslands in Tyrol (Austria). Erdkunde 67(4):325–343
Wilhelm F (1975) Schnee- und Gletscherkunde. Berlin
Wohl E (2004) Flood. In: Goudie A (ed) Encyclopedia of geomorphology. Routledge, London, pp 378–380
Wöhrer-Alge M (2013) Landslides management in Austria with particular attention to hazard mapping and land use planning. In: Margottini C, Canuti P, Sassa K (eds) Landslide Science and Practice, vol 7. Springer, Berlin, pp 231–237

Sven Fuchs is a senior scientist at the Institute of Mountain Risk Engineering at the University of Natural Resource and Life Sciences in Vienna, Austria. His research and teaching interests include high-mountain geomorphology, mountain hazard risk management, the study of coupled human–environment systems, climate change adaptation, and vulnerability assessment for natural hazards. His research has taken him to mountain regions of Europe, Southeast Asia, Central Africa, and countries of the former Soviet Union.

Martin Wenk works for the Federal Water Engineering Administration, a department of the Austrian Federal Ministry for Agriculture, Regions and Tourism in Vienna. His work and research is focused on the management of natural hazards and the assessment of related risks, with particular emphasis on the use of geoinformatics and cartography.

Margreth Keiler is professor at the Department of Geography, University of Innsbruck and director of the Institute for Interdisciplinary Mountain Research at the Austrian Academy of Sciences. She is head of the joint working group Coupled Human-Landscape Systems: Risk and Resilience at both institutions. Her research focus is on sediment connectivity, risk analysis and risk management under global change and multi-hazard risks in mountains, with regional expertise in the European Alps, in mountain areas of Central and Southeast Asia, and in Africa.

Geoheritage, Geotourism and Landscape Protection in Austria

7

Horst J. Ibetsberger and Christine Embleton-Hamann

Abstract

The variability of Earth's surface materials, forms and physical processes is crucial for sustaining ecosystems and their services, as it influences fauna, flora and land use, and via land use the development of specific cultural elements of a region. Many of Austria's nature protection areas display and interpret these abiotic attributes of nature, namely special landforms, landscapes or geological features. Nature protection under Austria's federal system has one notable weakness, namely that the legislation and implementation of nature protection falls under the jurisdiction of its provinces. Consequently, nine individual protection laws, differing in detail and quality, exist. Only protection categories are somewhat consistent. Geoheritage in Austria is conserved in six national parks, 47 nature parks, three UNESCO Global Geoparks and one national geopark. The last category – geoparks – is unfortunately not included in Austria's nature protection laws. Geotourism has the goal to raise the awareness of abiotic nature attributes, together with promoting a profound understanding of Earth sciences and the conservation of geodiversity. Geotourism and geoheritage are two sides of the same coin. Once a geoheritage site or park becomes a visitor magnet, its importance and value will rise, while geotourism depends on geoheritage and its protection. Geotourism must not be confused with adventure and activity tourism, in which geoheritage is occasionally used as a backdrop for activity setups. Challenges for the management of Austria's parks and sites are manifold. On the one hand, there is the goal to increase visitor numbers, and on the other hand, any negative impact of intensified visitation on the geoheritage itself must be prevented. As always, scarcity of funds is a pervasive problem, not only for maintenance of the existing infrastructure, but more importantly for the development of best practice education in order to make every visitor a multiplier in the protection of Austria's unique and beautiful places.

Keywords

Nature protection in Austria • Geoheritage • Geotourism • Geopark • Geotope • Thematic trail

7.1 Introduction

Austria's major landscape regions correspond, in whole or in part, to its main geological units. They comprise (i) the Bohemian Massif, (ii) the Alps, and (iii) the Alpine forelands and foothills together with a number of tectonic basins (cf. Chapter "Geomorphological Landscape Regions of Austria"). The granite and gneiss highlands of the Bohemian Massif lie north of the Danube and take up about 10% of the national territory. The Austrian Alps, as part of the European Eastern Alps, are characterized by a mountainous and high Alpine relief and cover 63% of the country. They can be divided into the Northern Alps, the Central Alps and the Southern Alps (whereby the Austrian share in this most southern unit is quite small). The hilly countryside and the plains of the third landscape region consist of the Alpine foothills and the Alpine and Carpathian foreland in the north, the Alpine foreland in the southeast and some large basins, such as the Vienna and the Klagenfurt basins. The total area of these low relief landscape region amounts to 27% of the national territory.

These landscape regions hold a number of specific landforms, old and new rocks, internationally important sites of stratigraphy together with many cultural sites that arose

H. J. Ibetsberger (✉)
GeoGlobe Häupl & Ibetsberger OG, Statzenbachgasse 5, 5202 Neumarkt am Wallersee, Austria
e-mail: horst.ibetsberger@geoglobe.at

C. Embleton-Hamann
Department of Geography and Regional Research, University of Vienna, Universitätsstr. 7, 1010 Vienna, Austria

C. Embleton-Hamann (ed.), *Landscapes and Landforms of Austria*, World Geomorphological Landscapes,
https://doi.org/10.1007/978-3-030-92815-5_7

from the exceptionally long use of the natural resources of the Alpine region (Feitzinger et al. 2003; Stöllner and Oeggl 2015). These topics are interpreted in the visitor centres, thematic trails and special exhibitions of Austria's various "Parks" (see Sect. 1.2). The contemporary issue of climate change has triggered a new topic that is currently being addressed with specific exhibitions (e.g. in the Hohe Tauern National Park or at the natural monument Dachstein giant ice cave), thematic trails (e.g. in the Mürzer Oberland Nature Park) or climate action days [e.g. organized by the Ore of the Alps UNESCO Global Geopark (Ibetsberger 2020)].

7.1.1 Geotourism and Geoheritage

Several definitions of "geotourism" have been published. While compatible in general, they differ in their respective focus on the various aspects of geotourism (Megerle 2008).

The first aspect refers to the central topic of geotourism. Newsome and Dowling (2010) defined geotourism as a form of tourism that specifically focuses on landscape and geology. On the basis of a simple "ABC model", in which A stands for the abiotic, B for the biotic and C for the cultural attributes of a given environment, they argued that information on and interpretation of B and C is widespread, while A receives little attention and consequently is hardly known and appreciated by the public. On the other hand, it is precisely A, the variability of Earth's surface materials, forms and physical processes that is crucial for sustaining ecosystems and their services, as it influences the biotic elements and the land use / cultural development of a given area (Newsome and Dowling 2018).

Secondly, the goal of geotourism is defined as promoting the conservation of geodiversity and an understanding of Earth sciences through appreciation and learning (Newsome and Dowling 2010, 2018). For geoheritage sites as the centres of visitation, this constitutes the need for scientifically sound and region-specific information, be it in the form of geotrails and viewpoints, visitor centres, guided tours, etc.

The third aspect addresses the "tourism" part in geotourism and defines it as form of sustainable tourism development that takes the well-being and economic viability of local communities and their residents into account. The goal is that local businesses, e.g. dealing with interpretation, tours, accommodation or gastronomy, work together with planning authorities and community groups to promote and provide a distinctive, authentic visitor experience (Dowling 2009).

All in all, the key principle of geotourism is that of environmental and economical sustainability. Since the natural environment underpins geotourism, it is essential that it is protected and conserved. Thus, the operators must ensure that the type, location and level of touristic use do not harm the geoheritage and its surrounding areas. At the same time, economic viability is important, as without worthwhile economic benefits the local communities will lose interest in managing and protecting their geoheritage (Dowling 2009).

The definition of geotourism naturally excludes any forms of event and activity tourism that use rocks, landforms, fossils, etc., as a backdrop for raising tourist attractiveness and visitor numbers (Newsome and Dowling 2010).

The relationship of geotourism and geoheritage is one of interdependencies (Newsome and Dowling 2018). Geotourism by definition depends on geoheritage and its protection. On the other hand, interested visitors coming from outside the region in order to see a specific geoheritage site will raise its importance and value. Although an increasing number of tourists provides potential risks, it is likely to be a valuable mechanism for public appreciation of geoheritage, provided a careful management is in place.

7.1.2 Nature and Geoheritage Protection in Austria

Austria has a total area of 83,858 km^2, of which only 37% are amenable to settlement. The reason for this is that Austria has the highest share of the Alps of all Central European countries. Nearly 50% of the Austrian population lives in mountainous regions with little space for settlement. These numbers highlight the gap between extensive unspoiled natural landscapes in high mountain regions (with the exception of glacial ski areas) and the stress on the ecosystem in densely settled regions. Examples for the latter are the inner Alpine valleys of the rivers Inn, Salzach and Enns, or the steadily growing agglomerations in the Alpine forelands, e.g. in the surroundings of the cities Innsbruck, Linz, Graz and Vienna. Soil sealing (caused by building houses, roads and other infrastructure) in Austria amounts to 15–20 ha per day (BOKU 2017). More than 5% of the habitable surface of Austria is sealed! Agriculture further consumes large areas. This means that in the habitable regions of Austria, biodiversity and geodiversity are under severe pressure. Protected areas are therefore of enormous ecological, social, cultural and scientific importance.

In Austria, legislation and implementation of nature conservation areas fall under the jurisdiction of its autonomous provinces. Consequently, there are nine different nature conservation laws, which differ in detail and quality. Consistency only exists with respect to protection categories. These are national parks, nature conservation areas, landscape conservation areas, nature parks and natural monuments. In addition, protection of geoheritage and the landscape also takes place in the four geoparks of Austria, but unfortunately without statutory background. This means that any kind of intervention is possible in geoparks. The inclusion of geoparks in Austria's nature conservation law is therefore an important issue for the future.

Some protected areas fall into several categories at once. The Styrian Eisenwurzen, for example, has been a nature park with a protected landscape area since 1996, a geopark in the Global Geopark Network since 2004 and an UNESCO global geopark since 2015. For a map of all Austrian "parks", see Fig. 7.1.

National Parks: To date six of Austria's most ecologically valuable regions are national parks according to International Union for Conservation of Nature (IUCN) category II protected areas. Human impact in these regions was small to start with and with the declaration as national park nature conservation became an absolute priority. Renunciation of any economic use on at least 75% of the area is a prerequisite. Other objectives and tasks pursued in the national parks are scientific research, education and recreation based on the experience of unspoiled nature (Platzgummer 2013).

The Austrian Federal Constitution identifies nature conservation as a responsibility of the provinces. When a province participates in a project of national importance, such as the establishment of a national park, a state treaty is concluded between the federal government and the respective provincial government clarifying the competences. Financing is jointly provided by the federal government, the province and the national park municipalities.

Nature Conservation Areas and Landscape Protection Areas: Nature conservation areas are devoted to preserving biological diversity and protecting endangered species of flora and fauna. In the context of this book, however, the category "landscape protection area" is of higher interest, as it aims to protect and preserve landscapes of high diversity, uniqueness and beauty (Fig. 7.2). As such landscapes are valuable assets for recreation and tourism, provincial governments can declare them as protected areas. This protection category comprises 258 areas in Austria (as of 2017, Federal Environment Agency). Financing and management of landscape protection areas is shared between the provincial government and the municipalities.

A **Natural Monument** is a protected natural formation (e.g. a tree, cave, gorge, waterfall) and is mainly of local or regional importance. No intervention or alterations may be made that could impair the existence or appearance, the peculiarity, the characteristic imprint or the scientific and cultural value of the natural monument. Some of Austria's national monuments have been protected since the early 1930s, when the first nature conservation laws were implemented. The designation is carried out by a decision of the district administrative authority. Financing is the responsibility of the provincial government and the municipalities.

It is difficult to tell the total number of natural monuments in Austria, as they are a matter of the provinces, which define and record them in different ways. All natural

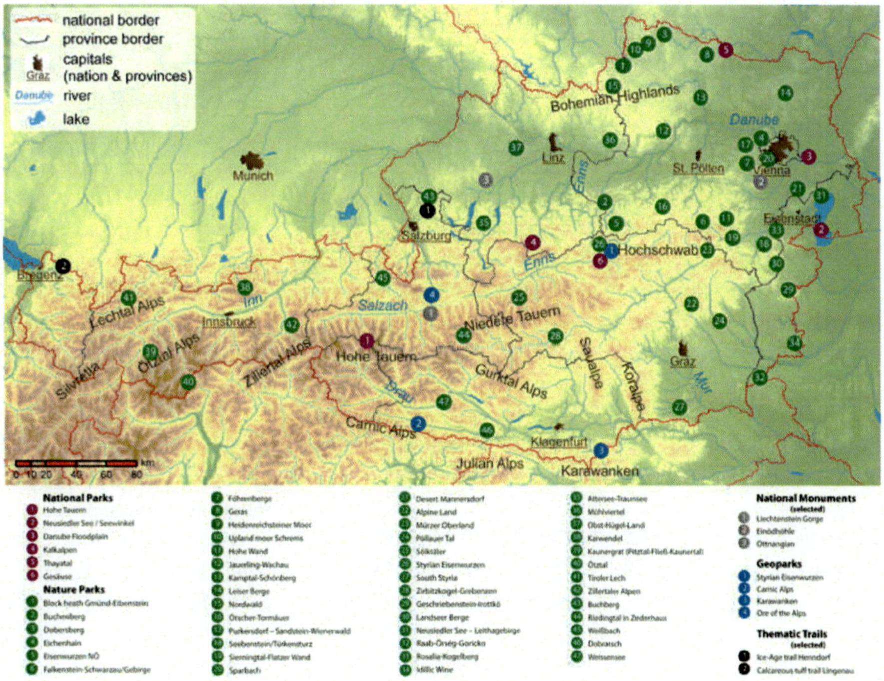

Fig. 7.1 Distribution of national parks, nature parks and geoparks in Austria. Further located are the positions of national monuments and thematic trails that are described in more detail in Sect. 1.2. Graphics: GeoGlobe, 2020

Fig. 7.2 Landscape protection area Paarseen – Schuhflicker – Heukareck in the Ore of the Alps UNESCO Global Geopark. *Photo* TVB Schwarzach – St. Veit

monuments are marked with the emblem of the province. The majority of them is devoted to flora and fauna. Hofmann (2000) reported the results of a project of the Austrian Geological Survey to identify all natural monuments in Austria that preserve abiotic features of geological, geomorphological or archaeological importance (geotopes). This survey arrived at a number of 612 sites. As of 2020, Hofmann expects a maximal change of this number by 1% (personal communication).

Nature Parks: This is a special designation awarded to nature conservation or landscape protection areas by the provincial government. What is new with respect to the categories listed above is the idea of protecting a cultural landscape and promoting sustainable economic development of rural areas alongside landscape and nature conservation. This is, for instance, achieved by conscientious husbandry that preserves the habitats of a wide range of species, or the development of high-quality "Nature Park" culinary specialties that sell at a good price.

As of 2020, Austria has 47 nature parks with a total area of about 5000 km^2 (https://www.naturparke.at/ueber-uns/) (Marek and Neffe 2013). Financing is mainly provided by the province and the municipalities.

Geoparks are a label of UNESCO. The "UNESCO Global Geopark" status is tied to strict obligations, and the certification is awarded for a four-year period only. After a positive evaluation, it can be prolonged for a further four years. The aspects primarily checked in this "revalidation process" are the quality of geotope protection and the touristic/educational concepts of the park.

In order to become an UNESCO global geopark, the area must have features or landscapes of international geological/geomorphological significance, which is assessed and acknowledged by professional scientists at Global Geopark Network (GGN) and UNESCO. After qualification, the sites must be managed with a holistic concept of protection, education and sustainable development. This implies that the people living in the area are to be involved in all decisions and that a sustainable economic plan for the local communities is to be established (http://www.unesco.org/new/en/natural-sciences/environment/earth-sciences/unesco-global-geoparks/). For the task of education, Earth scientists and universities help with the preparation of explanatory materials and service facilities. In addition, the geopark cooperates with schools and local partners such as tourist associations and outdoor providers (Allmrodt et al. 2015; Hess et al. 2016).

Thus, UNESCO geoparks have two things in common with nature parks in Austria. On top of nature conservation, education and recreation, they (i) protect a settled and therefore cultural landscape and (ii) care for the sustainable development of the local communities. What is unfortunately different is that the certification as "UNESCO Global Geopark" does not imply a protection by Austrian law.

7.2 Places to Visit in Order to Get to Know Austria's Geoheritage

7.2.1 National Parks

Steppes, original and alluvial forests, smooth valley surfaces, transverse valleys, rugged limestone mountains and glaciers characterize the landscapes of the Austrian national parks. The first phase of their implementation was not easy. They were caught in the area of tension between nature conservation and economic interests.

The internationally most conspicuous protest was certainly the occupation of the Danube floodplain near Hainburg in the winter of 1984/85, which eventually forestalled the construction of a hydroelectric power station there. It was the first heavy protest action of environmentalists in Austria and simultaneously a clear signal for the need of a political rethinking in terms of environment and democracy. Eleven years later (1996), the **Danube Floodplain National Park** was founded as the third of Austrian's national parks. Covering 9300 ha along the river Danube in Vienna and Lower Austria, the Danube Floodplain National Park is one of the largest virtually untouched wetlands in Central Europe (Fig. 7.3) (cf. Chapter "The Danube Floodplain National Park: A Fluvial Landscape with Expiration Date?" that is devoted to this national park).

The **Hohe Tauern** National Park is the oldest and most famous national park in Austria. It is situated in the Hohe Tauern range with peaks reaching more than 3000 m asl, and is shared by the provinces of Salzburg, Tyrol and Carinthia (Steyrer et al. 2011). The national park, which covers an area of 185 600 ha, was founded 1981 in Carinthia, 1984 in Salzburg and 1992 in Tyrol, and is the largest nature conservation area in the Alps. It includes all altitudinal levels of the Alps. All valleys are formed as glacial troughs. The visitor centres in Mittersil (Salzburg), Mallnitz (Carinthia) and Matrei (Eastern Tyrol) provide comprehensive information. They are devoted to selected issues like landscape development, current climate change and in connection with this, permafrost and the melting of glaciers. For some of the geomorphological highlights of the Hohe Tauern National Park, see Chaps. 24, 25 and 27.

More than 60 educational trails with information panels – partly interactive – have been installed. The most popular ones for geotourists are: Pasterze Glacier Trail, Tauernfenster Geotrail, Glacier educational trail Obersulzbachtal (Lieb and Slupetzky 2013), Glacier educational trails Ödenwinkel- and Sonnblickkees, Geomorphological trail to the Glorer Hütte (Stingl 2004) and Glacier educational trail Innergschlöß (Schlosser 2006) (https://hohetauern.at/de/besuchen/themenwege.html).

As an example among many, it is worth picking out the glacier educational trail Stubacher Sonnblickkees on the Salzburg territory of the national park. This 6-km-long trail starts at the mountain hut "Rudolfshütte" (2304 m asl) and runs to the Stubacher Sonnblickkees (Kees = glacier; ches = cold, ice) (Fig. 7.4). Educational panels inform about the sculpturing of the high-mountain landscape through glacial and fluvial erosion, and the downwasting of the glacier since the Little Ice Age glacier maximum around 1850. A good overview of the features that can be seen along the stops of this trail is provided by the following list of panel headings: (i) The glacier in 1850; (ii) The lateral moraine of 1850; (iii) An erratic boulder; (iv) Position of the glacier snout in 1880; (v) Glacier advance of 1920; (vi) The glacier stream, (vii) Glacial polish and glacial stria; (viii)

Fig. 7.3 Danube Floodplain National Park – banks of gravel near Orth an der Donau. *Photo* Wolfgang Schruf

Fig. 7.4 Mountainous area at the mountain hut Rudolfshütte (below the centre of the photo) with the artificial reservoirs of Weißsee (left) und Tauernmoossee (right) in the Hohe Tauern National Park. *Photo* H. J. Ibetsberger

Position of the glacier snout in 1950; (ix) Glacier measure mark; (x) The actual glacier front – a new lake emerges; (xi) Glacially polished boulder and (xii) View over the Stubacher Sonnblickkees (Slupetzky et al. 2011). This trail is easy to reach and offers an excellent opportunity for on-site learning.

The four other national parks of Austria are **Neusiedler See – Seewinkel** in Burgenland/Hungary (cross-border, 9064 ha – founded 1996, partly addressed in Chapter "Lake Neusiedl Area: A Particular Lakescape at the Boundary between Alps and Pannonian Basin"), **Kalkalpen** in Upper Austria (20,825 ha – founded 1997), **Thayatal** in Lower Austria/Czech Republic (cross-border, 1330 ha – founded 2000, see Chapter "Deeply Incised Valley Meanders of the Bohemian Massif") and **Gesäuse** in Styria (11,300 ha – founded 2002, partly addressed in Chapter "Gesäuse: River Gorge, Limestone Massifs and Sediment Cascades") Environmental education programmes – especially for younger visitors are a central topic in all six Austrian national parks (http://www.nationalparksaustria.at/en/pages/default.aspx).

7.2.2 Nature Parks

Three of the 47 Austrian nature parks are described in Chapter "Fluvial Geomorphology and River Restoration: Tiroler Lech Nature Park" (Tiroler Lech Nature Park), Chapter "The Upper Ötz Valley: High Mountain Landscape Diversity and Long Research Tradition" (Ötztal Nature Park) and Chapter "Dobratsch: Landslides and Karst in Austria's Southernmost Nature Park" (Dobratsch Nature Park). In the following two others will be briefly introduced.

Fig. 7.5 "Schwammerling", the symbol of the Mühlviertel Nature Park, is a spheroidally weathered granite. *Photo* H. J. Ibetsberger

The **Mühlviertel Nature Park** in Upper Austria is situated in the old gneiss and granite rocks of the Bohemian Massif. The hilly, small-structured landscape is strewn with giant granite boulders. The most impressive one is named "Schwammerling", because it looks like a mushroom (Fig. 7.5). Numerous smaller and bigger boulders – erosion remnants from the tropical climate during the Neogene – can be found everywhere amidst colourful meadows. Steep forested slopes with boulder fields accompany rivers and creeks. The nature park was installed in 1996 and encompasses an area of 1046 ha. Many trails lead through the park, past the granite boulders with information panels explaining their origin. A stone park near the open-air museum Großdöllnerhof exhibits rocks from all different geological units of Upper Austria. The main emphasis lies, however, on plutonic rocks such as granites and granodiorites.

Fig. 7.6 Exhibition of erratic boulders in the Nature Park Buchberg. *Photo* H. J. Ibetsberger

Another nature park, this one situated in the province of Salzburg, is the **Nature Park Buchberg**. Installed in 2009, it is a small park with an area of 35 ha only. Buchberg is the name of a local peak in the Alpine foreland with a summit elevation of 801 m asl. It belongs to the Flysch zone and consists of marls, silt- and sandstones that bear remnants of ichnofossils. During the LGM, the Buchberg was surrounded by the ice of the Salzach Glacier up to an altitude of 760 m asl. Only the top 40 m of its summit protruded over the ice surface. The viewing platform built underneath the summit provides a spectacular view over more than 120 summits of the Alps. This view together with the small area of the park gave rise to the park logo "small mountain with a great view". A circular thematic trail leads from the restaurant Alpenblick to the summit and back, where information panels along the path inform about geology, geomorphology, the Quaternary Salzach Glacier, forests and prehistoric settlements. The nature park also has an exhibition of erratic boulders that the Salzach Glacier transported to this area (Fig. 7.6).

7.2.3 Natural Monuments (Geotopes)

Austria's geotopes comprise a substantial number of gorges. Two spectacular examples, namely the Ragga slot and the Garnitzen Gorge in Carinthia, are described in detail in Chapter "Gorges and Slots—Three Examples of Geomorphosites in Western Carinthia".

In Salzburg, the **Liechtenstein Gorge** is a magnet for tourism. Situated about 5 km south of St. Johann im Pongau (Fig. 7.1), it stretches in a northeast-southwest direction over a distance of about 4 km. Long, spectacular walkways and bridges provide easy access. The gorge is composed of dark-grey to light-grey crinoidal limestones, belonging to the so-called Matrei-Nordrahmen nappe system, one of the uppermost tectonic elements of the Tauern window. Over thousands of years, the waters of the Großarler Ache (a tributary of the Salzach River) cut into these strongly tectonized rocks. The overhanging walls are up to 300 m high and in places only two metres apart (Fig. 7.7). The gorge was developed by the Pongau Alpine Association in 1875 and declared a Natural Monument in 1942 (Ibetsberger and Hilberg 2017). Nowadays, the natural monument attracts more than 200 000 visitors per year.

In order to mirror some of the other topics of the 612 abiotic natural monuments in Austria, the characteristics of two further sites are outlined in the following.

The **Einödhöhle** (solitude cave) is situated on the southern slopes of the peak Pfaffstättner Kogel north of Baden in Lower Austria (Fig. 7.1). The natural monument is a fossil sea cave that formed at the western edge of the Vienna basin during the Neogene. Developed in dolomites, it has two larger underground halls of 4 m height. The floor is covered by fine-grained sediments and cave debris. Figure 7.8 shows the two entrances, which are connected by a circular cave trail. In 1925, the Einödhöhle was developed as a show cave by the engineering battalion Klosterneuburg. Even electric light was installed, not least because of the high visitor frequency, which in the late 1920s reached more than 30,000 people per year. During World War II, it became a hiding place for the residents of the surrounding villages. In 1941, the cave was declared a Natural Monument. Today it is freely accessible and, equipped with a torch, it can be visited at one's own risk.

Another famous natural monument is the **Ottnangium** in Upper Austria (Fig. 7.1). This is the reference locality of a regional geological stage (c. 18.1–17.2 Ma) within the Neogene development of the Paratethys. A typical sediment of the Early Ottnangian is the so-called Schlier (an

Fig. 7.7 Natural monument Liechtenstein Gorge fascinates with its narrow passages and potholes. *Photo* M. M. Häupl

argillaceous marl). It was deposited in the flexural Molasse Basin of the Paratethys sea, which at that time was still connected to the Atlantic and therefore of marine character. The Ottnangian formation is famous for its biodiversity. In the Schlier outcrops of the natural monument, many fossils were preserved, either as trace fossils, shells or internal casts. Shells of mussels and snails are especially frequent. The place name "Ottnang" not only found its way into the geological timescale, but also appears in the names of many fossils, originally found here. Examples are the mussel "Thyasira ottnangiensis", the snail "Calliostoma ottnangiensis" or the benthic foraminiferes "Amphicoryna ottnangensis" and "Sigmoilopsis ottnangensis". In 1989, the site was declared a natural monument, and in 2014 new scientific information panels were installed.

7.2.4 Geoparks

Austria has three UNESCO Global Geoparks: Styrian Eisenwurzen (Styria), Ore of the Alps (Salzburg) and Karawanken (Carinthia and Slovenia), and one national geopark, the Carnic Alps (Carinthia) (Hejl et al. 2017, 2018).

Styrian Eisenwurzen UNESCO Global Geopark: The nature and geopark Styrian Eisenwurzen is located in the northern part of Upper Styria at its borders with the provinces of Lower- and Upper Austria. The park covers an area of 586 km^2 and spreads over the municipalities of Altenmarkt, Landl, St. Gallen and Wildalpen.

Fig. 7.8 Natural monument and former show cave Einödhöhle, located near Baden in the province of Lower Austria, is a wave-cut cave. *Photo* H. J. Ibetsberger

The nature and geopark Styrian Eisenwurzen is located entirely within the region of the Northern Limestone Alps with their Mesozoic rocks. One part of the Geo-Centre in Gams is dedicated to the sediments and fossils of the Cretaceous and the following Palaeogene period in and around Gams (municipality of Landl). An approximately one centimetre thick, dark layer on one rock marks the end of the Mesozoic. This layer has its origin in volcanic eruptions and shows traces of a large meteorite. It has not only been found in Gams, but in many other places worldwide. Scientists use this evidence as a sign marking the extinction of many animals, especially dinosaurs (Gulas and Kollmann 2018).

Gams is generally known as the "GeoVillage" of the park. The village hosts the GeoCentre and the GeoWorkshop and is the starting point of an interesting thematic trail, named "GeoTrail", linking two of the natural monuments of Austria. It is an easily accessible trail of about 5 km with 48 information points that address a number of geomorphological processes and landforms and the geology of the region. Designed in a family-friendly way, the trail includes several interactive stations. Starting at the centre of Gams, it follows the Gams River to the North Gorge, passes the entrance of Kraus Cave and leads back to the village.

With a total length of 350 m, the *Kraus Cave* is one of Central Europe's largest gypsum caves and one of the earliest show caves of Austria. Its uniqueness lies in the common occurrence of countless small and sparkling gypsum crystal dripstones, some of which are still growing in length (Fig. 7.9). Sulphurous thermal water transformed the original limestone into gypsum. These waters are still "escaping" from a spring at the bottom of the gorge, where they are trapped between concrete walls. Franz Kraus established the cave as a show cave, which was later named after him. It was opened on 28 March 1882. The Kraus Cave was the first cave in the world that provided electric lighting, but the technical equipment decayed after a short time.

Another attraction of the geopark is the *Wasserloch Gorge* near the village of Palfau. In this gorge, the waters of a powerful karst spring, named Wasserloch, rush over five huge rock steps down to the Salza Valley 300 m below. The enormous spring cave is the biggest aquiferous cave in Styria. An easily accessible walkway with wooden steps and bridges guides the visitor through the 900 m of the gorge.

Ore of the Alps UNESCO Global Geopark (see Chapter "The Ore of the Alps UNESCO Global Geopark (Salzburg) Geosites and Geotourism"): The UNESCO global geopark with its 212 km^2 is situated 50 km south of Salzburg within a mountainous area and includes the communities Bischofshofen, Mühlbach am Hochkönig, Hüttau and St. Veit im Pongau (Ibetsberger et al. 2018).

It extends from the Northern Calcareous Alps to the Central Alps and includes the Greywacke Zone. The Hochkönig mountain range in the Northern Calcareous Alps has an altitude of 2941 m asl (Fig. 7.10) and includes a permanent glacier, namely the "Übergossene Alm" (covered mountain pasture). The mountainous Greywacke Zone region is characterized by pastures and forests extending up

Fig. 7.9 Small gypsum crystals and stalactites, sparkling in the light. *Photo* S. Leitner

Fig. 7.10 Sunrise at the peak of Hochkönig (2941 m asl) with the "covered mountain pasture" glacier in the Ore of the Alps UNESCO Global Geopark. *Photo* H. J. Ibetsberger

to an altitude of 2000 m asl. The Salzach Valley, which crosses the geopark, is deeply incised into the soft rocks of the Greywacke Zone. The relative relief within the geopark is about 2400 m.

The geopark area has been continuously populated since 5300 BP. During the Bronze Age, the region was one of the most important sites for copper mining in Europe (Feitzinger et al. 2003; Ibetsberger 2015; Eibner 2016). In the Middle Ages, mining activities were extended and included gold, iron, lead and zinc. Since the 1970s all mining activities have ceased, but the mines still exist and some of them were developed to serve as fascinating show mines.

The transboundary **Karawanken UNESCO Global Geopark** is located between the two >2000 m high peaks Petzen and Koschuta. It is characterized by a rich geological diversity at the border region of the Alps and the Dinarides. 14 communities, 9 in Austria and 5 in Slovenia represent the park. It extends over an area of 1067 km^2.

The Karawanken are a relatively young mountain range of the Southern Limestone Alps on the border between Slovenia and Austria. With a total length of 120 km in an east–west direction, the mountain chain is one of the longest ranges in Europe. Geographically and geologically, the Periadriatic fault (PAL) divides it into the higher Western Karawanken and the lower Eastern Karawanken. The PAL is a major line within the plate tectonic evolution of the Austrian Alps, as it separates the Adriatic (Apulian) tectonic plate from the Eurasian plate.

Tectonic movements during the formation of the Karawanken together with Quaternary glaciation, karstification and erosional processes are responsible for the unique relief of the geopark. Its northern part within the Eastern Karawanken is mostly characterized by a hilly and mountainous landscape with valleys along the rivers Drau, Meža, Mislina and Vellach. To the south, along the Austrian/Slovenian border and in Slovenia, the Western Karawanken provide the highest peaks of the park area, namely Obir/Hochobir (2139 m asl), Košuta/Koschuta (2136 m asl) and Peca/Petzen-Kordeževa glava (2126 m asl).

The Karawanken Geopark is centred on geological topics and specializes in two themes, namely mines and caves. In the past, lead and zinc were mined in the Triassic Wetterstein and Partnach Formations of the park area. The highlight of the Slovenian side of the geopark is the show mine in Mežica. In AD 1870, the search for lead and zinc also triggered the discovery of the Obir flowstone cave in Eisenkappel/Austria, which is now the centre of the Austrian side of the geopark. The cave is located at an altitude of 1078 m asl, has a constant temperature of 8 °C and features actively growing stalagmites and stalactites (Poltnig et al. 2018).

A prominent endeavour of the Karawanken UNESCO Global Geopark is the environmental education of children, developed in close cooperation with local schools and kindergartens. Within the "Geopark young researchers" programme, children experience the nature and get to know

Fig. 7.11 Geoadventure for kids. *Photo* Geopark Karawanken Archiv

the connections between geology, geomorphology, fauna and flora. Qualified geopark guides supervise the educational activities, which take place both outdoors and indoors at the visitor centres (Fig. 7.11).

Carnic Alps National Geopark: The former UNESCO global and now national geopark covers an area of some 830 km^2 on both sides of the Gail Valley in the province of Carinthia at the border to Italy. It is a mountainous area with altitudes up to almost 2800 m asl. Since the beginning of the nineteenth century, this mountain area has been famous for its fossiliferous Palaeozoic rock sequence in the Carnic Alps and its Triassic succession in the Gailtal Alps. The latter are separated from the former by the structurally important Periadriatic fault (PAL), marking the different plate tectonic evolution of the Southern and Northern Alps. Both the Variscan and the Alpine orogeny affected the region, resulting in a landscape in which carbonate rocks and shales of different ages occur closely together (Krawanja-Ortner and Schönlaub 2018).

The geopark is equipped with a visitor centre and eight thematic trails on the Austrian side that communicate a rock history spanning almost 500 million years. Hikers can admire hundreds of metres high vertical cliffs, waterfalls, gorges (two of the gorges are described in Chapter "Gorges and Slots: Three Examples of Geomorphosites in Western Carinthia"), mountain lakes, fossilized remains of marine animals and imprints of land ferns and even silicified trees.

One of the highlights of the park is, for instance, the lake Wolayer Geotrail. In the surroundings of lake Wolayer, the oldest megafossil-bearing rocks of the whole Alpine chain occur. The rocks and their fossils developed in an ancient sea that embraced great parts of Europe during the earlier part of the Palaeozoic era until the Variscan orogeny. Brought together by later tectonic movements, shallow and deep realms of this ocean are today in juxtaposition in the lake Wolayer area, which makes it one of the 100 geologically most important regions of the world (Fig. 7.12). From a geomorphological point of view stop 6 (lake Wolayer, 1960 m asl) provides an overview of a perfect cirque landscape. The basin of the cirque lake is 14 m deep. Pollen found at the bottom of it indicates that it became ice-free 10 000 years ago. The lake is fed by underground springs from the surrounding fans and limestone cliffs.

7.2.5 Thematic Trails

The first thematic trails in Austria were installed in the mid 1960s. They mainly focused on biology and forests (Eder and Arnberger 2007). Today the variety of thematic paths is boundless. Resl (2017) carried out a survey of Austrian "Geotrails" that deal with the abiotic aspects of nature. He listed about 150 trails, of which 37 are mainly devoted to geomorphological topics, a further 34 mainly to geological topics, while the remaining 79 provide interdisciplinary information including biological or cultural topics.

The high number of interdisciplinary trails is no surprise, as much of Austria's territory is a cultural landscape. It encompasses even the high altitudinal levels of the Alps, where the first Alpine inhabitants lived, as the densely wooded valley bottoms were threatened by flooding and

Fig. 7.12 Lake Wolayer is located in a deep cirque that Pleistocene glaciers carved into the Seewarte mountain (2595 m asl). The middle part of the cirque backwall contains a reef core that developed in a shallow part of the Devonian sea. *Photo* H. P. Schönlaub

rock avalanches, and carried malaria. Precious ores were another attraction for the prehistoric settlement of the Alps. Important sites of archaeology, early mining, and local cultural or biological significance are therefore omnipresent, and frequently addressed in the stops of educational trails.

Most of the trails primarily devoted to geomorphological or geological topics focus on one special theme. Thus, geomorphological trails in Austria are clearly dominated by information on glacial processes, glacial landforms and Quaternary development (35%) (Weingartner 2007). Around 10% of the geomorphological trails focus on either gorge formation, periglacial forms (especially rock glaciers) or mass movements (especially "Bergstürze"). At the end of the list, there are three more topics, each addressed on two geotrails, namely fluvial processes, karst phenomena and hoodoos in unconsolidated tills. The preferred geomorphological topics of the 79 interdisciplinary geotrails mirror the situation described above: glacial processes and landforms head the list, followed by information on gorge development (cf. Resl 2017, table in the annex).

In the following, two trails are put forward: the first one belongs to the group of specialized geomorphological trails, and the second one to the group of interdisciplinary trails.

Ice-Age Trail Henndorf: The educational trail in the Alpine foreland of Salzburg province (Fig. 7.1) was installed in 1999 as the first nature trail that primarily focused on the traces of Quaternary glaciation (Ibetsberger and Häupl 2013). With the help of panoramic illustrations on panels, the visitor immediately recognizes the manifold glacial forms in the surrounding nature. In 2020, the trail was extended to include children's adventure stations, which were created especially for the purpose of environmental education (www.eiszeitrundweg.at; www.iceaetsch.at).

Calcareous Tuff Trail Lingenau: This 2.5-km-long trail explores and explains a unique natural monument in Austria's westernmost province of Vorarlberg (Fig. 7.1). The top attraction of this site are scenic sinter formations (Fig. 7.13). The five information panels of the trail provide comprehensive information on the formation of calcareous tuff and sinter deposits, but also address neighbouring topics, as for instance the interaction of vegetation and lime-loaded waters, or the historical use of the deposit as a construction material.

7.3 Management and Challenges of Geotourism

A pervasive management problem are the maintenance costs of repairing paths, signposts, information boards and adventure stations. Rock fall, weathering, snow pressure, falling trees, economic activities (e.g. lumbering), not to forget acts of vandalism, will damage the infrastructure that therefore needs to be repaired each year. Respective funds from the federal or provincial government are only available for officially protected areas (national parks, nature parks,

Fig. 7.13 Detail shot of the sinter cascade at Lingenau. Photo: Gemeinde Lingenau

nature conservation areas, landscape protection areas, natural monuments). In addition, the Alpine Club covers the maintenance costs for a few high Alpine trails. Looking at Resl's compilation (Resl, 2017, supplement 2), it becomes clear that only 35% of Austria's thematic trails can rely on the availability of maintenance funds.

This means that a large number of thematic trails, established by communities and tourism associations in the past are in jeopardy. Many of them generated the initial expenditure of around 50 000 Euros for installing the thematic trail with the help of earlier versions of European Union programmes like LEADER or INTERREG. On the opening day of a newly installed thematic trail, a wise and foreseeing major summarized the future challenges with the words "decay begins today!". Maintenance costs will arise that need to be covered by the community. In addition, local authorities will certainly change over time and responsibility for the trail might be handed over to a person, who considers trail maintenance as a minor matter within his/her duties. Under such circumstances, the trail could fall into disrepair. Professional builders of educational trails in Austria are therefore aiming to include annual services into the original contract of establishing a thematic trail.

Problems arising from insufficient funding can trigger developments that clearly harm the ideas of geotourism and geoheritage. For communities without access to state or province funding, the proposals of large investors, such as cable car owners, for financing regional tourism development by using local geoheritage as a backstage, is almost irresistible (Goroncy 2005; Kief 2008). Therefore, Austria is by no way free of aberrations, as demonstrated by the following two examples.

The motto of the Triassic Park near Waidring (Tyrol) is the Triassic age of the rock, together with the flora and fauna, especially the dinosaurs, of that geological period (https://www.triassicpark.at/en/). Figure 14 suggests that there is more scope for adventure than for scientific education in this park.

Another example is the natural monument Dachstein giant ice cave in Obertraun (Upper Austria). Due to its impressive ice formations, the cave attracts up to 180,000 visitors annually. Recently, a light and music show was installed in the cave. It features the cave bears "Ben and Boris" announcing supreme adventure lying ahead for the takers of the guided tour. The only connection of this spectacle with reality is the discovery of cave bear bones within the cave.

As geotourism is a knowledge-based tourism, best practice in education is another challenge for the management. The original principle of conceptualising a thematic trail or visitor centre was "learning by reading". This initial paradigm was replaced by the guiding principle "learning by doing", and subsequently by "learning through sensory experience". The required, sometimes elaborate stations for these modern approaches are, however, susceptible to failure. Especially in outdoor locations, simple, solid structures that can be put back into operation by minor repairs are preferable. Another issue is the design of information panels.

Fig. 7.14 Triassic Park near Waidring attracts families and children. *Photo* H. J. Ibetsberger

Fig. 7.15 Info panels should be a part of the whole and not impair the harmony of a landscape. *Photo* H. J. Ibetsberger

These should be discreet and should not disturb the scenic value of a landscape (Fig. 7.15). In Austria, the Alpine Club allows only ground-level markings on the high Alpine thematic trails under its management (Essl 2008).

Quality interpretation affords the involvement of Earth scientists, practitioners in geological heritage and a cooperation with respective university research programmes. Mao et al. (2009) conducted a questionnaire survey of potential geotourists and found that they are very discerning. They seek accurate and scientifically based information. Another point emerging from this study was that they wish for comprehensive website information, in order to study it in advance.

7.4 Conclusions

Nature protection in Austria encompasses a wide range of abiotic features. The formation of special landforms and scenic landscapes, the value of Earth materials, (e.g. rocks, minerals, fossils), together with the major tectonic processes acting in the past and in the present are explained at several natural monuments, in six national parks, 47 nature parks, three UNESCO global and one national geopark. Resulting from the presence of people, using the rich natural resources of the Alps since prehistoric times, Austria's abiotic geoheritage sites also include many important archaeological and old mining sites.

Thematic trails, exhibitions and geo-activities in visitor centres provide Earth education and geo-interpretation. Best practice in education is attempted, but sometimes compromises are necessary, as optimal education principles like "learning by doing" or "learning through sensory experience" require elaborate stations that are susceptible to failure and need constant upkeep. However, in order to win over younger visitors to the purposes of bio/geodiversity and nature conservation, it is essential to develop education programmes that include elements of fun and action.

All geoheritage parks and sites need to attract visitors, for two reasons. Firstly, national parks and UNESCO global geoparks are under obligation to educate the lay public in Earth sciences and raise the general awareness for nature conservation. The more visitors they reach, the better their performance. Secondly, many of Austria's geoheritage sites lie within permanently settled areas, where another issue becomes important, namely tourism as a source of income for the local communities. From the inherent need to increase visitor numbers a real management challenge arises, as this issue may lead to negative impacts on the geoheritage itself.

The only answer to this dilemma is to rely on the specific form of geo- and ecotourism with the attendant concepts of nature conservation, education and sustainable economic development of the local communities. The opposite form of tourism is adventure and event tourism (Dowling 2013, Fig. 7). In some cases, adventure parks use geoheritage themes as backdrop for staging an artificial "Walt Disney" world with the only goal of providing fun, action and attracting as many visitors as possible. This does not mean that these two opposite ends of tourism forms are without overlaps. Asian cultures demand some staging, even in geoparks, while some adventure parks also offer suitable Earth education programmes.

Lastly, the scarcity of funds adds to the management problems. Once suitable infrastructure is in place, such as a carefully kept path, information panels, signposts, etc., repair works will become necessary on an annual basis. Respective funds from the federal or provincial government are available for national parks, nature parks and natural monuments. Geoparks and communities that previously managed to set up a geoheritage trail with EU funds are, however, left out of the system.

References

Allmrodt N, Brauner S, Reyer S, Röhling HG (eds) (2015) GeoTop 2015. Geoparks – Brückenschlag zwischen Wissenschafts-Anspruch und Tourismus-Effekten. Stuttgart

Dowling RK (2009) The growth of global geotourism. In: Neto de Carvalho C, Rodrigues J (eds) New challenges with geotourism. Proceedings of the VIII. European Geoparks Conference Idanha-a-Nova, 14–16 September 2009 (Portugal), pp 24–30

Eder R, Arnberger A (2007) Lehrpfade - Natur und Kultur auf dem Weg. Grüne Reihe des Lebensministeriums 18:1–263

Eibner C (2016) The copper ore of Mitterberg. Eur Geoparks Mag 13:53

Essl J (2008) Freizeitsport im Einklang mit der Natur – Erfolgreiche Besucherlenkungs- und Besucherinformationsprojekte. Essl J (ed) Good Practices der Besucherlenkung im Alpintourismus. Alpine Raumordnung Nr. 34 / Fachbeiträge des Österreichischen Alpenvereins: 47–52 Innsbruck

Feitzinger G, Ibetsberger H, Vetters W (2003) Bergbau und Rohstoffe im Land Salzburg. Neumarkt

Goroncy F (2005) Spiel und Geschichten in Erlebniswelten. Tourismus Beiträge zu Wissenschaft und Praxis 6, Lit Verlag Berlin

Gulas O, Kollmann H (2018) The nature and geopark Styrian Eisenwurzen. In: Hejl E, Ibetsberger H, Steyrer H (eds) UNESCO Geoparks in Austria. Natur- u. Kulturerlebnisführer der Universität Salzburg, vol 5, pp 137–173, Pfeil Verlag, München

Hejl E, Ibetsberger H, Steyrer H (eds) (2017) UNESCO Geoparke in Österreich. Pfeil Verlag, München, pp 1–169

Hejl E, Ibetsberger H, Steyrer H (eds) (2018) UNESCO Geoparks in Austria. Pfeil Verlag, München, pp 1–177

Hess V, Rascher J, Zellmer H (eds) GeoTop 2016 Kultur.Wert.Stein Verantwortung und Chancen für Geoparks. Stuttgart

Hofmann T (2000) GAIAs Sterne. Ausflüge in die geologische Vergangenheit Österreichs. Grüne Reihe des Bundesministeriums für Umwelt, Jugend und Familie 12, Graz

Ibetsberger H (2015) Copper so close to the heavens. Eur Geoparks Mag 12:52

Ibetsberger H (2020) Together for our future - climate protection is environmental protection. Eur Geoparks Mag 17:27

Ibetsberger H, Häupl M (2013) Der Henndorfer Eiszeit-Rundweg - Geoedutainment pur. In: Angetter D, Hubmann B, Seidl J (eds) Geologie und Bildungswesen 12. Berichte der Geologischen Bundesanstalt, vol 103, pp 63–66, Wien

Ibetsberger H, Hilberg S (2017) Liechtensteinklamm - deep gorge and mystic springs. Eur Geoparks Mag 14:48

Ibetsberger H, Feitzinger G, Steyrer H (2018) UNESCO-Geopark Ore of the Alps. In: Hejl E, Ibetsberger H, Steyrer H (eds) UNESCO Geoparks in Austria. Natur- u. Kulturerlebnisführer der Universität Salzburg, vol 5, pp 8–46, Pfeil Verlag, München

Kief A (2008) Thematisierte Erlebnis- und Konsumwelten. Igel Verlag, Fluch oder Segen für Anbieter und Konsumenten

Krawanja-Ortner G, Schönlaub HP (2018) UNESCO-Geopark Karnische Alpen. In: Hejl E, Ibetsberger H, Steyrer H (eds) UNESCO Geoparks in Austria. Natur- u. Kulturerlebnisführer der Universität Salzburg, vol 5, pp 82–136, Pfeil Verlag, München

Lieb GK, Slupetzky H (2013) Naturkundlicher Führer Gletscherweg Obersulzbachtal. Innsbruck-Graz

Mao I, Robinson AM, Dowling RK (2009) Potential geotourists: an Australian case study. J Tourism 10(1):71–80

Marek H, Neffe E (2013) Naturparke in Österreich. Leopold Stocker Verlag, pp 1–260

Megerle H (ed) (2008) Geotourismus: Innovative Ansätze zur touristischen Inwertsetzung und nachhaltigen Regionalentwicklung. Stuttgart

Newsome D, Dowling RK (2010) Geotourism: the tourism of geology and landscape. Goodfellow Publishers, Oxford

Newsome D, Dowling R (2018) Geoheritage and Geotourism. In: Reynard E, Brilha J (eds) Geoheritage. Elsevier, pp 305–322

Opaschowski HW (1995) Freizeitökonomie: Marketing von Erlebniswelten. Leske + Budrich Opladen

Platzgummer S (2013) Österreichs Nationalparks. Falter Verlag, 1–256 http://www.nationalparksaustria.at/fileadmin/pdf_s/NPA_AP/NP_Guide_Broschuere_web.pdf

Poltnig W, Bedjanič M, Hartmann G, Fajmut Štrucl S, Varch Ch, Rojs L, Weissenbacher A, Vodovnik P, Glaser F, Herlec U and Klaura F (2018) UNESCO-Geopark Karawanken / Karavanke. In: Hejl E, Ibetsberger H, Steyrer H (eds) UNESCO Geoparks in Austria. Natur- u. Kulturerlebnisführer der Universität Salzburg, vol 5, pp 47–81, Pfeil Verlag, München

Resl F (2017) Geomorphologische Lehrpfade in Österreich. Diploma Thesis, University of Vienna

Schlosser H (2006) Gletscherweg Innergschlöß. Naturkundlicher Führer zum Nationalpark Hohe Tauern. Innsbruck

Slupetzky H, Slupetzky N, Wiesenegger H (2011) Gletscher und ihre Spuren beim Stubacher Sonnblick. In: Steyrer HP, Hejl E, Ibetsberger H, Schrott L (eds) (2011) Hohe Tauern. Naturkundliche Exkursionen. Natur- und Kulturerlebnisführer der Universität Salzburg, vol 2, pp 86–103; Pfeil Verlag, München.

Steyrer HP, Hejl E, Ibetsberger H, Schrott L (eds) (2011) Hohe Tauern. Naturkundliche Exkursionen. Natur- und Kulturerlebnisführer der Universität Salzburg, vol 2, pp 1–160, Pfeil Verlag, München.

Stingl H (2004) Geomorphologischer Lehrpfad. Landschaften um die Glorer Hütte. Eichstätt - Innsbruck

Stöllner T, Oeggl K (eds) (2015) Bergauf Bergab - 10.000 Jahre Bergbau in den Alpen. Bochum, pp 1–627

UNESCO Global Geoparks (2017) Accessed 20th May 2020. http://www.unesco.org/new/en/natural-sciences/environment/earth-sciences/unesco-global-geoparks/

Weingartner H (ed) (2007) Lehrpfad Hallstätter Gletscher. Linz, pp 1–127

Horst J. Ibetsberger is a self-employed geoscientist and scientific director of the company GeoGlobe, which specialises in the conception and implementation of geo- and ecotourism projects. He is also scientific head of the Ore of the Alps UNESCO Global Geopark. In addition, he is a lecturer of physical geography at the Department of Geography & Geology, Salzburg University.

Christine Embleton-Hamann is a retired professor at the Department of Geography and Regional Research at the University of Vienna. Her main interest is in alpine environments. Within this field she focusses on human–environment interactions with research topics like human impact on geomorphic processes, assessment of the scenic quality of landscapes, and geomorphological hazards. A second set of interest concerns the communication of geomorphological knowledge to a broader audience, in the pursuit of which objective she has written a well-received textbook on geomorphology. She is a former President of the Austrian Research Association on Geomorphology and served on the Executive Boards of the IAG and several IAG and IGU Working Groups.

Part II

Geomorphic Hotspots of High Scenic Quality and/or High Scientific Interest

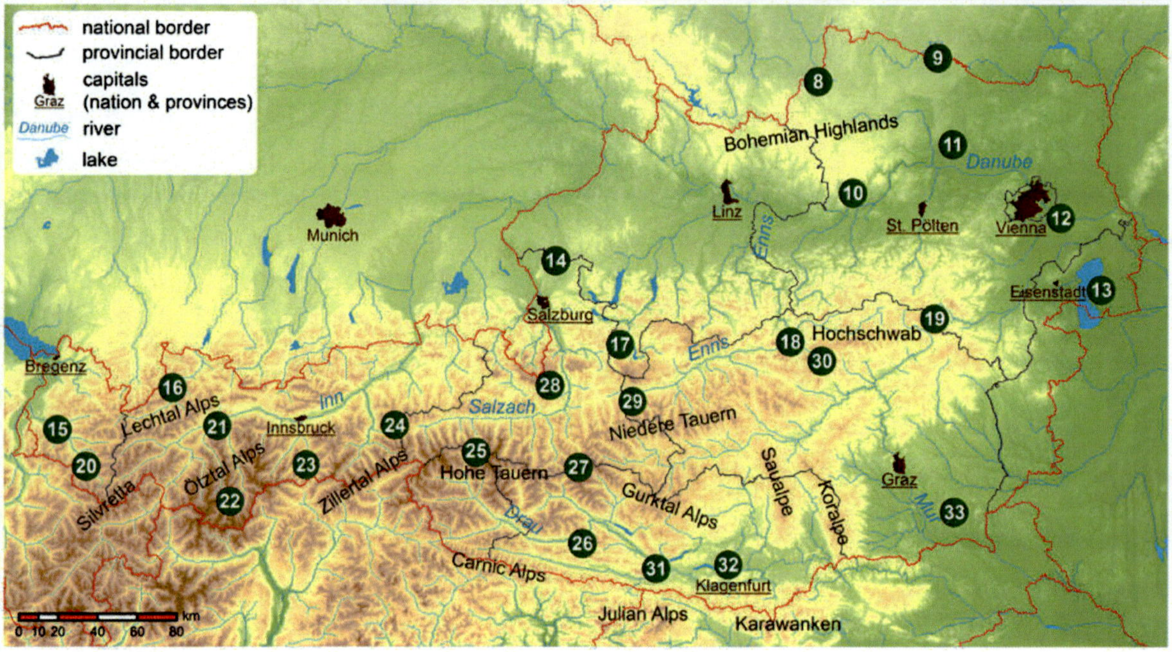

Location or region - main geomorphological focus

- 8 Northwestern Waldviertel - granite tors
- 9 Waldviertel Highlands - deeply incised valley meanders
- 10 Wachau World Heritage Site
- 11 Loess areas of Lower Austria - loess landforms, palaeosols
- 12 Danube Floodplain National Park
- 13 Lake Neusiedl and Seewinkel area – formation of large and small shallow lake basins
- 14 Alpine foreland - landforms and sediments of the Pleistocene Salzach and Traun glaciers
- 15 Walgau region - mass movements
- 16 Tiroler Lech Nature Park - fluvial geomorphology and river restoration
- 17 World Heritage Site Hallstatt-Dachstein / Salzkammergut
- 18 Gesäuse Gorge - landscape development, sediment cascade in the Johnsbach Valley
- 19 Rax massif - karst features and palaeosurfaces
- 20 Montafon region - landscape development, impact of hydropower generation
- 21 Western Tyrol - three „Bergsturz“ landscapes
- 22 Upper Oetz Valley - glacial and periglacial processes and landforms
- 23 Gschnitz Valley - Gschnitz end moraine and Alpine Lateglacial
- 24 Hohe Tauern National Park, Krimml Valley - waterfalls
- 25 Großglockner and Pasterze Glacier - past and future variations of Austria's largest glacier
- 26 Western Carinthia - gorges and slots
- 27 Hohe Tauern National Park, Dösen Valley - rock glaciers
- 28 Ore of the Alps UNESCO Global Geopark
- 29 Schladminger Tauern range - cirques
- 30 Erzberg area - impact of mining on geomorphologic processes, man-made landforms
- 31 Nature Park Dobratsch - landslides and karst
- 32 Klagenfurt basin - landscape development of the region
- 33 Styrian basin - volcano remnants

Location of the study areas in Chaps. 8–33 of the second part

Granite Tors of the Waldviertel Region in Lower Austria

8

Piotr Migoń, Aleksandra Michniewicz, and Milena Różycka

Abstract

The Waldviertel region in the northern part of Lower Austria hosts numerous localities of granite tors which add to the diversity of landscape developed through protracted long-term deep weathering and regolith stripping. They are varied in size and shape. Castellated tors tend to occur at higher elevations, whilst bouldery tors and monolithic boulders, isolated or in clusters, are common within gently rolling upland surfaces. A rich suite of surface microforms includes weathering pits, flared slopes, karren, rills, and pseudobedding. The area of the Kogelsteine near Eggenburg is a unique geosite at the eastern margin of Waldviertel, with the history of deep weathering, stripping, burial, exhumation, and further development in subaerial conditions.

Keywords

Waldviertel • Granite • Tors • Weathering • Exhumed landforms • Geoheritage

8.1 Introduction

The northern part of Austria is geologically and geomorphologically different from the remaining part of the country which belongs to the Alpine range or its forelands. The Mühlviertel region in Upper Austria and the Waldviertel region in Lower Austria are parts of the Bohemian Massif—a vast tectonic unit in the centre of Europe of protracted geological history that can be traced back to the Precambrian and dominated by crystalline basement rocks of igneous and metamorphic origin. In consequence, geomorphic landscapes of the Bohemian Massif are very different from those in other parts of Austria in terms of both landform inventory and the sequence of events which shaped regional relief. In Waldviertel (Fig. 8.1), except for the westernmost mountainous part along the Austrian/Czech border and deeply incised river valleys with entrenched meanders such as Thaya and Kamp canyons (see Chapter 9), the overall morphology is rather subdued due to long-term denudation regime, only minor uplift in the Cenozoic, and spatially restricted presence of active fault lines. However, within this seemingly inconspicuous regional landscape of Lower Austria occurs a fine suite of medium-size and minor landforms, developed upon granites and orthogneisses, which is among the best of its kind in Central Europe. Among them, the most characteristic are tors—upstanding residual rock landforms once defined by Linton (1955, p. 470) as 'solid rock outcrops as big as a house rising abruptly from the smooth and gentle slopes of a rounded summit or broadly convex ridge'. In Waldviertel, tors occur in a considerable variety of shapes and sizes, hosting also a range of smaller weathering features, such as weathering pits, flutes, and flared slopes. In this chapter, a selection of most representative tors and associated minor landforms developed in granitic areas of Waldviertel will be presented and set in a wider context of regional landform evolution. More detailed presentation of selected sites, along with the discussion of related geoconservation and geotourism issues, can be found in Migoń et al (2018).

8.2 Geological and Geomorphological Setting

8.2.1 Basement Geology

The crystalline basement of Waldviertel consists of two main tectonic units, the Moldanubian Superunit in the west and the Moravian Superunit in the east (Suess 1903; Höck

P. Migoń (✉) · A. Michniewicz · M. Różycka
Institute of Geography and Regional Development, University of Wrocław, pl. Uniwersytecki 1, 50-137 Wrocław, Poland
e-mail: piotr.migon@uwr.edu.pl

C. Embleton-Hamann (ed.), *Landscapes and Landforms of Austria*, World Geomorphological Landscapes,
https://doi.org/10.1007/978-3-030-92815-5_8

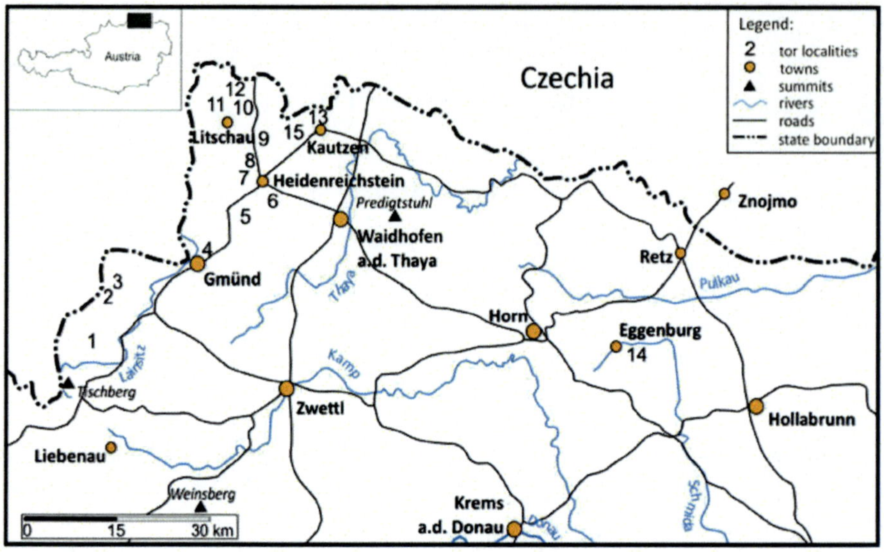

Fig. 8.1 Location map. Tors and other localities mentioned in text are numbered: 1—Nebelstein, 2—Mandlstein, 3—Dreilöcherstein, 4—Blockheide Nature Park, 5—Amaliendorf, 6—Hängender Stein, 7—Geyersteine, 8—Gugelhupfstein, 9—Kolomanistein, 10—Steingarten, 11—Graselstein, 12—Hutstein, 13—Hoher Stein, 14—Kogelsteine, 15—Skorpionstein

1999) (Fig. 8.2). The original ages of parent rocks are mainly Proterozoic/Early Palaeozoic, but these rock complexes were metamorphosed during the Variscan orogeny in the Devonian/Carboniferous. In the Moldanubian Superunit, gneisses dominate, with the subordinate occurrence of mica schist, amphibolite, marble, calcareous-silicate schist, quartzite, and granulite, testifying to medium-to-high grade metamorphism (cf. Schnabel 2002). The Moravian Superunit contains mainly different gneisses, mica schist, calcareous-silicate schist, marble, and granites (cf. Schnabel 2002). Moravian rocks were predominantly affected by low-grade metamorphism. Moravian granites form a large, SSW–NNE stretched body along the contemporary eastern border of the Bohemian Massif, known as the Thaya Batholith. Different variants of medium-grained, two-mica granites prevail. The age of the Thaya granite is about 580–600 Ma (Friedl 1997; Finger and Riegler 1999).

The Moldanubian gneisses and other metamorphic rocks were intruded by much younger granites during the Variscan mountain building in the Carboniferous. They build the large South Bohemian Pluton, which extends over 6000 km^2 and is exposed in both Austrian and Czech territory. Magmatic activity was multi-stage, occurring simultaneously with the orogenic deformations. The dominant granite type, Weinsberg granite, is coarse-grained, with large crystals of alkali feldspar. It shows ages between 330 and 325 Ma in the eastern and northern parts of the pluton, whereas in Mühlviertel, in the west and south-west of the pluton, it is about 322 Ma old (Gerdes et al. 2003; Finger and Schubert 2015). Eisgarn granite is a medium-grained, two-mica granite with

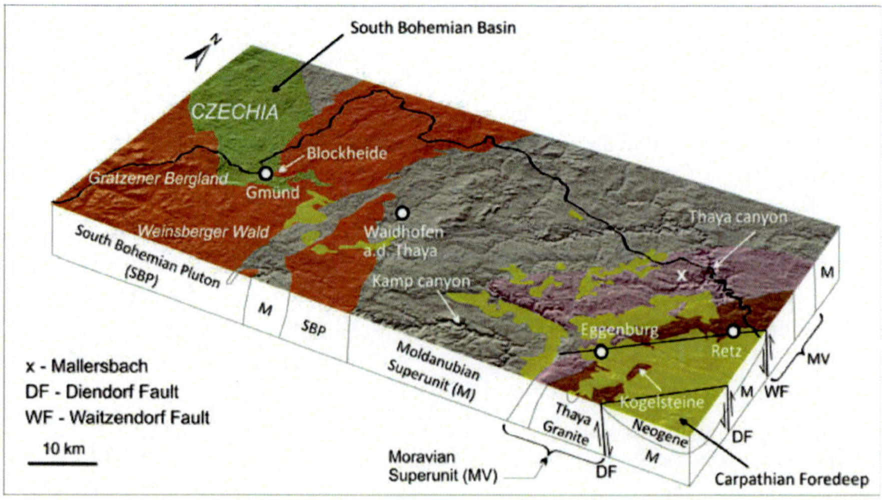

Fig. 8.2 General morphology and geological overview of Waldviertel. Green area in the north-western corner shows the extent of Cretaceous sediments in the South Bohemian Basin. Elevation data to build the model taken from SRTM 90 m dataset. Geology simplified after Höck and Roetzel (1996)

characteristic strip-like alkali feldspars, with an age of about 326–328 Ma (Friedl 1997; Koller 1999; Gerdes et al. 2003). Finally, fine-grained granites and granitoids like the Mauthausen granite or the Freistadt granodiorite intruded discordantly and much later, from 316 to 302 Ma (Gerdes et al. 2003).

Properties of Eisgarn granite clearly favour the origin of tors, which are abundant, whilst they are not so common in Weinsberg granite, except steep slopes in the mountainous terrain. Hence, most examples described in Sect. 3 are from the outcrop area of Eisgarn granite. Deformed granites of Thaya Batholith give rise to tors rather occasionally, but Kogelsteine (Sect. 5) are a notable exception.

8.2.2 General Relief

Waldviertel is best described as an upland characterized by average altitude around 500–600 m asl. The lowest part of the study area is located in the south-east and is connected with the Carpathian Foredeep, extending eastward from the Diendorf Fault and Waitzendorf Fault. The area with undulating morphology above the dissected fault scarp is characterized by isolated hills built of Thaya granite (inselbergs, see Roštinský and Roetzel 2005), up to 50–70 m high. The central part of Waldviertel, built of the metamorphic Moldanubian complex, has more rugged topography and includes a chain of hills located east of Waidhofen a. d. Thaya, which reach 718 m asl at the Predigtstuhl. Additionally, the eastern and central parts are dissected by deep fluvial valleys with incised meanders (see Chapter "2"). The western part of Waldviertel, coinciding with the South Bohemian Pluton, has a bipartite morphology. The area adjacent to the Moldanubian part has gently rolling topography with altitudes reaching 600 m asl, whereas the westernmost and south-western area (Gratzener Bergland (Novohradské hory), Weinsberger Wald) is characterized by mountainous topography with clear crest lines and moderately steep slopes, as well as the highest elevations of 1063 m asl at Tischberg and 1041 m asl at Weinsberg.

8.2.3 Two Contrasting Scenarios of Long-Term Geomorphic Evolution

The geomorphology of the southern part of the Bohemian Massif is a product of long degradation of the Variscan mountain chain, interrupted by several marine transgressions and regressions as well as episodes of relatively minor uplift (compared with other marginal parts of the Bohemian Massif). A regional surface of low relief was formed as early as the Permian and surface lowering continued in the Mesozoic. Deep weathering under warm and humid climate, followed by stripping of loose weathering products, were the main processes involved. Hence, an etchplain with scattered residual hills had formed. Being a surface of low elevation and low relief in the Cretaceous, it was likely subject to large-scale marine inundation or, like in the South Bohemian Basin and next to the town of Gmünd, to fluvial and lacustrine coverage (Huber 2003). In the beginning of the Cenozoic, surface lowering continued, with the significant role played by deep tropical weathering (Roetzel and Steininger 1999). Thick Palaeogene clayey, mainly kaolinitic weathering mantles on gneiss and granite survived in several places in the eastern and southern parts of Waldviertel and Mühlviertel (e.g. Mallersbach: Fig. 8.2, Schwertberg) (Wieden 1964; Jiranek et al. 1990). Moderate uplift of the Bohemian Massif in the Late Neogene and Pleistocene resulted in fluvial incision and the development of several canyon-like valleys in the eastern and southern parts of Waldviertel, but the tor-riddled upland surfaces continued to evolve rather slowly.

The story was more complicated along the eastern margin of the Bohemian Massif, along the boundary with the Carpathian Foredeep. Several marine transgressions and regressions in the Lower to Middle Miocene affected the area where the Thaya granite crops out. Here, geomorphic evolution of basement surfaces involved episodes of submergence, burial by marine sediments, followed by erosion and partial re-exposure (exhumation). The hilly granite morphology between Eggenburg, Retz and Znojmo in the Czech Republic is interpreted as an example of partially exhumed relief (Roštinský and Roetzel 2005).

8.3 Granite Tors and Boulders

8.3.1 Tor Typology

The definition of a tor by Linton (1955) gives no clue to the appearance of such outcrops. In fact, tor shapes are extremely diverse and characteristics of fracture patterns are recognized as the primary controls of what tors look like and explain why they are so different. Castle koppies (castellated tors) tend to be angular and develop where horizontal and vertical joint spacings are similar. If horizontal partings are closely spaced, tors are low but laterally extensive. Nubbins (bouldery tors) are heaps of rounded boulders, occasionally up to the height of 10 m. Pedestal rocks, sometimes evocatively called 'rock mushrooms', consist of a narrow lower part (stem) supporting a much larger cap. Certain outcrops may be singular monolithic compartments, as long as 10 m, hence transitional forms between boulders and tors. Finally, granite outcrops may take the form of a rock cliff, precipitous on one side and merging with the slope surface on the opposite side. Essentially each morphological type of tors is present in Waldviertel (Huber 1999) (Fig. 8.3).

Fig. 8.3 Representative tor types. **a** Castellated tor of Christophorusstein, **b** pedestal rock of Pilzstein, **c** monolithic boulders of Koboldsteine. Examples **a**–**c** come from the Blockheide Nature Park, **d** cliff at Hoher Stein in the vicinity of Kautzen

8.3.2 Castle Koppies

The best examples of castle koppies can be found in the most elevated, western part of the area, i.e. in Gratzener Bergland, but also in Weinsberger Wald. Among them is the summit tor of Mandlstein about 7 km north-west of Weitra—a large outcrop 95 m long, up to 15 m wide and up to 17 m high, with the top platform and vertical walls below (Huber 1999). It is dissected by subhorizontal and vertical joints. Some of the latter have been widened, resulting in the subdivision of the tor into three parts and separation of high pulpit rocks on its north-western side. Blocky talus is widespread around the tor. A rare feature is a natural rock tunnel, 1.5 m high, on the western side, developed through selective weathering acting along a vertical fracture. Another castle koppie crowns the summit of Nebelstein, which offers extensive views of the surrounding countryside.

Among many granite residuals in the Blockheide Nature Park near Gmünd is Christophorusstein, whose castellated outline rises above a gentle slope (Fig. 8.3a). It is more than 20 m long and 7 m tall, with a large group of weathering pits on the top platform. As within Mandlstein, some vertical joints have been widened and form narrow clefts. Notable, although much less known, is Graselstein about 2.8 km north of the town of Litschau. This tor is 5 m high, 20 m long, and is built of horizontal granitic plates with a thickness of 1 m. It is dissected by a wide cleft that passes into a tunnel under overhanging granite slabs.

8.3.3 Pedestal and Balanced Rocks

Pilzstein in the Blockheide Nature Park east of Gmünd constitutes one of the best examples of a 'mushroom' rock in Waldviertel. The tor reaches the height of 4.5 m and consists of two spherically shaped, massive blocks with the larger one (c. 40 m^2, 2 m thick) resting on the smaller one. The cap blocks are superimposed upon a thin and densely fractured stem with subhorizontal joint surfaces (Fig. 8.3b). A much wider stem relative to the cap width may be observed at Hutstein, which is a singular, 5 m high tor to the east of the village of Haugschlag, in the northernmost part of the region. A group of similarly shaped tors, up to 7 m in height and referred to as Geyersteine, rises from a valleyside of a small creek 4.1 km west from the town of Heidenreichstein. Their caps formed a continuous massive slab more than 2.5 m thick, now dissected by narrow fissures.

Balanced rocks (German: Wackelsteine) occur in a number of places in Waldviertel (Chábera and Huber 1995). Their characteristic feature is the presence of a large boulder which shows (or showed) ability to move if forced to do so. Several examples of such rocks can be found in the Blockheide Nature Park. The largest one, Wackelstein I, is 9.3 m long and 2.5 m high and located close to Christophorusstein, on a thematic granite trail. Another often visited example is a balanced rock (also named Wackelstein) within the cluster of large granite boulders located 0.8 km south-west from the village of Amaliendorf. This one is 3 m high and 5 m long, and rests on a flat pedestal. Its bottom part is convex downward, which results in its instability.

8.3.4 Monoliths and Boulder Clusters

Granite monoliths are abundant in the vicinity of Haugschlag village in the northernmost part of Waldviertel. They attain a length of 8 m and height of 3.5 m, are typically well rounded, and may have slightly undercut basal parts. In the same area, many convex rock platforms, barely protruding from the regolith-covered surface, are visible. The association of monoliths and platforms suggests that granite residuals have been exhumed from weathering mantles and the lower parts of granitic corestones are still in the subsurface. How separation of a rock mass into loose granite grus and unweathered compartments occurs can be seen in near-surface parts of some quarries (Fig. 8.4). Similar rock formations are present to the north-east of the town of Litschau, in the area aptly named Steingarten (Stone Garden). In the Blockheide Nature Park, the best examples of large granite monoliths are Teufelsbett (Devil's Bed) and Teufelsbrotlaib (Devil's Loaf of Bread), located c. 20 m from each other. The larger, elongated Teufelsbett is 4.2 m high and 10.6 m long, whereas Teufelsbrotlaib is more ball-shaped and reaches 4.2 m in height and 6.2 m in length.

Boulder clusters may be observed in many places in western Waldviertel, including the Blockheide Nature Park and the surroundings of Amaliendorf, c. 4 km north of the town of Schrems. Koboldsteine located in the north-western section of the Blockheide Nature Park represents a row of well-rounded monolithic boulders along a local N–S stretching terrain convexity, across the distance of approximately 50 m. Individual compartments are up to 5 m high and are separated by vertical fractures which in places turn into open fissures up to 2 m wide (Fig. 8.3c). Near Amaliendorf a cluster of large monolithic blocks is present in an open forest. The most popular among them is Wackelstein (see Sect. 3.3), but next to it other massive compartments, some as long as 8 m, with minor surface features such as weathering pits and flared slopes, can be seen.

8.3.5 Cliffs

Good examples of this type of tors are Gugelhupfstein in north-western Waldviertel, not far from the town of Heidenreichstein, and cliffs on Hoher Stein, in the vicinity of Kautzen. The former, located close to the road to Eisgarn, is c. 55 m long and represents a rocky section of the Räubersgraben Valley side. Its maximum height is 6 m, but in the northern part a 3 m high anvil-shaped pedestal rock rises above the upper rock platform. The main cliff, originated presumably from fluvial undercutting, is densely fractured and broken into large blocky compartments.

Below the summit of Hoher Stein, 2.8 km north-west of Kautzen, two adjacent tors, c. 30 m away from each other, emerge from the slope. The eastern cliff is 32 m long and up to 8 m high (Fig. 8.3d), the western one is 13 m and 7 m, respectively. They have been described as examples of frost-riven cliffs (Chábera and Huber 1999). Other rock cliffs interrupt smooth slopes below the summit of Nebelstein (1017 m) in Gratzener Bergland. In each case, dispersed blocky talus occurs in downslope direction from the cliffs.

8.4 Minor Surface Features

8.4.1 Flared Slopes

Flared slopes are basal concavities present at monolithic granite outcrops. They may be low and shallow (less than 0.5 m), but may also grow to gigantic dimensions of 10 m high and more (Twidale 1962). Their interpretation holds that they are former parts of a subsurface weathering front, which have become exposed after regolith was removed by erosion.

Fig. 8.4 Deep weathering of diorite in a quarry in Gebharts near Schrems. **a** Weathered zone with cores of unweathered diorite above solid rock, **b** onion-skin weathering and corestone development (photos courtesy of Reinhard Roetzel)

In Waldviertel, flared slopes are fairly common and provide evidence that tors and boulders have been excavated from grus weathering mantles (Huber and Chábera 1994). Best examples can be found in the Blockheide Nature Park next to Pilzstein tor, where two levels of concavities can be found on large monoliths, in the forest west of Amaliendorf, and in Steingarten north-east of Litschau (Fig. 8.5).

8.4.2 Weathering Pits

Weathering pits occur in many places, but two localities deserve particular attention. One is Christophorusstein in the Blockheide Nature Park (see Sect. 3.2), whose upper, nearly flat surface is dotted with pits of various dimensions. The largest one is 1.8 m long, 1 m wide and 0.3 m deep, but there are ten weathering pits altogether, including a group of four interconnected hollows, some holding water and some dry (Fig. 8.6a). An unusual cluster of weathering pits can be observed at Dreilöcherstein, near the village of Heinrichs bei Weitra. Three pits, 0.5–0.6 m across and up to 0.4 m deep, occur on a vertical surface of a boulder, indicating its 90° rotation when it slid down from an underlying pedestal (Fig. 8.6b). A large water-filled pit, 1.4 m across and 0.5 m deep, is present on a large boulder nearby.

Fig. 8.5 Flared slopes. **a** regular continuous basal undercutting at a solitary boulder in Steingarten near Schandachen, NE of Litschau, **b** double flared slope next to Pilzstein tor, Blockheide Nature Park

8.4.3 Karren and Rills

Several tors in Waldviertel host impressive karren and rill networks (Huber and Chábera 1993; Huber 1999). Gugelhupfstein is the best example, with more than 60 karren up to 3 m long, 50 cm wide and 28 cm deep (Fig. 8.7). Most of them are heavily overgrown by moss and lichen, and it is likely that biological weathering is a significant contributor to karren enlargement. Further localities are Teufelstein in the Steingarten area north-east of Litschau and Hutstein in Haugschlag, the latter typified by particularly deep karren, up to 56 cm into the rock surface. A fine dendritic pattern of rills has developed on the upper, sloping surface of Hängender Stein, south-east of Heidenreichstein, with the longest one being more than 4 m long.

8.4.4 Pseudobedding

Closely spaced (<0.5 m) (sub)horizontal partings in granites resemble bedding, typical for sedimentary rocks, but since they have not originated through deposition, the term 'pseudobedding' is applied. There are diverse views on their origin, but usually unloading of the rock mass and resultant stress release are implied. Partings are often parallel to the topographic surface, providing additional evidence of unloading. In Waldviertel, they are particularly clear in regularly jointed rock masses, whilst rounded granite monoliths hardly show similar features. Summit tors of Nebelstein and Mandlstein in Gratzener Bergland are good examples of these secondary discontinuities (cf. Chábera and Huber 1998).

8.5 Kogelsteine—Unique Geosite at the Eastern Margin of the Bohemian Massif

Kogelsteine (Rocks on Rounded Hilltop) is the name given to the largest and arguably most interesting assemblage of natural granite outcrops in the Austrian part of Thaya Batholith. It is located c. 3 km east of the town of Eggenburg, in the transitional zone between the Bohemian Massif and Carpathian Foredeep. Subject to marine transgressions in the Early Miocene, the area now consists of low granite elevations rising island-like from a flat terrain underlain by Lower Miocene sediments with Pleistocene loess on top (Figs. 2 and 8a) (Roštinský and Roetzel 2005). The summit of Kogelsteine rises to 336 m asl, whilst the relative height is about 25 m. This is mainly a treeless terrain, with scattered bushes and grassy surfaces surrounded by vineyards, hence of considerable appeal and scenic quality. Actually, these grasslands, representing a dry variety of Pannonian steppes developed on silica-rich substratum (*Silikattrockenrasen*),

Fig. 8.6 Weathering pits. **a** A large group on the top platform of Christophorusstein, Blockheide Nature Park, **b** displaced boulder with three pits at Dreilöcherstein near Heinrichs bei Weitra

Fig. 8.7 Overgrown karren at Gugelhupfstein tor

are of considerable botanical value, being home to rich biodiversity and hosting a range of rare and endangered species. The site is protected as a nature reserve (since 2009) and to sustain the existence of grasslands, periodic clearance of trees is organized which helps to keep granite outcrops exposed. In this way, geodiversity not only underpins biodiversity but benefits from biological conservation.

The area can be divided into three parts: the main central elevation, a minor elongated hill to the south-west, and the northern elevation of Fehhaube, named after an angular summit tor (Migoń et al. 2018). Kogelsteine are mostly rounded boulders, apparently in situ, scattered on gently inclined slopes (Fig. 8.8b). More angular shapes are occasionally present such as Wächter (Guardian), a 6-m-tall granite tower rising from the slope. Another angular exposure crowns the summit. Besides, low granite platforms and low steps (1–2 m high) occur on slopes, the latter clearly following joint faces. Boulders are either fully exposed or still partially embedded in grus weathering mantle. Some attain huge dimensions such as those on the top of Fehhaube which are up to 5 m long.

Some individual outcrops deserve special attention. On the south-facing slope of the main hill, the so-called Schwammerl (Mushroom), a 1.5 m high residual, resembles an anvil and shows deep basal undercutting, being an excellent example of a flared slope (Fig. 8.8c). In the southern part of Fehhaube hill, a large boulder 5.5 m long and 2.5 thick lies on the ground adjacent to a rounded pedestal, suggesting tor collapse in the past (Fig. 8.8d). A further feature of interest is a split boulder on the southern elevation. In the Wächter group, a few weathering pits occur, with the biggest one being 40 cm long.

Landforms and local geological context assist in inferring the origin and evolution of Kogelsteine. The hilly morphology on the Thaya Batholith is best explained as an etchsurface, developed due to selective deep tropical weathering and subsequent stripping of weathering products, in principle, prior to the Early Miocene marine transgression whose sediments buried and sealed the topographic surface again (Steininger and Roetzel 1999; Roštinský and Roetzel 2005). Later on, in response to regional uplift, these sediments were removed from the hilltops, but have remained in the topographic lows. Thus, hills were partially exhumed, and it is very likely that some outcrops and big loose boulders of Kogelsteine belong to the pre-Miocene landscape. Nevertheless, deep weathering likely continued, producing grus mantles, probably a few metres thick. Removal of grus exposed basal concavities on some boulders, including Schwammerl. At the same time, previously exposed granite tors were subject to mechanical disintegration and acquired angular shapes, particularly visible in the summit part of Kogelsteine. Thus, granite rocks of Kogelsteine are witnesses of complex geomorphic history involving deep weathering, stripping, burial, re-exposure, and surface weathering, spanning the period from the Eocene to the present.

8.6 Cultural Granite Landscape

There is a long history of association of people with granite landforms of Waldviertel. It is assumed that already in prehistoric time people were attracted by some remarkable granitic boulders and tors, giving them spiritual significance.

Fig. 8.8 Scenes from Kogelsteine outcrop near Eggenburg. **a** View from Kogelsteine to the north. Hills are built of Thaya granite, whereas intervening valleys and flatlands are underlain by Lower Miocene marine nearshore deposits, **b** tors and scattered boulders at the main elevation of Kogelsteine. 6 m tall Wächter on the right, **c** flared side of the Schwammerl boulder, **d** this elongated boulder near Fehhaube has probably collapsed from a pedestal seen to the left

In Christian times, surface weathering features such as weathering pits were interpreted as footprints of Jesus and saints, or pools where Saint Mary was bathing baby Jesus. On Kolomanistein near Eisgarn, a chapel was erected in the early eighteenth century on top of a granite residual, to commemorate an event from the life of St. Koloman, the former saint patron of Lower Austria. Numerous strangely shaped tors and boulders bear names related to the devil (German: Teufel), especially those with parallel karren, interpreted as scratches from devil's claws—a clear indication of ancient local beliefs in supernatural phenomena. Boulder assemblages themselves were the subject of folk tales too. For example, clustered boulders in the Blockheide area were told to roll down from a broken overloaded sack, into which the Creator collected them from the land. Even today, some granite tors are considered mystical places, such as Skorpionstein near Kautzen, famous for its numerous weathering pits and side hollows.

More recent history provides numerous examples of use of granite stones as building materials. Before mining in large quarries commenced, individual boulders were utilized as a source of material and many bear traces of exploitation. The so-called Lutherische Kirche in the Blockheide Nature Park was the name of a huge boulder that completely fell victim to industrial demand at the end of the nineteenth century (cf. http://www.vergangenes.gmuend.at/page.asp/-/193.htm). Various uses of granite and changing methods of exploitation through time are now presented along a thematic GranitKulTour trail in the Blockheide Nature Park, at nine stops enriched with interpretation panels and exhibits. In the same area, an open-air geological exhibition shows various types of granite present in Waldviertel.

8.7 Final Remarks

In terms of landform inventory and evolution of its geomorphological landscape, Waldviertel is an unusual part of Austria. Whilst its granite landscape in general may seem dull and monotonous, it is actually home to numerous distinctive tors and boulders which are important components of both natural and cultural heritage of the region. Tors occur in a variety of shapes, from 'classic' castellated, angular formations crowning hilltops, to apparently chaotic piles of rounded boulders, often of gigantic dimensions. Strangely shaped formations such as 'mushroom rocks', 'anvil rocks', and balanced stones are present at various localities throughout the region. Tors of Waldviertel record complex geomorphic history and landform evolution in changing environments, spanning the period since at least the Eocene. Whilst abundant flared slopes provide evidence of two- or multiphase excavation from weathering mantles, further development occurred in subaerial conditions, including cold environments of the Pleistocene. In addition, along the eastern margin of Waldviertel, burial by Lower Miocene marine sediments and subsequent exhumation of tors were involved.

From the geotourist point of view, good access to nearly all sites mentioned in this chapter (see Migoń et al. 2018 for details) is advantageous. Numerous locations have been developed for tourism, including trails and information panels, although the latter tell rather little about geological and geomorphological history. The Blockheide Nature Park near Gmünd is a popular outdoor recreational area, with thematic trails focused on granite use and biodiversity, but geomorphology is sadly a missing theme. However, this exciting granite landscape remains under-researched, and our knowledge about its long-term evolution is still rather sketchy, especially regarding the western part where Miocene sediments are missing.

Acknowledgements PM wishes to thank Karl-Heinrich Huber for generous supply of papers describing granite localities in the region and Reinhard Roetzel for introduction to sites along the eastern border of Waldviertel during a memorable field trip in 2002, literature advice, photo sharing, and corrections made to the early version of this chapter. Suggestions from Marie-Françoise André are also gratefully acknowledged. Last but not least, we thank the editor, Christine Embleton-Hamann, for invitation to contribute to this volume and help in gathering necessary cartographic materials.

References

Chábera S, Huber KH (1995) Pilzfelsen und Wackelsteine in Granitoiden des Südböhmischen Plutons. Sbor Jihočes muz v Čes Budějovicích. Přír Vědy 35:5–20

Chábera S, Huber KH (1998) Pseudoschichtung (pseudobedding) in Granitoiden des Südböhmischen Plutons. Sbor Jihočes muz v Čes Budějovicích. Přír Vědy 38:5–17

Chábera S, Huber KH (1999) Beispiele kryogener Verwitterungs- und Abtragungsformen im Eisgarner Granit. Sbor Jihočes muz v Čes Budějovicích. Přír Vědy 39:5–17

Finger F, Riegler G (1999) Der Thayabatholith und der kristalline Untergrund des Weinviertels. In: Roetzel R (ed) Arbeitstagung der Geologischen Bundesanstalt 1999, 3.-7. Mai 1999, Retz, Geologische Bundesanstalt Wien, pp 23–31

Finger F, Schubert G (2015) Die Böhmische Masse in Österreich: was gibt es Neues? Abh Geol Bundesanst 64:167–179

Friedl G (1997) U/Pb Datierungen an Zirkonen und Monaziten aus Gesteinen vom österreichischen Anteil der Böhmischen Masse. Dissertation University Salzburg

Gerdes A, Friedl G, Parrish RR, Finger F (2003) High-resolution geochronology of Variscan granite emplacement—the South Bohemian Batholith. Journ Czech Geol Soc 48:53–54

Höck V (1999) Der geologische Bau des Grundgebirges. In: Steininger FF (ed) Erdgeschichte des Waldviertels, 2nd ed. Waldviertler Heimatbund, Horn – Waidhofen/Thaya, pp 37–60

Höck V, Roetzel R (1996) Geologische Übersichtskarte des Waldviertels und seiner Randgebiete 1:200.000. In: Steininger FF (ed) Erdgeschichte des Waldviertels, 1st edn. Waldviertler Heimatbund, Horn – Waidhofen/Thaya

Huber KH (1999) Zum Formenschatz der Granitverwitterung und –abtragung im nordwestlichen Waldviertel. In: Steininger FF (ed) Erdgeschichte des Waldviertels, 2nd edn. Waldviertler Heimatbund, Horn – Waidhofen/Thaya, pp 113–132

Huber KH (2003) Some field observations and remarks on the Gmünd beds of the northwestern Waldviertel region (Lower Austria). Jahrb Geol Bundesanst 143:543–566

Huber KH, Chábera S (1993) Nové locality žlábkových škrapů v žulách centrálního masívu moldanubického plutonu (New discovery of granitic karren in the South Bohemian Pluton). Sbor Jihočes muz v Čes Budějovicích. Přír Vědy 33:5–14

Huber KH, Chábera S (1994) Úpatní výklenky na žulových balvanech moldanubika (Basal flared slopes on Moldanubian granite boulders). Sbor Jihočes muz v Čes Budějovicích. Přír Vědy 34:5–16

Jiranek J, Müller HW, Schwaighofer B (1990) Genetic types of the kaolin deposits in the Bohemian Massif. In: Minarikova D, Lobitzer H (ed) Thirty years of geological cooperation between Austria and Czechoslovakia, Federal Geological Survey, Vienna, pp 212–225

Koller F (1999) Plutonische Gesteine. In: Steininger FF (ed) Erdgeschichte des Waldviertels, 2nd edn. Waldviertler Heimatbund, Horn – Waidhofen/Thaya, pp 25–36

Linton DL (1955) The problem of tors. Geogr J 121:470–487

Migoń P, Różycka M, Michniewicz A (2018) Conservation and geotourism perspectives at granite geoheritage sites of Lower Austria. Geoheritage 10:11–21

Roetzel R, Steininger FF (1999) Älteres Tertiär. In: Steininger FF (ed) Erdgeschichte des Waldviertels, 2nd edn. Waldviertler Heimatbund, Horn – Waidhofen/Thaya, pp 75–78

Roštinský P, Roetzel R (2005) Exhumed Cenozoic landforms on the SE flank of the Bohemian Massif in the Czech Republic and Austria. Z Geomorph NF 49:23–45

Schnabel W (ed) (2002) Geologische Karte von Niederösterreich 1:200 000. Geol Bundesanst, Wien

Steininger FF, Roetzel R (1999) Jüngeres Tertiär. In: Steininger FF (ed) Erdgeschichte des Waldviertels, 2nd edn. Waldviertler Heimatbund, Horn – Waidhofen/Thaya, pp 79–88

Suess FE (1903) Bau und Bild der Böhmischen Masse. In: Diener C, Hoernes R, Suess FE, Uhlig V (eds) Bau und Bild Österreichs, Tempsky-Freytag, Wien – Leipzig, pp 1–322

Twidale CR (1962) Steepened margins of inselbergs from north-western Eyre Peninsula, South Australia. Z Geomorph NF 6:51–69

Wieden P (1964) Kaolinlagerstätte Mallersbach. Mitt Geol Gesell Wien 57:169–179

Piotr Migoń is professor of Geography at the University of Wrocław, Poland. Among his main research interests are granite landforms and he published a research monograph on this topic (*Granite Landforms of the World* Oxford 2006). Besides, he works on hillslope processes, mainly landslides, geomorphic evolution of sandstone terrains, geoheritage and geotourism. He is a long-standing member of IAG Executive Committee and is the series editor of 'Geomorphological Landscapes of the World'.

Aleksandra Michniewicz is a geographer and teaching assistant at the University of Wrocław, Poland. Her main field of interest is geomorphology of rock outcrops in the West Sudetes, Poland, where she investigates them using a combination of field mapping, geomorphometry and geophysics.

Milena Różycka is a geographer and lecturer at the University of Wrocław, Poland. Her main research interests include morphotectonics of the Sudetes range, geomorphometry of landslide and sandstone terrains, application of statistics in geomorphology, and geotourism.

9 Deeply Incised Valley Meanders of the Bohemian Massif

Ronald E. Pöppl, Reinhard Roetzel, and Doris Riedl

Abstract

The marginal eastern parts of the Bohemian Massif in Austria are characterized by incised valley meanders, which have formed in response to tectonic uplift and river channel incision starting before c. 5 Ma. The most impressive examples of incised valley meanders are located in the Waldviertel Highlands along the Thaya, Kamp and Krems rivers. In this chapter, some hotspot examples with canyon-like incised meanders and related other landforms with a special focus on the Thayatal National Park are presented. The Thayatal National Park is the smallest out of six Austrian national parks and is connected to the larger Czech Podyjí National Park. In the Thayatal National Park, an impressive example of a double bend meander can be found at Umlaufberg and Ostroh, caused by highly resistant orthogneisses in the western meander neck, which forced the river to flow around. At some locations, the Thaya Valley is cut into the Thaya granite and Biteš gneiss, and characterized by steep valley sides, castellated rock outcrops, boulder fields and block streams. Boulder fields and block streams developed by unloading of rocks in the Late Pleistocene and were accentuated during the Holocene by gravitational slope processes. Along the Kamp and Krems valleys, deeply incised valley meanders similar to those found along the Thaya Valley have formed. In the lower courses of the Kamp and Krems valleys, well-preserved Pleistocene loess profiles with several pedocomplexes (e.g. the Stiefern loess sequence in the Kamp Valley) can be found, serving as indicators of landscape evolution and palaeoclimatic changes. Moreover, along the Krems River impressive cave systems can be observed, such as the “Gudenushöhle”, which is considered to be one of the earliest human settlements in Austria.

Keywords

Valley meanders • Rock towers • Boulder fields • Pedocomplexes • Thayatal–Podyjí National Parks • Gudenus cave

9.1 Introduction

While the central parts of the Bohemian Massif are mainly characterized by rolling hills, gentle valleys, and broad and flat ridges, its south-eastern marginal parts are typified by deeply incised valley meanders, which have formed due to tectonic uplift and river incision starting at ca. 5 Ma. The most impressive Austrian examples are the Thaya, Kamp and Krems valleys in the Waldviertel region. In this chapter, some hotspot examples with canyon-like incised meanders and related other landforms with a special focus on the Thayatal National Park are presented. In Sect. 2, the geographical and geological setting of the study region, as well as the geomorphic evolution of deeply incised valley meanders are presented. Section 3 highlights hotspot examples of valley incision and other related landforms. In Sect. 4, special consideration is given to the cultural and natural values, tourism attractiveness and environmental problems of the Thayatal National Park region.

R. E. Pöppl (✉) · D. Riedl
Department of Geography and Regional Research, University of Vienna, Universitätsstr. 7, 1010 Vienna, Austria
e-mail: ronald.poeppl@univie.ac.at

R. Roetzel
Geological Survey of Austria, Neulinggasse 38, 1030 Vienna, Austria

C. Embleton-Hamann (ed.), *Landscapes and Landforms of Austria*, World Geomorphological Landscapes,
https://doi.org/10.1007/978-3-030-92815-5_9

9.2 Geographical and Geological Setting

9.2.1 Location, Climate, Hydrology and Land Use History of the Study Sites

The Waldviertel Highlands are located in the north-west of Lower Austria, at the border between Austria and Czechia (Fig. 9.1). The Waldviertel is characterized by continental climate, with mean annual temperatures of c. 7 °C and a mean annual precipitation between 500 and 700 mm, decreasing from west to east (Harflinger 1999). Maximum precipitation rates occur between April and September, which is also reflected in river runoff regimes that are further influenced by snowmelt processes. The main river systems of the Waldviertel Highlands drain into the Danube. These are mainly the Thaya and the Kamp rivers (see Fig. 9.1). The Kamp River as well as the smaller Krems River directly flow into the Danube, while the Thaya River (in Czech Dyje)—which is for long stretches in the Thayatal National Park the boundary river between Austria and Czechia—joins the Morava River flowing along the Austrian/Slovakian boundary, which then drains into the Danube. While the Krems and Kamp rivers solely drain Austrian territory, the larger part of the Thaya River catchment area is located in Czechia (Fig. 9.2). In Table 9.1, basic characteristics of the Thaya, Kamp and Krems river systems are shown.

The land cover of the Waldviertel region is mainly characterized by (commercial) coniferous forests ("Waldviertel" means "Forest Quarter") and non-irrigated arable land. However, in many areas—such as in the Thayatal National Park—a significant portion of near-natural vegetation remained. The main reason for the natural integrity of the Thayatal National Park lies in the political history of the region. After World War II, today's national park area

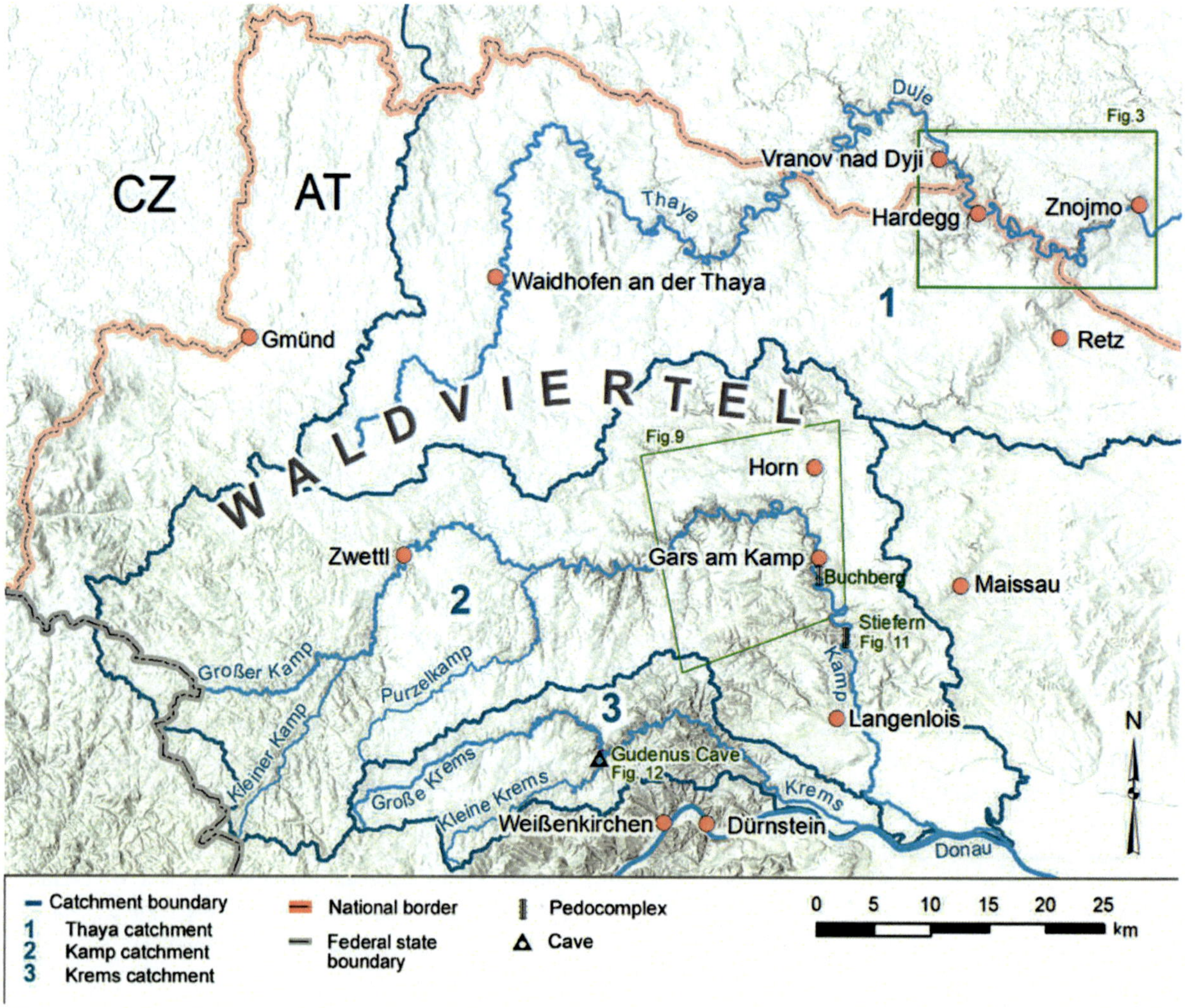

Fig. 9.1 Overview map of the study region. AT = Austria, CZ = Czechia

Fig. 9.2 Incised meanders of the Thaya River in the Bíteš gneiss in the western part of the bilateral cross-border national parks Thayatal and Podyjí. Photo: R. Roetzel

Table 9.1 Basic characteristics of the Thaya, Kamp and Krems river systems (BMLFUW 2015)

Characteristics	Thaya	Kamp	Krems
Total length (main stream)	235 km	153 km	81 km
Elevation (source)	410 m asl	920 m asl	850 m asl
Elevation (mouth)	147 m asl	180 m asl	180 m asl
Catchment area	13 419 km^2	1753 km^2	351 km^2
Mean discharge; in brackets: percentage of the catchment area recorded at the gauging station	~40 m^3/s (94%)	~9 m^3/s (85%)	~2 m^3/s (87%)

straddled the Iron Curtain. In the once-closed border strip, many endangered and rare animal and plant species could survive. The peripheral position of the national park area further resulted in a lack of infrastructure, missing trade and negative population development, however acting as a retreat for wildlife, which in other regions had to give way to modern agriculture and various construction projects.

9.2.2 Geology and Geomorphic Evolution

The Waldviertel Highlands belong to the tectonic unit of the Bohemian Massif. They are made up of igneous and medium- to high-grade metamorphic rocks, which were part of the Variscan mountain range, and developed during the Variscan orogeny in the Devonian and Carboniferous, about 400 to 300 Ma ago. The Bohemian Massif in this region consists of two main tectonic blocks: the structurally lower Moravian Superunit in the east and the tectonically higher Moldanubian Superunit in the west (Suess 1903, 1912). During the late phase of the Variscan orogeny, the Moldanubian Superunit was thrusted onto the Moravian Superunit along flat thrust planes. At the eastern and south-eastern margin of the Waldviertel Highlands, deeply incised canyon-like river valleys (rivers Thaya/Dyje, Pulkau, Kamp, Krems) show unique cross

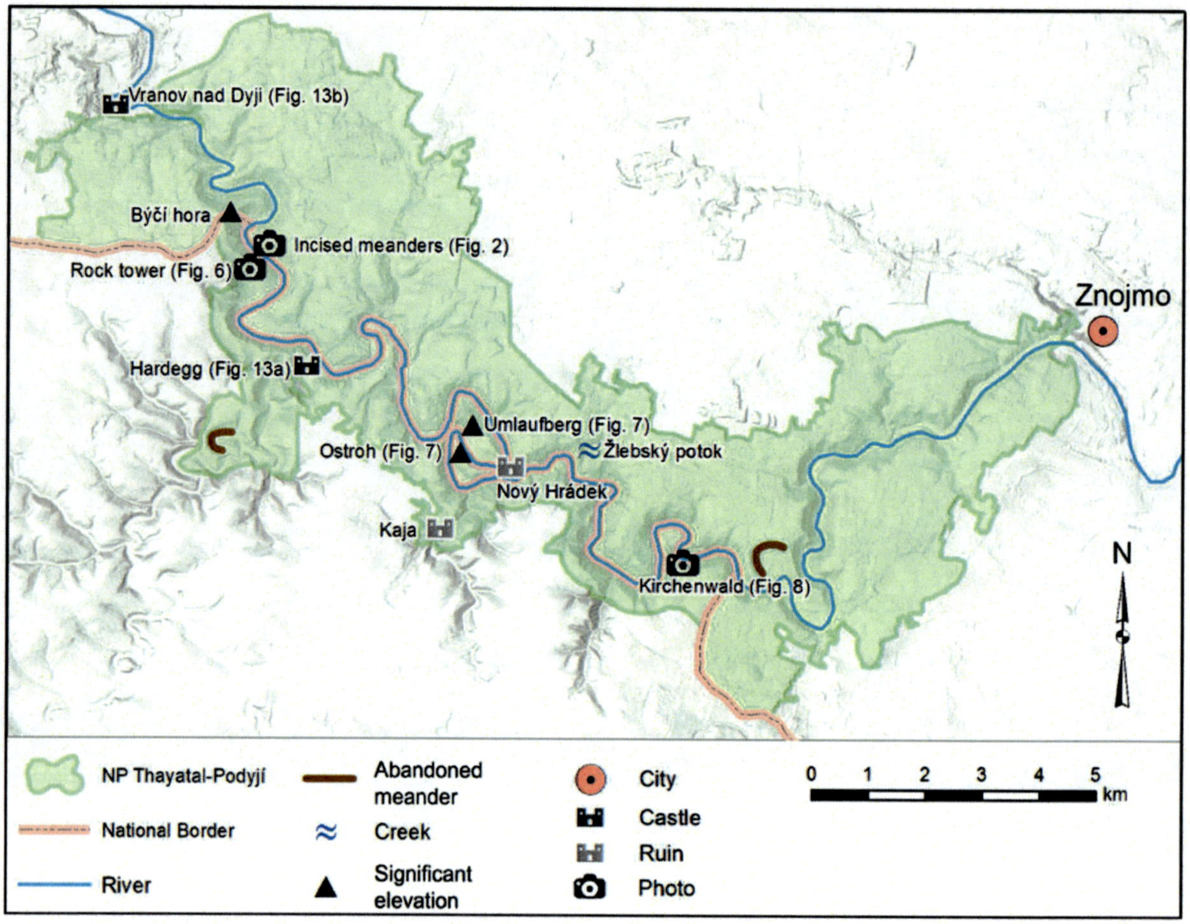

Fig. 9.3 Thayatal and Podyjí National Parks along the Austrian—Czech border

sections through Moravian and Moldanubian rocks (Figs. 9.2 and 9.3).

Moravian rocks are best exposed in the Thaya Valley (Fig. 9.4; Roetzel et al. 2005; Roetzel 2010; Kirchner 2016). In the Moravian Superunit in the Thaya Valley Upper Proterozoic plutonic rocks, like the ca. 600 Ma old Cadomian Thaya granite and Biteš gneiss (Scharbert and Batík 1980; Finger and Riegler 1999), together with other igneous rocks, like the Weitersfeld gneiss and Therasburg gneiss, alternate with even older metasediments, such as paragneiss, mica schist, quartzite, marble and calc-silicate gneiss (Batík 1992; Roetzel et al. 2004, 2005).

Further south-west the Kamp and Krems valleys similarly open spectacular insights into the Moldanubian Drosendorf and Gföhl nappe-systems. Where the valleys are developed in the Gföhl nappe-system, igneous rocks such as Gföhl gneiss, granulite, Wolfshof syenite gneiss or dioritic gneiss occur together with amphibolite, paragneiss, mica schist, marble, ultrabasic rocks, quartzite and graphitic quartzite. The tectonically lower Drosendorf nappe-system is characterized by the Dobra gneiss and a variegated alternation of paragneiss, quartzite, marble, calc-silicate gneiss and graphite (Fuchs and Matura 1976; Matura et al. 1983; Fuchs et al. 1984a, b; Fuchs and Fuchs 1986; Thiele et al. 1991; Schnabel 2002). The Dobra gneiss shows in parts an age of around 1.4 Ga and is one of the oldest rocks in Austria (Gebauer and Friedl 1994; Klötzli et al. 1999; Lindner et al. 2021).

The area in the wider surroundings above the Thaya Valley is marked by a planation surface, which most probably was formed during the Mesozoic to Palaeogene. Especially in the Eocene, tropical weathering produced thick weathering crusts of laterite and kaolin, which are preserved today only in small erosional remnants or as redeposited kaolinitic sands (Roetzel et al. 2005). On the planation surface and the marginal slope of the Waldviertel Highlands, marine to brackish sediments of the Lower (Eggenburgian—Ottnangian) and Middle Miocene (Badenian) are also preserved. During the early Miocene marine transgression, a number of different lithological formations of sand, gravel and clay, partly rich in fossils, were deposited in depressions

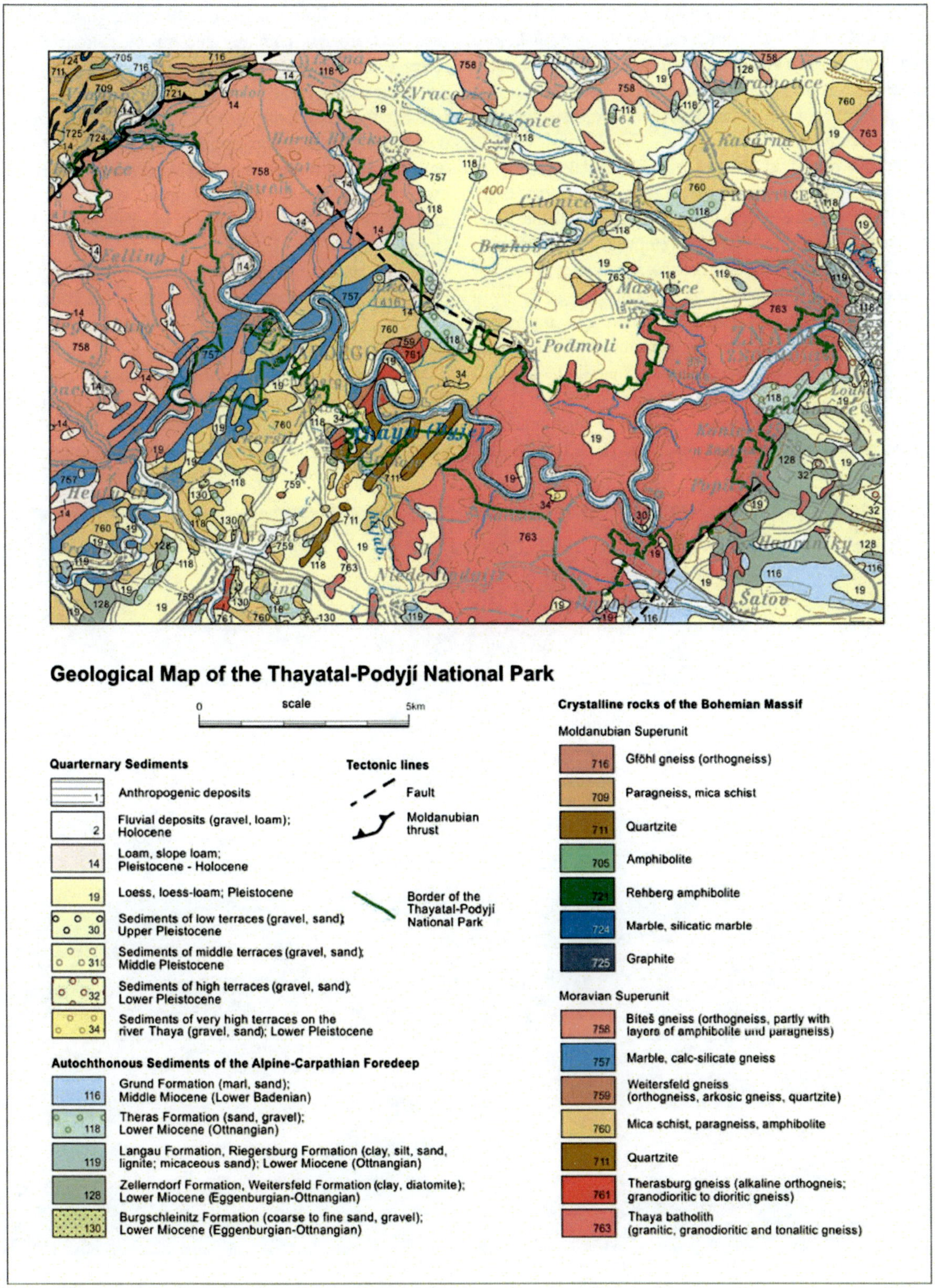

Fig. 9.4 Geological map of the Thayatal–Podyjí National Parks (Roetzel 2010, modified)

between crystalline elevations (Roetzel et al. 1999; Roštínský and Roetzel 2005). In the east, on the marginal slope of the Waldviertel Highlands, south-east of the prominent Diendorf fault zone, remains of the Middle Miocene sea with marls and sands of the Grund Formation were spared from erosion (Roetzel et al. 2005).

After the retreat of the sea and an initial erosion of meandering river systems into the Upper Miocene (Pannonian) deposits, the epochs of the Pliocene and the Early Pleistocene were characterized by a stepwise downcutting of river channels, thus giving rise to the valley meander development. This incision was probably triggered by a continuing movement of Alpine units towards the north, causing a significant rise of the Bohemian Massif and a simultaneous lowering of the drainage system in the foreland (Brzák 1997; Roetzel et al. 2005).

Two types of valley meanders may be distinguished. Ingrown meanders have a steep undercut slope on one side and a gentle slip-off slope on the other side. Entrenched meanders are less common and are characterized by narrower valleys with steep and symmetrical side slopes. With respect to valley meander origin, there is a scientific debate, whether the meandering pattern needs to be initiated in an overlying sediment cover (superimposed origin), or may also develop by uplift only (antecedent origin). As normally all overlying sediments have been stripped, this aspect of valley meander formation still remains to be elucidated (Sala 2004; Hooke 2013).

In the case of the valley meanders at the margin of the Bohemian Massif, there is no doubt about the critical role of uplift, but traces of an old sediment cover may also be found (Fig. 9.5).

In the Thaya Valley of the Thayatal–Podyjí National Parks, remains of fluvial gravel accumulations are located on both the Austrian and Czech side of the valley (Ivan and Kirchner 1994; Havlíček and Holásek 1995; Brzák 1997;

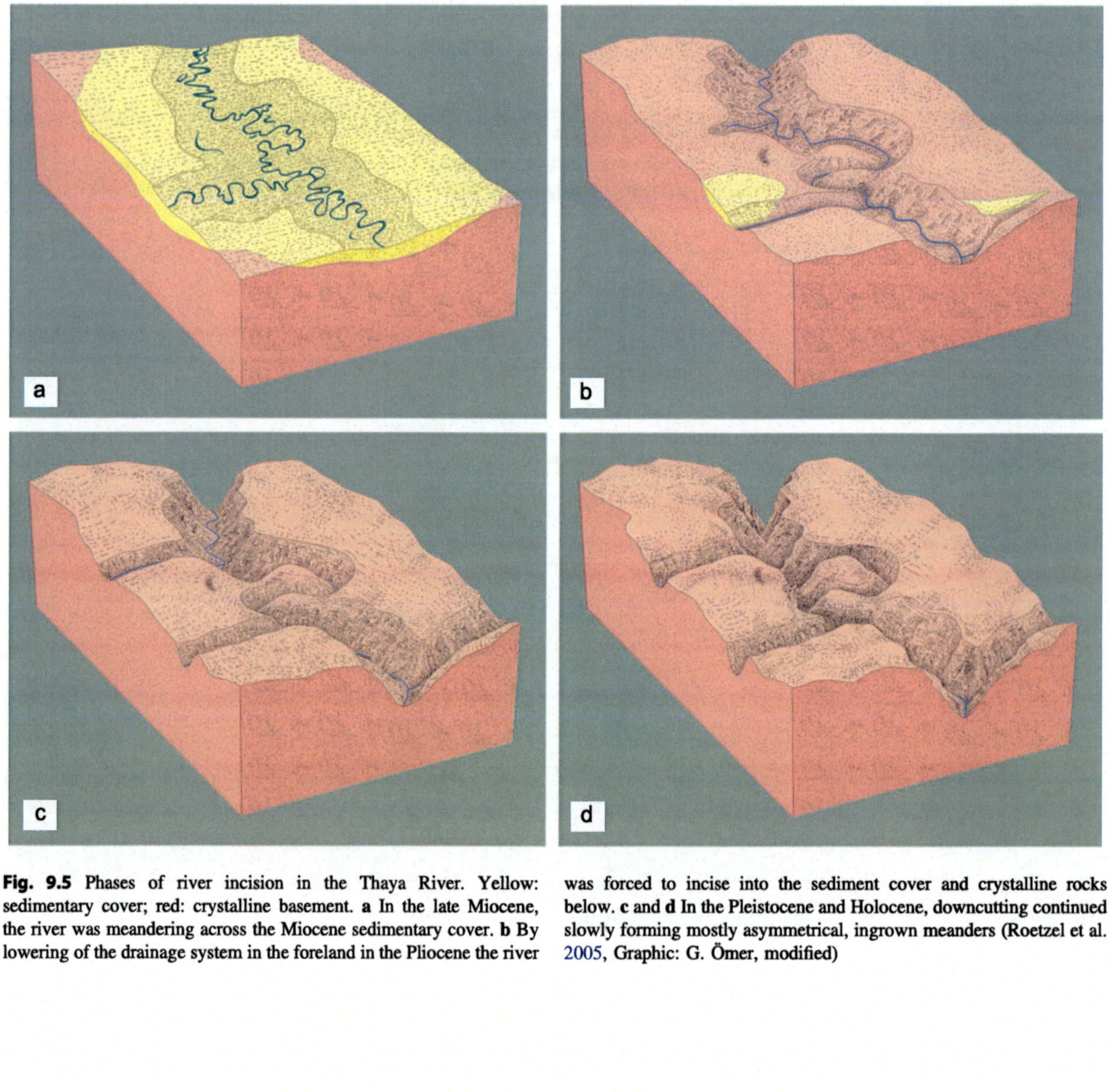

Fig. 9.5 Phases of river incision in the Thaya River. Yellow: sedimentary cover; red: crystalline basement. **a** In the late Miocene, the river was meandering across the Miocene sedimentary cover. **b** By lowering of the drainage system in the foreland in the Pliocene the river was forced to incise into the sediment cover and crystalline rocks below. **c** and **d** In the Pleistocene and Holocene, downcutting continued slowly forming mostly asymmetrical, ingrown meanders (Roetzel et al. 2005, Graphic: G. Ömer, modified)

Havlíček 2002; Roetzel et al. 2005). In Austria, the highest position of sandy gravels is recorded 110–135 m above the present Thaya River, proving the former course of the river on top of the planation surface at the onset of valley incision in the Pliocene. Inside the Thaya Valley, only few remains of fluvial gravel are preserved at different levels, confirming the incision of the river during the Pleistocene (Roetzel et al. 2004, 2005). Fluvial terraces along the Kamp River are reported mainly from the lower course of the river (Steininger 1969; Roetzel 2021), whereas along the Krems River only few gravel accumulations are known.

Loess and loess-loam cover extensive areas on the planation surface above the Thaya Valley, but are quite rare inside the valley. They occur locally in east- and southeastward positions in the lee of high scarps. Palaeosols from warmer and more humid periods of the Pleistocene are rarely preserved in the loess inside the valley and the surrounding areas (Havlíček and Holásek 1995; Havlíček 2002; Roetzel et al. 2005). Solifluction deposits accumulated in morphological depressions and on foothills during the Pleistocene by downslope creep of thawed, water-saturated, clayey-silty-sandy-stony masses on top of permafrost. The recent continuous incision of the river is documented by several erosional steps that can be found in the Holocene valley floor sediments (Roetzel et al. 2005).

9.3 Deeply Incised Valley Meanders: Hotspot Examples

9.3.1 Deeply Incised Valley Meanders and Related Landforms in the Thaya Valley

The incised meanders in the Thaya Valley in the Thayatal–Podyjí National Parks are mostly ingrown meanders, but entrenched types occur as well. In the highly resistant Bíteš gneiss occurring in the west, the valley meanders are rather rectangular than arched and relatively symmetric (entrenched type, Kirchner 2016). In the area below the Stierwiesberg (Býčí hora, 536 m asl) lies the deepest section of the valley, with a valley depth of about 220 m. On the valley slopes rock towers, tors, crags and ridges are common (Fig. 9.6).

Less resistant metamorphic rocks, like paragneiss, mica schist, quartzite, marble and calc-silicate gneiss characterize the central section of the Thaya Valley between Hardegg town and the Žlebský potok creek. The meanders show characteristic asymmetric transverse profiles (ingrown type) (Ivan and Kirchner 1994; Kirchner 2016). The depth of the valley varies between 120 and 160 m. The gentle relief in this part corresponds to the increased weathering susceptibility of the rocks. Even the overall relief reflects the different resistance of lithological units, with saddles developed in paragneiss and mica schist and elevations in quartzite, marble and calc-silicate gneiss. In this part of the valley, the most beautiful example of a double bend meander is located at Umlaufberg and Ostroh (Fig. 9.7). It is related to the occurrence of highly resistant rocks like Weitersfeld and Therasburg gneisses in the western meander neck, which forced the river to flow around (Poeppl 2007) (Fig. 9.4). The shape of the meanders in the Thaya Valley seems to depend on the weathering resistance of the rocks. Perfect meanders are most common in less resistant rocks such as paragneiss and mica schist, whereas the Bíteš gneiss and the Thaya granite often host deformed meanders (Brzák 1997; Poeppl et al. 2011). Overall, the downcutting of the river seems to have occurred at a less rapid pace, with both lateral as well as vertical erosion, also proved by a few abandoned meanders in the Thaya Valley (Fig. 9.3).

In the east, the Thaya Valley is cut into the Thaya granite and is characterized by steep valley sides, rock formations, boulder fields and block streams (Ivan and Kirchner 1994, 1998; Roetzel et al. 2005; Kirchner 2016). The meanders of this 120–160 m deep valley are both symmetric (entrenched) and asymmetric (ingrown) (Brzák 1997; Kirchner 2016). Through continued incision of the rivers, rocks became unloaded and gravitationally disaggregated. During cold periods of the Pleistocene, tors and steep cliffs formed by mechanical weathering and efficient downslope removal of debris. Boulder fields and block streams developed by unloading of rocks and frost wedging in the Late Pleistocene and were accentuated during the Holocene by gravitational slope processes (Fig. 9.8).

9.3.2 Deeply Incised Valley Meanders and Other Landforms in the Kamp and Krems Valleys

Along the Kamp (Fig. 9.9) and Krems valleys deeply incised meanders have developed in a similar way to those found along the Thaya Valley. West of Rosenburg Castle the Kamp Valley also hosts a double bend meander. The first bend turns south and surrounds a rising, crowned by the ruin "Ödes Schloss". The river then turns north and curves around the Umlaufberg (352 m asl), before it finally continues to the east (Fig. 9.10). Downstream from Rosenburg meandering sections alternate with straight sections in the final north–south directed river course of the Kamp.

In contrast to the Thaya Valley in the Thayatal–Podyjí National Parks, well-preserved Pleistocene loess profiles occur in the lower courses of the Kamp and Krems valleys. Such loess deposits often contain pedocomplexes (fossil soils), which are indicators of landscape evolution and palaeoclimatic changes. Typical Upper to Middle Pleistocene loess sequences with several pedocomplexes are located, among others, in Stiefern (Fig. 9.11; Verginis 1993; Sprafke 2016) and Buchberg in the Kamp Valley. The Late

Fig. 9.6 Rock tower in the about 600 Ma old Bíteš gneiss in the western part of the Thayatal National Park in Austria. Photo: R. Roetzel

and Middle Pleistocene (c. 781,000–130,000 BP) were characterized by several alternations of warm and cold climatic periods. During warmer periods, pedogenetic processes led to the development of organic soil layers, which for instance occur at different levels in the loess outcrop at Stiefern (Fig. 9.11). The upper, strongly developed palaeosol in Stiefern is correlated with the Paudorf palaeosol of MIS 5 (Riss-Würm) interglacial, the last major interglacial period before the Holocene (Sprafke 2016) (for a detailed discussion of palaeosols, see Sect. 3 of "Sunken Roads and Palaeosols in Loess Areas in Lower Austria: Landform Development and Cultural Importance").

In the so-called Kremszwickel region, i.e. the area at the confluence of the rivers Große Krems and Kleine Krems, impressive cave systems with 40 caves, like the Gudenushöhle, Eichmaierhöhle and Schusterlucke, together with rock shelters like Teufelsrast occur (Nagel and Rabeder 1991; Mayer et al. 1993). These caves developed in karstified

Fig. 9.7 Panoramic view of the double bend meander surrounding Umlaufberg and Ostroh in the national parks Thayatal and Podyjí. Photo: NP Thayatal_www.aufsichten.com. Insert: aerial photo of the entire double bend meander. Umlaufberg is situated in the lower right, Ostroh in the upper left, the centre of the panoramic view is marked in red. Photo: BEV, modified from Roetzel et al. (2005)

Fig. 9.8 Boulder field from Thaya granite in the Kirchenwald in the eastern part of the Thayatal National Park. Photo: NP Thayatal_www.aufsichten.com

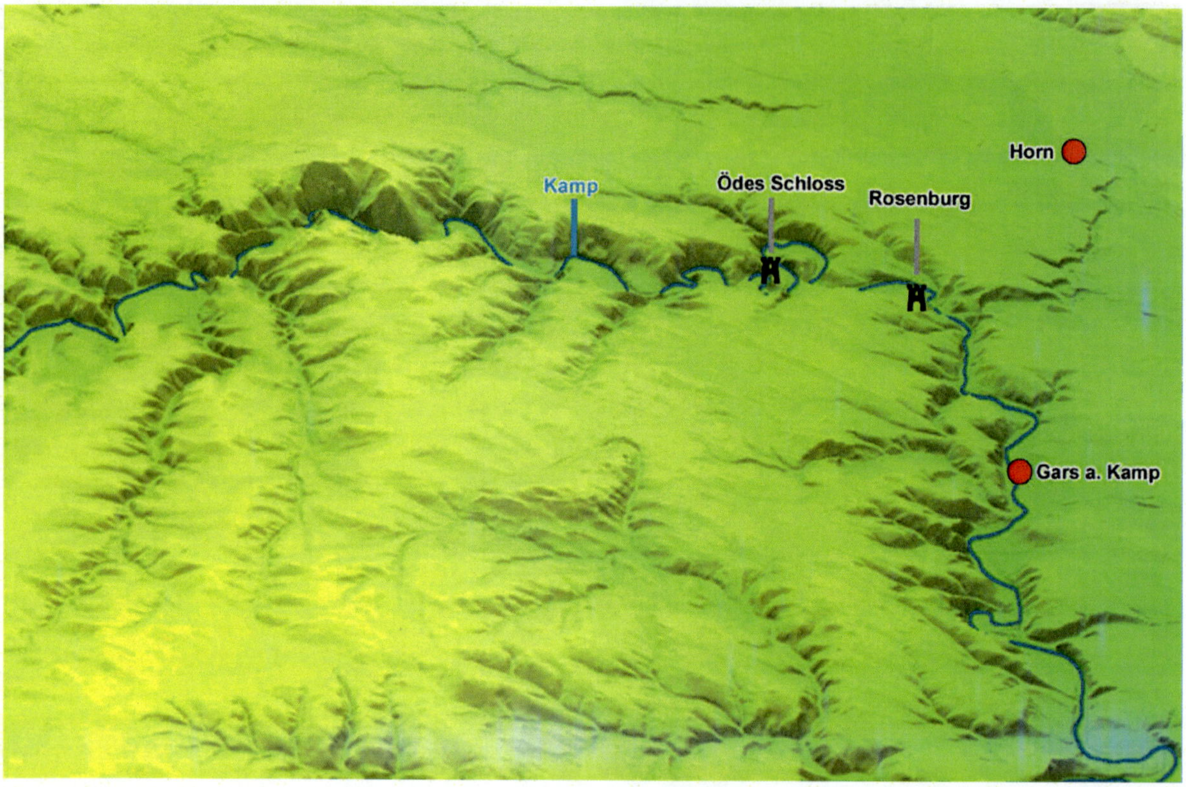

Fig. 9.9 Oblique view of the deeply incised meanders along the Kamp Valley

marbles intercalated with amphibolites of the Gföhl nappe-system (Matura et al. 1983; Matura and Heinz 1989). The Gudenushöhle is located c. 15 km WNW of the town of Krems (see Fig. 9.1), on the right bank of the Kleine Krems River, directly underneath Hartenstein Castle. The cave has a total length of 30 m, a maximum height of 3.7 m and a mean width of 4 m and comprises three surface openings (Mayer et al. 1993; Knobloch 2012) (Fig. 9.12). It is considered to be one of the earliest human settlements in Austria. Since 1883, more than 10,000 artefacts have been discovered, among which the oldest are dated to the Middle Palaeolithic (Mousterian, Neugebauer-Maresch 1995, cf. Mayer et al. 1993). In some caves, a species-rich fauna was found (Woldřich 1893), which in the Schusterlucke dates to the Early Würmian (Nagel and Rabeder 1991), indicated by an Uranium date of 115,000 + 9800/−8800 BP (Wild et al. 1989).

9.4 Cultural and Natural Values, Tourism Attractiveness and Environmental Problems of the Thayatal National Park

On 1 January 2000, the Thayatal National Park was established as an IUCN National Park, which is defined as "large natural or near natural areas set aside to protect large-scale ecological processes, along with the complement of species and ecosystems characteristic of the area, which also provide a foundation for environmentally and culturally compatible, spiritual, scientific, educational, recreational, and visitor opportunities" (IUCN 1994). The Thayatal National Park is the smallest out of six Austrian national parks and covers an area of 1330 ha, of which 1260 ha are nature zones and 70 ha are nature zones with management interference for the protection of ecosystems, while the buffer zone comprises less than one hectare. The Thayatal National Park is connected to the larger Czech Národní Park Podyjí, which has a total area of 6620 ha. The Thaya River constitutes the common border between both natural parks over ca. 25 km.

Due to specific climatic conditions, i.e. the national park Thayatal is located on a distinctive climate border (see Sect. 2.1), continental and central European flora and fauna are interspersed, which results in particularly high diversity of species in a small area. Moreover, the aspect changes constantly along the winding river bends, and the geological habitat factors also differ on a small-scale basis. In both national parks, 1288 plant species have been identified so far. In the Austrian part, over 100 bird species could be identified (more than 150 in both national parks), and almost 80 of them breed in the Thaya Valley. Furthermore, in 2007

Fig. 9.10 Incised meander of the Kamp River. View from the ruin "Ödes Schloss" to the west. Photo: R. Roetzel

the return of the extremely shy wild cat was recorded in the national park for the first time in 30 years in Austria. The only bigger settlement in the Thayatal National Park is the town Hardegg. With ca. 90 inhabitants, the municipality of Hardegg is the smallest "city" in Austria, but still well known because of the beautiful Hardegg Castle built around 1145 (Fig. 9.13a). The Národní Park Podyjí extends over a river length of 45 km between Vranov and Znojmo. From 1930 to 1934, upstream of the city of Vranov a hydropower station was built, which now constitutes a significant ecological problem (Fig. 9.13b). The intermittent operation of the plant at peak hours of power demand induces subdaily flow variations. During times of electricity production, the flow rate increases from at least 3 m^3/s to 30–45 m^3/s, which further alters the temperature regime and affects sediment dynamics and thus river bed sediment texture and habitats of the Thaya River (Poeppl et al. 2011).

9.4.1 Conclusions

The Thaya, Kamp and Krems valleys, located in the marginal eastern parts of the Bohemian Massif, exhibit a range of particular geomorphological features, of which the highlight are perfectly developed valley meanders. They started to form with river channel incision in response to tectonic uplift of the Bohemian Massif/lowering of the base level at ca. 5 Ma. Both the incised and the less common entrenched meander types can be found. Details in the shape seem to depend on rock resistance. In mechanically very strong rocks, like for instance the Bíteš gneiss, the shape of the valley meanders is rectangular rather than arched, and the entrenched type dominates.

Spectacular valley meanders, including a double bend meander, together with typical granitic weathering landforms like castellated rock outcrops, tors and boulder fields

Fig. 9.11 Late to Middle Pleistocene loess sequence in the former brickyard of Stiefern, with two visible palaeosols (**a**, **b**). The upper, well-developed palaeosol (**a**) is correlated with the Paudorf palaeosol of MIS 5 interglacial (Sprafke 2016). Photo: R. Roetzel

constitute the scenic attractions of the Thayatal and Podyjí National Parks. This cross-border national park is also characterized by a high biodiversity, due to several reasons, namely (i) its location on a distinctive European climate border, (ii) the constantly changing aspect along the winding Thaya Valley bends that further enhance the development of ecological niches and (iii) the survival of many rare species in the closed "Iron Curtain" border strip after World War II.

Additional sites of scientific value in the region are the well-preserved Pleistocene loess profiles with several pedocomplexes along the Kamp and Krems rivers, serving as indicators of landscape evolution and palaeoclimatic changes. Along the Krems River, important cave systems are located, out of which the "Gudenushöhle" is considered to be one of the earliest human settlements in Austria.

Fig. 9.12 Main entrance of the Gudenushöhle in the Krems Valley, located in marble, intercalated with amphibolite. Photo: F. F. Steininger

Fig. 9.13 a Hardegg Castle (Photo: R. Roetzel); b Vranov Dam (Photo: NP Thayatal, P. Lazarek)

References

Batík P (1992) Geologická Mapa Národního Parku Podyjí 1:25.000. *Český geologický ústav,* Praha

BMLFUW (2015) Hydrografisches Jahrbuch von Österreich 2013, 121. Band – Daten und Auswertungen. Bundesministerium für Land- und Forstwirtschaft, Umwelt und Wasserwirtschaft (ed), 976 pp

Brzák M (1997) Entwicklung des Thayatales im Nationalpark Podyjí (Tschechische Republik) aufgrund der Morphographie und Schotteranalyse. Mitt Österr Geograph Ges 139:261–275

Finger F, Riegler G (1999) Der Thayabatholith und der kristalline Untergrund des Weinviertels. In: Roetzel R (ed) Arbeitstagung der Geologischen Bundesanstalt 1999, 3.-7. Mai 1999, Retz, Geologische Bundesanstalt Wien, pp 23–31

Fuchs G, Fuchs W (1986) Geologische Karte der Republik Österreich 1:50.000 36 Ottenschlag. Geologische Bundesanstalt, Wien

Fuchs G, Kupka E, Steininger F (1984a) Geologische Karte der Republik Österreich 1:50.000 20 Gföhl. Geologische Bundesanstalt, Wien

Fuchs W, Grill R, Matura A, Vasicek W (1984b) Geologische Karte der Republik Österreich 1:50.000 38 Krems. Geologische Bundesanstalt, Wien

Fuchs G, Matura A (1976) Zur Geologie des Kristallins der südlichen Böhmischen Masse. Jahrb Geol Bundesanst 119:1–43

Gebauer D, Friedl G (1994) A 1.38 Ga protolith age for the Dobra orthogneiss (Moldanubian Zone of the southern Bohemian Massif, NE-Austria): evidence from ion-microprobe (SHRIMP) dating of zircon. J Czech Geol Soc 39:34–35

Harflinger O (1999) Die klimatischen Eigenschaften Niederösterreichs mit besonderer Berücksichtigung des Pannonikums. Arbeitstagung Geologische Bundesanstalt 1999, Retz-Hollabrunn, Beiträge: pp 106–110

Havlíček P (2002) NP Podyjí: Revizní kvartérně-geologický výzkum a mapování v letech 2001 a 2002. Zprávy o Geologických Výzkumech v roce 2002:71–73

Havlíček P, Holásek O (1995) Bericht 1994 über geologische Aufnahmen im Quartär des Kamptales auf Blatt 21 Horn. Jahrb Geol Bundesanst 138(3):481–482

Hooke J (2013) River Meandering. In: Wohl E, Schroder J (eds) Treatise on geomorphology, vol 9. Academic Press, San Diego, pp 260–288

IUCN (1994) The World Conservation Union Guidelines for protected area management categories. In IUCN 1994 Gland, Switzerland

Ivan A, Kirchner K (1994) Geomorphology of the Podyjí National Park in the southeastern part of the Bohemian Massif (South Moravia). Moravian Geographical Reports 1/1994(2), pp 2–25

Ivan A, Kirchner K (1998) Granite Landforms in South Moravia (Czech Republic). Geogr Fis Din Quat 21:23–26

Kirchner K (2016) The Dyje Canyon-like Valley: geomorphological landscape of Deep Valley at the Eastern part of the marginal slope of Bohemian Massif. In: Hradecký J, Pánek T. (eds) Landscapes and Landforms of the Czech Republic, Springer, pp 233–247

Klötzli U, Frank W, Scharbert S, Thöni M (1999) Evolution of the SE Bohemian Massif based on geochronological data—a review. Jahrb Geol Bundesanst 141(4):377–394

Knobloch G (2012) Erdgeschichte(n) zwischen Krems und Kamp. Living Edition, Sankt Leonhard

Lindner M, Dörr W, Hauzenberger CA, Reither D, Finger F (2021) In search of the oldest rock of Austria: The Hauergraben Gneiss, a 1.40 Ga old mafic quartz-monzonitic inlayer in the Dobra Gneiss (Drosendorf Unit, Bohemian Massif) as a new candidate. Austrian J Earth Sci 114:29–45

Matura A, Heinz H (1989) Geologische Karte der Republik Österreich 1:50.000. Erläuterungen zu Blatt 37 Mautern: 65 pp (Geol. Bundesanst.)

Matura A, Fuchs W, Grill R (1983) Geologische Karte der Republik Österreich 1:50.000 37 Mautern. Geologische Bundesanstalt, Wien

Mayer A, Raschko H, Wirth J (1993) Die Höhlen des Kremstales. Supplement of "Die Höhle" 33, pp 1–52

Nagel D, Rabeder G (eds) (1991) Exkursionen im Pliozän und Pleistozän Österreichs. Österr Paläont Ges, VI + 44 p

Neugebauer-Maresch C (1995) Altsteinzeit im Osten Österreichs. Wiss. Schriftenreihe Niederösterr 95/96/97: 98 pp

Poeppl RE, Bauer B, Keiler M, Glade T (2011) Die Thaya im Nationalpark Thayatal - eine flussmorphologische Analyse auf verschiedenen räumlichen Skalenebenen. Wiss Mitt Niederöst Landesm 21:209–220

Poeppl RE (2007) Die Hydrogeographie der Thaya innerhalb der Grenzen des "Nationalparks Thayatal" unter besonderer Berücksichtigung der Flussmorphologie und der Inselvegetation. Unpublished diploma thesis, University of Vienna, Austria, 124 pp incl. CD- Rom

Roetzel R (2010) Geologie und Geomorphologie im Nationalpark Thayatal-Podyjí. Wiss Mitt Niederösterr Landesmus 21:35–66

Roetzel R (2021) Bericht 2017–2019 über geologische Aufnahmen auf Blatt 21 Horn. Jahrb Geol Bundesanstalt 161:124–140

Roetzel R, Fuchs G, Batík P, Čtyroký P, Havlíček P (2004) Geologische Karte der Nationalparks Thayatal und Podyjí. Geologische Bundesanstalt, Wien
Roetzel R, Mandic O, Steininger FF (1999) Lithostratigraphie und Chronostratigraphie der tertiären Sedimente im westlichen Weinviertel und angrenzenden Waldviertel. In: Roetzel R (ed) Arbeitstagung der Geologischen Bundesanstalt 1999, 3.-7. Mai 1999, Retz, Geologische Bundesanstalt Wien: pp 38–54
Roetzel R, Fuchs G, Havlíček P, Übl Ch, Wrbka Th (2005) Geologie im Fluss. Erläuterungen zur Geologischen Karte der Nationalparks Thayatal und Podyjí. Geologische Bundesanstalt, Wien
Roštínský P, Roetzel R (2005) Exhumed Cenozoic landforms on the SE flank of the Bohemian Massif in the Czech Republic and Austria. Zeitschr Geomorph, N F 49(1):23–45
Sala M (2004) Valley meander. In Goudie A (ed), Encyclopedia of geomorphology, vol 2. Routledge, London and New York, pp 1090
Scharbert S, Batík P (1980) The age of the Thaya (Dyje) Pluton. Verh Geol Bundesanst 1980:325–331
Schnabel W (ed) (2002) Geologische Karte von Niederösterreich 1:200000. Geologische Bundesanstalt, Wien
Sprafke T (2016) Löss-Paläoboden Sequenzen in Niederösterreich als komplexe Archive quartärer Klima- und Landschaftsveränderungen. Dissertation, University of Würzburg
Steininger FF (1969) Bericht 1968 über Aufnahmen im Tertiär und Quartär auf Blatt 4555 (Horn). Verh Geol Bundesanst 1969:A69–A73
Suess FE (1903) Bau und Bild der Böhmischen Masse. In: Hoernes R, Suess FE, Uhlig V, Diener C (eds) Bau und Bild Österreichs, Tempsky-Freytag, Wien-Leipzig, pp 1–322
Suess FE (1912) Die moravischen Fenster und ihre Beziehung zum Grundgebirge des Hohen Gesenkes. Denkschr k Akad Wiss, Math-Naturwiss Kl 88:541–624
Thiele O, Boroviczény F, Erich A, Exner Ch, Kupka JE, Schwaighofer B (1991) Geologische Karte der Republik Österreich 1:50.000 19 Zwettl. Geologische Bundesanstalt, Wien
Verginis S (1993) Lößakkumulation und Paläoböden als Indikatoren für Klimaschwankungen während des Paläolithikums (Pleistozäns) mit ausgewählten Beispielen aus Niederösterreich. In: Neugebauer-Maresch C, Altsteinzeit im Osten Österreichs. Wiss. Schriftenreihe Niederösterr 95/96/97: pp 13–30
Wild E, Steffan I, Rabeder G (1989) Uranium series dating of fossil bones. Inst Radiumforsch Kernphys, Progess Rep 1987/88:53–56
Woldřich JN (1893) Reste diluvialer Faunen und des Menschen. Denksch k Akad Wiss, Math-Naturwiss Cl 60:565–566

Ronald E. Pöppl is a senior lecturer at the Department of Geography and Regional Research at the University of Vienna. His primary research interests include connectivity in fluvial systems, fluvial geomorphology, soil erosion, human impacts and human–environment interactions, river and catchment management.

Reinhard Roetzel is a sedimentologist and mapping geologist, who worked up to 2020 at the Geological Survey of Austria. His research focuses mainly on geological surface mapping of Cenozoic sediments and crystalline rocks as well as sedimentology and sedimentary petrology of Paleogene, Neogene and Quaternary sediments on the Bohemian Massif and in the Alpine–Carpathian Foreland Basin of Lower and Upper Austria.

Doris Riedl is a lecturer at the Department of Geography and Regional Research at the University of Vienna. Her primary research interests are hydrological networks and GIS.

Check for updates

Wachau World Heritage Site: A Diverse Riverine Landscape

10

Doris Riedl, Reinhard Roetzel, Ronald E. Pöppl, and Tobias Sprafke

Abstract

The Wachau is a famous and picturesque riverine landscape along the Danube River in Lower Austria. It is a valley incised into the crystalline rocks of the Bohemian Massif. Middle Miocene (Badenian) sediments in the eastern part between Krems and Spitz indicate a pre-existent fjord-like bay of the Badenian Sea. Fluvial sediments in valleys north of the recent course of the Danube show a different, more northerly course of the Palaeo-Danube in the late Miocene. In the late Pliocene and Pleistocene, the Danube deeply incised into the southeastern margin of the Bohemian Massif along its recent course, presumably triggered by uplift caused by the northward moving Alpine units. Due to the influence of a warm Pannonian climate, the Wachau is suited for wine and apricot production. Since the Middle Ages terraces with simple rock walls, which are now a characteristic feature of the valley, have been constructed for easier cultivation. The Wachau has been inhabited since Palaeolithic times. Several important artefacts like sculptures of women found in Stratzing and Willendorf document the early human habitation that led to a continuous development until today. Along the valley historical buildings from Roman times, the Middle Ages, Renaissance and Baroque can be found. In 2000, the Wachau became a UNESCO World Heritage Site to protect the unique combination of cultural and natural sites. The World Heritage Trail connects the 13 municipalities of the Cultural Landscape Wachau and encompasses 20 ruins and castles, monasteries and the Wachau wine region. Natural hazards threatening the World Heritage Site are mainly floods and rockfalls.

Keywords

Bohemian Massif • Danube Valley • Terrace cultivation • Palaeolithic sites • Floods • Rockfalls

10.1 Introduction

About 80 kms west of the city of Vienna, the Wachau, a culturally and geomorphologically fascinating landscape, extends over a distance of approximately 35 kms along the steep slopes of the Danube River valley. The region has been inhabited for millennia. In 2000, this land between the famous Melk Abbey and the city of Krems was added to the list of UNESCO World Heritage Sites as "Wachau Cultural Landscape". Since 2010, the "Wachau World Heritage Trail" has run along both sides of the Danube, providing insights in the natural and cultural marvels of the Wachau area. 180 km of trail are divided into 14 branches that lead to 20 castles, ruins, monasteries and the Jauerling (960 m)—the highest hilltop—and countless orchards, vineyards and small villages.

The Wachau is one of the deeply incised valleys at the southeastern margin of the Bohemian Massif, which was formed due to tectonic uplift and river channel incision starting at ca. 5 Ma. This contribution highlights the special geological, geomorphological and cultural features of the Wachau region. Section 2 of this chapter will provide an overview of the geographical, geological and climatic setting of the study region, and its geologic and geomorphic evolution. Section 3 gives an overview of the cultural history of the region. Finally, an outline of the most recent geomorphic hazards, i.e. floods and rockfalls, is presented (Sect. 4).

D. Riedl (✉) · R. E. Pöppl
Department of Geography and Regional Research, University of Vienna, Universitätsstr. 7, 1010 Vienna, Austria
e-mail: doris.riedl@univie.ac.at

R. Roetzel
Geological Survey of Austria, Neulinggasse 38, 1030 Vienna, Austria

T. Sprafke
Institute of Geography, University of Bern, Hochschulstrasse 12, 3012 Bern, Switzerland

C. Embleton-Hamann (ed.), *Landscapes and Landforms of Austria*, World Geomorphological Landscapes,
https://doi.org/10.1007/978-3-030-92815-5_10

10.2 Geographical Setting

The Wachau is a landscape along the river Danube through an incised valley between the cities of Melk and Krems in the Federal Province of Lower Austria (Fig. 10.1). The name already showed up in the Middle Ages. In AD 830 King Ludwig the German confirmed the monastery Altaich (Germany) as a possession of "Uuahouua " (Rentschler 2012), which evolved to "Wachova" and later to "Wachau".

The Wachau region includes the river valley of the Danube as well as the adjacent hills of the southern Waldviertel and the Dunkelstein Forest. In the Wachau, the Danube cuts through the southern part of the Austrian granite and gneiss highlands, tectonically belonging to the southeastern marginal zone of the Bohemian Massif. Its highest elevation is the Jauerling (960 m asl) in the Waldviertel and the Dunkelstein Forest on the right bank of the Danube River reaches up to 725 m asl, whereas the Danube Valley lies between 213 m asl (at Melk) and 203 m asl (at Krems).

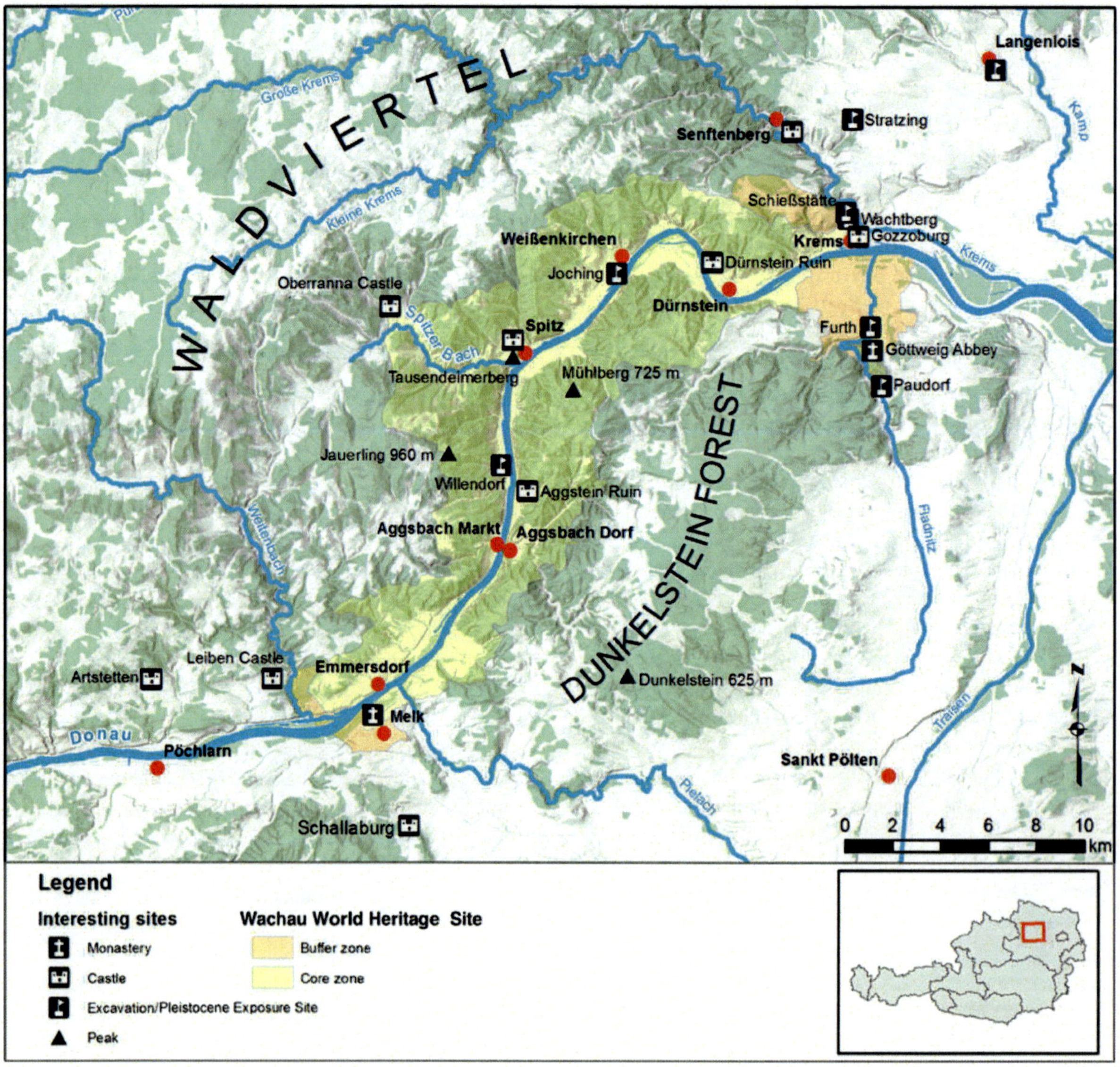

Fig. 10.1 Overview of Wachau area displaying the designated buffer and core zones of the Wachau World Heritage Site (Credits for Background map: USGS, NGA, NASA, CGIAR, N Robinson, NCEAS, NLS, OS, NMA, Geodatastyrelsen, GSA, GIS User Community)

10.2.1 Climatic Conditions

The Dunkelstein Forest as well as the southern Waldviertel with their higher altitudes are characterised by cold winters and moderately warm summers, while the Danube Valley itself with its lower altitude exhibits warmer and dryer (i.e. continental) climatic conditions reaching from the east into the Wachau and the hills of the Waldviertel. The climatic diversity of the region is shown in Table 10.1. The average temperature in Krems is nearly 3 °C higher compared to the Jauerling, further exhibiting fewer days of freezing. Table 10.1 further shows the continentally influenced Pannonian climate (Nagl 1983) in the east of the Wachau Valley as being characterised by less precipitation and more hot days and days of freezing over the year compared to the farther southwestern location (Melk) with a slightly more moderate Atlantic climate (Nagl 1983).

The mild Pannonian climate in combination with the variegated geological substrate also leads to a colourful floral and faunal diversity in the Wachau, especially on dry grasslands and extensively cultivated farmlands (Gamerith 2003).

From the heights of the Dunkelstein Forest and the Waldviertel, cool and wet air masses flow through small forested stony valleys (i.e. the "Wachauer Gräben") into the relatively warm Wachau Valley. The winegrowers in the Wachau see these continuous air circulations as one of the main reasons for the aromatic diversity in their wines. The mild climate in the Wachau Valley is not only favourable to the grapes of today's many vineyards, but as can be seen below, was an advantage for the early settlers in the region too. First permanent settlements go back to prehistoric time, Roman remains from around the second to the fifth century are visually identifiable in Mautern and towns like Dürnstein. Weißenkirchen and Emmersdorf exhibit historical buildings dating back to the twelfth century.

10.2.2 Basement Geology and Rock Landforms

The Danube in the Wachau Valley cut into crystalline rocks of the Bohemian Massif, which are part of the tectonic Moldanubian Superunit. It consists of medium to high grade metamorphic rocks, which were formed during the Variscan Orogeny in the Devonian and Carboniferous, about 400–300 Ma.

The river is incised into igneous and metamorphic rocks of the Drosendorf and Gföhl nappe-systems. Around Dürnstein and Weißenkirchen as well as in the vicinity of Aggsbach and at the left bank north of Melk the Gföhl gneiss of the Gföhl nappe-system is dominant (Fig. 10.2). Especially in the surroundings of Dürnstein and west of Weißenkirchen, numerous localities with tors can be found in the Gföhl gneiss (Fig. 10.3).

This granitic gneiss is accompanied by amphibolite, paragneiss, migmatitic and dioritic gneiss, and mica schist. Especially along the right banks of the Danube River, around the ruin of Aggstein, tors occur in migmatitic amphibolite and Gföhl gneiss (Fig. 10.4). The frequent occurrence of tors in this area may be associated with tectonic pre-shaping by the Diendorf fault system.

A variegated alternation of rocks of the Drosendorf nappe-system, like marble, calc-silicate gneiss, paragneiss and migmatitic gneiss makes up the western part of the river valley between Spitz and Willendorf. Granulite with ultramafic intercalations pinch out at the southern bank of the Danube south of Aggsbach Markt (Fig. 10.2) (Fuchs and Matura 1976, Matura et al. 1983, Fuchs et al. 1984, Schnabel 2002).

The Wachau area is located in the prominent NE-SW trending Diendorf fault zone. South of the Danube, in the Dunkelstein Forest, the master fault of the Diendorf fault zone divides the Gföhl gneiss in the north from granulite with intercalated ultrabasic rocks in the south (Fig. 10.2). Along this sinistral strike-slip fault, the rocks in the southeastern block have been moved towards northeast since the Permian by approximately 25 km (Scharbert 1962; Schermann 1966).

Table 10.1 Selection of climate parameters for the stations Melk, Krems and Jauerling

	m asl	$\sim t$ °C	Freezing/ice days	Summer/hot days	$\sum$ p mm
Melk	240	9.0	83.1/21.9	56.4/10.7	594.4
Krems	207	9.4	93.5/21.6	57.8/12.9	515.7
Jauerling	952	6.5	120.4/49.6	10.1/0.5	729.4

Source http://www.zamg.ac.at/fix/klima/oe71-00/klima2000/klimadaten_oesterreich_1971_frame1.htm). $\sim t$: daily average temperature, **freezing days**: $t < 0$, **ice days**: $t_{max} < 0$, **summer days**: $t_{max} \geq 25$, **hot days**: $t_{max} \geq 30$, $\sum p$ yearly precipitation

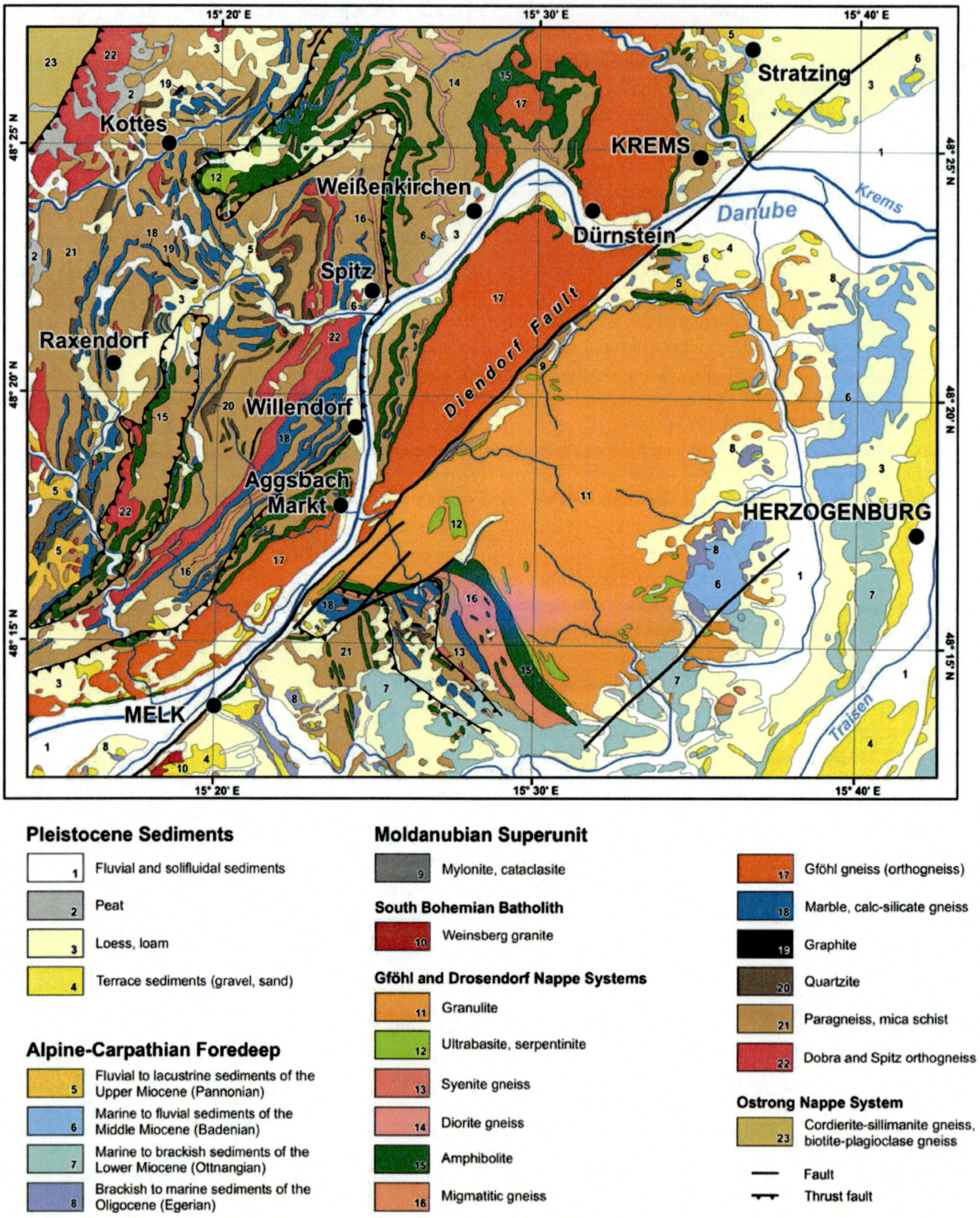

Fig. 10.2 Geological map of the Wachau (from the Geological Map of Lower Austria 1:200000—Schnabel 2002, modified)

Fig. 10.3 Tors in Gföhl gneiss north of the Dürnstein ruin in the Wachau Valley. Photo: W. Gamerith

10.2.3 Development of the Wachau Danube Valley

Together with its subsidiary faults, the Diendorf fault zone was also important for the development of the Danube Valley. Between Melk and Aggsbach Dorf (Figs. 10.2 and 10.5) and at the end of the Wachau near Krems, the Danube flows along the Diendorf master fault, whereas from Spitz to Dürnstein, the river runs along a second parallel fault. Further, NW-SE and N-S trending tectonic lineaments cause the valley segments between Aggsbach and Spitz and around Dürnstein.

Between Krems and Spitz Middle Miocene (Badenian), marine sediments can be found, indicating an ingression of the Badenian Sea into a valley-like depression (Bayer 1927; Papp 1952; Fuchs 1974). In Spitz, the elevation of the Tausendeimerberg (Figure 10.1), located at the confluence of the Spitzer Bach and the Danube, protected fossiliferous Badenian sands and clays from erosion. It can therefore be assumed that the lower Wachau section was in the Middle Miocene a fjord-like bay of the Badenian Sea (Steininger and Roetzel 1999). The wider valley morphology and the gentler slopes additionally confirm likely the existence of a former embayment.

West of Spitz a Miocene Palaeo-Danube can be traced north of the current Danube Valley. Palynologically dated gravelly, sandy and clayey fluvial sediments (Fuchs et al. 1990) in the present valleys of Weitenbach and Spitzer Bach (Figure 10.1) indicate the course of the Upper Miocene (Pannonian) Danube (Nagl and Verginis 1989; Fuchs et al. 1990) (Fig. 10.6)

Similarly to the Thaya, Kamp and Krems Rivers (see Chap. "Deeply Incised Valley Meanders of the Bohemian Massif"), in the late Pliocene and during the Quaternary, the Danube in the Wachau region deeply incised into the margins of the Bohemian Massif. This was probably triggered by a continuing movement of Alpine units towards the north causing a significant rise of the Bohemian Massif and a simultaneous lowering of the drainage system in the foreland (Brzák 1997; Roetzel et al. 2005).

Remnants of Pleistocene river terraces (Matura et al. 1983; Matura and Heinz 1989) as well as Pleistocene loess deposits are mainly preserved in the wider Wachau valley between Spitz and Krems.

Fig. 10.4 Tors in Gföhl gneiss and migmatitic amphibolite in the vicinity of the Aggstein ruin. Photo: R. Roetzel

10.2.4 Quaternary Environment

The Wachau was never directly affected by glaciation during the glacial epochs of the Quaternary. However, the different climatic conditions influenced the upstream morphodynamics of the Danube. During cold periods, the Wachau was part of the periglacial tundra area that extended between the alpine glaciers and the Eurasian Ice Sheet. Huge masses of sediment were produced by glacial and fluvial erosion as well as mechanical weathering due to freeze-thaw cycles. Small particles were blown mainly by strong westerly to northwesterly winds from the barren deposits of glacier forelands and glaciofluvial terraces. The mineral dust accumulated in areas with tundra or steppe vegetation and formed loess deposits (Frank and Rabeder 1997, Nigst et al. 2014).

The loess deposits in the Wachau region are variable in their thickness and age and important subjects of scientific studies. Numerous palaeosols are exposed in loess deposits from early to late Pleistocene in Krems-Schießstätte, Paudorf, Göttweig, Joching (Fig. 10.7) or Stratzing (for site locations see Fig. 10.1) (e.g. Fink 1976; Terhorst et al. 2011, 2013; Thiel et al. 2011; Sprafke et al. 2013). These are proof of the frequent oscillations of cold and warm periods during the Pleistocene (Sprafke 2016). Upper Pleistocene loess in the Wachau and Krems region has revealed several world famous archaeological findings, witnessing early human settlement in the area (Sect. 3.1), e.g. the Venus of Willendorf (Nigst et al. 2014), Venus of Galgenberg (Neugebauer-Maresch 1995) or the Krems-Wachtberg infant burials (Einwögerer et al. 2006).

10.3 Cultural Landscape Wachau

10.3.1 Archaeological Heritage

The settlement history in the Wachau goes back to the Palaeolithic (Fig. 10.8). Around 1900, an Upper Palaeolithic settlement was discovered at the Hundssteig in Krems in the course of loess quarrying. The area has been inhabited by humans since around 35 000 BP (Neugebauer-Maresch 2008), maybe even as far back as 43,500 BP (Nigst et al. 2014). Archaeological excavations in Senftenberg in the Krems Valley also showed an Aurignacian (i.e. early Upper Palaeolithic, see Fig. 10.8) settlement.

Fig. 10.5 View from ruin Aggstein towards southwest, showing the upstream Danube Valley. The Danube River, first running straight along the Diendorf fault between Melk (at the valley entrance in the background) and Aggsbach Dorf (below the ruin), is bending south of Aggsbach along a N-S trending fault to the north, while the Diendorf fault is continuing straight into the Dunkelstein Forest. Photo: R. Roetzel

In 1988, a 32,000-year-old statue made of local greenish amphibolite slate was found on Galgenberg near Stratzing during the excavation of a regularly used Aurignacian shelter. Because of its dancing pose, it was named "Fanny of Galgenberg" (small photo inserted in Fig. 10.8) after the famous Viennese dancer Fanny Elßler (Neugebauer-Maresch, 1995).

In Willendorf, a small village in the Wachau on the left bank of the Danube River right opposite Aggstein Ruin, excavations by scientists from the Museum of Natural History Vienna, unearthed an eleven cm high red chalk coloured female statuette of oolitic limestone that became famous as the Venus of Willendorf. It dates from the Gravettian period and was handcrafted about 29,500 years ago (small photo inserted in Fig. 10.8) (Antl-Weiser 2008).

At the beginning of the Neolithic period, settlements started to spread over the highlands of the Waldviertel. Danubian Linear Pottery Culture, an early Neolithic culture in Middle Europe, is proven in the area. With it the change from the nomadic life of hunters and food gatherers to the sedentary life of a food producing and domestic animals, breeding community is induced (Ruttkay et al. 1976).

10.3.2 Vines and Apricots: Development of a Picturesque Agricultural Landscape

The Celts already had a simple wine culture, but the Roman emperor Domitian passed an edict in AD 80 that prohibited viticulture north of the Alps. Only when the emperor Probus revoked the edict in AD 280 did viticulture start to flourish in the Wachau area. At the end of the fifth century, the Roman period in the Wachau ended. The region was confronted with the commotions during the migration of Germanic tribes, viticulture stagnated and only expanded again after Charles the Great evicted the Avars.

In the ninth century, Bavarian and Salzburgian monks came to the area and began to cultivate the land (Stadler

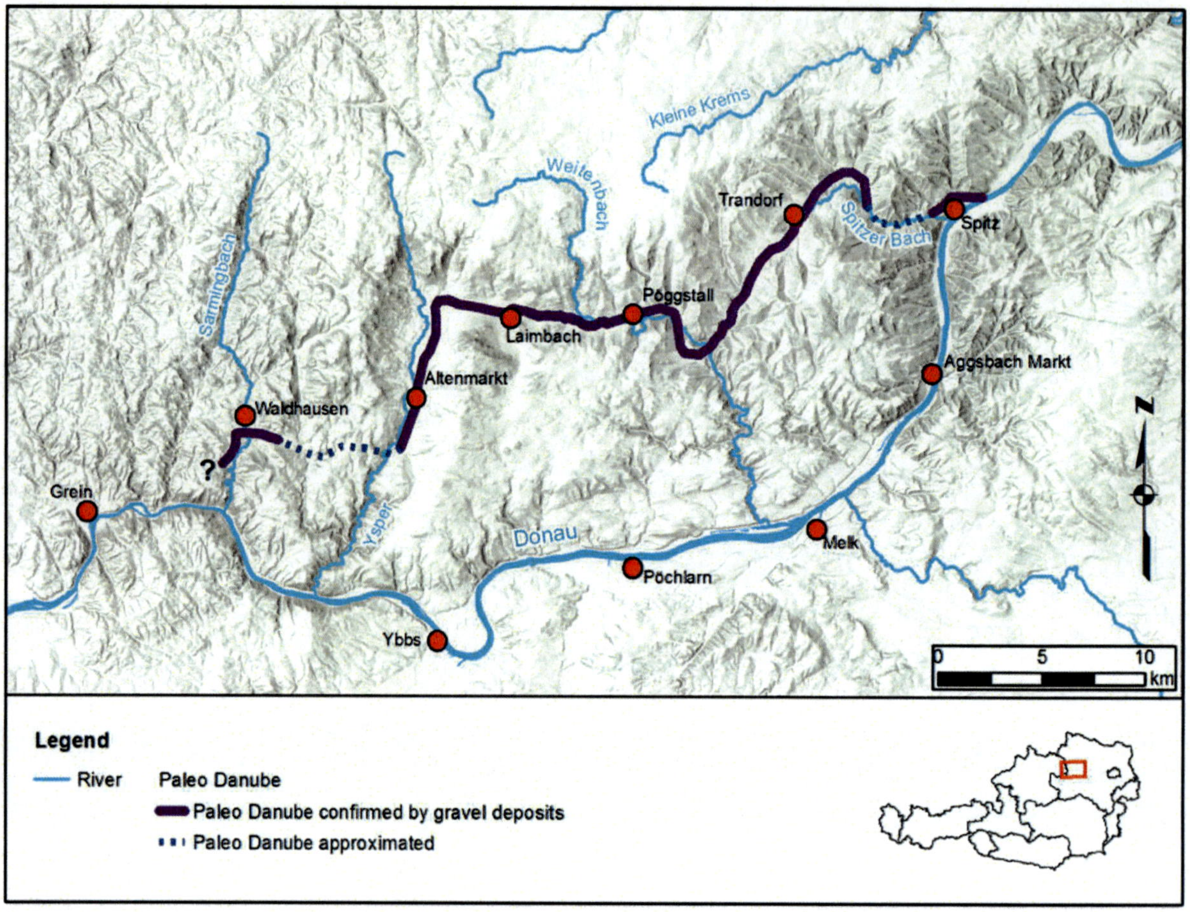

Fig. 10.6 Assumptive course of the Palaeo-Danube (Credits for Background map: USGS, NGA, NASA, CGIAR, N Robinson, NCEAS, NLS, OS, NMA, Geodatastyrelsen, GSA, GIS User Community)

1997). They planted new vineyards and established terraces to prepare flat areas, which are part of today's landscape pattern (ICOMOS 2000). The terraces are protected against slippage by stone walls, produced as dry masonry without mortar (Fig. 10.9). After devastating rainfalls in August 2002, many walls started sliding, especially the modern concrete walls. Since then, construction and repair of the old dry masonry has again intensified (Wachaumagazin 2003). Additionally, the walls provide a secondary benefit: they store heat, which has positive effects on the vines and apricots and provides shelter for lizards and other animals.

In the seventeenth century, labour shortage and increasing wages resulted in a decrease of viticulture. On the upper hillslopes forests advanced and in the valleys alternative cultures were attempted. The remaining areas turned to dry grass and shrublands. In the eighteenth century, hillside viticulture was promoted in ecologically optimal regions. Released areas were turned into pastures and in the upper stretches of the Wachau wine growing was stopped (ICOMOS 2000).

At the end of the nineteenth century, the appearance of Phylloxera, a small sap sucking insect, which was introduced from North America, stopped the rise of wine growing in the Burgenland and other wine areas. The Wachau winegrowers moved to higher yielding products. This was the opportunity for the apricot production (Fig. 10.10). In the 1950s, apricot growing in the Wachau was at its zenith. Since the mid-1950s, viticulture advanced again, but apricot farming in the Wachau is still an important commercial factor and the flowering apricot trees are a famous tourist attraction. (ICOMOS 2000). The Arbeitskreis Wachau (working group Wachau), which was installed in 1972 to prevent the construction of a hydro power plant Wachau, emphasises in its mission statement the essential role of wine growing for the region and promotes apricot farming

Fig. 10.7 Loess site of Joching: in a vineyard near Joching Middle Pleistocene loess with the last interglacial pedocomplex is overlain by Upper Pleistocene loess (Terhorst et al. 2011; Thiel et al. 2011). Photo: R. Roetzel

because of its importance for the local economy and landscape (Arbeitskreis Wachau 2016).

10.3.3 Protection and Management of the Cultural Heritage

From 1700 onwards, artistic and architectural monuments that are among the most important examples of Austrian Baroque were built or rebuilt in the Wachau. These include the Melk Abbey, the Canons' Abbey in Dürnstein (Fig. 10.11) and the Göttweig Abbey.

In the year 2000, the World Heritage Committee of the UNESCO decided on the basis of two cultural criteria (ii and iv) to inscribe the Wachau including the Abbeys Göttweig and Melk and the historic town centre of Krems as Cultural Landscape to the World Heritage List:

"*Criterion (ii): The Wachau is an outstanding example of a riverine landscape bordered by mountains in which material evidence of its long historical evolution has survived to a remarkable degree.*

Criterion (iv): The architecture, the human settlements and the agricultural use of the land in the Wachau vividly illustrate a basically mediaeval landscape that has evolved organically and harmoniously over time" (WHC 2000).

The inscription to the World Heritage List was only one stage along the way to protect the unique cultural landscape of the Wachau. Already in 1955, the "protected landscape area Wachau" was founded, which is one of 29 protected areas in Lower Austria. Its areal delimitation corresponds to the Wachau World Heritage Site (see Fig. 10.1). The Nature Park Jauerling—Wachau—already exists since 1972. The forms of protection intend to preserve characteristic designed cultural landscapes and a natural, healthy habitat for humans. In 2011, additionally, the European Protected Areas Wachau and Wachau-Jauerling were established under the Birds Directive of 2011 in accordance with the European Habitat Directive of 2009.

In 2010, a new hiking trail, the "World Heritage Trail Wachau", was opened. Along the 14 trail sections, 20 fortresses and castles, cultural highlights like the Abbeys Göttweig and Melk together with historical town centres (e.g. Dürnstein) alternate with typical Wachau landscapes like vineyards, apricot gardens, forests and feathergrass meadows. Several vantage points provide distant vistas and good views of the Danube Valley.

10.4 Natural Hazards in the Wachau

The Danube Valley Wachau and the Danube Floodplain National Park (see Chap. "The Danube Floodplain National Park—A Fluvial Landscape with Expiration Date?") are the last two free flowing reaches of the Danube in Austria. Many towns in the Wachau are located on the banks of the Danube and thus exposed to a high flood risk. A minor but peculiar hazard is rockfall, because the different rocks of the Bohemian Massif are not normally prone to this type of mass movements.

10.4.1 Floods

The most damaging floods in the Wachau are summarised in Table 10.2. The latest major flood (with a HQ > 100 discharge) struck the upper Danube basin in June 2013 and caused heavy damage along the Danube and numerous tributaries (Blöschl et al. 2013). Figure 10.12 depicts the rise and fall of this flood wave in the Wachau with discharge

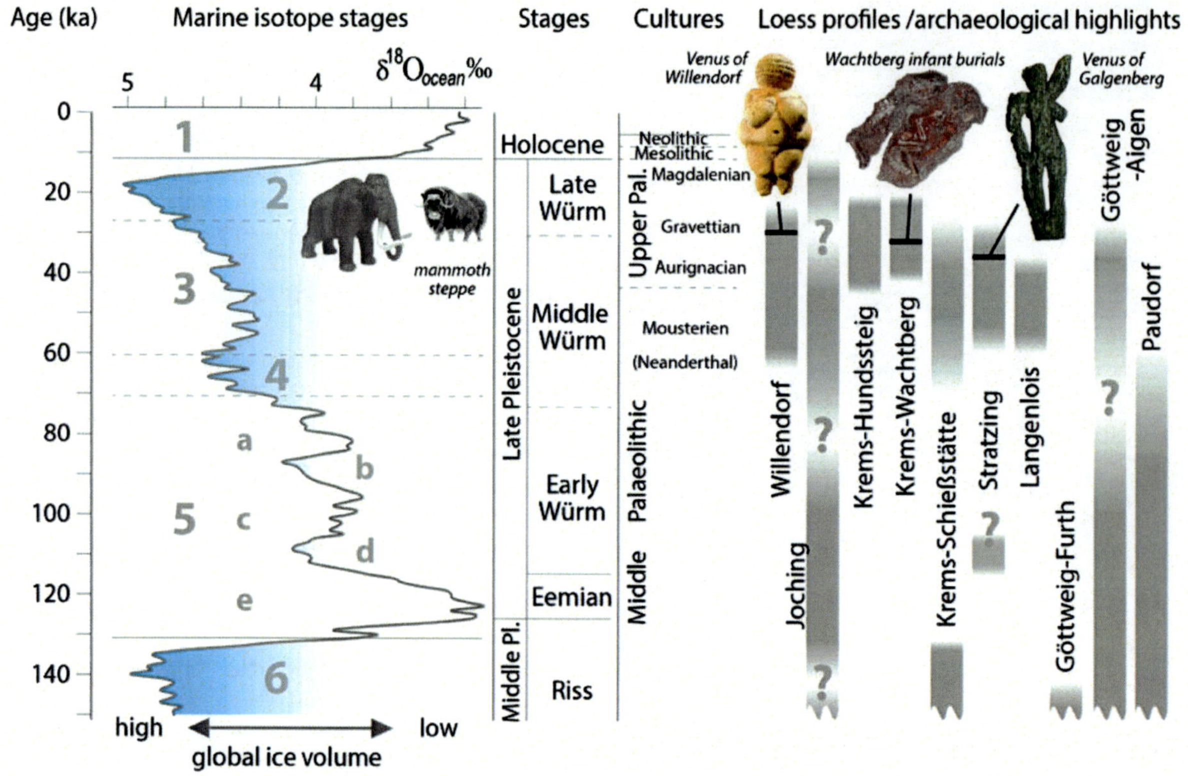

Fig. 10.8 Late Palaeolithic findings and loess deposits of the last 150,000 years in the Wachau and Krems region. Comparison with the marine oxygen isotope record, which indicates variations in the global ice volume (Lisiecki and Raymo, 2005). Redrawn and updated after Wessely and Draxler (2006). Grey bars indicate age range of loess deposition and pedogenesis recorded in the loess profiles and question marks probable erosional gaps; age information of loess deposits, archaeological cultures, Würm stratigraphy and marine isotope stages compiled from various sources (Einwögerer et al. 2006; Haesaerts et al. 2007; Neugebauer-Maresch, 2008; Nigst et al. 2014; Preusser, 2004; Rasmussen et al. 2014; Shackleton et al. 2003; Sprafke, 2016; Thiel et al. 2011; Photos: "Fanny" Venus of Galgenberg, Venus of Willendorf by Don Hitchcock; infant burial by Thomas Einwögerer)

indicated in blue and water level above the 2010 average level indicated in red.

The June 2013 flood came hard on the heels of another HQ 100 flood only 11 years earlier in August 2002. The 2002 flood hit the entire upper Danube basin and exceeded all previous HQ 100 events in death toll, inundation area and economic damage. It is still referred to as "the century flood" in the media (Blöschl et al. 2013).

In terms of flood protection, the Danube flood of 1991 with a damage of ~€6 Mio. led to the decision to install mobile flood protection structures costing a total amount of € 12 Mio. These structures proved to be very effective, for example, during the HQ 100 event on 3 June 2013 in Krems-Stein (Fig. 10.13).

10.4.2 Rockfalls

In 2002 and 2009, high volume rockfalls from the mining faces of two abandoned quarries located within the steep slopes of the Danube Valley caused severe damage to road and railway. Preparation factors of these mass movements were partly natural, arising from discontinuities within the tectonically highly fractured, fragmented and weakened rock mass along the Diendorf fault zone (see Fig. 10.2). However, the main source of slope instability was the former quarrying activity (Rachoy & Scheikl 2006; Laimer & Müllegger 2012; Müllegger 2013).

In the first quarry near Spitz, extreme undercutting of the footslope triggered the failure of a huge rock mass in 1961.

Fig. 10.9 Dry masonry walls in the Wachau near Arnsdorf. Photo: D. Riedl

Fig. 10.10 Vineyard with flowering apricot trees near Arnsdorf. Photo: D. Riedl

Fig. 10.11 Abbey in Dürnstein with its characteristic blue and white bell tower. Photo: R. Roetzel

Table 10.2 Historic Danube floods at the gauging station Kienstock/Wachau (HQ = flood return period, RHHW = highest flood level) (*source* http:www.noel.gv.at/wasserstand/static/stations/207357/station.html)

Date	Runoff (m3/s)	Annuality
4.6.2013	11.450	HQ 100
14.8.2002	11.305	HQ 100
23.3.2002	8.588	HQ 17
4.8.1991	9.647	HQ 26
2.7.1975	8.800	HQ 19
13.7.1954	10.200	HQ 35
17.9.1899	11.200	HQ 100
2.8.1897	9.900	HQ 28
4.1.1883	8.520	HQ 17
4.2.1862	10.500	HQ 50 (ice jam)
1787	11.900	HQ 100
1501	14.000	RHHW
1402	10.500	HQ 30

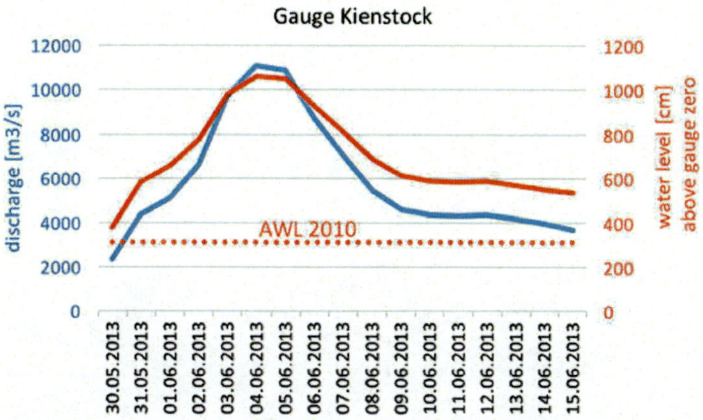

Fig. 10.12 Gauging station Kienstock during the flood in June 2013 (*source* https://www.bmlfuw.gv.at/dam/jcr:bb100102-eac3-4b51-8d8d-f40b94e5d546/Jahrbuch_2013_Datenteil.pdf)

Fig. 10.13 Mobile flood protection structures along the river banks of the Danube in Krems-Stein during the HQ 100 event on June 3rd 2013 in Krems-Stein. Photo: R. Schmid

Fig. 10.14 Gföhl gneiss in the Biratalwand in an abandoned quarry close to Dürnstein town after a huge rock mass fall in July 2009. Photo: R. Roetzel

After a second failure in 1984, geotechnical monitoring of the mining face was installed and revealed ongoing movements. In 1996, the quarry was closed and remedial measures to stabilise the former quarry faces were planned. But before, they could be implemented the above mentioned rockslide of 2002 with an estimated volume of 60 000 to 80 000 m^3 occurred. The second quarry in Dürnstein was already closed in 1903. In order to remove potentially unstable parts of the abandoned quarry, 65 000 m^3 of rock were blasted away in 1909. The residual rock face was destabilised and rockfall activities continued, culminating in the event of 2009 with a total volume of approximately 15,000 m^3 in 2009 (Laimer & Müllegger 2012; Fig. 10.14 and 15).

10.5 Conclusion

The Wachau is a beautiful, diverse riverine landscape with a strong anthropomorphic influence starting in prehistoric times and continuing during neolithic, celtic and roman times until today. Its unique combination of diverse geological development and history, cultural landscape and carefully managed building ensembles forms a worthwhile place to live, is an attraction for tourists and deserves a place in the list of the UNESCO World Heritage Sites. Today, in an association called "Arbeitskreis Wachau", people work together to protect the Wachau area and give guidelines for its further management. The main hazard for the Wachau

Fig. 10.15 Rock mass fall from the Biratalwand at Dürnstein in July 2009, destroying the track of the Wachau railway. Photo: R. Roetzel

landscape is flooding, induced by its nearness to the narrow free flowing Danube stretch. Another peculiar threat is rockfall which occurs anthropogenically induced in the otherwise rather stable rocks of the area.

References

Antl-Weiser W (2008) The anthropomorphic figurines from Willendorf. Wiss. Mitt. Niederösterr. Landesmuseum 19:19–30

Arbeitskreis Wachau (2016) Mission of Arbeitskreis Wachau. http://www.arbeitskreis-wachau.at Accessed 10 Aug 2016

Bayer J (1927) Entdeckung von Ablagerungen der I. Mediterranstufe in der Wachau. Verh Geol Bundesanst 1927:107–110

Blöschl G, Nester T, Komma J, Parajka J, Perdigão RAP (2013) The June 2013 flood in the Upper Danube basin, and comparisons with the 2002, 1954 and 1899 floods. Hydrol Earth Syst Sci 17:5197–5212. https://doi.org/10.5194/hess-17-5197-2013

Brzák M (1997) Entwicklung des Thayatales im Nationalpark Podyjí (Tschechische Republik) aufgrund der Morphographie und Schotteranalyse. Mitt Österr Geograph Ges 139:261–275

Einwögerer T et al (2006) Upper Paleolithic Infant Burials. Nat 444:285

Fink J (ed) (1976) Exkursion durch den österreichischen Teil des nördlichen Alpenvorlandes und den Donauraum zwischen Krems und Wiener Pforte. Erweiterter Führer zur Exkursion aus Anlaß der 2. Tagung der IGCP-Projektgruppe "Quaternary glaciation in the northern hemisphere". Mitt. Kommission Quartärforschung österr. Akad. Wiss. 1, 113 p

Frank C, Rabeder G (1997) Klimageschichte des österreichischen Plio-Pleistozäns. In Döppes D, Rabeder G (eds) Pliozäne und pleistozäne Faunen Österreichs. Mitt. der Komm. für Quartärforschung der öst. Akademie der Wissenschaften 10:375–380

Fuchs G, Matura A (1976) Zur Geologie des Kristallins der südlichen Böhmischen Masse. Jahrbuch Geol Bundesanst 119:1–43

Fuchs G, Roetzel R, Heinz H (1990) Geologische Karte der Republik Österreich 1:50.000. Erläuterungen zu Blatt 36 Ottenschlag. 64 p. (Geol. Bundesanst.)

Fuchs W (1974) Bericht 1973 über Aufnahmen auf den Blättern Mautern (37), Krems (38) und St. Pölten (56) sowie über Vergleichsbegehungen auf den Blättern Melk (54), Tulln (39), Deutsch-Wagram (41) und Wien (59). Verh Geol Bundesanst 1974:A47–A50

Fuchs W, Grill R, Matura A, Vasicek W (1984) Geologische Karte der Republik Österreich 1:50.000 38 Krems. Geologische Bundesanstalt, Wien

Gamerith W (2003) Wachau. Lebensräume einer Kulturlandschaft. Verlag Tyrolia, 232 p

Haesaerts P, Borziac I, Chirica V, Damblon F, Koulakovska L (2007) Cadre stratigraphique et chronologique du Gravettien en Europe centrale. Paleo - Revue d'archéologie préhistorique 19:31–52

ICOMOS (2000) Nomination Wachau Cultural Landscape http://whc.unesco.org/uploads/nominations/970.pdf. Accessed 10 Aug 2016

Laimer HJ, Müllegger M (2012) Geomorphological and geotechnical causes of anthropogenic-induced rock-mass falls in the Wachau-Danube Valley (Bohemian Massif, Lower Austria). Geogr Ann Ser B 94:157–174

Lisiecki, LE and Raymo, ME (2005) A Pliocene-Pleistocene stack of 57 globally distributed benthic $\delta^{18}O$ records. Paleoceanography 20: PA1003

Matura A, Fuchs W, Grill R (1983) Geologische Karte der Republik Österreich 1:50.000 37 Mautern. Geologische Bundesanstalt, Wien

Matura A, Heinz H (1989) Geologische Karte der Republik Österreich 1:50.000. Erläuterungen zu Blatt 37 Mautern. 65 p. (Geol. Bundesanst.)

Müllegger M (2013) Beispiele aus Niederösterreich Felssturz Dürnstein/ Wachau – Das schwere Erbe der Biratalwand. Berichte Geol Bundesanst 100:18–20

Nagl H (1983) Klima- und Wasserbilanztypen Österreichs – Versuch einer regionalen Gliederung mit besonderer Berücksichtigung des außeralpinen Raumes. Geographischer Jahresbericht aus Österreich 40:50–72

Nagl H, Verginis S (1989) Die Morphogenese der Wachau. Versuch einer neuen Deutung. Geogr Jahresber Österr 46 (1987):7–30

Neugebauer-Maresch C (1995) Die Statuette vom Galgenberg bei Stratzing. In: Kühnel H, Schönfellner F (eds) Tausend Jahre Krems. Ein Jubiläumsbuch, Wien, p 83 ff

Neugebauer-Maresch C (2008) Krems-Hundssteig - Mammutjägerlager der Eiszeit. Ein Nutzungsareal paläolithischer Jäger- und Sammler (innen) vor 41.000–27.000 Jahren. Mitteilungen der Prähistorischen Kommission 67, VÖAW

Nigst PR, Haesaerts P, Damblon F, Frank-Fellner C, Mallol C, Viola B, Götzinger M, Niven L, Trnka G, Hublin J-J (2014) Early modern human settlement of Europe north of the Alps occurred 43,500 years ago in a cold steppe-type environment. Proceed Nat Acad Science 111(40):14394–14399

Papp A (1952) Zur Kenntnis des Jungtertiärs in der Umgebung von Krems a.d.Donau (N.-Ö.). Verh Geol Bundesanst 1952:122–126

Preusser F (2004) Towards a chronology of the Late Pleistocene in the northern Alpine Foreland. Boreas 33(3):195–210

Rachoy Ch, Scheikl M (2006) Anthropogenic Caused Mass Movements and their Impact on Railway Lines in Austria. In: Marui H et al. (eds) Disaster Mitigation of Debris Flows, Slope Failures and Landslides, Proceed Interpreavent Int Symp, Nigata Japan, pp 639–643

Rasmussen SO, Bigler M, Blockley SP, Blunier T, Buchardt SL, Clausen HB, Cvijanovic I, Dahl-Jensen D, Johnsen SJ, Fischer H, Gkinis V, Guillevic M, Hoek WZ, Lowe JJ, Pedro JB, Popp T, Seierstad IK, Steffensen JP, Svensson AM, Vallelonga P, Vinther BM, Walker MJC, Wheatley JJ, Winstrup M (2014) A stratigraphic framework for abrupt climatic changes during the Last Glacial period based on three synchronised Greenland ice-core records: refining and extending the INTIMATE event stratigraphy. Quatern Sci Rev 106:14–28

Rentschler D (2012) Marken und Markgrafen im früh- und hochmittelalterlichen Reich. Eine vergleichende Untersuchung vorwiegend auf der Basis von Königsurkunden und anderen „offiziellen Quellen“. Dissertation, Universität Stuttgart

Roetzel R, Fuchs G, Havlíček P, Übl Ch, Wrbka Th (2005) Geologie im Fluss. Erläuterungen zur Geologischen Karte der Nationalparks Thayatal und Podyjí. Geologische Bundesanstalt, Wien

Ruttkay E, Wessely G, Wolff P (1976) Eine Kulturschicht der ältesten Linearbandkeramik in Prellenkirchen, p. B. Brück, Niederösterreich. Ann. Naturhist. Museum Wien, 80:843–861

Scharbert HG (1962) Die Granulite der südlichen Böhmischen Masse. Geol Rundsch 52:112–123

Schermann O (1966) Über Horizontalseitenverschiebungen am Ostrand der Böhmischen Masse. Mitt Ges Geol Bergbaustud Wien 16 (1965):89–103

Schnabel W (ed) (2002) Geologische Karte von Niederösterreich 1:200000. Geologische Bundesanstalt, Wien

Shackleton NJ, Sánchez-Goñi MF, Pailler D, Lancelot Y (2003) Marine Isotope Substage 5e and the Eemian Interglacial. Global Planet Change 36(3):151–155

Sprafke T, Terhorst B, Peticzka R, Thiel C (2013) Paudorf locus typicus (Lower Austria) revisited—the potential of the classic loess outcrop for Middle to Late Pleistocene landscape reconstructions. Quatern Science Journ 62:59–72. https://doi.org/10.3285/eg.60.1.06

Sprafke T (2016) Löss in Niederösterreich - Archiv quartärer Klima- und Landschaftsveränderungen. Würzburg University Press, Würzburg, 272 p

Stadler C (1997) Die Landschaftsveränderungen der Wachau im Spiel der Vegetation: ein Beitrag zur Landschaftscharakteristik. Diplomarbeit, Wien

Steininger FF, Roetzel R (1999) Jüngeres Tertiär. In: Steininger FF (ed) Erdgeschichte des Waldviertels, Schriftenreihe Waldviertler Heimatbund 38:79–88

Terhorst B, Thiel C, Peticzka R, Sprafke T, Frechen M, Fladerer FA, Roetzel R, Neugebauer-Maresch C (2011) Casting new light on the chronology of the loess/paleosol sequences in Lower Austria. Quatern Science Journ 60:270–277. https://doi.org/10.3285/eg.60.2-3.04

Terhorst B, Kühn P, Damm B, Hambach U, Meyer-Heintze S, Sedov S (2013) Paleoenvironmental fluctuations as recorded in the loess-paleosol sequence of the Upper Paleolithic site Krems-Wachtberg. Quat Int 351:67–82. https://doi.org/10.1016/j.quaint.2013.03.045

Thiel C, Buylaert JP, Murray AS, Terhorst B, Tsukamoto S, Frechen M, Sprafke T (2011) Investigating the chronostratigraphy of prominent palaeosols in Lower Austria using post-IRSL dating. E&G Quaternary Science Journal 60(1):137–152

Wachaumagazin (2003) Reisen in einer der schönsten Kultur-Landschaften Europas

Wessely G, Draxler I (2006) Pliozän und Quartär. In: Wessely G (ed), Niederösterreich. Geologie der Österreichischen Bundesländer), pp 235–252

WHC World Heritage Committee (2000) Report of the United Nations educational, scientific and cultural organisation convention concerning the protection of the world cultural and natural heritage, World Heritage Committee, 24th session, Cairns Australia.

Doris Riedl is a lecturer at the Department of Geography and Regional Research at the University of Vienna. Her primary research interests are hydrological networks and GIS.

Reinhard Roetzel is a sedimentologist and mapping geologist, who worked up to 2020 at the Geological Survey of Austria. His research focuses mainly on geological surface mapping of Cenozoic sediments and crystalline rocks, as well as sedimentology and sedimentary petrology of Paleogene, Neogene and Quaternary sediments on the Bohemian Massif and in the Alpine–Carpathian Foreland Basin of Lower and Upper Austria.

Ronald E. Pöppl is a senior lecturer at the Department of Geography and Regional Research at the University of Vienna. His primary research interests include connectivity in fluvial systems, fluvial geomorphology, soil erosion, human impacts and human–environment interactions, river and catchment management..

Tobias Sprafke is a postdoctoral researcher at the Institute of Geography, University of Bern, interested in terrestrial records of quaternary climate, soil and landscape evolution. His main research topic is the interplay between dust deposition, soil formation and slope processes. He has extensive experience in loess-palaeosol sequences of Austria, Germany, Poland and Kazakhstan and works on pedogenesis in Central Europe, Brazil and Cameroon.

Sunken Roads and Palaeosols in Loess Areas in Lower Austria: Landform Development and Cultural Importance

11

Helene Petschko, Tobias Sprafke, Robert Peticzka, and Heinz Wiesbauer

Abstract

Loess, a light-yellowish sediment composed of silt-sized material, takes up a large proportion of the northeast of Lower Austria and is full of surprises of cultural and scientific importance. On the one hand, the presence of loess results in a particular set of landforms such as loess dells, gullies, sunken roads and sinkholes of both natural and anthropogenic origin. On the other hand, the many outcrops of loess with several metres of height still reveal unprecedented insights in the climatic and settlement history of Lower Austria. Both, sunken roads and palaeosols may be seen as silent witnesses of past landscape changes but on a very different time scale. Originating from a natural gully erosion process, sunken roads evolved from gullies by their transformation in access paths leading to agricultural fields. Many driving forces influenced the vertical and lateral erosion of sunken roads, which showed increasing erosion rates of up to 15–30 cm per year until the mid-twentieth century. Nowadays, many sunken roads have disappeared and many remaining ones are paved or protected from further erosion and are transformed to cellar lanes (sunken roads with wine cellars dug next to each other into the loess wall). Results from analysing the ages of palaeosols found in Lower Austrian loess profiles are internationally important. Still to the day, with upcoming new dating methods and the chance for more detailed analysis, new insights into the stratigraphy, the related ages, palaeorelief and climate oscillations are found in close collaborations of geographers and archaeologists. Many internationally important archaeological artefacts and presumably the oldest loess of Europe were found in Lower Austria.

Keywords

Gully erosion • Cellar lanes • Natural and anthropogenic landscape change • Palaeoclimate change • Oldest loess of Europe

11.1 Loess and Loess Landscapes in Lower Austria

Loess is a light-yellowish Quaternary sediment, which is widespread in the earth's temperate climate zone; it is composed of silt-sized mineral powder and smaller amounts of sand and clay that is aggregated to fragile blocks (Pécsi and Richter 1996). As loess is the parent material for the world's most productive soils, since millennia its landscapes are strongly affected by human activity. Human impact is responsible for specific landforms like sunken roads, and in sufficiently thick deposits, human presence is documented by Palaeolithic artefacts or even fireplaces from our early ancestors. We start with an overview of the strong influence of loess on the geomorphology in Lower Austria. Landforms and landscapes of large areas in the Austrian lowlands along the Danube are related to the unique loess properties and distribution. In the following sections, we give details on the evolution, characteristics and cultural importance of sunken roads in Lower Austria (Sect. 3), and a summary of past and recent studies on local palaeosols, highlighting the importance of loess research in Lower Austria for the reconstruction of climate and landscape change of the last glacial and interglacial oscillations (Sect. 4).

H. Petschko (✉)
Department of Geography, Friedrich Schiller University Jena, Löbdergraben 32, 07743 Jena, Germany
e-mail: Helene.petschko@uni-jena.de

T. Sprafke
Institute of Geography, University of Bern, Hallerstrasse 12, 3012 Bern, Switzerland

R. Peticzka
Department of Geography and Regional Research, University of Vienna, Althanstraße 14, 1090 Wien, Austria

H. Wiesbauer
Freelance Civil Engineer for Landscape Planning and Maintenance, Börsegasse 7/14, 1090 Wien, Austria

C. Embleton-Hamann (ed.), *Landscapes and Landforms of Austria*, World Geomorphological Landscapes,
https://doi.org/10.1007/978-3-030-92815-5_11

11.1.1 Origin, Characteristics and Distribution

Loess is, in general, considered a deposit of windblown dust, although there is ongoing discourse on the validity of this simple definition (Sprafke and Obreht 2016). The Austrian loess formed during the cold stages of the Pleistocene in tundra to cold steppe environments, mainly during the most significant Alpine glaciations. Various sources of the material were involved and several phases of glacial, fluvial, slope and aeolian transport occurred before their final deposition. Next to a hemispheric component of fine dust, the majority of silt is derived from regional sources and to a large part was brought in by rivers (Smalley et al. 2009). Close to the Alps, a large amount of material came from glacier forefields. However, close to Krems, a coarse local bedrock component in the loess sediments points to a significant local to regional contribution to the deposits (Sprafke et al. 2013).

Despite the various sources of loess material, there are specific characteristics related to weak cementation processes after deposition of silt particles (Smalley and Marković 2014). In general, the mineral composition of loess is variable and partly depending on the geology found in the source region. As in most loess regions, Lower Austrian loess also has a dominating quartz component followed by feldspar and other silicates including clay minerals (Pécsi and Richter 1996). Carbonate contents in the Krems region are 20–25%, which is above the average and related to the proximity of the Calcareous Alps. Carbonate and clay are responsible for the weak cementation of loess, but it is unclear if it is related to pedogenesis or diagenesis (Sprafke and Obreht 2016).

Loess is relevant for agriculture and terroir of Austria's famous wine, the "Grüner Veltliner". It is also an important building ground with specific geotechnical properties. It forms rather stable, steep outcrop walls. However, the presence of running water easily erodes the silt-dominated, loosely cemented loess (Pécsi and Richter 1996) and the very erodible soils derived from loess (Poesen 1993).

Loess in Austria is restricted to the lowlands around the Alps and 90% of the Austrian Loess is found in Lower Austria (Wessely and Draxler 2006). The maximum thickness is 35–40 m at Krems a.d. Donau, at the eastern exit of the Wachau Valley of the Danube (Fig. 11.1; Fink and Kukla 1977).

A distinct feature of the Austrian loess belt is the strong gradient in weathering intensity from the west to the east, related to a strong decrease in precipitation from 700 to 900 mm to around 500 mm along this transect (Fink 1956, Hydrographic Service of Lower Austria 2011). The Bohemian Massif acts as main climate shed that reduces Atlantic moisture in the east. The present-day soil cover and the characteristics of the loess and its palaeosols reflect this climatic gradient. Fink (1956) differentiated the humid loess landscape (from the German boarder until St. Pölten) from the dry loess landscape (east of Krems to the eastern border; Fig. 11.1) and termed the region around Krems Fink a transition zone. However, the distinct characteristics of the loess profiles around Krems are mostly related to the specific morphological conditions at the eastern margin of the Bohemian Massif (Terhorst et al. 2015).

Luvisols, which are decalcified weathered forest soils with clay translocation, are characteristic for the humid loess landscape. The "brown loess", as widespread in the most humid parts of this region, lost its light pigmentation by decalcification (Fig. 11.1). Typical light-yellowish calcareous loess can be widely found in the dry loess landscape (Fink 1954). There, the present-day soil cover shows mainly humus rich, weakly weathered soils, typical for steppe environments of continental climate. The map of loess distribution also differentiates "sand loess", which is mainly related to the proximity of the sediment source (Fig. 11.1). It can mainly be found close to rivers grading into aeolian sands close to the Danube near Vienna and east of the Morava River (see Fig. 11.1 in Fink and Nagl 1979)).

11.1.2 Loess Landscapes and Their Morphology

In general, the morphology of loess landscapes depends on the thickness and stratigraphic differentiation of the loess cover as well as the (palaeo-) climatic conditions (Pécsi and Richter 1996). The Austrian loess landscapes as defined by Fink (1954, 1956, 1961) are regions covered by loess rather than landscapes developed within thick loess deposits such as, e.g. the Chinese or Middle Danube loess plateaus. In Austria, the loess cover predominantely is a thin veneer (of a few decimetres to maximal one metre thickness), partly mixed with local bedrock in the undulating hilly lowlands around the Alps, smoothing the pre-existing morphology. However, in specific morphological positions, several metres of loess thickness can be observed, especially in the dry loess landscape. Dominant loess landforms are trough systems with dry dells, dry valleys and loess covered terraces (syngenetic) or sinkholes, loess falls, gullies and ultimately steep slopes where slides occur (postgenetic; Leger 1990). Thus, the Austrian loess regions have their very own aesthetics: gently rolling hills that interchange with areas of several metre high yellowish cliffs, often close to river incisions. Human activity has had an important influence on this scenery by, for example, terracing for agriculture, creation of road cuts, development of sunken roads, also known

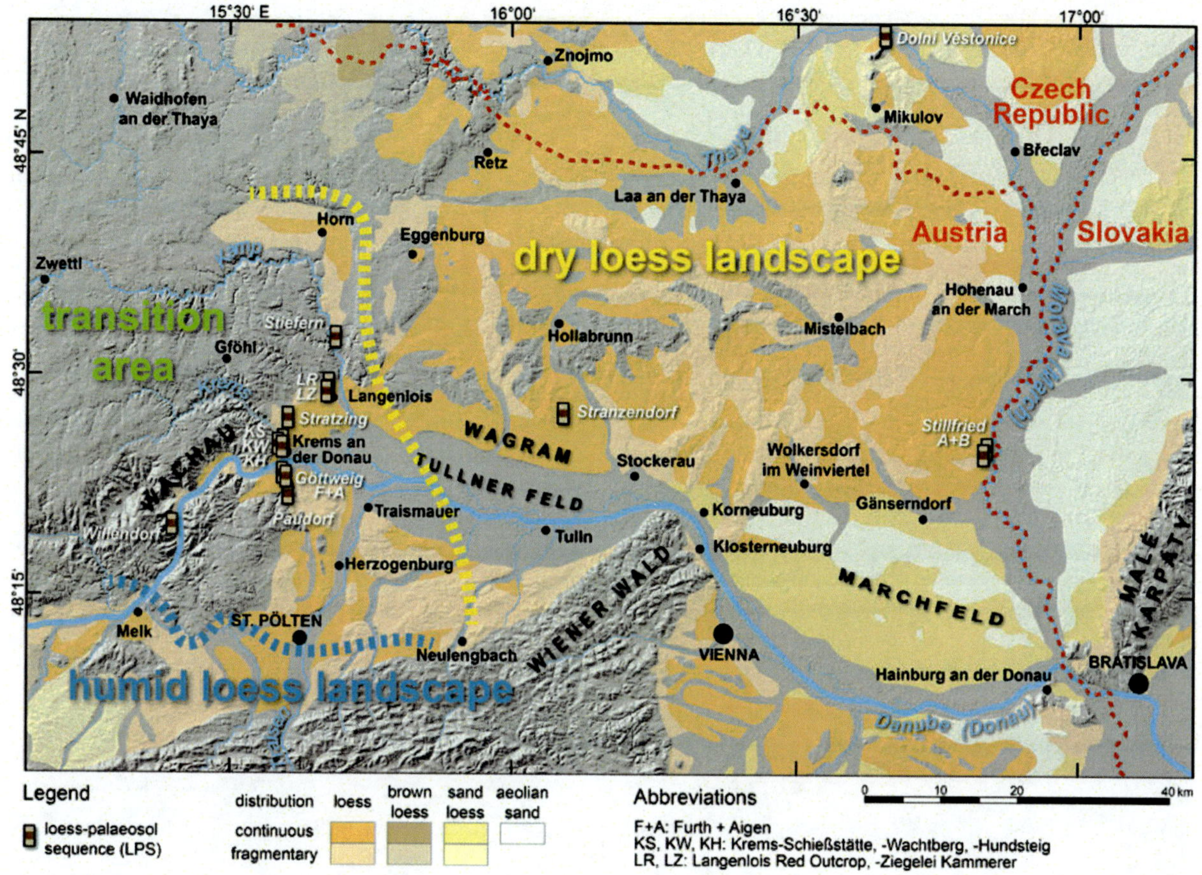

Fig. 11.1 Loess landscapes after Fink (1956) and locations of important loess–palaeosol sequences. Map adapted from Sprafke (2016), based on Fink and Nagl (1979), Fink (1965), Loishandl-Weisz and Peticzka (2005) and Shuttle Radar Topography Mission data of the US Geological Survey (USGS 2014)

as sunken lanes, hollow lanes or hollow roads, exploitation of loam pits and creation of loess cellars.

Loess-derived soils with a low organic matter content (<2%) of western and eastern Europe rank among the most susceptible soils for water erosion worldwide (Poesen 1993). Accordingly, the soils developed on loess in Lower Austria show many erosional landforms such as sinkholes and gullies (Schwenk 1992). In Lower Austrian loess, sinkholes are often related to artificial hollows due to excavations (cellar tubes) rather than natural hollows below the soil surface. Human-induced topography or land use changes reportedly affected the slope hydrology, causing erosion within cellar tubes and ultimately the collapse in form of sinkholes showing at the surface (Schwenk 1992). The large gullies of the wine-growing region “Wagram”, a distinct terrace landscape feature of the Danube River (Fig. 11.1), are silent witnesses of historical soil erosion. Local inhabitants transformed many of the original gullies into paths to access agricultural fields located on the higher terrain of the Wagram with machinery (Strauss and Klaghofer 2006). Over the years, the combination of unpaved paths and high erosion rates led to the formation of sunken roads (Krissl 1982).

11.2 Sunken Roads

11.2.1 Landform Development

Sunken roads are unpaved, dirt roads incised in the natural level of the surrounding landscape with height differences between the base of the road and the surrounding landscape ranging from sub-metre to up to 50 m (in China or the Ukraine Leger 1990; David et al. 2011). Actively used sunken roads typically have a U-shaped cross section with steeply inclined walls (up to 60°) in the upper part and a gentler slope (20–35°) at the lower part, where loose material from the top accumulates (Leger 1990). Deepening of the sunken road base is caused by sheet, rill and gully erosion, while soil falls and slides occur on the walls of the sunken roads causing lateral erosion. Those instabilities are caused by expanding cracks along the walls influenced by water erosion and frost processes (Leger 1990). Inactive sunken roads transform into gullies with a V-shaped cross section with less inclined walls.

The development of sunken roads is strongly influenced by natural and anthropogenic driving forces determining the formation of concentrated surface-water flow and local erosion processes. These include geotechnical properties and local topography of loess, type and presence of vegetation, the occurrence of intense rainfall events and anthropogenic influences on the local vegetation and soil surface and its loosening and sealing. These factors are of importance locally and in the entire catchment of the sunken road (Farres et al. 1993; Wiesbauer and Zettel 2014).

Sheet, rill or gully erosion occurs in areas where the shear forces of the concentrated surface-water flow overcome the resistance of the soil surface to particle detachment. Surface-water flow is collected and concentrated within natural linear landscape elements or anthropogenic linear features such as plough furrows, parcel borders or access roads (Farres et al. 1993). Additionally, soil compaction underneath tractor wheel lanes caused by the weight of the machinery contributes to faster surface-water runoff and to faster erosion once the compacted layer is eroded (Farres et al. 1993). However, not only modern agricultural machines contribute to sunken road formation, but sunken roads are also known to have evolved along mediaeval roads, and a strong influence of wheels and hoofs from ox or horse drawn carts by loosening the material and providing linear depressions for surface-water flow concentration is reported (David et al. 2011; Wiesbauer and Zettel 2014). Perhaps because of the possible variable age of sunken roads, the question of what initiated the development of sunken roads, artificial paths or natural gully erosion, is answered differently in national and international literature (Krissl 1982; Farres et al. 1993; Strauss and Klaghofer 2006). However, bank gullies have been used as access paths to agricultural fields or vineyards, and both erosion and anthropogenic impact played an important role in the continuous evolution of sunken roads (Farres et al. 1993; Wiesbauer and Zettel 2014). Likewise, the usage of sunken roads as access roads prohibits the growth of a dense vegetation cover that could protect the soil against erosion. Therefore, the formation of deep gullies incising the bottom of an unpaved sunken road is possible (Fig. 11.2).

Fig. 11.2 Gully erosion at the base of a sunken road in Groß Schweinbarth in Lower Austria. Photo: H. Wiesbauer

Besides the condition of the road surface itself, the potential evaporation and retention of surface-water flow in the catchment also influences gully erosion and sunken road formation. As loess provides perfect conditions for crop and wine cultivation, many catchments show extensive agricultural use. The crop type and distance between the vineyard lanes have a large influence on exposed bare soil and hence runoff production and erosion susceptibility (Krissl 1982). Besides the vineyard design and type of crop growing in the catchment, other anthropogenic influences on sunken road formation in Lower Austria are documented (Krissl 1982). As sheet, rill and gully erosion affected the unpaved sunken roads frequently, after heavy rainfall events, the local inhabitants needed to repair the road base to regain access to their fields and vineyards. At the same time, the agricultural fields and vineyards showed high losses of sediment and up to 1.5 m deep gullies may have formed. The material for filling these areas of soil loss often came directly from the sunken roads. Starting from the road base, the people dug caves inside the walls of the sunken roads which were close to the field. As vertical erosion was still going on over the years, these caves can nowadays be found at many different levels of the sunken lane wall. Therefore, they may be used as a proxy for the sunken road incision speed (Krissl 1982).

While sunken roads in loess are very famous because of their exceptional depth compared to the surrounding area, e.g. in China, Ukraine, Hungary (Leger 1990; David et al. 2011), similar landforms can be found worldwide within different lithologies such as Flysch, e.g. in Poland (Rejman and Rodzik 2006) or in Lower Austria (Bell et al. 2012), or Sandstone, e.g. in the Greensand of Wiltshire (Boardman 2013). All sunken road landforms have in common that they influence the local drainage system and erosional processes substantially. As sunken roads were used as the shortest and fastest connection between the farm and the field, concentrated runoff and muddy flooding of villages located nearby occurred frequently (Krissl 1982; Boardman 2013). In this context, their hydrological role has often been studied to analyse the increasing connectivity between agricultural areas and villages or rivers and to improve the design of protection measures (Evrard et al. 2007; Boardman 2013).

11.2.2 Characteristics of Sunken Roads in Lower Austria (Weinviertel)

One prominent area with the formation of bank gullies is the Wagram north of the Danube River in Lower Austria with elevation differences of around 30 m between the terrace base and bank (Fig. 11.3). Simultaneously, this elevation difference is the maximum depth of sunken roads found in Lower Austria. This area around and north of the Wagram is a good example illustrating the importance of high intensity rainstorm events triggering erosion processes, as the mean annual precipitation is as low as 550 mm (period 1971–2000, Hydrographic Service of Lower Austria 2011).

Given the terrace form of the Wagram, many sunken roads started out as ephemeral gullies (as defined by Poesen 1993) on top of the terrace. With ongoing gully erosion, some of these crossed the terrace bank and, as a result, transformed to bank gullies eroding the Wagram locally (Fig. 11.3, e.g. at Kirchberg am Wagram).

Consequently, ephemeral gullies and deeply incised bank gullies are present in this area. The Wagram has the highest density of sunken roads per map sheet of the Austrian topographic map. Here, more than 51 roads per sheet (=per 520 km^2) were counted (Fig. 11.4). A hotspot of sunken road density is Fels am Wagram (Fig. 11.3). Here, in a smaller sample area (20 km^2 instead of 520 km^2), a sunken road length density of 0.5 km/km^2 was recorded in 1995 (Wiesbauer and Zettel 2014). In total, the map in Fig. 11.4 shows an area of about 5504 km^2 with around 950 sunken roads visible on the topographic map. Of these sunken roads, 90% are located in loess, and their delimiting slopes show a slope angle of up to 90° (Wiesbauer and Zettel 2014).

Sunken roads in Lower Austria are remnants of historic (medieval) land use change when major deforestation of loess areas took place to provide larger areas for agriculture. Because of the low forest cover, an increase of soil erosion rates was observed and the first sunken roads evolved (Wiesbauer and Zettel 2014). Starting from around 1850 until the mid-twentieth century, a further acceleration of the erosion processes was observed due to land use changes, and sunken roads and their erosion became a big problem for land owners and villages in Lower Austria (Krissl 1982). The main observed land use change was the increasing parcel size and changes in the design of vineyards. Influenced by the avoidance of diseases (vine pest "Reblaus") and the increased use of machinery, the vine was cultivated with higher growth and more distance between the vine stocks (Krissl 1982). Furthermore, crops like corn, sunflower or sugar beets were cultivated more frequently (Wiesbauer and Zettel 2014). Due to a reduction of the vegetation cover near the soil surface, these areas were exposed to faster surface-water runoff and erosion. In 1956, the first technical measure against vertical and lateral erosion in sunken roads, and muddy flooding from these roads was installed (Krissl 1982). Historically, stones or wooden planks were used to pave the road to protect it from runoff erosion. Furthermore, the width of the ox or horse cart wheels was regulated in 1819 to avoid deep incision of the wheels in the road base (Wiesbauer and Zettel 2014). In modern times (twentieth and twenty-first century), the pavement with asphalt and installation of a paved drainage channel next to the road was recommended to control the runoff and erosion (Krissl 1982). In unused sunken roads and in-between vine stocks,

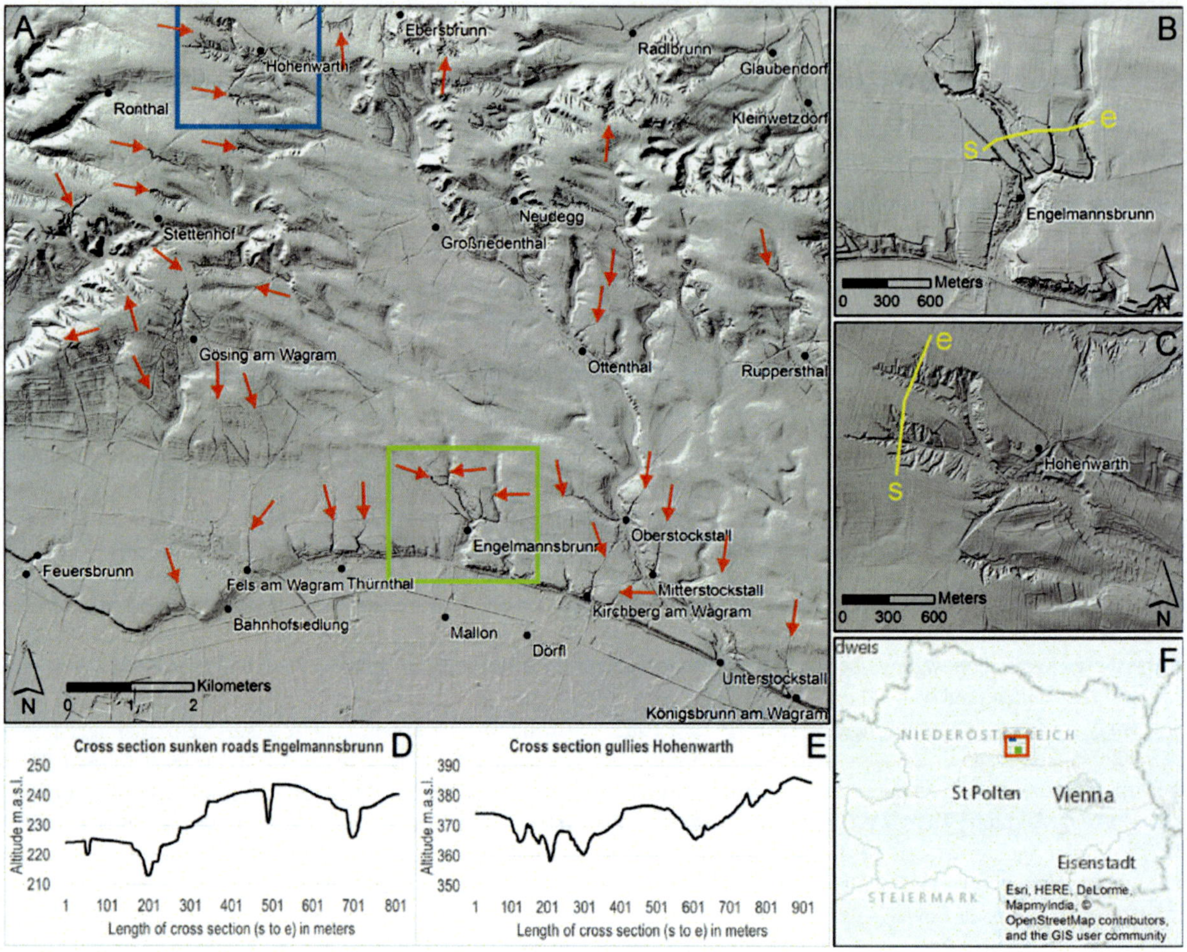

Fig. 11.3 **a** The distinct terrace landscape feature of the Danube River in the southern part of this figure, between Feuersbrunn, Fels am Wagram, Thürnthal, Kirchberg am Wagram and Königsbrunn am Wagram, is referred to as "Wagram". In the area along and north of the Wagram, sunken road and gully topography is visible on a hillshade of the 1 m × 1 m DTM. Red arrows indicate the head or location of the most prominent sunken roads or gullies. **b** and **c** Sunken roads close to Engelmannsbrunn and gully system close to Hohenwarth. The yellow lines show the location of the cross sections drawn from s to e. **d** and **e** Cross section of the sunken roads close to Engelmannsbrunn. and of the gullies close to Hohenwarth. *Source* Own graph obtained within esri ArcMap 10.2 from the 1 m × 1 m DTM. **f** Location of the study area in Lower Austria. The red outline shows the extent and location of the overview on sunken roads and gullies north of the Wagram (Fig. 11.3a). The green outline shows the location and extent of Fig. 11.3b. The blue outline shows the location and extent of Fig. 11.3c. *Data source* DTM: provided by the Provincial Government of Lower Austria, Land Niederösterreich 2016; Location of sunken roads or gullies: own mapping. Own graph based on the World light grey canvas basemap provided by esri ArcMap (Esri, HERE; DeLorme, MapmyIndia, © OpenStreetMap contributors, and the GIS user community)

the authorities recommended the growth of vegetation, and often gravel was used to protect the sunken road base (Fig. 11.5). In some sunken roads, even torrent protection measures such as check dams had to be installed (Krissl 1982).

Additionally, with the increase of the parcel size and as an erosion protection measure, smaller gullies were filled with soil material and "erased" from the landscape during land consolidation measures starting from the late 1950s (Fig. 11.6; Krissl 1982; Wiesbauer and Zettel 2014). These measures were also applied to sunken roads. In some areas of Lower Austria, a decrease of sunken roads by around 60–90% between 1915 and 1995 was observed. Comparably, a reduction by 0.8 km/km^2 was estimated in the surroundings of Fels am Wagram (20 km^2, Wiesbauer and Zettel 2014).

The question of the dimension of erosion rates in sunken roads in Lower Austria is difficult to answer as systematic measurements are missing, and gullies are often filled up quickly by local farmers. However, there are some proxies and estimates on erosion rates available based on

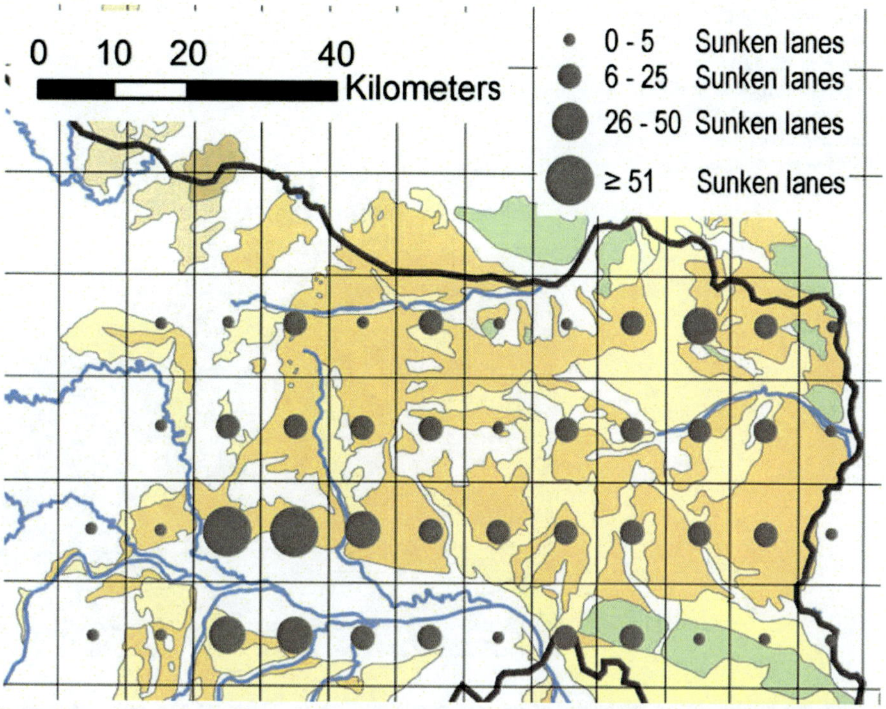

Fig. 11.4 Sunken roads in Lower Austria as observed in 1995 on the Austrian Map 1:50.000 (ÖK50), not including sunken roads in the forest. In the background, the distribution of loess is given (cf. Fig. 11.1). The thin black lines give the limits of the single ÖK 50 map sheets. The thick black line shows the boundary of Lower Austria. Source: translated from Wiesbauer and Zettel (2014), basemap: Loishandl-Weisz and Petizcka (2005)

Fig. 11.5 Typical vegetated and revitalised sunken road in loess close to Langenlois (refer to Fig. 11.1 for the location of Langenlois). Photo: H. Wiesbauer

observations of locals or measurements after major erosion events (Krissl 1982). Based on these proxies, Krissl (1982) estimated a vertical erosion rate of around 15–30 cm per year. Estimating the lateral erosion rate is more difficult, and no estimates for Lower Austria are available.

11.2.3 Cultural Importance of Sunken Roads

Sunken roads are not only important from a geomorphological point of view but also from a cultural perspective as they portray landmarks of national culture. Many sunken

Fig. 11.6 Filling of a sunken road in 1949 in Hippersdorf (close to Königsbrunn am Wagram as located in Fig. 11.3) to increase parcel size and prevent further gully erosion. *Source* Abt. Güterwege, Provincial Government of Lower Austria

roads are known to be remnants of ancient roads and paths (Fig. 11.7). It was possible to locate the "Rittsteig" in a sunken road close to Fels am Wagram. This path was probably already used during the Bronze Age and during Roman and medieval times (Csendes 1969; Papp 1993).

Within sunken roads, cellar lanes can often be found close to villages and vineyards. These are roads along which, on one or both side-slopes, wine press houses and cellars have been built and cellar tubes have been dug inside the loess slope (Fig. 11.8). These cellar lanes are a famous part of the cultural landscape of Lower Austria since the eighteenth century. The small houses, which form the entrance to the cellar, have been used for storing the wine press and other equipment. Furthermore, they have been hosting guests for tasting wine and local meals. Due to their visual appeal and good wine and food, many cellar lanes are still used for events to present the local vintners and their products (Wiesbauer and Zettel 2014).

Besides the cellar lanes and the enjoyment that comes along above the ground, the cellars themselves are of

Fig. 11.7 In the centre, a cellar lane with wine press houses and trees along a loess wall is visible. Additionally, loess terraces (in the centre behind the cellar lane) and sunken roads (on the right) were pictured in this painting from 1769 showing the Rohrendorfer Gebling close to Krems an der Donau, Lower Austria. *Source* Mayer 1769 in Wiesbauer and Zettel 2014

Fig. 11.8 Cellar lane in Wagram ob der Traisen (close to Traismauer) around 1925. Note the ox drawn cart, the incised tracks on the soil surface of the sunken road, the steep loess slopes, and the press houses built right next to the walls covering the entrance to the cellar tubes. *Source* ONB/Vienna, Picture Archives 57.515-B

cultural and touristic importance. Due to the elongated and arched shape of the cellar caves, they are commonly called cellar tubes. Underneath entire villages located in wine-growing regions extensive networks of connected cellar tubes can be found (e.g. Kirchberg am Wagram, Retz). These long but narrow underground cellar tubes, sometimes directly starting from loess outcrop walls, are very typical in this area. The digging in loess is comparably easy and the porosity of the material offers good ventilation for storing food and wine. Still to this day, these cellars are used for their original purpose. Additionally, guided tours, sometimes combined with wine tastings, are offered for tourists to explore the cellar tube maze.

Furthermore, educational trails on the topics of loess, its relation to the Ice Age, sunken roads and archaeological discoveries can be found in the area around Krems and the Wagram (Wessely and Draxler 2006). For example, west of Göttweig in a sunken road called Zellergraben, an educational trail on loess and wine (including wine cellars) was installed as this sunken road was declared a natural landmark in order to protect this landform (Fig. 11.9e, Wiesbauer and Zettel 2014).

11.3 Palaeosols in the Wagram Area Between Krems and Stockerau

Within the Weinviertel between Krems and Stockerau, the Tullnerfeld close to the Danube is distinct as it consists of the Holocene and last glacial Danube terrace with no significant loess cover. Looking from these wide plains (c. 180 m asl) to the north, there are the loess covered hills of the Western Weinviertel rising 100–200 m higher than the Tullnerfeld (Fig. 11.1). The boundary within these landscapes is formed by the prominent terrace of the Wagram with several metres of loess cover and its characteristic morphology. Occasionally reddish, brownish or greyish horizons can be traced within the yellowish loess cliffs.

Loess formation is not a continuous process but is related to climatic changes and ecosystem adaptations, phases of weathering and soil formation, with erosion or redeposition interrupting the deposition of dust and the formation of loess. In specific morphological positions, this can lead to the formation of thick loess–palaeosol sequences (LPS). These can record fluctuations of stadials and interstadials within a glacial period in high-resolution (Antoine et al. 2009; Terhorst et al. 2015) or long-term trends within several glacial-interglacial oscillations (Fink and Kukla 1977; Bronger 2003).

The Lower Austrian loess is one of the most intensely studied loess regions worldwide. The best conditions to study the stratigraphy of the widespread loess sediments and the palaeosols enclosed therein were maybe 50–100 years ago, when numerous loam pits were still in operation. The few and rather random outcrops of today cannot easily be integrated into a bigger picture of loess stratigraphy and chronology. Thus, their palaeosols are currently not investigated, although this would be helpful for reconstructing ancient palaeoreliefs and studying morphological response

Fig. 11.9 **a**: Stillfrieder Komplex at its type locality. This pedocomplex formed during marine isotope stage (MIS 5) and consists of the last interglacial Cambisol (c. 125–116 ka) superimposed by three humic horizons and loess of the early glacial period c. 116–75 ka). Photo: R. Peticzka; **b**: Mammoth skull within a Late Palaeolithic occupation horizon (c. 32 ka) at the excavation of Krems-Wachtberg in 2010. Photo: T. Sprafke; **c**: During the construction of a pipeline in the Eastern Weinviertel in 2011, a Stillfrieder Komplex following a different palaeotopography was discovered. Photo: U. Simon (Austrian Academy of Sciences); **d**: The MIS 5 pedocomplex (Paudorf soil) within loess sediments exposed at the sunken road of Göttweig-Aigen. It is a polygentic palaeosol typical for the transition area (cf. Sprafke et al. 2014). Photo: T. Sprafke; **e**: The Göttweiger Verlehmungszone at the type locality at the Zellergraben, west of Göttweig dating to the Middle Pleistocene attracts scientists since more than 100 years. Note the information board on the right side, belonging to an educational trail on loess and wine. Photo: I. Shorkunov; **f**: A series of Early Pleistocene brown loam palaeosols at the outcrop of Krems-Schießstätte. The two reddish horizons in the middle are part of the Kremser Komplex. Photo: T. Sprafke; **g**: The red loam of Stranzendorf that formed in the oldest loess of Europe, covered by Early Pleistocene loess. Photo: R. Peticzka

especially to climatic changes. The focus of present research is on the re-evaluation of some of the "classic" locations (Fig. 11.1) with the help of modern dating techniques (Sprafke 2016).

The outcrops of Krems-Schießstätte and Stranzendorf expose old Pleistocene loess. Several reversals of the earth's magnetic field recorded within these LPS indicate that these are among the oldest loess deposits in Europe, reaching back to the Pliocene, i.e. more than 2.6 Ma (Fig. 11.1; Fink 1979; Pécsi and Richter 1996). Younger palaeosols in the outcrops of Stiefern, Göttweig, Stillfried or Paudorf allow for comparison of regional palaeoenvironments when global temperatures were slightly warmer or cooler—a topic important in the present global change discussion (Sprafke 2016). In addition, the youngest loess deposits are widely known. Excavations in the last glacial loess of Austria reveal no strongly developed palaeosols, but archaeological discoveries gave important insights in the life of humans during the Late Palaeolithic. World famous is the Venus von Willendorf discovered in 1908 (Haesaerts et al. 1996). New discoveries like the Fanny vom Galgenberg in 1988 (Neugebauer-Maresch 1993) or the Wachtberg infant burial in 2005 (Einwögerer et al. 2006) raise the hope that there are many more archaeological secrets hidden in Austrian loess (see also Chapter "The Walgau—A Landscape Shaped by Landslides").

Naturally, structural and colour changes within loess related to palaeosol formation are not as evident for the public compared to prehistoric figurines, bones (Fig. 11.9b) and stone tools within the material that appeared during archaeological digs. Archaeology was often an important trigger for loess and palaeosol research in Lower Austria. Today, the collaboration with archaeologists to reconstruct palaeoenvironmental conditions during the last glacial period is among the most important tasks for loess researchers (Haesaerts et al. 1996; Hambach 2010; Terhorst et al. 2014).

The city of Krems with its almost 40-m-thick loess deposits is located at the core of loess research history. A large excavation for the construction of a flood-control dam at the beginning of the twentieth century did not only reveal thousands of archaeological discoveries, but also a 3–4 m thick, strongly weathered palaeosol complex (Fink 1976). This "Kremser Komplex" was presented by A. Penck in 1903 as one proof that there was not a single Ice Age (monoglacialism), but several glaciations accompanied by loess deposition, interrupted by interglacials with soil formation (polyglacialism).

A few years after the discovery of the Venus von Willendorf (in 1908), Archaeologist J. Bayer described a weathered horizon of 1–2 m thickness in a sunken road west of Göttweig which he termed "Göttweiger Verlehmungszone" (Bayer 1912)—Göttweig loam horizon—possibly Austria's most famous palaeosols (Fig. 11.9E). It was interpreted to represent a warm period within the last glacial, during the time of the Aurignacien culture of the late Palaeolithic.

In the 1930s, Götzinger tried to create a comprehensive loess–palaeosol stratigraphy for Lower Austria. At the outcrop of Paudorf, 7 km south of Krems, he found a 0.5–1-m-thick palaeosol, the "Paudorfer Boden" (Paudorf soil) in loess sediments above a more weathered horizon correlated to the "Göttweiger Verlehmungszone" (Fig. 11.9D, Götzinger 1935). Götzinger assumed that the "Göttweiger Verlehmungszone" was the palaeosol of the last interglacial period. The weaker developed "Paudorf soil" on top represented a single interstadial during the last glacial phase, whereas the "Kremser Komplex" was correlated to the penultimate interglacial (Götzinger 1936), the long interglacial defined by Penck and -Brückner (1909).

The 1950–60s are the classical era of loess research in Austria, with F. Brandtner and J. Fink investigating almost 100 outcrops in Lower Austria, to find palaeosols comparable to the ones in Paudorf, Göttweig and Krems (which are all located in the transition area after Fink 1956). For Brandtner, both the Göttweig and Paudorf palaeosols represented interstadial soils of the last glacial and the Krems palaeosol the last interglacial (Brandtner 1954, 1956). Fink used the at that time innovative method of radiocarbon dating of charcoal more convincingly and did not only defend Götzingers previous notion but made Paudorf, Göttweig and Krems key terms for the international Quaternary community that represent important periods during the Quaternary, namely the most prominent interstadial and the last two interglacials (Fink 1956, 1961). One important LPS that was established at that time in the dry loess landscape (Fink 1956) was the *Stillfrieder Komplex*, a 2-m-thick stack of a brown weathered horizon correlated to the last interglacial, superimposed by three humic horizons of the early glacial period (Fig. 11.9a,c; Fink 1954, 1962).

In the 1970s, marine stratigraphy emerged as a new continuous archive for Quaternary climate fluctuations (Shackleton 1967) and Kukla (1977) made important correlations to classical loess profiles. However, in this context, it was revealed that the stratigraphic continuity of the famous Lower Austrian outcrops in the transition area of the loess landscapes was only an assumption, not a reality: Specific mollusk species below the Paudorf soil proved that it represented the last interglacial; thus, the Göttweig soil has a much older age (Fig. 11.9d, e). The detection of the Brunhes–Matuyama magnetic reversal (c. 0.78 Ma) far above the Kremser Komplex revealed that this palaeosol complex formed more than a million years ago, during Early Pleistocene (Fink 1976). It became increasingly clear that the location of the loess profiles on slopes made it necessary to take into account erosion and redeposition processes in their interpretation (Sprafke 2016).

The 1980–90 s only saw a few studies on Austrian loess (e.g. Rabeder and Verginis 1987; Verginis 1993) with interesting results related to the application of the optically stimulated luminescence dating method to the loess (Zöller et al. 1994). In the last years, studies on Lower Austrian loess were intensified and new stratigraphic concepts appeared based on detailed reevaluations of the classic outcrops and luminescence dating (Thiel et al. 2011a, b, Sprafke et al. 2013, 2014; Terhorst et al. 2015).

The current model for Lower Austrian loess stratigraphy describes a last interglacial (Eemian) brown forest soil without clay translocation (Cambisol) that formed under increased Mediterranean influence in the continental part for the dry loess landscape and the transition area, i.e. from Krems to the Morava River (Fink 1962; Sprafke 2016). The difference with the present-day soil is probably related to the Eemian being c. 2 °C warmer on a global scale and different palaeoclimatic conditions in Lower Austria. In contrast, a last interglacial Luvisol, comparable to the recent surface soil (Terhorst 2013; Terhorst et al. 2015), is characteristic for the humid loess landscape.

The brown loess palaeosols that can be found in several outcrops of the Wagram and Krems region indicate that during the Middle Pleistocene several interglacials of a more Mediterranean character occurred. One famous example is the aforementioned Göttweiger Verlehmungszone. The Kremser Komplex and other palaeosols in the Krems-Schießstätte outcrop indicate these climatic conditions also for the interglacials of the older Pleistocene (Fig. 11.9f, Sprafke 2016). One outcrop with a very distinct palaeoenvironmental history is located at Stranzendorf in the Weinviertel (Fig. 11.9g). Below several metres of loess and palaeosols, a 3–4-m-thick strongly weathered red clay can be observed. This clay probably developed during the earliest periods of the Pleistocene under subtropical climate conditions in the presumably oldest loess of Europe (Fink 1979).

11.4 Summary

The loess region of Lower Austria has much to offer not only the picturesque landscape and the good wine. There are world famous archaeological discoveries, well-resolved records of the last glacial climate fluctuations, long records of landscape response to glacial–interglacial oscillations and the presumably oldest loess of Europe. Sunken roads testify human–environment interactions in historical times. Many of them have old wine press houses forming the entrance to wine cellars dug deep into the loess that are of cultural and touristic importance.

References

Antoine P, Rousseau DD, Moine O, Kunesch S, Hatté C, Lang A, Tissoux H, Zöller L (2009) Rapid and cyclic aeolian deposition during the Last Glacial in European loess: a high-resolution record from Nussloch Germany. Quat Sci Rev 28(25–26):2955–2973

Bell R, Petschko H, Röhrs M, Dix A (2012) Assessment of landslide age, landslide persistence and human impact using airborne laser scanning digital terrain models. Geogr Ann Ser Phys Geogr 94:135–156

Bayer J (1912) Chronologie des Temps quaternaires. In: Commont V (Ed): Congrès international d'anthropologie et d'archéologie préhistoriques: compte rendu de la XIVe session. Genève, 145–164

Boardman J (2013) The hydrological role of "sunken lanes" with respect to sediment mobilization and delivery to watercourses with particular reference to West Sussex, southern England. J Soils Sediments 13:1636–1644

Brandtner F (1954) Jungpleistozäner Löß und fossile Böden in Niederösterreich. Eiszeit Gegenw 4–5:49–82

Brandtner F (1956) Lößstratigraphie und paläolithische Kulturabfolge in Niederösterreich und in den angrenzenden Gebieten (zugleich ein Beitrag zur Frage der Würmgliederung). Eiszeit Gegenw 7:127–175

Bronger A (2003) Correlation of loess-paleosol sequences in East and Central Asia with SE Central Europe: towards a continental Quaternary pedostratigraphy and paleoclimatic history. Quatern Int 106:11–31

Csendes P (1969) Die Straßen Niederösterreichs im Früh- und Hochmittelalter. Verlag Notring, Wien

David L, Ilyes Z, Baros Z (2011) Geological and geomorphological problems caused by transportation and industry. Cent Eur J Geosci 3:271–286

Einwögerer T, Friesinger H, Händel M, Neugebauer-Maresch C, Simon U, Teschler-Nicola M (2006) Upper Palaeolithic infant burials. Nature 444:285–285

Evrard O, Bielders CL, Vandaele K, van Wesemael B (2007) Spatial and temporal variation of muddy floods in central Belgium, off-site impacts and potential control measures. CATENA 70:443–454

Farres P, Poesen J, Wood S (1993) Soil Erosion Landscapes. Geogr Rev 6:38–41

Fink J (1954) Die fossilen Böden im österreichischen Löß. Quartär 6:85–108

Fink J (1956) Zur Korrelation der Terrassen und Lösse in Österreich. Eiszeit Gegenw 7:49–77

Fink J (1961) Die Gliederung des Jungpleistozäns in Oesterreich. Mitteilungen Der Geologischen Gesellschaft in Wien 54:1–25

Fink J (1962) Studien zur absoluten und relativen Chronologie der fossilen Böden in Österreich. II. Wetzleinsdorf Und Stillfried. Archaeologia Austriaca 31:1–18

Fink J (1965) The Pleistocene in Eastern Austria. The Geological Society of America, Special Papers 84, 179–199

Fink J (1976) Exkursion durch den österreichischen Teil des nördlichen Alpenvorlandes und den Donauraum zwischen Krems und der Wiener Pforte (= Mitteilungen der Kommission für Quartärforschung der Österreichischen Akademie der Wissenschaften 1). ÖAW, Wien, p 113

Fink J, Kukla GJ (1977) Pleistocene climates in central Europe; At least 17 interglacials after the Olduvai Event. Quatern Res 7(3):363–371

Fink J (1979) Stand und Aufgaben der österreichischen Quartärforschung. Innsbrucker Geographische Studien 5:79–104

Fink J, Nagl H (1979) Quartäre Sedimente und Formen (1:1 000 000). In: Österreichische Akademie der Wissenschaften (ÖAW). Kommission für Raumforschung (ed) Atlas der Republik Österreich 1:1,000.000, Nr. II/6. Freytag-Berndt & Artaria, Wien

Götzinger G (1935) Zur Gliederung des Lößes. Leimen- und Humuszonen im Viertel unter dem Manhartsberge. Verhandlungen Der Geologischen Bundesanstalt (wien) 8–9:126–132

Götzinger G (1936) Das Lößgebiet um Göttweig und Krems an der Donau. In: Götzinger G (ed) Führer für die Quartär-Exkursionen in Österreich. Geologische Bundesanstalt, Wien, pp 1–11

Haesaerts P, Damblon F, Bachner M, Trnka G (1996) Revised stratigraphy and chronology of the Willendorf II sequence, Lower Austria. Archaeologia Austriaca 80:25–42

Hambach U (2010) Palaeoclimatic and stratigraphic implications of high resolution magnetic susceptibility logging of Würmian loess at the Krems-Wachtberg Upper-Palaeolithic site. In: Neugebauer-Maresch C, Owen LR (ed) New Aspects of the Central and Eastern European Upper Palaeolithic - methods, chronology, technology and subsistence. (= Mitteilungen der Prähistorischen Kommission 72). Österreichische Akademie der Wissenschaften, Wien, 295–304

Hydrographischer Dienst des Landes Niederösterreich (Hydrographic Service of Lower Austria) (2011) Wasserstandsnachrichten und Hochwasserprognosen Niederösterreich. http://www.noel.gv.at/Externeseiten/wasserstand/wiskiwebpublic/maps_N_0.htm?entryparakey=N. Accessed 2 Mar 2011

Krissl H (1982) Beschleunigte Gully-Entwicklung in den letzten Jahrzehnten in den Lössgebieten Ost-Österreichs. Beitr Zur Wildbacherosions- Lawinenforschung 144:115–124

Kukla GJ (1977) Pleistocene land-sea correlations I. Europe. Earth-Science Reviews 13(4):307–374

Land Niederösterreich (2016) NÖ Atlas, http://atlas.noe.gv.at/. Accessed 09 May 2016

Leger M (1990) Loess landforms. Quat Int 7–8:53–61

Loishandl-Weisz H, Peticzka R (2005) Vom Winde verweht – Die Sedimente und Böden im Verbreitungsgebiet der niederösterreichischen Kreisgrabenanlagen. 143–145. In: Daim F, Neubauer W (eds.): Zeitreise Heldenberg. Geheimnisvolle Kreisgräben. – Katalog zur Niederösterreichischen Landesausstellung, Berger, Horn und Wien

Neugebauer-Maresch C (ed, 1993) Altsteinzeit im Osten Österreichs (= Wissenschaftliche Schriftenreihe Niederösterreich, 95/96/97). NÖ Pressehaus, St. Pölten/Wien, 96 p

Pécsi M, Richter G (1996) Löss: Herkunft - Gliederung - Landschaften (= Zeitschrift für Geomorphologie, N.F., Supplementband 98). Bornträger, Berlin & Stuttgart, p 391

Penck A, Brückner E (1909) Die Eiszeiten in den nördlichen Ostalpen (= Die Alpen im Eiszeitalter 1). Tauchnitz, Leipzig, p 393

Papp H (1993) Altwege Nach Eggenburg. Waldviertel 42:1–13

Poesen J (1993) Gully typology and gully control measures in the European loess belt. In: Wicherek S (ed) Farm land erosion in temperate plains environment and hills. Elsevier, Amsterdam, pp 221–239

Rabeder G, Verginis S (1987) Die plio/pleistozänen Lößprofile von Stranzendorf und Krems (Niederösterreich). Griechische Geographische Gesellschaft Band B, 285–306

Rejman J, Rodzik J (2006) Poland. In: Soil Erosion in Europe. John Wiley & Sons, Ltd, pp 95–106

Schwenk H (1992) Massenbewegungen in Niederösterreich 1953–1990. Jahrbuch der Geologischen Bundesanstalt. Geologische Bundesanstalt, Wien, pp 597–660

Shackleton NJ (1967) Oxygen isotope analyses and Pleistocene temperatures re-assessed. Nature 215:15–17

Smalley IJ, O'Hara-Dhand K, Wint J, Machalett B, Jary Z, Jefferson I (2009) Rivers and loess: The significance of long river transportation in the complex event-sequence approach to loess deposit formation. Quatern Int 198:7–18

Smalley IJ, Marković SB (2014) Loessification and hydroconsolidation: There is a connection. CATENA 117:94–99

Sprafke T, Terhorst B, Peticzka R, Thiel C (2013) Paudorf locus typicus (Lower Austria) revisited—the potential of the classic loess outcrop for Middle to Late Pleistocene landscape reconstructions. Quat Sci J (e&g) 62(1):59–72

Sprafke T (2016) Löss in Niederösterreich—Archiv quartärer Klima- und Landschaftsveränderungen. Würzburg University Press, Würzburg, p 272

Sprafke T, Obreht I (2016) Loess: Rock, sediment or soil—what is missing for its definition? Quat Int 399:198–207

Sprafke T, Thiel C, Terhorst B (2014) From micromorphology to palaeoenvironment: The MIS 10 to MIS 5 record in Paudorf (Lower Austria). CATENA 117:60–72

Strauss P, Klaghofer E (2006) Austria. In: Soil Erosion in Europe. John Wiley & Sons, Ltd, pp 205–212

Terhorst B (2013) A stratigraphic concept for Middle Pleistocene Quaternary sequences in Upper Austria. Quat Sci J (e&g) 62(1):4–13

Terhorst B, Kühn P, Damm B, Hambach U, Meyer-Heintze S, Sedov S (2014) Paleoenvironmental fluctuations as recorded in the loess-paleosol sequence of the Upper Paleolithic site Krems-Wachtberg. Quatern Int 351:67–82

Terhorst B, Sedov S, Sprafke T, Peticzka R, Meyer-Heintze S, Kühn P, Solleiro Rebolledo E (2015) Austrian MIS 3/2 loess-palaeosol records-Key sites along a west–east transect. Palaeogeogr Palaeoclimatol Palaeoecol 418:43–56

Thiel C, Buylaert JP, Murray AS, Terhorst B, Hofer I, Tsukamoto S, Frechen M (2011) Luminescence dating of the Stratzing loess profile (Austria)—testing the potential of an elevated temperature post-IR IRSL protocol. Quatern Int 234:23–31

Thiel C, Buylaert JP, Murray AS, Terhorst B, Tsukamoto S, Frechen M, Sprafke T (2011) Investigating the chronostratigraphy of prominent palaeosols in Lower Austria using post-IR IRSL dating. Quat Sci J (e&g) 60(1):137–152

USGS (2014) Shuttle Radar Topography Mission, 1 Arc Second Non Void Filled (~30 m resolution), http://earthexplorer.usgs.gov/ Accessed 2014

Verginis S (1993) Paläoböden und Klimaschwankungen während des Jungpleistozäns am Beispiel des Profils Stiefern, NÖ. Griechische Geographische Gesellschaft Band B, 242–251

Wessely G, Draxler I (2006) Pliozän und Quartär. In: Wessely G (ed) Geologie der österreichischen Bundesländer – Niederösterreich. Geologische Bundesanstalt, Wien, pp 235–252

Wiesbauer H, Zettel H (2014) Hohlwege und Lössterrassen in Niederösterreich. Abteilung Naturschutz des Amtes der Niederösterreichischen Landesregierung, Wien

Zöller L, Oches EA, McCoy WD (1994) Towards a revised chronostratigraphy of loess in Austria with respect to key sections in the Czech Republic and in Hungary. Quat Geochronol (Quat Sci Rev) 13:465–472

Helene Petschko is a postdoc researcher at the Friedrich Schiller University Jena, Germany. Her research interests include terrain analysis and natural hazards, particularly landslides, their mapping, modelling, understanding of involved uncertainties, relationships with land use and climate change and presentation of research results, all aimed towards the improved preparedness of society.

Tobias Sprafke is a postdoctoral researcher at the Institute of Geography, University of Bern, interested in terrestrial records of quaternary climate, soil and landscape evolution. His main research topic is the interplay between dust deposition, soil formation and slope processes. He has extensive experience in loess-palaeosol sequences of Austria, Germany, Poland and Kazakhstan and works on pedogenesis in Central Europe, Brazil and Cameroon.

Robert Peticzka is assistant professor at the Department of Geography and Regional Research at the University of Vienna. His field of expertise is in soil science with a research focus on palaeopedology, aeolian sediments and soil physics.

Heinz Wiesbauer is a self-employed professional landscape planner in Vienna, Austria. He has written several books about nature conservation, river ecology and landscape history. In recent years, his research has also focused on the biology of wild bees, wasps and other insects. He is interested in the ecology, behaviour and pollination of wild bees.

12 The Danube Floodplain National Park: A Fluvial Landscape with Expiration Date?

Severin Hohensinner and Ronald E. Pöppl

Abstract

Situated between the European capitals Vienna and Bratislava in the Vienna Basin, the Danube Floodplain National Park covers one of the largest remaining floodplains in Central Europe. Here, the Danube River is still free flowing and forms the lifeline of the park. Prior to channelization in the nineteenth century, the Danube can be referred to as a high- to medium-energy variant of anabranching rivers that comprises both braiding and meandering elements. Between 1726 and 1817, main channel(s) and lotic side arms made up 85–95% of the total water bodies in the floodplain, pointing to the crucial role of river dynamics. Large shares of the fluvial landforms were permanently renewed by lateral bank erosion, avulsion of new channels or reoccupation of abandoned arms. River training programmes and flood protection projects in the nineteenth century severely truncated the system-inherent potential for channel adjustments. The consequences are the comprehensive stabilization of formerly dynamic fluvial landforms and the missing renewal of riverine habitats. Human interventions in the upstream Danube sections and lacking bedload influx also affected the morphological development in the national park. Today, the success of habitat restoration in such a channelized river reach is constrained by several factors. Locally, the requirements for unhampered navigation and the protection of the hinterland against floods are the most important concerns. Remote impacts, such as severely truncated bedload transport, restrain the recovery of the original river-typical fluvial processes and channel dynamics.

Keywords

Danube River • Alluvial • Historical • Anabranching • Channelization • Restoration

12.1 Geographical and Geological Setting

Although the classical subdivision of the river into the Upper, Middle and Lower Danube (Fig. 12.1) is justified from the geomorphological point of view, this approach reflects only a very generalized perspective of the Danube system. In particular, it neglects the broad range of morphologically differently developed river patterns, especially along the upper and also middle course. For example, the Upper Danube formerly also developed distinct meander bends in Germany, a process that normally would be expected mostly along the lower course of the river. Compared to the generally higher slopes at the Upper Danube (0.4–1.4‰), in such meandering reaches the river showed channel gradients of only 0.2–0.3‰ (Sommerhäuser et al. 2003; Jungwirth et al. 2014). The Austrian Danube section belongs to the upper course and today features a mean slope of 0.43‰ after river channelization.

The 40 km long Danube Floodplain National Park is located between Austria's capital Vienna and the mouth of the Morava River on the border to Slovakia, just upstream of the transition from the Upper to the Middle Danube (Figs. 12.1 and 12.2). Towards the north, the up to 4 km wide wetlands are delimited by the extensive plains of the Marchfeld region in Lower Austria, whereas to the south the boundary is formed by the fault lines of the Vienna Basin as well as by the risers of Pleistocene river terraces. Due to a dynamic rise and fall of water levels—sometimes up to 7 m

S. Hohensinner (✉)
Institute of Hydrobiology and Aquatic Ecosystem Management, University of Natural Resources and Life Sciences Vienna (BOKU), Gregor-Mendel-Str. 33, 1180 Vienna, Austria
e-mail: severin.hohensinner@boku.ac.at

R. E. Pöppl
Department of Geography and Regional Research, University of Vienna, Universitätsstr. 7, 1010 Vienna, Austria
e-mail: ronald.poeppl@univie.ac.at

C. Embleton-Hamann (ed.), *Landscapes and Landforms of Austria*, World Geomorphological Landscapes,
https://doi.org/10.1007/978-3-030-92815-5_12

Fig. 12.1 Location of the Danube Floodplain National Park (black arrow) within the Danube River Basin (modified after Tom Gonzales 2011, Wikipedia Map Workshop)

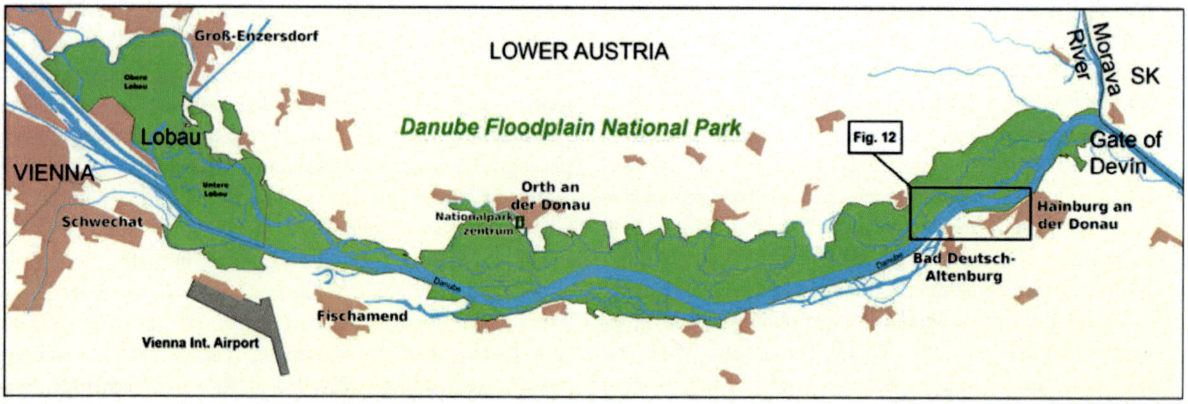

Fig. 12.2 Overview map of the Danube Floodplain National Park (Lobau: upstream section of the national park, SK: Slovakia; source: Anitagraser 2008, Wikimedia Commons, adapted)

—the wetland landscape is constantly being reformed, creating diverse and ever-changing habitats for numerous (endangered) plant and animal species. These include more than 800 vascular plant species, more than 30 mammal species, 100 breeding bird species, 8 reptile species, 13 amphibian species and c. 60 species of fish (http://www.donauauen.at).

Directly downstream of the national park, at the so-called Gate of Devín at the Austrian/Slovakian border, the Danube River enters the Pannonian Plains and the channel slope drops significantly to 0.1 ‰ (Figs. 10.2 and 10.3). As a consequence of the reduced slope, the Danube deposited huge volumes of bedload material after the Last glacial period between Bratislava and Györ. This led to a large inland delta with extensive sediment accumulations, which is expressed in the historical German names for *Žitný ostrov Island* (in Slovakia) and *Szigetköz* (in Hungary), namely, *Große und Kleine Schüttinsel* (Large and Small Accumulation Island).

Fig. 12.3 Danube Floodplain National Park close to the city of Hainburg, with the downstream Gate of Devín in the background (photo: F. Kovacs, 2003)

Due to the large tributary Inn River at the German/Austrian border, with its high sediment load, the Danube's flow regime changes from pluvial-nival to glacial-nival (Lászloffy 1967). The downstream Austrian Danube shows a sequence of four canyon sections through the Bohemian Massif that are interrupted by five alluvial sections. The last downstream alluvial section, which also includes the national park, is located in the Vienna Basin. Prior to river regulation, the break-through sections typically featured side bars, some mid-channel bars and in wider sections also small backwaters (Hohensinner 1995). In the alluvial sections, the Danube developed large anabranching channel networks due to the highly fluctuating flow and sediment fluxes of several large tributaries. Before the construction of dams and reservoirs, bedload transport (gravel, sand) in Vienna probably amounted to 350,000–500,000 m^3 annually, and the suspended load to 6–7 million tonnes (Penck 1891; Schmautz et al. 2000). As a consequence of the individual geological setting and the hydromorphological conditions, the alluvial sections along the Austrian Danube differed considerably (Hohensinner 2008). Owing to the nearby confluences of the large alpine tributaries Inn, Traun and Enns, the upstream floodplains in the Eferding Basin, Linz Basin and Machland Basin showed more intensive fluvial dynamics than the downstream Tulln and Vienna basins. The upper basins are characterized by distinct downstream canyon sections that constricted the flood conveyance severely and led to significant backwater effects in the upstream floodplains (today this backwater effect is reduced due to river regulation and local canyon widening). Apart from that, the downstream basins show greater lateral extension, which facilitated the evolution of wider floodplain systems. In the Vienna Basin the absence of distinct river terraces north of the Danube additionally favoured the development of a spacious river landscape.

The Vienna Basin can be referred to as a typical pull-apart basin that developed between the Bohemian Massif, the Alps and the Carpathians by tectonic subsidence in the Miocene, 17–8 million years ago (Grupe and Jawecki 2004). Within 9 million years, it subsided by up to 5500 m and was filled with marine, limnic, terrestrial and fluvial sediments. Until c. 12 million years ago, the area of today's national park was covered by the Paratethys Sea and later by the Pannon Lake, which primarily featured a brackish and later on a freshwater environment (Magyar et al. 1999; Fig. 12.4a). After the Pannon Lake had started to retreat towards the east 9.5 million years ago, the Urdonau (Palaeo Danube) extended its course in order to reach the remaining

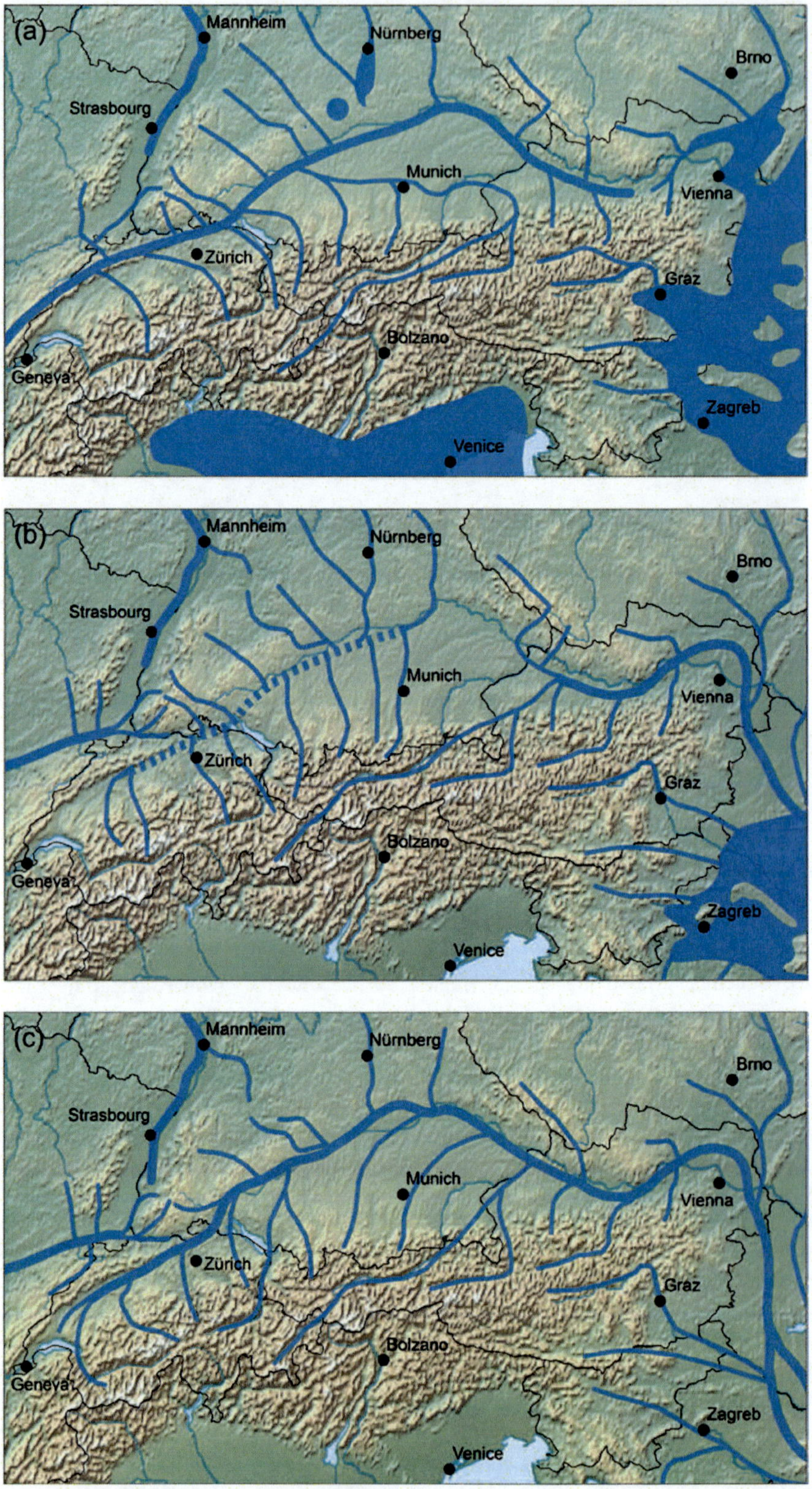

Fig. 12.4 **a** Palaeo Danube and Paratethys Sea 14 million years ago, **b** Palaeo Danube and Pannon Lake 9 million years ago and **c** Palaeo Danube 4.5 million years ago (the national park is located directly east of Vienna; modified after Jungwirth et al. 2014; base map: Natural Earth 2011)

lake (Fig. 12.4b). Until the Late Pliocene, approximately 3 million years ago, the Palaeo Danube flowed further north and crossed today's course in the national park in north-south direction (Fig. 12.4c).

At the beginning of the Pleistocene the braided Danube River entered the Vienna Basin probably already via its present gap and started to develop a succession of gravel terraces, which are dominant features of the contemporary landscape. River terrace development in the Vienna Basin was controlled by Pleistocene climatic fluctuations and tectonic processes (Fink and Majdan 1954; Salcher and Wagreich 2010). Based on outcrops, excavations and borehole data, six river terrace units in the urban area of Vienna can be distinguished at different levels of altitude; they developed between c. 600,000 (Laaerberg Terrace) and 10,000 BP (Prater Terrace) (Küpper 1968; Fink 1976; Table 12.1; Fig. 12.5). These fluvial gravel terraces—which are partly overlain by younger loess deposits—clearly delimit the recent floodplain of the Danube Floodplain National Park towards the south (Figs. 12.6 and 12.9).

Table 12.1 River terraces above the recent alluvium (Zone of recent meanders) within the urban area of Vienna (after Brix 1970; Fink 1976; Summesberger 2011)

Name of Terrace	Mean elevation (asl)
Laaerberg Terrace	250 m
Wienerberg Terrace	220 m
Arsenal Terrace	200 m
Mittel Terrace	185 m
Stadt Terrace/Gänserndorfer Terrace	175 m
Prater Terrace	160 m

12.2 Fluvial Morphology Prior to Channelization

According to historical sources from the eighteenth and nineteenth century, prior to channelization the alluvial Danube sections in Austria featured transitional channel patterns between braiding and meandering river types. The channel system comprised mostly one or two dominant branches as well as numerous secondary channels and oxbows. Large gravel bars on the river banks (point bars), but also mid-channel bars, were typical elements in the wider main river arm(s). Many small, dynamic islands were equally characteristic as large, stable islands that often showed the same elevation as the adjacent floodplain area. The islands separated individual branches that could evolve rather independently from each other. Some arms featured straight or sinuous courses, others even developed meandering channel patterns (Fig. 12.6). From this perspective, the historical Danube River is best described as a gravel-dominated, laterally active anabranching river (cf. Desloges and Church 1987, and Nanson and Knighton 1996).

What did the morphological configuration of the Austrian Danube River look like before the eighteenth century? Based on historical maps, a reconstruction of former fluvial morphology is only possible for the Viennese Danube from the early sixteenth century onwards (Hohensinner et al. 2013a, b). Geodata and written sources, on the other hand, enable drawing conclusions on the potential configuration of river sections before this period. Accordingly, the Danube floodplains can be understood as complex fluvial landforms that had developed in the Holocene by repeated channel incision and intense floodplain denudation in combination with subsequent sediment accumulation. Phases of amplified

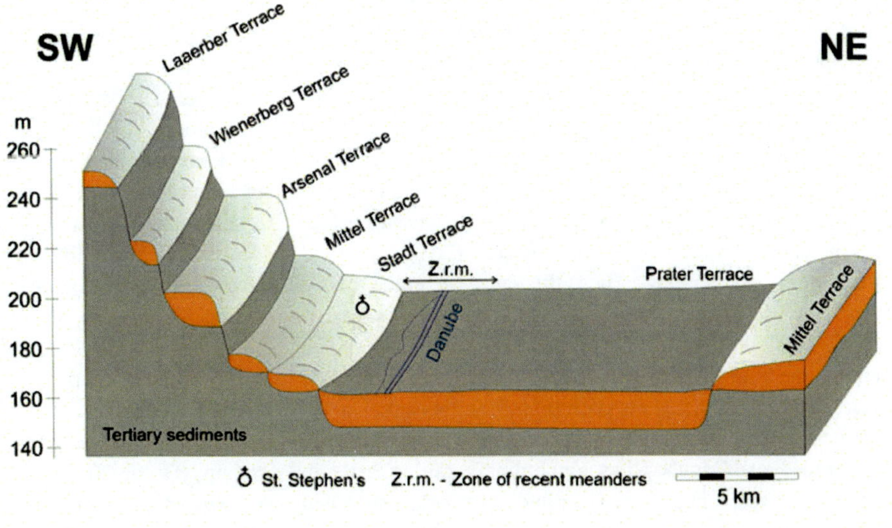

Fig. 12.5 Schematic illustration of the river terraces in the city area of Vienna (after Fink 1976; adapted from Voit and Embleton-Hamann 2011). Due to neotectonic subsidence the youngest terrace (Prater Terrace; cf. Peresson 2006), which formed c. 60,000–10,000 BP, is located only c. 3–4 m above the recent channel system (Zone of recent meanders) and is therefore also subject to episodic flooding

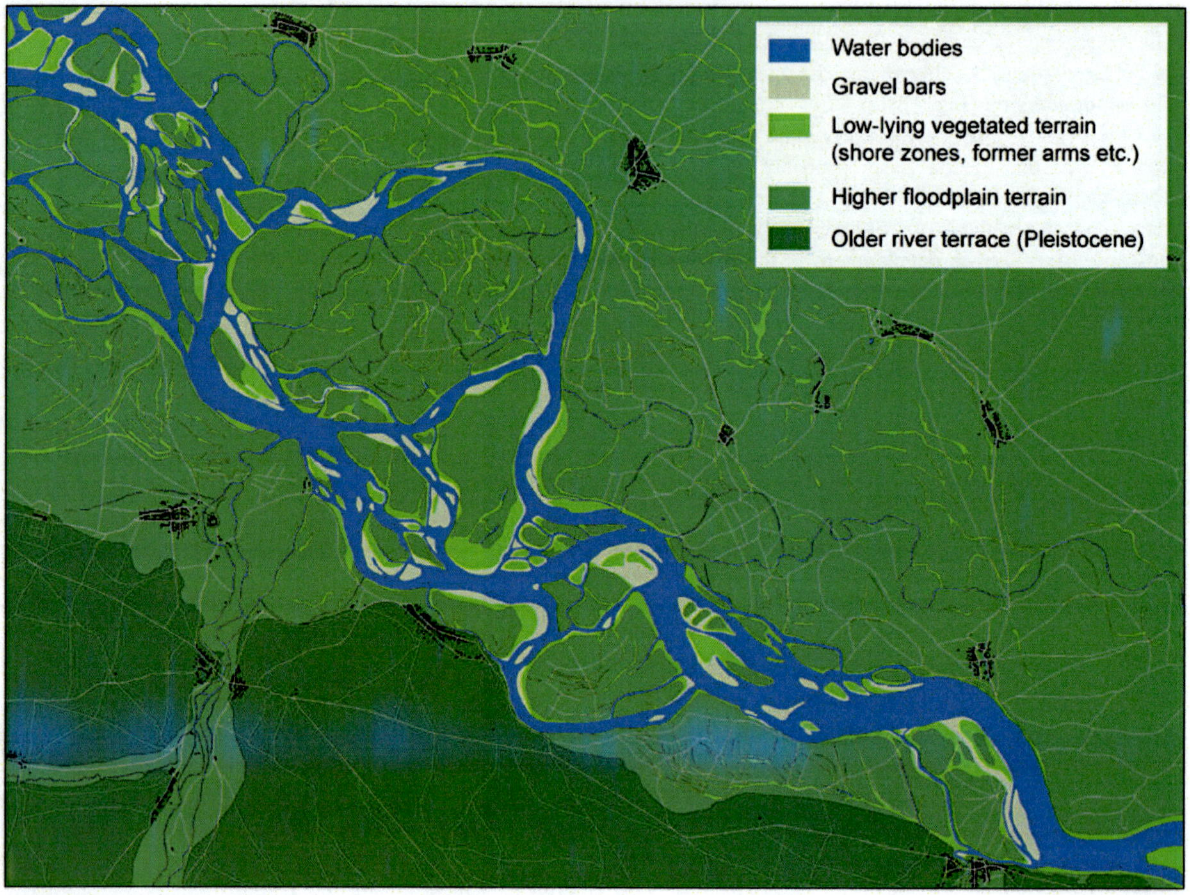

Fig. 12.6 Danube river landscape in the Lobau downstream of Vienna 1726 (upstream part of Danube Floodplain National Park; Eberstaller-Fleischanderl et al. 2004)

fluvial disturbances are evident in the Austrian Danube at 2600–2000 BP (Iron Age, Roman Times) and around 1500 BP (Late Antiquity/Early Middle Ages; Kohl 1991; Demek et al. 2008). During such phases, the anabranching river probably showed a tendency towards braiding, while a shift towards meandering may have occurred during periods of less intense fluvial activity. Climate-driven changes of the hydrological regime and, consequently, of river patterns were superimposed by human interferences in the whole basin. Large-scale land use changes, i.e. woodland clearings during Roman Times, no doubt led to increased surface runoff and sediment release that boosted fluvial dynamics and intensified braiding. Thus, the question posed above about the former channel patterns cannot be answered easily. In Vienna and north of the national park, the spatial arrangement of settlements founded in the High Middle Ages as well as remnants of former Danube arms in the eighteenth century provide references for the configuration of the medieval Danube River. Accordingly, during the High Middle Ages, large meandering channels stretched northwards far beyond the floodplain that was formed during the modern period (Slezak 1948). During the Late Middle Ages and the early modern period (c. AD 1300–1600), a phase of amplified fluvial dynamics and channel incision finally led to the evolution of the contemporary Danube floodplains.

In the national park, two geomorphological "hot spots" were crucial for the evolution of fluvial landforms. At the upstream end in the Lobau floodplain, where the tectonic subsidence ultimately reached 5500 m below the current terrain surface in the so-called *Schwechat Tief* (Schwechat subsidence), a significant reduction in channel slope was observed prior to channelization (Lorenzo 1819; see Lobau in Fig. 12.2). Here, the floodplain is one of the broadest along the Austrian Danube River. This is thought to be due to the drop in channel slope, which—as also historically reported—amplified the fluvial dynamics (deposition and remobilization of sediments) and fostered the formation of ice jams (Holub 2012). The second "hot spot" was located at the downstream end of the national park, where the large

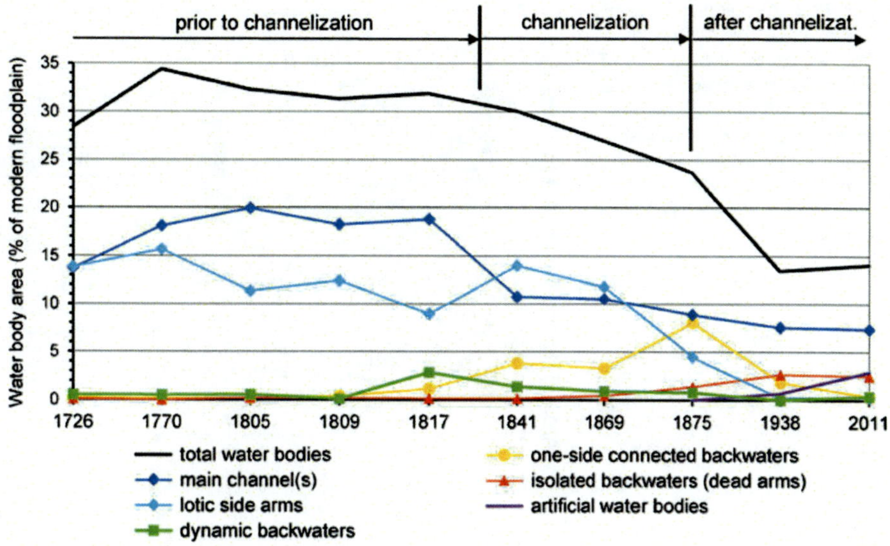

Fig. 12.7 Water body types in the Lobau between 1726 and 2011 (area shares in % of the modern floodplain formed in the last 500 years)

tributary Morava River discharges just upstream of the Gate of Devín (Fig. 12.2). Here, lateral channel constriction resulted in reduced flood conveyance capacity. In combination with the inflow of the tributary, this caused backwater effects associated with increased fluvial dynamics (Streffleur 1851). Prior to channelization, between 1726 and 1817, main channel(s) and almost permanently flowed-through side arms amounted to 85–95% of total water bodies in the contemporary floodplain, which was formed under the climatic-hydrological conditions of the modern period since approximately AD 1500 (Fig. 12.7). Consequently, the predominantly lotic habitat conditions primarily favoured rheophilic species; different types of backwaters were much less frequent (Hohensinner and Jungwirth 2009).

The evolution of such a river landscape was affected by several factors. Besides the highly fluctuating alpine flow regime with high sediment loads, these also included flow constrictions in the form of ice jams or large woody debris. When an ice jam suddenly disintegrated, the high bed shear stress caused new channels to be incised into the floodplain terrain (avulsion) or led to the reoccupation of abandoned arms (Pasetti et al. 1850; Streffleur 1851). Furthermore, flows between mean water and bankfull level (approximately 1-year flood) resulted in substantial lateral channel migration up to 30 m per year. Due to the different forms of channel adjustment and floodplain inundation (active overflow, backwater flooding, seepage inundation), such floodplains featured a great variety of depositional processes, each associated with specific sediment fractions. In the long-term, however, such medium-energy non-cohesive floodplains are considered to be in a dynamic equilibrium of aggradation and erosion processes (cf. Nanson and Croke 1992). In the Lobau, annually 1.6% of the area of the vegetated floodplain terrain were eroded between 1805 and 1817. At the same time, new vegetated terrain of equal size developed (Fig. 12.8).

Thus, within a few decades, large shares of the floodplain terrain were permanently renewed and high shares of morphologically young terrain were typical. In the case of the Lobau with a medium age of approximately 205 years, 25% of the terrain were younger than 35 years (Hohensinner et al. 2008). In comparison, the extremely dynamic floodplain in the eastern Machland 160 km upstream from Vienna (Lower/Upper Austria) showed significantly higher turnover rates, and, accordingly, a considerably younger floodplain terrain. Here, prior to channelization, approximately 2.5% of the vegetated floodplain area were both eroded and regenerated every year. Roughly estimated, fluvial dynamics in the national park downstream of the Lobau probably ranged between that of the Machland and the Lobau. Due to the high frequency of disturbances, the potential natural riparian vegetation (omitting human influences) along the main river corridor primarily consisted of pioneer vegetation, purple willow scrub and diverse softwood communities (white willow, black poplar). More remote sites with less disturbances on older floodplain soils were dominated by transition forms towards hardwood, and the oldest sites by mature hardwood communities (oak-elm woodland; Drescher and Egger 2013).

12.3 Human Modifications of the River Landscape

In the area of the national park, first hydraulic measures were already implemented in the 1770s or even earlier. These were local levees to protect the villages directly north of the alluvial zone against floods (Holub 2012). Probably due to

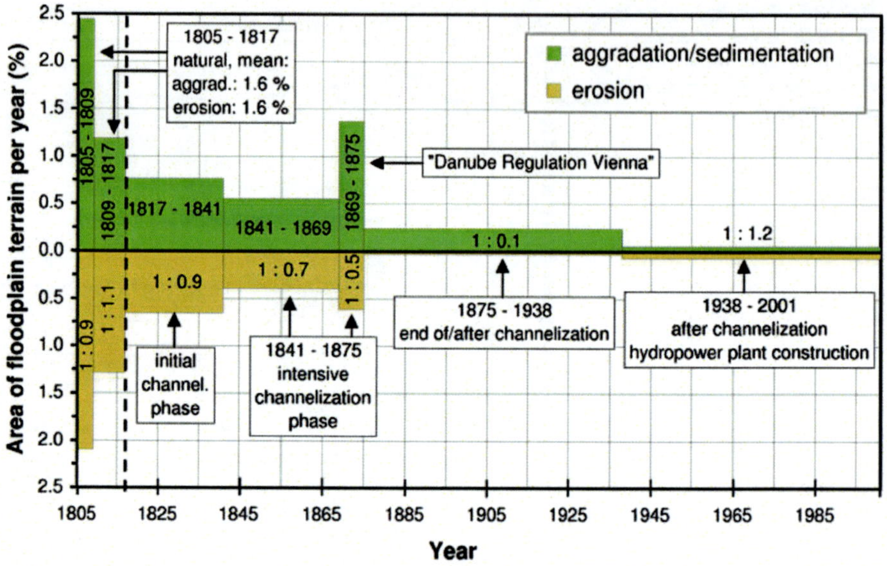

Fig. 12.8 Annual turnover rates in the Lobau 1805–2001 (aggradation and erosion in % of the area of the vegetated terrain in the modern floodplain); the ratio between aggradation and erosion is indicated (modified after Hohensinner et al. 2008, and Lair et al. 2009, with permission from Elsevier)

the severely increasing flood activity of the Danube River at that time, whole systems of local flood protection dikes had been installed by 1817 (for flood history see Hohensinner et al. 2013a; Hohensinner 2015). The era of river regulation started in 1836, when a two kilometre long cut off was created in order to shorten a river bend close to Fischamend and Schönau. In the following years, training walls and embankments were constructed at several sites along the main channel to improve navigation and prevent ice jam formation (Pasetti et al. 1850). Simultaneously, a main branch was cut off by a closure dam in the Lobau. While the piecemeal river training measures were continued in the national park, the Viennese Danube section was systematically channelized between 1870 and 1875. After termination of the Viennese regulation programme, the Marchfeld flood protection levee was extended further downstream, thereby separating large parts of the floodplain from the river. Finally, in 1902, the protection dike reached the Morava River at the downstream end of the national park. Since then, the floodplain areas behind the dike were never again inundated. The one exception is the Lobau, where a flood in 1892 breached the new levee at several sites, leading to the construction of the Schönau backwater levee along the margins of the Lobau floodplain in the same year. Already around 1880, another severe problem became apparent: Huge volumes of sediments were deposited in the river, resulting in a heightening of the river bed by up to 1.5 m in today's national park (Wiletal 1897; Schmautz et al. 2002). The material stemmed from the already channelized upstream river sections in Vienna and the Tulln Basin. This forced the waterway authorities to expedite their efforts and, finally, in 1912 the channelization programme for both mean flow and low flow was completed (Fig. 12.9). Due to the artificial narrowing of the channel, the aggraded material was eroded in the following decades and the river bed gradually incised. Between c. 1870 and the second half of the twentieth century, the flood level, and thus the sedimentation level, were higher than before and thereafter. This specific hydromorphological situation led to increased accretion of suspended load in the floodplain during floods and boosted terrestrialization in the cut off water bodies. Ongoing deposition of sediments on the one hand and channel incision on the other hand have led to an increasingly vertical decoupling between the river level (water/groundwater table) and the floodplain terrain.

The various forms of human interference are reflected by the altered composition of the water bodies in the Lobau (Fig. 12.7). In the course of the first channelization measures between 1817 and 1841, parts of the main channel were transformed into lotic side arms that, in turn, later were cut off at their upstream ends. While the lotic side arms significantly declined until 1875, one-side connected backwaters considerably increased. Prior to channelization, all backwaters together only made up 5–15% of the total water bodies. In contrast, in 1875 the different types of backwaters showed a larger expansion than the main channel, indicating an entirely unnatural configuration. As a consequence of the progressive channelization and associated terrestrialization processes, the area of one-side connected backwaters significantly declined until 1938, while isolated backwaters (dead arms) slightly increased (Fig. 12.7). Today, total water bodies show less than half of their former spread and originally typical side arms have vanished. Instead, isolated and one-side connected backwaters prevail.

The hydraulic structures, i.e. channel cut offs, embankments and flood protection dikes, implemented in the

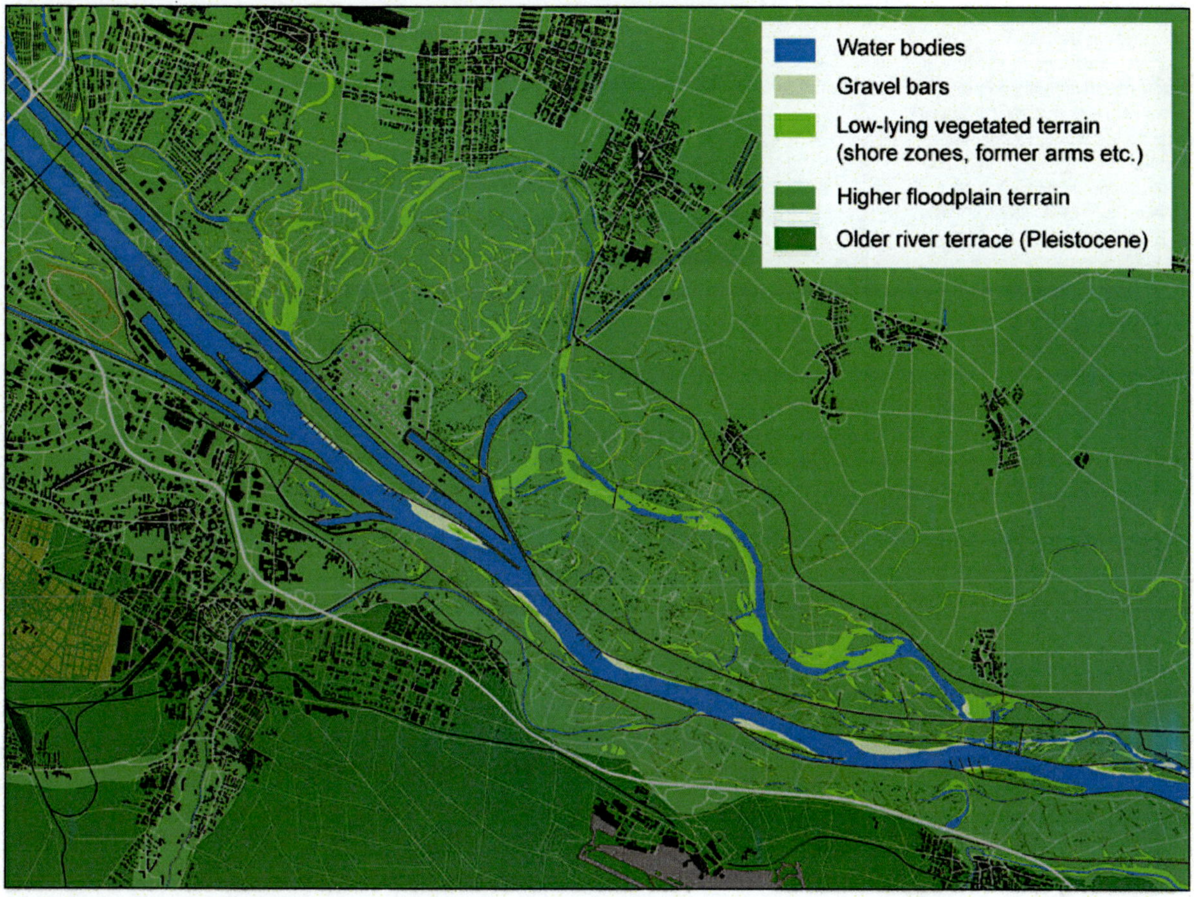

Fig. 12.9 Danube river landscape in the Lobau downstream of Vienna 2001 (upstream part of Danube Floodplain National Park; Eberstaller-Fleischanderl et al. 2004)

nineteenth century significantly constrained dynamic turnover processes (Fig. 12.8). During the early regulation phase between 1817 and 1869, both aggradation and erosion processes were reduced to a similar extent. As a consequence of the Viennese Danube regulation programme, lateral erosion was almost totally inhibited, whereas aggradation continued. Finally, in the second half of the twentieth century, also sedimentation rates substantially declined (the erosion indicated in Fig. 12.8 refers to the artificial excavation of harbours and a flood bypass). The human interferences also altered the morphological age structure of the river landscape (Fig. 12.10).

Before the first channelization measures between 1809 and 1817, the mean weighted age of all aquatic and terrestrial landforms within the modern floodplain was approximately 90 years. While the mean age gradually increased to 105 years in 1869, it was artificially reduced again during the Viennese regulation programme. Since then, a linear increase in the morphological age structure of the river landscape up to 200 years is clearly evident. Missing fluvial dynamics and ongoing ageing of the floodplain soils fundamentally altered site conditions for the successional processes of the riparian vegetation. Even without direct human interventions, i.e. forest management, this results in a significantly modified configuration of the potential natural vegetation (Drescher and Egger 2013).

12.4 Current Situation and River Restoration

In 1996 the IUCN-approved Danube Floodplain National Park was implemented. According to the IUCN, national parks (category II) are: "*large natural or near natural areas set aside to protect large-scale ecological processes, along with the complement of species and ecosystems characteristic of the area, which also provide a foundation for environmentally and culturally compatible, spiritual, scientific, educational, recreational and visitor opportunities*" (IUCN 1994). Within an area of over 9300 hectares (65% wetlands, 15% meadows and 20% water bodies) the reserve aims to

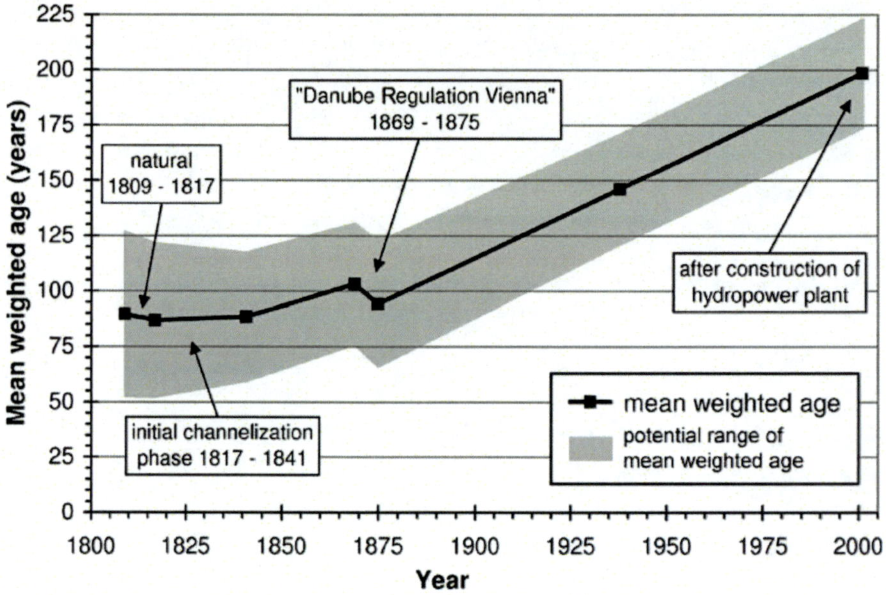

Fig. 12.10 Impact of human activities on the mean morphological age of the fluvial landforms in the Lobau floodplain between 1809 and 2001 (values refer to the modern floodplain formed since approximately AD 1500; modified according to Hohensinner et al. (2008), and Lair et al. (2009), with permission from Elsevier)

protect the last remaining major wetland environment in Central Europe. Here, the Danube River flows freely for c. 38 km, forming the national park's lifeline.

Since 2003, an ambitious river restoration project has been implemented along the Danube section in the national park. Coordinated by *viadonau* (i.e. the leading international waterway operator in the Danube region), the project is part of a larger European transportation infrastructure project (TEN-T) aimed at improving waterway conditions along the Danube River. Simultaneously, the project must fulfil a range of legal requirements and conditions, such as the EU Water Framework Directive, various European conservation directives and national park laws. Within the Danube Floodplain National Park, the project involves comprehensive ecological restoration and revitalization measures. These include the prevention of further channel incision by

Fig. 12.11 Lateral channel erosion in a restored section of the Danube River (i.e. removal of ripraps) in the Danube Floodplain National Park opposite Hainburg. At this location, the 100-year flood event in 2013 caused river bank retreat of 15–20 m. Photo: I. Reiter (2015)

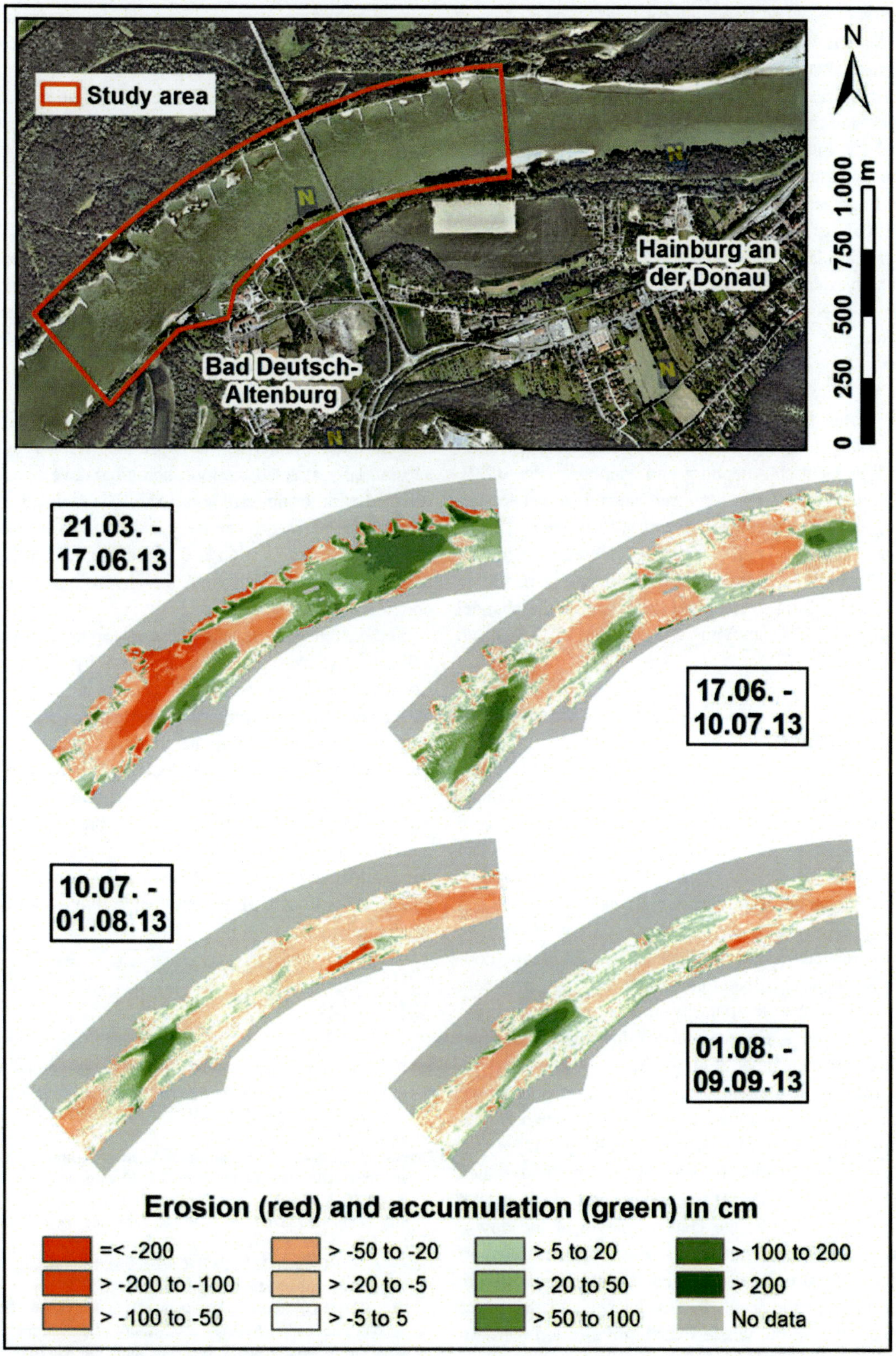

Fig. 12.12 Channel bed dynamics along the Danube River section between Bad Deutsch-Altenburg and Hainburg from March to September 2013 (data source: viadonau). Increased rates of scouring and aggradation were caused by the 100-year flood event in May/June 2013

granulometric river bed improvement (introduction of coarser bed material), restoration of hydrological connectivity and sediment transport (e.g. by reconnecting artificially cut off side arms), restructuring of river banks (e.g. the removal of ripraps), and the adaptation of groynes to be ecologically optimized for low flow conditions. However, the requirements for unhampered navigation and the protection of the hinterland against floods constrain the scope of action and the potential success of such restoration programmes. Moreover, the largely reduced bedload transport due to numerous barriers (i.e. hydropower plants) along the upstream river course and tributaries restrict the restoration of the original river-type specific habitats and fluvial processes.

From 31 May to 6 June 2013, a 100-year flood event occurred along the Danube River. It caused significant channel changes along a newly restored river section between Bad Deutsch-Altenburg and Hainburg (Fig. 12.2). In this Danube section, flood-induced lateral channel erosion caused river bank retreat of several metres (Fig. 12.11). Moreover, channel bed scouring increased locally (Fig. 12.12). Mobilized gravel-sized sediments were shown to migrate in the form of river bed dunes across the channel bottom (Seidel 2014), moving at velocities of 2.5–9 m/h (Gmeiner et al. 2016).

12.5 Conclusion

At first glance, the alluvial sections of the Austrian Danube River prior to channelization in the nineteenth century seem to show similar channel patterns. Indeed, these sections can be clearly identified as high- to medium-energy variants of anabranching channel styles that comprise both braiding and meandering elements. Nonetheless, they show significant differences related to the presence of (a) upstream confluences of large, bedload-rich tributaries, (b) upstream alluvial sections that temporarily functioned as retention areas for floods or sediments and (c) downstream channel constrictions (Hohensinner et al. 2008).

In the case of the Danube Floodplain National Park, two geomorphological "hot spots" defined the framework conditions for the evolution of the river landscape: (1) a significantly reduced channel slope at the upstream end, which potentially amplified fluvial dynamics and thus overall floodplain width and (2) the Gate of Devín at the downstream end, which constricted flow and sediment transport during floods. Next to local spatial constraints, also upstream channel changes, large-scale climate changes and alterations of the land cover in the whole river basin affected the configuration of fluvial landforms in the national park. As long as the controlling factors (climate, discharge, sediment supply, etc.) remain unchanged, a dynamic equilibrium or quasi-equilibrium can be assumed for river landscapes (Leopold and Wolman 1957). In such systems, erosion and aggradation processes are approximately balanced, leading to permanent morphological renewal of distinct parts of the floodplain. Because the environmental framework conditions, i.e. climate and land cover in the catchment, have been changing throughout the Holocene, the Danube River oscillated between more pronounced braiding and amplified meandering channel patterns.

River training programmes and flood protection projects in the nineteenth century have severely truncated the system-inherent potential for channel adjustments. The consequences are the comprehensive stabilization of formerly dynamic fluvial landforms and the missing regeneration of riverine habitats. The history of the Danube River shows that—besides direct river engineering measures—human interventions in the upstream river sections also substantially affected the morphological development in the national park. Long-term legacies of channelization, such as vertical decoupling of the river and floodplain levels due to ongoing channel incision and sedimentation during floods, have significantly changed the hydrological conditions above and below ground.

Today, the success of habitat restoration in such channelized river reaches is constrained by several factors. Locally, the requirements for unhampered navigation and the protection of the hinterland against floods are the most important concerns. Remote impacts, such as severely truncated bedload transport, restrain the recovery of the original river-typical fluvial processes and channel dynamics. Although the importance of a basin-wide sediment management to prevent further bed degradation and improve habitat diversity has been internationally recognized by river and waterway authorities, an integrative solution of that challenge is still pending.

References

Brix F (1970) Der Raum von Wien im Lauf der Erdgeschichte. In: Starmühlner F, Ehrendorfer F (eds) Naturgeschichte Wiens 1 (Lage, Erdgeschichte und Klima). Verlag Jugend und Volk, München, pp 27–190

Demek J, Drescher D, Hohensinner S, Schwaighofer B (2008) The Geology and Geomorphology of Floodplains. In: Klimo E, Hager H, Matic S, Anic I, Kulhavý J (eds) Floodplain Forests of the Temperate Zone of Europe. Kostelec nad Cernými lesy, Lesnická práce, Prague, pp 11–38

Desloges JR, Church M (1987) Channel and floodplain facies in a wandering gravel-bed river. In: Ethridge FG, Flores RM, Harvey MD (eds) Recent Developments in Fluvial Sedimentology. Special Publication Number 39. Society of Economic Paleontologists and Mineralogists, Tulsa, Oklahoma, pp 99–109

Drescher A, Egger G (2013) Wiener Holz—Vegetation. Report for the project "Genug Holz für Stadt und Fluss? Wiens Holzressourcen in dynamischen Donau-Auen" from Karl-Franzens-University Graz and eb & p Umweltbüro GmbH Klagenfurt commissioned by the Institute of Hydrobiology and Aquatic Ecosystem Management, BOKU Vienna

Eberstaller-Fleischanderl D, Hohensinner S, Jungwirth M (2004) Donau 1726–2001. Flussmorphologische Entwicklung der Donau im Wiener Teil des Nationalparks Donau-Auen 1726–2001 (Bereich Lobau, Strom-km 1924,4—1907, 6). Final report for MA 49 Forstamt u. Landwirtschaftsbetrieb, City administration of Vienna, 59 p

Fink J (1976) Exkursion durch den österreichischen Teil des nördlichen Alpenvorlandes und den Donauraum zwischen Krems und Wiener Pforte. Mitteilungen der Kommission für Quartärforschung der österreichischen Akademie der Wissenschaften, Vienna

Fink J, Majdan H (1954) Zur Gliederung der pleistozänen Terrassen des Wiener Raumes. Jahrb Geol Bundesanst 97:211–249

Gmeiner P, Liedermann M, Haimann M, Tritthart M, Habersack H (2016) Grundlegende Prozesse betreffend Hydraulik, Sedimenttransport und Flussmorphologie an der Donau. Österreichische Wasser- U. Abfallwirtschaft 68(5–6):208–216

Grupe S, Jawecki C (2004) Geomorphodynamik Der Wiener Innenstadt. Fundort Wien 7:14–31

Hohensinner S (1995) Bilanzierung historischer Flussstrukturen im Oberen Donautal als Grundlage für die Revitalisierung des ehemaligen Altarmes bei Oberranna. Master thesis, University of Natural Resources & Life Sciences Vienna, Austria

Hohensinner S (2008) Rekonstruktion ursprünglicher Lebensraumverhältnisse der Fluss-Auen-Biozönose der Donau im Machland auf Basis der morphologischen Entwicklung von 1715–1991. Ph.D. Thesis, University of Natural Resources & Life Sciences Vienna, Austria

Hohensinner S (2015) Historische Hochwässer der Wiener Donau und ihrer Zubringer. Materialien zur Umweltgeschichte Österreichs, vol. 1, Centre for Environmental History Vienna, 59 p

Hohensinner S, Eberstaller-Fleischanderl D, Haidvogl G, Herrnegger M, Weiß M (2008) Die Stadt und der Strom - Historische Veränderungen der Wiener Donau-Auen seit dem 18. Jahrhundert. Abhandlungen Der Geologischen Bundesanstalt 62:87–93

Hohensinner S, Jungwirth M (2009) Hydromorphological characteristics of the Danube River – the historical perspective. Österreichische Ingenieur- U. Architekten-Zeitschrift 154(1–6):33–38

Hohensinner S, Lager B, Sonnlechner C, Haidvogl G, Gierlinger S, Schmid M, Krausmann F, Winiwarter V (2013a) Changes in water and land: the reconstructed Viennese riverscape 1500 to the present. Water History 5(2):145–172

Hohensinner S, Sonnlechner C, Schmid M, Winiwarter V (2013b) Two steps back, one step forward: Reconstructing the dynamic Danube riverscape under human influence in Vienna. Water History 5 (2):121–143

Holub C (2012) Rekonstruktion der historischen hydromorphologischen Eingriffe an der Donau im Wiener und Tullner Becken. Master thesis, University of Natural Resources and Life Sciences Vienna and University of Vienna, 106 p

Jungwirth M, Haidvogl G, Hohensinner S, Waidbacher H, Zauner G (2014) Österreichs Donau. Landschaft - Fisch – Geschichte (Austria's Danube. Landscape – Fish – History). Institute of Hydrobiology & Aquatic Ecosystem Management, BOKU Vienna, 420 p, ISBN:978-3-900932-20-6

Kohl H (1991) Die Veränderungen der Flussläufe von Donau, Traun und Enns seit prähistorischer Zeit. Forschung zur Geschichte der Städte u. Märkte Österreichs 4:1–9

Küpper H (1968) Wien. Geologie der Österreichischen Bundesländer in kurz gefassten Einzeldarstellungen. Verhandlungen der Geologischen Bundesanstalt, Bundesländerserie, 206 p

Lair GJ, Zehetner F, Fiebig M, Gerzabek MH, van Gestel CAM, Hein T, Hohensinner S, Hsu P, Jones KC, Jordan G, Koelmans AA, Poot A, Slijkerman DME, Totsche KU, Bondar-Kunze E, Barth JAC (2009) How do long-term development and periodical changes of river-floodplain systems affect the fate of contaminants? Results from European Rivers. Environmental Pollution 157(12):3336–3346

Leopold LB & Wolman MG (1957) River Channel Patterns: Braided, Meandering and Straight. Geological Survey Professional Paper, 282-B:39–85

Lorenzo C (1819) Nieder Oesterreichische Donau-Stromkarte, M: 1:7200, surveyed 1816-1817, published 1819, 69 maps, Provincial Library of Lower Austria, Sign. B II 82

Magyar I, Geary DH, Müller P (1999) Paleographic evolution of the Late Miocene Lake Pannon in Central Europe. Palaeogeogr Palaeoclimatol Palaeoecol 147:151–167

Nanson GC, Croke JC (1992) A genetic classification of floodplains. Geomorphology 4:459–486

Nanson GC, Knighton AD (1996) Anabranching rivers: their cause, character and classification. Earth Surf Proc Land 21:217–239

Pasetti F et al (1850) Die Regulirung der Donau und der Bau einer stabilen Brücke über dieselbe bei Wien. Sonderabdruck Aus Der Allgemeinen Bauzeitung 15:41–137

Penck A (1891) Die Donau. Schr Naturwissen Verbr Naturwissen Kennt Wien 31:1–101

Peresson H (2006) Neotektonik. In: Wessely G (ed) Geologie der österreichischen Bundesländer: Niederösterreich. Geologische Bundesanstalt, Vienna, p 416

Reiter I (2015) Geomorphologische Veränderungen in einem deregulierten Flussabschnitt der Donau in Hainburg nach dem Hochwasserereignis 2013. Eine Erhebung mittels Kartierung im Gelände und GIS-Analysen. Bachelor thesis, Department of Geography and Regional Research, University of Vienna, Austria, 45 p

Salcher BC, Wagreich M (2010) Climate and tectonic controls on Pleistocene sequence development and river evolution in the Southern Vienna Basin (Austria). Quatern Int 222:154–167

Schmautz M, Aufleger M, Strobl T (2000) Wissenschaftliche Untersuchung der Geschiebe- und Eintiefungsproblematik der österreichischen Donau. Report for Verbund – Austrian Hydro Power AG (AHP), Vienna

Schmautz M, Aufleger M, Strobl T (2002) Anthropogene Einflussnahme auf die Flussmorphologie der Donau in Österreich. Österreichische Ingenieur- U. Architekten-Zeitschrift 147(5–6):171–178

Seidel C (2014) Der Einfluss des Hochwasserereignisses 2013 auf die Beschaffenheit der Flusssohle der Donau in einem Teilabschnitt des Pilotprojekts Bad Deutsch-Altenburg. Bachelor thesis, Department of Geography and Regional Research, University of Vienna, Austria, 68 p

Slezak F (1948) Historische Veränderungen der Donaustromlandschaft im Tullner und Wiener Becken. Dissertation, University of Vienna, 157 p

Sommerhäuser M, Robert S, Birk S, Hering D, Moog O, Stubauer I, Ofenböck T (2003) Developing the Typology of Surface Waters and Defining the Relevant Reference Conditions. Activity 1.1.6 Final report. UNDP/GEF Danube Regional Project, University of Duisburg-Essen, Germany and University of Natural Resources & Life Sciences Vienna, Austria

Streffleur V (1851) Einiges über Wasserstands- (Pegel-) Beobachtungen und deren Aufzeichnungen. Sitzungsberichte der Kaiserlichen Akademie der Wissenschaften. Mathematisch-naturwissenschaftliche Classe 7 (1):745–756

Summesberger H (2011) Vom Tropenmeer zur Eiszeittundra – 250 Millionen Jahre Wiener Stadtgeschichte. In: Berger R, Ehrendorfer F (eds) Ökosystem Wien – Die Naturgeschichte einer Stadt. Böhlau, Vienna, pp 61–87

Voit K, Embleton-Hamann C (2011) Die Terrassenlandschaft von Wien – Zeugnis einer kalten Vergangenheit. In: Embleton-Hamann C, Keiler M, Teufl I (eds) Umweltstadtführer Wien – Einblicke in die Natur einer Großstadt. Böhlau, Vienna, pp 27–34

Wiletal I (1897) Die Veränderungen des Donau-Laufes im Wiener Becken. Alt-Wien – Monatsschrift für Wiener Art und Sprache 6:48–51, 65–68

Severin Hohensinner, senior scientist and lecturer at the Institute of Hydrobiology and Aquatic Ecosystem Management (IHG), University of Natural Resources and Life Sciences Vienna (BOKU), investigates historical fluvial landscapes and past morphodynamic processes. His studies contribute to the identification of historical habitat conditions in riverine ecosystems. He is involved in several applied interdisciplinary projects in aquatic ecology, river landscape planning, river restoration and environmental history.

Ronald E. Pöppl is a senior lecturer at the Department of Geography and Regional Research at the University of Vienna. His primary research interests include connectivity in fluvial systems, fluvial geomorphology, soil erosion, human impacts and human–environment interactions, river and catchment management.

13 Lake Neusiedl Area: A Particular Lakescape at the Boundary Between Alps and Pannonian Basin

Erich Draganits, Michael Weißl, András Zámolyi, and Michael Doneus

Abstract

The Lake Neusiedl area is a unique lakescape, situated at the geodynamical and geomorphological boundary between the Alps, Carpathians and the Pannonian Basin, and therefore represents an important transition zone concerning terrain, climate, vegetation, fauna and cultures. We use geomorphological as well as geological data, topographical and historical maps plus historical charters to reconstruct the palaeohydrology of Lake Neusiedl and document dramatic landscape changes, especially in the last centuries. The present-day hydrological conditions of and processes at Lake Neusiedl are very different from those in the past. Virtually, all historical maps before 1780 show the Ikva River, Répce (Rabnitz) River and Kis-Rába River discharging into the connected Neusiedlersee/Hanság area that possessed a natural outlet, the Rábca River. The documented episodic variation of the water levels of Lake Neusiedl between desiccation and highest flood levels is c. 4.2 m, affecting enormous areas in this extremely low-relief region, with a huge impact on the landscape, fauna and vegetation, human settlement patterns, land use and communication routes—which should be considered in regional archaeological and historical interpretations. The numerous shallow lakes and presently dry basins in the Seewinkel originally formed as thermokarst lakes during early Lateglacial permafrost degradation after the end of the Late Glacial Maximum (LGM).

Keywords

Little Hungarian plain • Lake Neusiedl • Seewinkel • Hydrology • Airborne laser scanning topography • Thermokarst lakes

13.1 Introduction

Lake Neusiedl (German: *Neusiedler See*, Hungarian: *Fertő-tó*) area (Fig. 13.1) is the westernmost extension of the Little Hungarian Plain (Hungarian: *Kisalföld*) and characterized by low-lying, low-relief landscapes, dominated by Quaternary fluvial and lacustrine sediments, deposited on top of upper Pannonian (=Tortonian, upper Miocene) sediments (Fuchs and Schreiber 1985). This region is situated at the geodynamical and geomorphological boundary between the Alps, Carpathians and the Pannonian Basin (Székely et al. 2009) and represents an important transition zone concerning terrain, climate, vegetation, fauna and culture (Oberleitner et al. 2006; Korner 2008; Fally and Kárpáti 2012; Molnár et al. 2012; Boros et al. 2013). Grounded on the region's unique landscape and natural importance, a part of this area was declared a transnational Neusiedler See-Seewinkel National Park in 1992 and UNESCO World Heritage Site in 2001 (Fally and Kárpáti 2012).

Visualization of high-resolution digital terrain models (DTM) with 1 × 1 m resolution, derived from airborne laser scanning (ALS) measured in April 2010, is essential for the geomorphological investigation of this low-relief area (Figs. 13.1 and 13.2) (Doneus and Briese 2006; Mandelburger et al. 2009). In this low-relief landscape, even moderate lake level variations impact large areas. Thus, they are well visible in historical maps, which, therefore, represent invaluable sources for hydrological reconstructions (Csaplovics 2005; Draganits et al. 2008). As elevation information in maps and documents is affected by the difference in the Austrian and Hungarian levelling system, the latter being

E. Draganits (✉) · M. Weißl
Department of Geology, University of Vienna, Josef Holaubeck - Platz 2, 1090 Wien, Austria
e-mail: erich.draganits@univie.ac.at

A. Zámolyi
OMV Austria Exploration & Production GmbH, Protteser Str. 40, 2230 Gänserndorf, Austria

M. Doneus
Department of Prehistoric and Historical Archaeology, University of Vienna, Franz-Klein-Gasse 1, 1190 Wien, Austria

C. Embleton-Hamann (ed.), *Landscapes and Landforms of Austria*, World Geomorphological Landscapes,
https://doi.org/10.1007/978-3-030-92815-5_13

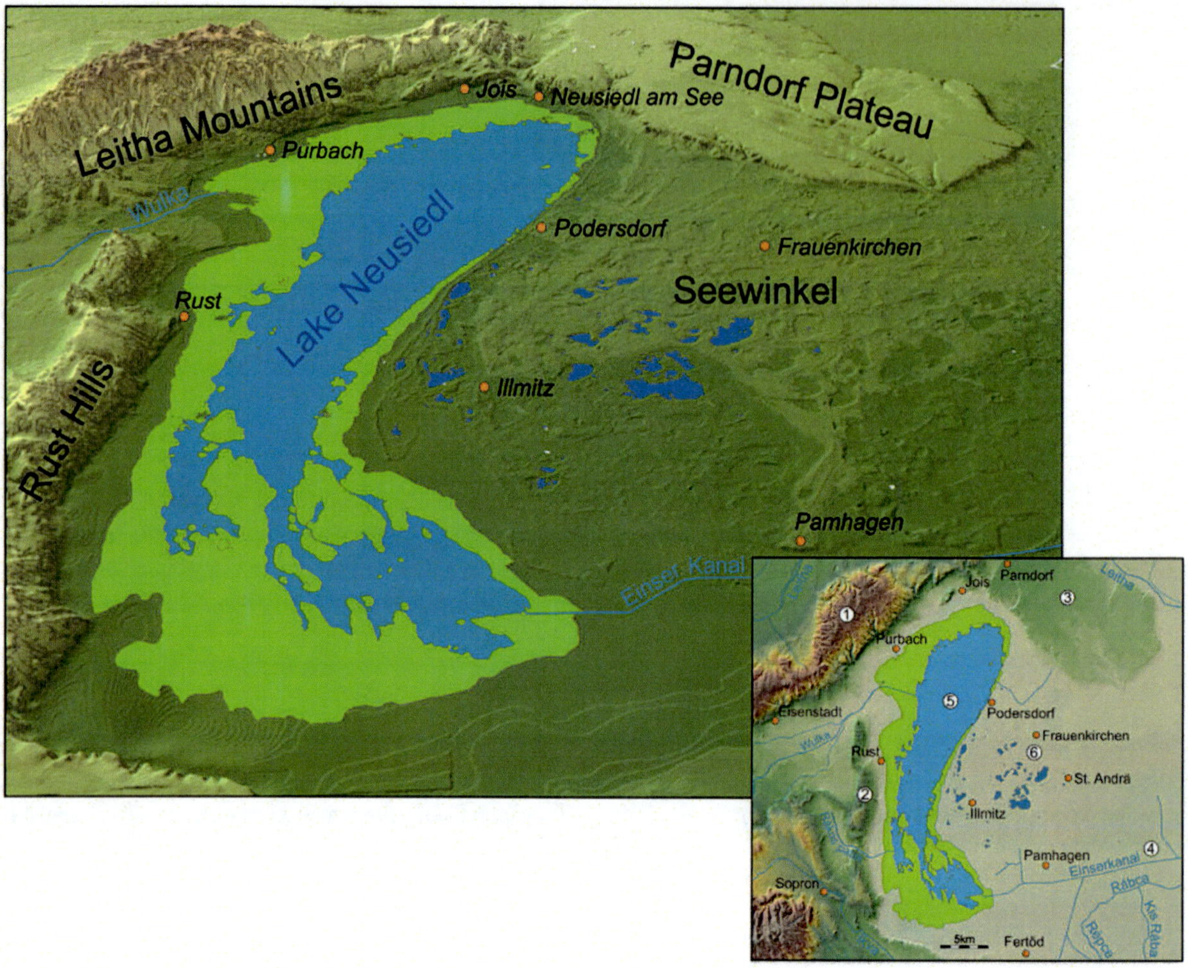

Fig. 13.1 Overview of the topography of the Lake Neusiedl area. Geomorphological elements are labelled in the order, they are mentioned in the text. Lake Neusiedl: light blue; reed area: light green; shallow lakes of the Seewinkel: blue. Oblique view is 8-times vertically exaggerated. Topographic data are a combination of 10 × 10 m DTM in the Austrian part and lower resolution data for the Hungarian portion

49.6–60.6 cm lower (Höggerl 2013), all altitude values were transformed to the Austrian system.

Lake Neusiedl, located at the political border of Austria and Hungary, is the most prominent geomorphological element of this area and thus receives most attention in this context. The previous studies about the lake are summarized in Sauerzopf (1959a), Löffler (1974, 1979), Fally and Kárpáti (2012) and Wolfram et al. (2014).

13.2 Geodynamical Setting

The boundary between the Eastern Alps, Carpathians and the Pannonian Basin represents a complex deformation zone, which has been active since the Early Miocene (Székely et al. 2009; Zámolyi et al. 2016). The tectonic processes in this area result from the collision of the Adriatic and Eurasian plates and are dominated by lateral escape and extensional collapse. Geomorphologically, significant tectonic activity in this area is indicated by seismicity (Tóth et al. 2007; Häusler et al. 2010), linear features related to normal- and strike-slip faults (Székely et al. 2009; Zámolyi et al. 2016) as well as tectonically influenced river courses (Zámolyi et al. 2010).

In the low-relief Little Hungarian Plain, hydrology is strongly influenced by active vertical crustal movements with highest recent subsidence values around 1–2 mm/year at the depocentre at Győr (Joó 1992). Long-lasting subsidence in this area is indicated by >8000 m thick Neogene sediments and >600 m thick Quaternary sediments (Scharek 1991). This subsidence causes tilting of the Lake Neusiedl area towards the east, and consequently, the thickness of

Fig. 13.2 Aerial photo from south of Podersdorf towards south-southwest, showing (left to right) the lakescape of the Seewinkel, the shore-parallel ridge (*Seedamm*) with some tree cover, followed by the brownish reed area and finally the southern part of Lake Neusiedl towards the west; The largest shallow lake in the foreground is the Oberer Stinker See. Photo: A. Ziegler, 20.12.2014

Pannonian sediments increases from zero at the Leitha Mountains to >2000 m below the Hanság south of the Seewinkel (Fuchs and Schreiber 1985). Similarly, Quaternary sediments thicken from virtually zero at Lake Neusiedl to >20 m below the Hanság (Tauber 1959).

13.3 Geomorphological Units

13.3.1 Leitha Mountains

The Leitha Mountains (Fig. 13.1, No. 1) are a NE-SW trending hilly landscape, rising above the surrounding low-relief areas from around 118 to 484 m asl. Despite their relatively low altitude, they represent a considerable weather divide, contributing to lower precipitation in the Lake Neusiedl area. Geologically, the Leitha Mountains represent a tectonic horst comprising Lower Austroalpine schists, gneisses and amphibolites overlain by low-grade metamorphic Triassic quartzites and dolomites (Pistotnik et al. 1993; Spahić and Rundić 2015). At the rim of the Leitha Mountains, these metamorphic rocks are covered by Badenian to Sarmatian (=Langhium to Serravallium, Middle Miocene) clastic sediments and limestones (Pistotnik et al. 1993; Wiedl et al. 2014).

13.3.2 Rust Hills

The Rust Hills (Fig. 13.1, No. 2) form a narrow N-S trending ridge with altitudes between 118 and 283 m asl at the western margin of the Lake Neusiedl. They comprise schists, gneisses and amphibolites of the Lower Austroalpine tectonic unit that are almost completely covered by Karpatian to Sarmatian clastic sediments and limestones (Fuchs 1965; Pistotnik et al. 1993). The Rust Hills are a tectonic horst, bordered by N-S trending normal faults, which affects both the Neogene cover and the basement rocks (Fuchs 1965; Spahić and Rundić 2015).

13.3.3 Parndorf Plateau

The Parndorf Plateau (Fig. 13.1, No. 3) represents a relatively even surface comprising Pannonian clastic sediments with a thin Pleistocene fluvial cover left by the Danube. It is elevated about 25–45 m above the surrounding lowland and its surface gently dips from northwest with around 184 m to some 144 m asl in the southeast. In the northwest, the Parndorf Plateau has a narrow connection with Danube terraces of the Vienna Basin, whilst in all other directions, it shows distinct slopes (Fig. 13.1). The Parndorf Plateau is dissected by dominantly northwest-southeast oriented dry valleys, which possibly represent periglacial features. Zámolyi et al. (2016) summarized arguments for a tectonic origin of the Parndorf Plateau.

13.3.4 Hanság/Waasen (Former) Wetlands

The Hanság depression (Fig. 13.1, No. 4) is an extremely flat area south of the Seewinkel Plain with an altitude below 117.5 m asl. Before drainage, this area was an extensive alder wetland (Fally and Kárpáti 2012) in the southeastern continuation of Lake Neusiedl. During episodic high water levels, this area formed a continuous L-shaped lake with Lake Neusiedl. In many places, a <0.5 m thin cover of peat is still preserved (Löffler 2000).

13.3.5 Seewinkel

The Seewinkel (Fig. 13.1, No. 6) is bordered by Lake Neusiedl in the west, the Parndorf Plateau in the northeast and the Hanság in the south. The name Seewinkel (*Winkel* = English: angle, corner) possibly originates from the former larger, L-shaped Lake Neusiedl with the Seewinkel in the corner of the 'L'. This roughly 300 km^2 large area is one of the flattest regions of Austria with less than 17 m relief variation. It contains the lowest area of Austria (c. 113 m asl), and its highest point is a small mound northwest of Frauenkirchen with the top at 130 m asl, which rises >4 m above the surrounding surface and is a (probably Iron Age) burial mound (Lindinger 1996).

13.4 Lake Neusiedl

The present-day (!) conditions of Lake Neusiedl are

(i) Largest lake of Austria: 321 km^2 (143 km^2 open water, 178 km^2 reed area; calculated for 116.50 m asl, Bácsatyai et al. 1997).
(ii) Exact lake level adjustment since 1965 (Wasserportal Burgenland 2016).
(iii) End of June 2016 lake level is 115.66 m asl (Wasserportal Burgenland 2016).
(iv) The average lake level of the last 10 years is 115.63 m asl, which is higher than in the period 1965–2006 (115.47 m asl) (Wasserportal Burgenland 2016).
(v) Average depth is only around 1.4 m (Heine et al. 2014).
(vi) Catchment area is 1120 km^2 (Herzig 2014).
(vii) On average, during the period 1965–2012, precipitation contributed 76% of the lake water and river inflow 24%; lake water is reduced by evaporation (89%) and drainage through the artificial outlet 'Einser Kanal' (Hungarian: *Hanság-főcsatorna*) (11%) (Maracek and Kubu 2014).
(viii) Chemically (Knie 1959), the lake belongs to subsaline lakes (Hammer 1986).
(ix) Lake Neusiedl is an endoreic lake and commonly called 'steppe lake' (Sauerzopf 1959a; Löffler 1974).

13.4.1 Origin of Lake Neusiedl

In contrast to most lakes in the Circum-Alpine area, which occupy glacially overdeepened valley floors, Lake Neusiedl lies outside the formerly glaciated areas (van Husen 1987), and therefore, other formation processes must have been involved. The lake is very shallow, and in contrast to deeper lakes, only less than 0.7 m thick lake sediments are deposited at the bottom (Heine et al. 2016), because even moderate wind generates waves that erode the lake floor.

Tauber (1959) and Löffler (1979) discussed various hypotheses of the formation of Lake Neusiedl. Hassinger (1905) proposed that the depression of Lake Neusiedl was formed by erosion by the Danube. This hypothesis was rejected because of the lack of any substantial gravel on top of the Pannonian sediments at the lake bottom of Lake Neusiedl (Küpper 1957). Formation by deflation has already been rejected by Wiche (1951). Finally, Szádeczky-Kardoss (1938) as well as Küpper (1957) suggested a tectonic origin of Lake Neusiedl depression. Tectonic activity, probably in a Horst-Graben type deformation style (Spahić and Rundić 2015), has been supported in several recent studies (Székely et al. 2009; Häusler et al. 2010; Zámolyi et al. 2016; Loisl et al. 2018). The steep eastern side of Hackelsberg, next to Jois, is most probably a normal fault, juxtaposing metamorphic rocks in the west besides Holocene Lake sediments in the east (Fig. 13.3a).

13.4.2 Age of Lake Neusiedl

Surprisingly, the age of Lake Neusiedl is still uncertain. So far, no well-defined geochronological age exists, but an age estimate is between 12 and 14 000 years, based mainly on ostracods (e.g. Löffler 1990). The only areas with thicker lake sediments are in reed areas (Löffler 1990) and around the present lake in areas up to about 117.5 m asl (Szontagh

Fig. 13.3 **a** Steep eastern slope of Hackelsberg, probably representing a normal fault, juxtaposing Lower Austroalpine metamorphic rocks against Holocene lake sediments. Note the extreme flatness of the former inundation area of Lake Neusiedl up to more or less the fault scarp. Approximate location of profile in Fig. 13.3b is indicated by red arrow. Photo taken from northeast of Winden am See towards northeast: Photo: E. Draganits, 24.9. 2006. **b** Trench exposing 3.5 m lithostratigraphy above the ground water, east of Jois (N47°57′14.1″; E016°47′30.8″; ±5 m). Below c 1 m of Chernozem soil are silty lake marls with hardly any indication of bedding except for two horizons, one with quartz pebbles and another one with Unionidae shells. Photo: E. Draganits, 19.11.2005

1904; Draganits et al. 2007). There is lake water with high content of suspended sediment flooded the surrounding of the lake, and later, during slow lake level drop, sediment was trapped in the vegetation. A construction trench within the former lake east of Jois exposed two horizons with Unionidae shells (Fig. 13.3b). The shells have been radiocarbon dated and yield an uncalibrated age of 12,340 ± 70 years BP (1σ confidence interval) (Table 13.1). As the radiocarbon age is not calibrated and the investigated trench did not reach the base of the lake sediments, this age represents a minimum age.

13.4.3 Palaeohydrology of Lake Neusiedl

There are many detailed studies about the limnology of Lake Neusiedl, including climate variations, precipitation and evaporation measurements, etc., (Sauerzopf 1959a; Löffler 1979; Eitzinger et al. 2009; Soja et al. 2013; Wolfram et al. 2014). We use geological (e.g. lake sediments), geomorphological (e.g. ALS DTM, palaeo shorelines, ice-push ridges), topographical and historical maps as well as historical charters to evidence dramatic changes of the lake, especially in the last centuries.

Table 13.1 Radiocarbon data from Unionidae shells from the lake section at Jois (Fig. 13.3b)

Sample	Laboratory number	$d^{13}C$*[‰]	^{14}C-age* [BP]
ED05/JOIS/2b[a]	LTL3911A	−5.5 ± 0.1	12,340 ± 70

[a]Unionidae bivalve shells (Fig 27_4); CEDAD radiocarbon facility in Brindisi

*1σ confidence interval

Some major hydrologic interventions for flood control and reclamation of wetland areas are responsible for the present-day conditions of the lake. The earliest preserved record of intervention into the lake's hydrology is a charter from June 5th 1568 archived in the Austrian *Finanz-und Hofkammerarchiv* in Vienna, documenting a considerable reduction of the lake level caused by a diversion of the Rabnitz River (Anonymous 1568). Probably, the most important intervention was the building of a dam road between Pamhagen and Fertőd (Fig. 13.1), which was finished in 1780 (Fig. 13.4, Korabinszky 1804). At the beginning, there existed some passages for water, which were closed later, cutting off the lake from important tributaries including Ikva and Répce Rivers.

The so called 'Einser Kanal' was connected with the lake in 1909 (Hicke 1996) and used to drain water from the lake. Since 1965, the Einser Kanal is used to remove water only if necessary—thus raising the average lake level (Wasserportal Burgenland 2016).

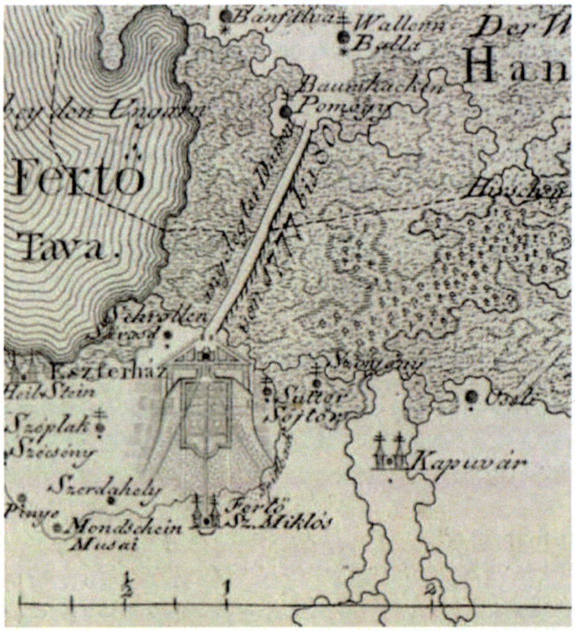

Fig. 13.4 Historical map showing the dam road between Pamhagen and Fertőd and the year of completion (Korabinszky 1804, Table XXIV; Digital Collections of the University Library Regensburg, W 02/8 5392)

Daily lake level measurements in Austria started in Neusiedl am See on June 1st 1930 (HDÖ 1938). Since then, the minimum lake level was 114.50 m asl (July 1949), the maximum value reached 116.08 m asl (May 1941) and the mean value is 115.57 m asl (Wasserportal Burgenland 2016). Lake levels before 1930 have been reconstructed mainly using maps and historical charters (Sauerzopf 1959b; Kopf 1963; Kiss 2009–2010). Kiss (2009, 2010) rightly questioned reconstructions based on terms like lake/river/bog in charters, leaving only few clear indications of the lake level before about 1784.

Larger lake extents than at present are clearly indicated by historical topographic maps (Fig. 13.5 and Csaplovics 2005) and lake deposits (Draganits et al. 2007). The investigated lake sediments east of Jois (Figs. 13.1 and 13.3b) comprise c. 3.5 m thick greyish, silty marl covered by Chernozem soil (Fig. 13.3b). Based on the content of characteristic authigenic carbonate minerals (Neuhuber et al. 2015), the silty marl unequivocally represents lake sediments. The investigated section is situated close to the western termination of an extremely flat area, in geomorphological continuity with the lake. Thus, the top of the section (117.65 m asl) is very close to the maximum inundation level of the lake during episodic flood events. What are the explanations for such high water levels of Lake Neusiedl (Figs. 5 and 6)? The lake's palaeohydrology cannot be understood based on its present conditions (e.g. Sauerzopf 1959a; Löffler 1979; Wasserportal Burgenland 2016), but only in the context of its entire palaeohydrological history. This includes its entire catchment prior to the construction of the dam road between Pamhagen and Fertőd in 1780 and the later closure of the dam's passages, its former natural outlet, the Rábca River, debouching into the Mosoni-Duna River at Győr (Fig. 13.7) (see Kopf 1963).

A combination of several factors may have contributed to episodically higher lake levels in the past (Fig. 13.6) including (i) climate variability (Eitzinger et al. 2009), (ii) less or no artificial drainage, (iii) changes in topography due to active vertical movements (Joó 1992) and (iv) the former larger catchment. It is important to remember that the maximum flood levels at the Rába gauge at Győr range between 115.85 and 115.96 m asl (altitude corrected for the difference in Austrian and Hungarian levelling systems) (http://www.edukovizig.hu/map/layout.html). Flooding in this area reportedly caused afflux of the Répce (Rabnitz) River and Kis-Rába River and consequently flooding of the Lake Neusiedl/Hanság lowland from the east (e.g. Kugler 1871). Virtually, all historical maps before 1780 show the Ikva River (383 km^2 additional catchment), Répce (Rabnitz) River (1268 km^2) and Kis-Rába River (6649 km^2) discharging into the connected Neusiedlersee/Hanság area

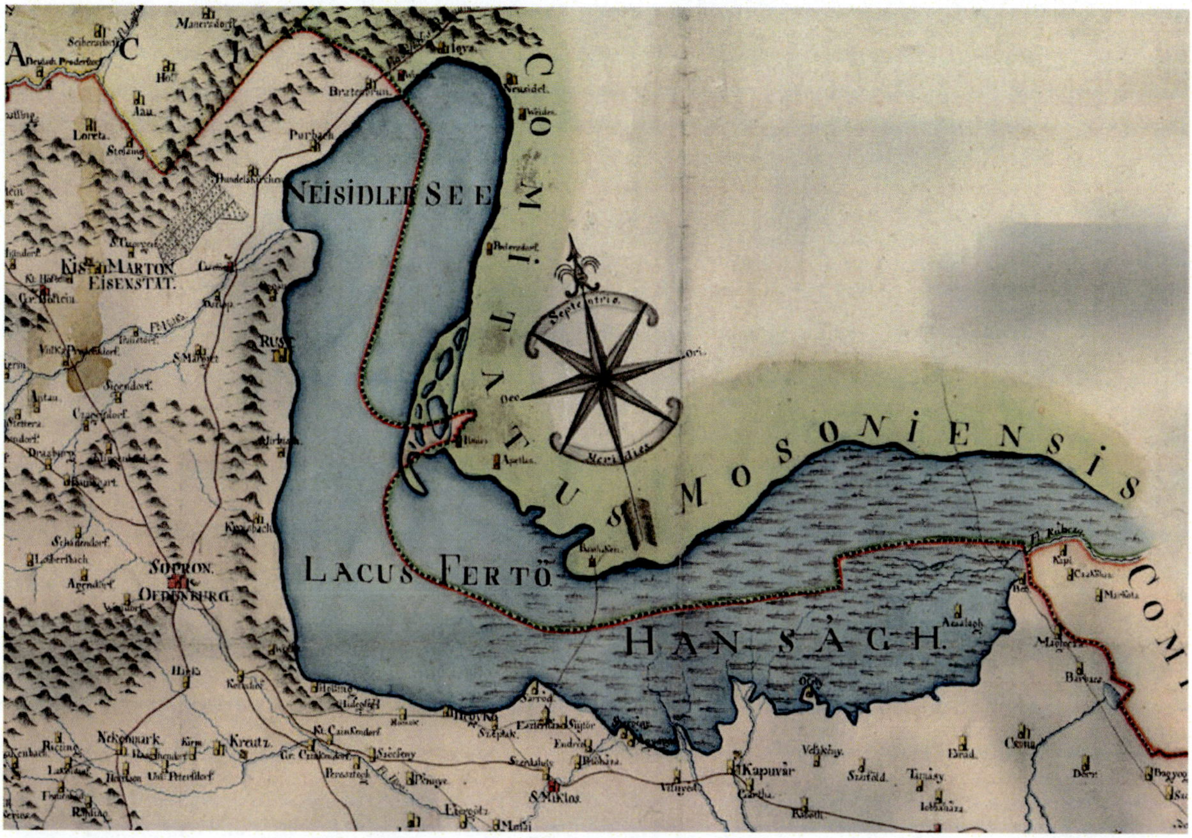

Fig. 13.5 Historical map from 1788 showing the intimate connection between Lake Neusiedl and the Hanság, shortly after the road between Pamhagen and Fertőd was completed (Hegedűs 1788, Országos Széchényi Könyvtár, TK 1614). Note that the Ikva, Répce and Kis Rába rivers drain into the connected Lake Neusiedl/Hanság system

(Fig. 13.7). Larger lake size and larger wetlands east of the lake may have also provided ideal habitats for malaria transmitting mosquitos (see Bruce-Chwatt and de Zulueta 1980; Wernsdorfer 2002).

Most historical maps show a natural outlet, the Rábca River (Fig. 13.5) (e.g. Zeller 1753; Hegedűs 1788), and consequently, Lake Neusiedl was unlikely an endoreic lake in most periods before 1780. Even in the period 1965–2012, when the Wulka River was its only major tributary, on average 11% of the lake's water was drained by the artificial Einser Kanal (Maracek and Kubu 2014). The much larger palaeohydrological catchment before c. 1780 also suggests a lower salt content of Lake Neusiedl compared to modern values, especially during flood periods. In this context, it is worth reconsidering the limnological basis for the categorization of Lake Neusiedl as 'steppe lake', a category lacking conclusive definition.

During 1865–1870, the lake has been more or less dry (Sauerzopf 1959b). Although there is a lot of evidence for exceptionally low precipitation before and during this period (ZAMG 2016), the maps of the second and third Military Survey of Austria, 1845–1846 and 1872–1881, respectively, already show a network of drainage channels tapping Lake Neusiedl. Therefore and based on the palaeohydrological reconstructions above, it seems very unlikely that Lake Neusiedl had '*about 100–200 dry periods since the lake came into existence*' as stated by Löffler (1990).

13.4.4 Geomorphological Features Related to Lake Neusiedl

Geomorphological features contribute considerably to our understanding of the palaeohydrology of the lake. Palaeo shorelines are well developed, especially in its northern and northwestern part. They are easily recognized in ASL digital terrain data, but many of them are also visible on the ground, for instance the palaeo shoreline north of Purbach (Fig. 13.8).

One of the most noticeable geomorphological features of Lake Neusiedl is a ridge (German: '*Seedamm*') that runs more than 22 km parallel to its eastern shore, closely

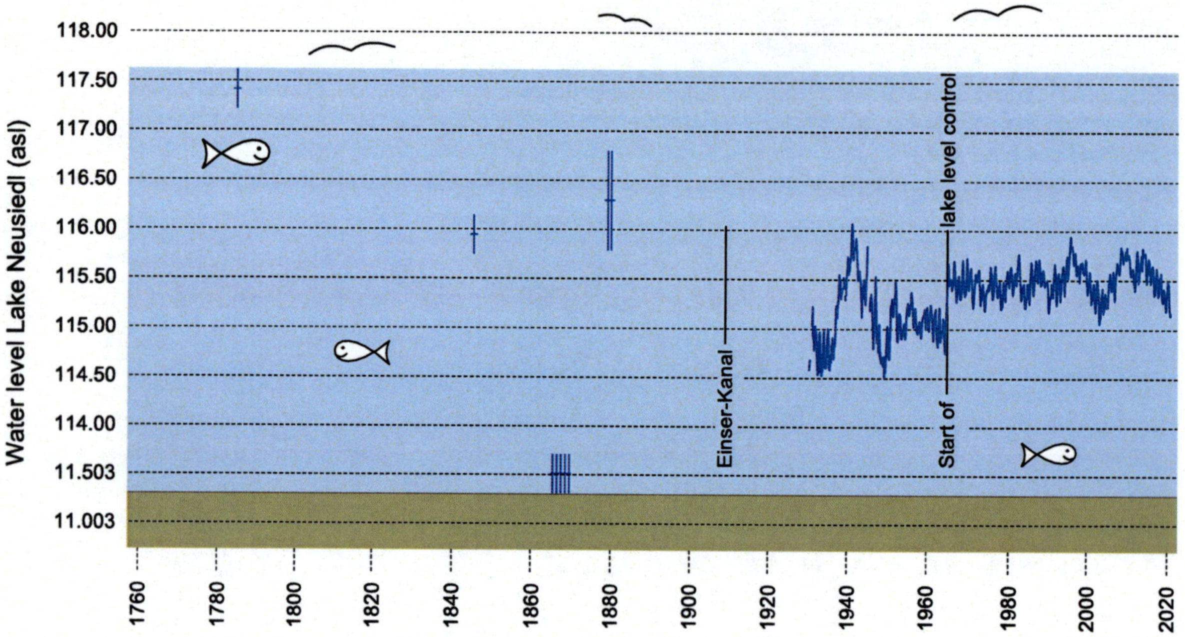

Fig. 13.6 Lake level variations reconstructed from the first Austrian land survey (1784), second Austrian land survey (1845/1846), historical sources (1865–1870) and third Austrian land survey (1872/1873). Modern daily water level measurement started in the Austrian part on June 1st 1930, exact lake level adjustment in 1965 (Wasserportal Burgenland 2016). Diagram is based on monthly values. Lowest lake bottom altitude from Heine et al. (2014)

following its outline (Figs. 13.9 and 13.10) (Bernhauser 1962). The structure comprises pebbly sand and is usually 100–150 m wide and 2.0–2.5 m high. It is generally accepted (e.g. Löffler 1974) that the ridge has been formed by ice-push (compare with Kelletat 1995 and Mahoney et al. 2004), which can occur when the lake is frozen and the dominant north-western winds push ice far onshore (Fig. 13.9a). Profiles across the structure indicate that this feature also formed at slightly higher lake levels than present. Already Bernhauser (1962) noticed the existence of at least one more similar ridge to the east. They are very well visible in the high-resolution ALS data (Fig. 13.10); however, their relationship with each other and with former lake levels is not completely clear and definitely deserves more research. Altitude data concerning palaeo lake levels and geomorphological features should be interpreted with care; active vertical crustal movements in this area are in the order of c. 1 mm/year (=1 m/millennium) (Joó 1992; Ruess and Mitterschiffthaler 2015), which may modify their altitude and spatial relationships, especially in this very low-relief landscape.

13.5 Seewinkel Lakescape

The Seewinkel is one of the flattest areas of Austria. Similar to the Parndorf Plateau, the Seewinkel is characterized by <25 m thick fluvial gravel and sand on top of Pannonian sediments (Tauber 1959; Häusler 2007). In some parts, thin layers of aeolian deposits have been described (Husz 1965). At present, the area is covered by anthropogenic steppe vegetation (Hungarian: *Puszta*), but originally, it most likely showed forest steppe conditions (Wendelberger 1950, 1955, 1987; Molnár et al. 2012).

The geomorphologically and ecologically most prominent elements of the Seewinkel are the numerous shallow lakes (Fig. 13.1) creating a particular lakescape (Fig. 13.2) of great beauty and importance (Löffler 1982). At present, only some 40 shallow lakes are preserved (Löffler 2000) but they were much more abundant in the past. Mid nineteenth century cadastral maps showed c. 150 shallow lakes in this area (Dick et al. 1994; Löffler 2000), which vanished due to artificial drainage and excessive extraction of groundwater.

In German, the shallow water bodies of the Seewinkel are either called '*See*' (lake) or '*Lacke*' (pond) (Löffler 2000, 2004). Similar water bodies have been called ephemeral lake, dry lake, playa, playa lake, saline lake, sabkha, salar, salina, salt flat, salt/clay pan perennial/seasonal astatic lake and soda lake (e.g. Briere 2000). However, neither playa, playa lake nor sabkha of Briere's (2000) definition are suitable because of the lack of arid climate conditions (e.g. Löffler 1957). Additionally, the lakes in the Seewinkel—except during very dry periods—have comparatively low salt contents (Krachler et al. 2012; Boros et al. 2013) and thus many of them classify as subsaline, only some as

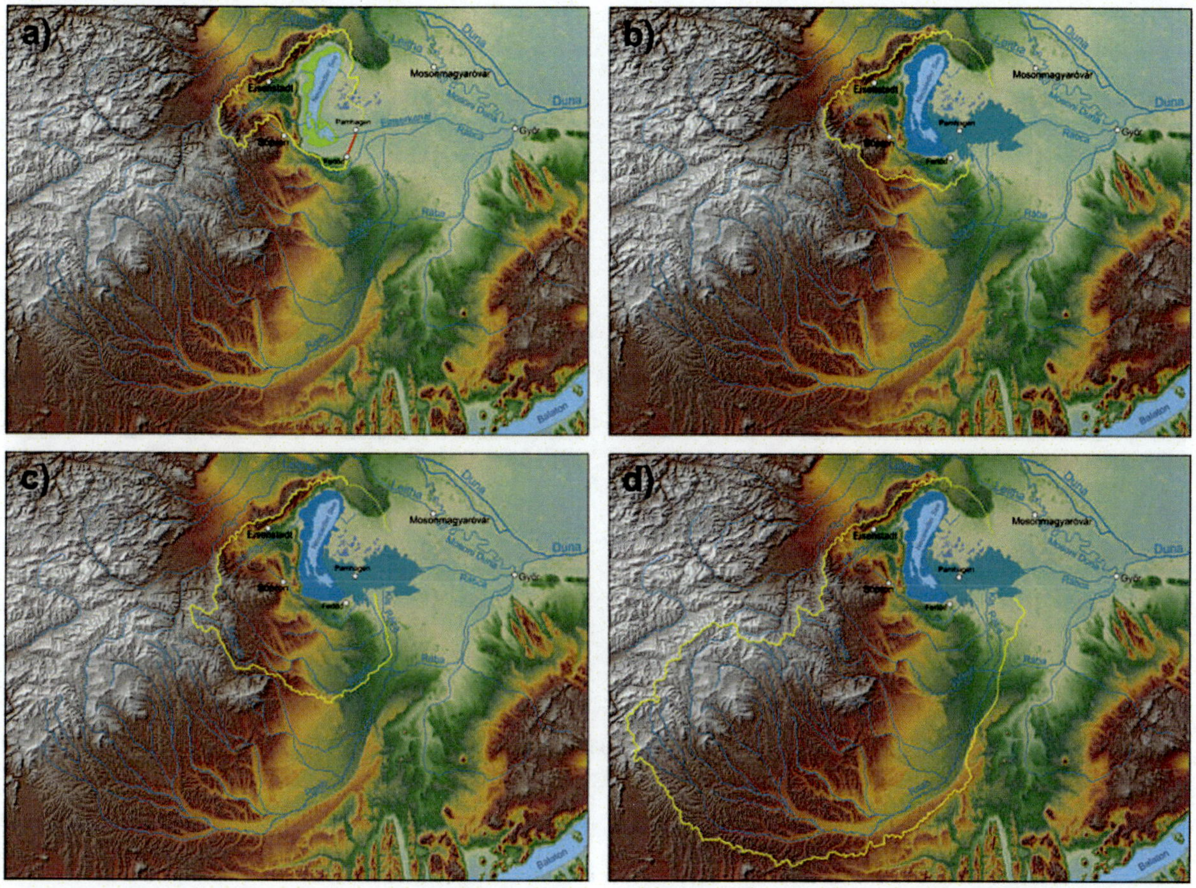

Fig. 13.7 Different watersheds (in yellow) of Lake Neusiedl; background DTM is height coloured (histogram equalize) with a 50% transparent hillshade (315° azimuth; 45° elevation). **a** Present-day catchment including only the Wulka River and some very small rivers in the Leitha Mountains (c. 1120 km^2). Current reduced catchment size results from the construction of the dam road between Pamhagen and Fertöd, drawn in red. Rivers draining into the lake as shown in historical maps increase the catchment of Lake Neusiedl additionally by **b** c. 383 km^2 (Ikva River), **c** c. 1268 km^2 (Répce River) and **d** c. 6649 km^2 by the Kis-Rába. Catchment areas calculated with SRTM 30 m DTM (http://earthexplorer.usgs.gov)

hyposaline lakes (e.g. Hammer 1986; Pinti 2011). Alkalinity and concentrations of specific anions are highly variable; pH-values are between 7.5 and 10.2 (Krachler et al. 2012; Boros et al. 2013). For all these reasons, we use the general term 'shallow lake' for the water bodies of the Seewinkel (see also Löffler 2004).

The largest of them measures 2 km in length (Fig. 13.10), but most are considerably smaller and even the largest are less than 1 m deep (Boros et al. 2013). The shallow lakes hardly have natural outlets, and one of their very characteristic features is an elevated salt content of the water, generally dominated by sodium carbonate (Na_2CO_3) (Krachler et al. 2012; Boros et al. 2014); many of them show perennial/seasonal astatic behaviour. The origin of the salt is a matter of long and still ongoing debate in the literature (Löffler 1957; Husz 1965; Krachler et al. 2000; Häusler 2007; Boros et al. 2013). Salt contents and compositions vary between the lakes and seasonally (Krachler et al. 2012), mainly depending on the amount of rain, groundwater influence, mineral precipitation and microbial activity (Löffler 1959; Krachler et al. 2000). Based on the hydrological properties, the shallow lakes can be divided into lakes completely depending on precipitation and surface runoff, lakes controlled by groundwater influx and mixed types (Steiner 2006). In all cases, permeability of the subsurface sediments is very important (Krachler et al. 2000; Steiner 2006).

13.5.1 Formation of the Shallow Lakes and Enclosed Depressions

One of the most important results of the ALS study was the discovery of more than 370 enclosed depressions in the

Fig. 13.8 Palaeo shoreline north of Purbach; photo taken towards west-northwest. The change in slope in this low-relief area is only gentle, and its visibility is enhanced by vineyard rows. The upper slope break is situated at 126 m asl, whilst the lower one is between 117.5 and 118 m asl, i.e. at the maximum flood levels of Lake Neusiedl. Photo: E. Draganits, 24.9.2006

Fig. 13.9 Ice-push processes. **a** Ice-push northwest of Podersdorf; even frozen soil on the lake shore is pushed by the wind forces; view towards northwest. Photo: E. Draganits, 28.1.2006. **b** Ice-push ridge west of Illmitz is c. 150 m wide and up to 2.5 m high; digital terrain model based on 10 × 10 m ALS DTM; visualization of colour-coded elevation (histogram equalize) (115.7–119.2 m) and 50% hillshade (315° azimuth; 45° elevation) above

Seewinkel and that these features are not restricted to the Seewinkel, but also exist in several other regions, including c. 25 in the Austrian part of the Hanság, in areas south and east of the Parndorf Plateau and even c. 13 on the plateau itself (Fig. 13.10). In Austria, a few more are found west of Lake Neusiedl near St. Margarethen, even more on a terrace between the Danube and Leitha Rivers and on a Danube terrace in the Northern Vienna Basin. Therefore, any

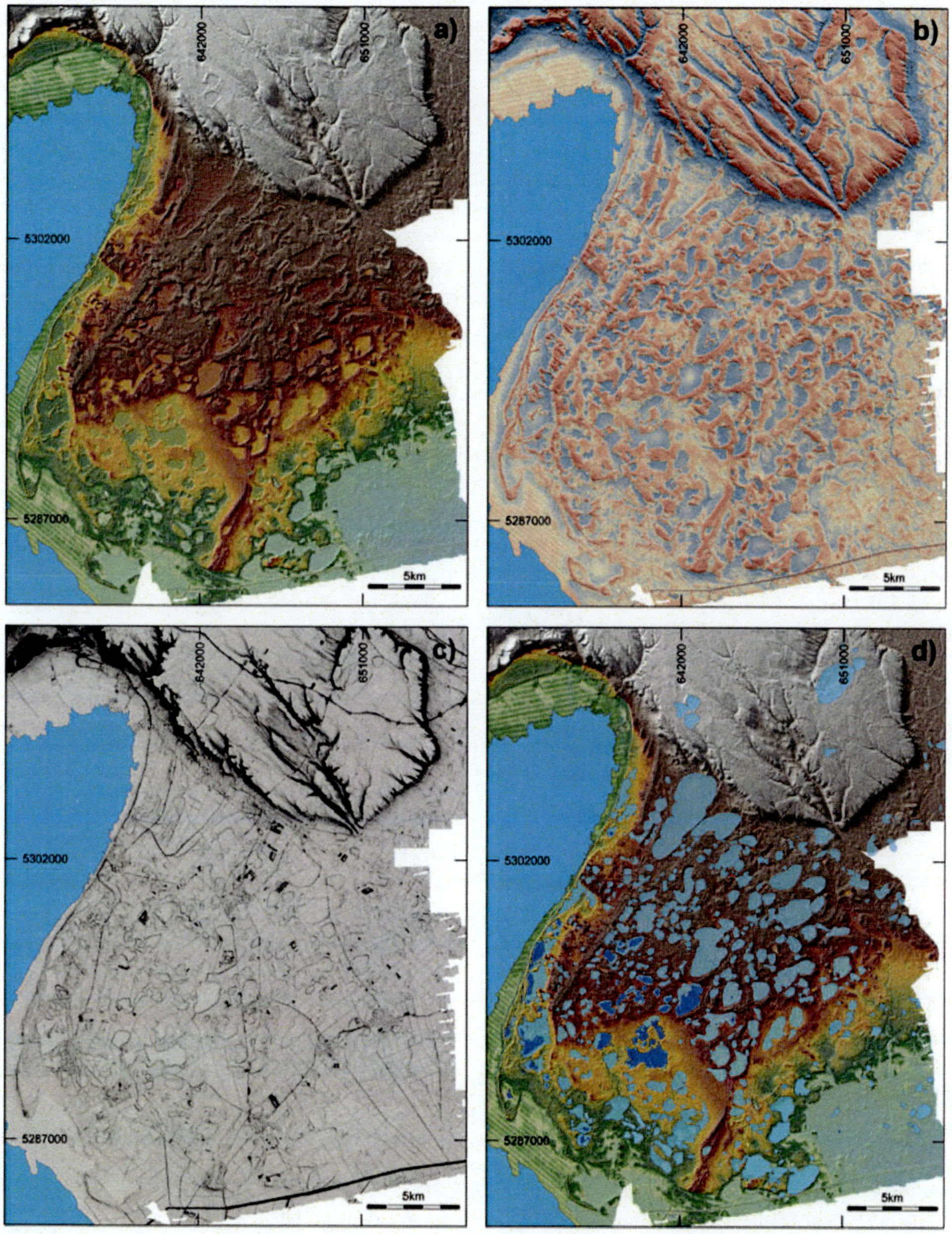

Fig. 13.10 Different visualizations of the digital elevation model (10 × 10 m ALS DTM), showing part of Lake Neusiedl, Parndorf Plateau and Seewinkel with the present-day extent of Lake Neusiedl. **a** Colour-coded elevation (histogram equalize) and 50% transparent hillshade (per cent clip) above. **b** Local relief model (LRM) (Hesse 2010) (1200 m kernel size, histogram equalize) 50% transparent above hillshade (315° azimuth; 45° elevation, histogram equalize) considerably increases the visibility of the shallow lakes. **c** Openness visualization (Doneus 2013) (900 m search radius) 50% transparent above slope map clearly showing the outline of the shallow lakes. **d** DTM visualization of Fig. 13.10a with interpreted mapping of shallow basins based on the previous visualizations in light blue and the presently existing lakes of the Seewinkel in blue. Note that the thermokarst basins are not restricted to the Seewinkel, but also occur southeast and on top of the Parndorf Plateau

explanation of the formation of these shallow basins (see also Boros et al. 2013) should also explain their distribution over different geomorphological units.

Riedl (1965) described several periglacial features from the Seewinkel, including ice-wedge pseudomorphs and deformed sediments, indicating a continuous permafrost regime. A recent review about permafrost extent during LGM shows the Lake Neusiedl area well within the continuous permafrost zone (Fábián et al. 2014; Vandenberghe et al. 2014).

Riedl (1965) suggested that the shallow lakes in the Seewinkel originally formed as pingos during glacial periods, a model hardly challenged in the past 50 years and still widely accepted (e.g. Löffler 2000). However, in recent reviews by French (2007) and Yoshikawa (2013), pingos are usually less than 0.6 km in diameter and are typically solitary features or appear in groups of only a few.

Both properties are in contrast with the up to 2.2 km long, and more than 370 basins in the Seewinkel (Fig. 13.10) visible in the high-resolution ASL survey. The basins are round to oval/kidney shaped with northeast-southwest oriented major axis. In some cases, smaller depressions seem to have merged—for example, south of Gols—to larger ones with sizes up to 2.2 km (Fig. 13.10). Several of the basins are surrounded by <3 m high ridges at their southeastern outline (Fig. 13.10), which probably formed by ice-push during northwestern winds, similar to the shore-parallel ridge (*Seedamm*) east of Lake Neusiedl. The appearance and properties of the basins closely resemble thermokarst lakes that develop by permafrost degradation and are very common in Arctic Canada and Northern Siberia (Bird 1967; Grosse et al. 2013; for comparison, see the area around the Kolyma River in Siberia: 69°00′00″N, 158°00′00″E). Already Székely et al. (2009) suggested that the basins of the Seewinkel may represent relict latest Pleistocene thermokarst lakes. High-resolution ALS DTM data proved extremely helpful for the interpretation of the enclosed depressions, which was more difficult in the past (see French and Demitroff 2001).

In contrast to pingos, which contain a core of relatively pure ice (Yoshikawa 2013), thermokarst lakes develop by thawing of pore ice of clastic sediments (French 2007; Grosse et al. 2013). Small, scattered natural depressions accumulate small amounts of water during warming periods. In these areas, the local ground thermal regime is disturbed by the high specific heat capacity of water, and consequently, a talik (unfrozen ground) forms underneath the water (Grosse et al. 2013). The positive feedback between lake growth and permafrost thawing results in basin growth; neighbouring basins may join to larger ones. Burn and Smith (1990) measured mean growth rates of 16 thermokarst lakes in the Yukon Territory of 0.7 m/year along their major axis, 0.5 m/year along their minor axis, and the maximum rate was 1.2 m/year. According to Edwards et al. (2016), even much lower growth rates are documented. The largest single thermokarst lake in the Seewinkel is more than 1.5 km long (Fig. 13.10), its growth, according to these data, may have taken a few millennia.

Another feature that the shallow depressions in the Lake Neusiedl area have in common with thermokarst lakes in permafrost areas in Northern Canada, and Siberia is the probable aeolian influence on their shape and growth. Recently, Sebe et al. (2015) presented evidence for strong northwestern winds impacting Eastern Austria during the Pleistocene. Modern observations in Northern Canada and Siberia indicate that thermokarst lakes are shaped by wind-generated waves and currents, which produce oval to kidney shaped lakes with their major axis perpendicular to the dominant wind direction (Fig. 13.10).

The thermokarst lakes in central and Eastern Europe probably developed by permafrost degradation during temperature increase at the end of the last glacial period. The thin cover of fine-grained overbank deposits and aeolian sediments in many parts of the Seewinkel area (Husz 1965) are very suitable for the formation of thermokarst lakes. The preservation of these features is probably linked to vertical tectonic movements leaving elevated areas with thermokarst features out of reach of subsequent fluvial destruction. In the case of Seewinkel, active subsidence of the Little Hungarian Plain around Győr (Joó 1992; Ruess and Mitterschiffthaler 2015) gradually shifted the Répce and Rába Rivers to the east (Scharek 1993), preserving these latest Pleistocene features until today.

So far, thermokarst lakes have been described almost exclusively from Arctic or high Alpine areas (Kääb and Haeberli 2001; Grosse et al. 2013). When the shallow lake basins formed by thermokarst degradation, these basins from Eastern Austria represent one of the first latest Pleistocene thermokarst lakes documented in central Europe.

Some basins visible in the ALS DTM are situated in the flooding zone of Lake Neusiedl (Fig. 13.10), for instance in the Hanság, but also east of the shore-parallel ridge (*Seedamm*). They resemble the other shallow basins in the Northern Seewinkel and Parndorf Plateau. Their relationship with the lake indicates that these basins are slightly older than Lake Neusiedl, as already concluded by Löffler (2000). Löffler (2000) divided the shallow lakes in the Seewinkel into (i) shallow lakes dammed by the ridge east of the lake, (ii) shallow lakes in the area of the Hanság and (iii) shallow lakes of central Seewinkel. Based on this study, this can be simplified by concluding that all of them formed as thermokarst lakes shortly before Lake Neusiedl started to exist—at least in its present extent. Later, some of the basins in the southern part were flooded episodically by the lake and some of them in the western part of the Seewinkel became adjacent to the later formed ridge (*Seedamm*) (Fig. 13.10). When the shallow lake basins formed by thermokarst processes during early Lateglacial permafrost degradation and before Lake Neusiedl came into existence (as shown by superimposition in the ALS data), their age is constrained between the end of the LGM at c. 19 000 years ago (van Husen 2011, Wirsig et al. 2016) and the formation of Lake Neusiedl around 12–14 000 years ago (Löffler 1990, this study).

13.6 Conclusions

The present-day conditions and processes of Lake Neusiedl strongly differ from conditions in the past. The earliest preserved record of modification of the lake's hydrological conditions is from 1568, followed by increasing drainage efforts and the building of a dam road between Pamhagen and Fertőd, finished in 1780, which subsequently cut off the lake from its most important tributaries and fundamentally changed its palaeohydrology.

Virtually, all historical maps before 1780 show the Ikva River, Répce (Rabnitz) River and Kis-Rába River discharging into the connected Neusiedlersee/Hanság area that possessed a natural outlet, the Rábca River. This palaeohydrological reconstruction of Lake Neusiedl and Hanság also implies a lower salt content of the water compared to modern values, especially during flood periods.

Therefore, it is not useful to compare the hydrological situation of Lake Neusiedl before 1780 (or even 1568) and after. Consequently, it is unlikely that Lake Neusiedl was an endoreic lake in most periods before 1780 and it is very unlikely that the lake had 'about 100–200 dry periods since the lake came into existence'.

The documented variation of the water level of Lake Neusiedl between desiccation and highest flood level is around 4.2 m. These variations affected enormous areas in this low-relief region, with a huge impact on the landscape, fauna, vegetation, human settlement patterns, land use, communication routes and even possible occurrence of malaria—which should be considered in regional archaeological, historical and biological interpretations. The dating of specific lake levels is an important challenge left for the future.

The numerous shallow lakes and presently dry basins in the Seewinkel originally formed as thermokarst lakes during permafrost degradation after the end of the LGM (after 19,000 years ago), and at least, some of them are slightly older than Lake Neusiedl as revealed by superimposition observed in the ALS DTM, which shows more than 370 enclosed depressions in the Seewinkel itself. The depressions of Eastern Austria represent one of the first latest Pleistocene thermokarst lakes documented in central Europe.

Acknowledgements High-resolution ALS data were provided by Amt der Burgenländischen Landesregierung. Bernhard Grasemann, Peter Barilits and Barbara Hodits contributed with their interest, local expertise and enthusiasm. Many thanks to Stephanie Neuhuber for fruitful discussion. We thank Martin Fera and Ulrike Fornwagner (Aerial Archive, Department of Prehistoric and Historical Archaeology) and Andreas Ziegler for extremely valuable aerial photos. Mathias Harzhauser determined the fresh water bivalves from the outcrop in Jois. Many thanks to Stephan Glatzel and Erwin Heine for their reviews.

References

Anonymous (1568) ÖStA/FHKA, Hoffinanz Ungarn, rote Nr. 17, Karton 37, fol. 346–352

Bácsatyai L, Csaplovics E, Márkus I, Sindhuber A (1997) Digitale Geländemodelle des Neusiedler See-Beckens. Wiss Arb Burgenland 97:1–53

Bernhauser A (1962) Zur Verlandungsgeschichte des burgenländischen Seewinkels (Geologisch-Bodenkundliche Auswertung der Ergebnisse der amtlichen Bodenschätzung der Gemeinden Illmitz und Apetlon—ohne Paulhof—1959/60). Wiss Arb Burgenland 29:143–171

Bird JB (1967) The physiography of Arctic Canada, with special reference to the area south of Parry Channel. John Hopkins Press, Baltimore, p 336

Boros E, Ecsedi Z, Oláh J (eds) (2013) Ecology and management of soda pans in the Carpathian Basin. Hortobágy Environmental Association, Balmazújváros, 551 p

Boros E, Horváth Z, Wolfram G, Vörös L (2014) Salinity and ionic composition of the shallow astatic soda pans in the Carpathian Basin. Ann Limn Int J Limn 50:59–69

Briere PR (2000) Playa playa lake sabkha: proposed definitions for old terms. J Arid Env 45:1–7

Bruce-Chwatt LJ, de Zulueta J (1980) The rise and fall of malaria in Europe: a historico-epidemiological study. Oxford University Press, Oxford, p 240

Burn CR, Smith MW (1990) Development of thermokarst lakes during the Holocene at sites near Mayo. Yukon Territory. Permafrost Perigl Proc 1(2):161–175

Csaplovics E (2005) Zur Topochronologie der Landschaft um den Neusiedler See bis zum Ende des 16. Jahrhunderts. Burgenländ Forsch 91:1–224

Dick G, Dvorak M, Grüll A, Kohler B, Rauer G (eds) (1994) Vogelparadies mit Zukunft? Ramsar-Gebiet Neusiedler See - Seewinkel. Umweltbundesamt, Wien, 356 p

Doneus M (2013) Openness as visualization technique for interpretative mapping of airborne Lidar derived digital terrain models. Remote Sens 5:6427–6442

Doneus M, Briese (2006) Digital terrain modelling for archaeological interpretation within forested areas using full-waveform laserscanning. In: Ioannides M, Arnold D, Niccolucci F, Mania K (eds) The 7th international symposium on virtual reality, archaeology and cultural heritage VAST, pp 155–162

Draganits E, Zámolyi A, Gier S, Hodits B, Exner U, Janda C, Grasemann B (2007) Neusiedlersee/Fertő Tó area (Austria/Hungary): minimum estimates of former lake levels. European Geosciences Union General Assembly 2007, Vienna, 15–20 April, Geophys Res Abstr 9:A-10052

Draganits E, Zámolyi A, Székely B, Timár G, Molnár G (2008) Reconstruction of the Neusiedlersee (Austria/Hungary) based on historical topographic maps from 1507 to present. European Geosciences Union General Assembly 2008, Vienna, 13–18 April, Geophys Res Abstr 10:A-08644

Edwards M, Grosse G, Jones BM, McDowell P (2016) The evolution of a thermokarst-lake landscape: late quaternary permafrost degradation and stabilization in interior Alaska. Sed Geol 340:3–14

Eitzinger J, Kubu G, Formayer H, Haas P, Gerersdorfer T, Kromp-Kolb H (2009) Auswirkungen einer Klimaänderung auf den

Wasserhaushalt des Neusiedlersees. Endbericht im Auftrag der Burgenländischen Landesregierung vom 15. Juli 2005. BOKU-Met Report 1, 81 p

Fábián SÁ, Kovács J, Varga G, Sipos G, Horváth Z, Thamó-Bozsó E, Tóth G (2014) Distribution of relict permafrost features in the Pannonian Basin. Hungary. Boreas 43(3):722–732

Fally J, Kárpáti L. (eds) (2012) Nationalpark Neusiedler See—Seewinkel. Fertő–Hanság Nemzeti Park. Monographische Studien über das Gebiet Neusiedler See und Hanság. Nationalparkverwaltung Fertő–Hanság, Szaktudás Kiadó Ház AG, Budapest, 450 p

French HM, Demitroff M (2001) Cold-climate origin of the enclosed depressions and wetlands ('spungs') of the pine barrens, southern New Jersey, USA. Permafrost Periglac Process 12:337–350

French HM (2007) The periglacial environment, 3rd edn. Wiley, Chichester, p 458

Fuchs R, Schreiber OS (1985) Das Pannonien im östlichen Burgenland (Seewinkel). In: Papp A (ed), M6 Pannonien (Slavonien und Serbien). Akadémiai Kiadó, Budapest, pp 68–72

Fuchs W (1965) Geologie des Ruster Berglandes (Burgenland). Jb Geol BA 108:155–194

Grosse G, Jones B, Arp C (2013) Thermokarst lakes, drainage, and drained basins. In: Shroder J (ed in chief), Giardino R, Harbor J (vol ed) treatise on geomorphology, vol 8, glacial and periglacial geomorphology. Academic Press, San Diego, pp 325–353

Hammer UT (1986) Saline lake ecosystems of the world. Junk, Dordrecht, p 616

Hassinger H (1905) Geomorphologische Studien aus dem inneralpinen Wiener Becken und seinem Randgebirge. Geogr Abh 8(3):1–205

Häusler H (2007) Geologische Karte der Republik Österreich 1:50000. Erläuterungen zu den Blättern 79 Neusiedl am See, 80 Ungarisch-Altenburg und 109 Pamhagen. Geologische Bundesanstalt, Wien, 88 p

Häusler H, Lenhardt W, Hammerl C (2010) Neotektonik. In: Häusler H (ed) Geologische Karte der Republik Österreich 1:50000. Erläuterungen zur geologischen Karte 78 Rust. Geologische Bundesanstalt, Wien, pp 97–99

Hegedűs JN (1788) Mappa Situationem Inclyti Comitatus Soproniensis cum omnibus eidem ingremiatis Urbibus, Oppidis, ac Possessionibus, non fecus Fluvius, Rivulis, Vüs Postalibus et Commercialibus, in üsdém que designates stationibus exhibens. ca. 1:111,550, Országos Széchényi Könyvtár, TK 1614, http://maps.hungaricana.hu/hu/1612

Heine E, Kogelbauer I, Prokoph A (2014) Loiskandl W (2014) Hydrographic surveying of the Steppe Lake Neusiedl—mapping the lake bed topography and the mud layer. Photogr Fernerk Geoinf 5:339–350

Heine E, Loiskandl W, Briese C, Eberhöfer C (2016) Csaplovics E (2016) Hydrografische Vermessung des Neusiedler Sees. Vermess Geoinf 1:9–24

Herzig A (2014) Der Neusiedler See—Limnologie eines Steppensees. In: Wöss E (ed) Süßwasserwelten. Limnologische Forschung in Österreich, Denisia 33, Kataloge des Oberösterreichischen Landesmuseums—Neue Serie, 163:101–114

Hesse R (2010) LiDAR-derived local relief models—a new tool for archaeological prospection. Archaeol Prospect 17(2):67–72

Hicke W (1996) Naturschutz im Burgenland. Teil II: Vom "nützlichen Vogelschutz" zum Europaschutzgebiet. 70 Jahre Naturschutzgesetzgebung im Burgenland. Amt der Burgenländischen Landesregierung, Eisenstadt, 85 p

Höggerl N (2013) Österreichische MGI-Gebrauchshöhen und deren Bezug zu Höhen abgeleitet aus GNSSMessungen. In: Hanke K, Weinhold T (eds) 17. Internationale Geodätische Woche Obergurgl. Wichmann, Berlin, pp 76–85

Husz G (1965) Zur Kenntnis der quartären Sedimente des Seewinkelgebietes (Burgenland—Österreich). Wiss Arb Burgenland 32:147–205

HDÖ—Hydrographischer Dienst in Österreich (1938) III. Das Mur- und Raab-Gebiet. Jb Hydrogr Zentralbureau Bundesministerium Land Forstw, NF 17. Jahrgang 1930, Hydrographisches Zentralbureau, Wien

Joó I (1992) Recent vertical surface movements in the Carpathian basin. Tectonophysics 202(2–4):129–134

Kääb A, Haeberli W (2001) Evolution of a high mountain thermokarst lake in the Swiss Alps. Arct Antarct Alp Res 33(4):385–390

Kelletat DH (1995) Atlas of coastal geomorphology and zonality. J Coastal Res Spec issue 13, 286

Kiss A (2009–2010) Fertő River—a low water level signal or something else? Chronica 9–10:66–77

Knie K (1959) Über den Chemismus der Wässer im Seewinkel und des Neusiedlersees. In: Sauerzopf F (ed) Landschaft Neusiedlersee: Grundriss der Naturgeschichte des Großraumes Neusiedlersee. Anläßlich des XIV. Internationalen Limnologen-Kongresses in Österreich 1959, Wiss Arb Burgenland 23:65–68

Kopf F (1963) Wasserwirtschaftliche Probleme des Neusiedler Sees und des Seewinkels. Österr Wasserwirtschaft 15(9/10):190–203

Korabinszky JM (1804) Atlas Regni Hungariae portatilis. Neue und vollständige Darstellung des Königreichs Ungarn auf LX Tafeln im Taschenformat. Ein geographisches Noth- und Hülfsbüchlein fürs gemeine Leben, Preßburg, Digital Collections of the University Library Regensburg, W 02/8 5392, urn:nbn:de:bvb:355-ubr16769-8

Korner I, Wrbka T, Staudinger M, Böck M (2008) Beweidungsmonitoring im Nationalpark Neusiedler See—Seewinkel Ergebnisse der vegetationsökologischen Langzeitmonitoring-Studie 1990 bis 2007. Abh Zool Bot Ges Österr 37:1–84

Krachler R, Krachler R, Milleret E, Wesner W (2000) Limnochemische Untersuchungen zur aktuellen Situation der Salzlacken im burgenländischen Seewinkel. Burgenländische Heimatblätter 62:3–49

Krachler R, Korner I, Kirschner A (2012) Die Salzlacken des Seewinkels: Erhebung des ökologischen Zustandes sowie Entwicklung individueller Erhaltungskonzepte für die Salzlacken des Seewinkels (2008–2011). Naturschutzbund Burgenland, Eisenstadt, 291 p

Kugler H (1871) Der Neusiedler-See. Verh. K.K. Geol Reichsanst 1871:99–100

Küpper H (ed) (1957) Erläuterungen zur geologischen Karte Mattersburg-Deutschkreutz 1:50000. Geologische Bundesanstalt, Wien, p 67

Lindinger V (1996) Die Geländeaufnahme der oberirdischen Bodendenkmale des Seewinkels, pol. Bez. Neusiedl am See. Proseminararbeit, Institut für Ur- und Frühgeschichte. Universität Wien, Wien, 111 p

Löffler H (1957) Vergleichende limnologische Untersuchungen an den Gewässern des Seewinkels (Burgenland) Verhandlungen der Zoologisch-Botanischen Gesellschaft in Wien 97:27–52

Löffler H (1959) Zur Limnologie, Entomostraken und Rotatorienfauna des Seewinkelgebietes (Burgenland, Österreich). Sitzungsberichte der Österr Akad Wiss Math-Nat Klasse, Abt 1, 168(4–5):315–362

Löffler H (1974) Der Neusiedlersee—Naturgeschichte eines Steppensees. Fritz Molden, Wien, p 175

Löffler H (ed) (1979) Neusiedlersee: The limnology of a shallow lake in Central Europe. Monogr Biolog 37, Junk, The Hague, 543 p

Löffler H (1982) Der Seewinkel: Die fast verlorene Landschaft. Niederösterreichisches Pressehaus, St.Pölten-Wien, p 160

Löffler H (1990) Paleolimnology of Neusiedlersee, Austria. In: Smith JP, Appleby PG, Battarbee RW, Dearing JA, Flower R, Haworth EY, Oldfield F, O'Sullivan PE (eds) Environmental history and palaeolimnology. Hydrobiologia 214:229–238

Löffler H (2000) Anmerkungen zur Bezeichnung der Gewässer des Seewinkels (Burgenland). Mitt Österr Geogr Ges 142:315–328

Löffler H (2004) The origins of lake basins. In: O'Sullivan PE, Reynolds CS (eds) The lakes handbook volume 1: limnology and limnetic ecology. Blackwell, Oxford, pp 8–60

Loisl J, Tari G, Draganits E, Zámolyi A, Gjerazi I (2018) High-resolution seismic reflection data acquisition and interpretation, Lake Neusiedl, Austria, NW Pannonian Basin. In: Németh B, Tari G, Radivojević D, Tomljenovic B, Krézsek C (eds) Special section: characterization of hydrocarbon and geothermal resource potential and carbon sequestration opportunities of the Pannonian Basin. Interpretation 6(1), pp SB77–SB97

Mahoney A, Eicken H, Shapiro L, Grenfell TC (2004) Ice motion and driving forces during a spring ice shove on the Alaskan Chukchi coast. J Glaciology 50(169):195–207

Mandelburger G, Höfle B, Briese C, Ressl C, Otepka J, Hollaus M, Pfeifer N (2009) Topographische Daten aus Laserscanning als Grundlage für Hydrologie und Wasserwirtschaft. Österr Wasser Abfallwirtschaft 61(7):89–97

Maracek K, Kubu G (2014) Der Neusiedlersee. Wasserwirtschaft Burgenland. http://wasser.bgld.gv.at/uploads/media/Poster_Der_Neusiedler_See.pdf

Molnár Z, Biró M, Bartha S, Fekete G (2012) Past trends, present state and future prospects of Hungarian forest-steppes. In: Werger MJA, van Staalduinen MA (eds) Eurasian Steppes. Ecological problems and livelihoods in a changing world. Plant and vegetation, vol 6, pp 209–252

Neuhuber S, Steier P, Gier S, Draganits E (2015) Radiogenic carbon isotopes in authigenic carbonate from Lake Neusiedl, Austria. European Geosciences Union General Assembly, Vienna, 12–17 April 2015, Geophysical Research Abstracts 17:EGU2015–15399

Oberleitner I, Wolfram G, Achatz-Blab A (eds) (2006) Salzlebensräume in Österreich. Umweltbundesamt, Wien, p 216

Pinti DL (2011) Soda lakes. In: Gargaud M, Amils R, Cernicharo Quintanilla J, Cleaves HJ, Irvine WM, Pinti DL, Viso M (eds) Encyclopedia of astrobiology. Springer, Heidelberg, pp 1524–1525

Pistotnik J, Herrmann P, Pascher GA (1993) 78 Rust. Geologische Karte der Republik Österreich 1:50000. Geologische Bundesanstalt, Wien

Riedl H (1965) Beiträge zur Morphogenese des Seewinkels. Wiss Arb Burgenland 34:5–28

Ruess D, Mitterschiffthaler P (2015) Rezente Höhenänderungen in Österreich abgeleitet aus geodätischen Wiederholungsmessungen. In: Hanke K, Weinold T (eds) 18. Internationale Geodätische Woche Obergurgl 2015, Wichmann, Berlin, pp 111–123

Sauerzopf F (ed) (1959a) Landschaft Neusiedlersee: Grundriss der Naturgeschichte des Großraumes Neusiedlersee. Anläßlich des XIV. Internationalen Limnologen-Kongresses in Österreich 1959. Wiss Arb Burgenland, 23, Burgenländisches Landesmuseum, Eisenstadt, 208 p

Sauerzopf F (1959b) Die Wasserstandsschwankungen des Sees. In: Sauerzopf F (ed) Landschaft Neusiedlersee: Grundriss der Naturgeschichte des Großraumes Neusiedlersee. Anläßlich des XIV. Internationalen Limnologen-Kongresses in Österreich 1959, Wiss Arb Burgenland 23:92–101

Scharek P (ed) (1991) The geological map series of the Little Hungarian Plain: Mosonmagyaróvár. Explanations. Hungarian Geological Institute, Budapest, 35 p

Scharek P (ed) (1993) The geological map series of the Little Hungarian Plain: Kapuvár. Explanations. Hungarian Geological Institute, Budapest, 39 p

Sebe K, Roetzel R, Fiebig M, Lüthgens C (2015) Pleistocene wind system in eastern Austria and its impact on landscape evolution. CATENA 134:59–74

Soja G, Züger J, Knoflacher M, Kinner P, Soja AM (2013) Climate impacts on water balance of a shallow steppe lake in Eastern Austria (Lake Neusiedl). J Hydrol 480:115–124

Spahić D, Rundić L (2015) Multi-phased normal faulting of the Eisenstadt-Sopron basin margins as a controlling factor over spatially confined coal mini-hydrocarbon play (East Austria). Carpathian J Env Sci 10(2):45–58

Steiner KH (2006) Hydrologie und Lackenhaushalt im Seewinkel. Salzlebensräume in Österreich. In: Oberleitner I, Wolfram G, Achatz-Blab A (eds) Salzlebensräume in Österreich. Umweltbundesamt, Wien, pp 59–70

Szádeczky-Kardoss E (1938) Geologie der Rumpfungarländischen Kleinen Tiefebene mit Berücksichtigung der Donaugoldfrage. Mitteilungen der berg- und hüttenmännischen Abteilung an der Königlich-Ungarischen Hochschule für Berg- und Forstwesen Sopron, 10, 444 p

Székely B, Zámolyi A, Draganits E Briese C (2009) Geomorphic expression of neotectonic activity in a low relief area in an airborne laser scanning DTM: a case study of the Little Hungarian Plain (Pannonian basin). Tectonophysics 474:353–366

Szontagh T (1904) Geologisches Studium des Fertő-Sees. Jahresber Königl Ungar Geol Anstalt 1902:206–211

Tauber AF (1959) Geologische Stratigraphie und Geschichte des Neusiedlerseegebietes. In: Sauerzopf F (ed) Landschaft Neusiedlersee: Grundriss der Naturgeschichte des Großraumes Neusiedlersee. Anläßlich des XIV. Internationalen Limnologen-Kongresses in Österreich 1959, Wiss Arb Burgenland 23:18–24

Tóth L, Mónus P, Zsíros T, Bus Z, Kiszely M, Czifra T (2007) Hungarian earthquake bulletin 1995–2007. http://www.georisk.hu. Accessed on 12 May 2016

van Husen D (1987) Die Ostalpen und ihr Vorland in der letzten Eiszeit (Würm). 1:500000, Geologische Bundesanstalt, Wien

van Husen D (2011) Quaternary glaciations in Austria. In: Ehlers J, Gibbard PL, Hughes PD (eds) Quaternary glaciations - extent and chronology. Closer Look, Developments in Quaternary Sciences, 15, Elsevier, Amsterdam, pp 15–28

Vandenberghe J, French HM, Gorbunov A, Marchenko S, Velichko AA, Jin H, Cui Z, Zhang T, Wan X (2014) The last permafrost maximum (LPM) map of the Northern Hemisphere: permafrost extent and mean annual air temperatures, 25–17 ka BP. Boreas 43:652–666

Wasserportal Burgenland (2016) http://wasser.bgld.gv.at/hydrographie/der-neusiedler-see.html

Wendelberger G (1950) Wald und Steppe am Neusiedlersee: Gedanken zu einer Wirtschaftsplanung am Neusiedler See. Burgenl Heimatbl 12(1):9–14

Wendelberger G (1955) Zur Frage der Waldlosigkeit der ungarischen Pußta. Burgenl Heimatbl 17(2):92–94

Wendelberger G (1987) Steppenheide und prähistorische Besiedlung am Westufer des Neusiedlersees. Wiss Arb Burgenland 75:285–294

Wernsdorfer WH (2002) Malaria in Mitteleuropa. Denisia 6:201–212

Wiche K (1951) Die Oberflächenformen des Burgenlandes. Burgenländische Landeskunde, pp 98–136

Wiedl T, Harzhauser M, Kroh A, Ćorić S, Piller WE (2014) From biologically to hydrodynamically controlled carbonate production by tectonically induced palaeogeographic rearrangement (Middle Miocene, Pannonian Basin). Facies 60:865–881

Wirsig C, Zasadni J, Christl M, Akçar N, Ivy-Ochs S (2016) Dating the onset of LGM ice surface lowering in the High Alps. Quat Sci Rev 143:37–50

Wolfram G, Déri L, Zech S (eds) (2014) Strategiestudie neusiedler see—phase 1. Studie im Auftrag der Österreichisch-Ungarischen Gewässerkommission. Wien-Szombathely, 246 p

Yoshikawa K (2013) Pingos. In: Shroder JF (ed) Treatise on geomorphology, vol 8. Academic Press, San Diego, pp 274–297

ZAMG (2016) HISTALP Österreich Winterbericht 2015/16 https://www.zamg.ac.at/cms/de/dokumente/klima/dok_news/dok_histalp/winterbericht-2015-16/oesterreich-klimareihen-1767-2016

Zámolyi A, Székely B, Draganits E, Timár G (2010) Neotectonic control on river sinuosity at the western margin of the Little Hungarian Plain. Geomorphology 122:231–243

Zámolyi A, Salcher B, Draganits E, Exner U, Wagreich M, Harzhauser M, Gier S, Fiebig M, Lomax J, Surányi G, Diehl M, Zámolyi F (2016) Latest Pannonian and quaternary evolution at the transition between Eastern Alps and Pannonian Basin: new insights from geophysical, sedimentological and geochronological data. Int J Earth Sci. https://doi.org/10.1007/s00531-016-1383-3

Zeller S (1753) Mappa Comitatus Soproniensis. Magyar Országos Levéltár, 51 × 36 cm, S 68 × No 0016, http://maps.hungaricana.hu/hu/18572

Erich Draganits studied geology and prehistoric archaeology at the University of Vienna, where he received his Ph.D. in geology in 2000 and his habilitation in 2008. He carries out research in various fields within the earth sciences, including sedimentology, geomorphology, geoarchaeology, applied geology and global change. Regionally, his study areas range from Southeast Asia, the Himalayas, Middle East, Mediterranean, Central Europe and Scandinavia to South America.

Michael Weißl studied archaeology and geology. He is a research scientist at the Department of Geodynamics and Sedimentology, University of Vienna. His research interests include Quaternary geology, geomorphology, and geoarchaeology. Current investigations focus on neotectonics in Central Europe.

András Zámolyi received a Ph.D. (2016) in geology from the University of Vienna. He conducted his dissertation on the Quaternary landscape evolution of the Little Hungarian Plain in close cooperation with the Eötvös University, Budapest. Before joining OMV he worked at the University of Vienna in industry research projects on the geodynamics of the Vienna Basin. His field of expertise includes active tectonics, structural geology, geomorphology, geostatistics and geomodeling.

Michael Doneus is landscape archaeologist with a focus on non-invasive archaeological prospection techniques and GIS. He is professor for landscape archaeology at the Department of Prehistoric and Historical Archaeology, University of Vienna, and key researcher at the Ludwig Boltzmann Institute for Archaeological Prospection and Virtual Archaeology.

14 Quaternary Landforms and Sediments in the Northern Alpine Foreland of Salzburg and Upper Austria

Johannes T. Weidinger, Horst J. Ibetsberger, and Joachim Götz

Abstract

This chapter highlights Quaternary landforms and sediments in the Northern Alpine foreland between Salzburg and Upper Austria—a region with a remarkable history of Quaternary research from the early pioneers in the twentieth century to the quantitative studies of today, making use of a range of modern (e.g., dating) techniques. Before any glaciers modified the region, the Mio-Pliocene landscape was dominated by massive accumulation of fluvial sediments originating from the still growing Alps in the hinterland to the Molasse basin in the foreland. Large parts of these deposits were later on eroded, reworked, and superimposed by glacial sediments during several Pleistocene cycles of glaciation and deglaciation. The sedimentary signatures of the regionally most important Salzach and Traun glaciers are exceptionally well preserved and can be traced from their inner-Alpine source areas, along their transport routes following the Salzach and Traun valleys, within several peripheral glacially shaped basins, and finally in the terminal moraines and adjacent outwash plains further north. In contrast to the Pleistocene Salzach Glacier that emanated from the Central Alps and formed a large piedmont glacier in the foreland, the branches of the Traun Glacier barely reached the foreland, because of a smaller ice accumulation zone in the Northern Calcareous Alps, predominantly on the high plateaus of Totes Gebirge and Hoher Dachstein. With the Traun-Enns-Platte and the Salzach-Inn confluence area, two important terrace systems north of the Quaternary ice margin developed. However, this unique landscape of glacial heritage constitutes a fragile and vulnerable system which needs to be protected.

Keywords

Pleistocene Traun and Salzach glaciers • Penck's model of the "glacial sequence," basins • Lakes • Moraines • Terraces

14.1 Introduction

Austria is well known for its spectacular mountains but also has plenty of hilly or even flat landscapes. One of these physiographically subdued areas is the Northern Alpine foreland with the former Molasse basin north of the Eastern Alps. The basin was filled with Neogene sediments, which are topped by (pre-) Quaternary deposits, and glacial landforms that developed within the past 700 ka. During the Quaternary, the Austrian Alps were repeatedly covered by ice with the development of complex glacier networks. Multiple ice streams followed the Alpine river network but also overflowed lower mountain passes and saddles. In the inner-Alpine area of Salzburg and Upper Austria, only the highest peaks, ridges, and crests above 2000–2500 m rose above the glacier surface during glacial maxima. These nunataks must have looked like islands in an ocean of ice.

Typical landforms of the so-called Glaziale Serie (glacial sequence, model depicted in Fig. 3.2 of Chap. 3), a model introduced by Penck and Brückner (1909), are frequently visible in formerly ice covered mountains and their fore-lands; however, between the Rivers Salzach (in the west) and Enns (in the east), this sequence of glacial and glaciofluvial landforms is exceptionally well developed and preserved. At the transition from the inner-Alpine main valleys toward the wide Alpine foreland, the ice streams

J. T. Weidinger (✉)
ERKUDOK© Institute in the Kammerhof Museums of Gmunden (K-Hof), Kammerhofgasse 8, 4810 Gmunden, Austria
e-mail: johannes.weidinger@gmunden.ooe.gv.at

H. J. Ibetsberger
GeoGlobe Häupl & Ibetsberger OEG, Statzenbachgasse 5, 5202 Neumarkt am Wallersee, Austria

J. Götz
Fachgruppe Geomorphologie, University of Bayreuth, Universitätsstraße 30, 95447 Bayreuth, Germany

C. Embleton-Hamann (ed.), *Landscapes and Landforms of Austria*, World Geomorphological Landscapes,
https://doi.org/10.1007/978-3-030-92815-5_14

spread laterally to form large piedmont glacier lobes. The easternmost of these semicircular ice fields was the Salzach Glacier. Further east, three channeled ice stream tongues of the Traun Glacier just reached the foreland (Fig. 14.1).

These glaciers formed distinct terminal moraines and adjacent outwash plains developed further north. The latter were dominated by permafrost and the deposition of loess due to dry and cold climatic conditions.

14.2 A Brief Review of Quaternary Research in the Alpine Foreland

The Northern Alpine foreland of Salzburg and Upper Austria is one of the World's classic locations and type localities for glacial morphology and Pleistocene stratigraphy.

It was as early as 1846, when the Austrian research pioneer Friedrich Simony—first Professor of Geography at

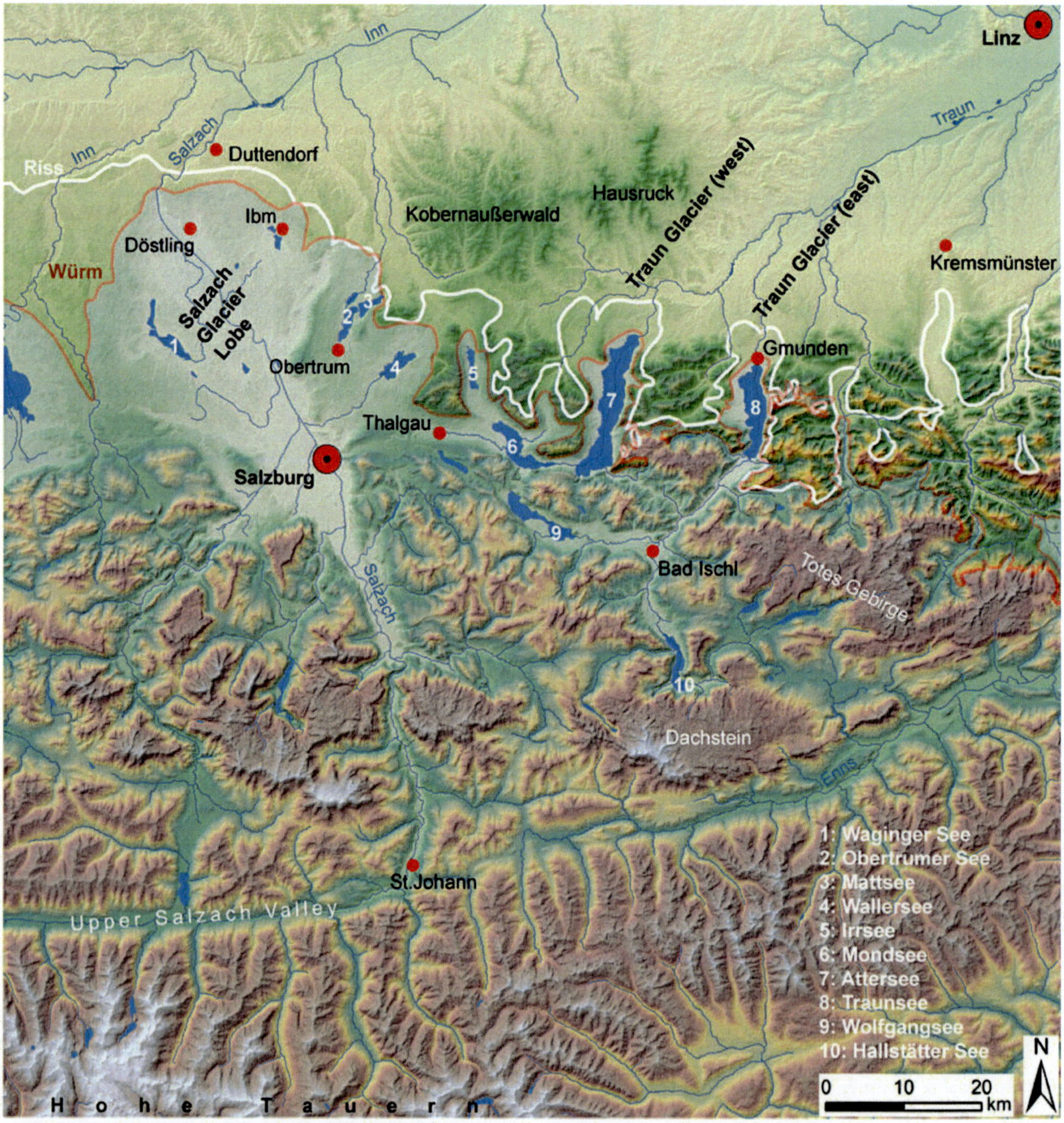

Fig. 14.1 Riss and Würm glacier extent in the northern foreland of the Eastern Alps between the Rivers Salzach, Traun, and Enns (from west to east) based on hillshaded elevation data (SRTM and ASTER data; 25 m resolution). Important place names mentioned in the text (villages, lakes, mountain ranges) are indicated

the University of Vienna—followed the revolutionary ideas of Louis Agassiz (1837) from Switzerland by finding geomorphologic and sedimentary features of a "pre-historic ice age" in the Hohe Dachstein and Salzkammergut regions of Upper Austria. After six decades of intensified investigations, the classical period of ice age research started, culminating in the book "Die Alpen im Eiszeitalter" (=The Alps in the ice age) by Penck and his student Brückner (1909).

Penck and Brückner based their research on a conceptual model of the sequence of landforms occurring at the terminus of an ice stream in the Northern Alpine foreland. They observed a regular arrangement of moraine ridges enclosing a till-covered basin, on the outside followed by an outwash cone that finally turns into a river terrace, and termed this morphosequence "*Glaziale Serie.*" Based on this model and the different weathering degrees of its landform members, they identified four major glacial periods, interrupted by warmer interglacial periods, and named them after Bavarian rivers Günz, Mindel, Riss, and Würm (from oldest to youngest).

No serious attempts to maintain the monoglacial hypothesis were made after this publication, as the broad outline of Penck and Brückner's work was accepted universally. It was not until the availability of radiocarbon chronology and extensive stacks of marine sediments in the 1970s that the fourfold classification of Pleistocene stratigraphy was seriously questioned. A few crucial publications in the 1980s established the new chronology of upwards of 50 so-called marine isotope stages (MIS) within the Quaternary. MIS 2, 6, 12, and 16 represent the glaciations Würm, Riss, Mindel, and Günz, and the odd numbers MIS 5, 9, and 11 mark interglacial periods like today (Wright 2000).

In the interwar period, Penck and Bückner's research in the Alpine foreland was carried on by G. Götzinger. He was not only a founding member of the International Union of Quaternary Research (INQUA) in Copenhagen 1928 but also elected president for the congress 1936. He guided the excursion of the 3rd INQUA-conference in 1936 through the Eastern Alps (Götzinger 1936a; b). In the 1950s, Ludwig Weinberger presented an excellent geomorphologic map of the area (Weinberger 1955). Toward the east, from the Lower Austrian foreland to the Vienna basin, Julius Fink intensified his work on terrace dating using loess-stratigraphy. In the second half of the twentieth century, methodological progress enabled a detailed analysis of the Quaternary sediments and landforms left by the Salzach and the Traun glaciers. The results are documented in the geological map of Austria 1:50,0000, sheets 47 (Rupp 2008), 49 (Krenmayr 1996; Kohl and Krenmayr 1997), 64 (Egger and van Husen 2003), 65 (Egger and van Husen 2014), 66 (Egger 1996), 67 (Egger 2007; Egger and van Husen 2007), and 69 (Egger and van Husen 2011).

14.3 Palaeorelief of the Miocene-Pliocene Landscape of Salzburg and Upper Austria

At the end of the Neogene, the morphology of the landscape was much different from today with large river systems originating from the still growing mountain range in the hinterland and transporting large quantities of alluvial sediments into the foreland Molasse basin. Remnants of these fluvial accumulations are preserved on top of two elevated hilly ridges of the Northern Alpine foreland located in the southwestern part of Upper Austria. They are called Kobernausserwald in the west and Hausruck in the east and are formed of 50–200 m thick gravel beds, which mainly came from the Hohe Tauern (Graul 1937; Rupp 2008; Fig. 14.2).

After the period of accumulation, which lasted until c. 8.5 Ma ago, the pre-Quaternary rivers dissected the gravel

Fig. 14.2 Reconstruction of the Miocene landscape in the Alpine foreland with the view toward south (naturalistic painting by Kobler). Upper left: close-up view of an outcrop of the alluvial gravel beds of the Hausruck Formation. Photo: J. T. Weidinger

layer and the Miocene bedrock underneath to successively form the drainage pattern that we find today. This landscape was overridden by the Pleistocene glaciers, which later on eroded and reworked the deposits and superimposed glaciogenic sediments.

14.4 Glacial and Glaciofluvial Sediments Along the Salzach Foreland Glacier

Glacial signatures left by the Quaternary Salzach foreland glaciers are extraordinarily well preserved in Bavaria, and particularly in Salzburg and Upper Austria, where remnants of all four terrestrial glaciations (Günz, Mindel, Riss and Würm) can be distinguished. However, since the evidence of the youngest Würm glaciation (115–11.7 ka BP) is best preserved, allowing for a palaeogeographic reconstruction in great detail, the extent, dynamics, and timing of the Salzach Glacier lobe during this time period are better understood than the situation at earlier glaciations.

The Salzach Glacier originated in the Hohe Tauern range, a high mountain range with peaks exceeding 3000 m in altitude, e.g., Großvenediger (3674 m), Ankogel (3250 m), and Sonnblick (3106 m). Multiple glaciers emerging from the valleys draining the Hohe Tauern range converged to one big ice stream in the east–west trending upper Salzach Valley. This ice stream had two separate outlets cutting through the barrier of the Northern Calcareous Alps. After merging again at the foot of the Alps, the ice mass of the Salzach Glacier developed into a piedmont glacier lobe that extended to a position c. 30 km north of the city of Salzburg (Fig. 14.1).

South of the city the Salzach Glacier carved out the trough and the main basin with an overdeepening of 420 m below the recent valley floor (Pomper et al. 2017). North of the city the ice spread out and the diverging ice streams eroded several further basins (e.g., the basins of lakes Waginger See, Obertrumer See and Wallersee). The whole northern foreland glacier front was bounded by a continuous terminal moraine ridge. To the north flysch mountains, protruded the ice surface as nunataks (Fig. 14.3).

A few data across the Northern Alpine foreland indicate that glacial carving into the Miocene bedrock predominantly occurred during the early glacial maxima, whereas during the later ones, accumulation dominated (e.g., Preusser et al. 2010). The same seems to be true for the Salzach foreland glacier. Glacial carving into the Miocene bedrock (composed of sand, silt, and marlstones) during the early glacial, maxima were followed by deposition of more than 100 m thick sediments, predominantly during the Riss glacial period (Salcher et al. 2015). In contrast, the sediment pile of the youngest Würm glaciation is minor but still holds the best scientific information.

Within the distal parts of the Würm Salzach foreland glacier, Penck and Brückner's model of the "*Glaziale Serie*" (see Sect. 2) were stratigraphically confirmed by Salcher et al. (2015). They observed the matching sediment record, namely a stacked succession of (i) early stage outwash (deposited during the glacier's advancing phase, so-called "*Vorstoßschotter*"), (ii) basal till (deposited during glacial

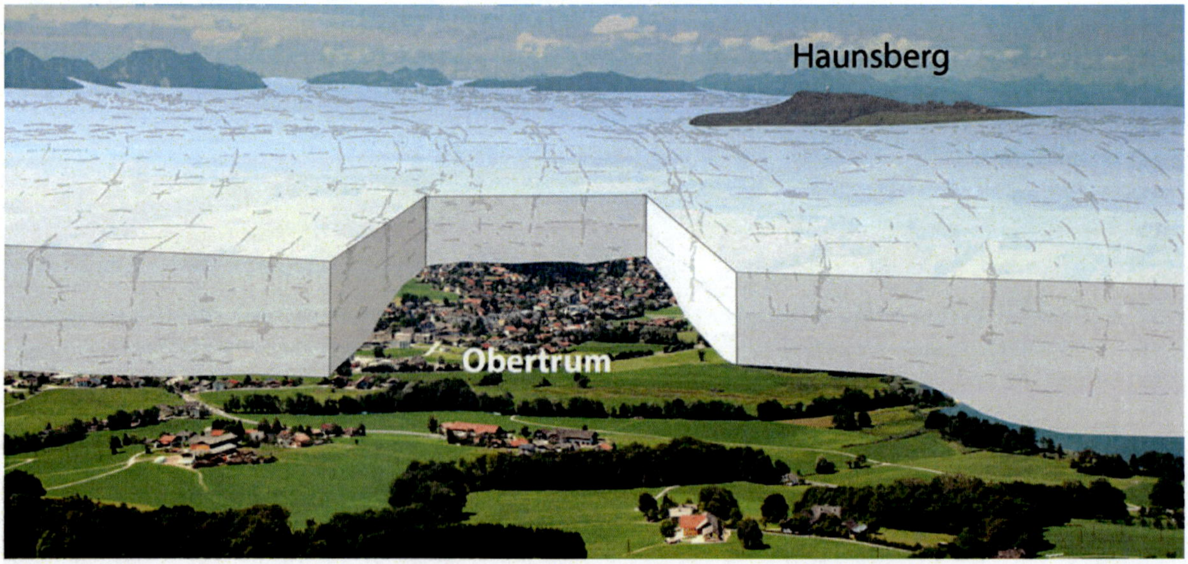

Fig. 14.3 Reconstruction of the Salzach foreland glacier in the Obertrumer Lake basin. The village of Obertrum (where it is located today) was covered by 300 m of ice. The flysch mountain Haunsberg (835 m) protruded above the ice surface as a nunatak. Flow direction of the glacier from left (south) to right (north). Photo and graphics: GeoGlobe

overriding), and (iii) late stage outwash on top (deposited at the end of the glacial cycle). An example outcrop located within the Würm terminal moraine shows the basal till confining massive and well-preserved sheetflood deposits, and the basal till is overtopped by later outwash (Fig. 14.4).

During deglaciation, a glacial lake between the retreating glacier front and the terminal moraines developed. But the largest and deepest post-LGM, glacial lake was the so-called Salzburg Lake in the main basin to the south (Figs. 14.5 and 14.6).

Fig. 14.4 Left: panoramic view of the outcrop (gravel pit near the village Döstling) within the LGM extent of the Salzach foreland glacier. Right: detailed sedimentary structures with interpretation. A1: later stage outwash, B: basal till, A2: early stage outwash (modified from Salcher et al. 2015). For position of the outcrop, see Fig. 14.1. View toward the east. Photos: J. Götz

Fig. 14.5 Salzburg basin at the end of the last glacial cycle (above) and today (below) (© Haus der Natur; museum Salzburg; Ice Age Exhibition). View toward the south. The former Salzburg Lake (above) was 35 km long and up to 10 km wide. The lake was filled with sediments in less than a few thousand years. Therefore, the landscape around the city of Salzburg (below) is almost flat, only protruded by few bedrock hills

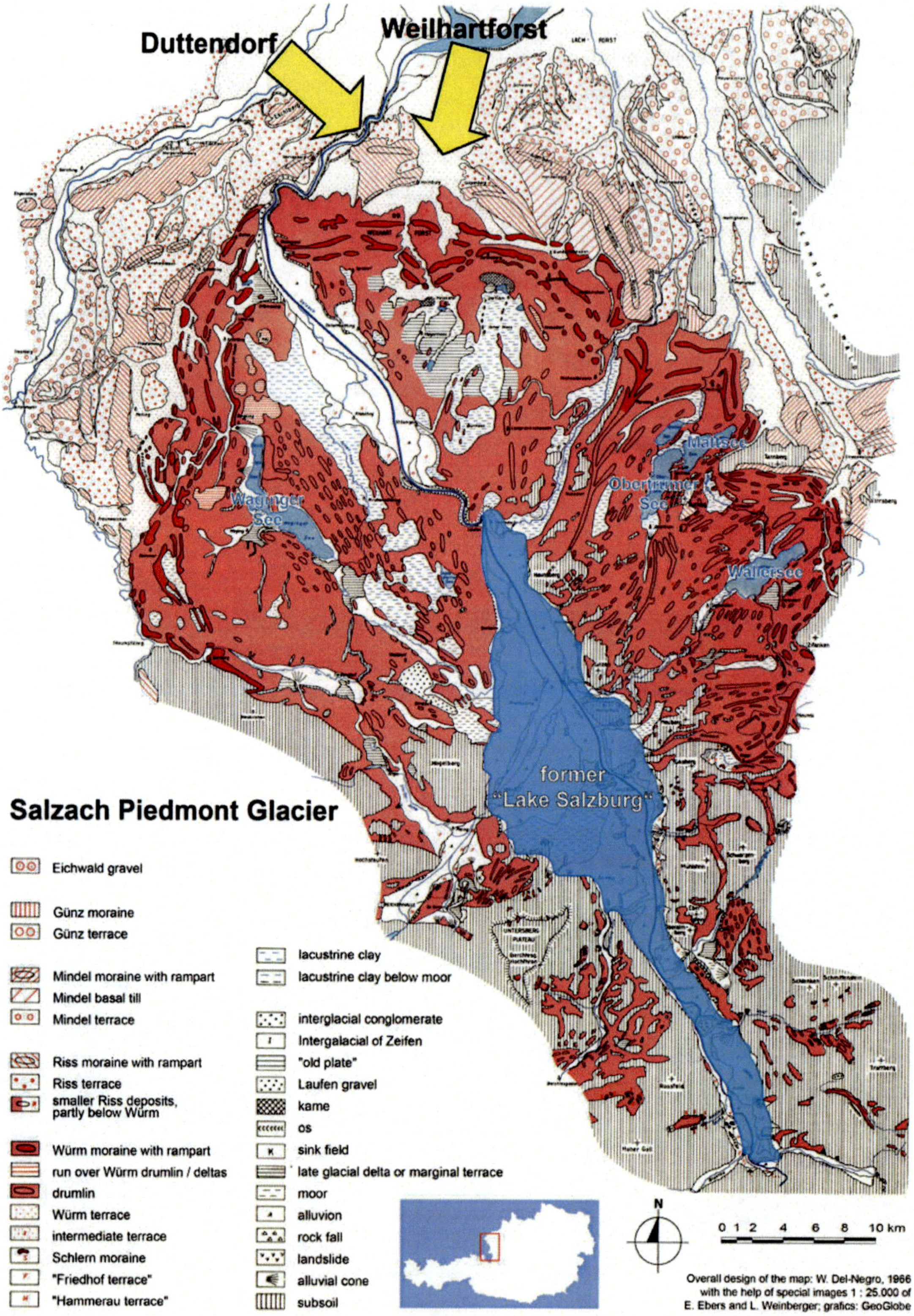

Fig. 14.6 Glacial and glaciofluvial landform assemblage left by the Salzach foreland glacier during the Würm. In the south, the extension of the former Salzburg Lake is indicated (Figure edited after Ebers et al. 1966)

In the area of the city of Salzburg, the lake was more than 250 m deep (Prey 1959; Ibetsberger et al. 2010); further south, the thickness of sediment fill reached 339 m (Van Husen 1979). However, glaciofluvial sediments rapidly accumulated and the lake disappeared before dense vegetation developed in the adjacent area. Since that time, the Salzach River cut through the sediments and established its postglacial river course. Comparable situations probably occurred after each of the four main glaciations.

Characteristic landforms left by the Würm Salzach Glacier advance and decay in the foreland are drumlins (Fig. 14.6, also middle and background of Fig. 14.11), which trace the direction of the diffluent ice flow typical for piedmont-type glacier lobes. While the drumlins on the western side of the Salzach foreland glacier are oriented from SE to NW, the eastern ones stretch from SW to NE. During deglaciation numerous typical ice decay, features developed within and close to the bounding terminal moraine ridge (Salcher et al. 2010). Ice-marginal (kame) terraces, eskers, and kettle holes are often well preserved and still observable in the region as typical examples of the ice age heritage. In the IBM basin (Upper Austria), accumulation of silty to clayey lake sediments caused the development of the largest peat bog complex in Austria, surrounded by several, still existing smaller lakes, and ponds embedded in a typical, so-called kame and kettle topography. The deepest parts of the terminal glacial basins are occupied by lakes (e.g., Waginger See, Abtsee, Obertrumer See, Mattsee, Wallersee in Fig. 14.6) that survived the last 19 ka and are today popular for swimming.

14.5 Glacial and Glaciofluvial Sediments Along the Traun Glacier

In contrast to the Quaternary Salzach Glacier that developed in central Alpine catchments (composed of crystalline rocks) and formed a large piedmont glacier in the foreland, the Quaternary Traun Glacier(s) originated in the Northern Calcareous Alps. The ice accumulation zone on the high plateaus of the Totes Gebirge and the Dachstein (2995 m asl) fed an ice stream flowing north through the historic salt mining region of Salzkammergut (described in Chap. 17). Due to the smaller catchment area, the glacier tongue(s) just about reached the edge of the Eastern Alps (Fig. 14.1). However, while moving north, the ice eroded a number of deep rock basins. An exploratory drilling in one of them revealed that it was at least c. 900 m deep, which is the greatest overdeepening so far known in the Alps (Van Husen and Mayr 2007). These overdeepened basins are still filled with lakes, although multiple deltas indicate different stages of recent infilling (e.g., Traunsee/Ebensee, Wolfgangsee/Zinkenbach, Hallstätter See/Obertraun). So, the whole area is well known for its beautiful scenery with more than 70 lakes in between rough mountain terrain. The main valley of the River Traun is the longest in Upper Austria (140 km).

Already the pioneers of research on the Quaternary of the Traun Glacier discovered that the basin of Bad Ischl—famous as the favorite place for the summer vacation of the Austro-Hungarian emperor, Franz Josef I—played a major role in the development of the different tongues of the Traun Glacier (Van Husen 1977). Here, the glacier divided into two main branches—the larger one flowed west toward Wolfgangsee, the other one northeast along the River Traun. Its western branch terminated in several individual glacier tongues.

14.5.1 The Western Branches of the Traun Glacier

Flowing from Bad Ischl toward the west, this branch followed the tectonic fault along the lake basin of Wolfgangsee. At the western end of the lake, a considerable part of the ice breached north over a low pass into the basin of lake Mondsee from where it further divided into four ice streams. Three of these headed north and northeast. They include the valley glacier that carved out the Attersee basin, which was additionally fed by local glaciers and ice overflowing from the eastern branch of the Traun Glacier. The second ice stream left a minor glacial basin east of Lake Attersee, and the third one eroded the lake filled Irrsee basin surrounded by terminal moraines (Fig. 14.1). The branches flowing straight west from the Wolfgangsee and Mondsee touched the Salzach Glacier. The confluence area of the LGM is marked by large kame terraces at Thalgau. During former glaciations, this confluence of the Salzach and Traun glaciers was located further north near the village of Strasswalchen (Riss) and near the Kobernausserwald und Hausruck during earlier glacial maxima (Fig. 14.1; Rupp 2008; Egger and van Husen 2003; Van Husen 1989; Van Husen and Egger 2014; Egger 1996, 2007).

At the center of these dividing branches, one of the most important outcrops for Quaternary stratigraphy in Austria is located—the so-called Mondsee-profile. It was first discovered during the construction of the A1-highway and has been much studied (Kohl 2000). The lacustrine sediments (photo in Fig. 3.10a in Chap. 3) and a delta complex of this outcrop are situated up to 50 m above the water level of Lake Mondsee and were formed in a greater ancient Lake Mondsee after the Riss glaciation (Van Husen 2000). The sequence was topped by the basal till of the Würm period. Palynological analyzes provided evidence for a warm Interglacial and two cold stadials interrupted by warmer interstadials. The sediment sequence, therefore, covers a period of about 50 ka, reaching from end of Riss to the middle of Würm (Drescher-Schneider 2000).

14.5.2 The Eastern Branch of the Traun Glacier

The eastern branch of the Traun Glacier followed the Traun Valley northwards along an important Miocene tectonic structure, the Königssee-Lammertal-Trauntal or KLT-fault (Decker 1994). Where this structure forms a so-called "horse-tail" (the fault splits into a fan of small fractures), the oscillating glacier left a prominent overdeepened basin. Seismic surveying in the area of Ebensee indicated its bedrock bottom in a depth of c. 400–450 m (Egger 2007). During Würm deglaciation, this basin was filled with (melt) water and the deepest lake of Austria—lake Traunsee—was born (Fig. 14.7). Lateglacial and Holocene sediment infill gradually raised its bottom, but the water depth of 190 m remains a national record.

Impressive Mindel, Riss, and Würm moraines are preserved at the northern rim of this lake. Penck and Brückner termed this landform assemblage a "Moraine-Amphitheater," highlighting the well-preserved glacial topography surrounding the city of Gmunden and adjacent areas to the north and west. Glacial and ice decay features of Gmunden town include at least two ridges of terminal moraines, glaciofluvial terraces (Kohl 2000), and the kettle hole lake Krottensee. The latter has a central, floating vegetation mat and is filled with lake and peat bog deposits. Infrastructure and regional planning around and within the former salt trading town Gmunden have always been influenced by this glacial topography—culminating in a contemporary railway project right through the inner town (Fig. 14.8).

14.6 Glaciofluvial and Alluvial River Terraces North of the Terminal Moraines of the Quaternary Foreland Glaciers

Research on Quaternary glaciofluvial and alluvial terraces in Austria advanced through the classical work of Julius Fink. In the course of his loess and palaeosoil studies (Fink 1956), Fink detected a regular terrace arrangement in the Northern Alpine foreland (Fig. 14.9a), which is confirmed by modern research. The oldest Günz-related terraces are always in the highest position, and the younger Mindel, Riss, and Würm terraces are not only lower in elevation but also stepwise

Fig. 14.7 Eastern branch of the Traun Glacier (reconstructed) formed the basin of lake Traunsee. Gmunden is positioned within the terminal Würm moraine. Mindel and Riss moraines are indicated in the background toward the north. Photo: I. Baron, edited by J. T. Weidinger

Fig. 14.8 Bird's eye view of the "Moraine-Amphitheater" of Gmunden town seen from the north, including reconstructed calving eastern Traun Glacier in the back, present water level of Lake Traunsee, bodies of dead ice in kettle holes and remnants of terminal moraines of the Würm glacial. Photo: M. Wojacek, edited by J. T. Weidinger

embedded into the older gravel accumulations (Van Husen and Reitner 2011a, b). Figure 14.9b shows the temporal position of these terraces in connection with the four glaciations and the development of global temperature. The widespread Riss and Würm terraces are named *Hochterrasse* (high terrace, deposited during Riss) and *Niederterrasse* (low terrace, deposited during Würm).

Between the Rivers Salzach and Enns two main systems of terraces can be distinguished:

(i) an older one in the east, in between the Rivers Traun and Enns, consisting of Günz and Mindel gravels, topped by loess and clay. This system is called Traun-Enns-Platte.
(ii) a younger one in the west, around the border of Bavaria and Upper Austria, near the confluence of the Rivers Salzach and Inn, with Riss and Würm gravels mainly exposed on the surface.

14.6.1 The Terrace System of the "Traun-Enns-Platte"

This Quaternary terrace system covers an area of ca. 1500 km^2—so it is in dimension larger than the earlier mentioned system and economically much more important. It has not only been interesting from a scientific point of view, but is a site for sand and gravel pits, valued in applied geotechnical sciences (Van Husen 1999). The *Weiße Nagelfluh*, a conglomerate composed of glaciofluvial gravels generated in a cold period between the Günz and the Mindel glacial stages (see photos in Fig. 3.8a–c in Chap. 3), originated from the Alm Valley and was deposited with decreasing thickness toward the Krems Valley in the east. This material has been mined since Roman times for building stones in a lot of quarries in the area around the market town of Kremsmünster (Kohl 1986, 2000; Van Husen and Reitner 2011a).

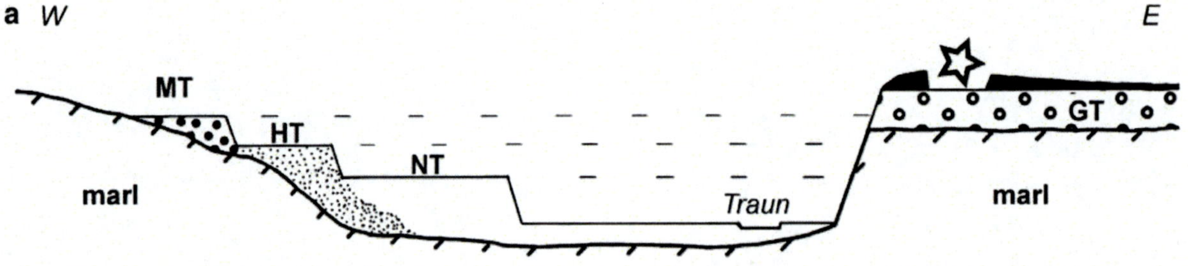

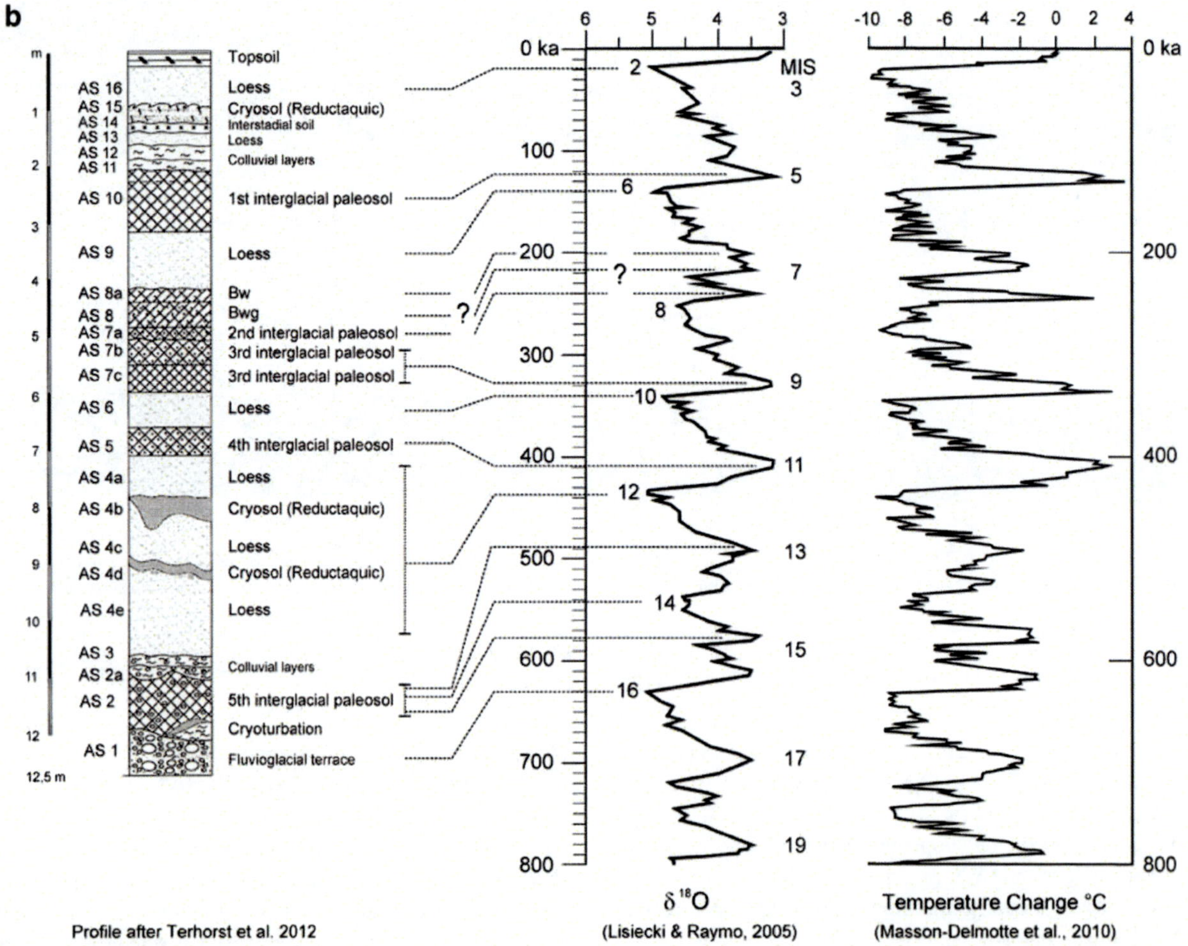

Fig. 14.9 a Sketch profile through the alluvial terraces in the Traun Valley shows the classical stratigraphy (not to scale): star marks position of the section in Fig. 14.9b, NT = Niederterrasse (Würm), HT = Hochterrasse (Riss), MT = Mindel terrace, GT = Günz terrace (after Van Husen and Reitner 2011a). **b** Loess sequence with palaeosoils from the brick yard Würzburger (left side) in correlation with oxygen isotopes distribution in the ocean and temperature development in the Antarctic ice shield for the past 800 ka

The growing need for water due to increasing construction of buildings and infrastructure, together with depletion and pollution of drinking water reserves put the Upper Austrian government more and more under pressure. From the 1960s to the 1970s, drinking water was explored here for the two largest cities of Upper Austria, Linz, and Wels (Kohl 2000) and the Traun-Enns-Platte became famous for its large ground water storage capacities. The Geologische Bundesanstalt (Geological Survey of Austria) compiled these features in a special hydro-geologic map (Schubert and Berka 2007). Additionally, the ground water situation of the western parts of Upper Austria was mapped by Baumgartner and Tichy (1981). Applied Quaternary hydrogeology was also the topic of a joint research program between the Upper Austrian and the Bavarian governments (Bayerisches Geologisches Landesamt 1981), focusing on the second most important terrace system.

14.6.2 Terrace Systems Near the Salzach-Inn-Confluence

In the central part of the Salzach foreland glacier, in front of the former glacier tongue and the terminal moraines, large outwash plains were deposited consisting of sand and gravel. This area with an extent of c. 90 km^2 is called the Weilhartforst. One of the gravel pits located at Duttendorf west of the Weilhartforst exposes a most helpful section for unraveling the history of terrace formation along the Salzach (see Fig. 14.1 and the yellow arrows in Fig. 14.6 for locating the Weilhartforst and Duttendorf).

The base of the loess profile of Duttendorf (Traub and Jerz 1975, Fig. 14.10) is built of gravel and sand of Neogene age. Above, a 25 m thick layer of Riss gravels was deposited. These gravels belong to the *Hochterrasse* (high terrace). A layer of loess accumulated above, which is yellow-gray colored, calcareous, bears concretions of iron and manganese oxides and shows rust patches. The thin gravel layer covering the loess was deposited during the LGM and is interpreted as the *Niederterrasse* (low terrace). Since the loess layer includes shells of the loess snails *Arianta* and *Trichia*, its deposition could be assigned to the LGM (21.65 ± 250 ka BP) by radiocarbon dating.

This loess layer marks the moment of the final accumulation of the lower terrace shortly before the LGM.

Recent luminescence data (Starnberger et al. 2009, 2011) indicate a continuous loess deposition during the LGM between 30 and 21 ka BP, and pollen and the mentioned fossils underpin a cold, arid steppe environment, correlating with other loess sections across Europe (Frechen et al. 2003). Starnberger et al. (2009, 2011) provide the first absolute, luminescence-based chronostratigraphy of the Duttendorf loess section and—along with other dated sites in the Eastern Alps—the first well-defined timing of the LGM in this part of the Alpine foreland. Although loess deposition was apparently continuous, there is evidence for a two-step oscillatory advance (Nunreuter- and Lanzinger phase) of the Salzach Glacier between 29 and 19 ka BP.

14.7 Glacial Heritage and Human Activity

Quaternary science is not only a matter of basic research but also of applied geology and geomorphology. Industrial minerals, such as gravel, conglomerate, clay, loess, phosphate, coal, and chalk, are mined in quarries leaving open pits behind that might be used for waste depositions or have to be revegetated and recultivated later on. Environmental problems such as ground water pollution force authorities to protect subsurface nature and landscape from degradation by rules and law.

Unfortunately, the (pre-) Holocene landscape is also the main target for constructing buildings for a society with rapidly growing economic and infrastructure needs. That is

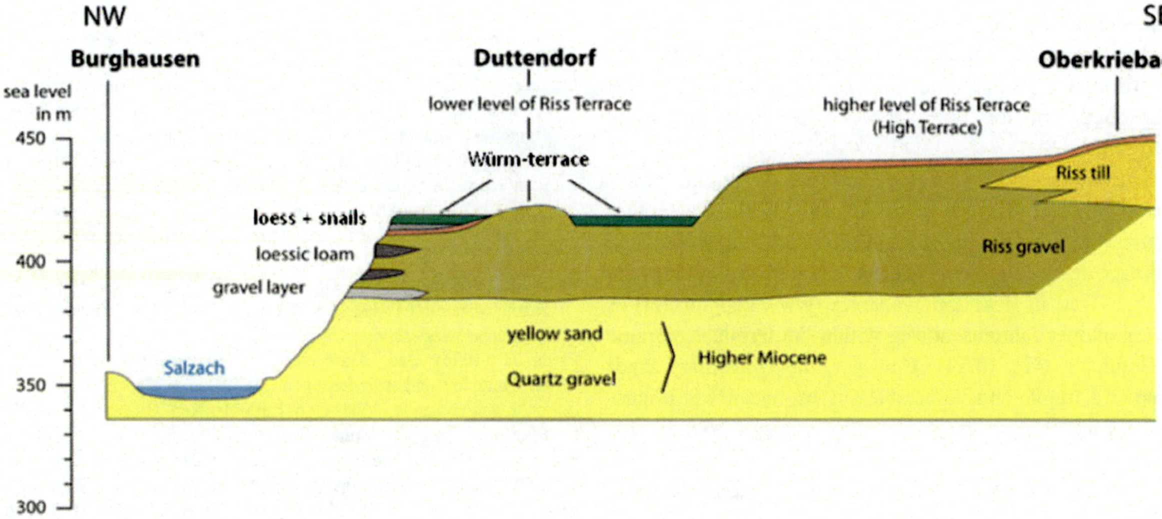

Fig. 14.10 Loess layer in *Duttendorf/Hochburg-Ach* is a typical example of a cold period sediment and a marker of the LGM (Graphics by GeoGlobe; after Traub and Jerz 1975)

Fig. 14.11 Left: view over the ground moraine landscape with drumlins in the county of Flachgau/Salzburg. The foreground of the picture shows some panels of the Ice Age Trail in Henndorf at Wallersee. Photo: H. J. Ibetsberger. Right: view over panels illustrating the Ice Age Trail Gmundner Jahrtausendweg toward the town of Gmunden positioned within the terminal moraines of LGM. Photo: J. T. Weidinger

why glacial landforms have suffered tremendously from landscape alteration and environmental degradation.

The area covered by the Salzach foreland glacier, including Salzburg, is one of the most popular residential areas of Austria. The county of Flachgau recorded a population increase of 4% between 1991 and 2001 and 5.8% between 2003 and 2012 (Amt der Salzburger Landesregierung 2013). Trade and commerce followed this trend. The result is that a former rural region developed into a "suburb" of the city of Salzburg. Each community opened its own business park. The transformation of agricultural land into an industrial estate destroys more and more of the natural environment and original landscape. The "footprint of the glacier" disappears between housing estates or production halls and warehouses. Drumlins were leveled, terraces excavated, and moraine landscapes transformed to golf courses. This development needs to be questioned, and the importance of geotopes should be strengthened.

Educational trails can help to increase the public awareness for intact ecosystems and/or still visible Quaternary features—and promising concepts focusing on the heritage of the ice age in Austria have been implemented, such as the Ice Age Trail in Henndorf/Wallersee (www.geoglobe.at) or the Gmundner Jahrtausendweg within the terminal moraine of Gmunden (Fig. 14.11). However, the Quaternary landscape is a fragile and vulnerable system and needs protection. Geomorphologists should play a major role in this context!

These are the main conclusions of this chapter that bear on questions of environmental awareness, economic development, and pedagogy.

References

Agassiz L (1837) Discours prononcé à l'ouverture des séances de la Société Helvetique des Sciences Naturelles, à Neuchâtel le 24 juillet 1837. Actes de al Soc Helv D Sc Nat, reunie à Neuchâtel le 24, 25 et 26 juillet 1837

Amt der Salzburger Landesregierung (2013) Bevölkerung Land Salzburg—Stand und Entwicklung 2013, Salzburg

Baumgartner P, Tichy G (1981) Geologische Karte des südwestlichen Innviertels und des nördlichen Falchgaus 1:50,000 und Erläuterungen. Amt der OÖ Landesregierung-Landesbaudirektion, Linz

Bayerisches Geologisches Landesamt (1981) Bericht über geologische und hydrogeologische Untersuchungen in Salzach-Saalach Talraum

Decker K, Peresson H, Faupl P (1994) Die miozäne Tektonik der östlichen Kalkalpen: Kinematik, Paläospannungen und Deformationsaufteilung während der "lateralen Extrusion" der Zentralalpen. Jahrbuch der Geol B-A 137/1:5–18. http://opac.geologie.ac.at/wwwopacx/wwwopac.ashx?command=getcontent&server=images&value=JB1371_005_A.pdf

Drescher-Schneider R (2000) Die Vegetations- und Klimaetnwicklung im Riß/Würm- Interglazial und im Früh- und Mittelwürm in der Umgebung von Mondsee. Ergebnisse der pollenanalytischen Untersuchungen. In: van Husen, Mondsee Klimaetwicklung im Riss/Würm Interglazial (Eem) und Frühwürm (Sauerstoffisotopenstufe 6–3) in den Ostlapen. Mitt. der Komm. f. Quartärforschung der ÖAW, Bd. 12, Verlag der Österr. Akademie der Wiss., 180p, Wien

Ebers E, Weinberger L, Del-Negro W (1966) Der pleistozäne Salzachvorlandgletscher. Veröffentlichungen Der Gesellschaft Für Bayrische Landeskunde 19–22:1–217

Egger H (1996) Geo. Karte der Rep. Österreich 1:50,000, Blatt 66-Gmunden und Erläuterungen (2007), 65 p, Geo B-A, Wien

Egger H, van Husen D (2007) Geo. Karte der Rep. Österreich 1:50,000, Blatt 67-Grünau im Almtal. Geo B-A, Wien

Egger H (2007) Erläuterungen zur Geo. Karte der Rep. Österreich 1:50,000, Blatt 67-Grünau im Almtal, 66 p. Geo B-A, Wien

Egger H, van Husen D, (2003) Geo. Karte der Rep. Österreich 1:50,000, Blatt 64-Strasswalchen und Erläuterungen (2009), 87 p. Geo B-A, Wien

Egger H, van Husen D (2011) Erläuterungen zur Geo. Karte der Rep. Österreich 1:50,000, Blatt 69-Grossraming, 117 p. Geo B-A, Wien

Fink J (1956) Zur Korrelation der Terrassen und Lösse in Österreich. Eiszeit Gegenw 7:49–77

Frechen M, Oches EA, Kohlfeld KE (2003) Loess in Europe—mass accumulation rates during the last glacial period. Quat Sci Rev 22:1835–1857

Götzinger G (1936) Führer für die Quartär-Exkursionen in Österreich—I. Teil, Geologische Bundesanstalt, 1–148

Götzinger G (1936) Führer für die Quartär-Exkursionen in Österreich—II. Teil, Geologische Bundesanstalt, 1–115

Graul H (1937) Untersuchungen über Abtragung und Aufschüttung im Gebiet des unteren Inn und des Hausruck. Mitt Geogr Ges München 30:179–259

Ibetsberger H, Jäger P, Häupl M (2010) Der Zerfall des Salzachgletschers und die nacheiszeitliche Entwicklung des Salzburger Gewässersystems aus der Sicht der Wiederbesiedlung der Salzburger Gewässer mit FischenGewässerschutz 14:7–54. http://www.geoglobe.at/DE/index.php?mact=News,cntnt01,detail,0&cntnt01articleid=69&cntnt01origid=67&cntnt01detailtemplate=Simplex_Detail&cntnt01returnid=67

Kohl H (1986) Die Weiße Nagelfluh als Bau-und Dekorationsstein. OÖ Heimatblätter 40(3–4):245–265

Kohl H (2000) Das Eiszeitalter in Oberösterreich. Schriftenreihe Des Oö Musealvereines - Gesellschaft Für Landeskunde 17:1–487

Kohl H, Krenmayr HG (1997) Erläuterungen zur Geol. Karte der Rep. Österreich 1:50,000, Blatt 49-Wels, 77 p. Geo B-A, Wien

Krenmayr HG (1996) Geo. Karte der Rep. Österreich 1:50,000, Blatt 49-Wels. Geo B-A, Wien

Penck A, Brückner E (1909) Die Alpen im Eiszeitalter. Leipzig

Pomper J, Salcher BC, Eichkitz C, Prasicek G, Lang A, Lindner M, Götz J (2017) The glacially overdeepened trough of the Salzach Valley, Austria: Bedrock geometry and sedimentary fill of a major Alpine subglacial basin. Geomorphology 295:147–158

Preusser F, Reitner JM, Schlüchter C (2010) Distribution, geometry, age and origin of overdeepened valleys and basins in the Alps and their foreland. Swiss J Geosci 103:407–426

Prey S (1959) Zwei Tiefbohrungen der Stieglbrauerei in Salzburg. Verhandlungen Der Geologischen Bundesanstalt 1959:216–224

Rupp Ch (2008) Geol. Karte der Rep. Österreich 1:50,000, Blatt 47 Ried im Innkreis, und Erläuterungen, 100 p. Geol B-A, Wien

Salcher BC, Hinsch R, Wagreich M (2010) High-resolution mapping of glacial landforms in the North Alpine Foreland, Austria. Geomorphology 122:283–293

Salcher BC, Starnberger R, Götz J (2015) The last and penultimate glaciation in the North Alpine Foreland: new stratigraphical and chronological data from the Salzach glacier. Quatern Int 388:218–231

Schubert G, Berka R (2007) Hydrogeologische Karte von Oberösterreich 1:200,000. Geol B-A, Wien

Starnberger R, Rodnight H, Spötl Ch (2011) Chronology of the last glacial maximum in the Salzach Palaeoglacier Area (Eastern Alps). J Quat Sci 26/5:502–510

Starnberger R, Terhorst B, Rähle W et al (2009) Paleoecology of quaternary periglacial environments during OIS-2 in the forefields of the Salzach glacier (Upper Austria). Quatern Int 198:51–61

Traub F, Jerz H (1975) Ein Lößprofil von Duttendorf (Oberösterreich) gegenüber Burghausen an der Salzach. Zeitschrfft Für Gletscherkunde Und Glazialgeologie 11(2):175–193

Van Husen D (1977) Zur Fazies und Stratigraphie der jungpleistozänen Ablagerungen im Trauntal. Jahrbuch der. Geol B-A 120:1–130

Van Husen D (1979) Verbreitung, Ursachen und Füllung glazial übertiefter Talabschnitte am Beispiel in den Ostalpen. Eiszeitalter und Gegenwart 29:9–22, Hannover

Van Husen D (1999) Geologisch-baugeologische Erfahrungen beim Bau des Eisenbahntunnels Lambach. OÖ Mitt Österr Geol Ges 90:137–154

Van Husen D (2000) Mondsee Klimaentwicklung im Riss/Würm Interglazial (Eem) und Frühwürm (Sauerstoffisotopenstufe 6–3) in den Ostlapen. Mitt. der Komm. f. Quartärforschung der ÖAW, Bd. 12, Verlag der Österr. Akademie der Wiss., 180 p, Wien

Van Husen D (1989) Geo. Karte der Rep. Österreich 1:50,000, Blatt 65-Mondsee. Geo B-A, Wien

Van Husen D, Mayr M (2007) The hole of bad Aussee. An unexpected overdeepened area in NW Steiermark Austria. Austrian J Earth Sci 100:128–136

Van Husen D, Egger H (2014) Erläuterungen zur Geo. Karte der Rep. Österreich 1:50,000, Blatt 65-Mondsee, 143 p. Geo B-A, Wien

Van Husen D, Reitner JM (2011a) Klimagesteuerte Terrassen- und Lössbildung auf der Traun-Enns-Platte und ihre zeitliche Stellung (Das Profil Wels/Aschet). Mitt Komm Qartärforsch Österr Akademie Der Wiss 19:1–12

Van Husen D, Reitner JM (2011b) Die Löss-Sequenz Wels/Aschet (ehemalige Lehrgrube Würzburger) – Ein Referenzprofil für das Mittel- und Jungpleistozän im nördlichen Alpenvorland (MIS 16 bis MIS 2). Mitt. der Komm. f. Quartärforschung der ÖAW, Bd. 19, Verlag der Österr. Akademie der Wiss., 70 p, Wien

Weinberger L (1955) Exkursion durch das österr. Salzachgletschergebiet und die Moränengürtel der Irrsee- und Attersee-Zweige des Traungletschers. Verh. Geol B-A SH D, Beiträge zur Pleistozänforschung in Österreich: 7–34. http://opac.geologie.ac.at/wwwopacx/wwwopac.ashx?command=getcontent&server=images&value=VS0004_007_A.pdf

Wright JD (2000) Global climate change in marine stable isotope records. In: Noller JS, Sowers JM, Lettis WR (eds) Quaternary geochronology: methods and applications. American Geophysical Union

Johannes T. Weidinger is head of the K-Hof, Kammerhof Museums of Gmunden. As a research earth and educational scientist he focuses on natural hazards with an emphasis on mass movements in high mountain regions, such as the Himalayas and the Alps. He is a guest lecturer at the Universities of Vienna, Salzburg, Linz/Upper Austria and Klagenfurt/Carinthia.

Horst J. Ibetsberger is a self-employed geoscientist and scientific director of the company GeoGlobe, which specialises in the conception and implementation of geo- and ecotourism projects. He is also scientific head of the Ore of the Alps UNESCO Global Geopark. In addition, he is a lecturer of physical geography at the Department of Geography and Geology, Salzburg University.

Joachim Götz is a research associate at the University of Bayreuth, Germany. As a mountain geomorphologist, his research and teaching interests include process-based monitoring studies focusing on drivers and controls of present-day sediment dynamics and the archive-based reconstruction of post- and paraglacial landscape evolution. His research in the Eastern Alps, the Nepalese Himalayas and the Argentinian Andes is based on a wide range of modern field, lab and modelling techniques.

The Walgau: A Landscape Shaped by Landslides

15

Stefan Steger, Elmar Schmaltz, Arie Christoffel Seijmonsbergen, and Thomas Glade

Abstract

Landslides of the slide-type movement are common on slopes of the federal state of Vorarlberg. This chapter focuses on landslides located in the Walgau region, where both shallow and deep-seated slope movements are widespread and leave a distinct geomorphic footprint on the hillsides. Landslide activity considerably increased since the Last Glacial Maximum (LGM) and still plays a substantial role in landscape evolution of the area. Several examples of this chapter highlight that the causes of those slope movements are manifold and a result of a complex interplay between various processes which act frequently at different spatial scales, but with varying intensity. The presented large, deep-seated, and small, frequently occurring landslides of the area are strongly influenced, respectively, caused by an interplay between the prevalent relief-rich topography, the local geological setting, regolith coverages, the humid climate, and recurrent land cover changes. Human activities are also known to directly impact the geomorphic dynamic of several recent landslide events. Major challenges for the future arise due to projected and already observable changes in land cover, climate, and human impact, which may modify the magnitude and frequency of the landslide process itself, but also the exposition of people, their properties, and the living environment.

Keywords

Landslides • Landform evolution • Walgau • Vorarlberg

15.1 Introduction

Vorarlberg is the westernmost federal state of Austria. In general, altitudes increase from northwest to south and range from 400 m asl at the plains of the Rhine Valley to over 3300 m asl at the southern lying peaks of the Silvretta mountain range. Gravity-driven degradation processes, such a landsliding, are very active throughout the mountainous and hilly parts of Vorarlberg and are considered among the most important processes that shape and reshape the landscape over time (Seijmonsbergen 1992).

Landslides can be defined as the downslope movement of soil, rock, or debris in response to the force of gravity (Crozier 1989; Glade et al. 2005). Within this chapter, the expression "landslide" is restricted to the slide-type movement according to Cruden and Varnes (1996). This specific landslide type describes those landslides, where a planar (translational slide) or concave (rotational slide) shaped shear plane separates the downslope moving material from the stable subsurface (Crozier 1989; Cruden and Varnes 1996).

Besides playing an essential role in landscape evolution (Crozier 2010; Bierman and Montgomery 2014), landslides also pose a serious threat to residents, private properties, and infrastructure (Bell and Glade 2004). Thus, they can be considered as hazards and even as risk when vulnerable infrastructure (e.g., buildings, roads) is exposed (Glade et al. 2005). A steadily increasing population density, the suburban spread, and the increased economic utilization of land substantially contribute to the ongoing trend that more and more people, their properties, and the economy are endangered by landslides in the Austrian Alps (Slaymaker and Embleton-Hamann 2009; Promper et al. 2014). An in-depth

S. Steger (✉)
Institute for Earth Observation, Eurac Research, Drususallee 1/Viale Druso 1, 39100 Bozen/Bolzano, Italy
e-mail: Stefan.steger@eurac.edu

E. Schmaltz · T. Glade
Department of Geography and Regional Research, University of Vienna, Vienna, Austria

A. C. Seijmonsbergen
Institute for Biodiversity and Ecosystem Dynamics, University of Amsterdam, Amsterdam, The Netherlands

C. Embleton-Hamann (ed.), *Landscapes and Landforms of Austria*, World Geomorphological Landscapes,
https://doi.org/10.1007/978-3-030-92815-5_15

investigation of these geomorphic phenomena is, therefore, also needed to develop sustainable strategies to reduce, or even prevent, future damage.

This chapter focuses on the characterization of landslides located in Vorarlberg with a regional focus on the Walgau region, an area characterized by widespread earth surface dynamics since the end of the last ice age (Ruff and Czurda 2008a; b). After presenting the contemporary environmental setting of the region and associated potential predisposing and triggering factors of landslides (Sect. 2), a short discourse on the role of landslides within the context of landscape evolution is given (Sect. 3). Section four examines several examples of deep-seated (Sect. 4.1) and shallow (Sect. 4.2) landslides, and subsequently embeds those processes within a multi-dimensional viewpoint (Sect. 4.3). This chapter concludes by highlighting future challenges in the context of landslide processes (Sect. 5).

15.2 Environmental Setting

The Walgau, a valley of the Ill River, is located in the western part of Vorarlberg (Fig. 15.1) at the geological boundary between the Eastern and the Western Alps (Friebe 2007). The corresponding alluvial plain is mainly filled with glaciofluvial gravels and alluvial sediments. The valley can generally be described as a "mixed-type valley" (i.e., due to its glacial and fluvial origin) with fluvial processes dominating the lower parts. Former glacial activity was responsible for a large portion of today's landforms, which have mainly been preserved at higher altitudes (De Graaff 1992; Ruff 2008a, b). The surrounding hillsides (Walserkamm, Großwalser Valley, Rätikon, Fig. 15.1) are affected by landsliding to varying degrees mainly due to their geological differences, which in turn co-determine regolith coverages and topography.

15.2.1 Geology

The geological setting highly influences the tendency of a specific slope to fail (Grelle et al. 2011). For instance, the dip of the bedding may act as a sliding surface while the thickness and properties of the overlying weathering mantle are also controlled by the geologic parent material (Crozier 1989; Schweigl and Hervás 2009). Consequently, the hydrogeological behavior in the deep subsurface, for example, the water pressure on potential sliding surfaces, is highly determined by the geological setting such as variation in lithology and the location of thrust zones and faults (Crozier 1989).

The Walgau Valley is located on the tectonic thrust zone between the Rhenodanubian Flysch to the north and the Northern Calcareous Alps to the south (Friebe 2004; Ruff 2008a, b). In general, four major lithological units can be distinguished (Fig. 15.1). The outermost northwestern part of the area mainly consists of continental marginal sediments (Helvetic and Ultrahelvetic), while the Walserkamm and the western part of the Großwalser Valley are underlain by sandstone sequences of the Flysch Zone. On the orographic left side of the Ill River, slopes consisting of Flysch sediments and carbonate rocks, e.g., the Rätikon mountains to the south, are prevalent as well as Penninic melange. A considerable portion of the area was formerly transformed by glacial processes. Thus, Quaternary glacial and postglacial deposits cover a large portion of the region (Ruff and Czurda 2008a; b).

The physical properties of the subsurface material considerably influence landslide activity of the region (Markart et al. 2007). In general, areas underlain by morainic deposits, and heterogeneous bedrock material with marls and sandstones such as in the Rhenodanubian Flysch or the Helvetic Zone, are most prone to shallow slope movements (Ruff 2008a, b; Ruff and Czurda 2008a, b; Zieher et al. 2016). Such smaller landslides are regularly triggered by a combination of heavy precipitation events and a long-term water supply (due to snowmelt for example), which leads to intensive water fluxes into the respective sediments and consequent increased pore water pressures in the slopes.

The Flysch Zone of Vorarlberg consists of interbedded sequences of sandstones, claystones, and marls. The deeply weathered marls and sandstones in the Walgau are partly interrupted by limestones and covered by till. Especially, the marly material within the Flysch is characterized by low permeability and better able to store infiltrated water than sandstone layers. In general, no clear transition between the overlying soil mantle and the underlying intensively weathered bedrock is discernible. The prevalent clayey layers are likely to promote landsliding due to their strong reaction to water-influx (for example the swelling of clays) (Heissel et al. 1967; Glade et al. 2001).

The marls of the Helvetic and Ultrahelvetic nappes are also highly prone to landsliding (Zieher et al. 2016). In the case of water saturation, changes of the consistency of the fine material and pore water pressure may promote shallow landsliding. The numerous cliffs and steep slopes in carbonate rocks of the Rätikon (Fig. 15.1) are regularly source areas of rockfalls (Seijmonsbergen 1992; Ruff 2008a, b). However, landslides of the slide-type movement can also be observed in these areas (cf. Sect. 4.1).

15.2.2 Land Cover and Socio-Economic Setting

The Walgau region has approximately 50,000 inhabitants in total while the majority of the population is spread across the

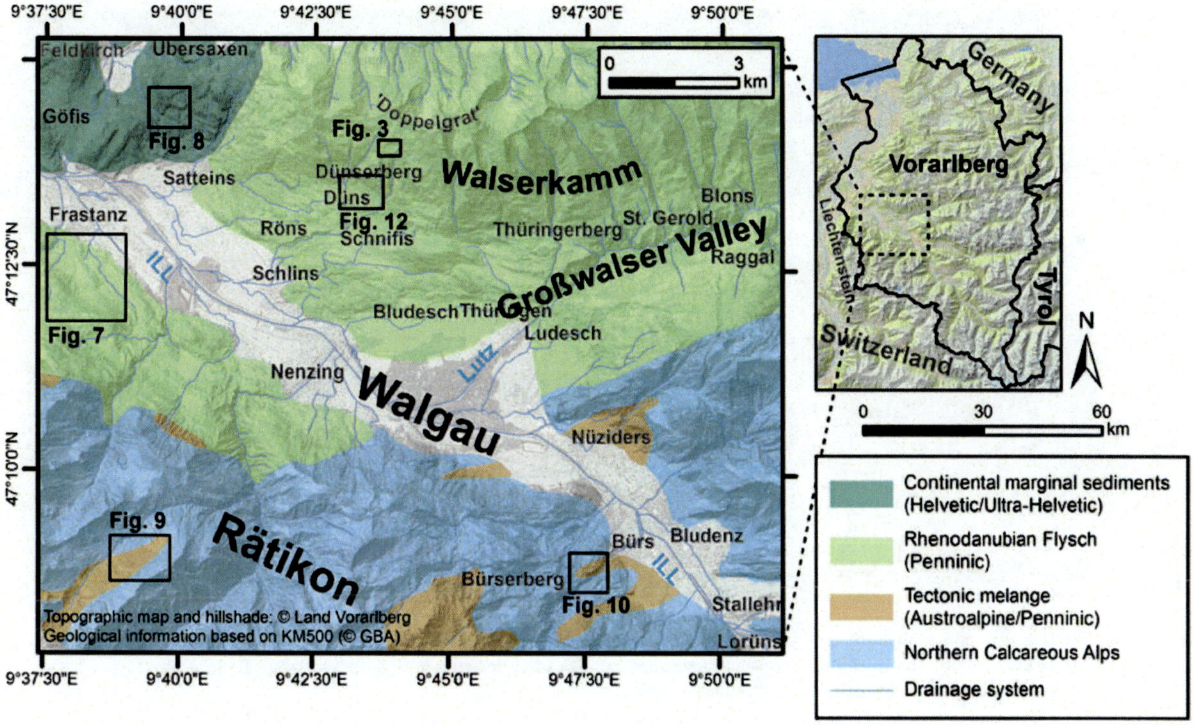

Fig. 15.1 Location of the focus area (right) and lithological overview (left). The black boxes refer to the locations of landslides addressed in the following sections

settlements at the valley bottom (Hoch and Rücker 2016). Agricultural and recreational areas are dominant on this plain. Smaller villages (e.g., Dünserberg) and dispersed single farms are located on the south-exposed slopes of the Walserkamm and connected by a dense network of roads and pathways (Fig. 15.2).

The prevalent Alpine pastures and meadows on the hillsides of the Walserkamm reflect the high importance of dairy cattle farming for the regional agro-economy. However, timber harvesting is in direct competition with areas used for dairy farming. Thus, a considerable portion of the slopes is also covered by managed coniferous and mixed forests (Schmaltz et al. 2017).

Smaller landslides occurring on the Walserkamm are more likely to interfere with infrastructure compared to the less-populated and mainly forested hillsides to the south of the valley. It is evident that the land cover of the region is closely related to its socio-economic setting and determines the exposure of elements at risk (e.g., houses, roads) potentially endangered by landslide events (Glade et al. 2005).

At the same time, land cover is also known to influence geomorphic processes and details on how vegetation affects slope stability are of high relevance to understand hillslope dynamics (Glade 2003; Marston 2010; Promper et al. 2015). In particular, woody vegetation is expected to reduce the tendency of a slope to fail due to hydrological and geomechanical effects (Sidle and Ochiai 2006), such as root reinforcement, and the ability of woody vegetation to extract water from the subsurface (Montgomery et al. 2000; Marston 2010). On the other hand, intensive grazing may contribute to an increased soil compaction preventing water from infiltrating into the soil (Markart et al. 2007).

In regions, where both branches (i.e., forestry and cattle farming) are of high economic importance, such as in the northern part of the Walgau (Fig. 15.2), human induced land cover changes were recognized during the last decades (Fig. 15.3). These changes are expected to influence the occurrence of shallow landslides in the Walgau region (Schmaltz et al. 2017).

Recent land use and climate change may increase the frequency of small landslide events, however, it is still under debate which factor has the higher influence on landslide occurrence (Glade et al. 2014; Promper et al. 2014). Since the land cover of an area can be managed directly by humans and is simultaneously considered to influence both, the landslide process itself, as well as potential consequences, hazard mitigation strategies focusing on land use management appear promising to prevent undesirable developments (Glade et al. 2005; Papathoma-Köhle and Glade 2013; Promper et al. 2015).

Fig. 15.2 Western part of the Walserkamm (cf. Fig. 15.1) with its patchy land cover distribution. The shown hillslopes are developed in Flysch, widely covered by glacial deposits, and are regularly affected by shallow landsliding (cf. Sect. 4.2). The red polygon refers to the location depicted in Fig. 15.12. Photo: E. Schmaltz

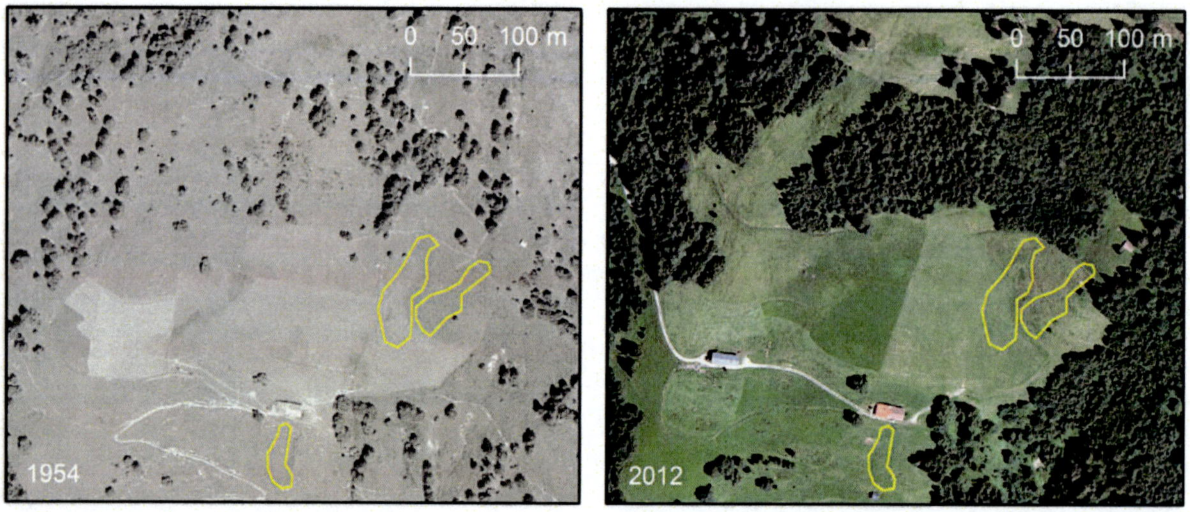

Fig. 15.3 Orthophotos (left: 1954, right: 2012) depicting considerable reforestation near Montanast on the Walserkamm, Vorarlberg. Note that the yellow polygons delimit landslides that occurred in 2005. By now, corresponding geomorphic features are not clearly visible in the field, mainly because of anthropogenic modifications such as slope leveling (Source of the orthophotos: © Land Vorarlberg)

15.2.3 Climatic Conditions

The climatic conditions in Vorarlberg favor landslide processes. The Walgau receives moisture throughout the year, in winter often as snow, in spring, summer, and autumn as rainfall (Fig. 15.4). In particular, the snowmelt season leads to an increase of soil moisture across the region and has to be regarded as an important pre-conditioning factor for landslide initiation. The regularity of these conditions can be depicted from the annual number of days with snow cover, which are 40 days with a snow cover depth of more than 1 cm in Feldkirch (438 m asl) and 93 days in Vandans (670 m asl) in the period 1981–2010. The actual triggering of landslides is, however, mostly associated with thunderstorm precipitation. The mean annual number of days with thunderstorms between 1981 and 2010 is 30 in Feldkirch

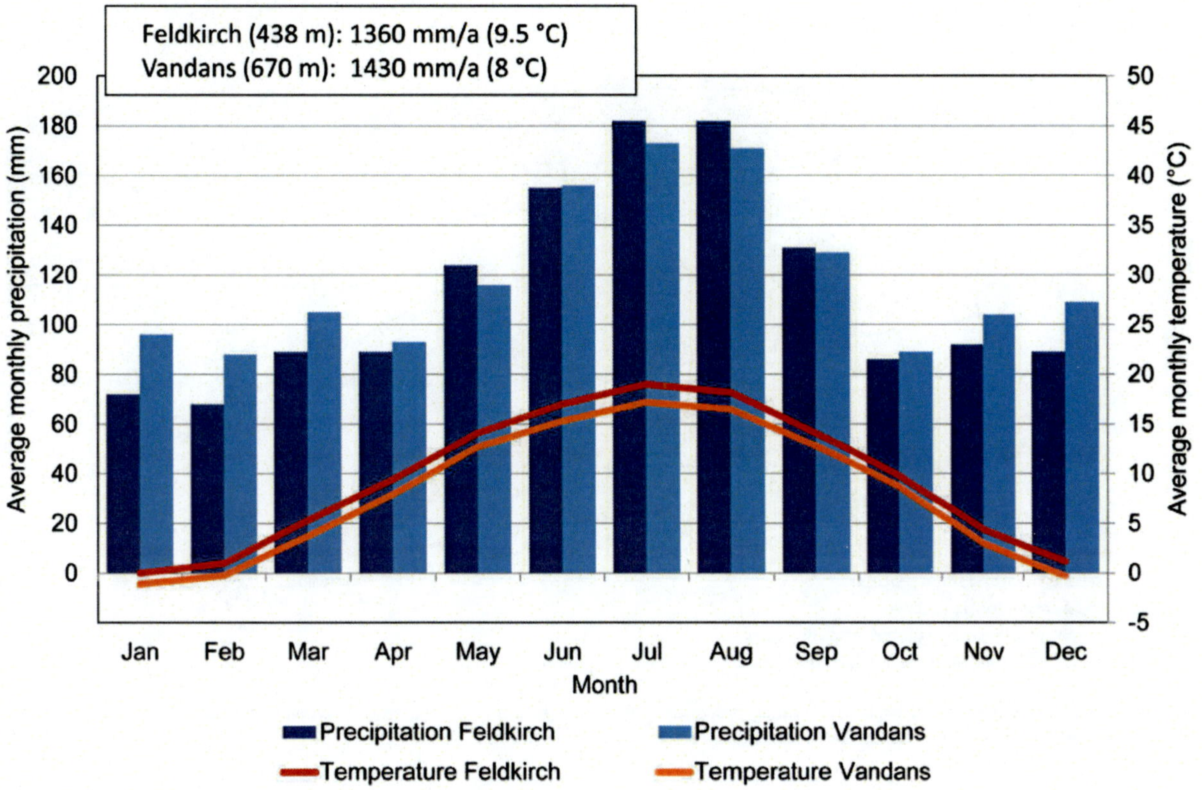

Fig. 15.4 Climate diagram of two stations representing the general climatic conditions (1981–2010) of the area. The two stations, Feldkirch and Vandans, are located close to the Walgau Valley (Data source: © ZAMG)

and 21 in Vandans. If a snowmelt period is followed by thunderstorms with strong precipitation, optimal landslide-triggering conditions are met (Zieher et al. 2016; Schmaltz et al. 2017).

15.3 Landscape Development and the Role of Landslides in the Walgau

The present landscape of Vorarlberg is the result of changing climate conditions since the LGM, which affected the relative intensity of geomorphological processes over time (Fig. 15.5). During maximum glaciation, the majority of the landscape was prone to subglacial and ice-marginal glacial processes. Fluvial deposition and ice-marginal glaciofluvial processes were widespread in Northern Vorarlberg and Southern Germany.

At the end of glaciation fluvial and slope processes became dominant in the valleys. During this period, optimal conditions for the development of low frequency, large deep-seated landslides occurred. Valley slopes were strongly eroded and steepened by glacial scour, while the loss of glacier mass promoted "tensional rebound." With a slight temporal lag, the permafrost disappeared from the slopes and increased their proneness to deep-seated landsliding. As major valleys were stripped from their valley bottom sediments, huge lakes formed in the Rhine (~500 m deep between Bregenz and Sargans) and Ill valleys (~200 m deep between Feldkirch and Bludenz). The combination of raised pore water pressure, slope steepening, and loss of support promoted deep-seated landslides on adjacent slopes (De Graaff et al. 1987), such as that of Flims and Tamins, of which sedimentological signatures were detected in Lake Constance (Schneider et al. 2004).

In the following non-glaciated period, the major valleys such as the Rhine and Ill valleys became filled with sediments which stabilized the adjacent slopes. Fluvial drainage patterns, which were disturbed during the glacial period, reestablished, while regrowth of vegetation as well as the development of an extensive soil cover lowered the probability of large deep-seated landslide events.

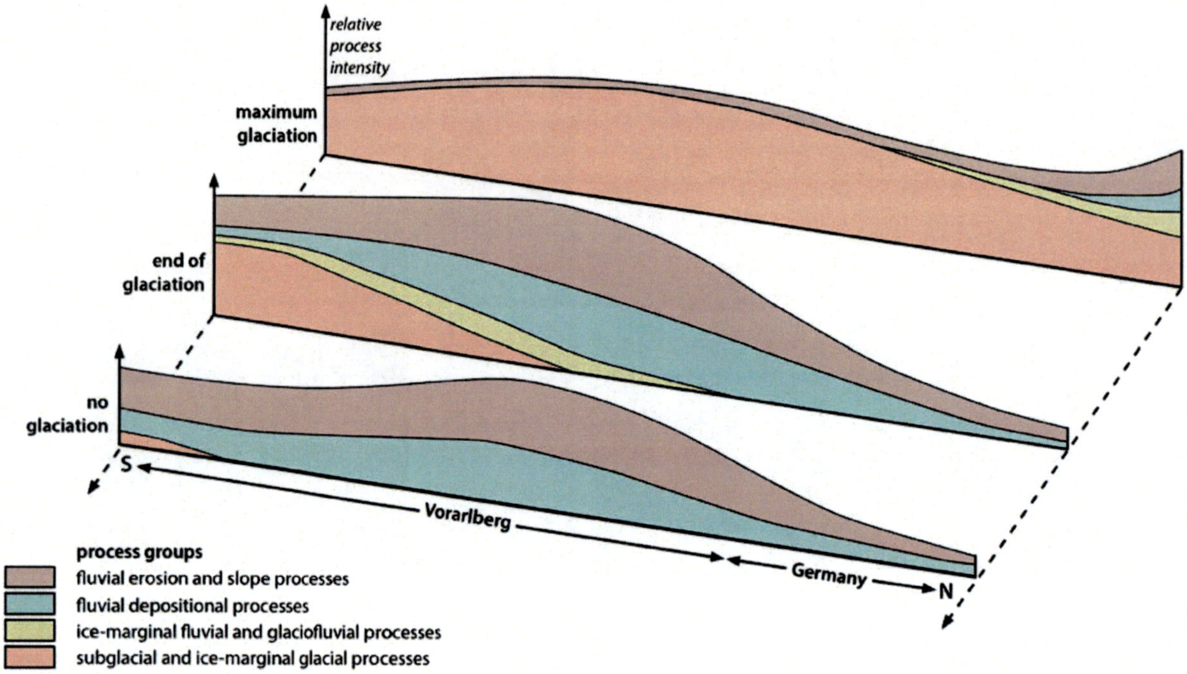

Fig. 15.5 Relative importance of landscape forming processes (y-axis) as a function of location (x-axis) and time (dashed line, depth-axis) in Vorarlberg and the adjacent German Alpine foreland (after De Graaff et al. 1987)

15.4 Historic and Recent Landslide Processes

Landslides are an important and dominant part of the natural scenery of the Walgau region. The following sections highlight examples of high-magnitude landslides (4.1), frequently occurring smaller slope movements (4.2), and the related multi-dimensional aspects (4.3) in this area.

15.4.1 Large Scale Landslides

In Vorarlberg many fossil, dormant and active large landslides have been recognized (Seijmonsbergen et al. 2005; Ruff and Czurda 2008a, b). Some developed already before the last glaciation, while others were active before or during the Lateglacial period (Seijmonsbergen et al. 2014; Ruff and Czurda 2008a, b). Active and relatively large landslides occurred near the villages of Doren (active periods since 1935), Riefensberg in 1984 and 1988, Rindsberg (Sibrätsgfall) in 1999 and Ebnit (long-term activity) (AdaptAlp 2010; Depenthal and Schmitt 2003; Jaritz and Marte 2013; Lindenmaier et al. 2008). In order to illustrate the variety of relatively large and deep-seated landslides in the Walgau area, we present four examples, which originate from different environmental conditions.

The first example portrays a deep-seated, dormant landslide located in the Flysch Zone around Frastanz (cf. Fig. 15.1). Corresponding geomorphic features, such as the large landslide deposits or the scar area, are still discernible in the landscape and occur between 475 and 1100 m asl (Figs. 6b and 7).

A comprehensive symbol-based geomorphic map (Fig. 15.7) specifically depicts individual landforms and their related sediments and processes within the wider surroundings of this landslide and allows to interpret the relative timing and former conditions under which those slope movement may have occurred (Seijmonsbergen 1992; Gustavsson et al. 2006). This deep-seated landslide as well as the surrounding area have been covered by a variety of Quaternary deposits, which form a key location for reconstructing the deglaciation history of Vorarlberg (De Graaff et al. 2007). Four landslide bodies (B1–B4) have been delineated, as have their corresponding scars. Field inspection of the landslide bodies revealed that they are partly overlain by subglacial till and ice-marginal fluvial deposits. The "dry valley" of Gampelün (code 1222) reflects a phase of the deglaciation in the Walgau when the Ill Glacier blocked the ice-free outlets of the Galina and Meng valleys. Their rivers were forced to follow an ice-marginal course along the Ill Glacier toward the northwest. Smaller Holocene streams followed the side scars of the landslide and deposited recent alluvial fans (code 2213) to the northwest of B1

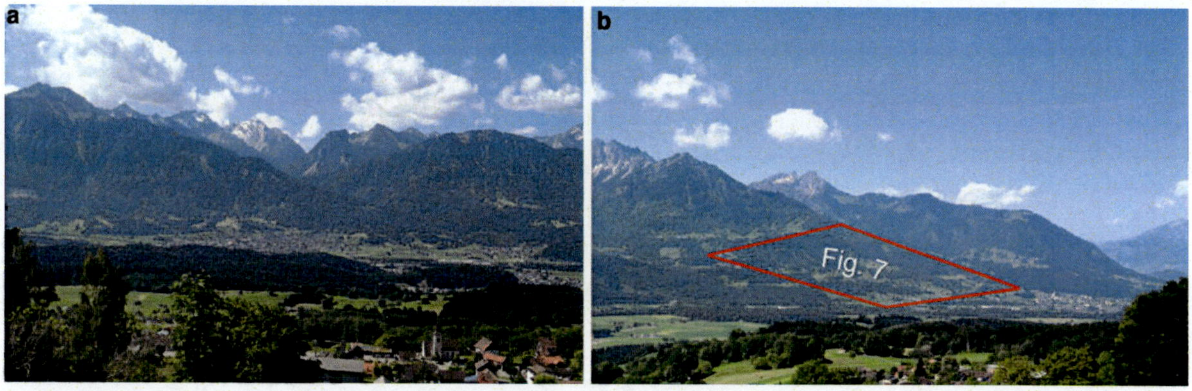

Fig. 15.6 View from Walserkamm (cf. Fig. 15.1) in south (**a**) and southwest (**b**) direction. The shown hillslopes are generally steeper and appear more rugged than on the opposite valley side (displayed in Fig. 15.2). The red polygon refers to the approximate position of the geomorphic map shown in Fig. 15.7. Photos: E. Schmaltz

and north of B4. The undisturbed terrain to the south of the landslide shows a sequence of moraine ridges and ice-marginal deposits (units 1221 and 1233), which were deposited 16,000–17,000 years ago (De Graaff et al. 2007). The terrace of Mariex in the north of the map is an ice-marginal terrace, formed during the waning stages of the Ill Glacier, which occupied the sediment free valley at that time (Seijmonsbergen 1992).

The second example for large-scale landslides is located northwest of the village of Satteins (cf. Fig. 15.1). The area is underlain by resistant folded Helvetic Schratten Limestone on top of non-resistant Drusberg marls. Large limestone blocks are dislocated on top of the marls and are separated by numerous tension fissures. Community archives provide data for a medieval local rock slide event at "Spiegelstein," which was initially detached from the steep cliffs between the years 1361 and 1371 at the onset of the Little Ice Age. The lake "Schwarzer See" was subsequently dammed by these landslide deposits. Recently, fallen blocks, freshly exposed bedrock surfaces, and tilted trees indicate that this landslide is still active. Geomorphic features like the scars, geological lineaments, and the dammed lake are the legacy of this major event (Fig. 15.8).

The third example for historic landslides of the region is a large landslide, dominated by slide-type movements that developed in the Gamp Valley (Fig. 15.9). The lower half of the southern valley slope is a tectonic half-window that exposes rocks of the Penninic Arosa Zone below the Austroalpine "Hauptdolomite" and gypsum bearing Raibler formations, separated by a tectonic thrust zone (Fig. 15.9b). Along the upper fault, a major approximately 50 m high landslide scar is exposed over a length of 2 km. Another striking landform is a collapse doline (see legend of the geomorphic map in Fig. 15.9a) which is 100 m in diameter and 50 m deep, located at approximately 1900 m asl on the water divide. This collapse doline is attributed to subsurface dissolution of thick gypsum layers, which have been stacked along tectonic thrust zones. Roof collapse resulted in a large depression (Seijmonsbergen 1992). The interpreted subsurface conditions have been indicated in the geological cross section in Fig. 15.9b.

Fourthly, near the village of Bürs, fissures up to 10 m wide, 500 m long, and over 100 m deep have developed in the Quaternary "Bürs" conglomerates near the mouth of the Alvier River. The conglomerates are interpreted as relicts of a cemented and raised alluvial fan which was deposited against the Ill Glacier (De Graaff and Seijmonsbergen 1993). Cementation of the conglomerate has been dated at 128 $\pm$ 10 ka by Ostermann et al. (2006). Locally, they are overlain by Lateglacial ablation till of the Ill Glacier and underlain by subglacial till (Ampferer 1908) and a palaeo bedrock surface developed in Raibler Formation. The tension fissures developed along two sets of joints, which are only observed on a 50 cm resolution 3D hill-shaded terrain surface from which the tree cover has been filtered (Fig. 15.10a).

15.4.2 Recent Shallow Landslides at the Walserkamm

Several areas in Vorarlberg, including the slopes of the Walserkamm, recently suffered from shallow slope movements. Considering the observed damages and aftermaths caused by these events, it appears that regions exhibiting complex hydrogeological conditions and recent anthropogenic transformations (e.g., land cover changes) were especially affected by these shallow landslides (Ruff 2008a; b; Haas et al. 2015; Schmaltz et al. 2016; Steger et al. 2020).

In general, the south-exposed slopes of Walserkamm show a lower inclination, and presumably more deeply weathered soils compared to their north-facing counterparts

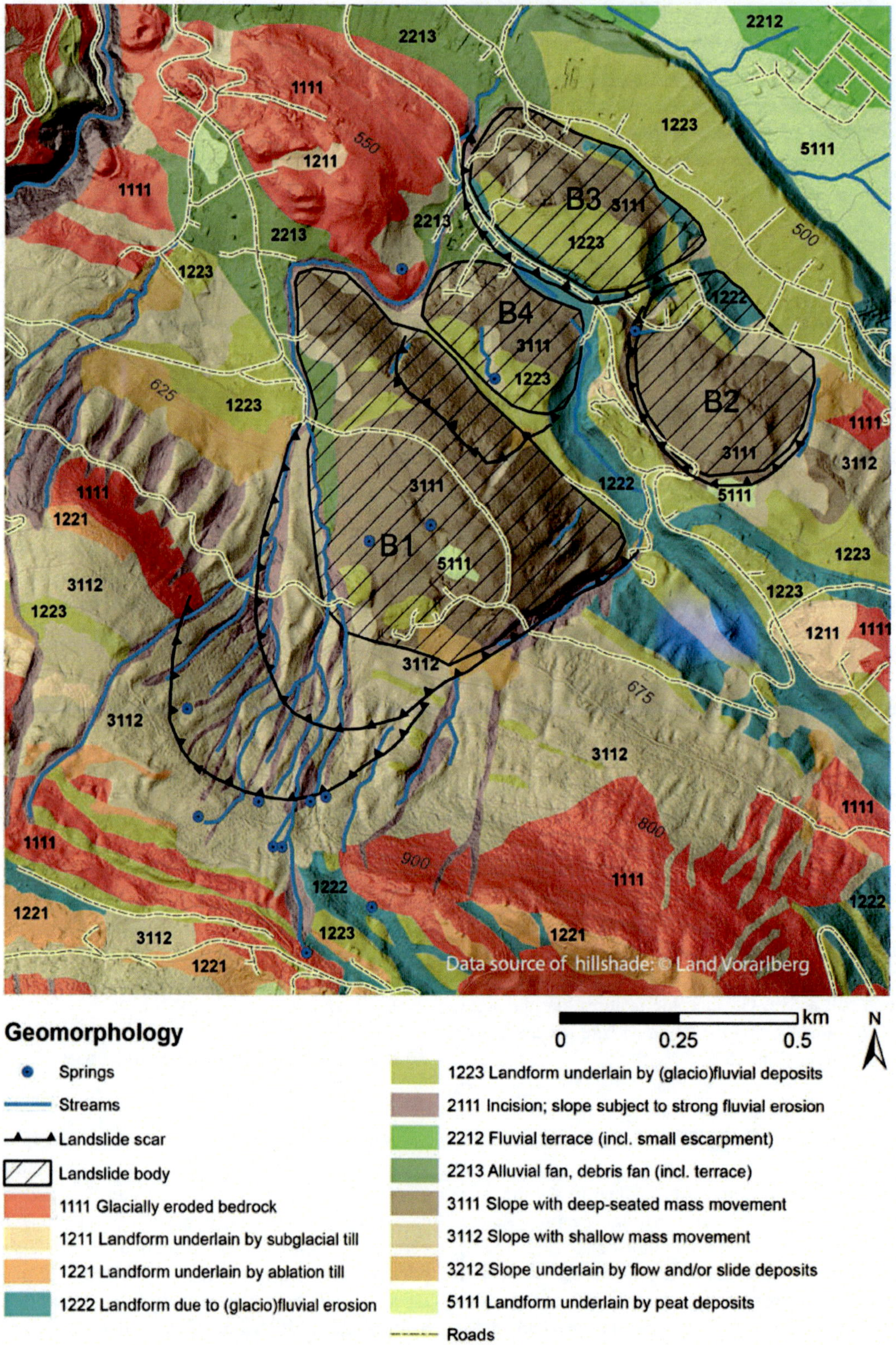

Fig. 15.7 Excerpt of the geomorphic map (1:10,000) sheet Gurtis showing the area south of Frastanz, southern Walgau, after Seijmonsbergen 1992 (location cf. Fig. 15.1). Landslide deposits of deep-seated landslides are denoted with B1–B4

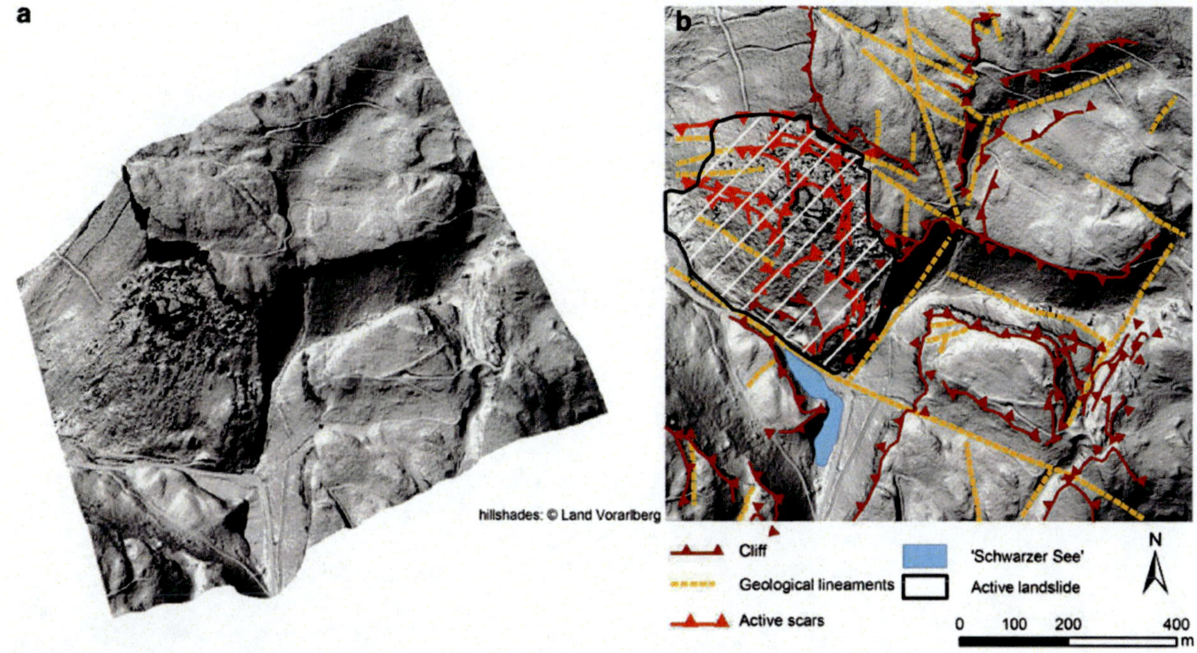

Fig. 15.8 Spiegelstein landslide and surroundings as visible on a hillshade presented in 3D (**a**) and with a plan view (**b**)

in the valley Laternsertal (Schmaltz et al. 2016; Zieher et al. 2016). Shallow landslides are common on these slopes (cf. examples in Fig. 15.11) and are mainly triggered by heavy rainfall events and strongly influenced by specific hydrogeological and topographical conditions, water availability, vegetation cover, and anthropogenic transformation as described in more detail in the following paragraphs (Ruff and Czurda 2008a; b; Schmaltz et al. 2017).

The hydrogeological conditions of the Walserkamm are characterized by the presence of the Flysch. Particularly, formations with a higher content of marl, like the Piesenkopf Formation, appear to be prone to slope failure under saturated or nearly saturated conditions (Seijmonsbergen 1992). A high amount of infiltrated water, for instance during snowmelt, leads to an increased water-influx into the bedrock material. This might cause a hydrological recharging effect of the deep-seated hydrogeological system (Geologist Walter Bauer, personal communication, October 14, 2015). Thus, the typical spring combination of high saturation of the underlying material combined with storm precipitation effectively promotes shallow slope failures (Haas et al. 2015; Zieher et al. 2016).

Comparing the distribution of vegetation with landslide occurrence, we found evidence that under similar topographical and lithological conditions, forested slopes of the Walserkamm appear to be less susceptible to shallow slope movements than their non-forested counterparts (Schmaltz et al. 2016, 2017). This was attributed to the rooting system, which contributes to an increased soil reinforcement and a decreased soil moisture (Sidle and Ochiai 2006; Meng et al. 2012). Additionally, trees are also known to promote evapotranspiration, while a considerable portion of rainfall is intercepted by the forest canopy (Ghestem et al. 2011; Meng et al. 2012). However, under certain conditions, trees might also contribute to a decreased slope stability as their roots can disturb the soil structure leading to higher infiltration capacities, while the weight of trees may add to increased downslope forces on steep terrain (Crozier 1989; Rickli et al. 2002; Sidle and Ochiai 2006; Marston 2010; Ghestem et al. 2011). Currently, ongoing investigations are aimed at gaining deeper insight into the linkage between vegetation cover and slope stability at the Walserkamm (Steger et al. 2015). Field observations by Markart et al. (2007) showed that simple correlations between landslide occurrences and vegetation types may not exist for the area due to a large variety of cultivation practices prevalent within each vegetation type.

Many areas of the south-exposed slopes are used for agriculture and therefore affected by anthropogenic transformations. Drainage channels were constructed in many areas to decrease soil moisture content and increase the stability of slopes. However, human interventions might also have a strong negative effect on slope stability. For instance, during the 2005 event, numerous landslides were triggered on or close to roads (Markart et al. 2007). Thus, we conclude that landslide activity on the Walserkamm is strongly related to an interplay of natural and anthropogenic factors.

Numerous shallow landslides during and after severe rainfall events in May 1999 and August 2005 are related to

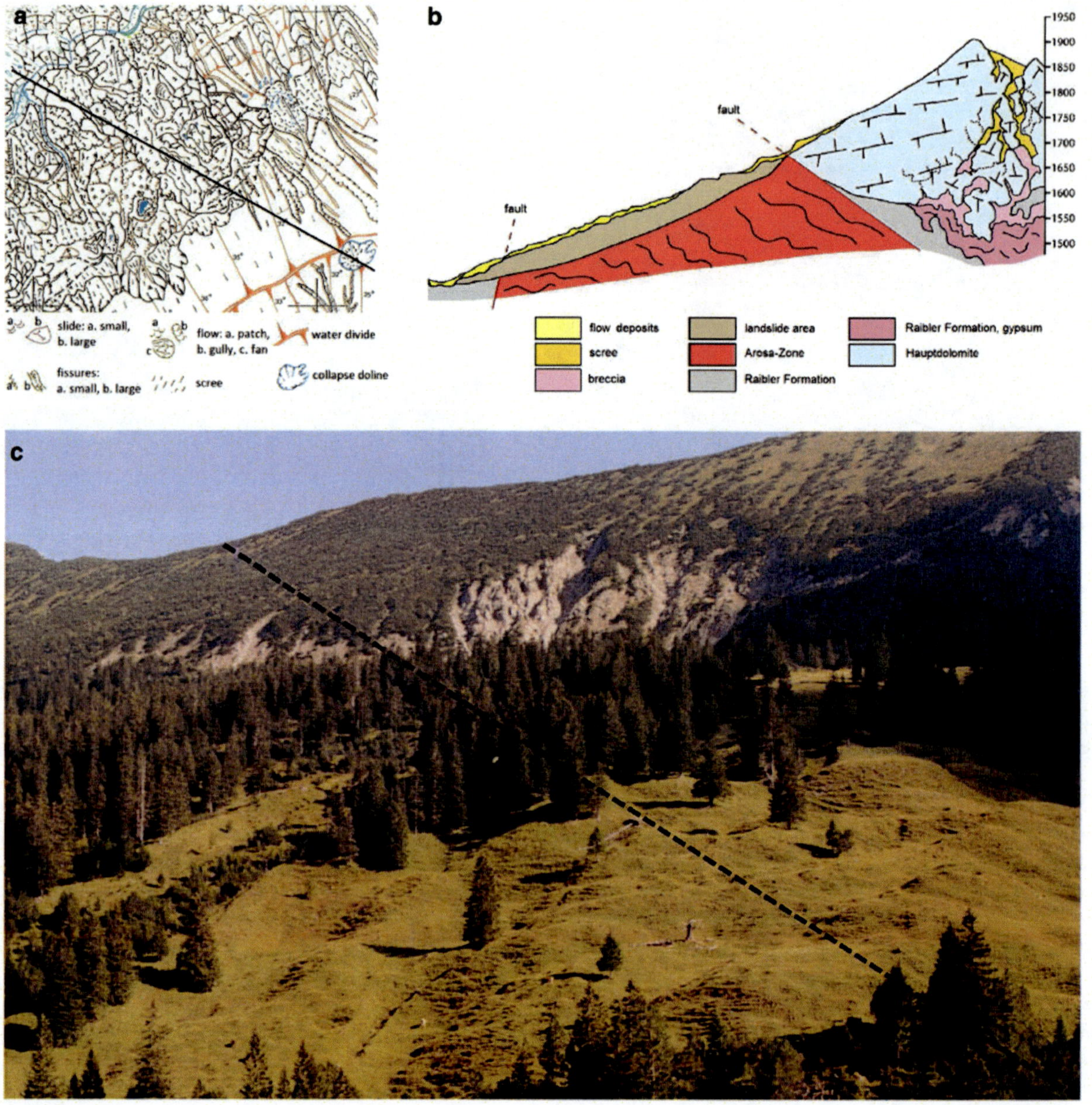

Fig. 15.9 Excerpt of a geomorphic map (**a**), geological cross section of the landslide area in the Gamp Valley (**b**), and corresponding photograph (**c**) (after Seijmonsbergen 1992). Photo: A. C. Seijmonsbergen

hydrometeorological conditions in the weeks before the respective landslide events. For instance, a large amount of snow was registered for the winter of 1998/99 with a mean temperature around the freezing point, leading to frequent freeze–thaw events. This promoted steady water infiltration and thus an increase of soil moisture content. Heavy rainfalls in the snowmelt season around the 22nd of May caused floods and triggered numerous shallow landslides (Haas et al. 2015). In August 2005, heavy rainfalls (Zieher et al. 2016; Haas et al. 2015) triggered new landslides in the areas of Montanast (Dünserberg, Fig. 15.1) and promoted reactivation of old landslides near Düns (Fig. 15.12). These rainfall events led to an oversaturation of the upper soil layers inducing consistency changes and shallow slope movements of the marl-rich material. In contrast to the event of 1999, the deeper subsurface was not affected by saturation and remained

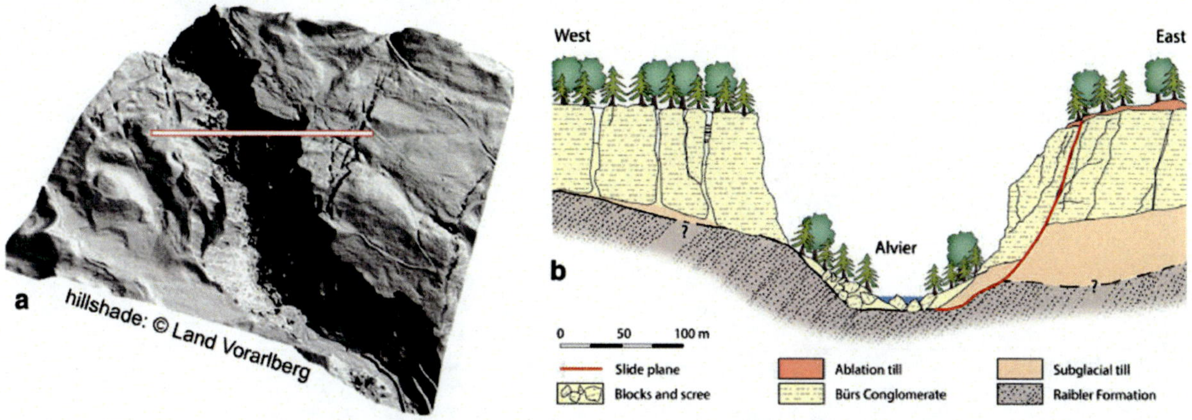

Fig. 15.10 3D-Bird's eye view (**a**) with location of cross section (**b**) indicated by a colored line. The cross section (**b**) of the lower Alvier River gorge near Bludenz shows tensional fissures developed in the Bürs conglomerates. Approximate height difference between the Alvier River and the upper conglomerate level is 80 m (after de Graaf and Seijmonsbergen 1993)

Fig. 15.11 Examples of typical shallow translational landslides on the Walserkamm. Landslide deposit northeast of Düns (cf. Fig. 15.12). Scars of shallow landslides on the steeper upper part of the hillside (**b**, **c**). Photos: E. Schmaltz

stable. Several landslides in this area are aligned along a transition between a steeper upper part and a flatter slope section (cf. potential lineament in Fig. 15.12). This spatial distribution supports the assumption of a locally strong relationship between landslide occurrence and hydrogeologically active lineaments. The drainage systems installed downslope of this transition may contribute to a confinement of the areal extent of these shallow movements. However, after the snowy winter of 2005/06, several parts of this area became unstable again and began to move (Geologist Walter Bauer, personal communication, October 14, 2015).

15.4.3 Landslides Within a Multi-dimensional Perspective

Landslides are a major agent within the landscape dynamics and the past, current, and future landscape development of the Walgau. But due to a number of reasons, great care is necessary when analyzing the causes of slope failure.

Firstly, landslides occur frequently over time, but with varying magnitudes and response times between trigger and event. For example, while large-scale and deep-seated landslides are prepared by long-term changes in

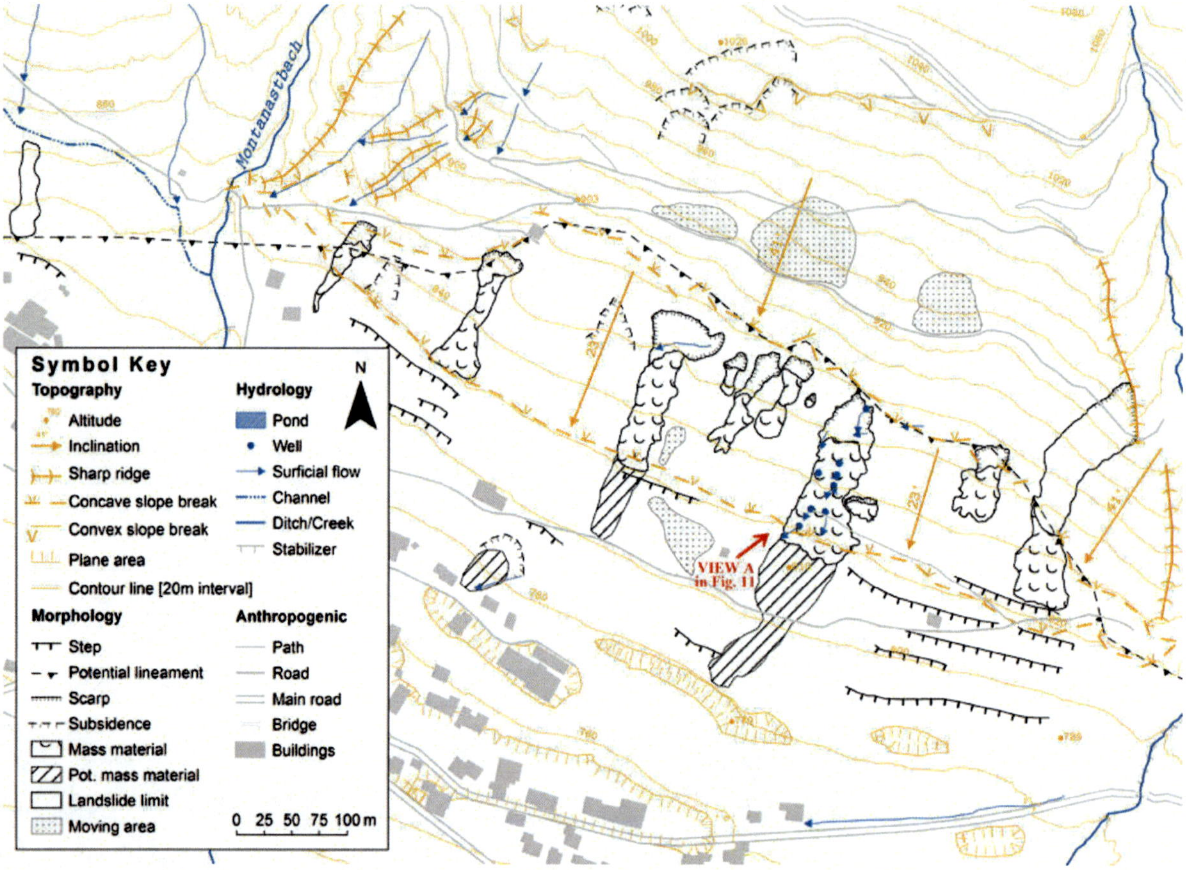

Fig. 15.12 Multiple landslides on the lower slopes of the Walserkamm. A complex hydrogeological setting as well as an alternation between marly sandstone flysch, limestones, and morainic sediments appear to promote translational landslides in the shallow soil mantle (after Schmaltz et al. 2016)

hydrogeological or climatic conditions, often related to glaciations (as explained in Sect. 4.1), small and shallow translational landslides respond quickly to intensive thunderstorm cells with high precipitation inputs as triggers (as shown in Sect. 4.2) (Zieher et al. 2016). In this case, landslide initiation can be related to an evident input. However, large landslides show long response times with possible large lag times between the changes of predisposing and triggering factors and the actual slope failure.

Secondly, one has to differentiate between first-time movements and reactivations. Most small landslides can be classified as first-time landslides which remove the available landslide mass from the slope system. In many cases, the material is delivered to the hydrological system where it is further distributed through debris flows or as sediment load in streams and rivers. In contrast to small landslides, large failures show different dynamics. Indeed, these large movements can occur instantaneously in the form of a "Bergsturz" (e.g., from Spiegelstein, Vorarlberg) (Oberhauser 1984). But many of these large slope movements creep with varying intensities for decades or even centuries (Czurda et al. 1983; Völk 2001), while these movements regularly do not involve the whole landslide mass, but only reactivated parts.

Thirdly, landslides might be nested and accordingly react to different landslide thresholds. On top of large-scale and creeping landslides, small landslides might occur regularly. It is obvious that these different landslides occur on a variety of spatial scales and are determined by different preparatory and triggering factors. However, they are all part of the landscape dynamics.

15.5 Future Challenges

15.5.1 Climate and Environmental Changes

The impact of environmental and climate changes on landslides is not only a global phenomenon (Glade and Crozier 2010; Promper et al. 2014, 2015). These changes are also recognizable in the regional conditions of Vorarlberg (Kromp-Kolb et al. 2014) where a increase of heavy rainfall events in the last 50 years was observed (Haas et al. 2015). Generally, the amount of rainfall and the intensity increased in winter, while summers became more and more dry (Matulla et al. 2004), with direct effects on certain geomorphic processes, particularly during snowmelt in spring. For example, the communities of Vorarlberg had to face more frequent and intense rainfalls in the last years (e.g., in 1999 and 2005) with increased damage, especially to the infrastructure (Haas et al. 2015). On the other hand, Orthophotos, since the early 1950s, provide evidence of a slowly rising timberline in parts of the region. This in turn may decrease erosional processes and slope movements in the upper parts of the Alpine area.

Strategies to cope with the anticipated hydroclimatic changes are recently under development (Haas et al. 2015). In the Eastern Alps, for instance, the establishment of protected areas and a reduced intensity of timber harvesting as one option within forest management may become more common (Slaymaker and Embleton-Hamann 2009) and lead to increased slope stability.

15.5.2 Human Impact and Socio-Economic Developments

Besides the already highlighted climate and environmental changes, humans modify the landscape to a great extent. For instance, the geomorphology of a populated area is regularly reshaped for agricultural purposes or for designing better building sites. Further direct interventions include the building of transport corridors, water and sewage pipelines, drainage systems, and reservoirs, the consequence of which is a changed slope hydrology. Agricultural use not only involves changes of the vegetation cover, but also the application of fertilizers, which affects the geophysical bonding of sediment particles (Glade 2003).

It is estimated that the population of the region Feldkirch will increase by 16.2% within the period 2010–2050, which will then increase the pressure on the environment through suburbanization (Hanika 2010). All these scenarios are expected to considerably intensify the exposure of humans and their properties to hazardous phenomena, but especially in the lower slope sections of the Walgau. Finally, we argue that a multi-dimensional evaluation of past, current, expected future environmental, and socio-economic developments is crucial for developing sustainable strategies to avoid future damages.

One day field trip: Several landslides presented within this chapter can be visited during a one day trip (for orientation refer to Fig. 15.1). When traveling by car from Feldkirch, we recommend visiting the landslide-dammed Schwarzer See near Satteins (Fig. 15.8). Afterward, slow moving shallow landslides along the described hydrogeological lineament can be seen northeast of Düns (Fig. 15.12). You can continue the trip to the top of the Walserkamm ("Dünser Älpele") for a panorama view over the entire Walgau. Several deep-seated landslides (e.g., Fig. 15.6b) can be seen from this position. From here, we recommend a walk (c.10 min in northeast direction) to the spectacular "Doppelgrat" (cf. Fig. 15.1) that relates to a large deep-seated landslide and contains several shallow and active slope movements.

References

AdaptAlp (2010) Working package 6 (WP6): platform for practitioners exchange program. http://www.alpine-space.org/2007-2013/uploads/tx_txrunningprojects/WP_Final_Report_02.pdf. Accessed on 20 July 2016

Ampferer O (1908) Glazialgeologische Beobachtungen in der Umgebung von Bludenz. Yearbook of the Geological Survey of Austria (GBA), 58/3

Bell R, Glade T (2004) Quantitative risk analysis for landslides—examples from Bíldudalur, NW-Iceland. Nat Hazard Earth Syst Sci 4(1):117–131

Bierman PR, Montgomery DR (2014) Key concepts in geomorphology. WH Freeman and Co Ltd., New York, p 500

Crozier MJ (1989) Landslides: causes, consequences and environment. Routledge, London/New York, p 272

Crozier MJ (2010) Landslide geomorphology: an argument for recognition, with examples from New Zealand. Geomorphology 120:3–15

Cruden DM, Varnes DJ (1996) Landslides types and processes. TRB special report. Nat Acad Sci 247:36–75

Czurda K, Jesinger, D (1983) Bergzerreißung und Talzuschub an der Flysch/Kalkalpen- Grenze (Hinteres Großes Walsertal, Vorarlberg). Jb. Vorarlbg. Landesmuseumsver. 125, Bregenz, pp 31–44

De Graaff LWS, De Jong MGG, Rupke J, Verhofstad J (1987) A geomorphological mapping system at scale 1:10 000 for mountainous areas (Austria). Z Geomorphol 31:229–242

De Graaff LWS (1992) Zur Morpho- und Chronostratigraphie des oberen Würm in Vorarlberg. Jahrb Geol Bundesanst 135:809–824

De Graaff LWS, Seijmonsbergen AC (1993) Die eiszeitliche Prozessfolge und Aspekte der jungquartären Talbildung und Hangentwicklung im Walgau (Exkursion D am April 1993). Jber Mitt Oberrhein Geol Ver 75:99–125

De Graaff LWS, De Jong MGG, Seijmonsbergen AC (2007) Landschaftsentwicklung und Quartär. In: Friebe JG (ed), Vorarlberg. Geologie der Österreichischen Bundesländer. Geological Survey of Austria (GBA), Vienna, Austria, pp 21–32

Depenthal C, Schmitt G (2003) Monitoring of a landslide in Vorarlberg/Austria. In: Proceedings of international FIG symposium on deformation measurements, 11th, Santorini (Thera) Island, Greece, pp 25–28

Friebe G (2004) Zur Geologie Vorarlbergs – eine Einführung unter besonderer Berücksichtigung verkarstungsfähiger Gesteine. – Vorarlberger Naturschau - forschen und entdecken. Dornbirn 15:19–39

Friebe G (2007) Geologie der Österreichischen Bundesländer Vorarlberg. Geologische Bundesanstalt, Wien, p 154

Ghestem M, Sidle RC, Stokes A (2011) The influence of plant root systems on subsurface flow: implications for slope stability. Bioscience 61:869–879

Glade T (2003) Landslide occurrence as a response to land use change: a review of evidence from New Zealand. CATENA 51:297–314

Glade T, Anderson M, Crozier MJ (eds) (2005) Landslide hazard and risk: issues, concepts and approach. Wiley Ltd, p 824. https://doi.org/10.1002/9780470012659.ch1

Glade T, Bell R, Dobesberger P, Embleton-Hamann C, Fromm R, Fuchs S, Hagen K, Hübl J, Lieb G, Otto JC, Perzl F, Peticzka R, Prager C, Samimi C, Sass O, Schöner W, Schröter D, Schrott L, Zangerl C, Zeidler A (2014) Der Einfluss des Klimawandels auf die Reliefsphäre. In: APCC (eds) Österreichischer Sachstandsbericht Klimawandel 2014 (AAR14). Austrian Panel on Climate Change (APCC), Verlag der Österreichischen Akademie der Wissenschaften, Wien, Österreich, pp 557–600

Glade T, Crozier MJ (2010) Landslide geomorphology in a changing environment. Geomorphology 120:1–2

Glade T, Kadereit A, Dikau R (2001) Landslides at the Tertiary escarpments of Rheinhessen. Germany. Zeitschrift Für Geomorphologie Suppl. 125:65–92

Grelle G, Revellino P, Donnarumma A, Guadagno FM (2011) Bedding control on landslides: a methodological approach for computer-aided mapping analysis. Nat Hazards Earth Syst Sci 11:1395–1409. https://doi.org/10.5194/nhess-11-1395-2011

Gustavsson M, Kolstrup E, Seijmonsbergen AC (2006) A new symbol-and-GIS based detailed geomorphological mapping system: renewal of a scientific discipline for understanding landscape development. Geomorphology 77:90–111

Haas W, König M, Pech M, Prettenthaler F, Prutsch A, Steininger KW, Themessl M, Wagner G, Wolf A (2015) Die Folgeschäden des Klimawandels in Österreich. Dimensionen unserer Zukunft in zehn Bildern für Österreich. Sonderheft des Klimaenergiefonds, p 51

Hanika A (2010) Kleinräumige Bevölkerungsprognose für Österreich 2010–2030 mit Ausblick bis 2050 ("ÖROK-Prognosen"). Endbericht zur Bevölkerungsprognose. Österreichische Raumordnungskommission (ÖROK), vol 1

Heissel W, Oberhauser R, Schmidegg O (1967): Geologische Karte des Walgaus 1:25,000. Wien (Geol B-A)

Hoch I, Rücker E (2016) Regionale Bevölkerungsprognose 2015 bis 2050. Amt der Vorarlberger Landesregierung, Landesstelle für Statistik, p 52

Jaritz W, Marte R (2013) 75 Jahre Sanierung Großhangbewegung Doren (Vorarlberg) - Ungleiches Ringen zwischen Natur und Technik? 19. Tagung für Ingenieurgeologie mit Forum für junge Ingenieurgeologen, München, pp 155–160

Kromp-Kolb H, Nakicenovic N, Seidl R, Steininger K, Ahrens B, Auer I, Baumgarten A, Bednar-Friedl B, Eitzinger J, Foelsche U, Formayer H, Geitner C, Glade T, Gobiet A, Grabherr G, Haas R, Haberl H, Haimberger L, Hitzenberger R, König M, Köppl A, Lexner M, Loibl W, Molitor R, Moshammer H, Nachtnebel HP, Prettenthaler F, Rabitsch W, Radunsky K, Schneider J, Schnitzer H, Schöner W, Schulz N, Seibert P, Stagl S, Steiger R, Stötter J, Streicher W, Winiwarter W (2014) Österreichischer Sachstandsbericht Klimawandel 2014: Synthese. Verlag der österreichischen Akademie der Wissenschaften, pp 67–131

Lindenmaier F, Zehe E, Wienhöfer J, Ihringer J (2008) Hydrological patterns and processes of a deep seated creeping slope at Ebnit, Vorarlberg. - Interpraevent 2008. Conf Proc 2:183–194

Markart G, Perzl B, Kohl B, Luizian R, Kleemayr K, Ess B, Mayerl J (2007) 22. und 23. August 2005—analyse von Hochwasser-und Rutschungsereignissen in ausgewählten Gemeinden Vorarlbergs. Naturgefahren und Landschaft. BFW-Dokumentation. Schriftenreihe des Bundesforschungs- und Ausbildungszentrums für Wald, Naturgefahren und Landschaft, Vienna, 5:48

Marston RA (2010) Geomorphology and vegetation on hillslopes: interactions, dependencies, and feedback loops. Geomorphology 116:206–217. https://doi.org/10.1016/j.geomorph.2009.09.028

Matulla C, Formayer H, Haas P, Kromp-Kolb H (2004) Possible climate trends in Austria in the first half of the 21st century. OeWAW 56:1–9

Meng W, Bogaard TA, Van Beek LPH (2012) The long-term hydrological effect of forest stands on the stability of slopes. In: Eberhardt et al (eds) Landslides and engineered slopes: protecting society through improved understanding, Taylor & Francis Group. London, pp 1631–1636

Montgomery DR, Schmidt KM, Greenberg HM, Dietrich WE (2000) Forest clearing and regional landsliding. Geology 28(4):311–314

Oberhauser R (1984) Exkursionsführer: Wandertagung 1986 der Österreichischen Geologischen Gesellschaft in Dornbirn mit Exkursionen in Vorarlberg und Tirol sowie mit Übertritten in die Schweiz und nach Liechtenstein. Geol Surv Austria (GBA), Vienna, p 130

Ostermann M, Sanders D, Kramers J (2006) 230 234 Th/U ages of calcite cements of the proglacial valley fills of Gamperdona and Bürs (Riss ice age, Vorarlberg, Austria): geological implications. Austrian J Earth Sci 99:31–41

Papathoma-Köhle M, Glade T (2013) The role of vegetation cover change for landslide hazard and risk. In: Renaud G, Sudmeier-Rieux K, Estrella M (eds) The role of ecosystems in disaster risk reduction. UNU-Press, Tokyo, pp 293–320

Promper C, Puissant A, Malet J, Glade T (2014) Analysis of land cover changes in the past and the future as contribution to landslide risk scenarios. Appl Geogr 53:11–19

Promper C, Gassner C, Glade T (2015) Spatiotemporal patterns of landslide exposure—a step within future landslide risk analysis on a regional scale applied in Waidhofen/Ybbs Austria. Int J Disaster Risk Reduct 12:25–33

Rickli C, Zürcher K, Frey W, Lüscher P (2002) Wirkungen des Waldes auf oberflächennahe Rutschprozesse. Schweiz Z Forstwes 153 (11):437–445

Ruff M (2008) Sind qualitative Analysen gut genug? – Erfahrungen aus Gefährdungsanalysen für Hangrutschungen in Vorarlberg/Österreich. AGIT Proc 2008:616–625

Ruff M, Czurda K (2008) Landslide susceptibility analysis at a regional scale—a qualitative approach in the eastern alps. Interpraevent Conf Proc 2:231–242

Schweigl J, Hervás J (2009) Landslide mapping in Austria. JRC scientific and technical reports. Centre Institute for Environment and Sustainability, Luxemburg, p 65

Schmaltz E, Steger S, Bell R, Glade T, Bogaard TA, Van Beek LPH, Wang D, Hollaus M, Pfeifer N (2016) Evaluation of shallow landslides in the Northern Walgau (Austria) using morphometric analysis techniques. In: Proceedings of the forth Italian workshop on landslides 2015, Naples (in press)

Schmaltz E, Steger S, Glade T (2017) The influence of forest cover on landslide occurrence explored with spatio-temporal information. Geomorphology 290:250–264

Schneider J-L, Pollet N, Chapron E, Wessels M, Wassmer P (2004) Signature of Rhine valley sturzstrom dam failures in Holocene sediments of Lake Constance, Germany. Sed Geol 169:75–91

Seijmonsbergen AC (1992) Geomorphological evolution of an alpine area and its application to geotechnical and natural hazard appraisal in the NW. Rätikon mountains and S. Walgau (Vorarlberg, Austria). Dissertation, University of Amsterdam, p 109

Seijmonsbergen AC, Woning MP, Verhoef PNW, de Graaff LWS (2005) The failure mechanism of a Late Glacial Sturzstrom in the Subalpine Molasse (Leckner Valley, Vorarlberg, Austria). Geomorphology 66:277–286

Seijmonsbergen AC, De Jong MG, de Graaff LWS, Anders NS (2014) Geodiversität von Vorarlberg und Liechtenstein—geodiversity of Vorarlberg and Liechtenstein, 41, Haupt Verlag AG, p 304

Sidle RC, Ochiai H (2006) Landslides: processes, prediction, and land use. Am Geophys Union. Water Res Monogr 18:312

Slaymaker O, Embleton-Hamann C (2009) Mountains. In: Slaymaker O, Spencer T, Embleton-Hamann C (eds) Geomorphology and global environmental change. Cambridge, pp 38–70

Steger S, Glade T, Hollaus M, Pfeifer N, Bogaard T, Van Beek R, Bell R (2015) Quantifying the effect of biomass and its change on landslide activity at regional scale. In: EGU EGU general assembly conference abstracts, p 13587

Steger S, Schmaltz E, Glade T (2020) The (f)utility to account for pre-failure topography in data-driven landslide susceptibility modelling. Geomorphology 354. https://doi.org/10.1016/j.geomorph.2020.107041

Völk HR (2001) Geomorphologie des Kleinwalsertals und seiner Gebirgsumrahmung. Landschaftsformen zur Eiszeit und Nacheiszeit unter Einbeziehung der geologischen Verhältnisse. Vorarlberger Naturschau, Dornbirn 10:7–96

Zieher T, Perzl F, Rössel M, Rutzinger M, Meißl G, Markart G, Geitner C (2016) A multi-annual landslide inventory for the assessment of shallow landslide susceptibility—two test cases in Vorarlberg, Austria. Geomorphology 259:40–54

Stefan Steger is a senior researcher at the Institute for Earth Observation, Eurac Research (Bozen/Bolzano, Italy). He received his Ph.D. degree in physical geography from the University of Vienna for his research in the field of statistically-based modelling of landslide phenomena. His interests include the detection, monitoring and modelling of gravity-driven natural hazards.

Elmar Schmaltz is a geomorphologist at the Federal Agency for Water Management in Austria and he received his Ph.D. degree at the University of Vienna. His main research interests are parameterisation strategies of landslide models and the influence of land cover on hillslope stability. His research is mainly located in the Austrian Flysch Zone in the Eastern Alps (Vorarlberg and Lower Austria), the Apennines in Italy, the badlands of Romania and the Loess Plateau in China.

Thomas Glade is professor for Physical Geography and he leads the research group on Geomorphic Systems and Risk Research at Vienna University, Austria. His research interests are geomorphology including theory, applied research, and human impact on landscapes; disaster risk reduction including natural hazard and risk research, early warning systems and risk governance; and aspects of Global Change. His current process foci are on landslides and soil erosion with study areas in Europe (e.g., Austria, Belgium, Germany, Romania) and further afield (e.g. China, New Zealand).

16 Fluvial Geomorphology and River Restoration: Tiroler Lech Nature Park

Martin Mergili and Ruginia Duffy

Abstract

Most major Alpine rivers have undergone extensive regulation in order to gain arable land and to control flooding in Alpine communities. The Tiroler Lech River is often considered the only near-natural river landscape in the northern Alps. Although it has undergone manifold alterations since the end of the nineteenth century—later than most other Alpine rivers—it is granted sufficient space to form a braided river landscape, with huge gravel banks and ecologically valuable riverine forests along a major part of its course. Also in these sections, longitudinal and transversal manmade structures containing the river are obvious features in the landscape. Those measures have resulted in gravel deficits and subsequent incision of the river bed, increased flood risk due to increased human activities on the floodplains, and have caused a threat to ecological diversity. These ramifications were recognized towards the end of the twentieth century and were later countered by restoration efforts, such as widening of the river bed, removal of retention dams in the tributaries, reconnecting detached lateral branches with the main river course, the management of target species, and by increasing public awareness.

Keywords

Braided river • Flood protection • Near-natural landscape • River regulation • River restoration

16.1 Introduction

Rivers—whether natural or managed—shape their valleys in various ways, mainly depending on the state of sediment transport, sediment equilibrium and flow patterns, as well as topographic and geological conditions, and—in the case of managed rivers—training structures. River bed incision and mass wasting, both in hard rock and soft rock, contribute to sediment load. When sediment mobilized upstream is in excess, deposition takes place and floodplains are formed. The sediment cover on floodplains is continuously being reshaped, especially following flood events. Hence, floodplains are variable, dynamic environments where rivers meander or become braided. Braided rivers are characterized by sand and gravel bars or islands that result from the river splitting off to form channels that then join, split again and re-join. Braided river systems in a near-natural state still exist in areas of the world where (i) human influence is limited or (ii) the resources for river regulation are lacking. Figure 16.1 illustrates some examples of remaining natural or near-natural braided river systems.

In Europe, humans have impacted river ecosystems from as far back as the late Stone Age, when the runoff of large lowland rivers was influenced by changes to the landscape in the catchment area. During the Roman Times, as agriculture increased, drastic changes were observed in the structure of lowland river plains. Consequently, the strong human influence on the flora and fauna of these systems is thought to have begun long before the eighteenth century, when the major civil engineering measures first took place (Müller 1996).

M. Mergili (✉)
Institute of Geography and Regional Science, University of Graz, Heinrichstrasse 36, 8010 Graz, Austria
e-mail: martin.mergili@uni-graz.at

M. Mergili
Institute of Applied Geology, University of Natural Resources and Life Sciences (BOKU), Peter-Jordan-Straße 82, 1190 Vienna, Austria

R. Duffy
Department of Environmental Geosciences, University of Vienna, UZA 2, Althanstraße 14, 1090 Vienna, Austria
e-mail: ruginia.duffy@my.jcu.edu.au

C. Embleton-Hamann (ed.), *Landscapes and Landforms of Austria*, World Geomorphological Landscapes,
https://doi.org/10.1007/978-3-030-92815-5_16

Fig. 16.1 Natural or near-natural braided river landscapes: **a** Macaulay River, New Zealand; **b** Tagliamento River, Italy; **c** Las Vueltas River, Argentina; **d** Surkhob River, Tajikistan. Photos: M. Mergili

Major alterations of the natural water and bedload regime of Alpine rivers first began in the Middle Ages. Since the nineteenth century, nearly all large river ecosystems in Europe have been altered by human activity (Müller 1996; Mohl 2004). This is because human impact was most intense on plains due to relief and steep slopes limiting the space available for permanent settlements and productive land. Hence, settlements in such areas are often in close proximity to main rivers and their tributaries and are susceptible to flood events that are a common occurrence in Alpine regions and have caused the largest economic damages of all natural hazards in the Alps (Agrawala 2007; Arlettaz et al. 2011; Cammerer and Thieken 2013).

The Lech River in Tyrol is often considered the only near-natural river in the northern Alps. However, the Lech River has also been influenced by various types of human activity. In this chapter, we discuss the regulation and restoration of the Lech River to highlight the shift from the earlier objectives to contain the river in a narrow bed to enable human activities nearby, to the newer objectives to restore near-natural conditions and maintain and restore biodiversity, realizing earlier failures. Firstly, we briefly review the basic concepts of river management (Sect. 2) before introducing the Lech River Valley and its regulation during the twentieth century (Sect. 3). We present and discuss the restoration efforts that have taken place in the beginning of the twenty-first century (Sect. 4) and lastly draw some conclusions (Sect. 5).

16.2 River Management: From Regulation to Restoration

Flood disasters and increasing pressure from human activities in Alpine valleys necessitated hydrological regulation measures (Konstanzer and Moritz 2008). River alterations are the result of river engineering of natural water courses, including river regulation, damming of floodplain areas, and river use for hydropower generation, as well as the removal of gravel. The containment and alteration of rivers has profoundly changed river morphodynamics, sediment transport behaviour, water regimes and, in turn, habitat dynamics, dramatically modifying conditions for aquatic and terrestrial wildlife. Human activity on rivers has also resulted in the enormous decline of freshwater biodiversity, loss of habitat connectivity, as well as the deterioration of natural flood retention capacity, deepening of the river bed, river bank erosion, and the lowering of groundwater tables (Mohl 2004; Habersack and Piégay 2007; Arlettaz et al. 2011).

These geomorphic, ecological and socio-economic problems have resulted in numerous projects being implemented to restore the natural river dynamics and connectivity of several rivers in Europe (Mohl 2004; Gurnell et al. 2016). These restoration efforts have been reinforced by the European Water Framework Directive (WFD) that aimed to ensure that rivers would reach good ecological status by 2015 (WFD 2000).

There are many dimensions and hence challenges to overcome in order to restore a river. The historical natural state of European river and floodplain ecosystems is needed as references to evaluate how the observed state of a river deviates due to human influence, and to direct the goals of restoration projects (Habersack and Piégay 2007; Müller 1996). Such reference rivers include the Fiume Tagliamento, one of the last large near-natural alpine rivers in Europe (Fig. 16.1b; Müller 1996). Understanding the historical natural state provides information on the geomorphic and ecological trends, adjustment conditions, the potential life span, recovery, and the diversity of recreated ecosystems. Moreover, the natural boundaries, dynamics, and processes, or long-term adjustments need to be considered when restoring a river. The restoration strategy should try to work with nature as much as possible and restore natural processes rather than simply desirable forms and habitats (Habersack and Piégay 2007; Kondolf et al. 2006). Such natural processes include sediment regime and sediment transport (Wohl et al. 2015). Adequate sediment supply is required for a sustainable river restoration project (Brooks and Brierley 2004). For example, paved river banks limit lateral erosion, adding to sediment deficit or exhaustion. Land use can also constrain possible restoration options, especially in areas where partial widening and lateral erosion are prohibited by settlements and infrastructure. Furthermore, if gravel bed thickness above the underlying finer material is less than 0.5 m and scouring occurs then bed breakthrough is a threat. This is particularly the case where river bed widening, a common restoration method to stop bed degradation by reducing transport capacity or shear stress, takes place (Habersack and Piégay 2007).

Undesired developments, such as sediment deficits, need to be recognized as early as possible in order for timely action to be taken using 'near natural' measures. Unfortunately, the current decisive criteria for achieving good ecological status outlined by the WFD are solely biological parameters, with hydromorphological elements supporting the biological elements. The WFD decisive criteria do not include abiotic parameters that can indicate the ecological status of the river prior to the biological parameters (Schiemer 1994; WFD 2000; Habersack and Piégay 2007). A more integrated concept was recently introduced with the REFORM (REstoring rivers FOR effective catchment Management) framework that includes river hydromorphology as a key element (Gurnell et al. 2016).

Once the undesired developments have been recognized, plans to restore a river can be implemented. While local restoration projects are beneficial in that one can communicate and demonstrate the efficiency of restoration propositions, they are very expensive and their cost-effectiveness has to be analysed. River restoration needs to be approached on a large scale; it cannot be managed like a state or region. Rivers need to be looked at in terms of hydrographic districts (catchment-wide river basins) and homogenous water bodies (Habersack and Piégay 2007). Brierley and Fryirs (2013) emphasize the importance to consider multiple scales (nested hierarchy). Hence, river restoration and management require international cooperation, especially considering that 66% of Europe's surface area lies in river basins that cross at least one national border (Nilsson et al. 2004).

Lastly, public relations have also become an important aspect of nature-conservation and restoration work (Konstanzer and Moritz 2008). Educating the public about the relations between the operation of a river and its delivery of valued amenities for ecosystems, the community, and the economy is critical (Norton 1998). This ensures that the public is equipped with the knowledge to make informed decisions about river management (Wohl et al. 2005).

16.3 The Lech River and Its History

16.3.1 General Characteristics

The Lech River is a 264 km long right-side tributary of the Danube and originates in the Lechquellengebirge near the Alpe Formarin in the province of Vorarlberg (Fig. 16.2; Meier 2002). It runs through the northwesternmost part of the province of Tyrol before draining northwards into the Danube, north of the city of Augsburg. The Tyrolean part of the river is referred to as 'Tiroler Lech'. Its catchment covers an area of about 1400 km^2 and is characterized by extreme differences in elevation over short distances (Fig. 16.3). While the highest summit in the catchment (Parseierspitze) peaks at 3038 m asl, the runoff gauge at Lechaschau in the lower section of the Tiroler Lech is located at 838 m asl. The long-term mean runoff measured at this gauge is 45 m^3 s^{-1} (Dobler et al. 2010). As the catchment is located near the northern edge of the Alps, it is characterized by a

Fig. 16.2 Lake Formarinsee, located close to one of the sources of the Lech River. Photo: R. Pöppl

considerable amount of precipitation all year round, with average annual precipitation between 1200 mm in the driest parts of the valley bottom and more than 2200 mm in some parts at high elevation (Mergili and Kerschner 2015).

While the uppermost reach of the Lech River—roughly its first 27 km—is predominantly incised into bedrock and displays a steep longitudinal gradient, floodplains several hundreds of metres wide dominate the valley from the village of Steeg downstream (Fig. 16.3). The present article focuses on the latter part, which can be subdivided into three subsections (Fig. 16.4; Sect. 4).

Figure 16.5 displays panoramic views of the Tiroler Lech which is often considered the last major river in the northern Alps in a semi-natural state (Mohl 2004; Cammerer et al. 2013). As stated by Konstanzer and Moritz (2008), the river is still the valley's most important 'land owner'. However, much of the river course is regulated in a way similar to other Alpine rivers (Konstanzer and Moritz 2008; Fig. 16.6a). In some major sections of the Lech, however, the transversal and/or longitudinal river training structures provide sufficient width to support a braided river system (Fig. 16.6b). As a result, the river course represents a sequence of alternating heavily regulated and near-natural sections (Fig. 16.6c). One of the most distinctive sections of the river is the 'string of pearls' between Stanzach and Forchach, where the generally wide floodplain is cut by a sequence of transversal structures (Figs. 16.5 and 16.7).

The braided river sections are characterized by broad areas of lowland riparian forest and huge gravel banks that span over 60 km and are up to 100 m wide in some parts (Fig. 16.8a). Due to erosion and deposition processes, the river and the surrounding river bed are constantly changing (Mohl 2004). The riparian forests include several species of willows (e.g. *Salix eleagnos*; *Salix daphnoides*) and alder (e.g. *Alnus glutinosa*; *Alnus incana*), as well as one of the few remaining large stands of the German tamarisk (*Myricaria germanica*; Fig. 16.8b) in the Alps. Dead wood from the forests is not removed by human intervention and is considered an integral part of the floodplain (Fig. 16.8c). Some parts of the river are declared as 'Natura 2000' area, an EU-wide network of nature protection areas (http://www.natura.org), due to its high biodiversity and the occurrence of numerous internationally endangered species within the dynamic braided river stretches (Mohl 2004; Pflüger 2004; Raven et al. 2007; Cammerer et al. 2013; Sect. 4). The high ecological value of the floodplain continues further north in the Bavarian portion of the river (Müller 1990, 1991).

Not only is the Lech River of environmental importance, but it is also of high scientific interest with regard to its

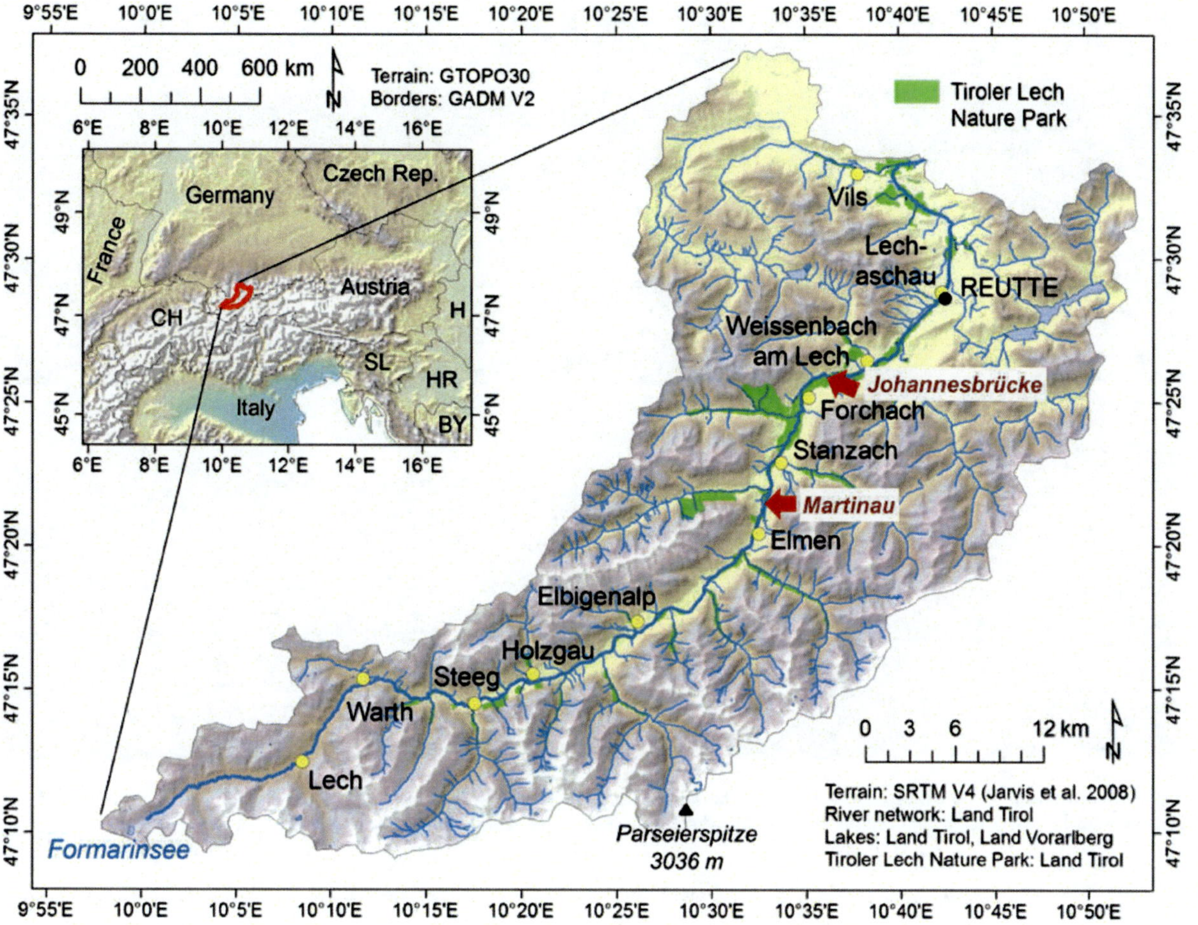

Fig. 16.3 Tiroler Lech and its catchment. Terrain representation in the detailed map derived from Jarvis et al. (2008)

Fig. 16.4 Longitudinal profile of the Tiroler Lech. Note the difference between the steep upper reach and the more gentle lower reach which is the main focus of the present article. The three subsections of the lower reach are further elaborated in Sect. 4

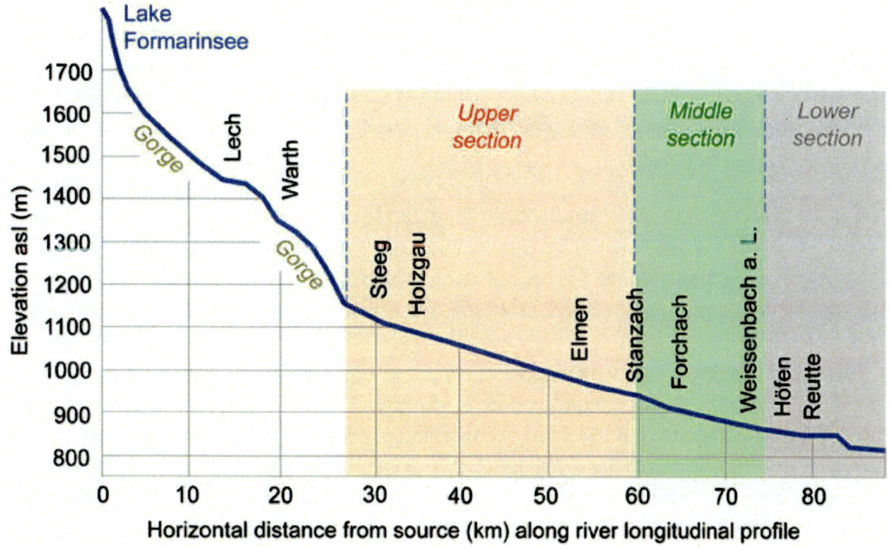

Fig. 16.5 Panoramic views of the middle section of the Tiroler Lech **a** between Stanzach and Forchach and **b** between Forchach and Reutte (Johannesbrücke Bridge in the left centre). See Fig. 16.3 for the location of the scenes. Photos: M. Mergili

fluvial sediment transport. Annual suspended sediment yield has been measured at a gauge at Füssen (southern Bavaria) for several decades, however without a significant trend and without a clear correlation with annual runoff in the period 1924–1990 (Walling 1997): high sediment yield relates to short-term peak discharge, which is not so much reflected in the annual runoff. Due to the availability of long-term data, the Lech River has been employed for sediment transport model calibration and validation (Diodato et al. 2012).

16.3.2 Flood Disasters and River Regulation

A timeline of the alterations to the Lech River and also of the restoration measures implemented at a later stage is illustrated in Fig. 16.9.

The Lechtal Valley was neglected from an economic point of view until the outgoing nineteenth century. Many people had to emigrate due to lacking economic opportunities and insufficient arable land (Meier 2002). River regulation measures—enabling the use of the valley bottom for agriculture, settlements and industry—therefore started much later than in other Alpine valleys so that the near-natural state of the floodplain was preserved. The construction of small-scale protection structures started in the last few decades of the nineteenth century, but they had to be replaced or repaired due to flood damage each spring.

Although average precipitation peaks in the summer months, the flow regime of the Lech River is governed by a combination of rainfall and snow melt. Lechaschau (Fig. 16.3) therefore experiences a runoff maximum in May and June, coinciding with the spring floods that occur annually (Tirol Atlas, 2001–2007). Some of the major historic floods in the Lech Valley were recorded during this time of the year. Catastrophic flood events are particularly likely when the snow pack in the mountains is thick and there are warm periods of prolonged, heavy precipitation that facilitate snow melt. The city of Augsburg located along the lower course of the river was heavily damaged by such a flood in June 1910. The next notable spring flood occurred soon after, in May 1912 (Meier 2002). Another major flood took place in May 1999 following an extraordinary snow fall period in February 1999. The above-average snow fall triggered the Galtür and Valzur avalanche disasters that resulted in 38 fatalities (Höller 2007) and indirectly caused a major flood event when a persistent cell of warm and moist air delivered heavy rainfall over the northern rim of the Alps and melted the snow pack. On 21 May, several

Fig. 16.6 Contrasting reaches of the Tiroler Lech: **a** valley bottom used as agricultural land near Holzgau, the river course contained to a narrow channel in the background; **b** braided river section between Stanzach and Forchach; **c** sequence of near-natural and heavily regulated portions of the Lech River between Weissenbach am Lech and Reutte, with near-natural parts in the foreground and in the background, and a channelized section (with some degree of restoration) in the middle ground. Photos: M. Mergili

Fig. 16.7 'String of pearls' between Stanzach and Forchach. This is the most distinctive section of the valley, where braiding of the river is constrained by a series of transversal structures. Photo: M. Mergili

Fig. 16.8 Braided river system in the 'string of pearls'. **a** Sediment banks located on the right side of the floodplain (view upstream) and channels and depressions filled with water. The lower terraces are regularly affected by floods (annual spring floods). The higher terraces are flooded less regularly and support riverine shrubs such as willows (*Salix* spp.); **b** shrubs of the German tamarisk (*Myricaria germanica*), which has become rare in the Alps, are commonly found on the floodplain of the Lech River; **c** dead wood, an integral part of the floodplain remains untouched by human intervention. Photos: M. Mergili

meteorological stations in the Lechtal Valley recorded their all-time maximum daily precipitation; Reutte experienced 212.5 mm of precipitation (Meier 2002). Two other major flood events took place in 2002 and 2005. However, in contrast, these events occurred in connection with wet periods in August.

Such flood disasters, along with increased pressure from human activities in the Lechtal Valley, necessitated hydrological regulation measures, including the construction of debris dams and channelization structures that severely narrowed the river bed in certain sections (Mohl 2004). A more systematic regulation programme, along with community-scale projects, was first launched in 1896 and extended after a flood in 1901. A general plan for river regulation was launched in 1907. This plan had to be revised after the flood events of 1910 and 1912. Technical measures, including longitudinal and transversal structures, as well as sediment retention structures, were largely completed by 1940. Thereafter, attention was directed towards complementary and additional structural measures, as well as

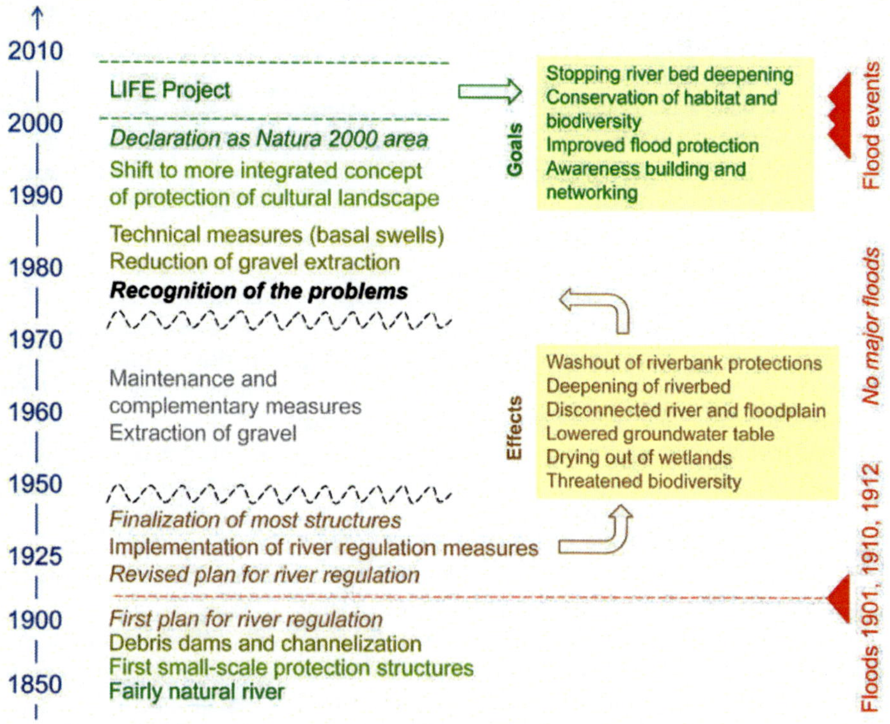

Fig. 16.9 Management of the 'Tiroler Lech' over time

towards the maintenance of the existing structures (Meier 2002).

Sediment-retaining dam constructions across major tributaries, channelization, and the growing exploitation of gravel from the Lech River bed, inadvertently resulted in sediment deficiencies and consequent river bed deepening (cf. Rinaldi et al. 2005). Sediment was removed or flushed downstream and was not replenished by sediment from the uplands. The lowering of the river bed level has consequently negatively impacted the fluvial system by disconnecting the rivers and the adjacent floodplain, further drying out fluvial wetlands, and by lowering the groundwater table (Mohl 2004; Konstanzer and Moritz 2008). Moreover, as a result of diminished flooding events and subsequent submersion of forests that were regularly affected during flooding events, numerous plant and animal species characteristic of gravel banks have declined, including the German tamarisk (*Myricaria germanica*; Fig. 16.8a), the pink-winged grasshopper (*Bryodema tuberculata*), and the little ringed plover (*Charadrius dubius*) (Mohl 2004).

In addition to these negative impacts, the settlement areas, to a certain extent, were also faced with the ramifications of river regulation that have caused problems for protective structures (washout of riverbank protections, etc.) (Konstanzer and Moritz 2008).

These ramifications were recognized as far back as in the 1970s and several ideas were developed to counter the vertical erosion and the associated consequences. The strategies shifted from erecting purely technical structures, such as transversal swells, basal ramps, and the reduction of commercial extraction of gravel, to a more integrated concept of the protection of the cultural landscape and the conservation of the Lech as a near-natural river and its protection as wildlife habitat (Meier 2002). These comprehensive ideas launched the LIFE project to 'direct' future development in the region (Konstanzer and Moritz 2008).

16.4 Restoration of the Lech River and Associated Challenges

The Lech River restoration project (LIFE; 00NAT/A/7053: Wildflusslandschaft Tiroler Lech) began in April 2001 and lasted until March 2007 (Konstanzer and Moritz 2008). The project budget totalled 7.82 million Euros, of which 49.5% was contributed by the European Union and the remainder by national and regional agencies. A precondition for the EU funding was that the restoration area had to be declared as a Natura 2000 area (the 'Tiroler Lech Nature Park'), despite the initial plan to declare it as a National Park. This precondition arose as National Park status brings with it major restrictions to hunting, to which local hunters strongly objected. This led to a lower-level protection status to be applied to the area.

The goals of the Lech River restoration project included (Lentner et al. 2003):

1. Conservation and restoration of the fairly natural, dynamic fluvial habitats,
2. Prevention of further lowering of the river bed and decrease of groundwater level,
3. Improvement of flood protection in accordance with environmental protection regulations,
4. Preservation of animals and plants that are listed by the EU as important, vulnerable or endangered,
5. Improvement of ecological awareness of the local people, and
6. Execution of a joint project with organizations from different fields of interest.

River restoration and engineering were the main measures employed to address the goals of the project. These primary measures focused on river morphology and bedload budget. To restore the characteristic habitats of free-flowing water while simultaneously providing flood protection, the river bed was widened over a length of 6 km where the demands of flood protection and the socio-economic conditions would allow. This was achieved by removing several constructions, such as in the vicinity of the Johannesbrücke (a bridge) and the hamlet of Martinau (Konstanzer and Moritz 2008; Figs. 16.3 and 16.5). Historical maps—mainly the Franciscaean Cadastre (1856) and a plan of the existing technical regulation measures (1925)—were used as a basis to reconstruct the historical river landscape (Haidvogl et al. 2011).

To restore the bedload budget of the river and to prevent further deepening of the river bed, unobstructed bedload transport was enabled by the removal of sediment-retaining dams (gravel barriers) on the tributaries, including the Hornbach and Schwarzwasserbach brooks dams. Furthermore, riverside waters were revitalized and linked up with their parent river (Konstanzer and Moritz 2008).

Public relations work was also undertaken and included communicating with the public, stakeholders, and opinion leaders, and encouraging their involvement, improving the ecological awareness of the local people, and execution of 33 public relations projects, including visitor platforms, themed trails, visitor management, and nature guide training (Konstanzer and Moritz 2008; Pflüger 2004).

The major challenge while restoring natural processes in the Lech River was the bedload balance. According to Lentner et al. (2003), a delicate bedload management balance has to be attained for the Lech River as the river has three different parts that have different bedload requirements (Fig. 16.4):

- Upper section (Steeg—Elmen; Fig. 16.6a): heavily regulated, straight course, stable river bottom,
- Middle section (Stanzach—Weißenbach; Fig. 16.6b): braided river; erosion processes, and
- Lower section (Höfen—Reutte—Weißhaus; Fig. 16.6c): unfavourable bedload deposition in the main settlement area; reduction of the sediment transport capacity, effective cross section and freeboard; multiple, ecologically disputed dredging operations.

In the upper and middle sections bedload is needed to maintain the highly dynamic, braided river type, whereas in the lower section a bedload surplus would be a severe problem (Lentner et al. 2003). Hence, well-coordinated measures were needed to fulfil all these requirements:

- Broad river widening measures for the upper and middle sections for simultaneous flood protection and river revitalization,
- Removal of some large sediment-retaining dams in major tributaries to improve the bedload balance in the main river (i.e. to counter the actual gravel deficit), and
- Construction of a big bedload trap in the lower section that improves the ecological situation and protects the main town in the district (Reutte; Fig. 16.3) from the possible bedload surplus induced by the measures in the upstream reaches (Lentner et al. 2003).

The ecological success of the measures implemented within the frame of the LIFE project is evaluated through a cross-disciplinary monitoring programme (Piegay et al. 2008). However, as to the knowledge of the authors, no comprehensive scientifically based assessment of the restoration success has been published until 2021. In general, the pertinent literature has become sparse from 2009 onwards, i.e. two years after the end of the LIFE project. As the targets of the European Water Framework Directive (WFD 2000) had to be met until the year 2015 and, most importantly, as another LIFE project (LIFE Lech, 2016 to 2021; Salchner 2020) has been completed, one may expect upcoming work in this direction. Still, the indicators of success have to be defined carefully, referring to a set of desired physico-chemical, hydromorphological and biological conditions. While the historical state of the river is sometimes used as a reference, this approach has been questioned as the historical state may be ill-defined and as some modifications to rivers may be irreversible (Piegay et al. 2008; Brierley and Fryirs 2016).

16.5 Conclusions

The Lech River, as well as most other Alpine rivers, has been subject to extensive regulation since the end of the nineteenth century. Towards the end of the twentieth century, ramifications of regulation of the Lech became apparent, including gravel deficits and subsequent incision of the river bed, increased flood risk on the floodplains, and threats to ecological diversity. This facilitated efforts to restore sediment equilibrium by promoting river and sediment continuity, by managing target species, and by increasing public awareness. Riparian systems functions are complex and require a multilateral approach that integrates many disciplines and points of view, such as those from geomorphologists, hydrologists, ecologists, and river engineers (Konstanzer and Moritz 2008). The key requirements for successful river restoration such as integrated ecological needs, morphological goals, flood protection, water resources, recreation and landscape aspects (Habersack and Piégay 2007; Chiari et al. 2008) were addressed in the LIFE project (2001–2007). Even though the restoration of the Lech River is broadly considered successful, a comprehensive, long-term evaluation of the restoration success is still needed.

References

Agrawala S (2007) Climate change in the European Alps: adapting winter tourism and natural hazards management, Organisation for Economic Cooperation and Development (OECD)

Arlettaz R, Lugon A, Sierro A, Werner P, Kéry M, Oggier P-A (2011) River bed restoration boosts habitat mosaics and the demography of two rare non-aquatic vertebrates. Biol Cons 144:2126–2132

Brierley GJ, Fryirs KA (2013) Geomorphology and river management: applications of the river styles framework. Wiley

Brierley GJ, Fryirs KA (2016) The use of evolutionary trajectories to guide 'moving targets' in the management of river futures. River Res Appl 32:823–835

Brooks AP, Brierley GJ (2004) Framing realistic river rehabilitation targets in light of altered sediment supply and transport relationships: lessons from East Gippsland Australia. Geomorphology 58 (1):107–123

Cammerer H, Thieken AH (2013) Historical development and future outlook of the flood damage potential of residential areas in the Alpine Lech Valley (Austria) between 1971 and 2030. Reg Environ Change 13:999–1012

Cammerer H, Thieken AH, Verburg PH (2013) Spatio-temporal dynamics in the flood exposure due to land use changes in the Alpine Lech Valley in Tyrol (Austria). Nat Hazards 68:1243–1270

Chiari S, Schmid F, Muhar A, Muhar S (2008) Recreational functions of rivers in Austria: an approach to the visitors' perspective. In: Raschi A, Trampetti S (eds) Management for protection and sustainable development 247–251

Diodato N, Gericke A, Bellocchi G (2012) Modelling the inter-annual variability of sediment yields: a case study for the upper Lech River. CATENA 97:12–19

Dobler C, Stötter J, Schöberl F (2010) Assessment of climate change impacts on the hydrology of the Lech Valley in northern Alps. J Water Clim Change 1:207–218

Gurnell AM et al (2016) A multi-scale hierarchical framework for developing understanding of river behaviour to support river management. Aquat Sci 78(1):1–16

Habersack H, Piegay H (2007) 27 River restoration in the Alps and their surroundings: past experience and future challenges. Devel Earth Surf Process 11:703–735

Höller P (2007) Avalanche hazards and mitigation in Austria: a review. Nat Hazards 43(1):81–101

Haidvogl G, Hohensinner S, Preis S (2011) Rekonstruktion historischer Flusslandschaften als Grundlage im Gewässermanagement—potential und limits. Oesterr Wasser Abfallwirtsch 63(9–10):174–182

Jarvis A, Reuter HI, Nelson A, Guevara E (2008) Hole-filled seamless SRTM data V4. International Centre for Tropical Agriculture (CIAT). Available from http://srtm.csi.cgiar.org, 2008, last access: 14 Apr 2016

Kondolf GM, Boulton AJ, O'Daniel S, Poole GC, Rahel FJ, Stanley EH, Wohl E, Bång A, Carlstrom J, Cristoni C (2006) Process-based ecological river restoration: visualizing three-dimensional connectivity and dynamic vectors to recover lost linkages. Ecol Soc 11(2):5

Konstanzer J, Moritz C (2008) Several landowners in a protected area: revitalisation of the tyrolean river Lech In: Arizpe D, Mendez A, Rabaça JE (eds) Sustainable Riparian Zones—a management guide 274–279. Generalitat Valenciana, España

Lentner R, Schletterer R, Moritz C (2003) LIFE-Project: Wildflusslandschaft Tiroler Lech

Meier IM (2002) Leben mit dem Hochwasser – Ausgewählte Hochwasserereignisse des 20. Jahrhunderts Im Tiroler Lechtal. Innsbrucker Jahresbericht 2001(2002):5–29

Mergili M, Kerschner H (2015) Gridded precipitation mapping in mountainous terrain combining GRASS and R. Nor Geogr Tidsskr 69(1):2–17

Mohl A (2004) LIFE River restoration projects in Austria. In: 3rd European conference on river restoration, pp 17–21

Müller N (1990) Die übernationale Bedeutung des Lechtals für den botanischen Arten-und Biotopschutz und Empfehlungen zu deren Erhaltung. Schriftenreihe Bayerisches Landesamt Für Umweltschutz 99:17–39

Müller N (1991) Veränderungen alpiner Wildflußlandschaften in Mitteleuropa unter dem Einfluß des Menschen. Augsburger Ökologische Schriften 2:9–30

Müller N (1996) River dynamics and floodplain vegetation and their alterations due to human impact. Large Rivers 9(3–4):477–512

Nilsson S, Langaas S, Hannerz F (2004) International river basin districts under the EU water framework directive: identification and planned cooperation. European Water Manage Online 2:1–20

Norton BG (1998) Improving ecological communication: the role of ecologists in environmental policy formation. Ecol Appl 8:350–364

Pflüger Y (2004) Value based decision making process for strategic visitor management in the Natura 2000 area Lech River Valley, Tyrol. Working Papers of the Finnish Forest Research Institute 2, 299–304

Piégay H, Naylor LA, Haidvogl G, Kail J, Schmitt L, Bourdin L (2008) Integrative river science and rehabilitation: European experiences. River futures: an integrative scientific approach to river repair. Island Press, Washington DC, pp 201–219

Raven P, Holmes N, Dawson H, Binder W, Muhlmann H (2007) River habitat survey in southern Bavaria and the Tyrolian Alps. Results from 2006. Report

Rinaldi M, Wyżga B, Surian N (2005) Sediment mining in alluvial channels: physical effects and management perspectives. River Res Appl 21(7):805–828

Salchner M (2020) LIFE lech - dynamic river system lech. Eco Mont 12(2):41–45.https://dx.doi.org/10.1553/eco.mont-12-2s41

Schiemer F (1994) Monitoring of floodplains: Limnological indicators

Tirol Atlas (2001–2007) Mittlere jährliche Abflussspende für Pegeleinzugsgebiete: Lechaschau. http://tirolatlas.uibk.ac.at/data/chart.py/index?lang=de;chart=3163;id=1311051. Accessed on 14 Mar 2016

Walling DE (1997) The response of sediment yields to environmental change. IAHS Publ 245:77–89

WFD (2000) Directive 2000/60/EC of the European Parliament and of the Council establishing a framework for the Community action in the field of water policy. Joint text approved by the Conciliation Committee provided for in Article 251.

Wohl E, Angermeier PL, Bledsoe B, Kondolf GM, Macdonnell L, Merritt DM, Palmer MA, Poff NL, Tarboton D (2005) River restoration. Water Resour Res 41(10):W10301. https://doi.org/10.1029/2005WR003985

Wohl E, Bledsoe BP, Jacobson RB, Poff NL, Rathburn SL, Walters DM, Wilcox AC (2015) The natural sediment regime in rivers: broadening the foundation for ecosystem management. Bioscience 65(4):358–371

Martin Mergili is professor for physical geography at the University of Graz, Institute of Geography and Regional Science. Before, he held positions as senior scientist at the Institute of Applied Geology, BOKU, and as post-doc at the Department of Geography and Regional Research, University of Vienna, Austria. His research focuses on high-mountain geomorphology, natural hazards, and GIS techniques for the simulation of landslide processes.

Ruginia Duffy graduated as an environmental scientist from the University of Vienna, Austria, where she specialised in geoecology. She has a keen interest in entrepreneurial and nature-based solutions to environmental challenges.

17 The World Heritage Site Hallstatt-Dachstein/Salzkammergut: A Fascinating Geomorphological Field Laboratory

Johannes T. Weidinger and Joachim Götz

Abstract

This chapter introduces the geomorphology of the World Heritage site Hallstatt-Dachstein/Salzkammergut. The historical significance of the region and its (geo) scientific implications are briefly summarized before focusing on various geological and geomorphological highlights within three selected areas. The Dachstein massif is addressed first. Based on a short tectonic history, the well-preserved 35 Ma paleosurfaces and the long-lasting formation of the complex karst cave system is illustrated. Further karst and karst water phenomena are described before highlighting recent glacier retreat and permafrost warming as a consequence of climate change. Within the section on the Gosau Valley, paleo-geographic and geologic basics, the Lateglacial history and paraglacial consequences, including deep-seated gravitational slope deformations and debris flows, are introduced. The glacial imprint in the root zone of the former Traun Glacier and the geomechanical system "hard on soft rocks" strongly affect the pattern and mechanics of potentially hazardous mass movements around Lake Hallstatt as illustrated by the Sandling, the Zwerchwand, and the active sediment cascade from the Plassen down to the village of Hallstatt. Some critical notes on, and future perspectives for the regional tourism are finally provided.

Keywords

Elevated plateaus • Karst • Caves • Glaciers • Fossil reef • Salt mining • Mass movements

J. T. Weidinger (✉)
ERKUDOK© Institute in the Kammerhof Museums of Gmunden (K-Hof), Kammerhofgasse 8, 4810 Gmunden, Austria
e-mail: johannes.weidinger@gmunden.ooe.gv.at

J. Götz
Fachgruppe Geomorphologie, University of Bayreuth, Universitätsstraße 30, 95447 Bayreuth, Germany

17.1 Introduction

Located east of Salzburg, along the Alpine foreland and the Northern Calcareous Alps to the peaks of the Dachstein Mountains, the Salzkammergut spans the federal states of Upper Austria, Salzburg, and Styria with the main river Traun, a right tributary of the Danube. The World Heritage Site Hallstatt-Dachstein/Salzkammergut (core/buffer zone) encompasses the southern, inner-Alpine part of the Salzkammergut around Lake Hallstatt and the Dachstein massif.

Regional geology and the tectonic history of the Salzkammergut are multifaceted. Sedimentary rocks of Permian to Neogene ages were several times dislocated during Alpine orogenesis piling up to a sequence of tectonic units. These slightly north-dipping nappes allochthonously rest on the crystalline rocks of the Bohemian massif with their overlying Molasse sediments, which are exposed further north. The uppermost nappe, mainly composed of thick-bedded Dachstein limestone, forms a barren plateau, which is only sparsely vegetated by Alpine flora. This plateau is on all sides bounded by almost vertical and up to 1500 m high rock walls. The surface of the lower nappe, the Hallstätter Zone, forming the footwall toward north, is much less elevated and hilly to hummocky in its relief. This densely forested area is composed of marls and evaporitic salt rocks, the so-called "Haselgebirge" (Mandl 2000).

Weather and climate are regionally and locally influenced by (i) the elevation of the mountains, (ii) an orographic effect, and iii) the surrounding lakes. Mean annual temperatures vary between 7.6 °C at the bottom of the valleys (e.g., Hallstatt village, 500 m asl) and 1.3 °C in the mountains (e.g., Krippenstein 2050 m asl). Due to the orographic effect, humid air from west and northwest is forced to precipitate north of the massif with high amounts of annual rainfall (e.g., 1922 mm/a at the Krippenstein). While free atmospheric wind circulation dominates on top of the mountains, relief influences the winds in the valleys. A famous example

C. Embleton-Hamann (ed.), *Landscapes and Landforms of Austria*, World Geomorphological Landscapes,
https://doi.org/10.1007/978-3-030-92815-5_17

is the Enns Valley with strong easterly winds (Weingartner 2008).

The geomorphology of the Salzkammergut is highly diverse as well and is characterized by a typical arrangement of rugged limestone cliffs on top of a much more gentle hilly to hummocky relief and a number of lakes in between. Quaternary glaciers originating from the Dachstein massif carved out deep U-shaped valleys with steep limestone flanks (e.g., Echern Valley), often ending up in large deltaic sediment bodies (e.g., Traun Delta), deposited into glacially overdeepened basins filled with residual lakes (e.g., Lake Hallstatt). Larger basins developed in areas of former glacier confluences and Lateglacial tongue basins, sometimes enclosed by morainic ridges (e.g., basins of Bad Goisern and Bad Aussee). Both situations (deltas, basins) are typical for the region and represent favorite settlement areas ever since (van Husen 1977).

Already 7000 years BP prehistoric settlers came to the Salzkammergut and discovered rich salt deposits, which never lost their economic importance until recent days (Kern et al. 2008). By mining and trading this mineral, the locals not only became rich and prospering, founding the archeologically famous settling areas, but were also affected by natural geomorphological processes that have increasingly become hazardous for them. Living and trading around today's villages of Hallstatt, Obertraun, Bad Goisern, and Gosau have thus been always closely connected to nature.

The term Salzkammergut originates from the Imperial Salt Chamber, the authority responsible for running the valuable salt mines in the region during the Habsburg Monarchy. As a consequence of the historico-cultural evolution, an inner and outer Salzkammergut can be differentiated. Whereas the inner Salzkammergut describes the salt bearing and thus mining areas adjacent to the Dachstein massif in the south, the outer Salzkammergut depicts the salt marketing and trading areas further north. There the town of Gmunden, today the regional center of the Salzkammergut, originally developed as an important salt trading town. This historical division, however, caused resentments between people of both areas partly persisting until present day!

Due to its variegated cultural history, the Hallstatt-Dachstein region officially became a UNESCO World Heritage site in 1999. Although this international certificate focuses on cultural highlights, nobody really excluded the spectacular landscape from protection providing the natural base for that kind of historic evolution (Jeschke 2002). In recent days, geosciences in general and geomorphological research in particular play a major role in the protection of cultural sites from natural hazards as well as in the popular education of both local inhabitants and tourists.

17.2 The Inner Salzkammergut—A Favorite Destination for Geoscientists

Due to its economic status and the prospection of salt mineral resources, the inner Salzkammergut region became a favorite destination and research focus of early geoscientists. The Czech Johann(es) Baptist Bohadsch in 1783, the Viennese physician Josef August Schultes in 1794, and the German geologist Christian Leopold von Buch in 1802 were the first geoscientists visiting the Salzkammergut region and describing a large variety of different natural phenomena (Lobitzer 2009; Lobitzer and Posmourný 2010). This first period of foreign investigations was followed by the times after the Viennese Congress in 1815—in German-speaking countries well known as the *Biedermeier*. This was a period of geoscientific awakening all over Europe and also in Austria. French geoscientists such as Ami Boué in 1829 as well as British ones, such as Sedgwick and Murchison in 1831, came to discover the Salzkammergut region for both geological and geomorphological research (Lobitzer 2001). In recent decades, regional research is diverse and manifold—the following map shows a selection of 14 geologic-geomorphological highlights in the inner Salzkammergut to which this chapter shortly introduces (Fig. 17.1).

17.3 The Dachstein Massif

Research activities mentioned above and the following popular scientific publications encouraged more people from all over Europe, but especially the rich ones from the Austro-Hungarian Monarchy, to visit the Salzkammergut region during their holidays. Mountain tourism therefore started quite early in the Dachstein massif, which includes not only the highest peak in the area (Hoher Dachstein; 2995 m asl), but also marks the border between Upper Austria, Styria, and Salzburg. Until today, the massif with its large plateau is a nature reserve and a remote place to stay and carry out research, with an extraordinary earth science history and a primeval landscape.

17.3.1 The Oldest Landscape of the Eastern Alps on Top of a Mountain

Although controversially discussed from geological and geomorphological points of view (Tollmann 1974), there is no doubt that remnants of paleosurfaces (elevated plateaus) that were originally generated in the Paleogene period, play a major role in shaping the "roof" of the Northern

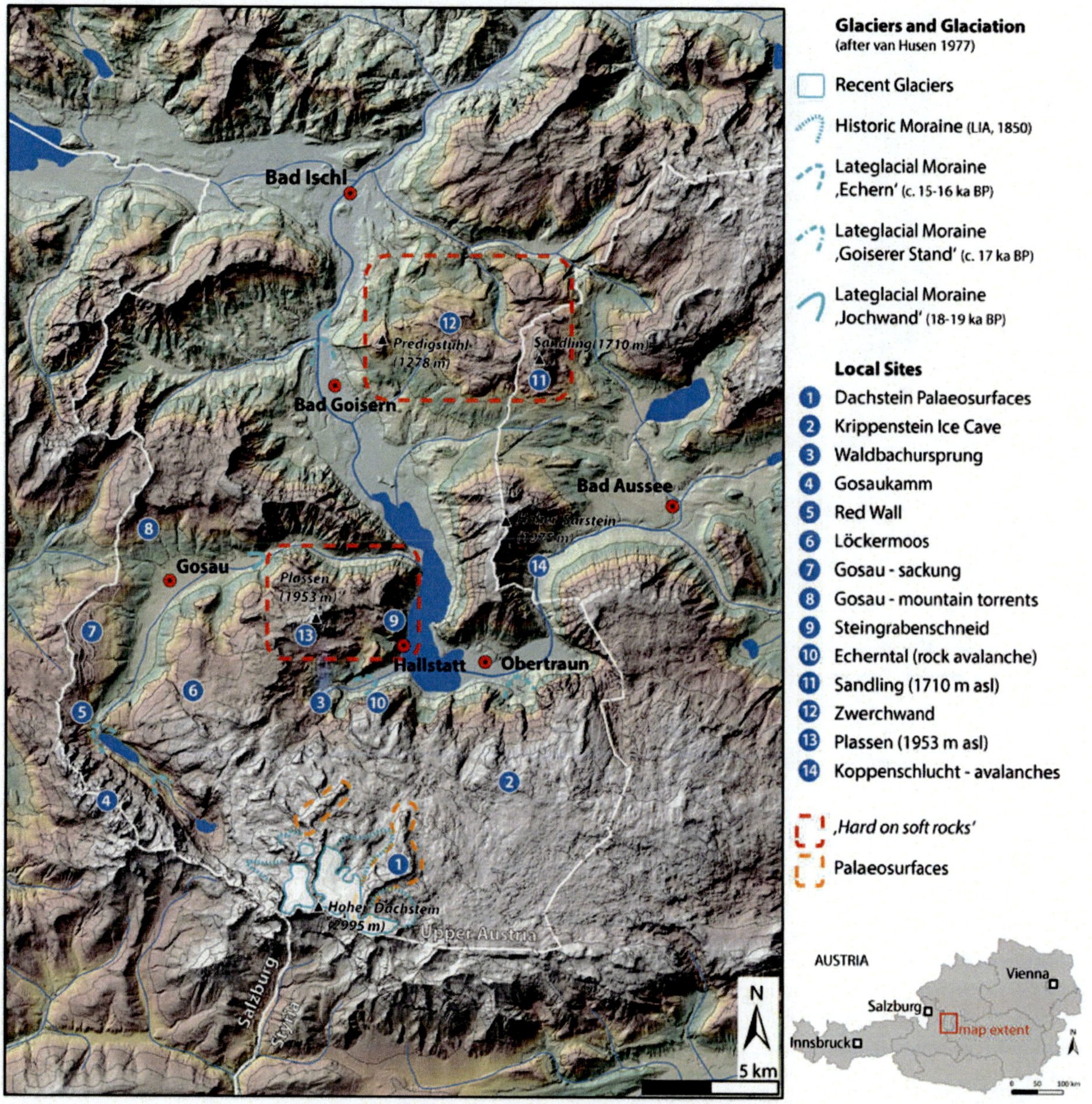

Fig. 17.1 Overview map of the Salzkammergut region with 14 geologic and geomorphologic highlights considered in this chapter (based on hillshaded elevation data; 10 m resolution; www.opendata.gv.at)

Calcareous Alps. As a remnant of an originally more than 10 000 km^2 large paleosurface, the Dachstein massif refers to one of these elevated plateaus, which formed during the late Eocene to the early Oligocene period c. 35 Ma BP (Frisch et al. 2001). During the entire Oligocene to the late Miocene (c. 31–10 Ma BP), this paleosurface has been sealed by a hundreds of meters thick pile of clastic sediments (the so-called *Augenstein-Formation*) transported by the rivers on their temporarily interrupted way from today's central Alps in the south toward and into the Molasse basin in the north. The Dachstein paleosurface became preserved under this sediment cover. Uplift within the last 10 Ma has caused dissection of the paleosurface and produced surface plateaus at various altitudes. (cf. Chap. 19, Sect. 19.2 and Fig. 19.3). After the erosion of the sediment cover and its final deposition in the Molasse basin, endogenous erosion of the massif by karstification of the Dachstein limestone exceeded exogenous surficial erosion. During the

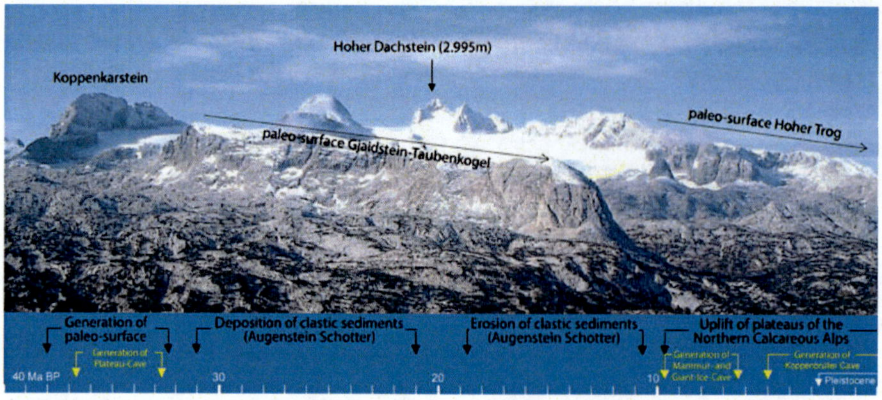

Fig. 17.2 View from Krippenstein (2108 m asl) toward SW over the paleosurfaces of the Dachstein plateau and summit areas. With an inclination of less than 10° the original 35 Ma old paleosurfaces dip toward north from the Hoher Gjaidstein (2794 m asl) to the Taubenkogel (2301 m asl) and in the area of the high pass called Hoher Trog (see also paleosurfaces and site 1 in Fig. 17.1)

Pleistocene the paleosurfaces morphologically survived as nunataks in a just slightly modified form (Fig. 17.2 and site 1 in Fig. 17.1).

17.3.2 Karst Geomorphology and Hydrology—Large Ice Caves and an Impressive Karst Spring

Haas et al. (2009) described paleokarst as a result of surficial and subsurface corrosion during the formation of the Dachstein limestone in the Upper Triassic (210 Ma BP)—perhaps the oldest karst phenomenon in the area. However, the main period of extensive corrosion of the cyclic and thick-bedded Dachstein limestone was the Neogene. Miocene tectonics and uplift of the Eastern Alps not only predetermined the exogenous shape of the 400 km^2 large Dachstein massif, but also influenced its endogenous structure (Frisch et al. 2003). During periods of tectonic quiescence, endogenous erosion and corrosion were most effective along permeable discontinuities in between the cycles of the thick-bedded Dachstein limestone and formed at least three major cave system levels, which are connected to each other by additional vertical systems. These well-known caves, which are extensively visited by tourists, include from the top of the mountain to the valley bottom.

(i) the oldest system, the so-called *Rumpfhöhlen* (= remnants of caves) system on the plateau, which developed between 37 and 33 Ma BP (site 2 in Fig. 17.1),
(ii) the *Mammut-Cave* and *Giant-Ice-Cave system* within the steep northern flanks of the mountain, which developed between 10 and 6 Ma BP, and the
(iii) youngest so-called *Koppenbrüllerhöhlen system* at the recent water level close to the bottom of the Traun Valley, which has been formed during the past 5 Ma.

Specific ventilation conditions with rising warm air during winter times and falling cold air in summer effectively cool the system and account for the formation of ice (Fig. 17.3) in the Giant Ice Cave, which is one of the regional sightseeing attractions.

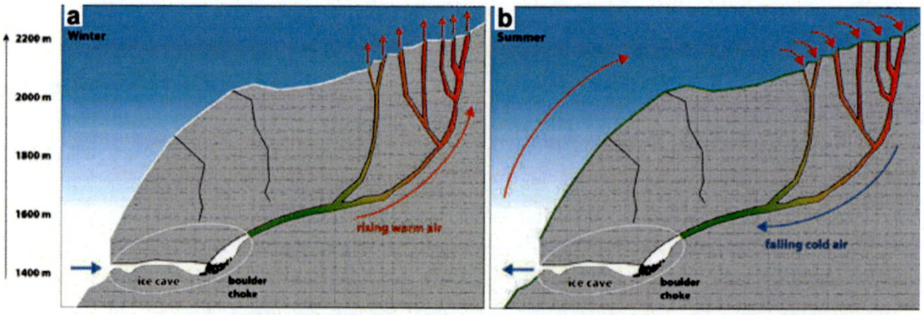

Fig. 17.3 Scheme of changing winter **a** and summer **b** air-circulation in the Giant Ice Cave system in the Dachstein massif. Specific ventilation conditions inside the mountain led to the formation of ice (site 2 in Fig. 17.1; modified from Gamsjäger 2010)

Fig. 17.4 Waldbachursprung is a famous, efficient karst spring in the Echern Valley close to the village of Hallstatt (site 3 in Fig. 17.1)

The extensive Dachstein karst cave system with a total length of 200 km is one of the largest in the Austrian Alps. In an area with high precipitation of 1800–2500 mm/a, this implies extensive subsurface runoff. Tracer experiments revealed the main flow direction of this subsurface karst water in the Dachstein massif toward north. The Waldbachursprung (=source of the river Waldbach) is a famous karst spring in the Echern Valley draining toward the village of Hallstatt (Fig. 17.4 and site 3 in Fig. 17.1). This important source for drinking water is highly vulnerable to contamination of waste waters generated by human activity on the plateau (Mandl 2014), especially during recent glacier retreat under climate warming.

17.3.3 The Dachstein massif—A Cradle of Research in Glacial Geomorphology

In 1840, Friedrich Simony, who became First Professor of Geography at Vienna University in 1851, came to the Dachstein region to study the mechanism of glacier movement and erosion as well as glacial geomorphology (Speta

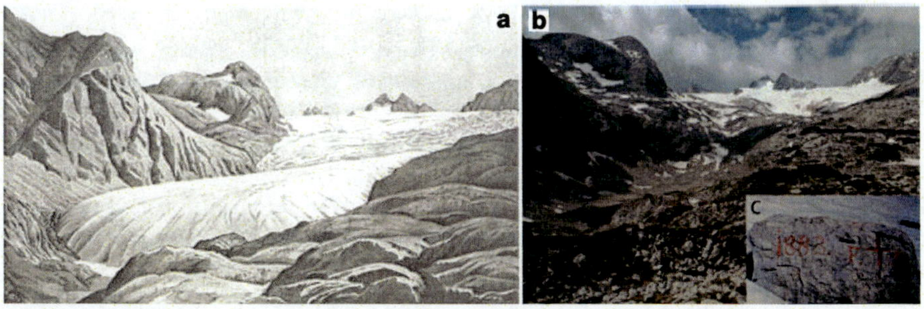

Fig. 17.5 **a** Hallstätter Glacier (former Karl's icefield) in 1840 (Simony 1889); view toward south. **b** The Hallstätter Glacier in 2014 after tremendous retreat from a similar viewpoint. **c** Stone carving by F. Simony marking the glacier retreat after the LIA close to today's Alpine Centre Simony Hut

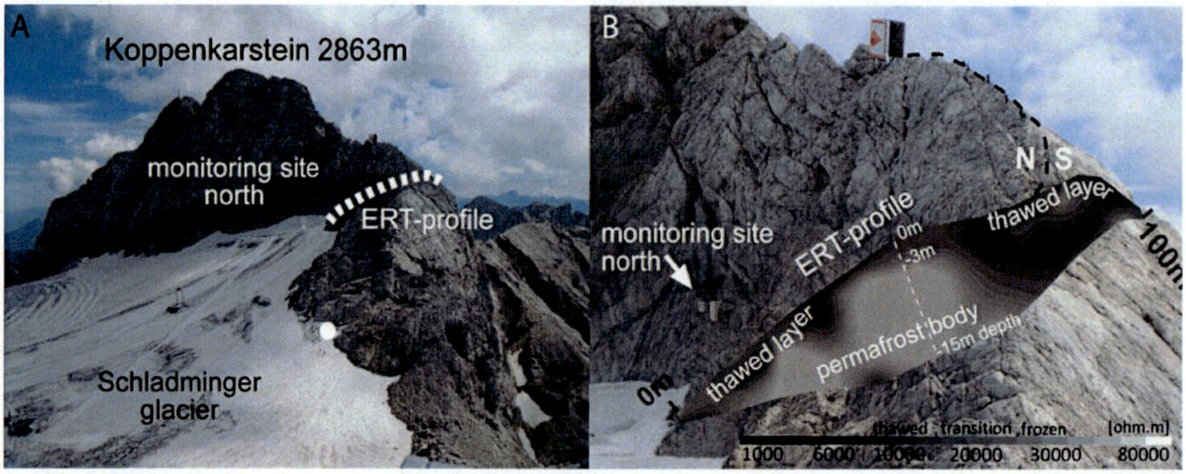

Fig. 17.6 **a** Permafrost monitoring site at the Koppenkarstein north face. **b** Electrical resistivity tomography including interpreted permafrost conditions (after Rode 2014; views toward east)

and Aubrecht 1996). His research on the Hallstätter Glacier, the largest of the five Dachstein glaciers, which lasted until the beginning of the twentieth century, has been fortunately carried out during a period of glacier advance—at the end of the well-known Little Ice Age (LIA). Friedrich Simony not only observed the local glacial maximum of this period close to 1850 (Fig. 17.5a) but also the rapidly retreating glacier afterward, marking different stages with simple stone carvings on erratic blocks (Fig. 17.5c). The negative trend in mass balance of the Dachstein glaciers has continued throughout the twentieth century until today, partly interrupted by short stagnations. Extraordinary warm and dry years, such as 2003 and 2015, speeded up this trend, and the complete disappearance of the largest glacier area in the Northern Calcareous Alps seems to be just a matter of time (Weingartner 2008).

As another consequence of climate change, permafrost warming and its potential degradation in rock walls is considered to constitute an increasing hazard in alpine environments due to enhanced rockfall activity potentially endangering infrastructure (Brommer et al. 2009). In the uppermost areas of the Dachstein massif, the interplay of permafrost dynamics, freeze thaw cyclicity, rock moisture distribution, weathering, and rockfall activity is recently investigated and monitored (Rode 2014; Schnepfleitner et al. 2016). Applied methods complement each other and include borehole temperature measurements, electrical resistivity tomography (ERT), terrestrial laser scanning, infrared photography, and rock moisture and temperature measurements. Exemplary results on ERT-based permafrost conditions at the Koppenkarstein test site are shown in Fig. 17.6 indicating a non-frozen active layer characterized by lower resistivities (dark gray) on top of a higher-resistive permafrost body underneath (light gray). The monitoring campaign will be continued in the future in order to better understand changing permafrost conditions as a consequence of climate warming.

17.4 The Gosau Valley

17.4.1 A Fossil Coral Reef and a Magnificent View

One of the most impressive places in the Salzkammergut region is the viewpoint from the late glacial moraine on the northern lakeside of the northernmost Vorderer Gosausee, visited by thousands of tourists each year. From there, or even better from the summit station of the Zwieselalm cable car, one can catch a glimpse of the still glaciated northwest face of the Dachstein massif (Gosau and Torstein glaciers), and it is easy to imagine how the glacier tongues, particularly during the Lateglacial, contributed to shaping the valley (Figs. 17.7a, b). The so-called Gosaukamm is visible to the west, a mountain ridge composed of several peaks, which are divided by faults and joints, and flanked by long run out cones of debris—derived from weathered Dachstein reef limestone (Fig. 17.7a; site 4 in Fig. 17.1). The transition of the two different limestone facies—cyclic bedded and unstratified reef limestone—with their different weathering behavior was described for the southern Dachstein plateau by Haas et al. (2010). Further down the valley, landscape morphology suddenly changes from rocky cliffs to a densely forested hilly region. This is the type locality of the Gosau Formation giving the landscape a completely different and smooth character (Fig. 17.8a).

Fig. 17.7 **a** Gosau Glacier during the early Late Glacial (possibly Steinach Stadial; locally known as Jochwand Stadium, 18–19 ka BP) filling up the Gosau Valley with the morphologically active Gosaukamm (site 4 in Fig. 17.1) in the background; view toward south. **b** View from the Zwieselalm toward southeast over the Vorderer Gosau Lake and the Dachstein massif. Glacier extents during the Lateglacial Daun Stadial (locally known as Echern Stadium, 15–16 ka BP, transparent white), the Egesen Stadial (locally known as Taubenkar Stadium, 11.5–12 ka BP, dashed white line), and the LIA at around 1850 (deep white) are illustrated. All glacier extents and ages after Plöchinger (1982) and Mandl et al. (2012). Photos: Weidinger (2012)

17.4.2 The Gosau Formation: The Cretaceous–Paleogene Boundary, a Giant Sackung, and Debris Flows

The bedded Gosau Formation developed in a shallow sea, which transgressed over a primary tectonic nappe structure of the Northern Calcareous Alps from the Upper Cretaceous to the Lower Eocene period (Wagreich 1988, 2003). Undisturbed stratigraphic sequences of these beds even bear the Cretaceous–Paleogene boundary in their upper parts, locally called the Zwieselalm Formation (Fig. 17.8b; site 5 in Fig. 17.1; Lahodynsky 2003). In these Flysch sequences, the turbiditic sediments of the Cretaceous to Eocene Gosau Formation contain layers with siliciclastic quartz grains from crystalline rocks of former nearby landmasses. These rocks are the source of the grindstone mining in Gosau, which has lasted at least during the past 400 years, but might date further back to prehistoric times (Lobitzer et al. 2010). Close to that mining area, the so-called Löckermoos is located (site 6 in Fig. 17.1). Pollen analysis of this peat bog allowed for reconstructing the vegetation and climate history in this montane to subalpine altitude from the Last Glacial Maximum (LGM, c. 20 ka BP) over the Lateglacial and Holocene period until modern times (Draxler 2014). Today, it is a nature reserve with a high diversity of protected peat flora and fauna.

Fig. 17.8 **a** Panoramic view toward north over the central part of the Gosau Valley. **b** The Zwieselalm Formation with its siliciclastic flysch material is highly susceptible to gully erosion as seen from the large Gosau Lake toward its most famous outcrop, the so-called Rote Wand (= red wall). Here, the Cretaceous–Paleogene boundary has been found (site 5 in Fig. 17.1) Photos: J. T. Weidinger

Fig. 17.9 **a** View from the skywalk "five fingers" close to the top of the Krippenstein (2108 m asl) toward the cascade of mass-wasting processes from the Plassen to the city of Hallstatt. The village Obertraun is located on a typical delta which has been deposited in Lake Hallstatt since Lateglacial times by the river Traun. **b** The basin of Bad Aussee southeast of Lake Hallstatt (seen from Sarstein) Photos: Weidinger (2012)

Except for the stratigraphically lower parts, Gosau rocks are mostly composed of sandstones, silts, marls, and slates, which all show weak mechanical behavior. This petrophysical predisposition and local high precipitation are the reasons for a combination of slow and deep-seated gravitational slope deformations and frequent debris flow activity in the entire Gosau Valley. One example is located in the northwestern part of the valley, where a large sackung from the Hornspitze (1433 m asl; site 7 in Fig. 17.1) is the source area for episodically active debris flows forming huge cones and fans in the valley bottom. But all other densely forested slopes of the Gosau Valley (Fig. 17.8a) are also home and source of creeks that episodically produce torrential debris flows, which are controlled by check dams (site 8 in Fig. 17.1). These processes present a severe hazard to local settlements and economy, particularly farming, skiing, and tourism infrastructure.

17.5 Around Lake Hallstatt to Bad Goisern and Bad Aussee

17.5.1 Glacial Imprint in the Root Zone of the Former Traun Glacier

The southern shore of Lake Hallstatt is surrounded by steep walls composed of the thick cyclic banks of Dachstein limestone. The geomorphic macro-relief was shaped by the Traun Glacier during the LGM and in Lateglacial stages (van Husen 1977), forming a deep trough that ended up in the accumulation of meltwaters—the recent Lake Hallstatt. The basin and village of Bad Goisern north of Lake Hallstatt (Fig. 17.9a) is surrounded by Lateglacial moraines of the so-called Goiserer (c. 14 ka BP) and Jochwand Stadial (16–17 ka BP; Fig. 17.1). To the southeast, the basin of Bad Aussee (Styria) is situated in the typical Salzkammergut scenery (Fig. 17.9b).

After Lateglacial glacier retreat, oversteepened rock walls were exposed. These started to mechanically react along dip-slope structures giving rise to numerous rock-, ice-, snow-, and debris avalanches, or mixtures between them. Some of these events generated famous landforms, as for example the Steingrabenschneid, a steep detachment scarp of a rock avalanche from 1652 (site 9 in Fig. 17.1; Weidinger 2012). More recent events, such as the rock avalanche in the Echern Valley from 2012 (site 10 in Fig. 17.1), regularly force authorities to adapt and mitigate in various ways (e.g., control structures, land use planning). However, hazardous rock and boulder falls or rock avalanches have been always recognized by locals—and Friedrich Simony already described the local profession of the so-called *Steinbewahrer* ("stone preserver") being responsible for securing mountain flanks from loose rock.

In the Holocene, the valley fills of Obertraun and Hallstatt-Lahn started to accumulate, most likely with the highest rates shortly after deglaciation. This assumption is based on the paraglacial concept (Church and Ryder 1972) describing the rapid adjustment of deglaciated, semi-stable landscapes through initially enhanced but successively declining rates of sediment reworking. The typical delta deposits are one of the rare flat areas in the region and therefore densely settled by the villages of Hallstatt-Lahn and Obertraun. The latter recently has become celebrated for the so-called Hallstein Mineral Water coming from a depth

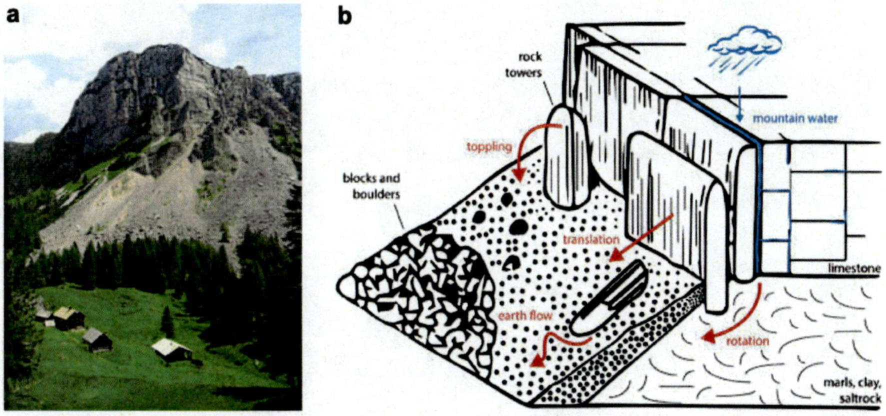

Fig. 17.10 a View of the mass movements from the Sandling (1710 m asl). b Sketch of the geomechanical system with mechanisms of dislocation (compiled after Moser et al. 2003)

of 214 m out of this hundreds of meters thick sediment body—an ideal aquifer for the karst water of the Dachstein massif beyond the rock walls. The water rises to the surface as an artesian spring at a constant temperature of 5.3 °C, ready for bottling and drinking.

17.5.2 The Geomechanical System "Hard on Soft Rocks" and Its Regional Geomorphological Relevance

Triassic to Jurassic stratigraphy and partly tectonics created a specific distribution and succession of lithologies in parts of the inner Salzkammergut, with soft and mechanically weak reacting rocks (such as salt and marls) underneath hard and rigid ones (mostly limestones). This setting of "hard on soft rocks" causes a variety of gravity-induced mass movements in the inner Salzkammergut. Areas specifically affected are shown in Fig. 17.1. Mass movements are manifold and follow typical sediment cascades with rockfall and gravitational toppling in the higher parts and subsequent earth, mud, and debris flows further down (Fig. 17.9). The latter preferentially occurs in materials which can easily be displaced during heavy precipitation events and might endanger villages or devastate forests and cultivated land (Moser et al. 2003; Wilson et al. 2003). The Sandling (1710 m asl) is a classic location for the system with multiple successions of rock avalanches (Fig. 17.10a and site 11 in Fig. 17.1).

The Zwerchwand (1341 m asl) east of Bad Goisern (site 12 in Fig. 17.1) is another example of this setting prone to mass movements. The rock wall (80 m high and 1 km long) is composed of Jurassic limestone and underlain by Haselgebirge, a mixture of clay, gypsum, and salt. Huge blocks and boulders scattered underneath the rock wall indicate multiple events, as for example in October 1978 (60,000 m^3), February 1980 (40,000 m^3), and March 1981 (30,000 m^3). These rock avalanches triggered further mass movements by either subsurface pore water mobilization or the effect of undrained loading (Wilson et al. 2003). An earth-, mud-, and debris flow with a velocity of 5 m/day and a volume of 14 Mm^3 started to move in February 1982 along the Stammbach Valley endangering the village of Bad Goisern. After several months, the complete mass movement could be stopped by intense mitigation measures including drainage of the moving mass.

A third example for the "hard on soft rocks" setting is located in the salt mining area between Hallstatt and the Plassen (1953 m asl)—in the heart of the World Heritage site (site 13 in Fig. 17.1). This risky cascade of complex mass movements is characterized by recurrent rock avalanches, earth- and debris flows. Two prehistoric events in the Bronze and late Iron Ages forced people to shift their housing and mining infrastructure (Kern et al. 2008). Numerous hazardous events have been reported by the local population from historic times as well (Weidinger 2012). To understand mass movements along the sediment cascade toward the village of Hallstatt, the endogenous structure of Plassen and its underlying rocks beyond the mining area need to be considered. Recent rockfall activity suggests lateral spreading of the limestone mass on top of soft, ductile, and petrophysically weak salt rocks, ending up in permanent morphodynamic activity (Melzner et al. 2015; Moser et al. 2003). The last huge rockfall at the southern flank of Plassen occurred in spring 2014 (Fig. 17.11a). In order to quantify the event and subsequent rockfall activity, the area has been regularly surveyed by terrestrial laser scanning (TLS) since September 2014. This monitoring allows detecting and quantifying morphological changes in detail (Fig. 17.11b). Based on the comparison of airborne laser scan (ALS) data acquired before the event in 2013, the first TLS dataset indicates an initial rockfall volume of c. 135,000 m^3. Post-event rockfall activity is considerable as well and continued until winter 2015/2016. An accumulation volume of c. 70,000 m^3 between September 2014 and September 2015 (Fig. 17.11c) was followed by c. 12,500 m^3 between

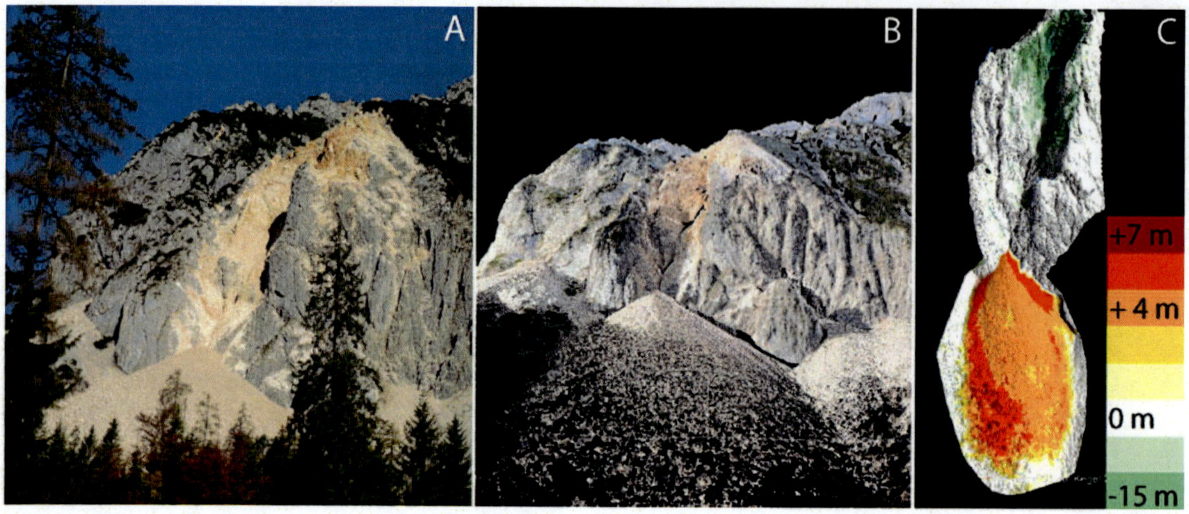

Fig. 17.11 **a** Recent rockfall event (spring 2014) at the southern flank of the Plassen, which is monitored since autumn 2014 using repeated TLS. **b** TLS-based 3D point cloud of the rockfall release area and the fresh rockfall deposits. **c** Surface change model between September 2014 and September 2015 indicating heavy post-event rockfall activity with a cumulative accumulation volume of c. 70,000 m^3 (site 13 in Fig. 17.1)

September 2015 and November 2015 indicating an average daily rockfall supply of 190–235 m^3.

17.5.3 The Koppenschlucht Bottleneck

The westernmost railway section of the Kronprinz Rudolf-Bahn provided the first north–south connection along Schärding, Attnang Puchheim, Stainach, Leoben, until Ljubljana and was completed in 1877 during the Austro-Hungarian Monarchy. The route was of particular importance for transporting salt and coal for the salines in the Salzkammergut. However, the Koppenschlucht section between Obertraun and Bad Aussee crossing the border between Upper Austria and Styria has always constituted a major bottleneck (site 14 in Fig. 17.1). Due to the deeply incised Traun Valley and the steep slopes between the Sarstein (1975 m asl) and the Dachstein massif, the route is endangered by multiple processes, including avalanches, flooding, and debris flows. While the first salt transport passed without difficulties in 1877, the route was closed due to snow avalanches already one year later for several days with over 60 pioneers in action. In 1896, the route needed to be closed for two months due to flooding and debris flows, and in the following year, the rail track was undercut and completely swept away at several places. After that, rails were rebuilt at a 20 m higher level. Protection measures against avalanches, two new bridges over the river Traun, and a tunnel crossing a section of Sarstein improved the situation, but in 1899 and 1920 the route was blocked again by flooding. However, despite a set of protection measures, the section needs to be occasionally closed until present times (the last time in 2009).

17.6 Tourism in the Inner Salzkammergut—Difficulties and Future Perspectives

Since the second half of the twentieth century changes in the economic structure of the inner Salzkammergut have forced authorities and local people to install and progressively develop tourism as a major source of income. Besides positive effects on economy, negative ones on nature can be exemplified at the Dachstein-Schladminger Glacier. Extensive infrastructure there includes not only classical skiing infrastructure, but also fancy "glacier experiences," such as a suspension bridge, the "stairway to nothingness," or the "ice palace" excavated into the glacier which in turn requests ice harvesting to keep the construction alive (Mandl 2014). Despite numerous problems in unifying economic development and nature conservation, a lot has been achieved with respect to the protection of the natural heritage of the inner Salzkammergut. The Dachstein massif and parts of adjacent areas correspond to different conservation areas, including a Natura 2000 region (Upper Austria), the *Landschaftsschutzgebiet Dachstein-Salzkammergut* and the *Naturdenkmal Dachstein-Südwand* (both Styria). Several educational paths have been successfully installed, and an integrated concept to establish a geopark is recently in planning (Weidinger 2013).

References

Brommer C, Phillips M, Keusen H-R, Teysseire P (2009) Bauen im Permafrost, Ein Leitfaden für die Praxis. Eidgenössische Forschungsanstalt für Wald, Schnee und Landschaft WSL: Birmensdorf:1–126

Church M, Ryder JM (1972) Paraglacial sedimentation: a consideration of fluvial processes conditioned by glaciation. Geol Soc Am Bull:3059–3071

Draxler I (2014) Archivwert der Moore im Dachsteingebiet. Gmundner Geostudien 5:41–55

Frisch W, Kuhlemann J, Dunkl I, Székely B (2001) The Dachstein paleosurface and the Augenstein formation in the Northern Calcareous Alps—a mosaic stone in the geomorphological evolution of the Eastern Alps. Int J Earth Sci 90:500–518

Frisch W, Kuhlemann J, Székely B, Dunkl I (2003) Die Hochplateaus in den Kalkalpen – 35 Millionen Jahre alte Landoberflächen. Gmundner Geostudien 2:181–189

Gamsjäger S (2010) (ed) 100 Jahre Dachsteinhöhlen - Beiträge zur Karst- und Höhlenkunde mit Schwerpunkt Dachsteinhöhlen. SPELDOK 21, Schwarzreiter, 23. Jahrgang, Sonderheft 2a

Haas J, Piros O, Görög Á, Lobitzer H (2009) Paleokarst phenomena and Peritidal beds in the cyclic Dachstein limestone on the Dachstein plateau (Northern Calcareous Alps, Upper Austria). Jahrb Geol Bundesanst 149(1):7–21

Haas J, Piros O, Budai T, Görög Á, Mandl WM, Lobitzer H (2010) Transition Between the Massive Reef-Backreef and Cyclic Lagoon Facies of the Dachstein Limestone in the Southern Part of the Dachstein Plateau, Northern Calcareous Alps, Upper Austria and Styria. Abhandlungen der Geol. B.-A. 65:35–56

Hager C (1992) Die Eisenbahnen im Salzkammergut. Wilhelm Ennsthaler Verlag, Steyr, pp 1–167

Jeschke HP (ed) (2002) Das Salzkammergut und die Weltkulturerbelandschaft Hallstatt-Dachstein/Salzkammergut. Grundlagenforschung, Kulturlandschaftspflegewerk und Monitoring, Bd. 1. In: Gesellschaft für Landeskunde – OÖ. Musealverein, I. Historische Reihe 13:1–392

Kern A, Kowarik K, Rausch A, Reschreiter H (eds) (2008) Salz – Reich, 7000 Jahre Hallstatt. Veröffentlichungen der Prähistorischen Abteilung des Naturhistorischen Museums Wien 2: 1–239

Lahodynsky R (2003) Die Kreide Tertiär-Grenze im Salzkammergut. Gmundner Geostudien 2:29–38

Lobitzer H (2001) Die frühe geologische Erforschung des Weltkultur- und Natur-Erbe Gebietes Hallstatt-Dachstein-Salzkammergut (Johann Bohadsch bis Carl Ferdinand Peters). Geschichte der Erdwissenschaften in Österreich 1, Graz

Lobitzer H (2009) Joseph August Schultes: Ein Pionier der geologischen Erforschung des Salzkammergutes vor 200 Jahren. Jahrbuch der Geol. Bundesanstalt 149/2+3:411–421

Lobitzer H, Lelkes-Felvári G, Ottner F, Svobodova M, Svábenická L (2010) Grindstone mining in Gosau—the classical locality of the Ressen Formation (Lower Campanian, Gosau, Upper Austria). Abhandlungen der Geol. B.-A. 65:169–179

Lobitzer H, Posmourný K (2010) Joahnn Baptist Bohadsch – Ein Pionier der naturwissenschaftlichen Erforschung des Salzkammergutes. Jahrbuch der Geol. Bundesanstalt 150/3+4:473–490

Mandl F (2014) Gletscherzustandsbericht 2012. Schladminger und Hallstätter Gletscher. Gmundner Geostudien 5:13–17

Mandl GW (2000) The alpine sector oft he Tethyan shelf—examples of Triassic and Jurassic sedimentation and deformation of the northern calcareous alps. Mitt Österr Geol Ges 92:61–77

Mandl G W, van Husen D & Lobitzer H (2012) Geologische Karte der Republik Österreich 1 : 50.000, Blatt 96-Bad Ischl Erläuterungen, 215p., Geologische Bundesanstalt, Wien

Melzner S, Ottowitz D, Pfeiler S, Moser M, Motschka K, Lotter M, Mandl GW, Rohn J, Otter J, Wimmer-Frey I (2015) A multidisciplinary methodology for the characterization of a large rock spread in the Northern Calcareous Alps (Eastern Alps). Geophysical Research Abstracts Vol. 17, EGU2015-10658-2

Mikulás R, Svobodová M, Svábenická L, Lobitzer H (2010) Ichnofossils of the Ressen Formation in Gosau (Campanian, Upper Gosau Subgroup, Upper Austria). Abhandlungen der Geol. B-A 65:155–168

Moser M, Lotter M, Meier H (2003) Giant Mass-movements in the area around hallstatt—geotechnic-kinematic-mechanism of movement. Gmundner Geostudien 2:343–352

Plöchinger B (1982) Geologische Karte der Republik Österreich 1 : 50.000, Blatt 95-St. Wolfgang und Erläuterungen, 74p., Geologische Bundesanstalt, Wien

Rode M (2014) Permafrost und Frostverwitterung am Dachsteinmassiv. Gmundner Geostudien 5:31–38

Schnepfleitner H, Kellerer-Pirklbauer A, Rode M (2016) Das Permafrostbohrloch "Koppenkarstein North Face" am Dachsteinmassiv: Ein österreichischer Beitrag zum Global Terrestrial Network for Permafrost (GTN-P). Joannea Geol. Paläont. 12, 28–36

Simony F (1889) Das Dachsteingebiet. Ein Geographisches Charakterbild aus den Österreichischen Nordalpen. Eduard Hölzel Verlag, Wien, pp 1–182

Speta F, Aubrecht G (1996) Ein Leben für den Dachstein – Friedrich Simony zum 100. Todestag. Stapfia 43, Katalog des OÖ Landesmuseums, NF 103, Linz: 1–355

Tollmann A (1974) Geologie im Umbruch. Mitt. Österr. Geograph. Ges. 116:53–78

van Husen D (1977) Zur Fazies und Stratigraphie der Jungpleistozänen Ablagerungen im Trauntal. Jahrb Geol Bundesanst 120(1):1–130

Wagreich M (1988) Sedimentologie und Beckenentwicklung des tieferen Abschnittes (Santon-Untercampan) der Gosauschichtgruppe von Gosau und Rußbach (Oberösterreich - Salzburg). Jahrbuch Der Geol Bundesanstalt 131(4):663–685

Weidinger JT (2012) Massenbewegungen im Salzkammergut – eine geologisch-geomorphologisch-kulturhistorische Betrachtung. Österreich in Geschichte Und Literatur Mit Geographie 56(1):77–95

Weidinger JT (2013) Die Dachstein-Klima-Arena. Unveröffentlichter Projektbericht zum Evaluierungsauftrag der geotouristischen Ressourcen im Gosautal-OÖ für die Naturschutzabteilung der OÖ Landesregierung und die Dachstein-Seilbahnen-Holding:1–178

Weingartner H (ed) (2008) Lehrpfad Hallstätter Gletscher. Ein Begleiter durch die Gebirgslandschaft am Dachstein. Atelier Tintifax, Wien, pp 1–123

Wilson AJ, Petley DN, Murphy W (2003) Down-slope variation in geotechnical parameters and pore fluid control on a large-scale Alpine landslide. Geomorphology 54:49–62

Johannes T. Weidinger is head of the K-Hof, Kammerhof Museums of Gmunden. As a research earth and educational scientist he focuses on natural hazards with an emphasis on mass movements in high mountain regions, such as the Himalayas and the Alps. He is a guest lecturer at the Universities of Vienna, Salzburg, Linz/Upper Austria and Klagenfurt/Carinthia.

Joachim Götz is a research associate at the University of Bayreuth, Germany. As a mountain geomorphologist, his research and teaching interests include process-based monitoring studies focusing on drivers and controls of present-day sediment dynamics and the archive-based reconstruction of post- and paraglacial landscape evolution. His research in the Eastern Alps, the Nepalese Himalayas and the Argentinian Andes is based on a wide range of modern field, lab and modelling techniques.

18 Gesäuse: River Gorge, Limestone Massifs and Sediment Cascades

Gerhard Karl Lieb and Oliver Sass

Abstract

Gesäuse is the name of a huge gorge the Enns River has cut into the Northern Calcareous Alps forming an impressive scenery with a relative relief of up to 1800 m. Besides a topographical overview, this chapter presents the most important facts on the interplay between lithology and relief and the impact of the Pleistocene glaciation on the region. Remnants of palaeosurfaces are discussed with respect to morphogenesis, resulting in a hypothesis on the development of the Gesäuse Gorge. Furthermore, we provide insight into ongoing research focussing on sediment cascades. The results show that in the picturesque dolomite landscape around the Johnsbach tributary river, rockwall retreat rates are exceptionally high. The debris is transported to the Johnsbach by episodic debris flows. There is a considerable human impact by former gravel mining and by undersized bridge openings, disconnecting some side valleys from the main river. Further interactions of geomorphology and socio-economic activities are discussed with a focus on natural hazards and nature protection (Gesäuse National Park). Finally, some selected future research agendas are outlined.

Keywords

Denudation chronology of the gorge • Glaciation • Palaeosurface • Sediment transport • Natural hazards

G. K. Lieb (✉)
Institute of Geography and Regional Science, University of Graz, Heinrichstrasse 36, 8010 Graz, Austria
e-mail: gerhard.lieb@uni-graz.at

O. Sass
Fachgruppe Geomorphologie, University Bayreuth, Universitätsstr. 30/GEO II, 95447 Bayreuth, Germany

18.1 Introduction

Gesäuse is the name of an impressive gorge in the Northern Calcareous Alps. The origin of the name comes from the roaring noise of the Enns River passing the cataract at the upper end of the gorge. The incision of the river and its tributaries has divided the Northern Calcareous Alps into separate mountain groups which are called Gesäuse Mountains (Gesäuseberge) in their entirety. Their striking appearance has been attracting attention since the beginning of the nineteenth century when painters found romantic motifs, and the first natural scientists began to explore the flora of the high mountains. In the second half of the nineteenth century, in particular after the opening of the railroad through the gorge, mountaineers (above all H. Hess) discovered the Gesäuse Mountains and due to the solid calcareous rocks gradually turned them into an arena for adventurous rock climbing.

The impressive scenery has always played an important role in creating the idea of protecting the Gesäuse Gorge and mountains which dates back to the 1950s. However, it took decades to implement this project, amongst other things because of the resistance of a power plant company planning a hydropower station just at the famous "Gesäuseeingang" site (Fig. 18.1), the construction of which was ultimately prohibited. Eventually, Gesäuse National Park was established in 2002 because of its suitability for being a high-quality protected area of international importance. With an area of 110 km^2, it comprises most of the Gesäuse Mountains and gorge as far as its natural value is not reduced by technical impacts (which is, for instance, the case in the section Kummerbrücke-Hieflau where river discharge is reduced because of the diversion of water to a power plant). The National Park is the youngest of Austria's six national parks, offering its visitors a wide range of environment-friendly activities.

From a geomorphological point of view, the Gesäuse serves as a good example of high mountains with a

C. Embleton-Hamann (ed.), *Landscapes and Landforms of Austria*, World Geomorphological Landscapes,
https://doi.org/10.1007/978-3-030-92815-5_18

Fig. 18.1 Enns Valley and Gesäuse mountains. View from the summit of Pleschberg towards south-east. The valley bottom east of Admont seems to be entirely locked by the Gesäuse Mountains which are divided into separate mountain groups by the gorge of the Enns River Photo: G. K. Lieb

predominant mountain chain character in the Northern Calcareous Alps. The typical processes and landforms of these mountains will be addressed in Sect. 3, especially with respect to sediment cascades, which are currently a research focus (Sect. 4). The topography of the gorge itself is of particular interest: Approaching the Gesäuse from the west, visitors use the road through the Enns valley with its wide, flat valley bottom. East of Admont, this broad valley is blocked by the Gesäuse Mountains towering above the valley like a mighty wall through which the Enns River finds its way in the narrow gorge of "Gesäuseeingang" (meaning gate of Gesäuse, Fig. 18.1), although the river could have flowed through the nearby situated low pass of Buchau.

18.2 Geographical and Geological Setting

The Enns Valley west of the Gesäuse is part of the Northern Longitudinal Valley Depression (cf. Chap. River and Valley Landscapes) which is prefigured by a major inner-Alpine fault, the so-called SEMP fault (cf. Sect. 3.2). All the rivers in this valley zone flow roughly W–E, parallel to the predominant direction of the neighbouring mountain ridges, before they turn northwards cutting through the Northern Calcareous Alps to finally reach the Danube River. The Enns River keeps its W–E direction within the Gesäuse Gorge, then turning north near Hieflau (Fig. 18.2). The incision of the Enns River and its tributaries has caused the separation of originally coherent calcareous massifs into individual mountain groups. These subareas (Buchstein, Reichenstein and Hochtor group) are marked in Figs. 18.1 and 18.2.

Another prominent relief feature is the saddle of Buchau. It represents a former valley bottom which was occupied until the upper Pleistocene. It forms the northern border of the Buchstein group and separates the calcareous mountain chain of Haller Mauern from the Gesäuse Mountains.

From a climatological point of view, the Gesäuse region is characterised by high annual precipitation of 1500 mm in the valleys and up to 2500 mm at the summits. Although most of the precipitation falls in summer, the winter climate is rich in snow with the consequence of significant avalanche hazard. Mean annual air temperatures vary between 7 °C in the valleys and slightly below 0 °C at the highest summits; annual variations are moderate due to the maritime character of the climate.

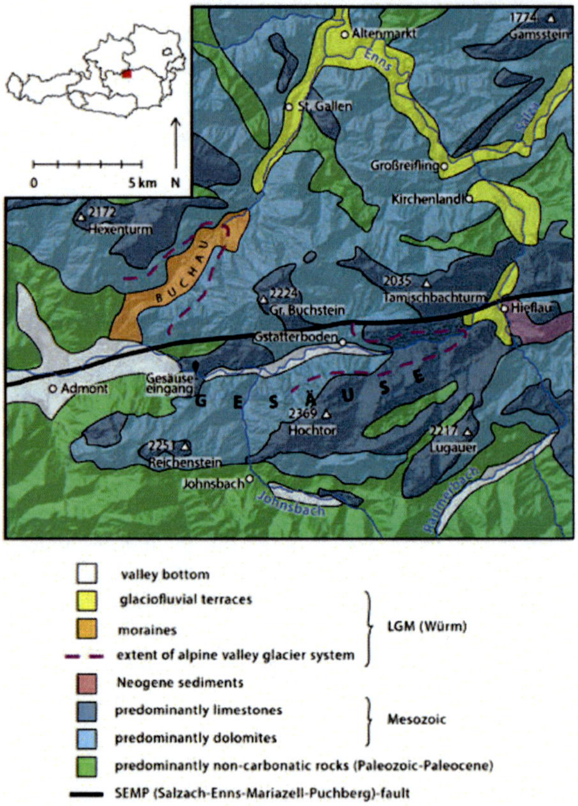

Fig. 18.2 Overview of topography and main lithological units (Rupp et al. 2011, simplified) of the Gesäuse mountains. The Gesäuse Gorge stretches from Gesäuseeingang in the W to Hieflau in the E. (Hillshade: Federal State Government of Styria, Board of Works—Geoinformation Staff Office; Cartography: V. Damm)

18.2.1 Lithology, Stratigraphy and Relief

The area shown in Fig. 18.2 belongs to two large lithological zones of the Eastern Alps, the Greywacke Zone built of Palaeozoic rocks in the south and the Northern Calcareous Alps predominantly built of Mesozoic rocks in the north. These two lithological units are characterised by completely different topographies because of the predominance of schists in the Greywacke and carbonatic rocks in the calcareous zone (Fig. 18.3). The Gesäuse Mountains belong entirely to the latter, which is why only the Mesozoic rock series will be considered in this section.

Table 18.1 gives a simplified overview of the Mesozoic stratigraphy of the Gesäuse Mountains; for further details, see Ampferer (1935), Büchner (1970), Kollmann (1983), in a supraregional context Flügel and Neubauer (1984) and Rupp et al. (2011). In most locations, the rocks occur in a concordant sequence with older rocks in lower and younger ones in higher positions. Landforms strongly depend on lithology—an aspect that already Geyer (1918), the author of the first regional study on geomorphology, had emphasised. From a geomorphological perspective, the rocks can be grouped as follows:

- Non-carbonatic rocks (predominantly Werfen and Raibl formation) form moderate mountainous relief without sheer rock walls.
- Dolomites (predominantly Wetterstein Dolomite) are prone to intensive erosion in some areas, e.g. in the lower Johnsbach Valley, which results in a picturesque "dolomite erosion landscape" (Lieb and Premm 2008) (Fig. 18.4). Weathering and erosional processes are highly active (Sect. 4).
- Limestones (predominantly Dachstein limestone) form steep rock faces and occur in different varieties, especially reef facies (Reichenstein group) and thickly bedded facies (other mountain groups). The different types of the limestone and the inclination of its layers result in diverse shapes of the summits (Fig. 18.5).

18.2.2 Quaternary Glaciation

During the Last Glacial Maximum (LGM), the Gesäuse was situated at the eastern margin of the Eastern Alpine complex of transection glaciers. The so-called Enns Glacier had large accumulation areas in the Central Alps west of the Gesäuse from where it flowed down the Enns Valley. About 15 km west of Admont, it branched into several glacier tongues so that the ice masses reaching the Gesäuse Mountains were already significantly reduced. When in contact with the Gesäuse barrier, the glacier formed two glacier tongues (outlined in Fig. 18.2):

- The main ice mass entered the Gesäuse Gorge, was also supplied by ice of local glaciers (e.g. the one coming down from Gr. Buchstein at Gstatterboden) and ended c.

Fig. 18.3 Inner Johnsbach Valley. View from the ascent to Reichenstein towards east. The landforms of the Greywacke Zone (right) contrast substantially with those of the Calcareous Alps (left) because of different lithologies Photo: G. K. Lieb

Table 18.1 Stratigraphic table of the Gesäuse Mountains (after Büchner 1970 and Rupp et al. 2011)

Period/Age	Numerical age (Ma)	Rock	Annotation
Jurassic	200	Fleckenmergel (marly limestone), crinoid limestone	Distinct in appearance because of dense vegetation cover, occurring in small areas only
Triassic/Rhaetian	204	Dachstein limestone (reef and banked limestone)	One of the main rocks building the summits
Triassic/Norian	217	Dachstein dolomite (light-grey dolomite)	Prone to erosion
Triassic/Carnian	228	Raibl formation (schists, sandstone)	Forming smooth topography, occurring in small areas only
Triassic/Ladinian	237	Wetterstein dolomite	One of the dominant rock types, erosional landscape, in small areas limestone
Triassic/Anisian	245	Gutenstein limestone and dolomite (dark-grey dolomites with limestone lenses)	Occurring in small areas only
Triassic/Early Triassic	251	Werfen formation (sandstones, schists, in small areas calcareous rocks)	Forming smooth topography
Late Permian	260	Präbichl formation (quarzite, sandstones, conglomerates)	Occurring in small areas only

Fig. 18.4 A part of the lower Johnsbach Valley ("Zwischenmäuer", meaning between the walls). Highly active erosional processes in dolomites have created a scenery of crests, rock towers and gullies. The landforms have stimulated imagination resulting in local legends and myths being told about single rock features Photo: G. K. Lieb

3 km west of the village Hieflau. The terminus position cannot be reconstructed with certainty because no morainic deposits were preserved on valley sides. Only the onset of glaciofluvial terraces further downvalley enables a rough estimate of the extent of the glaciation (van Husen 1968).

- The second glacier tongue crossing the Buchau pass left a well preserved sequence of terminal and lateral moraines in the saddle area. Because of its dense forest cover, this morainic relief is not well visible in the field. At its terminus, this glacier tongue was in contact with another local glacier descending from the Buchstein massif.

Fig. 18.5 Different shapes of summits depending on the varying character of the Dachstein limestone and the inclination of its layers. **a** Admonter Kaibling (reef facies, from SW), **b** Planspitz (asymmetric shape due to dip angle; from SW), **c** Hochtor (folded thick bedded limestone with incised glacial cirques; from E) Photos A, C: G. K. Lieb, B: O. Sass

Fig. 18.6 A fog stratum on a cold winter day resembles the surface of the LGM glacier branching into two glacier tongues when getting in contact with the Gesäuse mountains. View from the climb to a summit of the Haller Mauern to the south with the Reichenstein group in the background. Photo: G. K. Lieb

Figure 18.6 gives an impression of the LGM glaciation with an ice surface at c. 1000 m asl in the vicinity of Admont.

Besides these outliers of the Alpine valley glacier system, the Gesäuse Mountains were widely covered by local glaciers. The Buchstein glaciers have already been mentioned. Another example is a glacier tongue with its accumulation area in the upper parts of the Hochtor group reaching the inner part of the Johnsbach Valley with its terminus. In the well-developed cirques (Fig. 18.5c), small glaciers also existed during the cold phases of the Lateglacial period which is testified by respective moraine ridges.

In contrast to the Würm glaciation, the Gesäuse Mountains were part of the much larger transection glacier complex during the earlier Riss glaciation. In the lower Johnsbach Valley, for instance, (unpublished) geoelectrical profiles revealed a U-shaped valley floor underneath a c. 50 m thick valley fill. This U-shaped form is assigned to the erosion of a Riss glacier, as the Würm glaciation did not overprint this area according to our current knowledge. A reconstruction of both the Riss and the Würm glaciation prepared by D. van Husen can be found in Rupp et al. (2011). Due to the little morphological relevance of the older glaciation, no further emphasis is given to them in this chapter.

18.3 Morphogenesis of the Macrorelief Structures

This section focusses on the geomorphological development of the Gesäuse Gorge and the surrounding mountain groups which has been discussed by several authors since Geyer (1918). The significant change of relief which took place since the Neogene period is of special interest: Palaeosurface remnants testify to the existence of a former S–N drainage that was turned to a W–E one.

18.3.1 Remnants of Ancient Relief

Following the differentiation of the Eastern Alps developed by Frisch et al. (2000), the Gesäuse Mountains belong to the third “principal geomorphological domain” referred to as

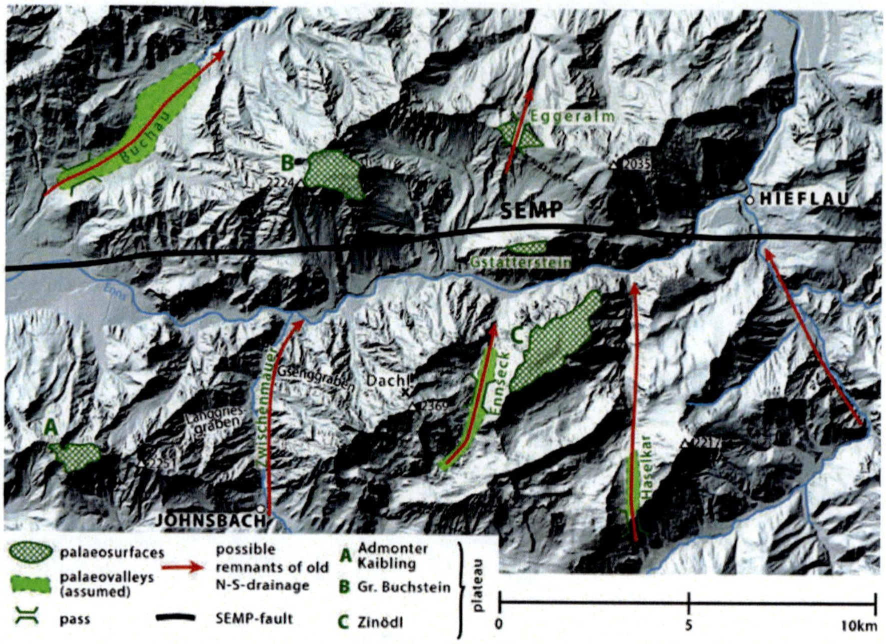

Fig. 18.7 Remnants of palaeosurfaces and past drainage patterns at both sides of the Gesäuse Gorge (hillshade: Federal State Government of Styria, Board of Works—Geoinformation Staff Office; Cartography: V. Damm). In addition, the sites mentioned in Sect. 4 are indicated

"high to intermediate relief with remnants of the Oligocene Dachstein palaeosurface". According to this hypothesis, this palaeosurface (its name is derived from the Dachstein massif c. 70 km further west) developed at the turn of the Eocene and Oligocene. Afterwards it was covered by fluvial deposits of the Augenstein Formation and uplifted not before the Late Miocene and Pliocene. It was this relatively late uplift that destroyed an older drainage system in which the rivers flowed roughly S–N, transporting sediments from the Central Alps (for details about the Dachstein palaeosurface and Augenstein Formation see Sect. 19.2 of Chap. "The Rax Karst massif: A Typical Plateau of the Northern Calcareous Alps?").

In the Gesäuse Mountains, some of these S–N trending valleys have been preserved: The depression at both sides of the Ennseck pass (1691 m asl, Figs. 18.7 and 18.8), which divides the Hochtor group into a western and an eastern part, is a good example. This palaeovalley was transformed by subsequent glacial processes, which carved two troughs into it —one declining to the north, the other one to the south—and by karstification inducing subterranean drainage and retrograde inclination because of karstic depressions. Ampferer (1935) and Kollmann (1983) assumed the continuation of this palaeovalley at the pass near Eggeralm (1442 m asl) in the Buchstein group on the other side of the Enns River. However, considering that lateral extrusion of the Eastern Alps moved the crustal blocks south of the SEMP some tens of kilometres eastwards (Frisch et al. 2000) this interpretation is no longer valid. Nevertheless, the fact that the current Enns River flows about 1000 m lower in W–E direction proves that the incision of the Gesäuse Gorge destroyed old drainage patterns and fundamentally changed river direction. Kollmann (1983) and Wiche (1951) also considered the direction of Haselkar and of the lower part of the Johnsbach Valley as legacies of ancient S–N-oriented valleys (Fig. 18.7).

Near Ennseck pass "Augensteine" deposited on the Dachstein palaeosurface can be found. Their occurrence in a valley incised 500–700 m in between the neighbouring summits is surprising. One explanation is that, similar to other mountain massifs of the Northern Calcareous Alps, Augensteine are not found in their primary location of sedimentation (which was at a higher level). Another explanation is given by Frisch et al. (2000) who pointed out that the palaeosurface was dissected by surface erosion up to several hundred metres deep at many locations. This process must have taken place at an early stage of relief evolution, i.e. before the incision of the Enns Valley, a conclusion which can already be found in Wiche (1951). Thus, today, only small remnants of the original palaeosurface are preserved, which form the three summit plateaus in Fig. 18.8 and are marked with A–C in Fig. 18.7.

18.3.2 Hypotheses on the Development of the Enns Gorge

The morphogenesis of the Gesäuse Gorge inevitably raises the question of how and when the old S–N drainage changed into a W–E one. The Enns Valley follows the SEMP fault, which in the Gesäuse Gorge does not run along the Enns River, but 1–2 km further north of it. This is clearly visible in the field by a sequence of saddles separating some rocky

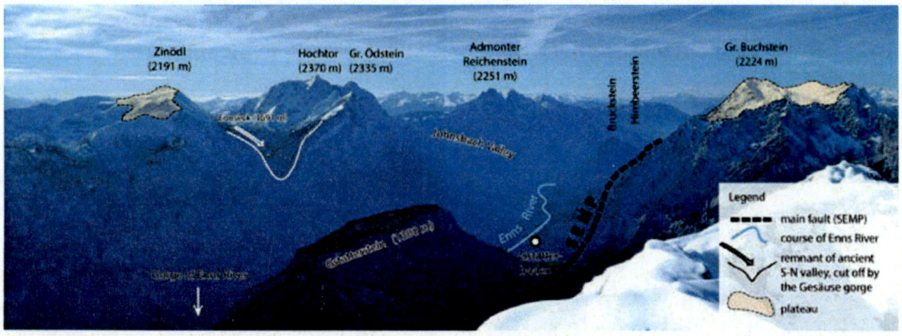

Fig. 18.8 The position of the SEMP (Salzach-Enns-Mariazell-Puchberg-) fault and some palaeorelief remnants at both sides of the Gesäuse Gorge. The ancient valley across the current Ennseck pass is described in the text. Its U-shaped cross profile reveals that it was glacially shaped during the Pleistocene. In the north, the valley ends abruptly high above the Gesäuse Gorge which was incised perpendicular to it at a later stage. View from the summit of Tamischbachturm to the south-east Photo: G.K. Lieb

summits (e.g. Gstatterstein, 1391 m) from the Buchstein group (Figs. 18.7 and 18.8). The SEMP is the primary reason for the W–E stretching valley because such a large fault zone attracts fluvial erosion. However, the question of why the strongest erosion did not occur directly at the fault but slightly south of it remains open. Possibly, secondary faults running parallel to the SEMP (as they were described a short distance to the west of Gesäuse by Keil and Neubauer 2011) have played a major role in this development.

Besides the SEMP fault, two other aspects could have facilitated the alteration of drainage direction: (1) the configuration of the Dachstein Limestone layer forms a syncline with the core of the depression running near to the present course of the Enns River (Kollmann 1983), and (2) there is an old subsidence area in the east near Hieflau where Miocene sediments ("Ennstal Tertiary" according to Flügel and Neubauer 1984) can be found and which also influenced other rivers to flow there (Fig. 18.2).

Based on these considerations, Kollmann (1983) developed the following model of drainage system evolution. As soon as the calcareous massifs started to uplift, underground karst water drainage developed. The water was successively directed to the area of Hieflau because of its low elevation. Thus, increasing amounts of water flowed W–E towards Hieflau, developing a cave river system which was later opened by the collapse of the caves and removal of loose material. This idea is supported by the fact that (because of intensive karstification of pure Dachstein Limestone) surface erosion is only of minimal importance. Frisch et al. (2000) highlighted this fact as a characteristic of this geomorphological domain. All rivers flowing on the surface (as it could have been the case with the Johnsbach River as soon as it had eroded the limestone above the dolomites) were directed eastwards, and finally—probably not before the Pleistocene—the former Enns River that was originally flowing across Buchau to the north-east changed its course.

The Enns Gorge was given its final shape during the last glaciation. The central part of the Gesäuse Gorge was formed as a glacial trough eroded in the dolomite rocks. In contrast, the short section of "Gesäuseeingang" (Figs. 18.1 and 18.2) is a narrow cataract in coarse boulders which fell down from the nearby limestone rock faces. South of it, the limestone monolith of Haindlmauer (1435 m asl) slid northwards due to lateral glacial undercutting (van Husen 1979). Its movement, together with torrential activity, blocked the Enns River leading to the formation of large lakes and nowadays to the shallow river gradient in the valley west of the Gesäuse Gorge (Fig. 18.1). This valley is a good example of an over-deepened glacial trough filled with lacustrine, biogenic and fluvial sediments up to 400 m in thickness (van Husen in Rupp et al. 2011).

18.4 Research on Current Geomorphological Processes

The impressive topography with a relative relief of almost 1800 m and intensive morphodynamic processes has attracted numerous geomorphological research activities in recent years. The backbone of the research infrastructure is the so-called Johnsbach Valley Platform (Strasser et al. 2013) which consists of a network of ten meteorological stations and three runoff gauges. In 2013–16, two third-party research projects on different parts of the geomorphological process cascade were carried out. The research mainly concentrated on the lower Johnsbach Valley ("Zwischenmäuer", Fig. 18.4) in which the S–N running course of the Johnsbach cuts through the dolomite erosional landscape. The location of the sites mentioned is indicated in Fig. 18.7.

Following the sediment cascade from the rockwalls down to the Enns River, the process chain starts with rockfall of different magnitudes originating from the Dachstein

Limestone and Wetterstein Dolomite outcrops. Whilst the Dachstein Limestone is prone to hazardous boulder and cliff falls (e.g. the Dachl rockfall in May 2016 which buried a frequently used Alpine trail; Lieb et al. 2017), the brittle dolomite mostly delivers large amounts of small-scale debris falls, which are highly important for the sediment budget. Terrestrial laser scan surveys supplemented by rockfall nets at the foot of the dolomite rockwalls in the Gsenggraben valley (Fig. 18.9) revealed very high back-weathering rates of c. 1 mm yr^{-1} which is 2–10 times higher than in previous studies on limestone rockwalls (e.g. Sass 2005). This finding underpins that the Wetterstein Dolomite is particularly prone to erosion.

Around the Zwischenmäuer area, the enormous sediment yield from the brittle Wetterstein Dolomite erosion landscape is temporarily deposited in narrow chutes and grooves of the rockwalls. Small debris flows and debris creep move the sediments into the tributaries of the Johnsbach Valley of which the impressive Gsenggraben and Langgriesgraben are the most important ones ("graben", literally meaning "trench", is a regional expression for a steep-sided tributary of a mountain valley). These tributaries do not usually show surface runoff because the water infiltrates into the thick debris fills. Thus, further transport is achieved by episodic debris flows during heavy precipitation events. Five valleys (including Langgriesgraben and Gsenggraben) were monitored by laser scanning and drone surveys over three years (Rascher and Sass 2016). The results show how sediments are routed from one section to the next before depleted channel sections are gradually being refilled again.

Fig. 18.9 Gsenggraben. **a** Overview of the catchment area, as seen from the head of the valley, looking towards west. **b** View of one of the investigated rockwalls showing the terrestrial laser scanner, the rockfall net and the cables of the sensor instrumentation Photos: A: G.K. Lieb, B: M. Rode

Despite the high morphodynamic activity, many of the side channels (e.g. Gsenggraben) are decoupled from the Johnsbach creek due to the effects of undersized bridge openings under the Johnsbach road (Fig. 18.10b) and former gravel mining. The former terrain of the mining company at Gsenggraben now provides a perfect opportunity to monitor how a heavily disturbed site is re-occupied by natural geomorphic process dynamics. The most important channel in terms of current sediment supply to the Johnsbach is the Langgriesgraben (Fig. 18.10a), which was also affected by gravel mining. Since mining was abandoned in 2009 due to national park policy, a light-coloured sediment slug (sensu Fryirs et al. 2007) has developed progressing slowly downvalley from the Langgries mouth.

In its uppermost reaches in the south-east of the valley, the Johnsbach had sufficient time to develop a typical step-pool morphology. Between Kölblwirt and the Johnsbach village a meandering transport reach is found; most of its sediments are deposited due to a damming rockslide mass of probably Lateglacial age. Due to the huge sediment input in the following Zwischenmäuer stretch, a succession of single thread and braided river sections has developed. Runoff measurements show that the Johnsbach is losing surface water into a karstic aquifer in this area. Thus, sediment transport capacity decreases downstream which probably further promotes debris accumulation. Therefore, the torrent runs on a gravel bed for the greater part of its course. Geophysical investigations indicate a valley fill of up to 50 m depth. The sediment yield at the outlet of the catchment is being monitored since 2015 at a bedload station (geophone sill) run by the University of Natural Resources and Life Sciences, Vienna.

18.5 Geomorphology and Socio-economic Activities

As in most regions of the Alps, there is a long and continuous history of human activities in the Gesäuse Mountains despite their rugged and inhospitable topography. The first

Fig. 18.10 Sediment transport in the Zwischenmäuer area. **a** View of the gravel bed in the Langgriesgraben in SW direction. **b** Example of a debris flow that was dammed by an undersized bridge opening and buried the only road to the village of Johnsbach Photo A: G. K. Lieb, B: J. Stangl (July 21, 2008)

period of intensified land use was as early as the Bronze Age when copper was mined in the Greywacke Zone south of the Gesäuse, driving people to lumber the forests of the Gesäuse Mountains in order to produce charcoal for ore smelting (Hasitschka 2005). These activities lasted until the nineteenth century and are still recognisable by remnants of charcoal piles and associated transport routes. Permanent settlements in the Middle Ages have been verified in the Johnsbach Valley and around Admont, whereas settlers and travellers avoided the Gesäuse Gorge itself. Traffic bypassed the gorge via the glacially widened and thus easily passable Buchau pass till 1872 when the railroad through the gorge was opened.

However, similar to the railroad through the Salzkammergut (described in Sect. 17.5.3 and Chap. "The World Heritage Site Hallstatt-Dachstein/Salzkammergut: A Fascinating Geomorphological Field Laboratory") this innovation encountered the problem of hazardous gravitational processes. Because of high relief, lithological and climatic conditions debris flows and avalanches are of particular significance (Fig. 18.11). For instance, 38 avalanche tracks may either reach the road or the railroad in the Gesäuse Gorge (Stangl 2009). The avalanche on 8 February 1924, which caused six fatalities near Hieflau railway station, was an example of a catastrophic event. This "hot spot" of natural hazards was only eliminated in 2005 by the completion of an avalanche tunnel for both road and railroad. Although a lot of permanent protection measures against avalanches and other gravitational processes have been taken, every few years extreme events force the authorities to close the traffic lines for safety reasons.

Debris flows and torrential activity yield great quantities of gravel to the Enns River and its tributaries. On the one hand, these processes directly amplify natural hazards and on the other hand provide the basis for quarrying of building materials at several locations. Quarrying of sediments, however, is incompatible with environmental interests and was therefore stopped after the Gesäuse National Park had been established in 2002. The striking geomorphological features as reflected by the impressive scenery of the Gesäuse have played an important role in establishing, promoting and managing of the national park. The sediment budget of Johnsbach and Enns, as

Fig. 18.11 The southern slopes of the Gr. Buchstein. Note the two prominent erosional channels, which also act as avalanche tracks. At their lower ends protection structures were erected in order to ensure the safety of the railroad. View from Hochtor summit towards north-west Photo: G. K. Lieb

of Alpine rivers in general, is not only of great interest for hazard protection and the demand of hydrological power plant suppliers, but also for river ecology and restoration (e.g. Habersack and Piégay 2008). The quantitative data on different sediment transport processes collected in recent years provide basic information for a sustainable sediment management strategy in the national park.

In order to experience natural features, the Gesäuse National Park offers a broad variety of educational sites, amongst them an interactive geological exhibition in Gstatterboden, and events. But, there is no educational trail which directly addresses geology or geomorphology. However, some trails allow visitors to study selected aspects discussed in this chapter. An example is the trail "Wilder John" which demonstrates river restoration and leads to the Gsenggraben (cf. Sect. 4). Another one is the educational trail through the Hartelsgraben Gorge which deals with the former charcoal production and the difficulties of former road construction in the steep relief of the Gesäuse Mountains.

18.6 Conclusions

The appearance of the Gesäuse Mountains contrasting with the plateau relief of the neighbouring mountain groups of the Northern Calcareous Alps (Totes Gebirge, Hochschwab) has been discussed in this chapter with special regard to lithology and geomorphological evolution since the Paleogene. However, the presentation was based on an exemplary approach omitting some aspects such as karstification. Karst is only mentioned in the context of the genesis of the Gesäuse Gorge although karst landforms are widespread and well developed.

Research is currently being carried out on catchment hydrology and sediment budgets. The interesting morphology, hydrology and hydrogeology (influenced by karst) and the excellent instrumentation give the Johnsbach catchment the potential to be one of the key sites of sediment budget research in Austria. This is emphasised by the very few catchments in which torrential sediment transport is quantified (e.g. in the Rofen Valley, a tributary of the Ötz Valley in Tyrol). The Gesäuse National Park and the nested Johnsbach catchment are LTSER sites (Long-Term Socio-economic and Ecosystem Research) in the LTER-Austria framework, and they are amongst the few sites in which geomorphological research plays a crucial role. The intensification of research since the establishment of the national park has shown that the Gesäsue mountains are a hot spot of biodiversity and endemic species—and there is still potential to investigate and better understand their geo(morpho)diversity.

References

Ampferer O (1935) Geologischer Führer für die Gesäuseberge. Geologische Bundesanstalt, Wien (with geological map 1:25.000)

Büchner KH (1970) Geologie der nördlichen und südwestlichen Gesäuse-Berge (Ober-Steiermark, Österreich). Dissertation, Univ. Marburg, p 118

Flügel HW, Neubauer F (1984) Erläuterungen zur Geologischen Karte der Steiermark 1:200.000. Geologische Bundesanstalt, Wien, p 127

Frisch W, Székely B, Kuhlemann J, Dunkl I (2000) Geomorphological evolution of the Eastern Alps in response to Miocene tectonics. Z Geomorph N.F. 44:103–138

Fryirs KA, Brierley GJ, Preston NJ, Kasai M (2007) Buffers, barriers and blankets: The (dis)connectivity of catchment-scale sediment cascades. CATENA 70:49–67

Geyer G (1918) Zur Morphologie der Gesäuseberge. Zeitschrift des Deutschen und Österreichischen Alpenvereins 49:1–32

Habersack H, Piégay H (2008) River restoration in the Alps and their surroundings: past experience and future challenges. In: Habersack H, Piégay H, Rinaldi M (eds) Gravel-bed rivers VI—from process understanding to river restoration. Elsevier, pp 703–738

Hasitschka J (2005) Gesäusewälder. Eine Forstgeschichte nach Quellen von den Anfängen bis 1900. Schriften des Nationalparks Gesäuse 1, Admont, p 120

Keil M, Neubauer F (2011) The Miocene Enns Valley Basin (Austria) and the North Enns Valley Fault. Austrian J Earth Sci 104(1):49–65

Kollmann W (1983) Hydrogeologische Untersuchungen in den nördlichen Gesäusebergen. Berichte der wasserwirtschaftlichen Rahmenplanung 66, Graz, p 299

Lieb GK, Hasitschka J, Kreiner D, Sass O (2017) Felsstürze im Gesäuse – zwischen Naturgefahr und Motor für Artenvielfalt. Im Gseis (Das Nationalpark Gesäuse Magazin). Sommer 2017:4–7

Lieb GK, Premm M (2008) Das Johnsbachtal – Werdegang und Dynamik im Formenbild eines zweigeteilten Tales. In: Kreiner D, Zechner L (eds.) Der Johnsbach. Schriften des Nationalparks Gesäuse vol 3, pp 12–24

Rascher E, Sass O (2016) Monitoring slope to channel coupling in an alpine catchment (Johnsbachtal, Austria) using multi-temporal terrestrial laser scanning. Ann Geomorphol 60(4). https://doi.org/10.1127/zfg_suppl/2016/0358

Rupp C, Linner M, Mandl GW (eds) (2011) Geologische Karte von Oberösterreich 1:200.000 Erläuterungen. Geologische Bundesanstalt, Wien, p 255

Sass O (2005) Spatial patterns of rockfall intensity in the northern Alps. Z Geomorph NF Suppl. 138:51–65

Stangl J (2009) Gravitative Naturprozesse im Nationalpark Gesäuse. Unpubl. master thesis, University of Graz, p 156

Strasser U, Marke T, Sass O, Birk S, Winkler G (2013) John's creek valley: a mountainous catchment for long-term interdisciplinary human-environment system research in Upper Styria (Austria). Environ Earth Sci 69:695–705

van Husen D (1968) Ein Beitrag zur Talgeschichte des Ennstales im Quartär. Mitteilungen der Gesellschaft der Geologie- und Bergbaustudenten 18:249–286

van Husen D (1979) Verbreitung, Ursachen und Füllung gazial übertiefter Talabschnitte an Beispielen aus den Ostalpen. Eiszeitalter und Gegenwart 29:9–22

Wiche K (1951) Zur Morphogenese der Gesäuseberge. Geographische Studien (J. Sölch-Festschrift), Wien, pp 203–224

Gerhard Karl Lieb is associate professor at the University of Graz, Institute of Geography and Regional Science. He has experience in physical geography of mountain regions with a special focus on permafrost, glaciers and geomorphic processes affected by global change as well as natural hazards. He is one of the two co-leaders of the national glacier monitoring programme of the Austrian Alpine Club. Besides those topics he works on regional geographies (Austria, Alps and Europe) and on didactics of the Austrian school subject geography and economics.

Oliver Sass holds a professorship for geomorphology at the University of Bayreuth and he worked for 11 years at the Institutes of Geography in Innsbruck and Graz. His fields of expertise are in alpine geomorphological processes, particularly rockwall weathering, talus sediments, fluvial processes and sediment budgets. He furthermore works on questions of man-environment interaction (e.g. natural hazards and wildfires).

19 The Rax Karst massif: A Typical Plateau of the Northern Calcareous Alps?

Christian Bauer and Lukas Plan

Abstract

The Rax massif is located at the border of the Austrian provinces of Styria and Lower Austria. The area is part of the Northern Calcareous Alps (NCA) and covers 104 km^2. The plateau is built up by a Triassic carbonate sequence with a thickness of more than 1 km. Similar to other plateaus of the NCA, the Rax massif indicates remnants of 'Tertiary' palaeosurfaces which carry Quartz-rich sediments of the Augenstein Formation. There is a long-lasting scientific controversy about the age and the evolution of these palaeosurfaces. Lichtenecker (Das Bewegungsbild Der Ostalpen. Naturwissenschaften 13:739–743, 1924) introduced the terms Augenstein-Landschaft and Rax-Landschaft. The latest model, introduced by Frisch et al. (Int J Earth Sci 90:500–518, 2001) proposed the term Dachstein-palaeosurfaces for karstified landscapes developed before the Augenstein sedimentation. The plateau area (approximately 27 km^2) is characterized by intense karstification including entire underground drainage. The morphogenesis of the epikarst is influenced by the presence of soil-cover and sparse vegetation. Due to dominantly subsoil development, karren are characterized by rounded forms. Closed depressions (dolines) represent the dominating surface karst feature. Based on airborne LIDAR data and intensive field mapping 914 dolines, with a summarized area of approximately 0.15 km^2, are delineated and their morphometric characteristics analyzed. Although the study area comprises a large number of caves, deep pits or large cave systems are unknown so far, which is in contrast to other karst massifs of the NCA. The Rax plateau also comprises the catchment areas of springs captured for the Viennese water supply.

Keywords

Rax • Karst • Rax-Landschaft • Dachstein-palaeosurface • Northern Calcareous Alps

19.1 Study Area

The Rax massif, together with the neighboring Schneeberg massif, comprises the easternmost karst plateaus of the Northern Calcareous Alps (NCA). It is situated at the border of the Austrian provinces of Styria and Lower Austria (Fig. 19.1). The outline of the Rax is well defined by deeply incised valleys or thalwegs, and based on the outline of the Austrian Cave Register, it covers an area of 104 km^2 (30 km^2 belonging to the province of Styria and 74 km^2 to Lower Austria; Stummer and Plan, 2002). While the stepped plateau of the Rax is located between 1350 and 2007 m asl, the surrounding valley bottoms lie at 500–700 m.

The NCA consists almost entirely of Mesozoic sediments. Unlike many other well-known plateaus like Totes Gebirge, Dachstein, or Tennengebirge, that are dominated by the Upper Triassic Dachstein limestone, the Rax plateau is almost entirely built up of carbonates of the Middle to Upper Triassic Wetterstein Formation. These shallow marine formation consists of several facies: (a) reef and reef breccia limestone, (b) massive to thick-bedded lagoonal limestone, and (c) dolomite. Only at the northern margin (Kleines Höllental) remnants of the Upper Cretaceous to Palaeocene Gosau Group occur that comprise clastic sediments. At the cliffs in the south and west well bedded limestone of the Middle Triassic Gutensteiner and Hallstätter Formation crop out. Below them, the Lower Triassic siliciclastic sediments of the Werfen Formation form more gentle slopes at the

C. Bauer (✉)
Institute of Geography and Regional Science, University of Graz, Heinrichstrasse 36, 8010 Graz, Austria
e-mail: christian.bauer@uni-graz.at

L. Plan
Karst and Cave Group, Natural History Museum Vienna, Burgring 7, 1010 Wien, Austria

C. Embleton-Hamann (ed.), *Landscapes and Landforms of Austria*, World Geomorphological Landscapes,
https://doi.org/10.1007/978-3-030-92815-5_19

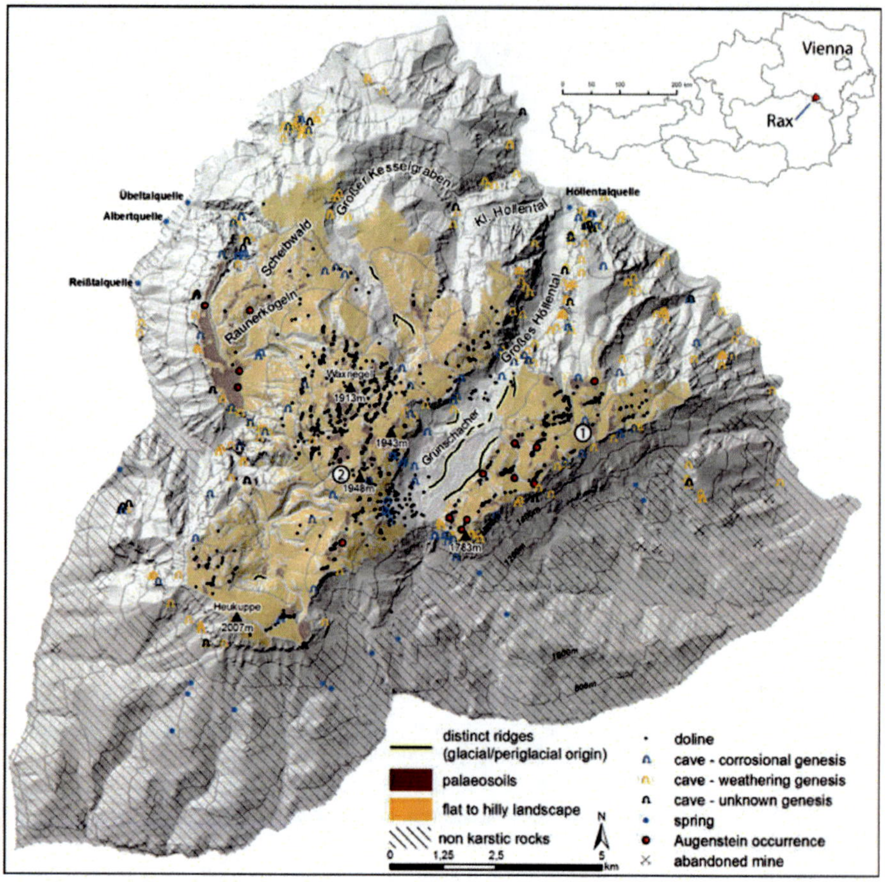

Fig. 19.1 The Rax massif as delineated by rivers and thalwegs. Encircled digits: (1) Grünschacher plateau and (2) elevated main plateau. The Augenstein occurrences are based on the 1:50.000 geological map (Mandl 2001)

surface and an aquiclude below the karst rocks that dip to the north. In the south, the area is bordered by the *Grauwackenzone* that consists of weakly metamorphic Palaeozoic rocks.

Parts of the plateau were affected by intense Pleistocene modifications (e.g., deep cuts into the plateau). This geomorphological impact is most obvious in the area of the Grünschacher, where glacial and periglacial processes formed distinct ridges, as well as small glacial cirques (Fig. 19.1). The elevated plateau areas do not show traces of significant modification.

The vegetation in the study area is characterized by strong variations due to altitude. The lower part of the Rax massif belongs to the montane zone and is covered by beech (*Fagus sylvatica*) and spruce (*Picea abies*) forests. Up to 1600 m, the slopes are covered with forests dominated by spruce and larch (*Larix decidua*). The subalpine zone extends from 1650 m to 1850 and is dominated by dwarf pines (*Pinus mugo*). Grassy vegetation persists even at the highest parts of the Rax plateau at 2000 m asl (Fig. 19.2).

19.2 Palaeosurfaces and Augenstein Formation

The Karst plateaus of the NCA are characterized by preserved 'Tertiary' planation surfaces (cf. Fig. 19.2a–c for the Rax massif) with a peculiar veneer of fluvial pebbles. Consisting mainly of quartz, these well-rounded pebbles (Fig. 19.2d) are called *Augensteine*. The etymology of this term is unclear. According to Bächtold-Stäubli and Hoffmann-Krayer (1987), the word has its origin in superstition and/or folk-medicine practice. Another explanation is that the term originated as analogy between eyes and the quartz-pebbles (Götzinger 1913a).

There is a long-lasting scientific controversy on the development of the palaeosurfaces and the allogenic Augenstein sediments in the NCA. There are three major theories about the type of formation dealing with geomorphological concepts as well as palinspastic reconstruction:

Fig. 19.2 Panorama-view of the Rax plateau. **a** View from Preiner Wand (1783 m) toward the west. The scarp between the two plateau-units is visible below Scheibwaldhöhe. Photo: C. Bauer. **b** View from Scheibwaldhöhe (1943 m) toward northwest showing a palaeosurface with flat to hilly landscape. Photo: C. Bauer. **c** View from Trinksteinsattel (1850 m) toward northeast (Grünschacher plateau). In contrast to the higher elevated main plateau, the lower Grünschacher plateau is densely vegetated. Photo: C. Bauer. **d** Well-rounded Augenstein pebbles, Dreimarkstein Photo: L. Plan

(1) Monophase formation. According to Lichtenecker (1924, 1926), the palaeosurfaces of the *Rax-Landschaft* are a result of a single (low-elevated) peneplained palaeosurface which was later disaggregated and tectonically uplifted at different rates. The terms *Rax-Landschaft* and *Augenstein-Landschaft* (Landschaft means landscape) describe different scenarios and were inconsistently used. Considering the geomorphologic situation of the palaeosurfaces in the context of the Augenstein sedimentation, the term Augenstein-Landschaft (Lichtenecker 1924) describes a landscape of a flat to moderate relief formed during the sedimentation of the Augenstein pebbles (syn-Augenstein Formation). In contrast, the term Rax-Landschaft (Lichtenecker 1924; 1926) describes the remnant hilly palaeosurfaces with differences in altitude of 200–300 m situated at high elevated karst plateaus after tectonic uplift (post-Augenstein Formation). According to this, Augenstein-Landschaft and Rax-Landschaft are successive landforms.

(2) Polycycling formation (e.g., Götzinger 1913b; Spreitzer 1951; Winkler-Hermaden 1957; Tollmann 1986). It is based on the concept of a *Piedmonttreppe* formed by pulses of uplift. Thus, the formation of a planation surface (peneplain) required periods of tectonic stability. Renewed tectonic uplift initiated the formation of a new planation surface. Although erosion increased, parts of the (subsequent) relict planation surface were preserved. The repeating sequence of tectonic uplift and tectonic quiescence formed different planation surface-levels of limited lateral extent.

(3) Polyphase evolution of the planation surfaces taking in account Miocene lateral extrusion of the Eastern Alps, accompanied by block segmentation of the Austroalpine mega-unit before uplift. The dismembered tectonic blocks were affected by different horizontal and —to a lesser extent—vertical movements (Frisch et al. 1998). These tectonic processes have resulted in a distinct decrease in (present-day) elevation from west to east (Frisch et al. 1998). According to this, differences in elevation of the palaeosurfaces reflect polyphase tectonics of dismembered blocks, rather than polycyclic (*Piedmonttreppe*) formation (Frisch et al. 2001). On the basis of DEM and sediment analysis as well as palinspastic reconstructions, Frisch et al. (2000) argued that the terms Rax-Landschaft and Augenstein-Landschaft as well as their concepts are no longer appropriate. Hence, he introduced the term *Dachstein-palaeosurface*—representing a pre-Augenstein geomorphologic situation at the Eocene Oligocene boundary (Frisch et al. 2001).

In the NCA, the tendency of decreasing elevation from west to east is even evident on single mountain blocks. Using the mountain group division of the Austrian Cave Register, Fig. 19.3 illustrates the morphometric pattern of

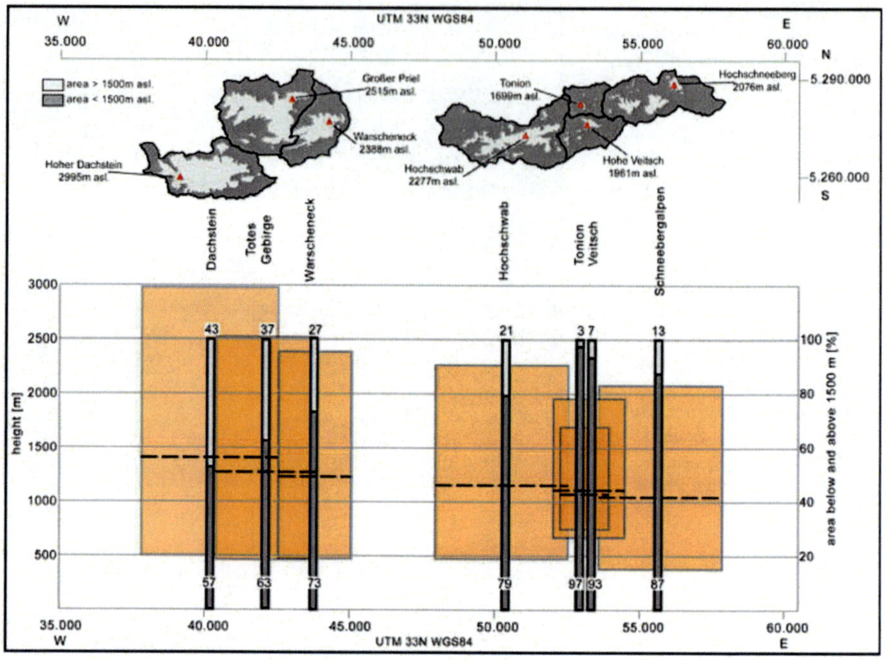

Fig. 19.3 Morphometric characteristics of selected karst plateaus of the NCA based on Austrian Cave Register boundaries. The orange rectangles represent the minimum and maximum elevation as well as the maximum W–E extensions of each mountain range. The dashed line indicates the mean elevation of each unit. The bar graph indicates the percentage of area above (light gray) and below (dark gray) 1500 m asl

seven karst areas between Dachstein and Schneeberg. The general decrease of maximum elevations from west to east is obvious. In the easternmost part of the NCA, the summits rarely exceed 2000 m asl (Rax, 2007 m; Schneeberg, 2076 m). According to Frisch et al. (2008), the present elevations of the palaeosurfaces range primary from 1500 to 3000 m, with a concentration between 1800 and 2500 m. This is in good accordance with the distribution of palaeosurfaces displayed in Fig. 19.3. Of the Schneebergalpen unit (including Schneealpe, Rax, and Schneeberg), 13% lies above 1500 m and host remnants of the Dachstein-palaeosurface. On the Rax massif, the Dachstein-palaeosurface ranges from 1400 in the NW (Scheibwald) to the summit at around 2000 m. The plateau of the Veitsch massif (7% area above 1500 m) also reveals remnants of the Dachstein-palaeosurface. In general, the percentage of the area above 1500 m as well as the elevation of the palaeosurfaces increases toward the west. This pronounced W–E trend is most significant in the mean elevations for each unit (dashed line in Fig. 19.3).

In general, if structural and lithological influences are negligible, the accumulation of horizontal cave passages indicates base-level controlled speleogenesis (e.g., Palmer 1987). Thus, stable base-level conditions (i.e., tectonic and climatic quiescence) for a longer period allow the development of distinct cave levels. Shifting elevation of the local relief due to tectonic uplift, glacial erosion, etc., initiates erosion and valley deepening that result in an abrupt base-level drop. For the NCA, the correlation of cave levels and surface morphology (e.g., palaeosurfaces) was analyzed by several authors (e.g., Riedl 1988; Haseke-Knapczyk 1989; Fischer 1990; Kuffner 1998). They distinguished three distinct cave levels from top to bottom: (1) *Ruin Cave Level*, (2) *Giant Cave Level*, and (3) *Spring Cave Level*. Frisch et al. (2001, 2002) correlate the formation of these three cave levels to (1) middle Eocene/early Oligocene, (2) Late Miocene, and (3) Pliocene/Quaternary. A recent study of the NCA addressed the issue of speleogenesis of several (epi)phreatic cave levels in the context of water tables. For the Totes Gebirge, Plan et al. (2009) examined if cave levels indicate different phases of speleogenesis within a height interval of a corresponding inclined water table. They concluded that cave levels are not readily correlated to the exact position of palaeo water tables. Moreover, particular cave levels are neither attributable to palaeo base levels over large distances nor to subsurface morphology. However, regarding the geomorphologic evolution of the NCA, the development of generalized cave levels—comprising different speleogenetic phases within large vertical extensions (several hundred meters)—as mentioned above, confirms the theory of a polyphase evolution, which was already postulated by Frisch et al. (2000).

The so-called *Augenstein Formation* (Frisch et al. 2001) represents remnants of allogenic sediments that were deposited on the palaeosurfaces (Dachstein-palaeosurface) of the NCA. Before the final uplift, the area of the NCA was modified by erosional processes during the Oligocene. Fluvial deposition occurred in the subsiding setting of the NCA, and the area was covered by up to several hundred meters of these fluvial deposits (Frisch et al. 1998, 2002). According to Frisch et al. (2001), the Augenstein Formation sealed the hilly low relief palaeosurfaces that formed in Late Eocene to

Early Oligocene time. The Augenstein sediments consist of low-grade metamorphic rocks dominated by quartz. They mainly derived from a Palaeozoic cover of the Austroalpine Crystalline, which today is preserved in a small strip along the south of the NCA and called *Grauwackenzone*. Rocks from the Austroalpine Crystalline are almost missing (Frisch et al. 1998, 2002). A change in the tectonic regime led to the so-called lateral extrusion of the Eastern Alps and the formation of significant E–W-striking fault systems (Ratschenbacher et al. 1991; Frisch et al. 2001). Subsequently, the formerly N-directed drainage across the NCA was interrupted and major E–W-trending rivers (Salzach and Enns valley) developed. Consequently, the supply of fluvial sediments was interrupted and in the NCA erosion became the dominating geomorphologic process. In all parts of the NCA, the elevated deposits were eroded or redeposited. Subsurface drainage of the karst plateaus conserved the palaeosurfaces and saved parts of the Augenstein Formation from erosion. Therefore, in most parts of the central and eastern NCA (between Steinernes Meer and Schneeberg), these 'Tertiary' palaeosurfaces (Dachstein-palaeosurface) have been preserved (Figs. 19.1 and 19.2). Present day occurrences of the Augenstein Formation are generally redeposited. According to Frisch et al. (2001) only very small remnants, most of them preserved in negative relief features, are considered to be in (para) autochthonous position. Reddish brown soils, often found on the karst plateaus of the NCA, are considered to be weathering products of the Augenstein Formation (Fig. 19.1). Their formation indicates intense weathering processes under (sub) tropical climate conditions.

As mentioned above, the Pleistocene glaciation had no significant impact on the Rax plateau. Thus, remnants of Augensteine as well as their weathering products (reddish brown soils) have survived in the highest parts of the karst plateau.

19.3 Karst Features

In contrast to other large karst massifs like Dachstein or Totes Gebirge, there are rather few works on surface karst features on the Rax massif. Only the morphological study of Baedeker (1922) contained some considerations, Fink et al. (2005) and Fink (2005) provide a brief description, some focused studies and an overview map based on field work in the 1990s. In 2004, within the framework of the KATER Projekt (KArst waTER research program; Kuschnig 2001) a detailed karst morphological mapping was initiated by the Vienna Waterworks. These karst morphological studies addressed the development of a database on karst features and morphological parameters influencing the vulnerability of infiltrating karst waters. Karst morphological field mapping resulted in a large and detailed dataset including a karst morphological map (1:10 000) of the Rax plateau (27 km^2), and quantitative data on karst features such as the spatial distribution, size, and shape parameters as well as phenomenological descriptions (Plan 2005).

19.3.1 Dolines

The presented analysis of dolines is based on two data sources: (1) For the Styrian part of the Rax massif, the available 1 m-spatial resolution airborne laser scanning digital terrain model (ALS-DTM) provides the only dataset for the delineation process. Large parts of the study area are densely vegetated with dwarf pine (*Pinus mugo*) and spruce. The effective canopy penetration of the ALS technique provides surface information of these areas in detail. (2) For the Lower Austrian part of the Rax, due to partial snow cover (especially in depressions) during ALS flight mission, ALS data were partly useless for karst feature delineation. The dataset was completed with an area wide field mapping of karst features (Plan 2005).

In order to make the two data sources comparable, the boundary extraction using the ALS data is based on the outermost closed contour line, a doline demarcation suggested by Rahimi and Alexander (2013). Including the glaciokarstic area of the Grünschacher, the Rax plateau covers an area of 27.3 km^2.

The minimum size of the analyzed features is set to 5 m^2. Within the ALS data, compound dolines with multiple nested contours are treated as single features. The combined database consists of 914 dolines with a total area of 0.15 km^2. A set of measurements as proposed by Bauer (2015) is applied to analyze the morphometric characteristics of doline shapes and relief attributes.

The vertical distribution of dolines (Fig. 19.4b) demonstrates that they are most frequent between 1550 and 1650 m and 1750–1850 m asl. These two maxima correspond with the two plateaus, Grünschacher (① in Fig. 19.1) and the more elevated main plateau (② in Fig. 19.1), respectively. Riedl (1966) argued that tectonics is the reason for the elevational difference between Grünschacher plateau and the elevated part. The reduced number of dolines within 1650–1700 m (cf. Grünschacher Fig. 19.1) is obvious. Large parts of this area have been subject to Pleistocene modifications including distinct ridges of glacial or periglacial origin.

Williams (1969) suggested simple spatial statistics to quantify karstification of a given area (Table 19.1). By calculating the relation of the investigation area and the total area of dolines, the pitting index indicates to which extent karst depressions dominate the surface. A pitting index value of 1 (complete pitting) implies that the whole surface is dissected by dolines. Figure 19.1 illustrates the spatial

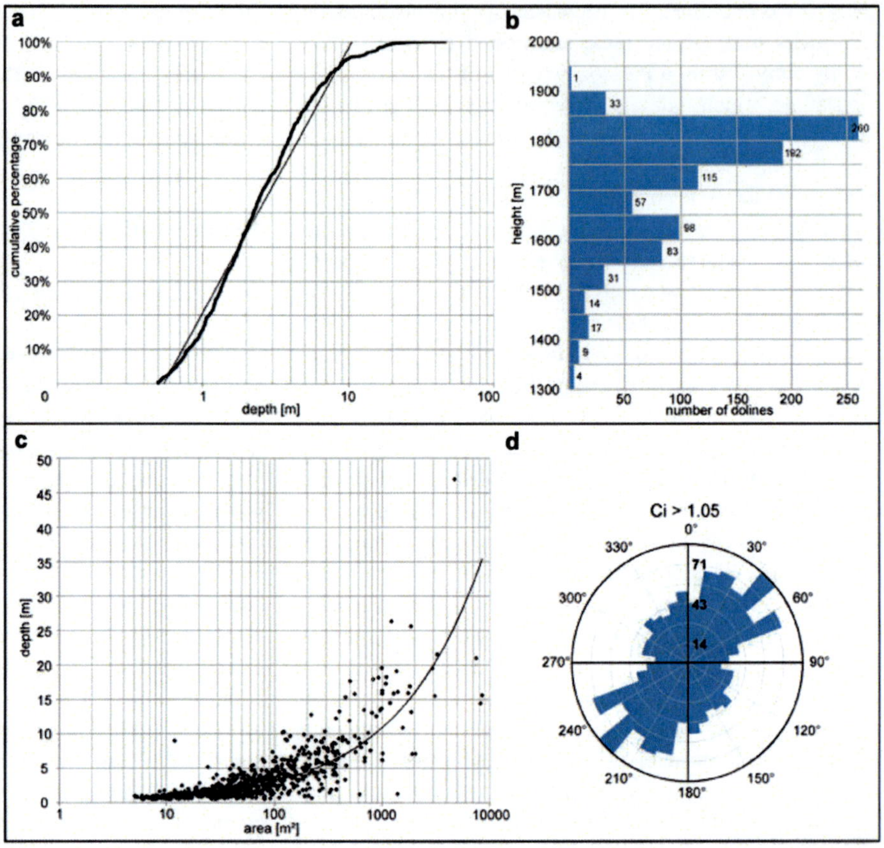

Fig. 19.4 Morphometric analysis of 914 dolines larger than 5 m^2. **a** Doline depth frequency distribution; **b** Histogram of doline elevation distribution; **c** Distribution of doline depth as a function of area. The area-to-depth ratio illustrates a controlled spread; **d** Symmetric, bidirectional rose-diagram of the longest transect of non-spherical (Ci > 1.05) dolines (n = 796)

Table 19.1 Spatial statistic of analyzed dolines

Indices	Formula	Values
Pitting index	$\mathrm{Rp} = \frac{\mathrm{Ak}}{\sum_i \mathrm{Ad}}$	183
Doline density [/km^2]	$\mathrm{Dd} = \frac{\mathrm{Nd}}{\mathrm{Ak}}$	33.5
Circularity index	$\mathrm{Ci} = \frac{\mathrm{Ad}}{\pi \times \left(2\frac{\mathrm{Ad}}{\mathrm{Pd}}\right)^2}$	

Ak = studied area of karst [m^2]; Ad = area of doline [m^2]; Nd = number of dolines; Pd = perimeter of doline [m]

distribution of dolines in the study area. Especially, large areas of in the western and northwestern plateau are not dissected by Karst depressions. Therefore, the pitting index of 183 indicates that the plateau surface is only partly (0.55%) shaped by dolines. In fact, the distribution of dolines is clustered. The second index is the doline density on the plateau area. This index relates the investigation area to the number of dolines as point shape elements. Compared to other quantitative karst investigations, the doline density of the Rax plateau (33.5 dolines/km^2) is low. For example, in the study of the Hochschwab massif (NCA) Plan and Decker (2006) found a doline density of 122.

As mentioned above, every depression matching the criteria is assumed as a single doline (i.e., compound depressions count only as one feature). This results in a large number of shallow features: More than 80% of all dolines are less than 5 m deep (Fig. 19.4a). Only 2.5% of all features are deeper than 15 m (maximum depth is 47 m). The doline depth distribution shows a strict log-normal distribution (R^2 = 0.96). The plotted area-to-depth ratio (Fig. 19.4c) illustrates a controlled spread of the features. The result shows a moderate relation (R^2 = 0.69). Assuming solution origin of the dolines (e.g., enhanced corrosion due to centripetal flow of the depression), deepening at the doline base is coupled with perimeter widening. Thus, the area-to-depth ratio rather reflects solution origin of the dolines. Outliers do not reflect depth as a function of area. In general, dolines with decoupled deepening (at the doline base) from perimeter widening are assumed to be of collapse origin. The morphometry of these features is characterized by steeper side walls and a smaller plan area (e.g., Ford and Williams 2007). It has to be considered that the present doline morphology is not attributable to a single process. Thus (1), varying dominant doline-forming processes, (2) polygenetic origin (e.g., Pleistocene glaciation), and (3) significant process rate variations due to changing climate conditions may have influenced doline morphometry. As mentioned below (see Sect. 3.3) caves, related to corrosion, developed primarily under vadose conditions. Considering also the lack of extensive phreatic caves (see below), outliers are rather pits that developed under vadose conditions and were opened or reshaped by (peri) glacial processes than classical collapse dolines.

The preferred elongations of doline long axes are considered to be a function of structural and tectonic constraints (e.g., Williams 1971; Pfeffer 2010). Firstly, the circularity of the features was investigated. The circularity index (Ci), calculated by the formula of Seale (2005), indicates the planimetric deviation of a doline shape from a perfect circle. Hence, a Ci of 1 represents a perfectly circular-shaped doline. Secondly, the direction analysis of dolines was performed by calculating the longest axis within each feature. To minimize the random effects of small and nearly circular dolines, sample features had to exceed a Ci > 1.05. The calculated elongations of dolines (Fig. 19.4d) show a preferred orientation between 5° and 65°, with a maximum at 45° (NE-SW). The Rax massif is dominated by a SSW-NNE striking fault system. Thus, the results demonstrate a good agreement between the preferred elongation of dolines and the regional tectonic lineations. Furthermore, the examined preferred orientations cannot be referred to the hypothesis of Schappelwein (1966), who considered a preferred doline elongation of E (NE)/W(NW) as a result of increased dissolution from snow accumulations in high mountain reliefs. According to Schappelwein, these snow accumulations are aligned by the prevailing (regional) wind directions.

Figure 19.5a, b illustrates that dolines are often aligned in small dry valleys. Some of these valleys are a result of structural constraints. Within the dry valley in Fig. 19.5a, b, five individual dolines are delineated. The elongation of the longest axis of these dolines (direction NW) corresponds with the direction of the dry valley (which is also indicated as 'assumed fault' on the geological map). The most distinctive and geomorphologically obvious system is the *Großes Höllental fault,* which displaced the elevated western (main plateau) and the lower eastern (*Grünschacher* plateau) part by some 200 m (Mandl 2006). This is also illustrated by the doline elevation distribution in Fig. 19.4b.

19.3.2 Karren

On the Rax plateau, small-scale karst features (Karren) are not as common as on other karst plateaus of the NCA (e.g., Totes Gebirge, Dachstein) that are dominated by limestone of the Dachstein Formation. According to the classification by Ford and Williams (2007), hydrodynamically controlled linear solution runnels are the predominant type of Karren on the Rax. They display rounded forms (*Rundkarren*), if developed subcutaneously and sharp rims (*Rinnenkarren*) if developed on bare slopes. As mentioned above, most parts of the plateau area are partly or entirely covered by soil and vegetation. Thus, the rounded forms are most frequent. Where *Rundkarren* are exposed due to soil removal, they may have been sharpened by direct rain water to form *Rillenkarren* (Fig. 19.5c).

19.3.3 Caves and Subsurface Drainage

The water drainage system of the plateau is controlled by the underlying non-karstic rocks that crop out in the south. The siliciclastics of the Werfen Formation act as an aquiclude layer below the Triassic carbonate sequence. As the contact dips to the north, the southern slopes of the Rax are characterized by several rather small springs at the contact of these rock types (Cornelius 1936; Fig. 19.1). The plateau area drains to a few large springs in the north and northwest. These springs can be merged into two groups, both of which are captured for the Vienna Water Supply (Stadler et al. 2008): (1) The *Höllentalquelle* (Fig. 19.1), which is strongly influenced by the *Großes Höllental* fault (Figs. 19.1 and 19.5), has a mean discharge of 149 l/s (from 2004 to 2007). It is supposed to drain the high plateau area, where the catchment area reaches from *Scheibwaldhöhe* (1943 m) to *Preinerwand* (1763 m). (2) Springs near *Hinternasswald: Übeltalquelle, Reißtalquelle, and Albertquelle* (Fig. 19.1). Compared to *Höllentalquellle,* these springs show a lower mean discharge (e.g., *Übeltalquelle*: MQ (2005–2007): 7 l/s; *Reißtalquelle*: MQ (2007): 90 l/s). Based on Oxygen-18 and Deuterium analysis, the catchment area of these springs is situated in lower parts of the plateau, at *Raunerkögeln* (1550 m) and *Scheibwald* (1447 m).

However, water balances of the infiltrating precipitation and the spring discharges suggest that a significant portion of the water directly flows as groundwater into the sediments of the Vienna Basin (Fink et al. 2005).

Up to now, according to the Austrian Cave Register, 314 caves were explored and documented in the Rax massif (Fig. 19.1). Thus, the study area is characterized by a high density of caves (2.9 caves/km^2), similar to other karst plateaus of the NCA (cf. Hochschwab: 3.2 caves/km^2; Plan and Decker 2006). In comparison to other massifs of the NCA,

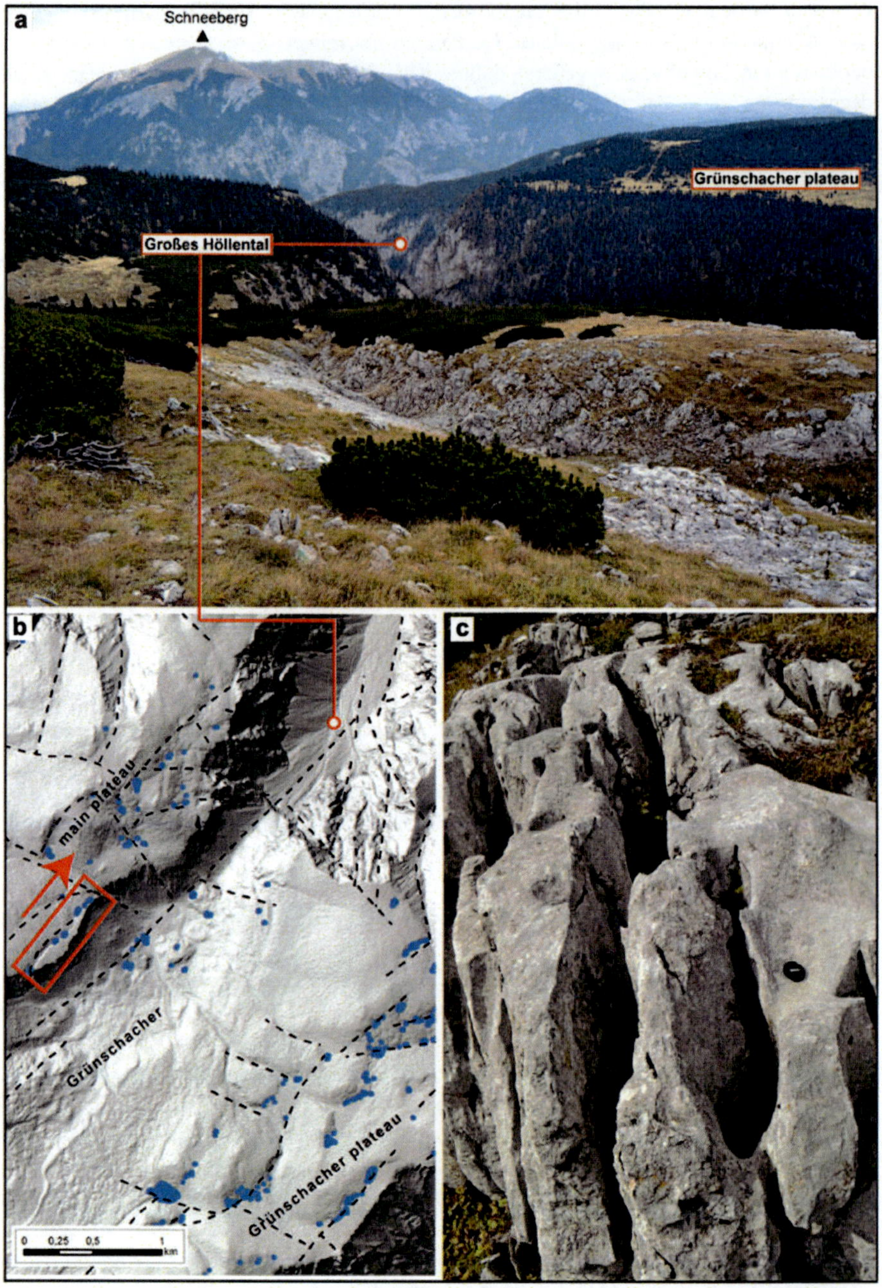

Fig. 19.5 Karst Features. **a** dry valley on the main plateau in the continuation of Großes Höllental. View toward NE. Photo: C. Bauer. **b** Hillshade visualization: The dashed line illustrates tectonic lineations based on the geological map (Mandl 2001). The blue polygons represent delineated dolines. The orange box indicates the structural valley of Fig. 19.5a. The Pleistocene/periglacial modification (distinct ridges) of the Grünschacher is obvious. **c** Assembly of *Kluftkarren*: Features that developed subsoil (*Rundkarren;* upper right) were exposed and have been successively reshaped (sharpened) by direct corrosion of rainwater (lower left). Lens cap provides scale Photo: C. Bauer

however, caves are very small, and karstic caves (where dissolution is the main genetic process) form the minority. So far, *Krampusschacht* (west of *Scheibwaldhöhe*) is the longest and deepest cave with a length of 412 m and a depth of 90 m (Fig. 19.6). Only 20 caves are longer and six caves are deeper than 50 m. Most of the karstic caves are pits or small canyons. Their morphology shows that they have been formed by dissolution under vadose conditions (Fig. 19.6). So far, the access to these caves and thus their possible continuation into greater depth is blocked by boulders (local breakdown) or perennial ice.

This, it is in contrast to most other karst massifs in the NCA where extensive cave systems of more than 100 km length and 1 km depth have been documented (Plan and Oberender 2016). Also, extensive cave levels that developed under phreatic or epiphreatic conditions are missing on the

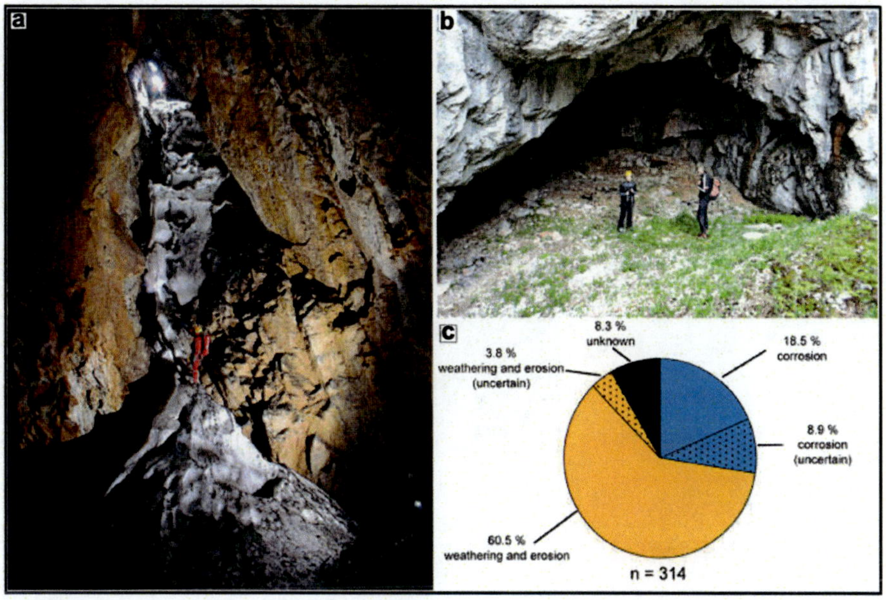

Fig. 19.6 a Snow cone at the bottom of the 60 m deep entrance pit of the *Krampusschacht*. Photo: L. Plan. **b** The *Übeltal-Halbhöhle*: Speleogenesis of the 17 m long shelter cave is dominated by frost weathering and erosion processes. Photo: E. Herrmann. **c** Distribution of cave types, according to dominant genetic processes (cf. Fig. 19.1)

Rax. So far, there is no explanation for this. The same holds true for the neighboring Schneeberg, which is also built up of a more than 1 km thick sequence of well karstified Triassic carbonates. Some small caves and dolines show fillings with reddish sandstones that could be the reason why old, phreatic caves are not accessible (Herrmann 2004).

A genetic classification of caves revealed that for almost two thirds of them—even though they have formed in limestone—the dominating speleogenetic processes are weathering and erosion (Oberender und Plan 2013). These caves mainly comprise shelter caves (Fig. 19.6b) or short ascending galleries where studies by Oberender and Plan (2015) suggest that frost weathering and erosion of the scree are the dominant processes. Other common non-karstic caves are crevice caves that formed due to gravitational mass movements near the cliff-tops.

19.4 Anthropogenic Modifications

The southern slopes of the study area are characterized by local landscape-forming remnants of iron mining activities from the second half of the eighteenth century until the end of the nineteenth century (Weber 1997). Most abandoned mines (Fig. 19.1) are situated in the Permian Präbichl Formation, transgressing on the Palaeozoic Grauwackenzone. The geomorphic remnants comprise (1) material excavations (e.g., collapse shafts, adit openings), (2) material accumulations (e.g., mine heaps) and to a lesser extent, (3) construction activities (e.g., roads, buildings). However, these activities have no geomorphic impact for the higher elevated karst plateau.

According to Lichtenecker (1926), some doline-like features may be attributable to anthropogenic modifications. He indicated negative relief features SW of the *Scheibwaldhöhe* (1943 m) as impact structures due to artillery shells from target practice during World War I. In fact, it is often difficult to differentiate between anthropogenic structures due to artillery impacts and natural geomorphic features (dolines) in an environment with pronounced karst relief. Additionally, due to natural modifications, the traces of impact structures of warfare origin (e.g., walls) diminish over time (almost 100 years), especially in an alpine environment. However, artillery shell fragments can be found on stone cairns in the area described by Lichtenecker (Fig. 19.7a). Moreover, near *Dreimarkstein* (1948 m) distinct linear features illustrate another anthropogenic impact on the epikarst. These structures are supposed to be a trench system of warfare origin (Fig. 19.7b, c).

The Rax plateau is characterized by intensive tourist use and therefore well-developed tourist infrastructure (including seven alpine huts and a cable car). Associated activities induce several indirect impacts/problems on the karst environment: (a) soil erosion due to trampling damage and impact on the epikarst, (b) waste disposal, (c) sewage treatment and sewage disposal. To prevent associated impacts on a significantly vulnerable karst aquifer, supplying drinking water to the city of Vienna (cf. Chap. "Karst Landscapes in Austria", Sect. 5.1) is still a major challenge in managing recreation activities in the study area.

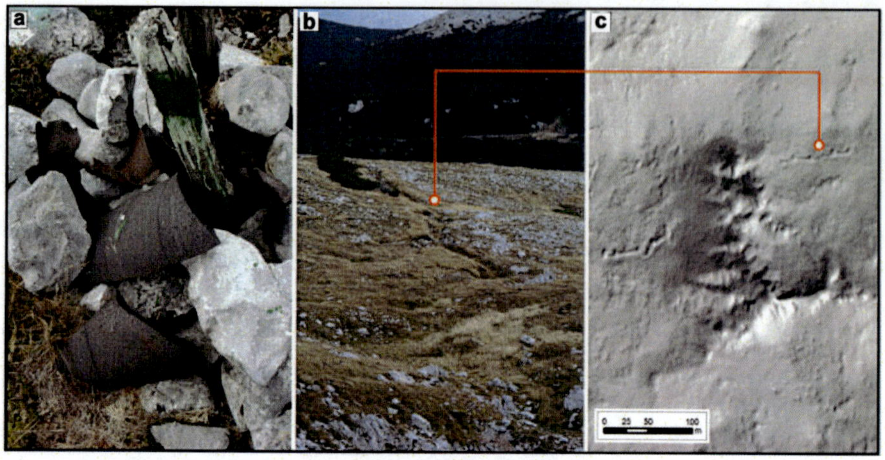

Fig. 19.7 Remnants of anthropogenic modifications of military origin. **a** Fragments of artillery shells. Photo: C. Bauer. **b** Possible trench near Dreimarkstein (cf. Fig. 19.1). Photo: C. Bauer. **c** Hillshade visualization

19.5 Conclusions

The NCA are characterized by high alpine karst plateaus built up by thick Triassic carbonate sequences, of which the most important karstified strata are limestones of the Dachstein and Wetterstein formations. The plateau systems are susceptible to karstification, and the drainage systems are mostly subsurface. The same holds true for the studied Rax plateau, which is almost entirely built up of carbonates of the Wetterstein Formation.

Karst massifs of the NCA host remnants of planation surfaces. These high elevated palaeosurfaces have long been recognized, and their origin has been discussed controversially. Especially on the Rax plateau, these conspicuously leveled areas are well developed. Even though parts of the Rax were affected by Pleistocene glaciations, the higher plateau area shows little glacial modifications. Thus, remnants of planation surfaces are well preserved. Lichtenegger (1924, 1926) even introduced the term *Rax-Landschaft* as locus typicus for elevated remnants of planation surfaces in the NCA. Recent concepts (e.g., Frisch et al. 2001) propose a polyphase evolution of the planation surfaces (the so-called *Dachstein-palaeosurface*) accompanied with block segmentation of the NCA before uplift.

The most obvious surface karst features of the Rax plateau are dolines. Their morphological analysis illustrates: (1) That doline distribution is clustered and large areas, especially in the NW, are not dissected by dolines; (2) Doline density on the Rax plateau is low, when compared to data from other karst massifs and (3) Doline morphology cannot be attributed to a single process like collapse or dissolution only. Karren are not as common as on the other karst plateaus, developed in the limestones of the Dachstein Formation. The drainage system of the Rax plateau is controlled by underlying non-karstic rocks (e.g., Werfen Formation) that dip to the north, which is the case for many massifs in the NCA. Consequently, the plateau area drains to a few large springs in the north and northwest which are captured for the Vienna Water Supply.

Caves show considerable differences to many other karst plateaus, especially in the central NCA (e.g., Dachstein, Totes Gebirge, Tennengebirge). These host-extensive (more than100 km long) cave systems that developed under (epi) phreatic conditions and were controlled by the base level, which resulted in three distinct cave levels. So far, there is no explanation for the lack of extensive phreatic caves in the Rax and neighboring massifs (Schneeberg and Scheealpe).

Acknowledgements The ALS data were kindly provided by the Federal State Government of Styria, Board of Works—Geoinformation Staff Office. Vienna Water financially supported the karstmorphologic field mapping.

References

Baedeker D (1922) Beiträge zur Morphologie der Schneebergalpen. Geographischer Jahresbericht Aus Österreich 12:5–100

Bauer C (2015) Analysis of dolines using multiple methods applied to Airborne Laser Scanning data. Geomorphology 250:78–88. https://doi.org/10.1016/j.geomorph.2015.08.015

Bächtold-Stäubli H, Hoffmann-Krayer E (1987) Handwörterbuch des deutschen Aberglaubens. de Gruyter, Berlin, New York

Cornelius HP (1936) Erläuterungen zur geologischen Karte der Rax. Geologische Bundesanstalt, Wien

Fink MH (2005) Das Karstrelief der östlichen Kalkhochalpen. In: Nagel D (ed) Festschrift für Prof. Gernot Rabeder. Mitt. d. Komm. f. Quartärforschung d. Österr. Akademie d. Wissensch. 14:17–27, Wien

Fink MH, Pavuza R, Stummer G (2005) Daten zur Karstverbreitung und Karstgefährdung in den östlichen Kalkhochalpen. Speldok 16. Verband Österreichischer Höhlenforscher, Wien

Fischer K (1990) Höhlenniveaus und Altreliefgenerationen in den Berchtesgadener Alpen. Mitteilungen Der Geographischen Gesellschaft München 75:47–59

Frisch W, Kuhlemann J, Dunkl I, Brügel A (1998) Palinspastic reconstruction and topographic evolution of the Eastern Alps during late Tertiary tectonic extrusion. Tectonophysics 297:1–15

Frisch W, Kuhlemann J, Dunkl I, Székely B (2001) The Dachstein paleosurface and the Augenstein Formation in the Northern Calcareous Alps—a mosaic stone in the geomorphological evolution of the Eastern Alps. Int J Earth Sci 90:500–518

Frisch W, Kuhlemann J, Dunkl I, Székely B (2008) Die geomorphologische Entwicklung der Ostalpen. Mitt Öster Geog Gesell 150:123–162

Frisch W, Kuhlemann J, Dunkl I, Székely B, Vennemann T, Rettenbacher A (2002) Dachstein-Altfläche, Augenstein-Formation und Höhlenentwicklung – die Geschichte der letzten 35 Millionen Jahre in den zentralen Nördlichen Kalkalpen. Die Höhle 53:1–34

Frisch W, Székely B, Kuhlemann J, Dunkl I (2000) Geomorphological evolution of the Eastern Alps in response to Miocene tectonics. Z Geomorphol 44:103–138

Ford D, Williams P (2007) Karst Hydrogeology and Geomorphology. Wiley, Chichester

Götzinger G (1913a) Zur Frage des Alters der Oberflächenformen der östlichen Kalkhochalpen. Mitteilung Der Geographischen Gesellschaft Wien 52:39–57

Götzinger G (1913b) Neue Funde von Augensteinen auf den östlichen Kalkhochalpenplateaus. Verhandl. Geol. Reichsanstalt 1913:61–65

Haseke-Knapczyk H (1989) Der Untersberg bei Salzburg. Die oberund unterirdische Karstentwicklung und ihre Zusammenhänge. Ein Beitrag zur Trinkwasserforschung. MaB Reihe, 15. Österreichische Akademie der Wissenschaften, Innsbruck

Herrmann E (2004) Höhlen in den Abstürzen des Scheibwaldes an der Westseite der Rax. Höhlenkundliche Mitteilungen Wien 60:5–9

Kuffner D (1998) Höhlenniveaus und Altflächen im Westlichen Toten Gebirge. Die Höhle supplement No. 53

Kuschnig G (2001) Das Karstforschungsprogramm der Wiener Wasserwerke. In: Mandl G (ed) Geologische Bundesanstalt Arbeitstagung 2001. Geologische Bundesanstalt, Wien, pp 213–219

Lichtenecker N (1924) Das Bewegungsbild Der Ostalpen. Naturwissenschaften 13:739–743

Lichtenecker N (1926) Die Rax. Geographischer Jahresbericht Aus Österreich 13:150–170

Mandl GW (2001) Geologische Karte der Repulbik Österreich 1:50.000 - Blatt 104, Mürzzuschlag. Geologische Bundesanstalt

Mandl GW (2006) KATER II (KArst waTER research program) - Geology of the Rax-Schneeberg-Region. Explanatory notes to the digital geological map of the Rax-Schneeberg-Region. Geological Survey of Austria, Wien. URL: http://www.ccwaters.eu/downloads/KATERII_Explanation%20Geological%20map.pdf

Oberender P, Plan L (2013) A genetic classification of caves in Lower Austria. In: Proceedings 16th international congress of speleology, Czech Republic, Brno, vol 3, pp 390–392

Oberender P, Plan L (2015) Cave development by frost weathering. Geomorphology 229:73–84

Palmer AN (1987) Cave levels and their interpretations. NSS Bull 49 (2):50–66

Pfeffer KH (2010) Karst: Entstehung - Phänomene – Nutzung. Borntraeger, Stuttgart

Plan L (2005) Karstwasserschutz und andere umweltrelevante Fragestellungen im Bereich Rax. Schneeberg und Hochschwab. Unpubl, Forschungsbericht der Wiener Wasserwerke

Plan L, Decker K (2006) Quantitative karstmorphology of the Hochschwabplateau. Z Geomorphol 147:29–54

Plan L, Oberender P (2016) Höhlen in Österreich. In: Spötl C, Plan L, Christian E (eds) Höhlen und Karst in Österreich. Oberösterreichisches Landesmuseum (Linz), pp 11–22

Plan L, Filipponi M, Behm M, Seebacher R, Jeutter P (2009) Constraints on alpine speleogenesis from cave morphology—a case study from the eastern Totes Gebirge (Northern Calcareous Alsp, Austria). Geomorphology 106:118–129. https://doi.org/10.1016/j.geomorph.2008.09.011

Ratschenbacher L, Frisch W, Linzer HG, Merle O (1991) Lateral extrusion in the eastern Alps, part 2: structural analysis. Tectonics 10:250–271

Riedl H (1966) Neue Beiträge zum Problem: Raxlandschaft – Augensteinlandschaft. Mitteilungen Der Österreichischen Geographischen Gesellschaft 108:98–109

Riedl H (1988) Neue Beiträge zur Geomorphologie der Nördlichen Kalkalpen. Geographischer Jahresbericht aus Österreich XLV:81–96

Rahimi M, Alexander EC (2013) Locating sinkholes in LiDAR coverage of a glacio-fluvial karst, Winona County, MN. In: Thirteenth multidisciplinary conference on sinkholes and the engineering and environmental impacts of Karst: Carlsbad, National Cave and Karst Research Institute, Symposium 2, pp 469–480

Schappelwein K (1966) Geomorphologische Untersuchungen in den nordöstlichen steirischen Kalkalpen (Östl. Hochschwab, Zeller Staritzen, Veitschalpe). Geographischer Jahresbericht Aus Österreich 31:95–131

Seale LD (2005) Creation, analysis, and evaluation of remote sensing databases for Pinellas Counts, Florida. Diploma thesis, University of South Florida, United States of America

Spreitzer H (1951) Die Piedmonttreppen in der regionalen Geographie. Erdkunde 5:294–305

Stadler H, Benischke R, Strobl E (2008) Hydrogeologie Schneeberg/Rax - Final Report. Institut für Wasser Ressourcen Management, Hydrogeologie und Geophysik, Graz

Stummer G, Plan L (eds) (2002) Speldok-Austria. Handbuch zum Österreichischen Höhlenverzeichnis. Speldok 10. Verband Österreichischer Höhlenforscher, Wien

Tollmann A (1986) Geologie von Österreich, vol 3. Franz Deuticke, Wien

Weber L (ed) (1997) Handbuch der Lagerstätten der Erze, Industrieminerale und Energierohstoffe Österreichs. Archiv für Lagerstättenforschung 19, Geological Survey of Austria, Wien

Williams PW (1969) The geomorphic effects of ground water. In: Chorley RJ (ed) Water, earth and man. A synthesis of hydrology, geomorphology and socio-economic geography. Methuen & Co Ltd, London, pp 269–284

Williams PW (1971) Illustrating morphometric analysis of karst with examples from New Guinea. Z Geomorphol 15:40–61

Winkler-Hermaden A (1957) Geologisches Kräftespiel und Landformung. Springer, Wien

Christian Bauer is senior lecturer for physical geography at the University of Graz, Institute of Geography and Regional Science. His research focuses on geomorphology and geomorphometry in karst terrains, applied geomorphological topics using GIS (natural hazards) as well as human-environment interactions. His interdisciplinary research experience involves different disciplines, including Geosciences, System Science and History.

Lukas Plan is a research scientist at the Karst and Cave Group at the Natural History Museum Vienna. As geologist, he mainly deals with speleogenesis, evolution of karst features and landscapes, karst hydrology, and tectonics. He has explored and surveyed many kilometres of caves predominantly in the Eastern Alps and was co-editor of a book on caves and karst in Austria.

Montafon: Geodiversity and Human Impact

20

Sven Fuchs, Margreth Keiler, and Martin Wenk

Abstract

The Montafon Valley, located in southwestern Austria, is—for a comparatively small region of approximately 560 km^2—characterized by a high diversity of geological and geomorphological features. The landscapes of the Montafon comprise landforms and associated processes common in high mountains, a deeply incised main valley of the Ill River as a result of both glacial erosion and fluvial deposition, and several headwater tributaries. Montafon is dominated by a variety of relict and recent mass movements. The high geodiversity is still a main resource for the region and ranges from ancient mining activities to the current use of the landscape for hydropower and tourism. Thus, the landscape is an interplay of natural geomorphology and anthropogenic impact. The chapter provides an excursion guide to the diverse landscapes and landforms of this region, following the Ill River from its lower reach to its source.

Keywords

High mountains • Geodiversity • Mass movements • Human impact on landscape

S. Fuchs (✉)
Institute of Mountain Risk Engineering, University of Natural Resources and Life Sciences (BOKU), Peter-Jordan-Str. 82, 1190 Vienna, Austria
e-mail: sven.fuchs@boku.ac.at

M. Keiler
Department of Geography, University of Innsbruck, Innrain 52f, 6020 Innsbruck, Austria

M. Keiler
Institute for Interdisciplinary Mountain Research, Austrian Academy of Sciences, Innrain 25, 6020 Innsbruck, Austria

M. Wenk
Federal Ministry for Agriculture, Regions and Tourism, Federal Water Engineering Administration, Marxergasse 2, 1030 Wien, Austria

20.1 Geographical Setting

The Montafon Valley is the southernmost valley of the Federal State of Vorarlberg, situated along the upper reaches of the Ill River between the Rätikon, Verwall and Silvretta mountain ranges in the central Eastern Alps (Fig. 20.1). The valley has a length of 39 kms and is located between 47° 08′ N (valley constriction of Lorüns, around 580 m asl) and 46° 50′ N (Piz Buin mountain, 3312 m asl) and between 9° 47′ W (upper Rellstal Valley) and 10° 09′ W (Verwall mountain range), following with the exception of lake Lünersee, a centripetal drainage pattern towards the Ill River. The Ill River drains the valley over the entire length, and together with the whole catchment, the river has been considerably modified for more than 100 years in order to meet the requirements of hydropower generation. The climate of the Montafon Valley is temperate with an intermediate position between sub-oceanic and sub-continental conditions. In the village of Schruns, the annual mean temperature is 7.4 °C and precipitation amounts to 1243 mm (Werner 2005).

The valley constriction of Mauren (760 m asl) is the divide between the Outer and the Inner Montafon Valley, and is the result of debris accumulation from the east (Zamangspitze, 2386 m asl) and a torrential fan from the west (Maurentobel, 47° 2′ 34″ N, 9° 56′ 39″ E). The northern Outer Montafon is characterized by alluvial fans deposited by the tributary rivers at their mouths into the main valley. Major tributary valleys are the Relltal and the Gauertal from the southwest, and the Silbertal from the northeast. The latter valley is the largest tributary valley in the Montafon with a length of about 20 km. The major villages in the Montafon are located on these fans since in the past settlement on alluvial fans has been relatively safe from mountain hazards in comparison with the frequent flood events affecting the main valley bottom. In contrast, the southern Inner Montafon is a narrow high Alpine valley with an almost missing valley floor, the only exception being

C. Embleton-Hamann (ed.), *Landscapes and Landforms of Austria*, World Geomorphological Landscapes,
https://doi.org/10.1007/978-3-030-92815-5_20

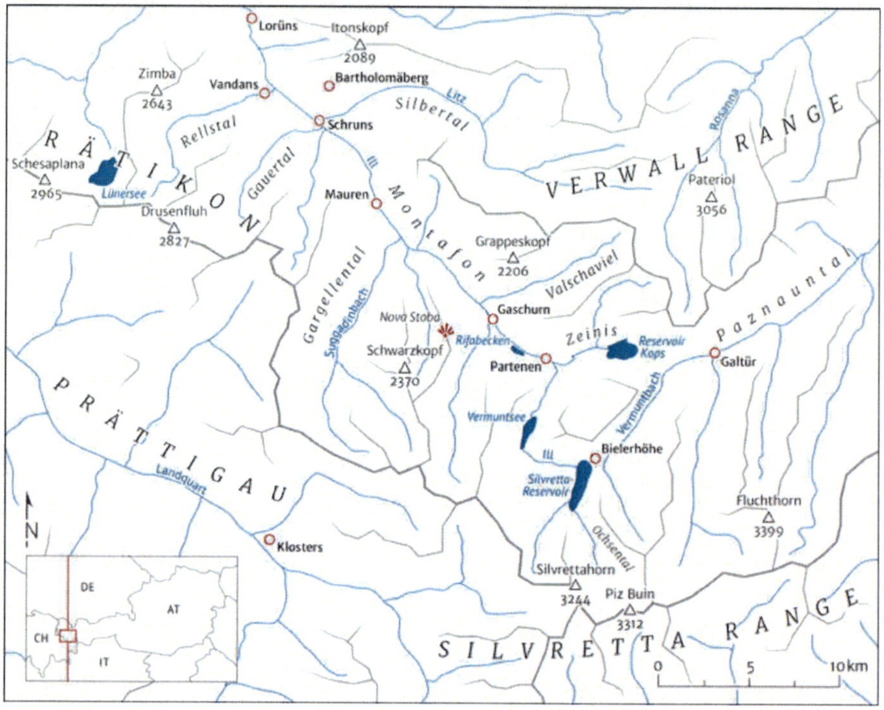

Fig. 20.1 Geographical overview of the Montafon Valley

the plain of Gaschurn where the Valschaviel Valley feeds into the main valley. At the village of Partenen (1051 m asl), the valley head of the main valley is located. The Ill River has its source at the Ochsental and the Vermunt glaciers (around 2800 m asl) in the Silvretta range. Its head waters flow north through the Silvretta reservoir (1990 m asl) and the Vermunt Valley with the Vermunt reservoir (1747 m asl) before the river enters the SE-NW trending Montafon Valley at Partenen. The highest elevation in Montafon is the Piz Buin (3312 m asl).

20.2 Geology

20.2.1 General Setting

Montafon is located at the border between the Eastern Alps and the Western Alps, which crosses the valley at Vandans (Friebe 2007) (Fig. 20.2). Three major mountain ranges meet here, the Rätikon in the west, the Verwall in the east and the Silvretta in the south. The northern and northwestern Montafon is dominated by limestones and dolomites, as well as subordinated shale and sandstones. These rocks are of Permian and Lower Cretaceous age and belong to the Lechtal nappe of the Northern Calcareous Alps (Oberhauser 2007). The dolomites form such major peaks as Davennakopf (1707 m asl) and Itonskopf (2089 m asl), and are partly overlain by the so-called Kössen Formation. A major coral reef is located at Itonskopf (accessible from 47° 6′ 58″ N, 9° 56′ 7″ E). The Raibler Formation underneath the dolomites contains gypsum deposits, which around the location Küngs Maisäss in the Silber Valley led to the development of impressive gypsum dolines (accessible with the Kristberg cable car followed by a 1.5-h walk, 47° 7′ 15″ N, 9° 57′ 21″ E). Southwards, in the Rellstal Valley and around the community of Bartholomäberg, the so-called Graywacke zone is present with a lateral extent of about 15 km. This zone is represented by mudstones, sandstones and marlstone.

From the village of Vandans, the entire Montafon until the Silvretta reservoir is part of the Silvretta crystalline and characterized by less metamorphic units such as phyllite gneiss, amphibolites and mica schist layers. The orographic western parts of the watershed are composed of Penninic units such as the Arosa zone, the Sulzfluh nappe and the Prättigau Flysch (Reithofer 1961; Fuchs 1984; Wolkersdorfer 2005; Friebe 2007). The morphology of the entire valley is influenced by the Quaternary glaciations, and as such, large parts of the surface are covered by glacial and fluvioglacial sediments of the last and youngest glaciation (Würmian).

Some of the tributary valleys are structurally controlled: the Rellstal Valley follows the border between the Limestone Alps and the Silvretta crystalline, and the SW/NE Gargellen fault zone can be even detected in satellite images as a line from Wasserstubental Valley (47° 6′ 4″ N,

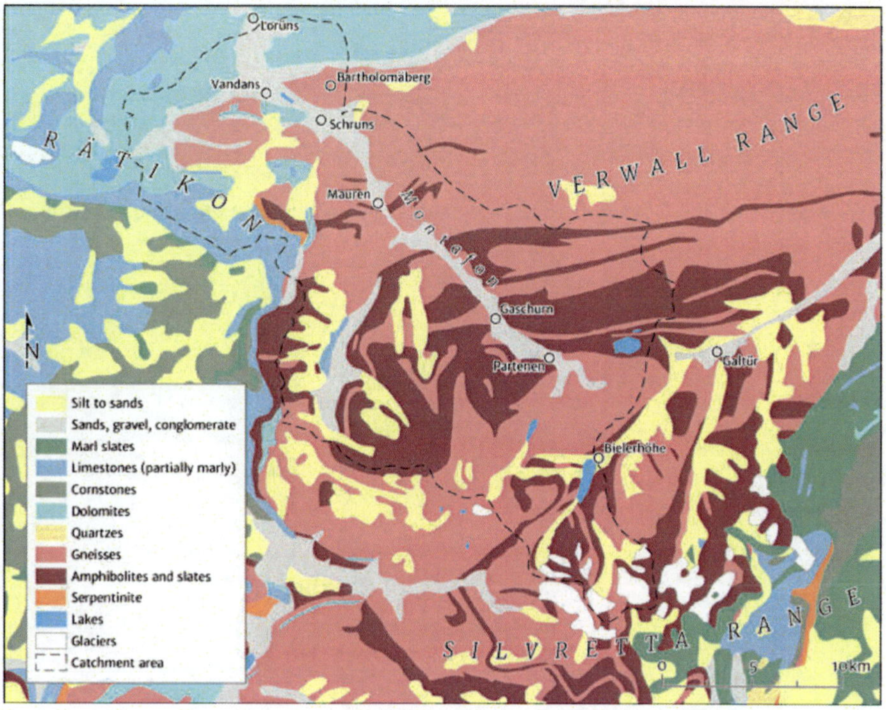

Fig. 20.2 Generalized map of the geology in Montafon Valley. *Data Source* Fuchs and Oberhauser (1990), Fuchs and Pirkl (1990), Geologische Bundesanstalt (2007) and Oberhauser et al. (2007)

10° 2′ 57″ E) through Gieslabach, Tramosabach and Gargellental Valley to the Schlappiner Joch (2202 m asl, 46° 55′ 31″ N, 9° 54′ 29″ E), a historic pass connecting the Montafon and the Swiss Prättigau region. The entire fault zone is part of the large lineament of Landwasser-Gargellen, spanning from Davos in Switzerland to the Kleinwalsertal in Austria.

20.2.2 Geological Thematic Trail

The geologic hiking trail in Montafon provides an overview of the northern geological units and was created in 1978 (Fig. 20.3). The loop trail starts at the church of the village Bartholomäberg (47° 5′ 30″ N, 9° 54′ 32″ E). With a total length of 16 km, equal to 4–5 h walking time, the path has 24 presentation boards illustrating the earth history of the Montafon Valley. The age, genesis and composition of rocks and minerals are explained, as well as the tectonics of the region. A complementary guidebook can be purchased at the tourism office of Bartholomäberg (Bertle 1979). Starting from the overview board at the parking lot next to the church (1087 m asl), the trail stretches in northwest direction to Rellseck (alpine hut, 1492 m asl) and passes four presentation boards addressing rock characteristics. At Rellseck, the path turns right and climbs in northeast direction along the Monteneu crest to the Wannaköpfle peak (2032 m asl). Passing to the right of Itonskopf, the peak of Alpilakopf (2078 m asl) will be reached. Like the entire crest, the peak is composed of dolomite and underlain by the gypsum of the Raibler Formation. The path continues to the limestone cliff at "Obere Wiese" which is of Jurassic age, and contains corals, shells and other fossils. Further, the trail surrounds the minor summit of the Schwarzhorn (1883 m asl) and after passing a number of lakes turns back to the starting point. At Knappa Gruaba (1340 m asl), anthropogenic geomorphology can be studied since in this region silver and copper mining activities date back to the ninth century, including the former tunnel mouths and the mine tailings. The historical St. Anna mine is located here, today a museum that re-opened in 2010 (Schaubergwerk Bartholomäberg, www.kristberg.at/Bergwerkstollen.htm).

20.3 Geomorphology

The northern summits of the Montafon are characterized by rugged relief developed in the white to light grey rocks of the Northern Calcareous Alps and of the Penninikum. Prominent examples include the Zimba (2643 m asl), the Drusenfluh (2827 m asl) and the Schesaplana (2965 m asl) (Fig. 20.4). In contrast, the Silvretta crystalline is characterized by broad and rounded summits due to the different weathering processes, which also provide the mountains with darker colours (Fig. 20.5). Examples include the Geißspitze (2334 m asl), the Kreuzjoch (2395 m asl) and the

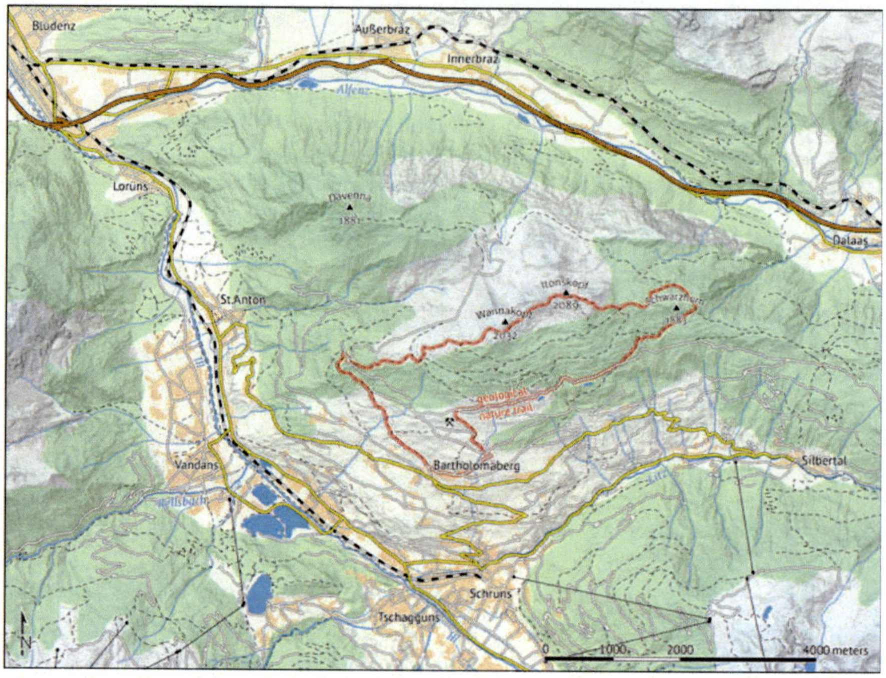

Fig. 20.3 Overview of the geological hiking trail in the Montafon Valley. *Data Source* OGD Vorarlberg (2016)

Zaferna (2277 m asl). As such, the topography of the Montafon mirrors the different weathering and geomorphological downwearing dictated by geology.

Due to the Würmian glaciation, but also throughout the entire Quaternary, the original surface relief shaped by the geological bedrock and fluvial processes was modified by glacial erosion (Aulitzky et al. 1994). As a result, only few V-shaped valleys remain, such as the one of the Balbier torrent (Fig. 20.6, five minutes by foot to the Balbier waterfall from Gortipohl, 47° 0′ 51″ N, 10° 0′ 14″ E).

Typical U-shaped valleys can be found as a result of glacial erosion, such as the Gauertal Valley. A possible hiking trail starts at Matschwitz (accessible from Golm via the cable car starting at the Latschau reservoir, middle station at 47° 04′ 10″ N, 9° 51′ 33″ E) and leads to the mountain pastures of Plazadels and Wachters Dieja. The

Fig. 20.4 View from Sennigrat (2289 m asl) in northwestern direction outwards the Montafon Valley. The communities in the valley bottom are Schruns and Tschagguns. In the middle ground, the reservoirs of Rodund (645 m asl) and Latschau (994 m asl) are located, and the background shows the limestone summits of Steinwandeck (1996 m asl), Zimba (2643 m asl) to Schesaplana (2946 m asl, right to left). On the left hand side of the main valley, the tributary valleys Gauertal and Rellstal can be seen. Photo: S. Fuchs

Fig. 20.5 View from Versettla/Nova Stoba (1992 m asl, 46° 58′ 40″ N, 9° 59′ 48″ E) in southeastern direction over the Inner Montafon Valley with its distinct U-shaped cross profile left by the Quaternary glaciers. The foreground shows the Rifabecken reservoir and the background the Kops reservoir. Photo: S. Fuchs

Fig. 20.6 Bablier waterfall. At the foot of the 17 m high waterfall, you can find a barbecue site. Photo: F. Müller, CC BY-SA 3.0, https://commons.wikimedia.org/w/index.php?curid=11937974

buildings on these pastures date back to the early seventeenth century, as proven by dendrochronological studies (Keiler et al. 2001). The entire orographic left slopes of Gauertal are sliding slowly, which results in visible tilting of some of the alpine huts in Plazadels. For Wachters Dieja, the movement is a result of recent torrential erosion undercutting the slope toe at the orographic left side of Gauerbach (Fuchs et al. 2001).

The entire Montafon is characterized by torrential fans. A textbook example is the torrential fan of Vand (orographic right of the Valschaviel Valley mouth, 46° 59′ 54″ N, 10° 1′ 25″ E), a tributary of Valschaviel torrent at Gaschurn (Fig. 20.7). The small torrent shows typical characteristics of a mountain watershed such as a steep and confined upper channel and a cone-shaped depositional area in the lowest part of the catchment filling the angle between the steep slopes and the floor of the main valley. The size of the fan is related to the rock fall area of Grappeskopf (2206 m asl). The rockfall area is a significant source of material through the process of frequent torrent events (debris flows and related processes). The torrent is anthropogenically modified by torrential mitigation measures in order to prevent flooding and debris flows from the upper part of the catchment.

Upstream to the SE, the Montafon Valley becomes narrower. As shown in Fig. 20.5, the community of Partenen as well as the Rifabecken reservoir, which is part of the Kops hydropower system, are located within the valley floor. Prominent features on the orographic right valley side are the valley shoulder of Tafamunt Maisäss (1478 m asl,

Fig. 20.7 Middle ground shows the Valschaviel Valley to the right and the torrential fan and watershed of Vand to the left. The picture is taken from the opposite slope at 46° 58′ 40″ N, 9° 59′ 48″ E. Photo: S. Fuchs

46° 58′ 33″ N, 10° 3′ 37″ E) and the adjacent rockfall area (Fuchs and Keiler 2003). The latter is caused by a gravitational deformation extending deep into the mountain ("Bergzerreißung Vesal", Reithofer 1961; Bertle 1992) and forming the debris cone together with the burden material from the tunnel system of the Kops hydropower reservoir. According to historical sources, the main period of rockfall activity can be dated back to 1892 (Loacker 1971). The horizontal structures on the lower part of the cone are artificial dams preventing the release of rocks towards the community of Partenen and were constructed as a result of the 1922 events (Fuchs and Keiler 2003).

The Kops reservoir is situated in the Zeinis Valley, which is a glacial hanging valley, carved out by a tributary Ice Age glacier that had a smaller ice volume than the main Ice Age Ill Glacier. Thus, only a shallow valley as compared to the deep U-shaped valley of the Inner Montafon was eroded, resulting in the present-day situation of the Zeinis Valley mouth "hanging" high above the main valley floor. In the background of Fig. 20.5, the Kops reservoir (1809 m asl) with its massive concrete dam is visible, surrounded by the peaks of the Silvretta massif. The orographic right side of the Zeinis Valley is accompanied by a prominent valley shoulder comparable to the Tafamunt Maisäss above Partenen. In this case, the formation of the terrace is favoured by the outcropping of less resistant mica schists. The small Wiegensee Lake is located there (46° 58′ 33″ N, 10° 5′ 28″ E) which is part of a 65 ha large raised bog protected under the Habitats Directive (formally known as Council Directive 92/43/EEC on the Conservation of Natural Habitats and of Wild Fauna and Flora).

An excellent viewpoint is the mountain restaurant of Nova Stoba (46° 58′ 47″ N, 9° 59′ 31″ E), accessible by the Silvretta Nova cable cars from Gaschurn to the mountain station of section Versettla II. From there, a few minutes' walk provides you with the panoramic view captured in Fig. 20.5. Another interesting feature is the inactive rock glacier on the northeastern slope of Schwarzkopf (2370 m asl) south of St. Gallenkirch at Alpe Nova (46° 58′ 17″ N, 9° 58′ 30″ E), which is clearly visible from the restaurant terrace (Fig. 20.8). Rock glaciers are sensitive indicators of climate change: when their internal ice content melts out, they stop moving downhill and become inactive. Satellite images such as provided by Google Earth depict rock glaciers very well. Active rock glaciers in the Montafon can be found in the inner Silbertal Valley around lake Dürrwald (47° 2′ 7″ N, 10° 4′ 4″ E). Visiting them affords a four-hour walk, starting and ending at the bus stop "Untere Dürrwaldalpe". For the outward journey, the bus can either be boarded at the car park at "Felsa" or at the valley station of Kristberg cable car (community of Silbertal).

The Litzbach draining the Silbertal Valley is a highly active mountain torrent (Fischer 1989) with an overall catchment area of 102 km^2. As a result of the 1965 flood events, a slit dam was constructed at 47° 4′ 17″ N, 10° 1′ 26″ E (Fig. 20.9). The events in the mid-1960s resulted in the reactivation of a large slump at the location "Bärenalpe" (with an erosion area of around 5 ha) and the successive mobilization of considerable amounts of loose gravel. With the snowmelt at the end of May to the beginning of June 1965, the discharge increased remarkably so that along the entire Litzbach course almost all bridges and

Fig. 20.8 Inactive rockglacier at the northeastern slope of Schwarzkopf (2370 m asl). *Source* Google Earth, 2016 Geoimage Austria

the valley road were destroyed. In order to protect the village of Schruns at the confluence of the Litzbach and the Ill rivers, in 1966 and 1967 a retention area and a slit barrier were built (Schilcher 1973). In the subsequent years, the high discharge repeatedly led to aggradation and the retention area had to be emptied using caterpillars. Between 1967 and 1971, more than 1 million m^3 of material was either removed from the retention area or re-mobilized and transported further downstream. Due to the high amount of gravel and sand and the lack of cohesive material, the re-mobilization turned out to be quite successful, in particular during periods of high discharge following snowmelt. Once empty, the retention volume available amounts to approximately 300,000 m^3.

The Silvretta Lake located in the valley head at Bielerhöhe (2030 m asl) is the highest reservoir in the Illwerke hydropower chain. Starting in 1938, the massive dams were constructed by war prisoners and forced labourers; in 1943, the lake was partly filled, and in 1951, the maximum water surface area was reached. The overall catchment of the lake amounts to 45 km^2 and the water volume is 36.6 million m^3. To the south of the western dam, glacially polished bedrock, consisting of amphibolite from the Silvretta crystalline, can be examined (46° 54 ′30″ N, 10° 5′ 24″ E), showing the

Fig. 20.9 Slit dam constructed after the 1965 flood events. The dam has a concrete core and the surface is covered with granite rock in order to better withstand abrasion. Photo: S. Fuchs

transition between different types of amphibolite, their minerals and structures of deformation (Bertle 2007). Further amphibolites along the path contain epidotes and garnets of 2 cm diameter with a dark selvage (the so-called korona structure resulting from pressures of 8–10 kbar and temperatures of 500–600 °C during the Variscan metamorphosis, Friebe 2007). The Silvretta Lake is a sediment trap, which can be seen at periods of low water level at its southern end. Here, the Ill debouches and because of the decelerating flow velocity deposits the entrained sediment in form of a delta.

Box: Hydropower Generation in the Montafon and its Impact on the Environment

Hydropower generation has a long tradition in the Montafon (Plitzner 1999). Already in 1895, a first hydroelectric power plant was established in Schruns, followed by another facility in Lorüns in 1901. Today, the Illwerke operate ten hydroelectric power plants in the Montafon Valley, which produce peak power for the domestic and international electricity market. Streams of the Silvretta mountain range and surrounding areas are captured and transferred in a widely ramified system of tunnels, pipes, channels and reservoirs to storage power plants (Fig. 20.10). If the entire system is considered, the hydroelectric turbines have a maximum power of 1812 MW and a power consumption of 1044 MW when pumping the water back. The electricity produced is being fed into the European power grid and predominantly sold to the long-term partner Energie Baden-Württemberg AG as well as to the local markets of Vorarlberg and Tyrol. The most visible components of the hydroelectric system are the four large reservoirs. In 1930, the Illwerke began the construction of Vermunt reservoir, later followed by the Silvretta reservoir at Bielerhöhe (completion 1951), the Lünersee (1959) and the reservoir Kops (1969). The Silvretta Hochalpenstraße, a toll road between the Montafon Valley and the Paznaun in the Federal State of Tyrol with the Bielerhöhe pass at an elevation of 2071 asl as highest point, was originally planned as construction road and is owned by the Vorarlberger Illwerke AG.

As shown in a review by Magilligan and Nislow (2005), hydropower has major impacts on river hydrology, primarily through changes in the timing, magnitude and frequency of low and high flows, ultimately producing a hydrologic regime differing significantly from the natural flow regime. Given the implementation of the EU Water Framework Directive in Austria, there is an ongoing discussion of the admissibility of hydropeaking, which introduces short-term changes from high discharge to very low water flow to the river. These issues are also connected to the legally required residual amount of water for a sustainable riverine environment and the resulting business restrictions.

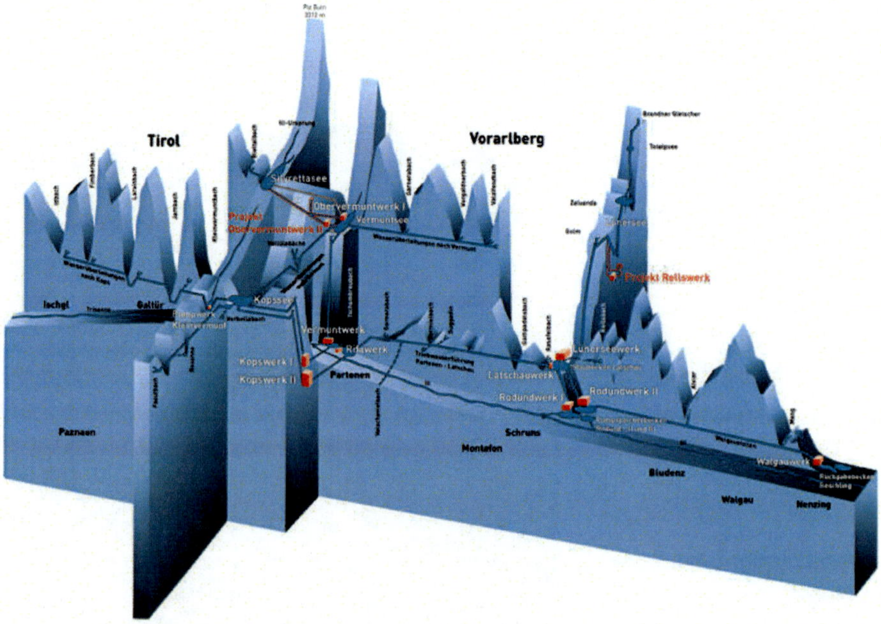

Fig. 20.10 Overview of the hydropower system of Illwerke. Reproduced with permission of Vorarlberger Kraftwerke Aktiengesellschaft

20.4 Concluding Remarks

The Montafon Valley in the SE of the Federal State of Vorarlberg shows a high geodiversity as a result of the geological setting at the border between the Western and the Eastern European Alps, the Würmian glaciation responsible for the main erosive features and the Holocene deposition of material. Moreover, the landscape has been repeatedly influenced by human activity, such as the long history of agriculture and mining during the last centuries and the artificial interventions necessary for hydropower generation since the outgoing nineteenth century. Over a length of approximately 40 km, the valley is characterized by high-alpine landscapes with a variety of associated sgeomorphologic features and plays an important role for tourism and recreation as well as landscape and nature conservation.

References

Aulitzky H, Heuberger H, Patzelt G (1994) Mountain hazard geomorphology of Tyrol and Vorarlberg, Austria. Mount Res Dev 14 (4):273–305

Bertle H (1979) Führer für den geologischen Lehrwanderweg Bartholomäberg. Schruns, Heimatschutzverein im Tale Montafon

Bertle H (1992) Angewandte Geologie in Vorarlberg auf Grundlage der Landesaufnahme - Rückblick und Einfluß von R. Oberhauser. Jahrbuch Der Geologischen Bundesanstalt 135(4):791–808

Bertle H (2007) Silvrettastausee: Silvrettakristallin. In: Friebe JG (ed) Geologie der österreichischen Bundesländer - Vorarlberg. Geologische Bundesanstalt, Wien, pp 143–144

Fischer A (1989) The control of pebbles in the Austrian Rhine basin. Historic development and practice. Mensuration, Photogrammétrie, Génie Rural 87 (1):17–20

Friebe JG (2007) Vorarlberg. Geologie der österrechischen Bundesländer - Vorarlberg. Geologische Bundesanstalt, Wien

Fuchs F, Oberhauser R (1990) Geologische Karte der Republik Österreich 1:50.000 - Blatt 170 Galtür. Geologische Bundesanstalt, Vienna

Fuchs F, Pirkl HR (1990) Geologische Karte der Republik Österreich 1:25.000 - Blatt 169 Partenen Ost und West. Geologische Bundesanstalt, Vienna

Fuchs G (1984) Erläuterungen zu Blatt 169 Partenen (Ost und West). Geologische Bundesanstalt, Wien

Fuchs S, Keiler M (2003) Naturräumliche Rahmenbedingungen. In: Keiler B, Pfeifer K, Rudigier A (eds) Die Maisäße auf Tafamunt. Montafoner Schriftenreihe 9. Heimatschutzverein im Tale Montafon, Schruns, pp 9–14

Fuchs S, Keiler M, Pfeifer K (2001) Naturräumliche Grundlagen. In: Keiler B, Pfeifer K (eds) Plazadels und Wachters Dieja. Maisäßsiedlungen im Gauertal, vol Montafoner Schriftenreihe 2. Heimatschutzverein im Tale Montafon, Schruns, pp 9–14

Geologische Bundesanstalt (2007) Geologische Karte von Vorarlberg 1:100.000. Geologische Bundesanstalt, Wien

Keiler B, König G, Pfeifer K (2001) Geschichte der Maisäßwirtschaft. In: Keiler B, Pfeifer K (eds) Plazadels und Wachters Dieja. Maisäßsiedlungen im Gauertal, vol Montafoner Schriftenreihe 2. Heimatschutzverein im Tale Montafon, Schruns, pp 15–20

Loacker H (1971) Berg- und Grundwasserverhältnisse im Illgebiet. Verhandlungen Der Geologischen Bundesanstalt 1971(3):441–449

Magilligan FJ, Nislow KH (2005) Changes in hydrologic regime by dams. Geomorphology 71(1–2):61–78

Oberhauser R (2007) Zur Geologie der Ost-Westalpen-Grenze anhand der geologischen Karte von Vorarlberg 1:100.000 mit dem zugehörigen Profilschnitt vom Bodensee ins Unterengadin. Geologie der österreichischen Bundesländer - Vorarlberg. Geologische Bundesanstalt, Wien

Oberhauser R, Bertele H, Bertele R (2007) Geologische Karte von Vorarlberg 1:100.000. Geologische Bundesanstalt, Vienna

Plitzner K (1999) Der Weg nach Süden! Oder doch nach Norden? Von den Anfängen der Elektrizitätswirtschaft in Vorarlberg bis zur Gründung der „Vorarlberger Illwerke“ bis in die 1930er Jahre. In: Maier H (ed) Elektrizitätswirtschaft zwischen Umwelt, Technik und Politik: Aspekte aus 100 Jahren RWE-Geschichte 1898–1998. Technische Universität Bergakademie Freiberg, Freiberg, pp 89–127

Reithofer O (1961) Bericht 1960 über Aufnahmen auf den Blättern Gaschurn (169) und Mathon (170) In: Bundesanstalt G (ed) Verhandlungen der Geologischen Bundesanstalt. Wien, pp A 58–A 59

Schilcher C (1973) Erfahrungsbericht über die Schlitzsperre im Litzbach in der Gemeinde Silbertal, Bez. Bludenz. Mitteilungen Der Forstlichen Bundes-Versuchsanstalt Wien 102:293–305

Werner R (2005) Klima und Wetter im Montafon. In: Rollinger JM, Rollinger R (eds) Montafon 1: Mensch, Geschichte, Naturraum. Stand Montafon, Schruns, pp 57–98

Wolkersdorfer C (2005) Geologische Verhältnisse des Montafons und angrenzender Gebiete. In: Rollinger JM, Rollinger R (eds) Montafon 1: Mensch, Geschichte, Naturraum. Stand Montafon, Schruns, pp 25–55

Sven Fuchs is a senior scientist at the Institute of Mountain Risk Engineering at the University of Natural Resource and Life Sciences in Vienna, Austria. His research and teaching interests include high-mountain geomorphology, mountain hazard risk management, the study of coupled human-environment systems, climate change adaptation, and vulnerability assessment for natural hazards. His research has taken him to mountain regions of Europe, Southeast Asia, Central Africa, and countries of the former Soviet Union.

Margreth Keiler is professor at the Department of Geography, University of Innsbruck and director of the Institute for Interdisciplinary Mountain Research at the Austrian Academy of Sciences. She is head of the joint working group Coupled Human-Landscape Systems: Risk and Resilience at both institutions. Her research focus is on sediment connectivity, risk analysis and risk management under global change and multi-hazard risks in mountains, with regional expertise in the European Alps, in mountain areas of Central and Southeast Asia, and in Africa.

Martin Wenk works for the Federal Water Engineering Administration, a department of the Austrian Federal Ministry for Agriculture, Regions and Tourism in Vienna. His work and research is focused on the management of natural hazards and the assessment of related risks, with particular emphasis on the use of geoinformatics and cartography.

21 Giant "Bergsturz" Landscapes in the Tyrol

Martin Mergili and Christoph Prager

Abstract

The line connecting the Zugspitze and the central Ötz Valley is characterized by a remarkably high density of large prehistoric landslides, with distinctive "Bergsturz" landscapes in their deposition areas. Those events north of the Inn Valley (e.g., Fernpass, Tschirgant) originated in carbonate rock units and those south of the Inn Valley in metamorphic rocks (e.g., Köfels). All events were geologically predisposed by complex fold-, fault- and fracture systems. The failed slopes mobilized substantial rock mass volumes and led to the accumulation of fluvio-lacustrine backwater sediments. Temporally, these early and middle Holocene landslides cluster with other events in the surrounding regions in the Alps. The landslide accumulation areas are characterized by rough debris and rather permeable terrain, and are therefore unfavourable for agricultural use. Distinctive forest ecosystems often dominated by Scots pine (*Pinus sylvestris*) are best adapted to these conditions. Whilst the deposition areas of large landslides were—and are still—obstacles for traffic and unfavourable for many types of land use, they are often perceived as appealing and scenic, and are therefore popular for recreational activities.

Keywords

Landslide • Rock avalanche • Rockslide • Backwater sediments • Land cover

21.1 Introduction

The Ötz Valley represents a major tourist destination in summer and even more so in winter. Many guests approach their holiday resorts straight from the north, crossing the Northern Calcareous Alps to the Inn Valley before entering the Ötz Valley. On busy days, their cars get stuck in traffic jams in a place known as Fernpass where the road is winding through an undulating, forested mass of blocky material. Some lake sites crowded with tourists are passed on the way. Having overcome this nuisance, observant travellers note another hilly mass of blocks spread over the entrance of the Ötz Valley, vegetated by sparsely developed pine forests. Several more rock masses are passed on the way through the Ötz Valley. Most impressive, between Umhausen and Längenfeld—nearby the village of Köfels—the road passes a steep gorge incised hundreds of metres deep into an undulating barrier of crushed rocks.

Travellers and scientists have thought about the origin of these rock masses—mainly the one of Köfels—for almost two centuries. Escher von der Linth (1845) interpreted the Köfels barrier as the result of a gravitational mountain collapse. The presence of fused rocks led to the hypothesis of volcanic activity (Pichler 1863) or a meteorite impact (Stutzer 1936; Suess 1937; Bond and Hempsell 2008). Whilst seismic activity was brought into discussion too (Trientl 1895), more recent studies mainly interpret the phenomenon as the deposit of a prehistoric giant landslide. The other rock masses all the way from the Fernpass to the central Ötz Valley are interpreted in similar ways, and a considerable amount of work has been published with regard to the characteristics, dynamics and timing of these events

M. Mergili (✉)
Institute of Geography and Regional Science, University of Graz, Heinrichstrasse 36, 8010 Graz, Austria
e-mail: martin.mergili@uni-graz.at

M. Mergili
Institute of Applied Geology, University of Natural Resources and Life Sciences (BOKU), Peter-Jordan-Straße 82, 1190 Vienna, Austria

C. Prager
alpS GmbH, Grabenweg 68, 6020 Innsbruck, Austria

ILF Consulting Engineers Austria GmbH, Feldkreuzstraße 3, 6063 Rum, Austria

C. Embleton-Hamann (ed.), *Landscapes and Landforms of Austria*, World Geomorphological Landscapes,
https://doi.org/10.1007/978-3-030-92815-5_21

(e.g., Ampferer 1904, 1939; Abele 1974, 1997b; Preuss 1974; Ivy-Ochs et al. 1998; Erismann and Abele 2001; Prager et al. 2006, 2007, 2008, 2009a, b, 2012; Prager 2010; Patzelt 2012a, b, Zangerl et al. 2021; Nicolussi et al. 2015; Ostermann and Prager 2014, 2016; Dufresne et al. 2016a, b, and references therein). Some of these events were characterized by long run-out distances of several kilometres, most likely favoured by the enormous landslide volumes, by dynamic disintegration and, crucially, by interaction with the saturated valley fills.

Compared to other regions in the Alps, the area between the Fernpass and the central Ötz Valley displays spatial clustering of large landslide deposits. Table 21.1 lists the major characteristics of the most notable landslides in this region, whilst Fig. 21.1 provides an overview of their spatial distribution. Other events such as Eibsee or Ehrwald have occurred in the vicinity (Abele 1974; Prager et al. 2008; Prager 2010; Ostermann and Prager 2016), but are not subject of the present chapter.

With this background, one may raise a number of questions. The following ones will be addressed—and as far as possible answered—within the next sections:

1. Why is there a spatial concentration of large prehistoric landslide events in this specific area—i.e., what were the geological and geomorphological factors predisposing the events?
2. What were the temporal patterns of occurrence, and can they be helpful to draw conclusions on the triggers of the mass movements?
3. Which types of habitats do the deposits of the large prehistoric mass movements support?
4. How do those events reflect themselves in the patterns of today's cultural landscape?

Before diving into the issue more deeply, some terms have to be defined. The term "landslide" is used throughout as a general description for all types of gravitational mass movements (Cruden 1991: 27: "*A landslide is the movement of a mass of rock, earth or debris down a slope*"). Concerning this work, reference is made to a practical and useful classification based on the type of material (rock, soil) and type of movement (slides, falls, topples, spreads and flows) of Cruden and Varnes (1996). The term "rock avalanche" describes rapid rockslide and "flow" processes, commonly larger than 1 million m^3 in volume, with a substantial degree of dynamic fragmentation and often displaying long run-out distances, even on only gently inclined valley floors (also referred to as "Sturzstrom"; e.g., Heim 1932; Hsü 1975). The German term "Bergsturz" is commonly applied to a variety of large, rapid rock slope collapses (Abele 1974) and may comprise major rock falls, rockslides and rock avalanches (see, e.g., Erismann and Abele 2001; Hungr and Evans 2004).

Table 21.1 Major characteristics of the landslides considered in the present chapter

Name	Age (ka)	Rock types	Volume (million m^3)	Travel distance (km)	Angle of reach (°)	Selected references
Fernpass	4.1–4.2	Dolostone, marl, limestone	1000[a]	15.5[b]	5.3[b]	Abele (1964, 1997b), Prager (2010), Prager et al. (2006, 2008, 2009a, 2012)
Tschirgant	3.7–3.5[c] 3.0–3.2	Dolostone, breccia, limestone, rauhwacke	100–125[a] 200–250[d]	6.2	12.7	Pagliarini (2008), Patzelt (2012a), Dufresne et al. (2016a; b), Ostermann and Prager (2016)
Haiming	3.7–3.5[c] 3.0–3.2	Dolostone, breccia, limestone	25–60[d]	2.4	13	Abele (1974), Patzelt (2012a)
Habichen	≥ 11.5	Orthogneiss	27[d]	2.9	18.2	Prager et al. (2008), Ostermann and Prager (2016)
Achplatte	≥ 3.6[e]	Orthogneiss	60[d]	2.2	24	Ostermann and Prager (2016)
Tumpen	≥ 3.6[e]	Orthogneiss	Multiple events	Multiple events	Multiple events	Poscher and Patzelt (2000), Ostermann and Prager (2014)
Köfels	9.5	Orthogneiss, subordinary paragneiss	3100[a] 4,000[d]	5.6	11–12	Prager et al. (2009b), Zangerl et al. (2021), Nicolussi et al. (2015)

[a]Release volume (estimation of scarp area)
[b]Complex accumulation path comprising two rock avalanche braches, data given for the longer (southern) branch
[c]According to Patzelt (2012a), the Tschirgant and Haiming rock avalanches comprise at least two events each; according to Ostermann et al. (2016), at Tschirgant only one accumulation event occurred at approx. 3 ka
[d]Deposition volume
[e]Minimum age of the backwater sediments (Poscher and Patzelt 2000)
See Fig. 21.1 for the spatial distribution of the events

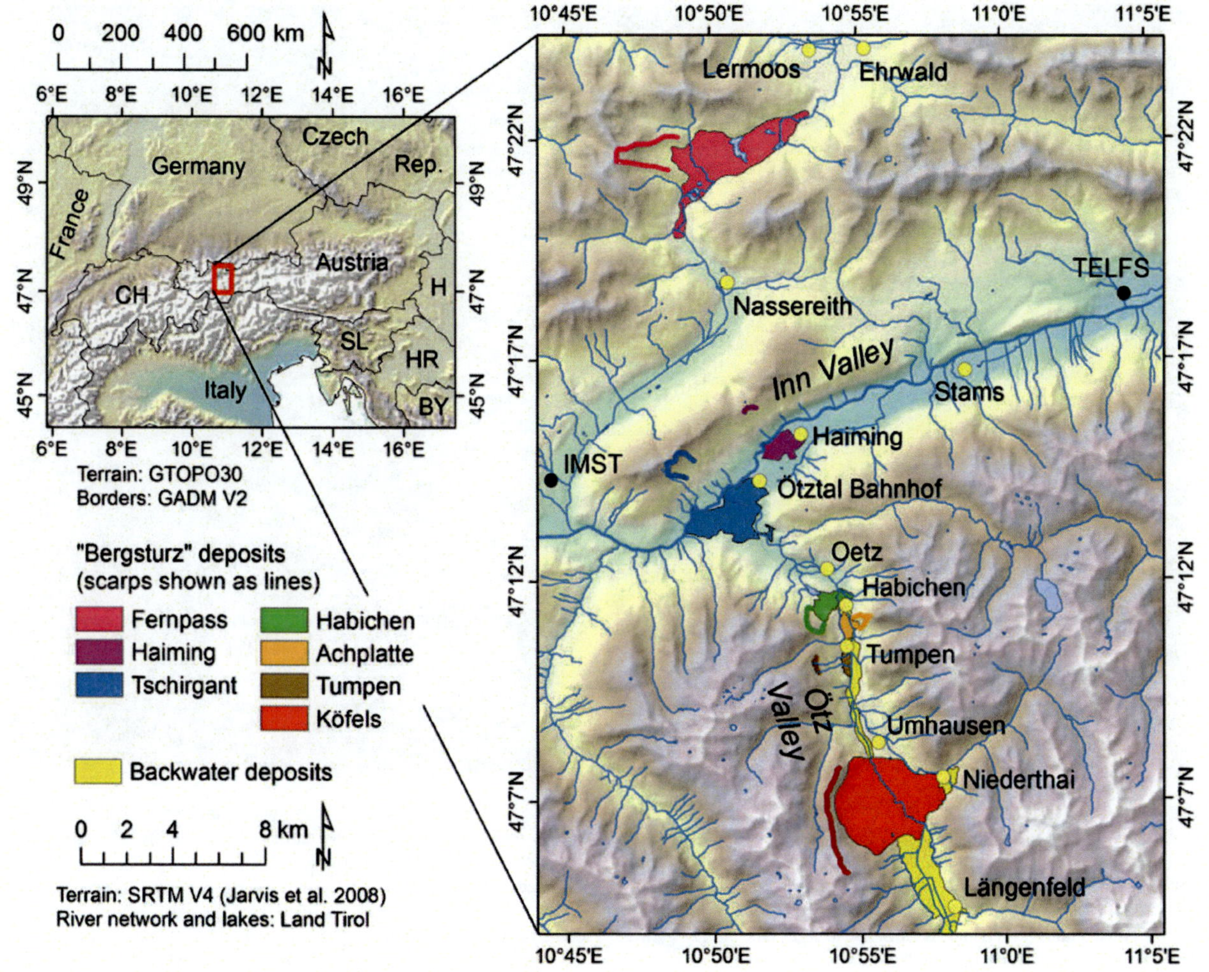

Fig. 21.1 The terrain representation in the right pane is based on Jarvis et al. (2008)

21.2 Geological Setting and Geomorphological Features

The area of interest is composed of Mesozoic sedimentary strata of the Northern Calcareous Alps (north of the Inn Valley) and the Palaeozoic metamorphic Ötztal basement complex (south of the Inn Valley) (Brandner 1980; GBA 2011). The Inn Valley is preconditioned by a major tectonic fault, facilitating glacial and fluvial erosion that shaped the landscape of today. The Ötz Valley, a northward discharging tributary to the W–E-trending Inn Valley, is deeply incised in the Ötztal and Stubai Alps, with a local relief of up to 2500 m.

The Ötztal Alps are made up of poly-metamorphic rocks of plutonic, volcanic and sedimentary origin (orthogneisses, amphibolites, paragneisses and micaschists; see Purtscheller 1978 for a more detailed account). In the medial and lower sections of the Ötz Valley, the landslide masses at Tumpen, Achplatte, Habichen and Köfels (Figs. 21.1 and 21.2c) mainly displaced various types of orthogneisses, embedded in less competent paragneiss units (Purtscheller 1978; GBA 2011; Ostermann and Prager 2016, and references therein). Layering and main foliation mainly trend in W–E direction and are cut by different fault and fracture systems which predisposed the bedrock slope failures. Also the Köfels rockslide, representing the largest known landslide of the Alps in crystalline rocks (Fig. 21.2a, b), was sourced mainly in orthogneisses and structurally predisposed by the coalescence of brittle fracture sets as well as by fault-related valley deepening (Prager et al. 2009b).

In contrast, the Fernpass, Tschirgant and Haiming events occurred in carbonate rocks, which detached from complexly folded and faulted thrust sheets of the Northern Calcareous Alps, the Lechtal nappe in the case of Fernpass and the Inntal nappe in the case of Tschirgant and Haiming

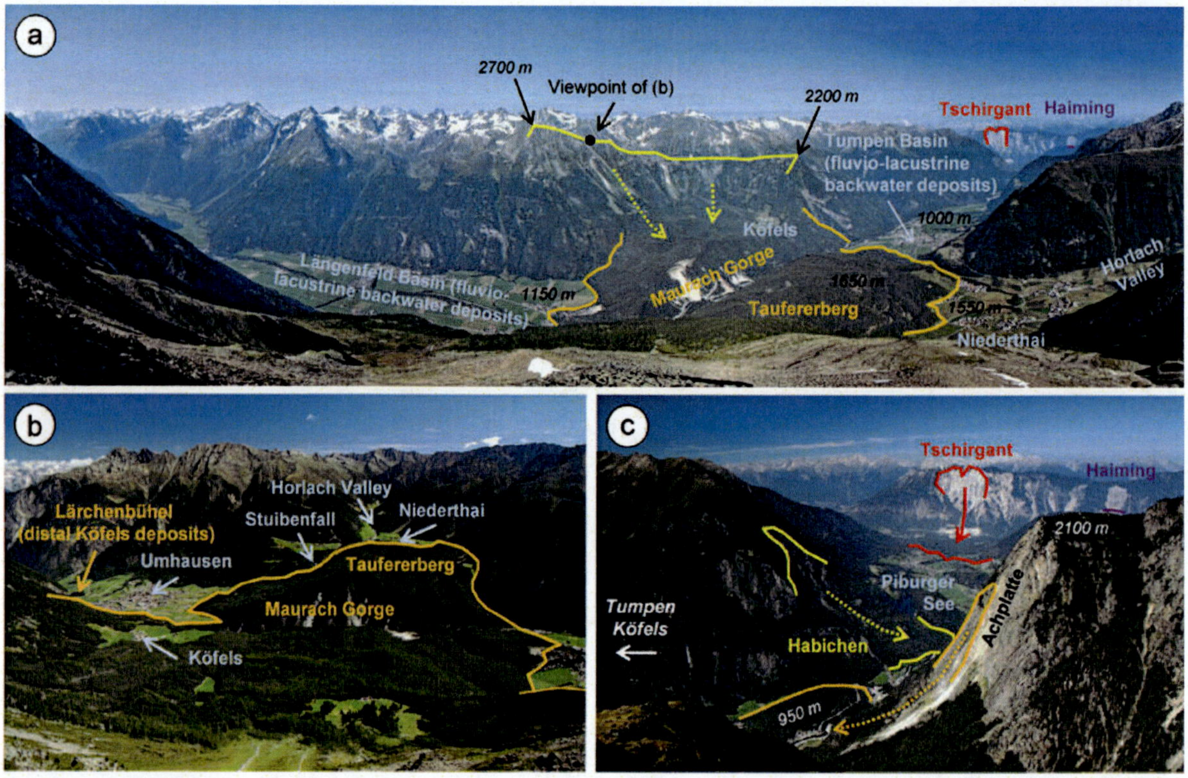

Fig. 21.2 Large landslides of the lower and central Ötz Valley. **a** Panoramic view of the central Ötz Valley with the Köfels rockslide. The left and central background shows the Ötztal Alps (metamorphic rocks), the Northern Calcareous Alps with the headscarps of the Tschirgant and Haiming rock avalanches are seen in the very right background. The river Ötz drains from left to right and the Längenfeld basin represents a backwater area dammed by the deposit of the Köfels rockslide (prior to the incision of the Maurach gorge). The village of Niederthai in the lower right centre of the image is built on a system of terraces of unclear origin. Viewpoint: Hemerkogel; **b** view of the Köfels rockslide from the headscarp; **c** Achplatte and Habichen rock avalanches near the outlet of the Ötz Valley. Lake Piburger See was dammed by the Habichen rock avalanche. The Tschirgant rock avalanche and the headscarp of the Haiming rock avalanche are seen in the background (Northern Calcareous Alps). All line symbols shown are approximate. *Photos* M. Mergili; line symbols modified after Zangerl et al. (2021), Ostermann and Prager (2016)

(Brandner 1980). These events were predisposed by the orientation of bedding planes and fracture systems, dipping out of the slopes and therefore providing preferentially oriented sliding planes (Pagliarini 2008; Prager et al. 2008, 2009a; Ostermann and Prager 2016). The scarp area of the Fernpass rock avalanche consists of dolostones, limestones and marls of the Seefeld Formation (Upper Triassic), several hundreds of metres thick (Prager et al. 2006, 2009a). The scarp area of the Tschirgant rock avalanche is mainly made up by dolostones, breccias and limestones of the Wetterstein Formation (Middle to Upper Triassic), which also represent the main landslide mass. At the lower sections of the failed slope, some weaker units of the lithologically heterogeneous Raibl Group (Upper Triassic) are encountered, including evaporitic strata. Geological field surveys indicate that karst processes may have contributed to the failure (Prager 2010; Prager et al. 2008, Dufresne et al. 2016a). Due to the extremely rugged terrain, the Tschirgant scarp area is still a rockfall-prone site and thus a source of debris flows, indicated also by local site names such as “Breitmure” or “Galgenmure” (the German word “Mure” means debris flow).

The distinctive geometry and topography of the Fernpass rock avalanche deposits indicate an unusual dynamic behaviour and have therefore attracted research for several decades (e.g., Ampferer 1904; Abele 1964, 1975, 1991; Prager et al. 2006, 2009a, 2012; Prager 2010). Figure 21.3 provides panoramic overviews of the scarp and parts of the accumulation area. The initial movement impacted the opposite slope and consequently split into two branches, propagating further in two different directions (Prager et al. 2006, 2012). The northern branch of the rock avalanche—representing the logical continuation of the initial movement—reached a maximum travel distance of at least 10.8 km in the fluvio-lacustrine plain of the Lermooser Moos. More remarkably, the southern part of the rock avalanche was

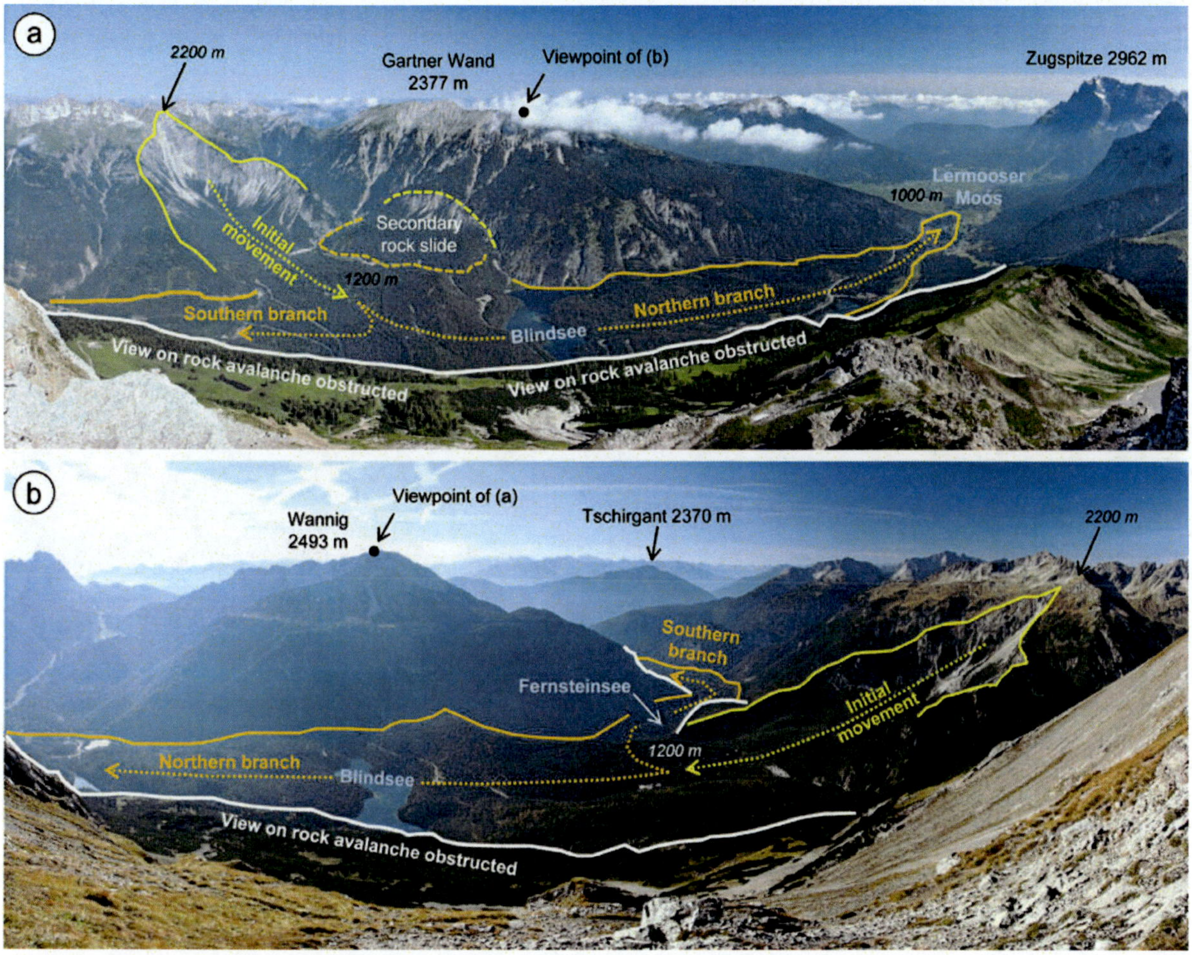

Fig. 21.3 Panoramic views of the Fernpass rock avalanche. **a** Viewpoint: Wannig; **b** viewpoint near Gartner Wand. Note the separation of the rock avalanche into a northern and a southern branch. Only the uppermost section of the southern branch is visible in (**b**). All line symbols shown are approximate. *Photos* M. Mergili; line symbols modified after Prager et al. (2006) and Prager (2010)

deflected by 140° and, after a further curious turn near Nassereith, reached a total travel distance of at least 15.5 km. The angles of reach were extraordinarily low, with 6.7° for the northern and even 5.3° for the southern branch. Prager et al. (2006) inferred that the high mobility of the southern part—despite a comparatively lower volume—was favoured by its propagation over saturated fine valley floor sediments, which were investigated by drillings and geophysical surveys. High pore water pressures could have developed as a response to the undrained loading, due to the inability of the groundwater to escape through the rock avalanche mass (because of the low permeability of the basal sliding zones). The initial rock avalanche deposit was subsequently reshaped by gravitational spreading and creeping, leading to the development of significant extensional structures, e.g., graben with several lakes in the proximal to medial accumulation areas (see also Abele 1972, 1997b), and so-called toma hills in the distal sections of the deposits (Prager 2010; Prager et al. 2012). According to the morphological definition by Abele (1974: 119), toma hills are "isolated, cone- to pyramidal- or roof-shaped elevations, predominately made up by rockslide debris and characterized by more or less planar hill slopes with constant inclination". The distal portion of the deposition area was also modified by fluvial processes and human activities.

Field surveys and seismic measurements indicate that the proximal Fernpass rock avalanche deposits are some hundreds of metres thick, and that the gravitational collapses of this thick accumulation ridge released the curiously deflected southern branch of the Fernpass rock avalanche (Prager 2010; Prager et al. 2012). Further local particularities of the Fernpass area include springs featuring very high discharge rates (e.g., the Loisach springs) and springs ranking among the most radioactive ones in North Tyrol. Most likely, the

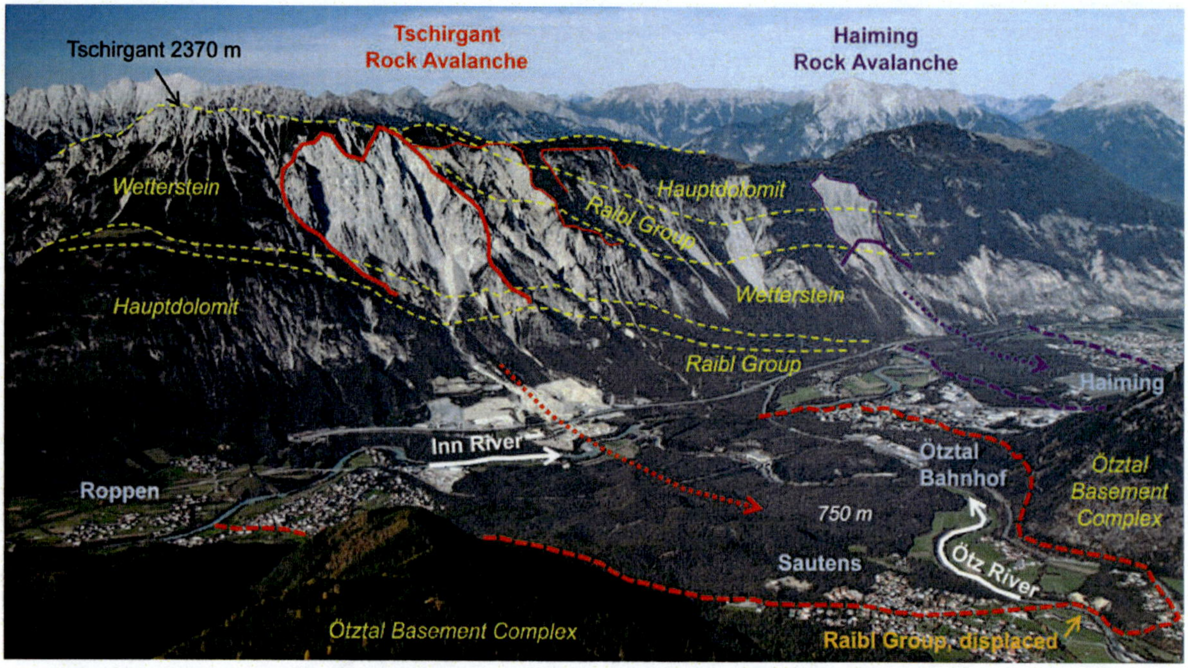

Fig. 21.4 Panoramic view of the Tschirgant and Haiming rock avalanches, with schematic delimitation of lithological units (strata of the Northern Calcareous Alps involved in the events; complex fold- and fault-systems here not depicted). Note that the headscarp of the Haiming rock avalanche is located approximately in the middle of the slope whilst the upward scarp represents an erosional feature. All line symbols shown are approximate. Viewpoint: Blose. *Photo* M. Mergili; line symbols modified after Prager et al. (2008), Patzelt (2012a), Ostermann and Prager (2016), Dufresne et al. (2016a)

known radon emanations in the Fernpass area (Krüse 1937) are to be attributed to intense dynamic fragmentation of the thick landslide rocks, i.e., organically-rich successions of the bituminous Seefeld Formation (for details, see Prager 2010 and references therein).

The Tschirgant rock avalanche (Ampferer 1904, Patzelt 2012a) travelled for a distance of 6.2 km, producing a hummocky landscape with a variety of ridges and depressions (Fig. 21.4). Thereby, the carbonate Wetterstein beds make up the main parts of the accumulation area, with different rocks of the Raibl Group encountered rather at the lateral and distal areas of the deposits (Dufresne et al. 2016a; b). The displaced Raibl beds are well exposed mainly along the banks of the river Ötz (Fig. 21.5, see also Patzelt 2012a). A further very particular feature of the Tschirgant deposit consists in the locally well-exposed basal shear zone. In some places, the substrata (i.e., pre-slide valley deposits) were injected into the landslide debris, and some intermingling took place (Patzelt and Poscher 1993; Prager 2010; Dufresne et al. 2016a, b). Further, the rock avalanche deposit impounded the rivers Ötz and Inn, and thus caused the accumulation of some minor backwater deposits, later eroded to terraces, e.g., in the Roppen and Ambach areas (Ampferer 1904; Ostermann and Prager 2014).

The Habichen, Achplatte and Tumpen rock avalanches (Fig. 21.2c; Heuberger 1975; Poscher and Patzelt 2000; Ostermann and Prager 2014, 2016) disintegrated into relatively coarse blocky masses during motion, resulting in blocks between sub-metre size and >10 m^3, lacking the finer components found in the Fernpass and Tschirgant rock avalanches. The Habichen rock avalanche cut off a drainage channel, and dammed the still existing lake Piburger See. Backwater sediments in the Ötz Valley indicate temporary lakes impounded by all three of the rock avalanches mentioned above.

The prominent Köfels event (Table 21.1 and Figs. 21.1, 21.2a, b) deposited a rather coarse rockslide mass which displays a broad variety of fabrics. The deposits are well exposed in the Maurach gorge, which has been incised by the river Ötz: here, some parts show fractured but rather intact rock slabs; others resemble a fault breccia, with large fragments interspersed in a matrix of crushed material. Other parts are full of fractures, but with lower contents of fine particles. Locally, fused rocks indicate intense shear deformation (frictionites, see below).

The Köfels slide blocked the Ötz Valley—and the outlet of the tributary Horlach Valley—with a dam height of several hundreds of metres. The resulting backwater deposits in the Längenfeld basin upstream reached a depth of up to

Fig. 21.5 Depositional area of the Tschirgant rock avalanche. **a** Fragmented dolostones and rauhwacken of the Raibl Group, accumulated in the distal part of the deposit of the Tschirgant rock avalanche (see Fig. 21.4 for the location; *Photo*: M. Mergili); **b** and **c** dynamic interaction of rock avalanche debris (RD) and fluvial gravels (FG). Bold arrows indicate the direction of rock avalanche propagation and thin arrows the direction of injection and back-thrusting (Prager 2010; Prager et al. 2012)

100 m (Ampferer 1939; Klebelsberg 1951; Heuberger 1966, 1975). Most likely, the surface of the backwater deposits did not exceed the present-day surface of the Längenfeld basin due to a sufficiently high hydraulic permeability of the Köfels deposits to allow a subsurface drainage (Ampferer 1939; Klebelsberg 1951). The origin of the terraced sediments at Niederthai (in the lower Horlach Valley; Fig. 21.6a) is not yet clear. Presumably a few tens of metres thick, they may relate to a pre-slide (possibly periglacial) origin and/or to the Köfels event (either as displaced valley fill and/or backwater sediments; Geitner 1999; Prager et al. 2009b and references therein; Jarman 2011). Temporary exposures in local construction pits show coarse rockslide material next to fine-grained sediments (Fig. 21.6b and c).

Fig. 21.6 Terraces of Niederthai are developed in a depositional mass of unknown origin, representing either displaced sediments from the Ötz Valley floor and/or backwater sediments (see Fig. 21.2 for the location). **a** Overview. Photo M. Mergili; **b** excavation pit at Niederthai, exposing coarse rockslide deposits (angular gneiss debris with clast- and matrix-supported fabric). Photo C. Prager; **c** fine terrace deposits in the same pit as shown in (**b**), adjacent to the east (poorly stratified sands with fining-upward trend). *Photo* C. Prager

Fig. 21.7 Frictionite (fused hyalomylonite) locally encountered at the Köfels rockslide. The photo was taken in the Maurach gorge, and the thickness of the frictionite in the photo is up to 3 cm. *Photo* C. Prager (Prager et al. 2007)

Formerly, in an abandoned gravel pit, un-stratified gravels and sands with an obscure fining-upward trend were locally exposed. Presumably, they represent sediments displaced/liquefied by the Köfels event.

The most distinctive feature of the Köfels rockslide, however, is the occurrence of fused rocks, i.e., "pumice" and glassy frictionites (Fig. 21.7) that led researchers to theories of volcanic activity (Pichler 1863) and a meteorite impact (Stutzer 1936; Suess 1937; Bond and Hempsell 2008). However, according to current knowledge, these frictionites originated from partial melting in the sliding zone (Preuss 1974, 1986; Erismann et al. 1977; Erismann and Abele 2001). Such frictionites are up to now known from a few sites worldwide (Weidinger and Korup 2009; Weidinger et al. 2014). Related to the intense rock mass fragmentation, radioactive springs discharge locally and very high radon gas concentrations are emitted from the deposits (Purtscheller et al. 1995).

21.3 Temporal Patterns and Possible Triggers

The previous decades have brought a dramatic increase of knowledge with regard to the timing of large landslides in the Alps. The ages summarized in Table 21.1 were derived by means of radiocarbon, surface exposure and/or U/Th dating (e.g., Ivy-Ochs et al. 1998; Prager et al. 2009a; Patzelt 2012a, Nicolussi et al. 2015; Ostermann et al. 2016).

In the Alps, several major rock slope failures occurred throughout the Holocene. However, compiled age determinations indicate two clusters in the temporal distribution: one around 10.5–9.4 ka BP and another one around 4.2–3.0 ka BP (Fig. 21.8; Prager et al. 2008). The first, early Holocene cluster includes the Köfels rockslide (±9.5 ka, Nicolussi et al. 2015) as well as other very large events such as Flims (±9.4 ka, Deplazes et al. 2007) and Kandertal (±8.6 ka, Tinner et al. 2005), along with others, e.g., the two Rinderhorn rock avalanches (±9.8 ka, Gräminger et al. 2016). For the Köfels rockslide, Nicolussi et al. (2015) constrained the age to 9527–9498 years (earlier radiocarbon dating of this wood fragment and exposure dating campaigns had indicated slightly higher ages; e.g., Ivy-Ochs et al. 1998). This age of the Köfels failure possibly overlaps with the range of the largest Alpine landslide, the Flims rockslide 130 km farther west and ranges close to the timing of some other major events cited above. The second temporal cluster, addressed as "Fernpass cluster" (Prager et al. 2008), includes the rock avalanches at Fernpass, Tschirgant and Haiming, as well as at Eibsee, Stöttlbach, Pletzachkogel and others (Fig. 21.8 and Table 21.1).

However, ideas about a possible common trigger, e.g., climate change and/or seismic loading, remain hypothetic so

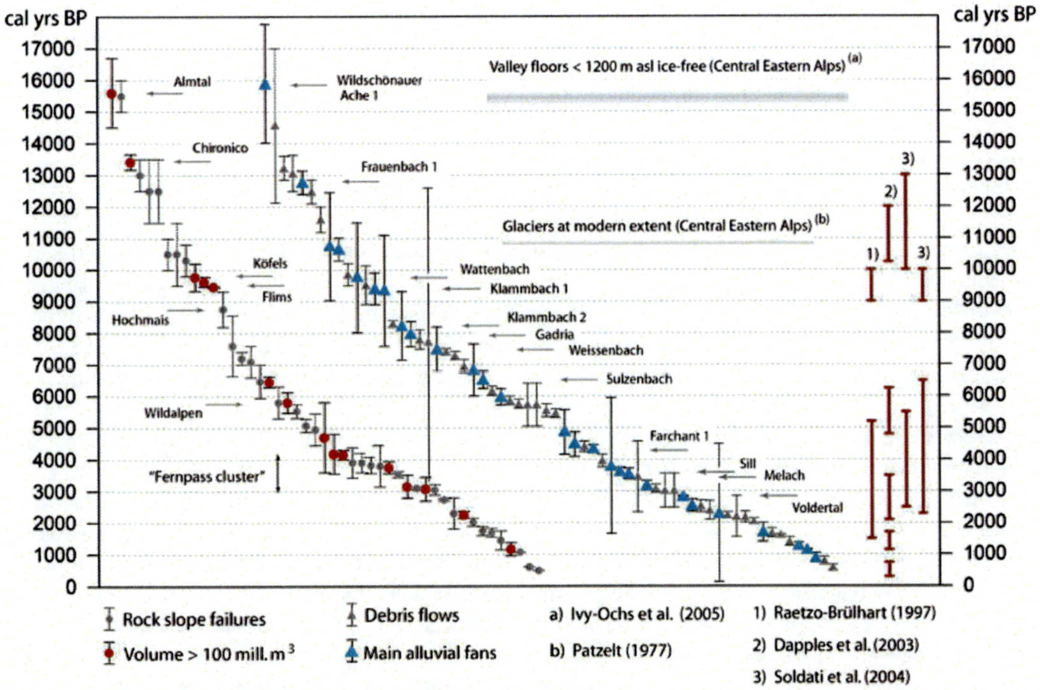

Fig. 21.8 Temporal distribution of fossil landslides in the Tyrol and its surrounding areas. Vertical axes: calibrated years BP, horizontal axes: dimensionless sequence of dated events (see Prager et al. 2008; Prager 2010 for details)

far, but represent an interesting field for further research. In general, the observed temporal distribution has led to various hypotheses with regard to the disposition (causes) and triggering of the events. Disregarding exotic and yet not validated theories such as a meteorite impact, the following factors (or a combination of them) may have been involved (Prager et al. 2008, and references therein):

- Slope dis-equilibrium resulting from glacial debuttressing, exposure of potential sliding planes and stress redistribution due to glacial erosion, the associated valley-deepening and subsequent retreat of the Pleistocene glaciers may have either directly triggered landslides, or at least reduced the stability of slopes (longer termed rock mass weakening and progressive failure).
- Dynamic loading induced by earthquakes may either have directly triggered events or led to progressive weakening and destabilization of the rock masses.
- A climatically humid period may have led to high groundwater levels and may have increased the pore water pressures, thereby fostering the growth of cracks and triggering hydromechanically coupled destabilization processes.

In the older literature, large Holocene landslides have often been attributed to glacier retreat, debuttressing and associated slope destabilization effects (e.g., Heuberger 1966, 1975; Abele 1969, 1972, 1997a). However, compiled age determinations indicate a time lag of several thousands of years between deglaciation and the actual occurrence of many of the landslides (Prager et al. 2008). In the long term, the propagation and coalescence of fractures may have triggered landsliding by reaching a stability threshold or by reducing the equilibrium far enough to facilitate landslide initiation even through a small trigger of different kind.

Increased earthquake activity, compared to other regions nearby, is measured in the surroundings of the area investigated (Inn Valley and Engadin Line), with some of the strongest earthquakes ever recorded in Austria (up to magnitudes 5.3; Drimmel 1980; Reiter et al. 2003). Long-term seismic weakening of fractured bedrock slopes may also trigger large failure events as proposed, e.g., for the Tschirgant and Haiming rock avalanches (Patzelt 2012a). On the other hand, historical records of large earthquake-triggered landslides are rare. The recorded seismic activity is limited at least in the Ötz Valley (ÖNORM B 4015: 2007), so that the occurrence of a sufficiently strong seismic event here is not indicated yet. Evidence of earthquake-triggered landslides in the past may be given, e.g., by systematic analyses and dating of disturbed lake sediments and speleothems (Becker et al. 2006; Strasser et al. 2013).

Similar ages of dated alluvial fans and debris flow deposits may indicate periods of increased water and/or debris supply in the catchment areas, possibly resulting in temporally increased mass displacements (Prager et al. 2008). Since such periods in the Holocene coincide with several dated landslides (Fig. 21.8), there might be a relationship between the large landslides in the investigation area and climatic conditions, even though this issue is disputed. Climate change might be an issue for slope stability since increased groundwater levels (and related pore water pressures) would lower the rock mass strength, especially in terms of weakened potential failure planes. For the Köfels rockslide, for example, Zangerl et al. (2021) proposed that an increased groundwater level and an internal friction angle of the basal shear zone of <27–30° would be necessary to enable failure. These assumptions are in accord with geotechnical data on fault and shear zone materials comparable to the Köfels rocks (e.g., Engl et al. 2008). However, although at Köfels several major sliding planes are well exposed (see, e.g., Prager et al. 2009a), the question in which way a rather continuous shear zone required for such a massive rock slope failure can have developed remains open. It has to be emphasized that simple, i.e., mono-causal explanations for the release of the large landslides are probably inappropriate. More likely, the long-term effects of crack growth and fracture propagation due to stress redistribution (post-glacial debuttressing), seismic activities and/or climate changes (varying groundwater/pore water pressure conditions) might control processes of slope destabilization. Thus, even minor triggers below significant thresholds may shift the equilibrium of a fractured bedrock slope far enough to promote failure and release even major landslides (Prager et al. 2008).

21.4 Ecological Implications

Whilst vegetation is largely able to cope with slow moving landslides ("creeping slopes"), large, rapid landslides leave behind pristine environments which have to be newly colonized. Therefore, such events may give ecologists the chance to study the succession of ecosystems. Thousands of years of succession have already taken place on the prehistoric landslides in the Tyrol. However, the land cover of large "Bergsturz" deposits still differs markedly from the surroundings. This phenomenon may be attributed to two closely connected reasons: the highly distinctive topography and substrate created by the sometimes turbulent motion, and, as a consequence, the unfavourableness of these environments for agricultural use. Whilst the floors of the Inn and Ötz valleys have almost completely been converted to agricultural or built-up areas, some of the large landslide deposits have remained in a near-natural state.

On the other hand, landslide deposits often consist of highly porous material, so that the habitats are dry due to the soil-related conditions, even when there is plenty of precipitation. This effect increases in limestone, where very slow soil formation and local karst effects further facilitate the drainage of water. The Scots pine (*Pinus sylvestris*) is well adapted to this type of condition. Whilst it is weak in competition with other tree species at more favourable sites, it can reasonably well survive in dry habitats. Kral (1989) underlines the relict character of the pine forests in the "Bergsturz" landscapes: in the Lateglacial and early Holocene, before immigration of more competitive tree species such as spruce, pines were equally important in various habitats, but later could only persist in unfavourable sites. On the deposits of the Fernpass (Starlinger 1992), Tschirgant (Mair 1997) and Haiming rock avalanches, *Pinus sylvestris* forms characteristic plant communities together with the heather *Erica carnea* in the undergrowth. These sparse stands of pine with an undergrowth of variable density are known as *Erico-Pinetum*, which also grows on sunny southerly exposed slopes in other parts of the Inn Valley and can shelter a high variety of species. The Fernpass deposit also hosts one of the easternmost populations of *Pinus mugo* ssp. *uncinata* ("Spirke"; Starlinger 1992), rare elsewhere in North Tyrol.

From an ecological point of view, the best studied site is the deposit of the Tschirgant rock avalanche (Fig. 21.9; Knoflach and Thaler 1994; Mair 1997; Oberhuber and Mayr 1998). The mean annual precipitation in that area is less than 800 mm. Oberhuber and Mayr (1998) emphasize the poor growth of *Pinus sylvestris* in years of reduced spring precipitation, when the water stress induced by the shallow soils with a high content of coarse material is accentuated. This phenomenon was observed for different types of habitats within the rockslide deposit. Besides the drought, the poor growth of the pines also points to the infertile soils, only weakly developed and often with thick acidic organic layers, on the nutrient-poor and weathering-resistant carbonates of the Wetterstein Formation (dolostones, breccias, limestones). The undergrowth is strongly differentiated according to the topographic position within the rockslide mass, with a very sparse cover on steep, southerly exposed slopes and a dense cover of mosses and *Vaccinium myrtillus*, remarkably an acidity indicator on carbonates, in sinks (Oberhuber and Mayr 1998). Only the most favourable places are suitable for spruce (*Picea abies*).

Fig. 21.9 Scots pine (*Pinus sylvestris*) forests on the deposits of the large landslides. **a** Stand of *Pinus sylvestris* on blocky material deposited by the Köfels rockslide; **b** *Erico-Pinetum* on the carbonate deposit of the Tschirgant rock avalanche; **c** *Erica carnea* in the undergrowth of an *Erico-Pinetum* on the same deposit. *Photos* M. Mergili

Also the crystalline "Bergsturz" deposits in the Ötz Valley—particularly the deposit of the Köfels rockslide—are partly covered by *Pinus sylvestris* forest, along with a species-poorer undergrowth adapted to the gneissic substrate (Mair 1997). Those areas where the soils can retain enough water are covered by spruce or larch forests broadly common in that area.

The large landslide-dammed backwater plains, e.g., the Tumpen and Längenfeld basins (Klebelsberg 1951; Heuberger 1975; Poscher and Patzelt 2000), are integral parts of the "Bergsturz" landscapes. However, they differ substantially from the actual rockslide or rock avalanche deposits not only in terms of their flat terrain surface, but also in terms of hydrology and ecology. The often fine-grained fluvio-lacustrine deposits have a varying but generally low hydraulic permeability (i.e., a comparatively higher capacity to retain water). Thus, they were originally swampy but are rather fertile due to replenishment of nutrients by the effects of flooding. Therefore, by means of water management measures such as river regulation and drainage, such areas have been converted into cultural landscapes used for agriculture and settlements and contrast sharply with the topographically rugged, forested "Bergsturz" deposits (Fig. 21.2a).

21.5 Implications for the Cultural Landscape and Society

The giant "Bergsturz" landslides have modified the landscapes of the affected areas. Mountain ridges were displaced, slopes completely reshaped, valleys blocked and fluvio-lacustrine backwater areas created upstream. Those landscapes were populated at least at the time of the Tschirgant rock avalanche (see Table 21.1), as indicated by a fire place at the base of the deposit (Patzelt 2012a), so that this event may have been observed by humans and may even have badly affected the society. It may be inferred that the Tschirgant rock avalanche induced logistic problems due to the interruption of old traffic lines. In general, the deposits of large landslides were and still are major obstacles for traffic and communication and as such even led to the development of certain types of borders: the Fernpass rock avalanche represents the border between the Bajuvarian and Alemannic language domains.

Comparable ramifications are observed in other cases such as the multi-phase Pletzachkogel rock avalanches (Patzelt 2012b) in the lower Inn Valley (former border between Tyrol and Bavaria, and maybe former border between Noricum and Raetia), or the Sierre landslide and the debris cone of Pfyn in Switzerland (German and French language border) (pers. comm. Gernot Patzelt). Whilst the Fernpass still represents some kind of obstacle for traffic, this is not so much the case for the "Bergsturz" landscapes at the entrance of and within the Ötz Valley. However, thick landslide debris can be relevant for the integrity of infrastructure facilities and make protection works necessary. In the steep Maurach gorge (Fig. 21.2), the Ötztal road B186 was sporadically affected by secondary rock falls originating from the thick Köfels deposits, and thus requiring some technical mitigation measures (retention walls). Not only traffic, but also other types of use are complicated. For example, coarse boulders and mega-blocks can hardly be excavated but may require drill and blast works to enable construction of foundations for roads and buildings.

Possible indirect negative effects of large prehistoric landslides may include less obvious health risks: the radon emissions from the Köfels deposit (Purtscheller et al. 1995) have led to increased rates of lung cancer in the surroundings, compared to the average for the province of Tyrol (Ennemoser et al. 1994). In this context, also the radioactive spring "Pseirer-Brünnl" leaking from the Köfels deposit has to be mentioned. Its emanation rate of 400 Bq/l and more (Krüse 1940) is the second highest among all springs in Northern Tyrol. Also, the emanation from the "Radiumquelle" (Krüse 1937), most likely leaking from the Fernpass deposits and/or the adjacent Seefeld beds, may be related to the intensive degree of rock fragmentation (Prager 2010).

In general, the "Bergsturz" deposits are potential pore aquifers and some are used for the extraction of groundwater (e.g., Loisach-Quellen and others; Prager 2010). They are also used for the commercial exploitation of mineral resources such as gravel at the Tschirgant, Haiming and Köfels deposits. With respect to the morphological setting, i.e., several 100-m-thick rockslide debris blocking the Ötz Valley and surface discharge occurring only through the fluvial Maurach gorge (Fig. 21.2a, b), the Längenfeld basin and the damming rockslide barrier were formerly investigated for a hydropower project (Ampferer 1939; Klebelsberg 1951). In a related investigation adit, surveying the pre-rockslide mouth of the Horlach Valley, a larch wood fragment was encountered between the rockslide debris and the bedrock underneath. The wood samples thus obtained enabled a quite early application of ^{14}C dating of the Köfels event (Heuberger 1966; Ivy-Ochs et al. 1998; Nicolussi et al. 2015).

As outlined above, the backwater plains are highly favourable for agricultural land use and for settlement areas due to their flat terrain surfaces and fine sediments. In contrast, the landslide deposits are highly unfavourable for these types of use. Nevertheless, the hummocky landscapes with the prevailing near-natural ecosystems—a kind of remaining wilderness—are often perceived as scenic and appealing, and therefore are of high recreational value. The deposits of the Köfels rockslide and the Tschirgant rock avalanche offer networks of hiking paths. Furthermore, the deposit of the Fernpass rock avalanche hosts a number of scenic lakes (Fig. 21.10a), one of which is used to generate income through

Fig. 21.10 Elements of "Bergsturz" landscapes appropriate to generate income through tourism: **a** Lake Fernsteinsee, privately owned with a hotel nearby (Fernpass rock avalanche); **b** Stuibenfall Waterfall (area of Köfels rockslide). *Photos* M. Mergili

tourism (adjacent view point, kiosk and hotel). Lake Piburger See (Habichen rock avalanche) represents a popular spot of recreation especially during summer time. The Stuibenfall Waterfall (Fig. 21.10b), having formed due to reshaping of the landscape by the Köfels rockslide, is the highest waterfall in the province of Tyrol. It provides the basis for local tourism, especially for a nearby restaurant and a snack station, and represents the perfect ambience for the "Ötzi village", a hiking trail and the spectacular Stuibenfall climbing path.

21.6 Conclusions

Several Holocene giant rockslides and rock avalanches, often referred to as "Bergsturz" landslides, have shaped the landscapes between the Fernpass and the central Ötz Valley. Some of the events occurred in carbonate rock, others in metamorphic crystalline rock. Ideas concerning the causal factors of the spatial—and, to some extent, also temporal—clustering of events still remain hypothetic. A sound geological and geotechnical understanding of the predisposing and triggering factors, and of the mechanisms involved in these "Bergsturz" landslides is essential, not only for reconstructing the past but also for anticipating possible hazards and risks. Due to the dimensions, the implications for society and the recreational importance of the "Bergsturz" deposits, these landscapes display a high (yet only partly deployed) potential for environmental education. They may serve as "field laboratories" for enhancing the awareness and understanding of geomorphological processes in mountain areas, but also of the implications of such processes for ecosystems and society at various spatial scales.

References

Abele G (1964) Die Fernpaßtalung und ihre morphologischen Probleme. Tübinger Geogr Studien 12:1–123

Abele G (1969) Vom Eis geformte Bergsturzlandschaften. Z Für Geomorphol NF Suppl 8:119–147

Abele G (1972) Kinematik und Morphologie spät- und postglazialer Bergstürze in den Alpen. Z Geomorphol NF Suppl 14:138–149

Abele G (1974) Bergstürze in den Alpen. Ihre Verbreitung, Morphologie und Folgeerscheinungen. Wiss Alpenvereinshefte 25:1–230

Abele G (1975) Fernpaß-Garmisch-Partenkirchen, 1.Teil: das Fernpassgebiet, Tirol, ein geographischer Exkursionsführer. Innsbrucker Geogr Studien 2:145–157

Abele G (1991) Der Fernpassbergsturz, eine differentielle Felsgleitung. Öster Geogr Ges Zweigverein Innsbruck Jahresber 1989(1990): 22–32

Abele G (1997b) Rockslide movement supported by the mobilization of groundwater-saturated valley floor sediments. Z Geomorphol NF 41:1–20

Abele G (1997a) Influence of glacier and climatic variation on rockslide activity in the Alps. In: Matthews JA, Brunsden B, Frenzel B, Gläser B, Weiß MM (eds) Rapid mass movement as a source of climatic evidence for the Holocene. Paläoklimaforschung, vol 19, Fischer, pp 1–6

Ampferer O (1904) Die Bergstürze am Eingang des Ötztales und am Fernpaß. Verh Geologischen Reichsanst 73–87

Ampferer O (1939) Über die geologischen Deutungen und Bausondierungen des Maurach Riegels im Ötztal. Geol Bauwesen 2:25–43

Becker A, Davenport CA, Eichenberger U, Gilli E, Jeannin PY, Lacave C (2006) Speleoseismology: a critical perspective. J Seismolog 10(3):371–388

Bond A, Hempsell M (2008) A sumerian observation of the Köfels' impact event: a monograph. Alcuin Academics

Brandner R (1980) Geologische und Tektonische Übersichtskarte von tirol, tirol-atlas, C1, C3. Univ.-Verlag Wagner, Innsbruck

Cruden DM (1991) A simple definition of a landslide. Bull Int Assoc Eng Geol 43:27–29

Cruden DM, Varnes DJ (1996) Landslide types and processes. In: Turner AK, Schuster RL (eds) Landslides investigation and mitigation. Transportation research board, US National Research Council. Special Report 247, Washington, DC, Chapter 3, pp 36–75

Deplazes G, Anselmetti FA, Hajdas I (2007) Lake sediments deposited on the Flims rockslide mass: the key to date the largest mass movement in the Alps. Terra Nova 19:252–258

Drimmel J (1980) Rezente Seismizität und Seismotektonik des Ostalpenraumes. In: Oberhauser R (ed) Der geologische Aufbau Österreichs, Springer, Berlin, pp 507–527

Dufresne A, Prager C, Bösmeier A (2016a) Insights into rock avalanche emplacement processes from detailed morpho-lithological studies of the Tschirgant deposit (Tyrol, Austria). Earth Surf Proc Land 41 (5):587–602

Dufresne A, Bösmeier A, Prager C (2016b) Rock avalanche sedimentology—case study and review. Earth Sci Rev 163:234–259

Engl DA, Fellin W, Zangerl C (2008) Scherfestigkeiten von Scherzonen-Gesteinen. Ein Beitrag zur geotechnischen Bewertung von tektonischen Störungen und Gleitzonen von Massenbewegungen. Bull Angew Geol 13(2):63–81

Ennemoser O, Ambach W, Auer T, Brunner P, Schneider P, Oberaigner W, Purtscheller F, Stingl V (1994) High indoor radon concentrations in an Alpine region of western Tyrol. Health Phys 67 (2):151–154

Erismann TH, Heuberger H, Preuss E (1977) Der Bimsstein von Köfels (Tirol), ein Bergsturz-"Friktionit'. Tschermaks Mineral Petrogr Mitt 24:67–119

Erismann TH, Abele G (2001) Dynamics of rockslides and rockfalls. Springer, Heidelberg, p 316

Escher von der Linth A (1845) Beiträge zur Kenntnis der Tyroler und Bairischen Alpen. Neues Jahrb Mineral Geognosie, Geol Petrefakten-Kunde 536–561

GBA (2011) Geofast 1:50.000, Zusammenstellung ausgewählter Archivunterlagen der Geologischen Bundesanstalt, Blatt 116—Telfs (Ausgabe 2011/04), 145—Imst (Ausgabe 2011/07), 146—Oetz (Ausgabe 2011/07), Geol Bundesanst Wien

Geitner C (1999) Sedimentologische und vegetationsgeschichtliche Untersuchungen an fluvialen Sedimenten in den Hochlagen des Horlachtales (Stubaier Alpen/Tirol)—Ein Beitrag zur zeitlichen Differenzierung der fluvialen Dynamik im Holozän. Münchener Geogr Abh Reihe B 32:1–247

Gräminger LM, Moore JR, Vockenhuber C, Aaron J, Hajdas I, Ivy-Ochs S (in press) Two early Holocene rock avalanches in the Bernese Alps (Rinderhorn, Switzerland). Geomorphology

Heim A (1932) Bergsturz und Menschenleben. Wasmuth, Zürich, p 218

Heuberger H (1966) Gletschergeschichtliche Untersuchungen in den Zentralalpen zwischen Sellrain und Ötztal. Wiss Alpenvereinshefte 20:1–126

Heuberger H (1975) Das Ötztal. Bergstürze und alte Gletscherstände, kulturgeographische Gliederung. Innsbrucker Geogr Studien 2:213–249

Hsü K (1975) Catastrophic debris streams (Sturzstroms) generated by Rockfalls. GSA Bull 86:129–140

Hungr O, Evans SG (2004) The occurrence and classification of massive rock slope failure. Felsbau 22:16–23

Ivy-Ochs S, Heuberger H, Kubik PW, Kerschner H, Bonani G, Frank M, Schlüchter C (1998) The age of the Köfels event. Relative, ^{14}C and cosmogenic isotope dating of an early Holocene landslide in the Central Alps (Tyrol, Austria). Z Gletscherk Glazialgeol 34:57–68

Jarman TR (2011) Sedimentary complexities associated with the Köfels rockslide (Tyrol, Austria). Geology and Physical Geography (Honours), 68 pp., University of Edinburgh

Jarvis A, Reuter HI, Nelson A, Guevara E (2008) Hole-filled seamless SRTM data V4. International Centre for Tropical Agriculture (CIAT), available from http://srtm.csi.cgiar.org, 2008, last access: 14 Apr 2016

Knoflach B, Thaler K (1994) Epigäische Spinnen im Föhrenwald der Ötztal-Mündung (Nordtirol, Österreich). Ber des naturwiss-medizinischen Ver Innsbruck 81:123

Kral F (1989) Pollenanalytische Untersuchungen im Fernpaßgebiet (Tirol): Zur Frage des Reliktcharakters der Bergsturz-Kiefernwälder. Verh Zoologischen-Botanischen Ges Öster 126:127–138

Krüse K (1937) Beiträge zur Kenntnis der Radioaktivität der Mineralquellen Tirols (VIII. Mitteilung mit einer Gesamtübersicht der bisherigen Untersuchungen). Jahrb Geol Bundes-Anstalt 87:41–56

Krüse K (1940) Beiträge zur Kenntnis der Radioaktivität der Mineralquellen Tirols (IX. Mitteilung und Schluß). Mitt Reichstelle Bodenforschung (N. F. Jahrb Geologischen Bundes-Anstalt) 1:69–80

Mair P (1997) Die Föhrenwälder der Bergsturzgebiete Tschirgant und Köfels (Tirol). Unpublished Diploma thesis, University of Innsbruck p 138

Nicolussi N, Spötl C, Thurner A, Reimer PJ (2015) Precise radiocarbon dating of the giant Köfels landslide (Eastern Alps, Austria). Geomorphology 243:87–91

Oberhuber W, Mayr, S (1998) Dendroklimatologische Untersuchung von Kiefernbeständen (Pinus sylvestris L.) auf Schuttmaterial des Tschirgant-Bergsturzes (Tirol). Ber Naturwiss-Med Ver Innsbruck 85:35–46

ÖNORM B 4015 (2007) Belastungsannahmen im Bauwesen – Außergewöhnliche Einwirkungen – Erdbebeneinwirkungen, Grundlagen Berechnungsverfahren. Österr. Normungsinstitut, Wien, pp 61

Ostermann M, Ivy-Ochs S, Sanders D, Prager C, Patzelt G (2016) Multi-method (^{14}C, ^{36}Cl, ^{234}U/^{230}Th) age bracketing of the Tschirgant rock avalanche (Eastern Alps): implications for absolute dating of catastrophic mass-wasting. Earth Surf Proc Land. https://doi.org/10.1002/esp.4077

Ostermann M, Prager C (2014) Major holocene rock slope failures in the inn-Ötztal valley region (Tyrol, Austria). In: Kerschner H, Krainer K, Spötl C (eds) From the foreland to the Central Alps. Geozon, pp 116–126

Ostermann M, Prager C (2016) Rock slope failures shaping the landscape in the Loisach-, Inn- and Ötz Valley region (Tyrol, Austria). Geo Alp 13:257–276

Pagliarini L (2008) Strukturelle Neubearbeitung des Tschirgant und Analyse der lithologisch-strukturell induzierten Massenbewegung (Tschirgant Bergsturz, Nördliche Kalkalpen, Tirol). Unpublished Master thesis, University of Innsbruck, pp 90

Patzelt G (2012a) Die Bergstürze von Tschirgant und von Haiming, Oberinntal, Tirol. Begleitworte Kartenbeilage. Jahrb Geologischen Bundesanst 152(1–4):13–24

Patzelt G (2012b) Die Bergstürze vom Pletzachkogel, Kramsach, Tirol. Jahrb Geol Bundesanst 152(1–4):25–38

Patzelt G, Poscher G (1993) Der Tschirgant-Bergsturz. Arbeitstagung 1993 der Geologischen Bundes-Anstalt, Geologie des Oberinntaler Raumes, Schwerpunkt Blatt 144 Landeck, Exkursion D: Bemerkenswerte Geologische und Quartärgeologische Punkte im Oberinntal und aus dem äußerem Ötztal, pp 206–213

Pichler A (1863) Zur Geognosie Tirols II. Die vulkanischen Reste von Köfels. Jahrb Geologischen Reichsanst Wien 13:591–594

Poscher G, Patzelt G (2000) Sink-hole collapses in soft rocks. Felsbau 18:36–40

Prager C, Krainer K, Seidl V, Chwatal W (2006) Spatial features of Holocene sturzstrom-deposits inferred from subsurface investigations (Fernpass rockslide, Tyrol, Austria). Geo Alp 3:147–166

Prager C, Zangerl C, Patzelt G, Brandner R (2008) Age distribution of fossil landslides in the Tyrol (Austria) and its surrounding areas. Nat Hazard 8(2):377–407

Prager C, Ivy-Ochs S, Ostermann M, Synal HA, Patzelt G (2009a) Geology and radiometric ^{14}C-, ^{36}Cl- and Th-/U-dating of the Fernpass rockslide (Tyrol, Austria). Geomorphology 103(1):93–103

Prager C, Zangerl C, Nagler T (2009b) Geological controls on slope deformations in the Köfels rockslide area (Tyrol, Austria). Austrian J Earth Sci 102(2):4–19

Prager C (2010) Geologie, Alter und Struktur des Fernpass Bergsturzes und tiefgründiger Massenbewegungen in seiner Umgebung (Tirol, Österreich). Unpublished Ph.D. thesis, University of Innsbruck, p 307

Prager C, Zangerl C, Poscher G (2007) Prominent mass movements in the Tyrol (Austria): the deep-seated Tschirgant, Tumpen and Köfels rockslides. Geo Alp 4/2007 (Sediment 2007):159–162

Prager C, Zangerl C, Kerschner H (2012) Sedimentology and mechanics of major rock avalanches: implications from (pre-) historic Sturzstrom deposits (Tyrolean Alps, Austria). In: Eberhardt E et al. (eds) Landslides and Engineered Slopes: Protecting Society Through Improved Understanding. Proceedings of the ISL NASL 2012. Taylor & Francis, Banff/Canada, pp 895–900

Preuss E (1974) Der Bimsstein von Köfels/Tirol: die Reibungsschmelze eines Bergsturzes. Jahrb Ver Zum Schutze Alpenpflanzen -Tiere 39:85–95

Preuss E (1986) Gleitflächen und neue Friktionitfunde im Bergsturz von Köfels im Ötztal, Tirol. Mater Tech 3:169–174

Purtscheller F (1978) Ötztaler und Stubaier Alpen. Sammlung Geologischer Führer 53:1–128, Bornträger

Purtscheller F, Pirchl T, Sieder G, Stingl V, Tessadri R, Brunner P, Ennemoser O, Schneider P (1995) Radon emanation from giant landslides of Koefels (Tyrol, Austria) and Lang Tang Himal (Nepal). Environ Geol 26:32–38

Reiter F, Ortner H, Brandner R (2003) Seismically active Inntal fault zone: inverted European rift structures control upper plate deformation. Mem Soc Geol Ital 54:233–234

Starlinger F (1992) Rotföhren- und Spirkenwälder am Fernpaß (Tirol). Tuexenia 12:67–91

Strasser M, Monecke K, Schnellmann M, Anselmetti FS (2013) Lake sediments as natural seismographs: A compiled record of late quaternary earthquakes in central Switzerland and its implication for alpine deformation. Sedimentology 60(1):319–341

Stutzer O (1936) Die Talweitung von Köfels im Ötztal (Tirol) als Meteorkrater. Z Dtsch Geologischen Ges 88:523–525

Suess FE (1937) Der Meteor-Krater von Köfels beim Umhausen im Ötztale, Tirol. Neues Jahrb Mineral Geol Paläontol 98–155

Tinner W, Kaltenrieder P, Soom M, Zwahlen P, Schmidhalter M, Boschetti A, Schlüchter C (2005) Der nacheiszeitliche Bergsturz im Kandertal (Schweiz): Alter und Auswirkungen auf die damalige Umwelt. Eclogae Geol Helveticae 98:83–95

Trientl A (1895) Die Bimssteine von Köfels. Tiroler Landeszeitung 6 (5)

von Klebelsberg R (1951) Das Becken von Längenfeld im Ötztal. Schlern-Schriften 77:399–422

Weidinger JT, Korup O, Munack H, Altenberger U, Dunning SA, Tippelt G, Lottermoser W (2014) Giant rockslides from the inside. Earth Planet Sci Lett 389:62–73

Weidinger JT, Korup O (2009) Frictionite as evidence for a large late quaternary rockslide near Kanchenjunga, Sikkim Himalayas, India. Implications for extreme events in mountain relief destruction. Geomorphology 103(1):57–65

Zangerl C, Schneeberger A, Steiner G, Mergili M (2021) GIS-based topographic reconstruction and geomechanical modelling of the Köfels Rock Slide. Nat Hazards Earth Syst Sci 21:2461–2483. https://doi.org/10.5194/nhess-21-2461-2021

Martin Mergili is professor for physical geography at the University of Graz, Institute of Geography and Regional Science. Before, he held positions as senior scientist at the Institute of Applied Geology, BOKU, and as post-doc at the Department of Geography and Regional Research, University of Vienna, Austria. His research focuses on high-mountain geomorphology, natural hazards, and GIS techniques for the simulation of landslide processes.

Christoph Prager is an engineering geologist at ILF Consulting Engineers Austria GmbH and has participated in several landslide projects as senior researcher. He has provided expert services to various engineering projects worldwide, with a strong focus on field surveys, site investigations and assessments of natural hazards (landslides).

22 The Upper Ötz Valley: High Mountain Landscape Diversity and Long Research Tradition

Thomas Geist, Clemens Geitner, and Kati Heinrich

Abstract

The upper Ötz Valley comprises Vent Valley, Gurgl Valley and their tributaries. It is one of the most intensively glaciated regions in Austria. The area gained worldwide attention when almost 5300 years old "Ötzi" was discovered here in 1991, showing how early a human presence and influence on the landscape must be expected. The main morphological shaping in this mountain area was done by the huge glaciers during the Pleistocene including all typical geomorphological features of the glacial and periglacial process systems. The glacier forefields of Gurgler Ferner and Rotmoosferner offer classic localities for glacier history investigation as well as appropriate chronosequences for ecological research. With Vernagtferner, Hintereisferner and Kesselwandferner, some of the best investigated glaciers in the Eastern Alps are located here. In historical times, several advances of Vernagtferner into the Rofen Valley caused the formation of an ice dam and the subsequent filling of a meltwater lake followed by an outburst with catastrophic impacts further downstream. As a consequence, systematic observations and investigations have a long tradition, both on the glaciers and in the periglacial areas. This includes also Hochebenkar rock glacier in the Gurgl Valley. In recent years, the upper Ötz Valley was used as a test site for remote sensing approaches for glacier monitoring and subsequent geomorphological analysis in the glacier periphery, mainly focusing on Airborne Laser Scanning (ALS). By the combination of new methodological approaches with traditional research knowledge, this study area has a strong potential to analyse recent changes and also to impart this knowledge to the public.

Keywords

Ötz Valley • Glacier history • Glacier forefield • Moraine • Rock glacier • Airborne Laser Scanning • Research history

22.1 Introduction

The upper Ötz Valley as described in this chapter comprises Vent Valley, Gurgl Valley and their tributaries. The entire Ötz Valley has a total catchment area of 894 km^2 and is the largest tributary valley of the Inn Valley. The elevation ranges from 675 m asl at the mouth to the highest peak, Wildspitze, at 3768 m asl (Gattermayr 2013). Delimiting the Ötztal Alps in the west and the Stubai Alps in the east, the catchment is one of the most intensively glaciated regions in Austria with a glacier area of approximately 103 km^2 in 2006 (Austrian glacier inventory, Institute of Atmospheric and Cryospheric Sciences, University Innsbruck, project leader: Michael Kuhn; see also Abermann et al. 2009). Around 1850, at the end of the Little Ice Age, glaciers extended over 210 km^2 (Patzelt 2010).

Around 50% of the Ötz Valley catchment is located above 2500 m asl (Patzelt 2010). Only an area of 130 km^2 is covered by forest, which is far below the potential forest area of 290 km^2 and demonstrates the strong deforestation activities that started more than 3500 years ago, mainly at higher elevations, to expand the grazing area. Only 3.6% of the catchment area is suitable for permanent settlements. Currently, they comprise about 15 000 inhabitants and yield about 3 million overnight stays per year (Province of the Tyrol 2015). Tourism plays an important role, especially in the winter season.

T. Geist (✉)
FFG Austrian Research Promotion Agency, Sensengasse 1, 1090 Vienna, Austria
e-mail: Thomas.Geist@ffg.at

C. Geitner
Institute of Geography, University of Innsbruck, Innrain 52f, 6020 Innsbruck, Austria

K. Heinrich
Institute for Interdisciplinary Mountain Research, Austrian Academy of Sciences, Innrain 25, 6020 Innsbruck, Austria

C. Embleton-Hamann (ed.), *Landscapes and Landforms of Austria*, World Geomorphological Landscapes,
https://doi.org/10.1007/978-3-030-92815-5_22

Scientific research involving many disciplines has been done in the Ötz Valley over a long period. This had been mainly triggered by several catastrophic outbursts of meltwater lakes in historical times in Rofen Valley (see Sect. 4.3). Heuberger (1975) and Patzelt (1996) give comprehensive overviews of such research activities. In recent decades, several protected areas of various categories have been designated. Since 2006, the Ötztal Nature Park serves as an umbrella organization for these protected areas.

In 1991, the area gained worldwide attention when an almost 5300 years old mummy ("Ötzi") was discovered in an ice patch at Tisenjoch (3208 m asl) on the border of Italy and Austria. It was not only the oldest archaeological find of this kind in Europe, it also served as a reminder of how early a human presence and influence on the landscape must be expected (Oeggl et al. 1997; Kutschera et al. 2014). These insights have stimulated archaeological as well as palaeoenvironmental Alpine research over the years (e.g., Bortenschlager and Oeggl 2000).

22.2 Geographical Setting

The upper Ötz Valley south of Zwieselstein (1470 m asl) as the lowermost point at the confluence of Venter Ache and Gurgler Ache covers an area of 363 km^2 or c. 40% of the entire Ötz Valley catchment (Fig. 22.1). It includes more

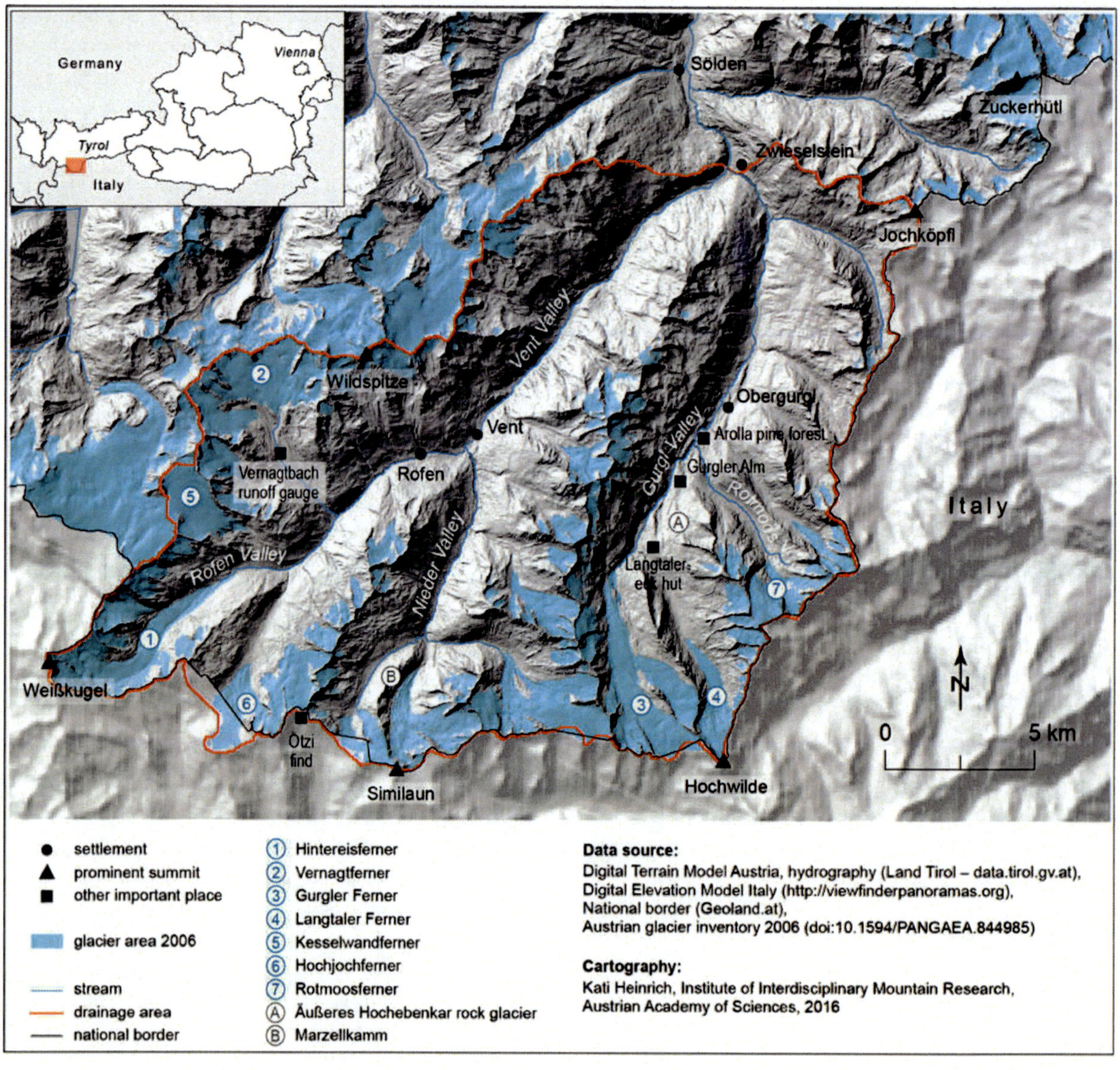

Fig. 22.1 Overview map of the study area with main locations

Fig. 22.2 Hintereisferner in the Rofen Valley with Langtauferer Spitze (3529 m asl, right) and Weißkugel (3739 m asl, centre), the second highest summit of the Ötztal Alps. In September 2013, the glacier was largely clear of snow, and thus, crevasses are easily recognizable. The isolated snow accumulation patches are remains of winter avalanches. Below Langtauferer Spitze, a typical cirque glacier is visible. *Photo* Prinz (2013)

than 15 peaks higher than 3500 m asl and more than 50 individual glaciers (Fig. 22.2).

To the south, the Ötz Valley head is surrounded by the main Alpine ridge, which forms a continental watershed with the Inn/Danube system to the north and the Etsch/Po system to the south. It is worth noting that this high Alpine ridge is not really situated in the middle of the Alps, but rather somewhat to the south. This clear asymmetry is also reflected in lengths (linear distance 50 km to the north and 15 km to the south) and gradients of the corresponding valleys, and also means that the levels of the valley floors differ markedly: the mouth of Ötz Valley lies at about 675 m asl and the corresponding Schnals Valley in the south joins the Etsch Valley at an altitude of about 550 m asl.

Geologically, the area belongs to the Ötztal-Stubai Complex and the Schneeberg Complex, both parts of the Austroalpine Nappe System. They are mainly composed of paragneiss and mica schists (Fig. 22.3), modified by three metamorphic events, Caledonian, Variscan and Alpine. As a special feature within the silicate rocks, the Schneeberg Complex contains phenocrysts of marble (Krainer 2010). Apart from solid rock, different unconsolidated sediments characterize the surface of large areas, documenting recent or past gravitational, fluvial or glacial processes.

The main morphological shaping in this mountain area was done by the huge glaciers during the Pleistocene. Respective landforms include erosion forms such as U-shaped and hanging valleys, cirques and roches moutonnées, but also various moraine types as typical accumulation features in former glaciated landscapes. During the period of the Last Glacial Maximum (LGM), a wide and thick ice field covered large parts of the Alps, whilst only a few nunataks towered above the glaciers. The former local elevation of the glacier surface can be reconstructed by mapping the trimline ("Schliffgrenze") in the rockface, which often clearly marks the boundary between glacier erosion with smooth surfaces and rough surfaces from frost weathering (Fig. 22.4).

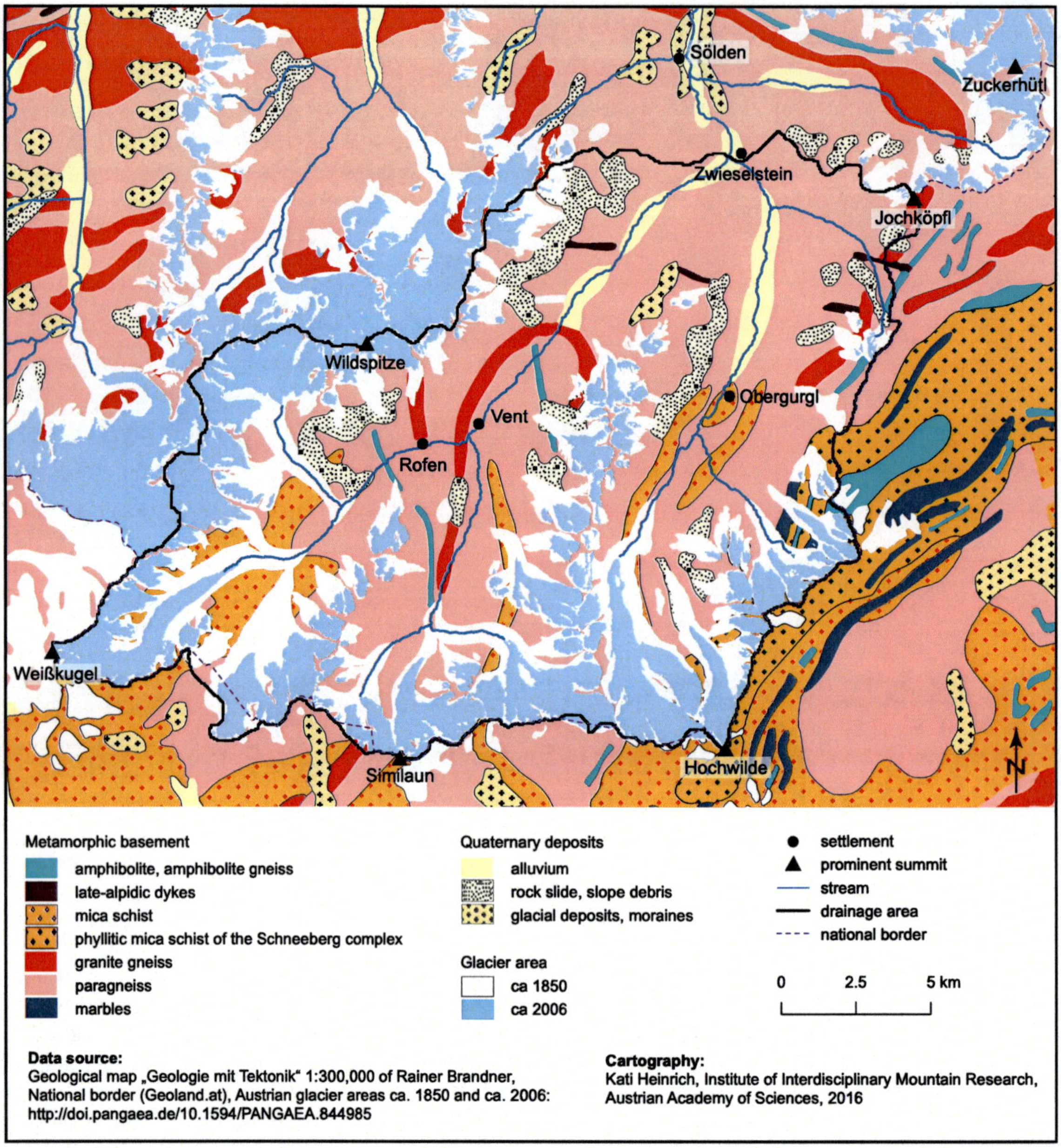

Fig. 22.3 Geological overview map with glacier extent around 1850 and 2006

Gravitational processes occur on the valley slopes, ranging from rock falls to some deep-seated rock slides. Other active processes are debris flows, fluvial erosion and snow avalanches. Moreover, the periglacial environment creates lobes formed by solifluction and earth hummocks. However, the most important permafrost-related forms in this area are rock glaciers.

Within the large Alpine LGM glaciation, the glacier of the Ötz valley region was a part of the huge Inn valley glacier system with its terminal moraines in the Alpine foreland near Rosenheim (Bavaria), approximately 250 km downwards of Obergurgl. After the LGM, the ice decayed remarkably fast, interrupted by a series of successively smaller glacier advances during the Lateglacial period. Within our study area, the last of those advances of the Ötz Valley glacier reached to a few kilometres south of Zwieselstein (Aulitzky et al. 1994). Generally, the next younger terminal moraines originate from the Little Ice Age (LIA), defined as a period extending from the Late Middle Ages to about 1850.

Situated in the Central Alps and thus surrounded by mountain ranges, the Ötz Valley is part of the continental inner-Alpine dry zone. This is highlighted by the fact that the average annual precipitation hardly increases from the mouth of the Ötz Valley into the Inn Valley (675 m asl) up to Vent (1895 m asl). At the weather station in Obergurgl (1907 m asl), the mean annual temperature is 2.8 °C, annual precipitation reaches 851 mm (with a significant summer maximum), and a snow cover of at least 1 cm has been recorded for nearly 200 days per year (period 1961–1990, Fischer 2013). Nevertheless, notable precipitation differences occur in the study area, which is typical for local climatic deviations in the Alpine region. For instance, in case of winds coming from the south, the cloud accumulation often reaches north beyond the Alpine ridge, causing precipitation in the Obergurgl region but not quite reaching further west and north.

In terms of vegetation, the area includes the montane, subalpine, alpine and nival altitudinal belts. The timberline, which marks the border between the subalpine and alpine elevation zones, is mostly established by Arolla pine (*Pinus cembra* L.). The Arolla pine forest of Obergurgl (1950–2100 m asl) covers an area of c. 20 ha and includes some individual trees more than 300 years old. Because of its uniqueness, it was designated a National Monument in 1963. Nevertheless, for thousands of years, the forests at the timberline ecotone were strongly affected by humans. In order to enlarge the pasture areas for cattle, the timberline has been lowered widely by about 300 m; currently, it is located at around 2000 m asl and has a mosaic-like structure, alternating with dwarf shrub heaths (Mayer and Erschbamer 2012).

As to the higher settlements, Rofen farmstead at an elevation of 2014 m asl is the highest permanent settlement in Austria (see Fig. 22.1). Furthermore, Obergurgl (1907 m asl) is the highest parish of the Eastern Alps. Both were established as permanent settlements with exclusively livestock farming during the High Middle Ages.

22.3 Long Research Tradition and Test Site for New Monitoring Approaches

The long research tradition in the study area includes meteorological, glaciological, geomorphological, dendrochronological, ecological and microbiological investigations, some continuing for more than 100 years. The Alpine Research Centre Obergurgl (AFO), established in 1951, constitutes an important field station of Innsbruck University. It is responsible for the monitoring programmes and research projects in Obergurgl, for research support, data provision and dissemination. This research centre was an ideal starting position for the international and interdisciplinary *Man and the Biosphere Programme* (MAB) in the 1970s. In 2000, the AFO monitoring programme was integrated into the international *Long Term Ecological Research Network* (LTER), whose sites range from the subalpine to the subnival zone (AFO n. d.). In recent years, a lot of this interdisciplinary scientific knowledge was also published for the interested public in an illustrated version (Koch and Erschbamer 2010, 2012, 2013; Schallhart and Erschbamer 2015).

In recent years, the upper Ötz Valley with the area of Hintereisferner and Kesselwandferner in particular was used as test site for several remote sensing approaches for glacier monitoring and subsequent geomorphological analysis in the glacier periphery, mainly focusing on Airborne Laser Scanning (ALS). ALS is an active remote sensing technique, which uses a laser beam to acquire 3D point data, representing the surface of the earth or objects on that surface. ALS data are widely used for topographic surveying and the generation of digital terrain models with high resolution and high accuracy (usually between 0.05 m and 0.20 m). Airborne laser scanning is well suited for remote mountain areas, because there is no need to obtain ground control points within the area of interest.

Due to the high precision and spatial resolution, ALS data and the derived Digital Elevation Models (DEMs) have been applied to different aspects of geomorphology in recent years, in particular for the calculation of morphometric variables. Van Asselen and Seijmonsbergen (2006) used ALS-based DTMs for expert-driven, semi-automated geomorphological mapping in Alpine regions and compared morphometric analyses of various geomorphologic units like fluvial terraces, talus slopes, rock cliffs or glacial landforms with a validation dataset.

The detection, quantification and monitoring of changes in surface morphology with the use of multi-temporal ALS are other widespread applications. Since 2001, ALS measurements have been carried out regularly in this area, with Hintereisferner and Kesselwandferner as the main targets, resulting in a unique data record of 18 ALS flight campaigns over the same area. Furthermore, a monitoring station with a Terrestrial Laser Scanner (TLS) was installed in 2017, allowing for a permanent observation of the Hintereisferner area.

The multi-temporal analysis of the data sets allows monitoring changes in the surface elevation. This was initially investigated for glacier volume changes and glacier dynamics by Geist (2005) and Geist and Stötter (2007). More recently, Abermann et al. (2010) and Bollmann et al. (2011) applied multi-temporal ALS data to detect and quantify periglacial processes. Sailer et al. (2012) demonstrate how ALS data series can be used for the detection and quantification of geomorphological processes in an Alpine periglacial environment. They quantify volumetric earth surface changes caused by dead-ice melting, fluvial erosion/deposition, rockfall activity, gravitational displacements and

Fig. 22.4 Distinct trimline in Rofen Valley at about 2700 m asl. It marks the local elevation of the LGM glacier surface. *Photo* G. Patzelt

permafrost degradation in glaciated, recently deglaciated and periglacial terrain (see Fig. 22.5). Furthermore, they aim to determine the time lapse between two ALS campaigns required to give significant results for different geomorphological processes, depending on their magnitude. As a recent example, Klug et al. (2017) verify that surface changes are attributed to permafrost thaw and thus can be used as a possible indicator for permafrost distribution.

Future investigations in the field of geomorphology based on ALS data will focus on process chains. Perfect examples are areas of debris redistribution, where the results of mainly fluvial and gravitational processes are analysed; but in reality, a complex system of cascading processes is involved: (1) glacial erosion, forming the trough-like shape of the valley, (2) the deposition of ground and lateral moraines, (3) the regressive fluvial erosion with subsequent gravitational denudation, (4) the reallocation of the unconsolidated debris towards the glacier, (5) the transport of material within the glacier dynamics and (6) dead-ice formation after glacier melt (Sailer et al. 2012). The quantification of these "geomorphological mass balances" is a challenging future task. The ALS technology has without doubt the potential to become an important tool in small-scale spatial monitoring and quantification of process-chain-related mass reallocations in high alpine terrain.

22.4 Key Sites

22.4.1 Gurgl and Rotmoos Valley—Glacier History and Glacier Forefields

In the Gurgl Valley, the glacier inventory of 2006 includes 34 individual glaciers. The Gurgler Ferner (the term *ferner* is a regional expression for glacier) is the largest, with almost 9 km^2 in 2006 (and thus the second largest glacier in the Ötztal Alps after Gepatschferner), followed by Rotmoosferner (c. 2.7 km^2), Langtaler Ferner (c. 2.6 km^2) and Gaisbergferner (c. 1.2 km^2). The other glaciers in the Gurgl valley catchment are much smaller. A comparison of the Austrian glacier inventories of 1969 and 2006 shows a reduction of the entire glacier surface in this area by 27% as a result of the recent global warming.

The Gurgler Ferner and the Rotmoosferner provide classic localities to reconstruct Holocene and modern glacier history. The Gurgler Ferner glacier forefield, defined by two lateral moraines of maximum Holocene glacier expansion, is located just below the Langtalereck hut. The cross section (Fig. 22.6) shows the two lateral moraines, the inner ridge (on the left side) marks the glacier extent around 1855 and the outer ridge marks an earlier glacier extent. The latter proves that Gurgler

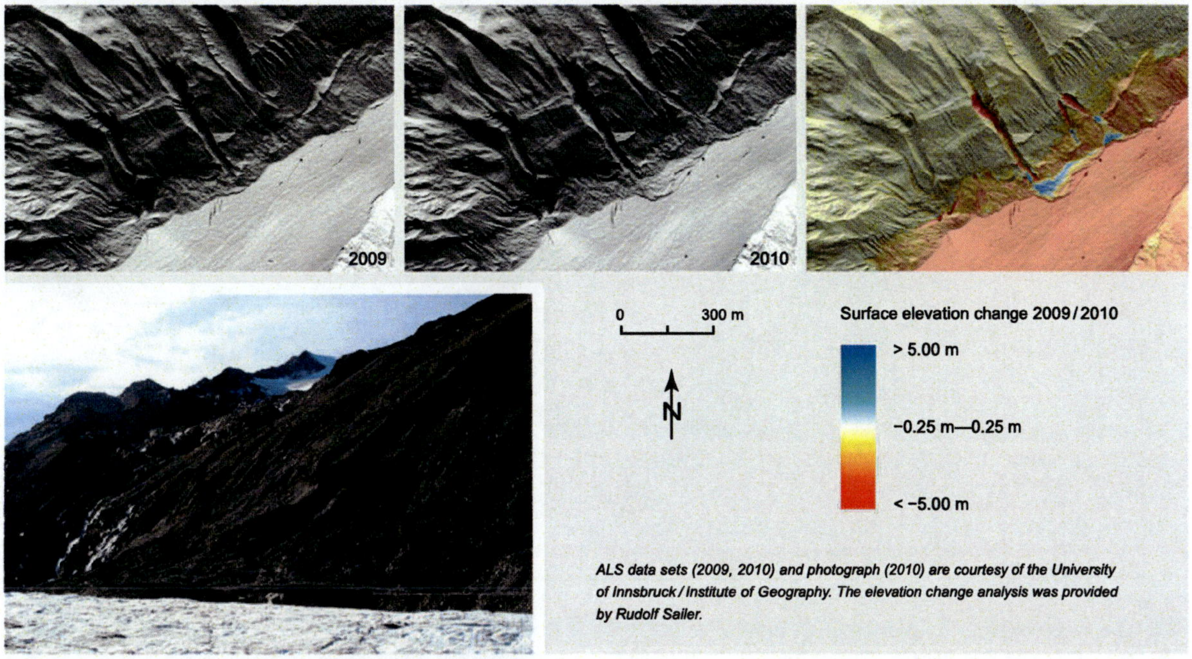

Fig. 22.5 Current rapid downwasting of glacier tongues leads to a significant lowering of the ice surface at glacier margins. Unconsolidated moraine material is piled up on the slopes and can be mobilized as debris flows after heavy rainfall. The figure shows the situation at the orographic left side of Hintereisferner before and after a debris flow event. The visualization of the digital elevation models of 2009 and 2010 (upper left and middle) and, even better, the elevation change analysis (upper right) illustrate the accumulation area of the debris flow sediments, superimposed on the glacier surface (blue). The analysis clearly shows the down-melt of the glacier surface, as well as the depletion of unconsolidated moraine material and erosion in debris flow channels (red). Stable slopes are pictured in grey. Lower left: view of the investigated valley slope. *Photo* R. Sailer

Ferner advanced to a maximum position in medieval times around AD1300. Dating of the peat covered by this moraine reveals that Gurgler Ferner had not advanced beyond this ridge for more than 7500 years (Patzelt 1986; Patzelt et al. 1996). Based on such field and laboratory methods, as well as on historical records, Patzelt et al. (1996, 2019) with a new and very comprehensive compilation of all picture documents, maps and field data) reconstructed a detailed Holocene chronology of Gurgler Ferner, which verifies glacier advances in the middle Holocene, in medieval times and during the Little Ice Age. This glacier history of Gurgler Ferner confirms the general pattern that during the Holocene glaciers in the Alps were never significantly larger than they were during the LIA. Therefore, the glacier expansion of the LIA provides a suitable reference for palaeoglaciological comparisons. Nevertheless, it should be mentioned that radiocarbon and dendrochronological dating of wood and peat remains in some other glacier forefields point to quite long-lasting periods within the early and mid-Holocene, when the glaciers were much smaller than today (e.g., Nicolussi et al. 2005).

The advances of Gurgler Ferner dammed the tributary Langtal Valley several times, causing the formation of a temporary lake with a maximum volume of c. 11.7 million m^3 (Patzelt et al. 1996). Contrary to the situation in Rofen Valley (see Sect. 4.3), this so-called Gurgler Ice Lake, occurring between the years 1716 and 1867, caused destruction within the cultural landscape only twice, because usually it drained slowly. Lake terraces and related sediments of this "Ice Lake" can be found near the confluence of Langtal and Gurgl valleys.

Rotmoos Valley, an eastern tributary valley to the Gurgl Valley, is a classic locality for Holocene glacier history as it contains a large mire (2260 m asl) that provided palynological and sedimentological evidence for the suggestion of a two-part Rotmoos glacier advance during a cool and humid period between c. 4200 and 3000 BC (Patzelt and Bortenschlager 1973; Bortenschlager 1984; Patzelt et al. 1996). Moreover, Rotmoos Valley represents the structure, dynamics and chronology of glacier forefields in an exemplary way. Therefore, it serves as a long-term investigation site for various ecological chronosequence studies (e.g., Erschbamer

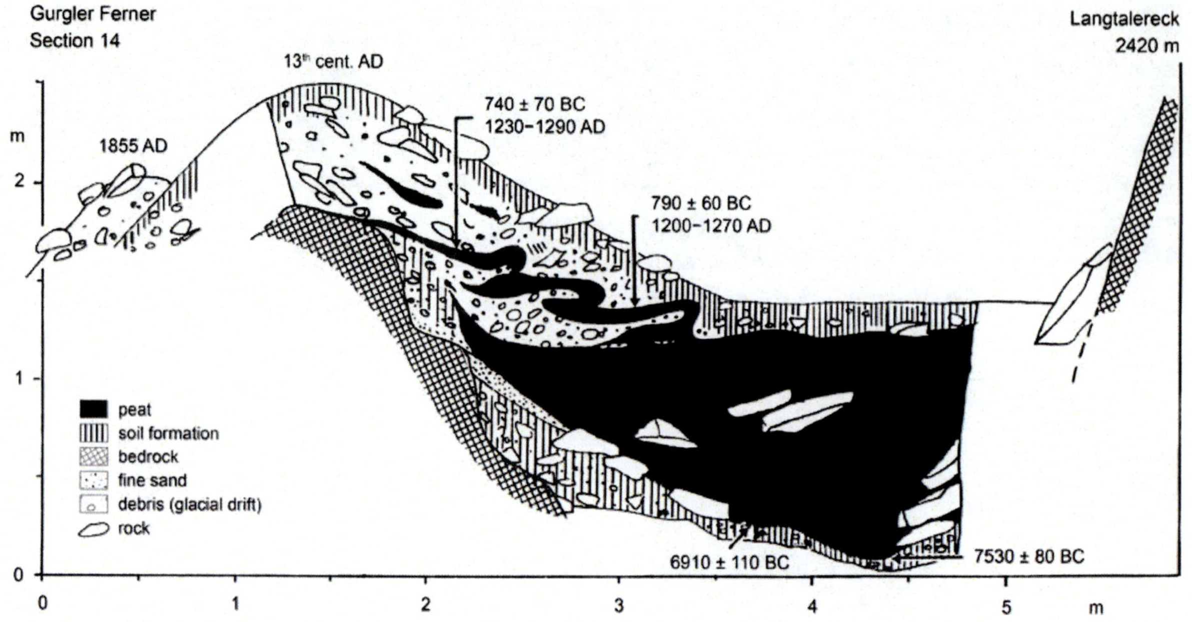

Fig. 22.6 Cross-section of moraine ridges of Gurgler Ferner, with peat accumulation and dating results (after Patzelt et al. 1996)

et al. 1999, 2008; Kaufmann 2002). In Fig. 22.7, the deposits of the LIA terminal moraine at about 2290 m asl are easily identifiable. They mark the maximum ice cover in the year 1858. The general retreat of Rotmoosferner during the last 150 years was interrupted by two minor re-advances around 1910/1920 and 1970. This last advance was favoured by a period of rather cool summers and mild but snowy winters in the 1960s and 1970s. However, since the mid-1980s, significant warming in summer and winter has led to rapid shrinkage of glacier tongues.

Shrinking glaciers imply expanding glacier forefields with a tendency to strong morphodynamic processes, redistributing the fluvioglacial deposits. Nevertheless, only a few years after deglaciation primary succession of plants and initial soil formation begin. According to Schwienbacher (2004), Fig. 22.8 illustrates the spatial distribution pattern of pH values of the uppermost horizons within the glacier forefield of Rotmoosferner. The eastern side reveals a link between the age of deposition and pH values, which can be explained by the state of soil formation. Nevertheless, considering the geological setting, pH values are unusually high, which must be due to the carbonate content of the Schneeberg Complex at the valley head mentioned above. On the western side, the level of pH values and their spatial distribution pattern are very different. Here, the recent dynamics of the glacier forefield is dominated by accumulation from the slopes, which are free from carbonates, and is mainly caused by debris flows and avalanches. Thus, the pattern of pH values reflects the high and small-scale diversity of morphodynamics, which provides the background for biological succession and diversity.

22.4.2 Gurgl Valley—Hochebenkar Rock Glacier

Active rock glaciers are widespread features in Alpine environments and are commonly regarded as geomorphological indicators of Alpine permafrost. They are lobate or tongue-shaped and composed of frozen, unconsolidated material that slowly creeps downslope (Barsch 1996). Within the catchment area of the Ötztal Ache, 421 rock glaciers were identified (135 are classified as active rock glaciers), covering an area of 30.5 km^2 (Krainer and Ribis 2012) (for rock glaciers also see Chap. "Rock Glaciers in the Austrian Alps: A General Overview with a Special Focus on Dösen Rock Glacier, Hohe Tauern Range").

Probably, the best investigated rock glacier location in the Austrian Alps is Äußeres Hochebenkar (= AHK rock glacier), c. 4 km SSW of Obergurgl and located in a small northwest facing cirque (see Fig. 22.1). It is a tongue-shaped rock glacier covering an area of 0.4 km^2, with a length of about 1600 m along the main flow line, expanding from an altitude of c. 2850 m asl down to c. 2350 m asl. It is characterized by a typical surface topography with transverse and longitudinal ridges and furrows, a steep front and steep lateral sides (Fig. 22.9). A detailed morphological characterization of AHK rock glacier is given by Vietoris (1972) and Haeberli and Patzelt (1982). Krainer (2015) states that morphology, high ice content and high surface flow velocities indicate that the rock glacier contains a massive ice core and probably developed from a debris-covered cirque glacier.

The history of systematic measurements and scientific investigations at AHK rock glacier dates back to the 1930s,

Fig. 22.7 Rotmoosferner and its glacier forefield. *Photo* Kautzky (2009). Former glacier front locations dated by Gernot Patzelt

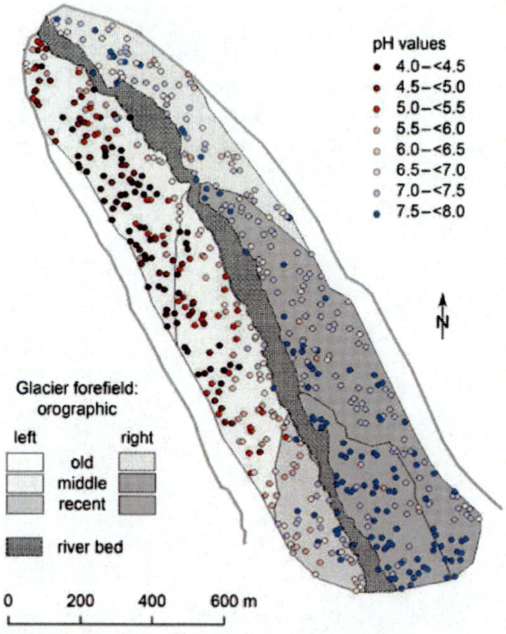

Fig. 22.8 Pattern of pH values of uppermost soil horizons within the glacier forefield of Rotmoosferner (570 samples, pH measurements in $CaCl_2$ suspension; according to Schwienbacher 2004)

establishing long-term records on surface flow velocities, front advance rates and surface elevation changes (e.g., Pillewitzer 1938; Vietoris 1972; Schneider and Schneider 2001). AHK is characterized by comparably high surface velocities in its lower part, especially below an edge in the terrain at 2570 m asl. Annual displacement rates of up to 6.6 m were recorded in the 1950s and 1960s and are considered the highest measured flow velocities of a rock glacier in the Alps (Schneider and Schneider 2001). In comparison, annual displacement rates of active rock glaciers are mostly in the order of centimetres to decimetres. In recent decades, changes in the velocity of AHK rock glacier showed a close correlation with changes in the mean annual air temperature at nearby weather stations. Above-average temperatures resulted in increased flow rates, whereas below-average temperatures caused a decrease in the flow rate (Krainer 2015).

In 2007, a new monitoring network was initiated at AHK rock glacier. Velocity measurements are now supported by the application of a Differential Global Positioning System (DGPS). Geological mapping is supported by orthophotos and ALS elevation models. The thermal regime of the rock glacier and the extension of the permafrost area are derived from patterns of the basal temperature of the winter snow cover (BTS). Ground penetrating radar (GPR) is used to determine thickness (maximum thickness is 49 m) and size of the ice body. The discharge of the rock glacier is studied quantitatively by automatic gauges, and regular chemical analysis is used to determine the water properties of the runoff. The results of recent field campaigns in the context of long-term monitoring are summarized by Nickus et al. (2015).

Recently, horizontal displacement rates as well as surface elevation changes of AHK rock glacier were calculated over different time spans using ALS data. The results clearly show a sharp distinction between areas with less activity and areas with high surface elevation changes, especially in the middle part of AHK rock glacier (Klug 2015). These areas correspond well with the description by Schneider and Schneider (2001) who assume that deep lateral cracks occur as a consequence of local sliding of the rock glacier on the bedrock and increased tension in the permafrost body. Regarding the entire area of the AHK rock glacier, an area-averaged mass loss is not detected. Therefore, the ice content seems to be well protected from surface energy input and significant ice melt did not occur between 2006 and 2010 when the ALS measurements were carried out (Klug 2015).

Fig. 22.9 Hochebenkar rock glacier (centre) with Langtalferner to the right and Rotmoos and Gaisberg valleys with small glaciers to the left. *Photo* Abermann (2009)

22.4.3 Rofen Valley—Large Glaciers and Natural Disasters in the Past

Vent Valley consists of the sub-catchments of Niedertal Valley and Rofen Valley, both joining in the village of Vent. Rofen Valley can be characterized as a typical high alpine valley in the Central Alps, with many features of glacial and periglacial geomorphological process systems. With Vernagtferner, Kesselwandferner, Hintereisferner and Hochjochferner, this area contains four of the best investigated glaciers in the Eastern Alps. At the end of the Little Ice Age (LIA), these glaciers and their tributaries covered a major part of the Rofen Valley (Fig. 22.3). Glacier retreat since the last LIA advance around 1850 was only interrupted by a few periods with minor glacier advances (e.g., 1920s). In recent decades, the melting of the glacier ice has increased dramatically, leading to ice-free and unstable moraine-covered slopes characterized by a variety of active geomorphological processes. In particular, the lateral moraines of the LIA advance are prone to fluvial erosion, gravitational displacements and combined processes like debris flows (Fig. 22.5). At Hintereisferner, the recent downwasting of the glacier tongue and the increased morphodynamic activity result in a gain of sedimentary deposits on both sides of the glacier tongue and the subsequent formation of dead-ice bodies (Abermann et al. 2010). Furthermore, in the glacier forefield, intensive fluvial geomorphodynamics can be observed.

Hintereisferner has the typical geometry of a valley glacier with a length of still several kilometres along the flow line, extending from an elevation of c. 3700 m down to c. 2450 m asl (Fig. 22.2). Neighbouring Kesselwandferner extends from c. 3500–2750 m asl. Most of this glacier consists of a plateau-like accumulation area ending in a short tongue. Kesselwandferner was a tributary glacier of Hintereisferner until c. 1935. Whilst Hintereisferner has steadily retreated since c. 1920, Kesselwandferner advanced significantly (>300 m) between 1965 and 1983. The different responses of these two adjacent glaciers, exposed to the same regional climatic conditions and variations, can be explained by topographical factors and individual dynamic response times (Kuhn et al. 1985). The glaciers have contrasting longitudinal profiles, with a relatively flat tongue and a steeper accumulation area for Hintereisferner and a large, flat accumulation area with a short and partly steep tongue for Kesselwandferner. Thus, Kesselwandferner has a greater mass gain in years with positive mass balance, which makes this glacier climatically very sensitive. This is expressed by short-term fluctuations of the glacier tongue.

Both glaciers have been intensively investigated for many decades. Observations on length variations of Hintereisferner began as early as 1847. First measurements of glacier velocity and ice thickness change date back to 1894 (Blümcke and Hess 1899) and have been continued up to the present (e.g., Span et al. 1997). The mass balance of Hintereisferner has been determined annually since 1953 (Kuhn et al. 1999). At Kesselwandferner surface elevation change, flow velocity, annual mass balance and length variations have been surveyed continuously since 1965. Since 2001, extensive multi-temporal ALS measurements have been carried out over the area (see Sect. 22.3).

Areas that were not covered by LIA glaciers have been exposed to conditions favourable for permafrost. This is the case for the Rofenberg area between Hintereisferner and Hochjochferner. This area is also characterized by considerable gravitational activities, e.g., rock falls.

One prominent geomorphological feature further down the Rofen Valley is the deep incision of the Rofen Gorge. Above the gorge, in the bedrock slope on the eastern side of the valley, the trimline is clearly visible as an indicator for the maximum glaciation during the LGM (Fig. 22.4). At the upper end of the gorge is the confluence area of meltwater streams coming from Hochjochferner, Hintereisferner and Kesselwandferner up-valley and the meltwater stream from Vernagtferner. This location is the reason why the Rofen Valley is one of the best investigated glaciological sites in the Alps.

In historical times, at least four advances of Vernagtferner into the Rofen Valley caused the formation of an ice dam at the confluence area and subsequent filling of a meltwater lake followed by an outburst, partly with catastrophic impacts further downstream: these occurred around 1600, 1680, 1770 and 1850. The advances and the glacial lakes were documented by drawings and paintings, one of them being presumably the first illustration of an Alpine glacier (Fig. 22.10). Vernagtferner advanced largely at the same time as other Alpine glaciers. However, these advances were exceptional in terms of magnitude and dynamics (Nicolussi 2013). As a consequence of this potentially dangerous situation, systematic observations and scientific investigations have a long tradition in this region, starting with the first complete photogrammetric survey of an Alpine glacier by Sebastian Finsterwalder in 1889. Since 1965, glacier mass balance has been measured continuously with the glaciological method. Since 1973 runoff, precipitation and evaporation data have been collected at the Vernagtbach gauge (Braun et al. 2013).

Directly above the confluence area on the orographic left side of Rofen Valley, a dense network of open fissures indicates the deep-seated mass movement at a place called Plattei. Here a slope the size of c. 10 ha, with an estimated volume of c. 2 million m^3 of material, is moving slowly downwards. An acceleration of the movement and a subsequent high-impact landslide event would block the mouth of Vernagtbach and the upper part of the Rofen Gorge (Patzelt 2013).

In the Marzellkamm area (see Fig. 22.1), another deep-seated mass movement has been observed for almost 15 years. Annual slope movements of more than 50 cm in

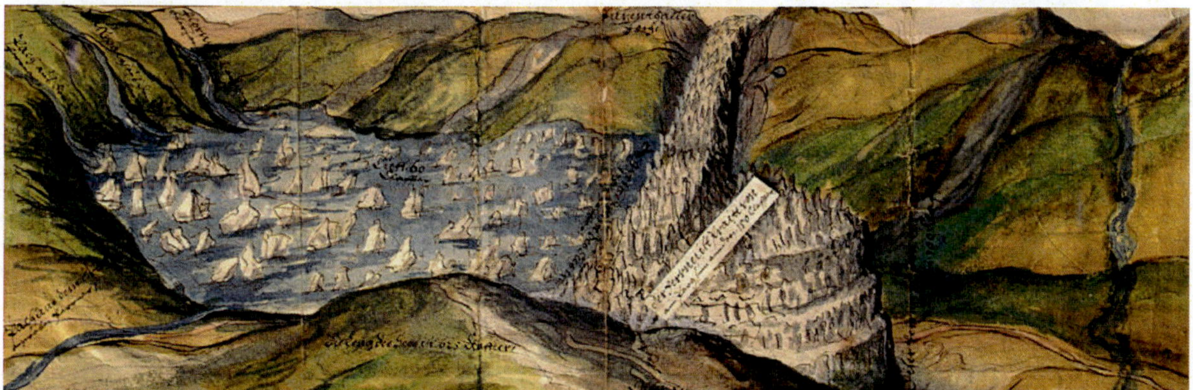

Fig. 22.10 Historical painting of Vernagtferner and the proglacial lake, 9 July 1601 (adapted from Abraham Jäger). This is the oldest known illustration of a glacier anywhere in the world (Nicolussi 1993)

places were measured. The reason is the rapidly retreating glacier below, which stabilized the slope for many years (Fey et al. 2015).

22.5 Additional Aspects High Alpine Landscape and Human Interaction

The upper Ötz Valley not only provides insights into a high diversity of geomorphological forms and processes, it also serves as an arena for human activities. Figure 22.11 illustrates how exposed this high-altitude living space is. However, wood remains, and charcoal particles and pollen spectra from a mire just below these buildings reveal the land use history of this recently treeless site. Located within the timberline ecotone during the early and mid-Holocene, the woodland was cleared as early as 4300 BC (and at least some timber was burnt) to enlarge the pasture areas (Vorren et al. 1993). These palaeoenvironmental results are confirmed by comparable findings from other sites in the region. Archaeological investigations at various sites in the upper Ötz Valley also indicate Mesolithic and Neolithic activities (Zanesco 2012). Apparently, land use intensified during the Bronze Age, as irrigation sediments of this period could be found (Patzelt et al. 1997). Consequently, there seems to be a general pattern for the oldest traces of mankind at high alpine elevations to be discovered above or close to the former timberline. That means that these very early human activities took place in the same part of the landscape that is nowadays preferred for recreation in the summer as well as in the winter season.

Fig. 22.11 Rockfall area near the Gurgler Alm, 2240 m asl. *Photo* Patzelt (1994)

22.6 Outlook

This section focused on high-altitude Alpine landscapes and their past and present morphodynamics. Currently, we face remarkable environmental shifts as a result of changes in climate and land use. As they are running in parallel, it is sometimes difficult to separate the influences from each other. At higher elevations, without vegetation cover, changes within the geomorphological system can be clearly assigned to climate change, which can also be confirmed by local climate data. Since the first meteorological measurements in Obergurgl in 1953, the mean annual air temperature has increased by 1.2 °C (Fischer 2013). However, the alpine and subalpine ecosystems are also exposed to changes in land use. Transformation of mostly agricultural areas into a scene relevant for tourism implies secondary effects which we will have to deal with in the future.

The research tradition in many disciplines and the valuable long-term observation records in the upper Ötz Valley make this area ideally suited as an "observatory" for monitoring landscape change in changing climate and land use conditions. In addition, the significant changes currently taking place (e.g., rapid downwasting of glacier tongues) have a high potential for environmental education in many aspects (e.g., excursion guides of Obergurgl region by Patzelt et al. 1996 and 2007). This is generally supported by the fully developed touristic infrastructure in this part of the Alps and in particular by the activities of the Alpine Research Centre Obergurgl.

References

Abermann J, Lambrecht A, Fischer A, Kuhn M (2009) Quantifying changes and trends in glacier area and volume in the Austrian Ötztal Alps (1969–1997–2006). Cryosphere 3(2):205–215

Abermann J, Fischer A, Lambrecht A, Geist T (2010) On the potential of very high-resolution repeat DEMs in glacial and periglacial environments. Cryosphere 4:53–65

AFO (=Alpine Research Centre Obergurgl) (n.d.) Alpine research in the region of Obergurgl. https://www.uibk.ac.at/afo/publikationen/pdf/a4-afo-broschuere-engl_final.pdf (20.04.2016)

Aulitzky H, Heuberger H, Patzelt G (1994) Mountain hazard geomorphology of Tyrol and Vorarlberg, Austria. Mt Res Dev 14(4):273–305

Barsch D (1996) Rockglaciers—Indicators for the present and former geoecology in high mountain environments. Berlin, p 331

Blümcke A, Hess H (1899) Untersuchungen am Hintereisferner. Z Dtsch Österr Alpenvereins 1(2):1–87

Bollmann E, Sailer R, Briese C, Stötter J, Fritzmann P (2011) Potential of airborne laser scanning for geomorphologic feature and process detection and quantification in high alpine mountains. Z Geomorphol Supplementary Issues 55(2):83–104

Bortenschlager S (1984) Beiträge zur Vegetationsgeschichte Tirols I. Inneres Ötztal und unteres Inntal. Ber Naturwiss-Med Ver Innsbruck 71:19–56

Bortenschlager S, Oeggl K (eds) (2000) The man in the ice. The iceman and his natural environment (Volume 4): Palaeobotanical results. Wien

Braun L, Reinwarth O, Weber M (2013) Der Vernagtferner als Objekt der Gletscherforschung. Z Gletscherk Glazialgeol 45(46):85–104

Erschbamer B, Niederfriniger Schlag R, Winkler E (2008) Colonization processes on a central Alpine glacier foreland. J Veg Sci 19(6):855–862

Erschbamer B, Bitterlich W, Raffl C (1999) Die Vegetation als Indikator für die Bodenbildung im Gletschervorfeld des Rotmoosferners (Obergurgl, Oetztal, Nordtirol). Ber Naturwiss-Med Ver Innsbruck 86:107–122

Fey C, Rutzinger M, Wichmann V, Prager C, Bremer M, Zangerl C (2015) Deriving 3D displacement vectors from multi-temporal airborne laser scanning data for landslide activity analyses. Giscience Remote Sens 52(4):437–461

Fischer A (2013) Klima und Gletscher in Obergurgl. In: Koch E-M, Erschbamer B (eds) Klima, wetter, Gletscher im Wandel. Alpine Forschungsstelle Obergurgl Band 3, Innsbruck, pp 55–72

Gattermayr W (2013) Das hydrographische regime der Ötztaler ache. In: Koch E-M, Erschbamer B (eds) Klima, wetter, Gletscher im wandel. Alpine Forschungsstelle Obergurgl Band 3, Innsbruck, pp 121–155

Geist T (2005) Application of airborne laser scanning technolog in glacier research. Ph.D. thesis, Institute of Geography, University of Innsbruck, Austria

Geist T, Stötter J (2007) Documentation of glacier surface elevation change with multi-temporal airborne laser scanner data—case study Hintereisferner and Kesselwandferner, Tyrol, Austria. Z Gletscherk Glazialgeol 41:77–106

Haeberli W, Patzelt G (1982) Permafrostkartierung im Gebiet der Hochebenkar-Blockgletscher, Obergurgl, Ötztaler Alpen. Z Gletscherk Glazialgeol 18(2):127–150

Heuberger H (1975) Das Ötztal. Bergstürze und alte Gletscherstände, kulturgeographische Gliederung. Innsbrucker Geogr Studien 2:213–249

Kaufmann R (2002) Glacier foreland colonisation: distinguishing between short-term and long-term effects of climate change. Oecologia 130:470–475

Klug C (2015) Blockgletscherbewegung im Äußeren Hochebenkar 1953–2010—eine Methodenkombination aus digitaler Photogrammetrie und Airborne Laserscanning. In: Schallhart K, Erschbamer B (eds) Forschung am Blockgletscher—Methoden und Ergebnisse. Alpine Forschungsstelle Obergurgl Band 4, Innsbruck, pp 135–158

Klug C, Rieg L, Ott P, Mössinger M, Sailer R, Stötter J (2017) A multi-methodological approach to determine permafrost occurrence and ground surface subsidence in mountain Terrain, Tyrol, Austria. Permafrost Periglac Processes 28(1):249–265

Koch E-M, Erschbamer B (eds) (2010) Glaziale und periglaziale Lebensräume im Raum Obergurgl. Alpine Forschungsstelle Obergurgl Band 1, Innsbruck, p 304

Koch E-M, Erschbamer B (eds) (2012) An den Grenzen des Waldes und der menschlichen Siedlung. Alpine Forschungsstelle Obergurgl Band 2, Innsbruck, p 222

Koch E-M, Erschbamer B (eds) (2013) Klima, wetter, gletscher im wandel. Alpine Forschungsstelle Obergurgl Band 3, Innsbruck, p 214

Krainer K (2010) Geologie und Geomorphologie von Obergurgl und Umgebung. In: Koch E-M, Erschbamer B (eds) Glaziale und Periglaziale Lebensräume im Raum Obergurgl. Alpine Forschungsstelle Obergurgl Band 1, Innsbruck, pp 31–52

Krainer K (2015) Blockgletscher: Einführung. In: Schallhart K, Erschbamer B (eds) Forschung am Blockgletscher—Methoden und Ergebnisse. Alpine Forschungsstelle Obergurgl, Band 4, Innsbruck, pp 9–32

Krainer K, Ribis M (2012) A Rock glacier Inventory of the Tyrolean Alps (Austria). Austrian J Earth Sci 105(2):32–47

Kuhn M, Markl G, Kaser G, Nickus U, Obleitner F, Schneider H (1985) Fluctuations of climate and mass balance: different responses of two adjacent glaciers. Z Gletscherk Glazialgeol 21:409–416

Kuhn M, Dreiseitl E, Hofinger S, Markl G, Span N, Kaser G (1999) Measurements and models of the mass balance of Hintereisferner. Geogr Ann Phys Geogr 81(4):659–670

Kutschera W, Patzelt G, Wild E-M, Jettmar B, Kofler W, Lippert A, Oeggl K, Pak E, Priller A, Steier P, Wahlmüller-Oeggl N, Zanesco A (2014) Evidence for early human presence at high altitudes in the Ötztal Alps Austria/italy. Radiocarbon 56(3):923–947

Mayer R, Erschbamer B (2012) Lärchen-Zirbenwälder und Zwergstrauchheiden. In: Koch E-M, Erschbamer B (eds) An den Grenzen des Waldes und der menschlichen Siedlung. Alpine Forschungsstelle Obergurgl Band 2, Innsbruck, pp 99–123

Nickus U, Krainer K, Thies H, Tolotti M (2015) Blockgletscherabflüsse im Äußeren Hochebenkar—Hydrologie, Wasserchemie und Kieselalgen. In: Schallhart K, Erschbamer B (eds) Forschung am Blockgletscher—Methoden und Ergebnisse. Alpine Forschungsstelle Obergurgl Band 4, Innsbruck, pp 117–134

Nicolussi K (1993) Bilddokumente zur Geschichte des Vernagtferners im 17. Jahrhundert. Z Gletscherk Glazialgeol 26(2):97–119

Nicolussi K (2013) Die historischen Vorstöße und Hochstände des Vernagtferners 1600–1850 AD. Z Gletscherk Glazialgeol 45(46):9–23

Nicolussi K, Kaufmann M, Patzelt G, van der Plicht J, Thurner A (2005) Holocene tree-line variability in the Kaunertal, Central Eastern Alps, indicated by dendrochronological analysis of living trees and subfossil logs. Veg Hist Archaeobotany 14:221–234

Oeggl K, Patzelt G, Schäfer D (eds) (1997) Alpine vorzeit in Tirol, Begleitheft zur Ausstellung. Innsbruck, p 65

Patzelt G (1986) Begleitworte zur Karte des Gurgler Ferners 1981. Z Gletscherk Glazialgeol 22(2):163–170

Patzelt G (1994) Former lakes dammed by glaciers and resulting floods in the Oetz Valley. Mt Res Dev 14(4):298–301

Patzelt G (1996) Modellstudie Oetztal—Landschaftsgeschichte im Hochgebirgsraum. Mitt Österr Geographischen Ges 138:53–70

Patzelt G (2010) Einleitung: Das Ötztal—topographische Kennzeichnung. In: Koch E-M, Erschbamer B (eds) Glaziale und periglaziale Lebensräume im Raum Obergurgl. Alpine Forschungsstelle Obergurgl Band 1, Innsbruck, pp 9–11

Patzelt G (2013) Das Vorfeld des Vernagtferners und seine Umgebung. Begleitworte zur Karte 1:10.000. Z Gletscherk Glazialgeol 45/46:25–39

Patzelt G (2019) Der Gurgler Ferner und seine nacheiszeitlichen Schwankungen. In: Patzelt G (ed) Gletscher—Klimazeugen von der Eiszeit bis zur Gegenwart. Hatje Cantz Verlag, Berlin, pp 123–170

Patzelt G, Bortenschlager S (1973) Die postglazialen Gletscher- und Klimaschwankungen in der Venedigergruppe (Hohe Tauern, Ostalpen). Z Geomorphol N.F Suppl-Band 16:25–72

Patzelt G, mit Beiträgen von Bortenschlager S, Poscher G (1996) Exkursion A1 Tirol: Ötztal—Inntal. Exkursionsführer DEUQUA 1996 Gmunden, Institut für Hochgebirgsforschung, Innsbruck, p 23

Patzelt G, Kofler W, Wahlmüller B (1997) Die Ötztalstudie—Entwicklung der Landnutzung. In: Oeggl K, Patzelt G, Schäfer D (eds) Alpine vorzeit in Tirol, Begleitheft zur Ausstellung. Innsbruck, pp 46–62

Patzelt G, Stötter J, Nicolussi K, Wastl M (2007) Indicators of past and present climate changes—excursion from Obergurgl to Gurgler Ferner. In: Kellerer-Pirklbauer A, Keiler M, Embleton-Hamann C, Stötter J (eds) Geomorphology for the future, joint-meeting of the commission on geomorphology of the Austrian geographical society and the IAG working group on geomorphology and global environmental change 2007, Conference Proceedings, Innsbruck, pp 13–22

Pillewitzer W (1938) Photogrammetrische Gletscheruntersuchungen im Sommer 1938. Z Erdkunde 9(19):367–372

Province of the Tyrol (2015) Regionsprofil Ötztal, Planungsverband 13, Statistik 2015. https://www.tirol.gv.at/statistik-budget/statistik/regionsprofile/plv13/ (20.04.2016)

Sailer R, Bollmann E, Hoinkes S, Rieg L, Sproß M, Stötter J (2012) Quantification of geomorphodynamics in glaciated and recently deglaciated terrain based on airborne laser scanning data. Geogr Ann Ser B 94:17–32

Schallhart N, Erschbamer B (eds) (2015) Forschung am Blockgletscher—Methoden und Ergebnisse. Alpine Forschungsstelle Obergurgl Band 4, Innsbruck, p 198

Schneider B, Schneider H (2001) Zur 60jährigen Messreihe der kurzfristigen Geschwindigkeitsschwankungen am Blockgletscher im Äußeren Hochebenkar, Ötztaler Alpen, Tirol. Z Gletscherk Glazialgeol 37(1):1–33

Schwienbacher E (2004) Populationsbiologische Studien an frühen Sukzessionsarten im Gletschervorfeld des Rotmoosferners (Ötztal, Tirol). Verbreitung und Standortsansprüche von *Artemisia genipi* und *A. mutellina*, Populationsstruktur von *A. genipi*. Diplomarbeit an der Universität Innsbruck

Span N, Kuhn M, Schneider H (1997) 100 years of ice dynamics of Hintereisferner, Central Alps, Austria, 1894–1994. Ann Glaciol 24:297–302

van Asselen S, Seijmonsbergen AC (2006) Expert-driven semi-automated geomorphological mapping for a mountainous area using a laser DTM. Geomorphology 78(3/4):309–320

Vietoris L (1972) Über den Blockgletscher des Äußeren Hochebenkars. Z Gletscherk Glazialgeol 8(1/2):169–188

Vorren K-D, Morkved B, Bortenschlager S (1993) Human impact on the Holocene forest line in the Central Alps. Veg Hist Archaeobotany 2:145–156

Zanesco A (2012) Zum archäologischen Fundbild in Obergurgl. In: Koch E-M, Erschbamer B (eds) An den Grenzen des Waldes und der menschlichen Siedlung. Alpine Forschungsstelle Obergurgl Band 2, Innsbruck, pp 75–98

Thomas Geist is an expert for earth observation at the Austrian Research Promotion Agency (FFG) and holds a Ph.D. degree in physical geography from the University of Innsbruck. He is interested in geomorphology, landscape evolution and climate and environmental change in (formerly) glaciated areas. His activities include adult education projects and teacher training, mainly in the field.

Clemens Geitner is associate professor at the Institute of Geography at the University Innsbruck. His research focuses on soil science, geomorphology, landscape ecology, landscape evolution, and geoarchaeology, mainly in the Alps. It combines field work, remote sensing, laboratory analysis and modelling approaches. His interests include interdisciplinary approaches to investigate human-environmental interactions with respect to sustainable land use and resource preservation.

Kati Heinrich is a cartographer and technician at the Institute for Interdisciplinary Mountain Research of the Austrian Academy of Sciences in Innsbruck, Austria.

The Moraine at Trins and the Alpine Lateglacial

23

Hanns Kerschner, Susan Ivy-Ochs, and Christian Schlüchter

Abstract

Since the early twentieth century, a prominent end moraine at the village of Trins in the Gschnitz Valley in the Tyrolean Alps serves as type locality for an alpine-wide lateglacial phase of glacier advances. Dating of the moraine with the terrestrial cosmogenic radionuclide ^{10}Be gives a stabilization age for the moraine in the range of 16.6 ± 1.4 ka. As the glacier at Trins and comparable glaciers elsewhere in the Alps advanced over ice-free terrain, the "Gschnitz Stadial" is the first glacier advance in the Eastern Alps, which is independent of the large system of dendritic valley glaciers and local ice caps of the Last Glacial Maximum. The large end moraine and lateral moraines at the type locality and at comparable glaciers in other valleys indicate that the glacier geometry remained stable for at least several decades or even a few centuries. At the type locality, the equilibrium line altitude was about 700 m lower than during the Little Ice Age. From the morphological and glaciological characteristics of the glacier and its moraines, a cold and dry climate can be inferred. Summers were 8–10 °C colder compared to the twentieth-century mean, and annual precipitation sums were reduced to about one-third of present-day amounts.

Keywords

Alpine Lateglacial • Gschnitz stadial • Glacier-climate modelling • Surface exposure dating

H. Kerschner (✉)
Institute of Geography, University of Innsbruck, Innrain 52, 6020 Innsbruck, Austria
e-mail: Hanns.Kerschner@uibk.ac.at

S. Ivy-Ochs
Laboratory of Ion Beam Physics, ETH Zürich, HPK H 25, 8093 Zürich, Switzerland

C. Schlüchter
Institut für Geologie, Universität Bern, Balzerstrasse 1-3, 3012 Bern, Switzerland

23.1 Introduction

Studying glacier advances of the Alpine Lateglacial gives valuable insight into the climatic conditions during the most recent period of worldwide deglaciation (Termination 1), as glaciers react immediately to changing climatic conditions by changes in their mass balance and surface geometry. Prolonged unidirectional changes lead finally to phases of glacier advance and the formation of landforms of glacial deposition like moraines or, on the other hand, to phases of glacier recession. On a longer timescale, the behaviour of glaciers is predominantly driven by climate, but the influence of climate is modified by the physical properties of ice dynamics. Hence, former glacier extents provide a climatic information filtered on a timescale of decades to perhaps a few centuries, depending on the size of the glacier and on its dynamical characteristics.

Here, we will discuss such a prominent phase of glacier advance during a rather early phase of the Alpine Lateglacial. It is largely based on papers by Ivy-Ochs et al. (2006a) and Kerschner et al. (2014), where the more technical aspects and suggestions for field trips can also be found.

23.2 The Stadials

The general ice decay after the Last Glacial Maximum (LGM) around ~ 26.5–21 ka (Monegato et al. 2007; Ivy-Ochs et al. 2008) left a highly dynamic landscape. It was characterized by the availability of large amounts of loose material from moraines and recently ice-free and unvegetated slopes with periglacial activity, large amounts of meltwater and gravitational mass movements on the valley flanks. As a consequence, considerable river aggradation, subsequent erosion and resedimentation on already ice-free valley floors, along stagnating ice bodies, into more or less short-lived lakes and in niches at the mouth of lateral valleys were typical for this period of "Early Lateglacial Ice Decay" (van Husen

C. Embleton-Hamann (ed.), *Landscapes and Landforms of Austria*, World Geomorphological Landscapes,
https://doi.org/10.1007/978-3-030-92815-5_23

1987; Reitner 2007, 2013) in the Eastern Alps. During this period, early large-scale mass movements started to develop (van Husen et al. 2007). Some glacier advances occurred, which, however, were not always caused by climatic reversals to a more glacier-friendly climate, but rather had their origin in the changing topography of the large downwasting glaciers and changing ice dynamics (Reitner 2005). In valleys with already independent local glaciers, they advanced to a climatically controlled "Lateglacial maximum position" (Reitner and Gruber 2014). This phase, which characterizes the onset of Termination 1 (Denton et al. 2010) in the Eastern Alps, was centred around 19 ka, when elsewhere in Europe and North America large ice sheets still existed (van Husen 2004; Klasen et al. 2007; Reitner 2007). It came to an end before 18 ka (van Husen 1977; Draxler 1987).

After a first stabilization of the landscape, the ice recession or stagnation of the ice extent in the alpine valleys was interrupted by several phases of subsequently smaller glacier advances. Since the days of Penck and Brückner (1901/1909), they are called "stadials" and are usually signified by more or less closely spaced sets of end moraines and lateral moraines. They were used to set up a moraine stratigraphy based on morphological criteria, equilibrium line altitude (ELA) and, moreover, the vertical distance ("ELA depression") between the former altitude of the ELA and a reference altitude like the ELA of the "Little Ice Age" (Gross et al. 1977). Implicitly, it is assumed that marked morphological boundaries represent important stratigraphic boundaries. During the time period since Penck and Brückner (1901/1909), the system became increasingly elaborate and detailed (e.g., Senarclens-Grancy 1958; Heuberger 1966; Mayr and Heuberger 1968; Heuberger 1968; Maisch 1982; Schoeneich 1998; Ivy-Ochs et al. 2006a, b; Heiri et al. 2014; Ivy-Ochs 2015). Some stadials seem to be well defined and established, while the evidence for other stadials is still patchy. A main and hitherto unresolved problem of the system is that no single valley with a complete and well-dated series of moraines is known to the present date.

Following Penck and Brückner (1901/1909) and Mayr and Heuberger (1968), the first documented independent glacier advances were "Bühl" and, considerably smaller, "Steinach". In the Eastern Alps, these classical stadials of "Bühl" and "Steinach" are today seen as part of the "Early Lateglacial Ice Decay" (Reitner 2007), while the first independent glacier advance is the "Gschnitz Stadial", which will be discussed here in more detail. In western Austria and in Switzerland, the main advance was followed by a series of increasingly smaller glacier advances (Clavadel/Senders and Daun: Maisch 1981, 1987; Kerschner and Berktold 1982; Müller 1982; Schoeneich 1998), which are missing farther to the east (Reitner et al. 2016). It lasted until the onset of the Lateglacial Interstadial (Bølling-Allerød Interstadial) at 14.7 ka. The interstadial caused a preliminary end to large glacier advances and was characterized by the rapid spread of tree growth and finally reforestation in areas that were previously covered by glaciers or occupied by tundra habitats. With the onset of the Younger Dryas cold phase (ca. 12.900–11.700 ka) around 12.9 ka, alpine glaciers advanced for the last time far beyond their Holocene extent and built a series of moraines with successively smaller glacier extents, which are summarized under the name of "Egesen" (Heuberger 1966; Ivy-Ochs et al. 2009). The moraines of the final Egesen advance seem to have been deposited around the Younger Dryas/Holocene transition (Ivy-Ochs et al. 2009; Moran et al. 2016). The Alpine Lateglacial, and particularly the Younger Dryas, was also a phase of widespread rock glacier development in areas where the ELA was too high to permit the existence of glaciers, but which were still affected by the presence of (discontinuous) permafrost (Kerschner 1978a; Frauenfelder and Kääb 2000; Reitner 2006). In some places, the first phase of glacier advances was followed by a second, drier phase of rock glacier development in previously glacierized areas (Sailer and Kerschner 1999).

At present, it seems that the sequence of moraines shows a more varied and detailed picture in the western part of Austria and in Switzerland than in the eastern part of the Austrian Alps, where it is much more restricted to the Gschnitz Stadial and the Egesen Stadial (Reitner 2013, 2015; Bichler et al. 2016; Reitner et al. 2016).

Absolute dating with minimum radiocarbon ages in combination with palaeobotanical studies was already on its way in the 1970s and gave a first insight into the temporal structure of the Alpine Lateglacial (Patzelt 1972). Since the late 1990s, surface exposure dating with terrestrial cosmogenic radionuclides (Gosse and Phillips 2001) helped to significantly refine the absolute chronology of lateglacial glacier advances. The method does not depend on the presence of organic material, which is usually missing at localities of interest. It gives absolute ages for the stabilization of the moraines and related landforms, and hence minimum ages for the responsible glacier advances. While surface exposure dating made it possible to constrain the age of moraines and glacier advances with increasing accuracy, it also gave rise to a number of new chronological and geomorphological questions.

23.3 The Moraine Landscape at Trins

In the village of Trins in the Gschnitz Valley, 30 km south of Innsbruck and about 150 km upstream from the LGM end moraines of the Inn Glacier system north of Rosenheim in Germany (Figs. 23.1 and 23.2), a beautifully developed end moraine crosses the valley at an altitude of 1200 m asl (Fig. 23.3). It is so prominent that it was already shown on the first geological map of Tyrol by Pichler (1859). Penck

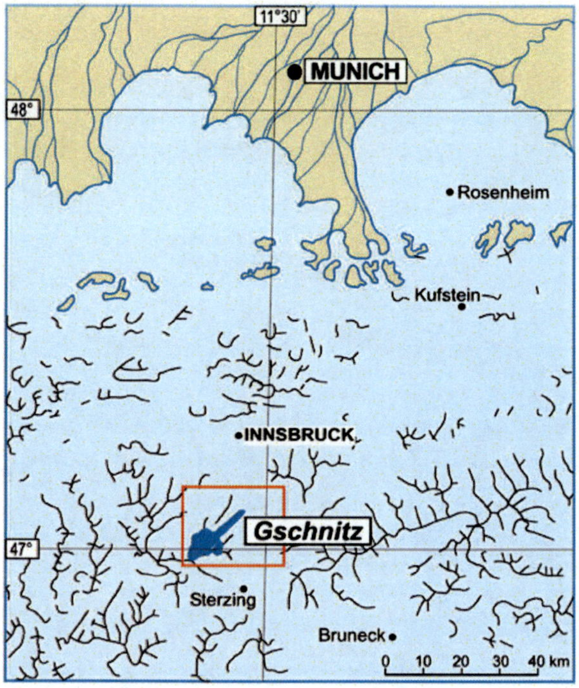

Fig. 23.1 Index map of the Gschnitz glacier. LGM glacier extent after van Husen (1987). Red line indicates outlines of Fig. 23.2

and Brückner (1901/1909) chose it as the type locality for their "Gschnitz Stadial". Since then, various authors have studied the glacial morphology of the Gschnitz Valley and its surroundings, defining and redefining the moraine stratigraphy of the valley or of the Alpine Lateglacial in general.

The valley itself is almost straight, running from the SW to the NE. Crystalline bedrock from the Ötztal-Stubai complex dominates in the headwaters of the valley, while in its middle part metamorphic carbonates (Brennermesozoikum) form the bedrock. In its outermost part, already beyond the glacier end discussed here, Palaeozoic schists prevail (Rockenschaub and Nowotny 2009). In the areas with carbonate and crystalline bedrock, the valley is a classically developed glacial trough with a parabolic cross section (Fig. 23.4). Near the headwaters of the valley, above the Lapponesalm, a 500-m-high step leads to the wide cirques, which formed most of the accumulation area of the Gschnitz glacier. Present-day glaciers, the remnants of Simmingferner, are confined to the highest peaks (Pflerscher Hochjoch, 3164 m; Schneespitze, 3173 m; all place names used in this chapter are entered in Fig. 23.2) in the SW. The regional mean of the Little Ice Age (LIA) equilibrium line altitude in the catchment was at 2630 m, while presently it runs across the highest parts of the glacier or is missing at all.

The moraine complex is built of an arcuate end moraine with lateral moraines on both sides of the valley (Fig. 23.5 and 23.6). At its lowest point, it is dissected and breached by the modern-day river. On the right hand side of the river, it consists of a single ridge which is about 30 m high and 100 m wide. The distal slope is in the order of 30–35° and does not show any signs of periglacial reworking. From the moraine, a small outwash terrace stretches downvalley for about 500 m. On the proximal side, moraine and glacier-transported landslide deposits extend about 100–200 m upvalley. An exposure opened by construction works on the distal side of the moraine showed a diamict of predominantly crystalline clasts and some Mesozoic marbles. Many showed striations and all of them were embedded in a densely packed clayey matrix. On top, the moraine is covered by much less densely packed cover beds with a higher content of coarse silt and sand, from which the modern soil developed (Damm and Terhorst in Kerschner et al. 2014). The sheer size of the end moraine suggests that the glacier geometry remained more or less stable for at least a few decades or perhaps even for a century or more.

On the left hand side of the modern river, the moraine consists of a set of two to three closely spaced subparallel ridges with interspersed kettle holes. On the proximal side, the large amount of sediment forms an undulating and somewhat terrace-like feature with the Krotenweiher dead ice hole. It extends about 1 km upstream and suggests that the lowermost left hand side of the glacier was thickly covered with crystalline landslide debris. The lateral moraines extend almost continuously for three kilometres on both sides of the valley up to an altitude of 1410 m. Together with some remnants above the village of Gschnitz at 1520–1540 m asl, they allow a detailed reconstruction of the glacier tongue for six kilometres (Fig. 23.7). The right lateral moraine in its lower part runs parallel to the hillslope at a distance of a few tens of metres to about 100 m (see Figs. 23.5 and 23.6). This is typical for less well-nourished glaciers in a dry climate, when the lateral component of glacier flow cannot fully compensate the ablation caused by the sensible and long-wave heat flux from the valley side. Higher up, a small river (Fallzambach) intersected the moraine and built a small cone against the former glacier side. Today, the cone ends about 60 m above the valley floor.

Minor lateral moraines on both sides of the valley show that downwasting of the glacier was interrupted by at least one smaller readvance, but the glacier size was not significantly smaller than during the maximum advance.

All over the end moraine and partly also along the lateral moraines, large subangular to subrounded gneiss boulders can be found, most of all on the left hand side of the river. In former times, large boulders were even more abundant. However, those which were mapped by Kerner von Marilaun (1890) in the cultivated areas have been removed since then.

Downvalley from the end moraine at Trins, some lateral moraines and basal till above ice-marginal deposits (Senarclens-Grancy 1958; Mayr and Heuberger 1968;

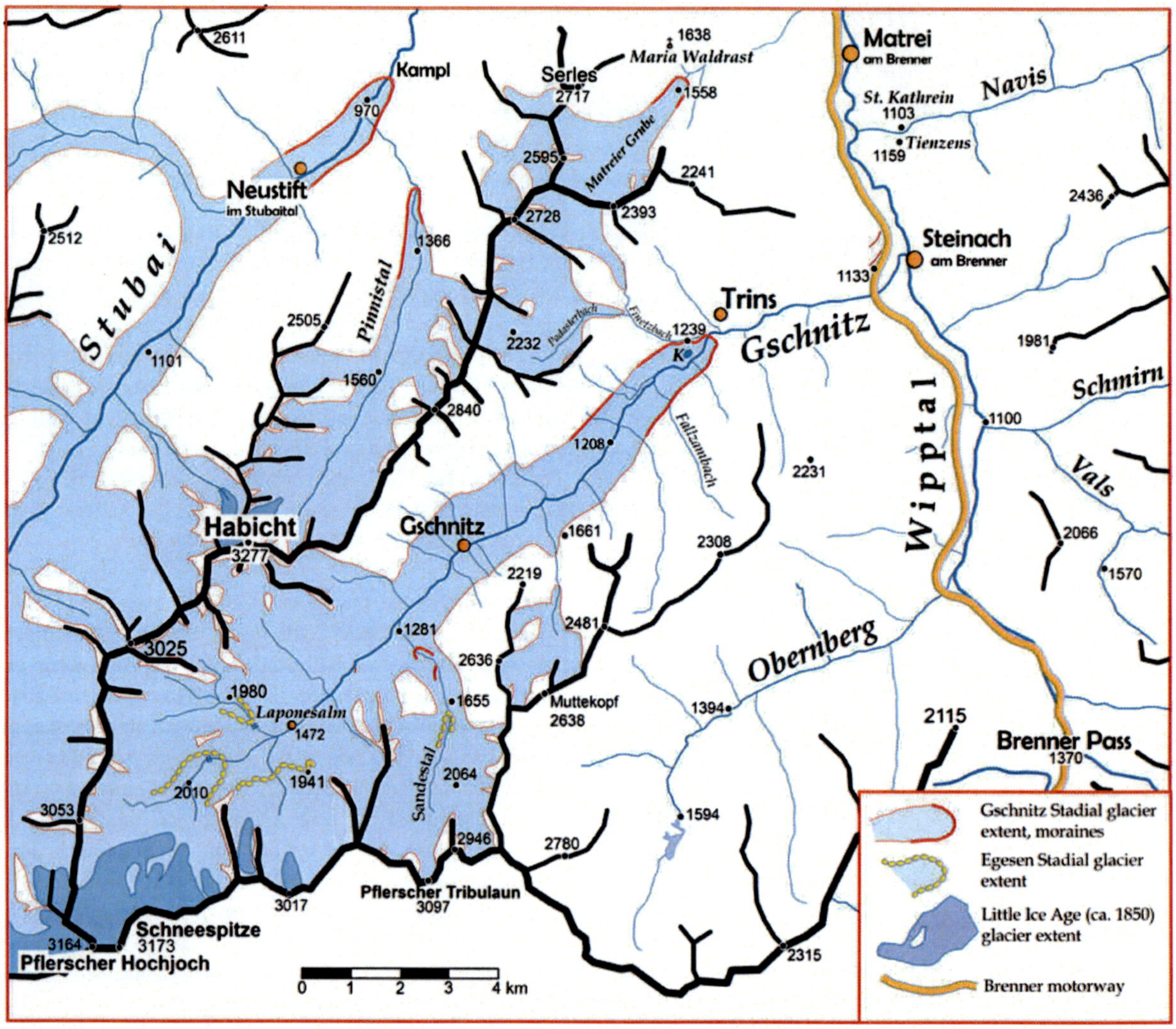

Fig. 23.2 Gschnitz Valley and surroundings. Glacier extents from field mapping and Digital Elevation Model. Gschnitz glacier extents in the catchments of Padasterbach and Finetzbach (west of Trins) and north of Muttekopf are very tentative because of scarce field evidence. Lateral moraines of the former "Steinach Stadial" are north of Pt. 1133 near the town of Steinach

Rockenschaub and Nowotny 2009) can be observed at the junction of the Gschnitz Valley with the main valley (Wipptal) to the SW of the town of Steinach (Fig. 23.2). They were assigned to a "Steinach Stadial" by Senarclens-Grancy (1958). However, associated end moraines and lateral moraines further down in the Wipptal are missing. The general morphological situation at the junctions between the Wipptal and its tributary valleys suggests the presence of a larger ice body in the main valley. Together with fine grained lake sediments in the lower part of the Gschnitz Valley and around the town of Steinach (Mayr and Heuberger 1968; Draxler et al. 2003), the abundant ice-marginal terraces and the dejection cones from the lateral valleys (Figs. 23.8 and 23.9) built along the downwasting glacier in the Wipptal are today seen as part of the Lateglacial Ice Decay phase.

Upvalley from the Trins moraine, a set of end moraines can be found at the mouth of the Sandestal tributary. Moraines of the Egesen Stadial were deposited at the upper end of the step near Lapponesalm and higher up in the cirques. Similar moraines can also be found in the Sandestal with a glacier end at 1630 m asl (Fig. 23.2).

23.4 The Age of the Gschnitz Stadial

Numerous radiocarbon ages can be used as minimum ages for moraines that were classified as Gschnitz based on morphological properties, position in the field relative to other moraine systems and with characteristic ELA depressions in the range of 600 to 700 m relative to the LIA ELA

Fig. 23.3 Gschnitz end moraine (1) at the village of Trins. Lateral moraines (2) can be seen in the centre of the photograph to the left and right. In the background, note parabolic shape of valley and step (3) above Lappones Alm. The lower end of the cloud (3) marks the approximate position of the Younger Dryas glacier snout. View towards SW. *Photo* Hanns Kerschner, Sept. 2003

(Ivy-Ochs et al. 2006a). While earlier authors assumed a Younger Dryas age of the moraine at Trins (summary in Mayr and Heuberger 1968), it became increasingly clear that the Gschnitz Stadial must be much older (Patzelt 1972). At a number of localities, mainly in Switzerland, it can be shown from minimum radiocarbon ages and pollen analysis that higher areas or even former accumulation areas of possible Gschnitz Stadial glaciers became ice free around the onset of the Bølling-Allerød Interstadial (e.g., Welten 1982, see also Ivy-Ochs et al. 2006b). More recently, this was confirmed with surface exposure dating (Dielforder and Hetzel 2014; Hippe et al. 2014).

At the type locality itself, minimum radiocarbon ages from the Krotenweiher peatbog did not give convincing results, as the bottom sediments date from the Early Holocene (Preboreal period) (Bortenschlager 1984; Patzelt und Sarnthein 1995). Because of the prominent position of the moraine within the moraine stratigraphy of the Alpine Lateglacial, it was among the first sites in the Alps which were selected for surface exposure dating with terrestrial cosmogenic radionuclides (^{10}Be, ^{26}Al; Gosse and Phillips 2001, Ivy-Ochs and Kober 2008). The great number of large boulders in apparently stable positions and the presence of quartz veins and big single quartz crystals on the boulders facilitated sampling, which took place in 1996 and 1998. The only possible problem encountered is potential tree growth on top of some boulders, perhaps already since the late Bølling-Allerød Interstadial, and subsequent toppling or spalling of boulders. The position of the dated boulders together with the ages is shown in Fig. 23.5. Five of them are located on the end moraine and two more on the upper part of the right hand lateral moraine. Recent advances in the dating method necessitated a recalculation of the original ages from Ivy-Ochs et al. (2006a) and produced a mean age for moraine stabilization of 16.6 to 16.9 ka $\pm$ 1.4 ka. The dating results of the six samples lie between 18.2 $\pm$ 0.7 and 13.4 $\pm$ 1.3 ka. Considering the youngest age and the oldest age as outliers, the spread of ages is in the order of 2000 years. Apart from other causes, this may indicate a prolonged morphological instability of the moraine due to the melt out of ice cores and post-depositional surface erosion.

This places the Gschnitz Stadial to the beginning of the "Mystery Interval" (Denton et al. 2006), to Greenland

Fig. 23.4 View of the Gschnitz Valley showing parabolic cross section. Lower parts of the bedrock slopes are partially masked by scree slopes. View towards SW. *Photo* Hanns Kerschner, June 2009

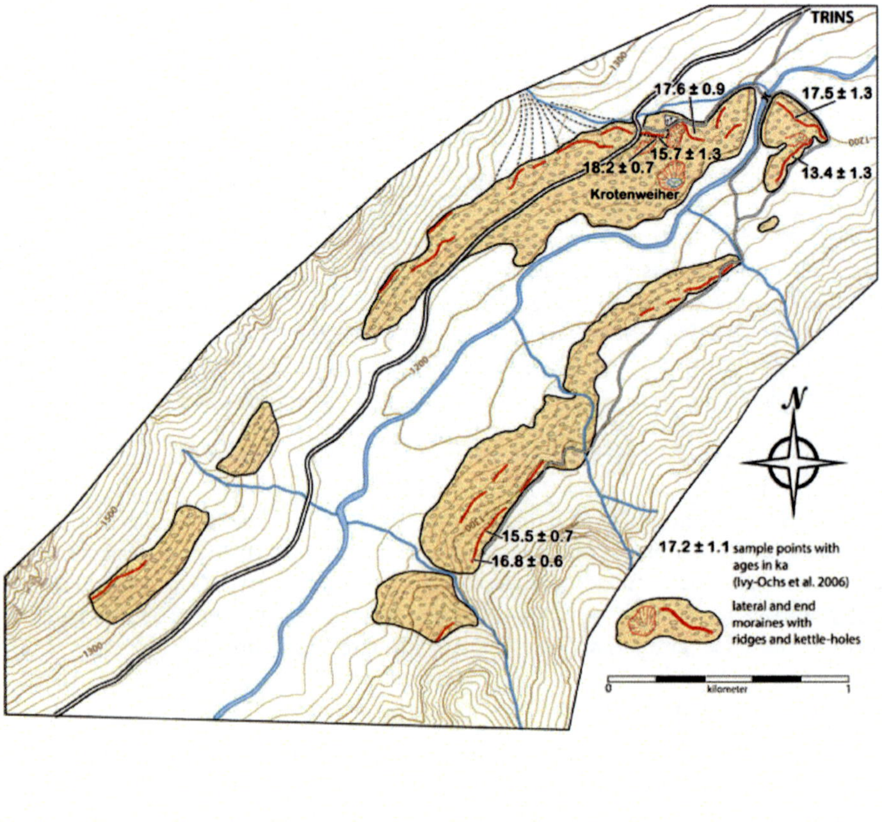

Fig. 23.5 Map of the moraine complex at Trins with ^{10}Be ages (modified from Kerschner et al. 2014). ^{10}Be ages were recalculated from raw data in Ivy-Ochs et al. (2006a) with version 3 of the "online calculators formerly known as the CRONUS-Earth online calculators" (http://hess.ess.washington.edu/math/v3/v3_age_in.html), see also Borchers et al. (2016)

Fig. 23.6 Ensemble view of the end and lateral moraines of the Gschnitz Stadial type locality. Due to their forest cover, the Gschnitz moraine ridges stand out prominently in this view. Remnants of modern glaciers are visible in the background. View towards SW. *Photo* Martin Kerschner, 14 Aug 2016

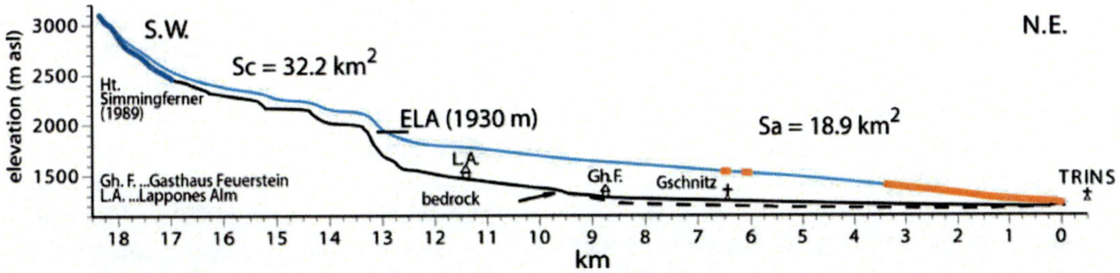

Fig. 23.7 Long profile of the Gschnitz Valley with the reconstructed Gschnitz Glacier surface entered in a blue/orange line (modified from Ivy-Ochs et al. 2006a). Orange markings indicate lateral moraines; Sc: surface of accumulation area; Sa: surface of ablation area; Gasthof Feuerstein is situated at Pt. 1281 in Fig. 23.2. Ht. Simmingferner is the modern glacier at the upper end of the valley

Stadial 2.1a (Rasmussen et al. 2014) and to the Heinrich 1 ice rafting event (Hemming 2004) in the North Atlantic ocean, which, according to Andrews and Voelker (2018), covered a period of approximately 1000 years from 17.7 ± 0.14 to 16.7 ± 0.2 ka. It led to a severe climatic downturn in the adjacent continental areas of Europe and possibly beyond. A similar age is reported for a distinct climate deterioration in the Tagliamento Basin in northern Italy, the Ragogna oscillation (Monegato et al. 2007). As the glacier advance occurred about 2 ka before the onset of the Bølling-Allerød Interstadial, there is time enough left for several further glacier advances before rapid warming at the onset of the Lateglacial Interstadial at 14.7 ka put a preliminary end to large glacier advances in the Alps.

23.5 Glaciological and Climatological Characteristics

The glacier was about 18 km long and had a surface area of 51 km^2. It was characterized by a long and rather flat tongue, which was separated from a wide accumulation basin by a

Fig. 23.8 Ice-marginal terrace at hamlet of Tienzens (between Matrei and Steinach) from the Lateglacial Ice Decay period in the central part of the northern Wipptal. The church of St. Kathrein at the mouth of Navis Valley (in the very left of the photograph) is on top of an end moraine (Heissel 1932; Magiera 2003). View towards SE. *Photo* Hanns Kerschner, 14 Aug 2016

step upvalley from today's Lappones Alm (Fig. 23.7). From the lateral moraines, the surface slope of the glacier tounge and the ice thickness can be determined. This allows the calculation of the basal shear stress along the glacier tongue, which depends on surface slope and ice thickness. It ranged from about 20 kPa near the front to 70 to 80 kPa farther upglacier. A basal shear stress in that range is typical for a glacier in a rather dry climatic environment. Present-day alpine glaciers usually show much higher basal shear stresses in the range of 100 up to 150 kPa. With a shear stress of 80 kPa, the remaining tongue between the present-day village of Gschnitz and the step at Lappones Alm could be reconstructed, using the theoretical long profile of a glacier tongue (Nye 1952; Kerschner 1978b). In the accumulation area, the ice thickness was also calculated with a shear stress of 80 kPa. These data allow to draw a contour map of the glacier surface. Its accuracy should be similar to that of a good glacier map from the early twentieth century. The glacier map serves as the basis for the calculation of the equilibrium line altitude (ELA) and for the modelling of simple glacier flow scenarios.

Calculated with the accumulation area method (Gross et al. 1977) and an accumulation area ratio of 0.63, the Gschnitz Stadial ELA is 1930 m asl, 700 m below the average Little Ice Age ELA. Other reasonable estimates of the Gschnitz ELA range between 1850 m and somewhat less than 2000 m.

Past glacier extents always provide mixed climatic information. In any case, a glacier-friendly climate is characterized by a positive mass balance due to a reduction of ablation, an increase in accumulation or a combination of both. While an increase in accumulation is always caused by more humid conditions and hence an increase of precipitation, ablation may be reduced by a number of factors, including a lowering of the sensitive heat flux due to lower summer temperatures, a reduction of the shortwave radiation balance by increasing cloud cover or by an increase in surface albedo, a reduction of the long-wave atmospheric radiation due to lower air temperatures, an increase of energy-intensive evaporation from the glacier surface, by a significantly shorter ablation season, a significant debris cover on the glacier surface or, most likely, a combination of these factors. However, almost all of these changes in energy fluxes can rather successfully be parameterized by changes in summer temperature (e.g., Ohmura et al. 1992). The advantage of glacier-derived palaeoclimatic information over biological proxies is its direct relation to commonly used climatic variables like annual precipitation sums and summer temperature, and also a fairly high precision. On the other hand, the glacial record is always discontinuous and filtered on a longer timescale.

Fig. 23.9 Ice-marginal terraces and alluvial cones built along stagnating ice masses in the lowermost part of the Wipptal. Summits of the southernmost chain of the Karwendel mountains in the background. View towards N. *Photo* Hanns Kerschner, May 1988

From the size and morphology of the moraine at Trins, it can be concluded that the glacier tongue has remained stable for decades to centuries. The nature of the Gschnitz Glacier as a stationary glacier has two advantages for modelling palaeoclimate: (i) it allows the use of glacier flow models and glacier-climate models that are based on steady-state conditions, and (ii) it can be assumed that ablation equalled accumulation. Thus, information on the net ablation of the Gschnitz Glacier from a glacier flow model would allow calculating net accumulation and in turn precipitation. Precipitation can then be used as input data for estimating temperature in a glacier-climate model.

Ablation was estimated with the help of a simple glacier flow model, assuming steady-state conditions (already Paschinger 1952; Kull and Grosjean 2000), for which the straight Gschnitz Valley with its nearly perfect parabolic cross section (Fig. 23.4) is ideally suited. The glacier flow model calculates the ice velocity from ice deformation and basal sliding. While the contribution of ice deformation to the overall velocity is rather small and mainly determined by the small shear stress, basal sliding is more difficult to assess. Therefore, several scenarios have been calculated with reasonable assumptions on the contribution of basal sliding to the velocity of the glacier tongue (Ivy-Ochs et al. 2006a).

Based on the reconstructed glacier geometry (glacier map, see above), the model calculates ice velocity and ice discharge across given cross sections between the ELA and the end moraine at Trins. The difference in ice flux between two cross sections, divided by the area of the enclosed glacier segment, is the net ablation for this particular segment. With each downvalley following cross section net ablation increases from zero at the ELA to a maximum at the glacier snout. The resulting data for all glacier segments allow calculating the change of net ablation with altitude, the ablation gradient. It is a useful variable for the climatic characterization of glaciers (Kuhn 1984). For the most reasonable scenario with a basal sliding rate of 70% of the ice velocity, it is calculated as -2.3 kg m^{-2} m^{-1} (Ivy-Ochs et al. 2006a), similar to that of present-day White Glacier on Axel-Heiberg-Island in the Canadian Arctic (Cogley et al.

1996). Finally, based on the reconstructed net ablation and the ELA (=1930 m asl, see above) of the Gschnitz Glacier, precipitation at 1930 m asl was calculated to be about 500 mm per year, which is one-third of present-day precipitation at this altitude. Lower and upper extremes for estimates of annual precipitation range around 300 and 900 mm a^{-1}.

To calculate summer temperature from precipitation, the glacier-climate model of Ohmura et al. (1992) was used. It relates summer temperature (June–August) and annual precipitation totals at the ELA of a worldwide sample of 70 glaciers in the form of a nonlinear regression equation. With precipitation as input data, it can be solved for summer temperature. During the formation of the Gschnitz Stadial moraine, summer temperature at the ELA was slightly negative, which is typical for an at least partly cold glacier. It was about 8–10 °C lower than the twentieth-century mean. Hence, at the end moraine at Trins, summer temperature should have been in a range of 4–6 °C. In the Inn Valley, it was between about 8 °C and 10 °C, just warm enough to maintain a shrub tundra. All in all, all qualitative and quantitative information derived from the glacier and the moraines shows that climate during the Gschnitz Stadial was considerably cold and very dry for a prolonged period.

In comparison, Younger Dryas precipitation in this area was about 10% less than today, and summer temperature was 3.5–4 °C lower than modern values (Kerschner et al. 2000; Kerschner and Ivy-Ochs 2008).

23.6 Gschnitz Moraines Elsewhere

Already Penck and Brückner (1901/09) found numerous moraines from glaciers showing the characteristics of the Gschnitz type locality at Trins. Unfortunately, they determined the "snowline depression" (ELA lowering) at a different locality as 600 m against "present" (i.e., early 1900s). This gave rise to considerable terminological confusion in later years (Ivy-Ochs et al. 2006a; Kerschner 1986), as an ELA lowering of 600 m against "present" is not sufficient for the glacier, which built the moraine at Trins. However, it fits to a smaller glacier extent, which is also documented by moraines at various places. Nevertheless, moraines which can be classified as "Gschnitz" are found at numerous localities all over the Alps (e.g., Patzelt 1975; van Husen 1977; Maisch 1981, 1982, 1987; Kerschner and Berktold 1982; Lieb 1987; Schoeneich 1998; Coutterand and Nicoud 2005; Reitner 2006, 2013, 2015; Federici et al. 2011; Reitner et al. 2016; Wirsig et al. 2016; Fabbri et al. 2018; Hofmann 2018; Gschwentner et al. 2020).

In the neighbouring valleys (Fig. 23.2), remnants of possible Gschnitz Stadial moraines can be mapped in the Stubaital near Neustift (Fig. 23.10; Kerner von Marilaun 1890; Mayr and Heuberger 1968), at the mouth of Pinnistal (Fig. 23.11) and in the Matreier Grube cirque near the monastery of Maria Waldrast. In the Obernberg Valley, a possible Gschnitz Stadial glacier extent is masked by an early Holocene landslide (Ostermann et al. 2012).

Fig. 23.10 Remnants of Gschnitz Stadial lateral moraines and ice-marginal terraces at Neustift im Stubaital. The moraines and related features are wide enough to accommodate farmhouses. Modern glaciers can be seen in the background to the very right. View towards S. *Photo* Hanns Kerschner, 13 Aug 2018

Fig. 23.11 View of the mouth of Pinnistal (to the left of Fig. 23.10). Mountains in the background are mainly built of Mesozoic carbonates. The arrow points to the position of the Gschnitz Stadial glacier end. *Photo* Hanns Kerschner, 13 Aug 2018

Large dendritic glaciers were still widespread in the Alps during the Gschnitz Stadial, especially in the highest parts of Switzerland and western Austria, for example in the upper Inn Valley and its tributaries, in the catchments of Vorderrhein and Hinterrhein, in the Reuss Valley, the Aare catchment or in the Rhone catchment. For various reasons, the field evidence in areas with very large glaciers is sparse at best, while smaller glaciers up to about one hundred square kilometres and cirque glaciers frequently developed large moraines. They are best preserved at altitudes lower than about 1500 m asl, while at higher altitudes they show a considerable post-depositional overprint by solifluction. Recently available LIDAR hillshades show that many smaller glaciers, which may be attributed to the Gschnitz Stadial, must have been partly covered by rockfall debris. Hummocky landscapes with an irregular pattern of small hills and depressions are frequently preserved within the lateral and end moraines, which grade into rock glaciers in a number of places (Steinemann et al. 2020).

However, as far as the assignment of undated moraines to the Gschnitz Stadial is concerned, there are considerable uncertainties involved. The calculation of ELAs of large dendritic glaciers with a complicated topography is often technically complex and may also lead to unreliable or not representative results, especially when the accumulation basin of the glacier extended across valleys with widely differing precipitation regimes. The same is true for the calculation of the reference elevation of the LIA ELA. All too often large areas in the accumulation basins of Gschnitz stadial glaciers were not covered by glaciers during the Holocene, and hence, LIA ELAs are missing. In such a case, numerical values for ELA depressions are only guesses. Where somewhat reliable field evidence is available, ELA depressions along the northern slope of the Alps may have been in the order of −800 m. In the driest part of the central Alps, −700 m similar to the type locality should be a reasonable value. In the south, closer to the Mediterranean Sea, the ELA rise from the Gschnitz Stadial to the LIA may have even been in the order of 1000 m (Kerschner and Ivy-Ochs 2008). If such a large value can be substantiated during future research, glaciers of considerable extent may have existed in the Southern Alps (Federici et al. 2011).

Deglaciation prior to and immediately after the Gschnitz Stadial glacier advance also led to the first development of rock glaciers down to altitudes of 1200 m asl, which formed in ice-free cirques as landforms characteristic for discontinuous alpine permafrost (Lieb 1987; Reitner 2006; Kellerer-Pirklbauer et al. 2012; Steinemann et al. 2020).

23.7 Conclusions

The Gschnitz Stadial was a significant multi-decadal to multi-century phase of glacier advance during the early Alpine Lateglacial. Wherever moraines are preserved, they are usually large and well developed. In many places, glacier tongues were partly debris covered, adding to the large amounts of sediment moved and finally deposited by those glaciers. In the central Tyrolean Alps, an ELA lowering of 700 m relative to the LIA is typical. Further to the north, the ELA may have been 800 m lower than the LIA, while in the south of the Eastern Alps the ELA lowering may have been even in the order of 1000 m. Results from a glacier flow model show that climate during the Gschnitz Stadial was rather dry and cold. Annual precipitation sums were in the order of one-third of modern amounts in the central Eastern Alps and summer temperatures were 8–10 °C lower than during the twentieth century.

Surface exposure dating with terrestrial cosmogenic radionuclides gives an age for moraine stabilization at the type locality in the Tyrolean Alps of about 16.6 to 16.9 ka. This supports earlier minimum ages from radiocarbon dating, which pointed to an age clearly older than the onset of the Lateglacial Interstadial (Bølling—Allerød) at 14.7 ka. It is also similar to the age of the Ragogna cold phase in the Tagliamento Basin in northern Italy. The ages suggest that the climatic downturn caused by the Heinrich 1 ice rafting event in the North Atlantic Ocean caused the glacier advance in the Alps.

References

Andrews JT, Voelker AHL (2018) "Heinrich events" (&sediments): a history of terminology and recommendations for future usage. Quatern Sci Rev 187:31–40

Bichler MG, Reindl M, Reitner J, Drescher-Schneider R, Wirsig C, Christl M, Hajdas I, Ivy-Ochs S (2016) Landslide deposits as stratigraphical markers for a sequence based glacial stratigraphy: a case study of a Younger Dryas system in the Eastern Alps. Boreas 45:537–551

Borchers B, Marrero S, Balco G, Caffee M, Goehring B, Lifton N, Nishizumi K, Phillips F, Schaefer J, Stone J (2016) Geological calibration of spallation production rates in the CRONUS-Earth project. Quat Geochronol 31:188–198

Bortenschlager S (1984) Beiträge zur Vegetationsgeschichte Tirols I. Inneres Ötztal und unteres Inntal. Ber des Naturwiss-Med Ver Innsbruck 71:19–56

Cogley JG, Adams WP, Ecclestone MA, Jung-Rothenhäusler F, Omanney CSL (1996) Mass balance of White Glacier, Axel Heiberg Island, N.W.T., Canada, 1960–91. J Glaciol 42:548–563

Coutterand S, Nicoud G (2005) Les stades de retrait du glacier de l'Arve entre le verrou de Cluses et l'ombilic de Chamonix au cours du tardiglaciaire (Vallée de l'Arve, Haute Savoie). Quaternaire 16:85–94

Denton GH, Broecker WS, Alley RB (2006) The mystery interval 17.5 to 14.5 kyrs ago. PAGES News 14:14–16

Denton GH, Anderson RF, Toggweiler JR, Edwards RL Schaefer JM, Putnam AE (2010) The Last glacial termination. Science 328:1652–1656

Dielforder A, Hetzel R (2014) The deglaciation history of the Simplon region (southern Swiss Alps) constrained by ^{10}Be exposure dating of ice-moulded bedrock surfaces. Quatern Sci Rev 84:26–38

Draxler I (1987) Zur Vegetationsgeschichte und Stratigraphie des Würmspätglazials des Traungletschergebietes. In: van Husen D (ed.) Das Gebiet des Traungletschers, O.Ö. Eine Typregion des Würm-Glazials. (Mitteilungen der Kommission für Quartärforschung der Österreichischen Akademie der Wissenschaften 7), pp 19–35

Draxler I, Ottner M, Rockenschaub M, Wimmer-Frey I (2003) Seetone im äußeren Gschnitztal. In: Rockenschaub M (ed) Geologische Bundesanstalt, Arbeitstagung 2003, pp 191–198

Fabbri SC, Buechi MW, Horstmeyer H, Hilbe M, Hübscher C, Schmelzbach C, Weiss B, Anselmetti FS (2018) A subaquatic moraine complex in overdeepened Lake Thun (Switzerland) unravelling the deglaciation history oft the Aare Glacier. Quatern Sci Rev 187:62–79

Federici PR, Granger DE, Ribolini A, Spagnolo M, Pappalardo M, Cyr AJ (2011) Last glacial maximum and the Gschnitz stadial in the maritime Alps according to ^{10}Be cosmogenic dating. Boreas 41:277–291

Frauenfelder R, Kääb A (2000) Towards a palaeoclimatic model of rock-glacier formation in the Swiss Alps. Ann Glaciol 31:281–286

Gosse JC, Phillips FM (2001) Terrestrial in situ cosmogenic nuclides: theory and application. Quatern Sci Rev 20:1475–1560

Gross G, Kerschner H, Patzelt G (1977) Methodische Untersuchungen über die Schneegrenze in alpinen Gletschergebieten. Z Gletscherk Glazialgeol 7:223–251

Gschwentner Ph, Kerschner H, Spötl Ch (2020) Lateglacial ice advance in the Kellerjoch region near Schwaz (Tyrol, Eastern Alps). Austrian J Earth Sci 113:211–227

Heiri O, Koinig KA, Spötl C, Barrett S, Brauer A, Drescher-Schneider R, Gaar D, Ivy-Ochs S, Kerschner H, Luetscher M, Moran A, Nicolussi K, Preusser F, Schmidt R, Schoeneich P, Schwörer C, Sprafke T, Terhorst B, Tinner W (2014) Palaeoclimate records 60–8 ka in the Austrian and Swiss Alps and their forelands. Quatern Sci Rev 106:186–205

Heissel W (1932) Quartärgeologie des Silltales. Jahrb Geol Bundesanst 82:429–468

Hemming SE (2004) Heinrich events: Massive late pleistocene detritus layers of the North Atlantic and their global climate impact. Rev Geophys 42 RG1005. https://doi.org/10.1029/2003RG000128

Heuberger H (1966) Gletschergeschichtliche Untersuchungen in den Zentralalpen zwischen Sellrain und Ötztal. Wissenschaftliche Alpenvereinshefte 20. Wagner, Innsbruck

Heuberger H (1968) Die Alpengletscher im Spät- und Postglazial. Eiszeit und Gegenw 19:270–275

Hippe K, Ivy-Ochs S, Kober F, Zasadni J, Wieler R, Wacker L, Kubik PW, Schlüchter C (2014) Chronology of Lateglacial ice flow reorganization and deglaciation in the Gotthard Pass area, Central Swiss Alps, based on cosmogenic ^{10}Be and in situ ^{14}C. Quat Geochronol 19:14–26

Hofmann FM (2018) Glacial history of the upper Drac Blanc catchment (Écrins massif, French Alps). Master thesis, Dept. of Physical Geography and Ecosystem Science, Lund University, 62 p. + Appendix

Ivy-Ochs S (2015) Glacier variations in the European Alps at the end of the last glaciation. Cuad de Invest Geográfica 41:295–315

Ivy-Ochs S, Kober F (2008) Surface exposure dating with cosmogenic nuclides. Eiszeit und Gegenw Quat Sci J 57:179–209

Ivy-Ochs S, Kerschner H, Kubik PW, Schlüchter C (2006a) Glacier response in the European Alps to Heinrich event 1 cooling: the Gschnitz stadial. J Quat Sci 21:115–130

Ivy-Ochs S, Kerschner H, Reuther A, Preusser F, Heine K, Maisch M, Kubik PW, Schlüchter C (2008) Chronology of the last glacial cycle in the European Alps. J Quat Sci 23:559–573

Ivy-Ochs S, Kerschner H, Maisch M, Christl M, Kubik PW, Schlüchter C (2009) Latest Pleistocene and Holocene glacier variations in the European Alps. Quatern Sci Rev 28:2137–2149

Ivy-Ochs S, Kerschner H, Reuther A, Maisch M, Sailer R, Schaefer J, Kubik PW, Synal HA, Schlüchter C (2006b) The timing of glacier advances in the northern European Alps based on surface exposure dating with cosmogenic ^{10}Be, ^{26}Al, ^{36}Cl, and ^{21}Ne. In: Siame LL, Bourlès BL, Brown ET (eds) In situ cosmogenic nuclides and quantification of geological processes. Geological Society of America, Special Paper 415, pp 43–60

Kellerer-Pirklbauer A, Lieb GK, Kleinferchner H (2012) A new rock glacier inventory of the eastern European Alps. Austrian J Earth Sci 105:78–93

Kerner von Marilaun F (1890) Die letzte Vergletscherung der Central-Alpen im Norden des Brenner. Mittheilungen der k.k. Geogr Ges Wien 33:307–332

Kerschner H (1978a) Paleoclimatic inferences from Late Würm rock glaciers, eastern central Alps, western Tyrol, Austria. Arct Alp Res 10:635–644

Kerschner H (1978b) Zur Rekonstruktion eines spätglazialen Gletscherstandes mit Hilfe eines rechnerisch ermittelten Zungenlängsprofiles. Z Gletscherk Glazialgeol 14:119–123

Kerschner H (1986) Zum Sendersstadium im Spätglazial der nördlichen Stubaier Alpen, Tirol. Z Geomorphol Suppl 61:65–76

Kerschner H, Berktold E (1982) Spätglaziale Gletscherstände und Schuttformen im Senderstal, nördliche Stubaier Alpen, Tirol. Z Gletscherk Glazialgeol 17:125–134

Kerschner H, Ivy-Ochs S (2008) Palaeoclimate from glaciers: examples from the Eastern Alps during the Alpine Lateglacial and early Holocene. Global Planet Change 60:58–71

Kerschner H, Kaser G, Sailer R (2000) Alpine Younger Dryas glaciers as paleo-precipitation gauges. Ann Glaciol 31:80–84

Kerschner H, Ivy-Ochs S, Terhorst B, Damm B, Ottner F (2014) The moraine at Trins—type locality of the Gschnitz Stadial/Die Trinser Moräne—Typlokalität für das Gschnitzstadium. In: Kerschner H, Krainer K, Spötl C (eds) From the Foreland to the Central Alps. Field trips to selected sites of Quaternary research in the Tyrolean and Bavarian Alps. Geozon, Berlin

Klasen N, Fiebig M, Preusser F, Reitner JM, Radtke U (2007) Luminescence dating of proglacial sediments from the Eastern Alps. Quatern Int 164/165:21–32

Kuhn M (1984) Mass budget imbalances as criterion for a climatic classification of glaciers. Geogr Ann 66A:229–238

Kull C, Grosjean M (2000) Late Pleistocene climate conditions in the north Chilean Andes drawn from a climate-glacier model. J Glaciol 46:622–632

Lieb GK (1987) Zur spätglazialen Gletscher- und Blockgletschergeschichte im Vergleich zwischen den Hohen und Niederen Tauern. Mitt der Österr Geogr Ges 129:5–27

Magiera J (2003) Quartäre Ablagerungen in einigen Nebentälern des Sill- und des Inntals. In: Rockenschaub M (ed) Geologische Bundesanstalt, Arbeitstagung 2003, pp 199–203

Maisch M (1982) Zur Gletscher- und Klimageschichte des alpinen Spätglazials. Geogr Helv 37:93–104

Maisch M (1987) Zur Gletschergeschichte des alpinen Spätglazials: analyse und interpretation von Schneegrenzdaten. Geogr Helv 42:63–71

Maisch M (1981) Glazialmorphologische und gletschergeschichtliche Untersuchungen im Gebiet zwischen Landwasser- und Albulatal. (Kt. Graubünden, Schweiz). In: Physische Geographie 3. Geographisches Institut der Universität Zürich, Zürich

Mayr F, Heuberger H (1968) Type areas of lateglacial and postglacial deposits in Tyrol, Eastern Alps. In: Richmond GM (ed) Glaciations of the Alps. Series in Earth Sciences 7. University of Colorado, Boulder, pp 143–165

Monegato G, Ravazzi C, Donegana M, Pini R, Calderoni G, Wick L (2007) Evidence of a two-fold glacial advance during the last glacial maximum in the Tagliamento end moraine system (Eastern Alps). Quatern Res 68:284–302

Moran A, Ivy-Ochs S, Schuh M, Christl M, Kerschner H (2016) Evidence of central Alpine glacier advances during the Younger Dryas-early Holocene transition period. Boreas 45:398–410

Müller HN (1982) Zum alpinen Spätglazial: Das Zwischenbergstadium. Z Gletscherk Glazialgeol 17:135–142

Nye JF (1952) A comparison between the theoretical and the measured long profile of the Unteraar glacier. J Glaciol 2:103–107

Ohmura A, Kasser P, Funk M (1992) Climate at the equilibrium line of glaciers. J Glaciol 38:397–411

Ostermann M, Sanders D, Ivy-Ochs S, Alfimov V, Rockenschaub M, Römer A (2012) Early Holocene (8.6 ka) rock avalanche deposits, Obernberg valley (Eastern Alps): Landform interpretation and kinematics of rapid mass movement. Geomorphology 171–172:83–93

Paschinger H (1952) Die spätglazialen Gletscher des Gschnitztales. Z Gletscherk Glazialgeol 2:35–57

Patzelt G, Sarnthein M (1995) Late glacial morainal arc at Trins in the Gschnitz valley/Tyrol—"Krotenweiher" peat bog. In: Schirmer W (ed) Quaternary field trips in central Europe, vol 2, pp 669–670

Patzelt G (1972) Die spätglazialen Stadien und postglazialen Schwankungen von Ostalpengletschern. Ber der Dtschen Botanischen Ges 85:47–57

Patzelt G (1975) Unterinntal—Zillertal—Pinzgau—Kitzbühel. Spät- und postglaziale Landschaftsentwicklung. In: Fliri F, Leidlmair A (eds) Tirol—ein geographischer Exkursionsführer (Innsbrucker Geographische Studien, 2). Geographisches Institut der Universität, Innsbruck, pp 309–329

Penck A, Brückner E (1901/1909) Die Alpen im Eiszeitalter, 3 vol., Tauchnitz, Leipzig

Pichler A (1859) Beiträge zur Geognosie Tirols. Z des Mus Ferdinandeum Innsbruck 3:1–232

Rasmussen SO, Bigler M, Blockley SP, Blunier T, Buchardt SL, Clausen HB, Cvijanovic I, Dahl-Jensen D, Johnsen SJ, Fischer H, Gkinis V, Guillevic M, Hoek WZ, Lowe JJ, Pedro JB, Popp T, Seierstad IK, Steffensen JP, Svensson AM, Vallelonga P, Vinther BM, Walker MJC, Wheatley JJ, Winstrup M (2014) A stratigraphic framework for abrupt climatic changes during the Last Glacial period based on three synchronized Greenland ice-core records: refining and extending the INTIMATE event stratigraphy. Quatern Sci Rev 106:14–28

Reitner JM (2015) Die Landschaftsentwicklung vom ausgehenden Neogen bis ins Quartär. In: Heinisch H, Pestal G, Reitner JM (eds) Erläuterungen zu Blatt 122 Kitzbühel. Geologische Bundesanstalt, Wien, pp 81–99

Reitner JM, Ivy-Ochs S, Drescher-Schneider R, Hajdas I, Linner M (2016) Reconsidering the current stratigraphy of the Alpine Lateglacial: Implications of the sedimentary and morphological record of the Lienz area (Tyrol/Austria). Eiszeitalter und Gegenwart. Quatern Sci J 65:113–144

Reitner JM, Gruber A (2014) Glacial dynamics and large pre-LGM rock-slides in the lower Inn valley and in the Brixen valley. In: Kerschner H, Krainer K, Spötl C (eds) From the foreland to the Central Alps. Field trips to selected sites of quaternary research in the Tyrolean and Bavarian Alps. Geozon, Berlin

Reitner JM (2005) Quartärgeologie und Landschaftsentwicklung im Raum Kitzbühel-St. Johann i.T.-Hopfgarten (Nordtirol) vom Riss bis in das Würm-Spätglazial (MIS 6–2). Dissertation, Universität Wien

Reitner JM (2006) Landschaftsentwicklung vom Neogen bis in das Quartär. In: Schuster R, Pestal G, Reitner JM (eds) Erläuterungen zu Blatt 182 Spittal an der Drau. Geologische Bundesanstalt, Wien

Reitner JM (2007) Glacial dynamics at the beginning of Termination I in the Eastern Alps and their stratigraphic implications. Quatern Int 164/165:64–84

Reitner JM (2013) The effect of climate change during the Lateglacial in the Hohen Tauern. In: 5th Symposium for Research in Protected Areas, Mittersill, pp 653–658, 10–12 June 2013

Rockenschaub M, Nowotny A (2009) Geologische Karte der Republik Österreich 1:50000, Blatt 148 Brenner. Geologische Bundesanstalt, Wien

Sailer R, Kerschner H (1999) Equilibrium line altitudes and rock glaciers in the Ferwall-group (Western Tyrol, Austria) during the younger Dryas cooling event. Ann Glaciol 28:141–145

Schoeneich P (1998) Le retrait glaciaire dans les vallées des Ormonts, de l'Hongrin et de l'Etivaz (Préalpes vaudoises), Institut de Géographie, Université de Lausanne, Lausanne

Senarclens-Grancy W von (1958) Zur Glazialgeologie des Ötztales und seiner Umgebung. Mitt der Geologischen Ges Wien 49:257–313

Steinemann O, Reitner JM, Ivy-Ochs S, Christl M, Synal HA (2020) Tracking rock glacier evolution in the Eastern Alps from the lateglacial to the early holocene. Quatern Sci Rev 241

van Husen D (1977) Zur Fazies und Stratigraphie der jungpleistozänen Ablagerungen im Trauntal (mit quartärgeologischer Karte). Jahrb Geol Bundesanst 120:1–130

van Husen D (1987) Die Ostalpen in den Eiszeiten. With map "Die Ostalpen und ihr Vorland in der letzten Eiszeit (Würm)" 1:500.000. Geol Bundesanst Wien

van Husen D (2004) Quaternary glaciations in Austria. In: Ehlers J, Gibbard PL (eds) Quaternary glaciations—extent and chronology part I: Europe. Elsevier, London

van Husen D, Ivy-Ochs S, Alfimov V (2007) Mechanism and age of lateglacial landslides in the calcareaous Alps: the Almtal landslide, Upper Austria. Austrian J Earth Sci 100:114–126

Welten M (1982) Vegetationsgeschichtliche Untersuchungen in den westlichen Schweizer Alpen: Bern-Wallis. Denkschriften der Schweiz Naturforschenden Ges 95:1–105

Wirsig C, Zasadni J, Ivy-Ochs S, Christl M, Kober F, Schlüchter C (2016) A deglaciation model of the Oberhasli, Switzerland. J Quat Sci 31:46–59

Hanns Kerschner is a retired professor at the Department of Geography of Innsbruck University. Working in many fields of geography, his scientific interests are focused around glacial geomorphology and chronology, and glacier-climate interaction.

Susan Ivy-Ochs is professor at the ETH Zürich and has a special interest in Quaternary geology and geomorphology. Her research focuses on unravelling the various processes that have shaped and are shaping the European Alps. Cosmogenic nuclides are a key tool.

Christian Schlüchter is a professor emeritus at the University of Bern. His professional activities are in glacial geology, ice-age stratigraphy and engineering geology. He has done extensive fieldwork worldwide on the extent of the Last Glaciation and he has supported the early development of advanced dating techniques.

24 The Krimml Waterfalls in the Hohe Tauern National Park

Erich Stocker

Abstract

The Krimml Waterfalls drop from a nearly 400 m high, multiple structured rock step in the Krimml trough valley near the border of the gneiss core of the Subpenninic thrust. They are the most famous of a series of waterfalls, concentrated in the upper end of the Salzach longitudinal valley exemplifying the type of well-preserved consequent waterfall. Holocene incisions in the hard rock bars and the valley step are relatively insignificant while the other two spectacular waterfalls in the neighbouring Sulzbach valleys, the Sulzbach Fall and the Gamseck Fall, are well developed erosional waterfalls. All the waterfalls show high discharge during the summer months giving spectacular waterfall views. Their initiation goes back to varying structural and geomorphic conditions as well as to the process systems which act in their whole drainage basins. Since the beginning of the nineteenth century, the Krimml Waterfalls have attracted nature lovers and alpinists and trails have been generated by the Alpine Club. In 1983, the Krimml Waterfalls were incorporated into the Hohe Tauern National Park. As one of the most visited sites, the waterfall area can also be considered an outstanding geomorphosite.

Keywords

Waterfall • Cascade • Consequent waterfall • Erosional waterfall • Knickpoint • Geomorphosite • Valley step

E. Stocker (✉)
Department of Geography and Geology, University of Salzburg, Hellbrunnerstraße 34, 5020 Salzburg, Austria
e-mail: er.stocker@aon.at

24.1 Introduction

Waterfalls play an important role in constituting natural monuments. The most famous waterfalls of Austria, by their peculiarity, their variety of landforms, their height and their diversity of stages of development, must be also appreciated as geomorphosites. For two centuries, the waterfalls near Krimml in the north-western part of the Venediger Mountains have been well known as a great natural wonder (Fig. 24.1). Although at that time geological and geomorphological research was just beginning, it is surprising that the unusual quality of the landscape scenery of the waterfalls of this area has been nearly neglected in geomorphological studies. Substantial analyses regarding geomorphic details such as geodiversity, classification, development and mechanisms are few.

The waterfalls in the Hohe Tauern, considered only as knickpoints in valley long profiles, have been used primarily to infer the history of the stages of landform development. The more recent focus on classifying landforms as natural peculiarities in terms of protection and preservation has moved waterfalls and gorges, connected with the acting processes and their diversity, in the centre of immediate geomorphological interest.

24.2 Geographical and Geological Setting

The mountain range of the Hohe Tauern constitutes the divide between the tributaries of the Salzach River on the northern side and the Drau River on the southern slope. Waterfalls generally are evenly distributed across the whole range (Fig. 24.2), though the most famous ones are located on the borders of the Großvenediger Mountains to the west of the Hohe Tauern.

The most important geological structure of the Hohe Tauern is the Tauern Window, the largest tectonic window of the Eastern Alps. It extends over a length of 160 km

C. Embleton-Hamann (ed.), *Landscapes and Landforms of Austria*, World Geomorphological Landscapes,
https://doi.org/10.1007/978-3-030-92815-5_24

Fig. 24.1 Krimml Waterfalls Lower Fall, view from the fall base on the edge of the Krimml basin, about 4 km from the mouth of the Krimmler Ache into the Salzach River. The catchment area above the waterfalls is about 110 km^2, the discharge is 5.5 m^3/s on average, during summer between 12 and 16 m^3/s. *Photo* Stocker (1993)

between the saddles of Brenner and Katschberg and permits an insight into the internal tectonic structure which lies below the Austroalpine thrust system. During the Paleogene (50–40 Ma), continental collision and tectonic compression led to a distinctive Austroalpine overthrusting from south to north. The Neogene longitudinal updoming, supported by processes of exhumation, generated a large anticline verging to the north. The Tauern Window reveals the interior structure below the Austroalpine: the basal Subpenninic thrust systems and their multiple covers, the Penninic thrusts and the Matrei-fault-block-zone (Fig. 24.3).

An overview of the whole Tauern Window shows that in the western part, especially in the Venediger Mountains, the Subpenninic thrusts (Venediger thrust system and Eclogite zone) are largely exposed. In particular, the Central Gneisses and the Cambrian Old gneisses are exceptionally widespread, while the covers up to the Mesozoic calcareous and dolomitic rocks are distributed primarily on the northwest border. Moreover, the Subpenninic units nearly adjoin to the northern frame of the tectonic window, which is marked by the Salzach fault line going from west to east. The Penninic covers, superimposed on the Subpenninic base, are widespread also around the gneiss cores of the central and eastern mountain groups of the Hohe Tauern (Geologische Bundesanstalt 2009).

In the course of the Miocene (21–12 Ma), the uncovering of the anticline was initiated by lag thrusts and transverse faults which led to a stretching of the mountain ranges along W-E or NW–SE directed fault lines, described as "lateral tectonic extrusion" by Frisch et al. (2000).

During the Neogene and Quaternary, the important fault lines (e. g., Salzach-Enns fault line, Isel- and Möll fault lines) gave rise to the development of longitudinal valleys which extend subparallel to the strike of the mountain chains. The rivers occupying these longitudinal valleys lie about 2.5 km below the accordant summits of the Hohe

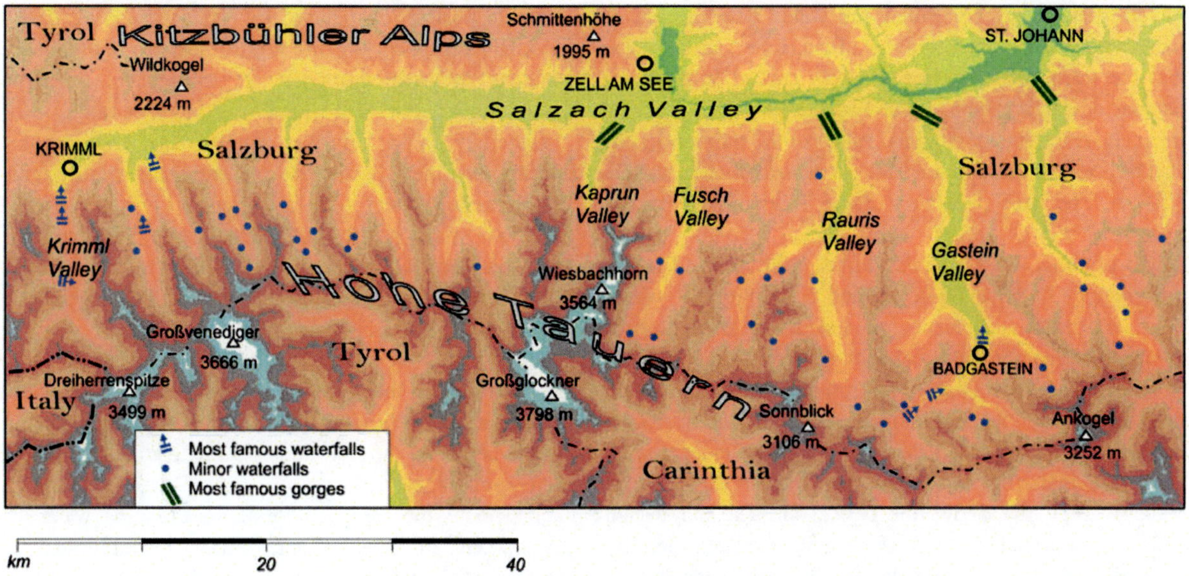

Fig. 24.2 Distribution of waterfalls in the northern Hohe Tauern shows that the most important falls are situated in the upper part of the Salzach River drainage basin. Waterfalls with a smaller discharge occur in nearly all parts of the mountain range. The most famous gorges occur along a strip of resistant limestones which cross the lower courses of the main confluents. Topographical base: Austrian Amap Fly

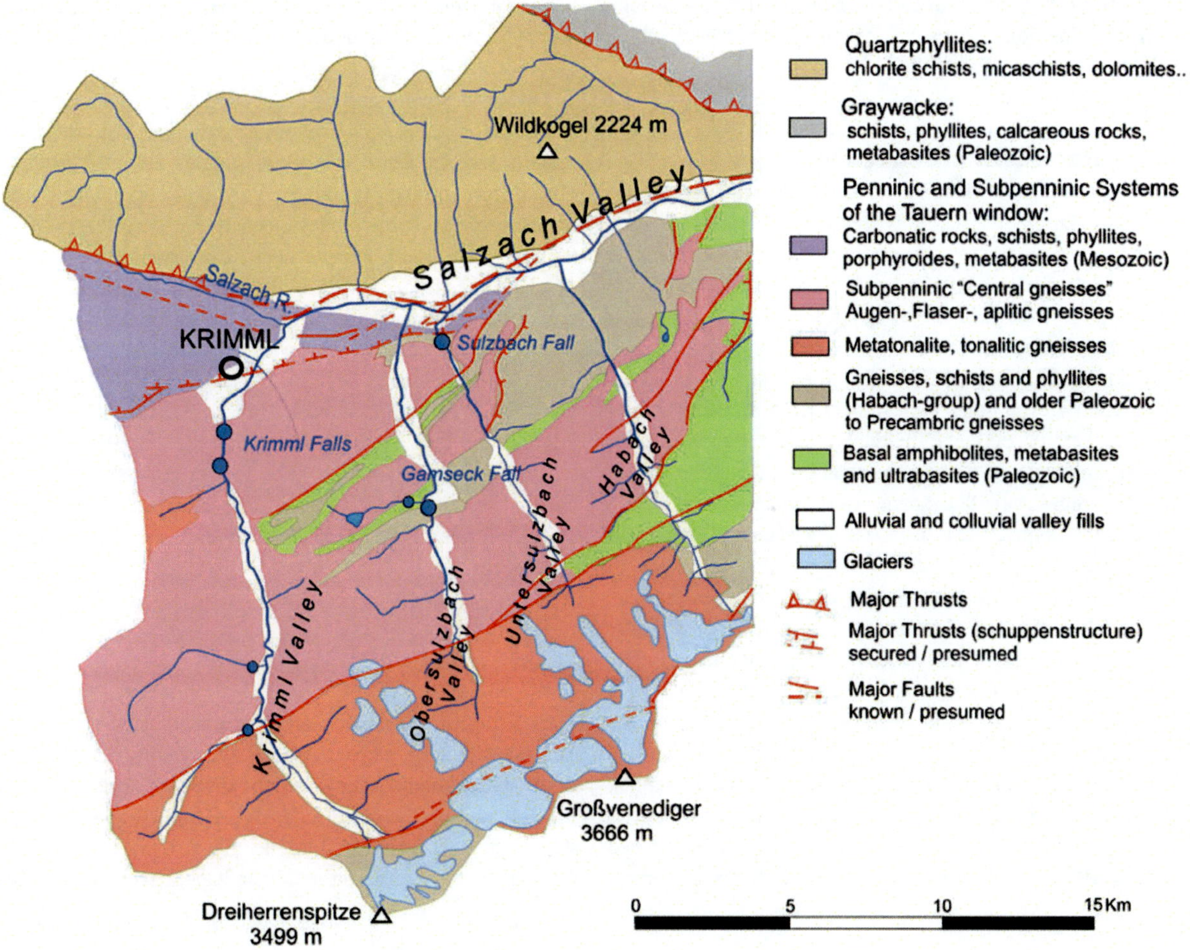

Fig. 24.3 Geological setting of the Krimml Waterfall area (Simplified after GBA 1979, 2005)

Tauern main chain, controlling the valley development of their tributaries. The northern side of the Hohe Tauern shows that the main tributaries are regularly spaced in a typical parallel drainage pattern, headed by the slope between the main chain and the longitudinal valley of the Salzach River.

The valley incision amounts to approximately 1500–2000 m. The deep dissection of the Tauern range results from a strong Quaternary uplift, which recently reaches values of about 0.5–1.2 mm/a. (Frisch et al. 2000). Erosion processes by rivers and glaciers, and denudation by weathering and mass movements, produced a high degree of diversity of geomorphological features corresponding to the structures and the climatic development. Glacial erosion decisively shaped the landforms of the Hohe Tauern. Many of their famous mountain tops are developed as pyramidal peaks (Dreiherrenspitze, on the watershed of the Krimml Valley). They rise above the typical cirques and trough valleys. Additionally, rock slope failures controlled by geological factors and periglacial conditions, created an outstanding variety of features of slope deformations and microfeatures of denudation. In comparison with the other main Tauern valleys, the trough valley basin with the river channel of Krimmler Ache is the most elevated. It stretches between 1800 and 1500 m altitude over nearly 14 kms with an average longitudinal gradient of only 2.5° before reaching the Upper Krimml Waterfall.

24.3 Waterfalls: Introduction and Peculiarities in the Krimml Area

Waterfalls are very steep, generally vertical drops in a river course (Schwarzbach 1967; Ford 1968; Young 2004). In analysing the waterfalls in Switzerland, the free fall of water as a characteristic is also emphasized by Schwick and Spichtig (2007); but additionally, they include cascades which follow each other immediately (in a distance shorter than the heights of the individual cascade steps). Rivers, crossing rocky steps, undergo gradual or sudden changes of their flow regimes (in dependence of the channel slope) from subcritical to supercritical turbulent (Morisawa 1985). River sections which are characterized by smaller cascades or supercritical turbulent flows during low water seasons can change into a shooting flow during floods in which singular cascades disappear in the general strong turbulences. The hydraulic flow of water, under these conditions, changes immediately, leading to a disruption and spraying of the water surface.

Many alpine waterfalls, in addition, show that the water, at least partially, is not dropping in a pure free fall, but is shooting over steeply descending rock surfaces. River bed sections which are characterized by rapid changes of slope angles or obstructions caused by perpendicular patterns of resistant bedrock ledges cause a further increase of turbulence, while the rapid water drops generate similar effects (break-up, developing spray) as in a typical waterfall (free fall).

Generally, a waterfall is divided into the crest-line (knickpoint), the fall-face and the base of the fall (Young 1985). The features of these three elements result from initial conditions which are controlled by the geological structure and preexistent relief features as well as by the subsequent fluvial and weathering processes which lead to the waterfall development.

The celebrity of waterfalls depends both on the height of the waterfall and the quantity of the dropping water. There are further attributes, which increase the degree of popularity: the height of the free falling water, distinctively developed cascades, typical erosional features like potholes and plunge pools, the compactness of the dropping water, the development of veils of spray, the structuring and the character of the water curtain (Figs. 24.1 and 24.4).

Most of the famous waterfalls in the Alps have respectable free fall-heights as shown in Switzerland (Schwick and Spichtig 2007). The most impressive waterfalls occur in rivers of higher orders. However, because of their high power of erosion, the originally existing waterfall steps are mostly incised. As spectacular as are the bigger waterfalls, they are as rarely conserved.

On the northern side of the Hohe Tauern, the density of waterfalls is about 0.04/km^2 both in the western and in the eastern part. In detail, however, smaller falls are arranged much more frequently on trough walls, especially in the upper parts of some valleys generating densities of 0.2–0.6/km^2 (Fig. 24.2). Furthermore, it is striking that the biggest and the most important waterfalls in the Hohe Tauern are concentrated in the vicinity or in contact with the exposed gneiss cores. However, they are very few in the areas of the Penninic System. Even during the Pleistocene, geological conditions (rock type, metamorphism, dip, jointing and fault systems) controlled the glacial excavation in size and differentiation. The former graded slopes and valleys become structured by steps and basins (Trenhaile 1979), forcing the Holocene river courses to adapt themselves.

24.3.1 The Krimml Waterfalls

The Krimml Waterfalls are associated with the most impressive valley steps of the Eastern Alps. The Krimml River (Krimmler Ache) which drains a catchment of 110 km^2 above the waterfalls, drops from the 400 m high step over a length of 1.2 km to the Krimml basin, about 4 km from its mouth into the Salzach River. The valley

Fig. 24.4 Krimml Waterfalls Left: the Upper Fall and the cascade section during high discharge. Right: detail with shooting flow between steps, the river bed conforms to the dip of the gneiss-cleavage. *Photos* Stocker (1993)

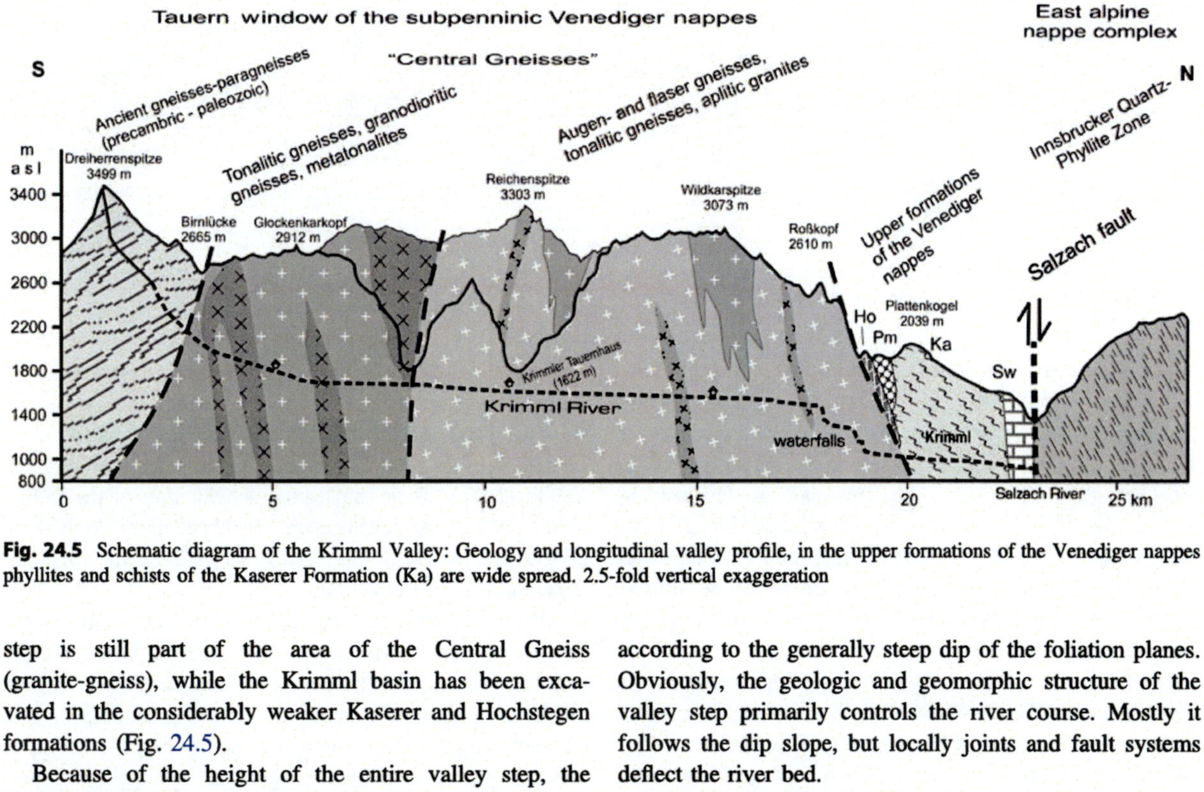

Fig. 24.5 Schematic diagram of the Krimml Valley: Geology and longitudinal valley profile, in the upper formations of the Venediger nappes phyllites and schists of the Kaserer Formation (Ka) are wide spread. 2.5-fold vertical exaggeration

step is still part of the area of the Central Gneiss (granite-gneiss), while the Krimml basin has been excavated in the considerably weaker Kaserer and Hochstegen formations (Fig. 24.5).

Because of the height of the entire valley step, the Krimml Waterfalls are classified among the most spectacular European waterfalls. A longitudinal profile, however, shows a mean slope of the step of merely 22°. Moreover, in the middle a broad flatter section divides the step exactly into two parts (Fig. 24.6). In detail, the slopes of the step are subdivided into a series of smaller rock scarps, arranged according to the generally steep dip of the foliation planes. Obviously, the geologic and geomorphic structure of the valley step primarily controls the river course. Mostly it follows the dip slope, but locally joints and fault systems deflect the river bed.

The Krimml Waterfalls therefore represent exemplarily the type of "consequent waterfalls" described by Schwarzbach (1967) and Ford (1968). Considering the modest Holocene downcutting of about 5–10 m, the fluvial development can be classified as extremely low, despite of the extraordinarily large discharge of the river. The top of the

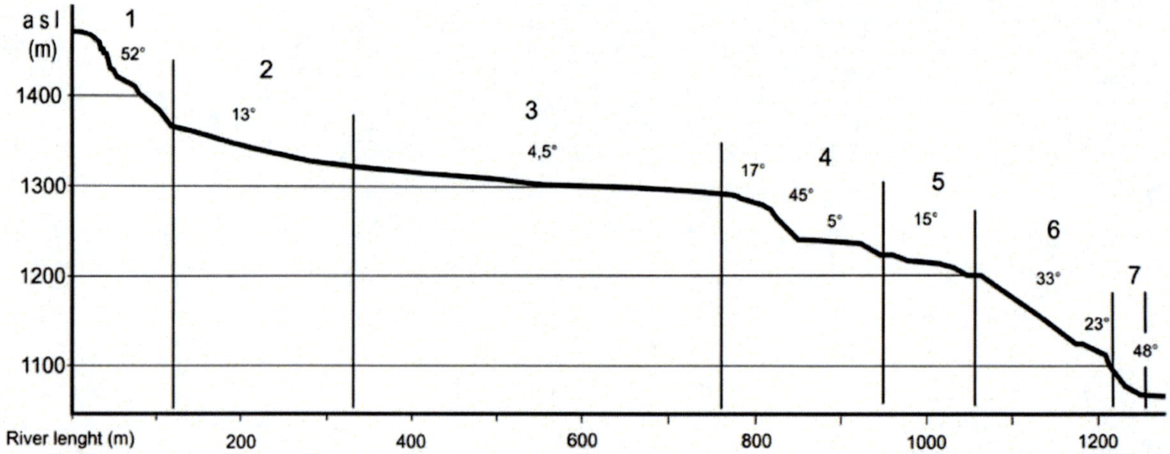

	1 Upper Fall	2 torrent	3 torrent	4 Middle Fall	5 gorge	6 cascades	7 Lower Fall
asl (m)	1465-1368	1368-1324	1324-1292	1292-1226	1226-1203	1203-1115	1115-1070
Height (m)	97	44	32	66	23	88	45
Length (m)	117	212	441	188	90	160	69
Average Slope	39,7°	11,7°	4,2°	19,3°	14,3°	27,9°	33,1°
width (m)		40	10	3-10	2-3	3-8	3-10
depth (m)	<5		1-2	3-10	3-5	5-7	<3
type	waterfall, cascade	anasto-mosed	torrent, braided	waterfall cascades	gorge, cascades	gorge, cascades	waterfall
top	convexity, knickpoint	debris cone boulders	boulder steps	convexity, knickpoint	convexity, knickpoints	knickpoints	knickpoint
base	bedrock platform			plunge pools			plunge-pool

Fig. 24.6 Longitudinal profile and channel sections on the valley step of the Krimml Waterfalls. Parameters: altitude, height, length, slope, width, depth of Holocene incision and waterfall characteristics

main waterfalls is generally convex in the long profile and ends downwards in an edge, favouring acceleration of flow velocity as well as the free fall of the water masses. In addition, the increasing incision leads to a narrowing of the lip. The convexities above the crest-line are favoured by general dip of the foliation planes. The Lower Fall shows a narrowing due to a diversion of the river course in the strike direction of the cleavage plains. The lip of the fall crosses the dipping gneisses and favours the development of a sharp edge so that a free fall is best developed.

Similarly, the fall-face of the Upper Fall is developed as a steep cascade (Fig. 24.4). Rock juts in the middle cause a division and widening of the 70 m high fall. The base of the fall is eroded and deepened by the dropping water into a V-shaped channel, in which the water masses converge, creating strong turbulences and veils of spray (Fig. 24.6). The cascading flow ends in a buttressed section and finally in an accumulation of coarse blocks.

The Middle Fall is developed as a cascade with shooting waterfalls and two plunge pools. The fall-face of the Lower Fall is about 55 m high. The changing structures from Augen to Flasergneiss, or steeply dipping cleavage and jointing planes originated by pressure release influence the formation of the dropping river course in detail (Schmidt and Steyrer 2011). They can lead to a widening of the falling water and a strong development of veils of spray. At the base of the free fall, a plunge pool is surrounded by a dam of coarse block accumulations.

One of the most scenic parts of the waterfalls is the cascade section between the Middle and the Lower Fall. It consists of a narrow gorge and a broader channel which conforms with the slope of the step of about 30°. The channel, incised from about 5 m to more than 10 m, is crossed by transverse rock ledges and equipped with four bigger cascades. During high discharge, the individual water curtains merge into each other, creating a shooting flow with a thick veil and clouds of spray (Figs. 24.1 and 24.4), especially during the summer months when the discharge is 10–16 m^3/s on average. The annual mean of discharge amounts to 5.6 m^3/s (Wiesenegger 2012).

Despite the generally high discharge, it is remarkable that features due to corrosion are few. Potholes of minor dimensions are mostly distributed above the waterfall crest and along the cascade section. The material around the bases of the waterfalls generally consists of sharp-edged coarse block fragments transported by rockfall. Steeply dipping cleavage and jointing due to pressure release favoured processes like toppling and the initiation of debris flows from lateral small torrential watersheds.

24.3.2 The Sulzbach and Gamseck Fall in the Adjacent Tauern Valleys

The two adjacent Salzach confluents (Ober and Untersulzbach) show, despite their similar geological conditions, remarkable waterfalls in very different positions (Fig. 24.3). The most spectacular is the Sulzbach Fall, situated on the 150 m step upstream of the outlet of the Untersulzbach, creating a hanging valley above the plain of the Salzach River. Like the other two valleys, the Untersulzbach Valley is a typical trough valley eroded into the resistant granitic gneisses which here nearly reach the Salzach fault line. However, the waterfall has been developed during the Holocene in the emerging meta-arcose-schists which are imbedded in a thin strip between the Salzach fault and the gneisses (Fig. 24.7). The Untersulzbach River is only slightly incised into the bottom of the trough valley; a stronger downcutting starts on the edge of the step (of the hanging valley) in the less resistant, nearly vertical dipping meta-arcose schists.

In contrast to the glacially rounded step profile at the mouth of the valley bottom, the river drops over a vertical fall-face of 50 m height and has a sharp crest-line. The free falling water masses are symmetrically widened, producing a beautiful water curtain with a thick mantle of spray. A well-formed plunge pool has developed at the base (Fig. 24.8).

The adjacent channel sections, especially those located downstream, are characterized by cascades with potholes and are incised in a gorge. On the right sidewall of the gorge, beginning 10 m beyond the current waterfall, several parts of ancient spillway surfaces are well conserved. They attest to the recent recession of the fall-face. It is significant that the waterfall has not been destroyed by incision; on the contrary, both the fall-crest and the base of the fall are accentuated. The waterfall therefore is not a relict of a primarily higher waterfall from the step of the hanging valley but it has been developed and enhanced during the Holocene into the relatively hard, vertically dipping meta-arcose schists and therefore it can be considered as a barrier fall (Young 1985).

In the Obersulzbach Valley, the floor of the glacial trough has a prominent 300 m step in its middle. Exactly at this position, a broad band of amphibolites, phyllites and schists, imbedded between areas built predominantly of Central Gneiss, crosses the valley (Fig. 24.3). The edge of the upper valley floor is based on a lens of aplitic granite, which provides a threshold at 1500 m asl. The valley step is characterized by a V-shaped cross profile. Favoured by the weaker phyllites and schists which occur at the base of the step, Holocene downcutting triggered headward erosion and the development of a gorge. The crest of the fall is rounded, the fall-face of 40 m shows a free fall (Fig. 24.8), and the base of the fall is formed by a plunge pool, followed downstream by further cascades with plunge pools. The development of the Gamseck Fall and further cascades has been focused on the position where resistant amphibolites cross the river bed. Therefore, their evolution was spatially restricted in comparison with the Sulzbach Fall. On the entrance to the waterfall, the catchment area is 55 km^2.

Glacial undercutting of the left side of the Obersulzbach Valley created two consequent waterfalls near the Gamseck Fall (Fig. 24.2). They result from two smaller hanging valleys: the Seebach River, which drains the cirques of the Seebach (5.6 km^2), and the Foißbach (2.1 km^2). The Seebach Falls consist of cascades and have a total height of

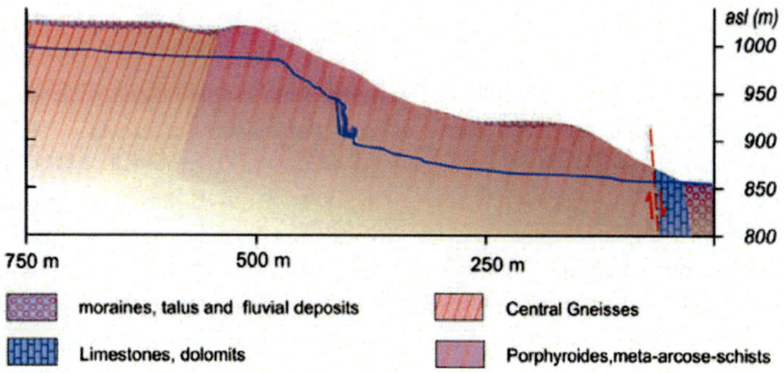

Fig. 24.7 Long profile and geological section of the Sulzbach Fall developed on the 150 m step of the hanging valley of the Untersulzbach River above the Salzach River valley bottom

Fig. 24.8 (left) Sulzbach Fall, a 50 m free fall, and (right) Gamseck Fall in the middle of the Obersulzbach Valley. (*Photos* Stocker 2011)

370 m. The Foißbach Falls show two separated falls of 50 m height each.

24.4 Initiation and Development of the Waterfalls in the Northern Hohe Tauern

Waterfall diversity is strongly connected with their initiation and development. Waterfalls, "where the discordance is primarily attributable to causes other than differing erodibility" (Ford 1968), are designated by Schwarzbach (1967) as *consequent waterfalls.* In alpine areas especially, the smaller waterfalls below cirque thresholds, from sidewalls of glacial troughs or hanging valleys and the bigger step-waterfalls in trough valleys can be assigned to this type. However, Holocene fluvial development could modify these waterfalls into *erosional waterfalls,* or *secondarily subsequent waterfalls* (Schwarzbach 1967), which are remodelled by fluvial action; such subsequent falls show different pathways of development, from accentuation to retreat and reduction in height.

In the high mountain area of the Hohe Tauern, waterfalls cover a broad range of phenotypes in a succession from consequent to well-developed erosional waterfalls. An evaluation regarding waterfalls on valley steps shows interdependencies between geological and petrographic structures and the geomorphic sculpting. Originally, smaller scarps imposed on a valley long profile as a result of a crucial change of geologic structure, may have been increased to accentuated rock steps by the Pleistocene valley glaciers (Embleton and King 1971). The 400 m high rock step of the Krimml Waterfalls clearly shows that its position marks the northern end of the granitic gneiss core (Central Gneiss). Nevertheless, the rock bar is characterized by glacial landforms as demonstrated by the U-shaped cross profiles and the widespread occurrence of roches mountonées. Therefore, one may suppose that the Pleistocene valley glaciers accentuated the step by overdeepening the weak phyllitic rocks which are confining the gneisses at the base of the falls.

Holocene rivers have had to adapt their river beds to the relict conditions. The geological and geomorphological parameters were decisive, whether the fluvial processes could substantially change their channels or not. The river bed in the reach of the Krimml Waterfalls, however, shows only insignificant incision, which is probably not only caused by the hard bedrock. The interactions of the process systems between the valley sides and the river, especially the extent of their coupling, can also play an important role (Harvey 1994).

Distinct from the other Tauern valleys, the low stream gradient of the Krimml Valley above the waterfalls generally reduces coarse sediment flux through the fluvial system, becoming visible in extended storage areas; the river gradient is only 2.5% over a length of 13 km, and only 1.7% over a length of more than 10 km (Fig. 24.9). A series of indicators speak for weak coupling between the slope and river systems through nearly the whole trough valley basin above the waterfalls (Stocker 2011). Wiesenegger (2012) described thunderstorm events from 1964 to 2002 when the Krimml River was dammed by debris flows impounding two lakes and leading to an interruption of the waterfall activity for two hours.

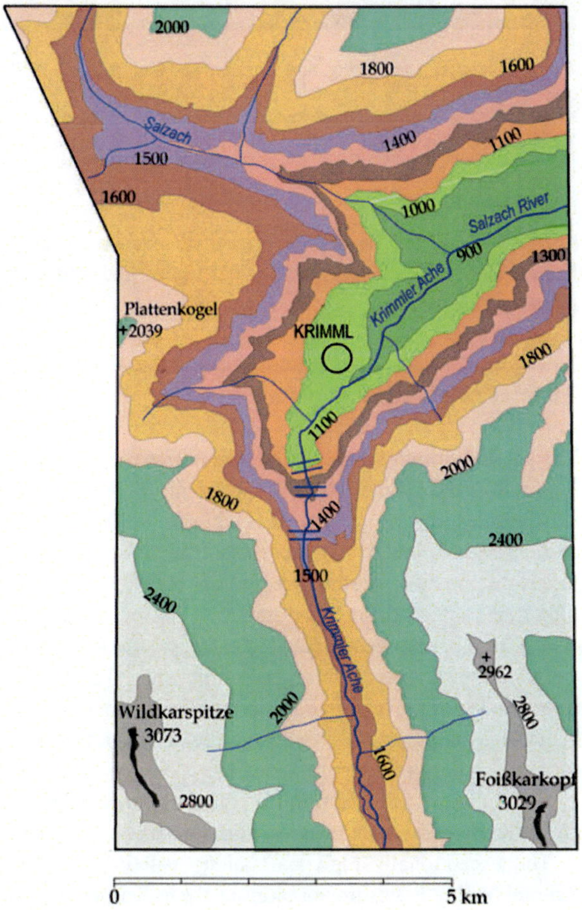

Fig. 24.9 Upper end of the Salzach River catchment shows that the bottom edge of the broad longitudinal valley is closed by a 400 m high step between 1100 and 1500 m asl (100 m contour lines). While the Krimmler Ache, the main river, drops over the step, creating the waterfalls, the smaller Salzach River from its source area (a third of that of the Krimml River) has incised the step. Above the step, both in the Krimml Valley and the middle of the Salzach source area between 1500 and 1800 m asl flat valley floors are largely extended (Contour lines based on ÖK 200 Austrian Amap Fly, BEV 2005)

The Krimml Valley mouth is a basin excavated by the glacier in soft mica schists and carbonatic phyllites. However, the question of which parameters can be recognized as the most substantial for the development of the Krimml Waterfalls has been answered contradictorily. Seefeldner (1961) refused to recognize the origin of the valley step as a consequence of glacial overdeepening. He explained the step by a succession of knickpoints (attributed to periods of rapid uplift) which extended headward, running into the resistant gneiss barrier. Furthermore, the position of the waterfall on the upper end of the whole Salzach drainage basin has to be taken into account. It must also be appreciated that the recent bottom of the Salzach longitudinal valley lies 400 m beneath an ancient valley system, visible both in the flat upper Krimml Valley and on the watershed of the Salzach River system (Fig. 24.10). The valley of the Salzach-headwater also drops in a 400 m high valley step to the local base level of about 900 m asl, situated in the Salzach longitudinal valley. This implies both the relevance of the tectonic depression by the Salzach fault line (a gradual lowering of the base level) and of the overdeepening by the Pleistocene Krimml Glacier.

In the upper Krimml Valley, it created two hanging valleys on the left side (with waterfalls) and, downstream the glacier induced the accentuation of the Krimml Waterfall step by overdeepening the Krimml basin. Leaving its valley, the Pleistocene Krimml Glacier produced a third hanging valley at the mouth of the Salzach River. In comparison with the Krimml drainage basin (130 km^2), the Salzach source area amounts only to a third (44 km^2) and has a watershed 1000 m lower than the Krimml catchment. The Salzach River, favoured by the presence of the Salzach fault line, sharply incised the valley step during the Holocene; possible former waterfalls are not preserved.

The Sulzbach Fall has developed at the 150 m step of the hanging valley of the Untersulzbach River. It can be classified as an example of an erosional waterfall. The drainage basin extends over 40 km^2 and the channel slope above the waterfall crest is 9.1% on average, nearly four times larger than that in the Krimml Valley. Similarly, the channel slope of the Obersulzbach River amounts on average to 4.9% (above the crest of the Gamseck Fall valley step). The trough valley above 1500 m is subdivided into four basins and steps. In the other Tauern valleys, the steps are generally situated much further upstream. However, supported by a strip of hard limestones which crosses the valley mouths, they show characteristic hanging valleys. During the Lateglacial and Holocene, river incision created the most famous gorges and slots of the Hohe Tauern.

Fig. 24.10 Upper section of the Krimml Valley at about 1580 m asl, 3 km upstream of the Upper Waterfall. The 13 km long bottom of the glacial trough has an average inclination of only 2.6%. Four flat basins occupy 74% of the length of the valley bottom. Between them rock thresholds form steps of only 5–7%. The river channel alternates generally between meandering, anastomosing and braided flow and has only short sections of a rocky river bed. The valley long profile of the Krimml Valley is unique in the whole mountain range of the Hohe Tauern. *Photo* Stocker (2011)

24.5 The Economic and Cultural Values of the Krimml Waterfalls—The Waterfall Areas as Geomorphosites

Since the beginning of the eighteenth century, the Krimml Waterfalls have attracted nature lovers and painters. In 1828, a first trail leading to a pavilion at the Lower Fall was constructed, and with the effort of the Alpine Club, a waterfall trail was opened in 1878. In 1912, the number of visitors increased to 11,000. However, based on the ideal natural conditions, plans for a hydroelectric mega-power station were developed. Their realization was averted only in the 1950s by public interventions. The award of the European Diploma for Conservation of Nature in 1967 was an important declaration for this natural monument. Finally, in 1983, the Krimml Waterfalls were incorporated into the Hohe Tauern National Park. Visitor numbers rose from 200,000 in the 1960s to more than 500,000. Today, with about 400,000 visitors per year, it is one of the most visited sites in the National Park (Haßlacher 2012). In recent decades, new infrastructures such as the House of Water exhibit and the Aqua-park have been built. Major topics such as the general importance of water and the medical application of the waterfall environment are shown.

The waterfall area as an outstanding example of the alpine heritage can be considered generally as an important "geosite" and "geomorphosite", based on its high scientific value. Its additional values (Reynard 2009) imply also cultural, historical, ecological, social/economic and aesthetic significance. These issues are presented extensively at the Visitor Centres, especially ecological topics and traditional alpine agriculture.

24.6 Conclusion

The parallel trough valleys of the Hohe Tauern range draining north to the Salzach River are differently stepped in their long profiles, due to geological conditions, stages of development and fluvial/glacial processes. The most spectacular waterfalls on the northern edge of the Großvenediger Mountains offer not only an attractive natural scenery but also a fascinating geomorphological research area. Thus, these waterfalls represent true geomorphosites. Their origin, however, is not yet fully understood and scientific statements about their initiation are often contradictory. Apart from the expected variety of the entire area, the waterfalls of the first three tributaries of the Salzach River show an especially high diversity of geomorphic features, initiation and development.

The exceptionally high and unique valley step of the Krimml Valley is positioned close to the valley mouth, and the associated spectacular waterfalls are situated near the village of Krimml. Originated by the contrast between the Central Gneiss and the confining weak phyllitic rocks, and probably accentuated by Pleistocene glaciation, the waterfalls show only weak fluvial transformation, conserving to a large extent an original aspect. This can be explained by the petrographic and structural resistance against erosional processes despite the high discharge. The second great peculiarity is the more than ten kilometres long upper Krimml Valley, developed as a glacial trough, situated above the waterfalls. As against all other Tauern valleys, its longitudinal profile shows a low stream gradient which reduces coarse sediment flux through the fluvial system, visible in extended storage areas. A series of indicators speak for low coupling of the slope and river systems in the whole trough valley basin above the waterfalls, with the consequence of a reduced capacity of river erosion along the waterfall steps. The waterfalls in the neighbouring valleys, especially the Sulzbach Fall, form perfect examples of how an erosional waterfall developed, supported by steeply dipping meta-arcose schists.

The value of the landscape at the borders of the Hohe Tauern National Park is increased by the absence of big power stations, preserving a natural experience of the dropping waters. Trails along the falls permit not only observations of the broad range of aspects from numerous lookouts but also entry to the spectacular alpine trough valleys and the high mountain area of the watersheds.

Acknowledgements I am particularly grateful to Prof. Michael Church for kindly reading through the manuscript, and for his constructive comments and helpful suggestions.

References

BEV (2005) Austrian amap fly. Version 4.0, Bundesamt für Eich- und Vermessungswesen. Wien

Embleton C, King C (1971) Glacial and periglacial geomorphology. Arnold, London

Fally W (1985) Die Krimmler Wasserfälle. In: Österreichischer Alpenverein (ed) Wasserfallweg Krimmler Wasserfälle. 3rd edn. Naturkundlicher Führer zum Nationalpark Hohe Tauern, vol 3, Österreichischer Alpenverein, Innsbruck, pp 20–26

Ford D (1968) Waterfalls. In: Fairbridge R (ed) Encyclopedia of geomorphology, Dowden. Hutchinson & Ross, Inc., Stroudsburg, Pennsylvania, pp 1219–1220

Frisch W, Székely B, Kuhlemann J, Dunkl I (2000) Geomorphological evolution of the Eastern Alps in response to Miocene tectonics. Z Geomorph N F 44:103–138

GBA (ed) (1979) Geologische Karte der Republik Österreich 1:50.000, Bl. 151 Krimml. Geologische Bundesanstalt, Wien

GBA (ed) (2005) Geologische Karte von Salzburg 1.200.000. Geologische Bundesanstalt, Wien

GBA (ed) (2009) Geologie der österreichischen Bundesländer. Salzburg 1:200.000. Erläuterungen. Geologische Bundesanstalt, Wien

Harvey A (1994) Influence of slope/stream coupling on process interactions on eroding gully slopes: Howgill Fells, Northwest England. In: Kirkby M (ed) Process models and theoretical geomorphology. Wiley, London, pp 247–270

Haßlacher P (2012) Die Bedeutung der Krimmler Wasserfälle für den Tourismus. In: Oesterreichischer Alpenverein (ed) Naturkundlicher Führer Wasserfallweg Krimmler Wasserfälle. 4rd edn. Naturkundlicher Führer zum Nationalpark Hohe Tauern, vol 3, Österreichischer Alpenverein, Innsbruck, pp 151–156

Morisawa M (1985) Rivers. Form and Process. Longman, London, New York

Reynard E (2009) Geomorphosites: definition and characteristics. In: Reynard E, Coratza P, Regolini-Bissig G (eds) Geomorphosites. Pfeil, München, pp 9–20

Schmidt R, Steyrer H (2011) Die Krimmler Wasserfälle—die höchsten Österreichs. In: Steyrer H, Hejl E, Ibetsberger H. Schrott L (ed) Hohe Tauern. Naturkundliche Exkursionen, Bd 2, Pfeil, München, pp 25–33

Schwarzbach M (1967) Isländische Wasserfälle und eine genetische Systematik der Wasserfälle überhaupt. Z Geomorph N F 11:377–417

Schwick C, Spichtig F (2007) Die Wasserfälle der Schweiz. AT Verlag, Aarau

Seefeldner E (1961) Salzburg und seine Landschaften. Eine geographische Landeskunde. Verlag. Das Bergland-Buch Salzburg-Stuttgart

Stocker E (1993) Zur Geomorphologie der Krimmler Wasserfälle. In: Österreichischer Alpenverein (ed) Fachbeiträge des Österreichischen Alpenvereins. Alpine Raumordnung, Nr. 7, Österreichischer Alpenverein, Innsbruck, pp 23–25

Stocker E (2011) The Krimml-waterfalls—a "geomorphosite" in the national park of the Hohe Tauern, Salzburg, Austria. In: University of Bucharest (ed) Proceedings of the international conference environment—Landscape—European identity, 2011 Annual Meeting of the Faculty of Geography, Nov 2011

Trenhaile A (1979) The morphometry of valley steps in the Canadian Cordillera. Z Geomorph N F 23:27–44

Wiesenegger H (2012) Die stürzenden wasser der Krimmler Ache. In: Oesterreichischer Alpenverein (ed) Naturkundlicher Führer Wasserfallweg Krimmler Wasserfälle. 4rd edn. Naturkundlicher Führer zum Nationalpark Hohe Tauern, vol 3, Österreichischer Alpenverein, Innsbruck, pp 115–125

Young R (1985) Waterfalls: form and process. In: Bremer H (ed) Fluvial geomorphology. In memoriam J. N. Jennings, Z Geomorph N F Suppl-Bd 55, Borntraeger, Berlin-Stuttgart, pp 81–95

Young R (2004) Waterfall. In: Goudie A (ed) Encyclopedia of geomorphology, vol 2, J-Z. Routledge, London, pp 1099–1101

Erich Stocker is a retired associate professor of physical geography at the University of Salzburg, Department of Geography and Geology. His research in Alpine and Mediterranean areas is focused on slope development, geomorphological mapping and monitoring of environmental change on alpine and subalpine slopes in the Eastern Alps.

Check for updates

Großglockner and Pasterze Glacier: Landscape Evolution at Austria's Highest Summit and Its Neighbouring Glacier System

25

Gerhard Karl Lieb and Andreas Kellerer-Pirklbauer

Abstract

The Großglockner-Pasterze Glacier area represents one of the most famous mountain sceneries of the entire European Alps. It has attracted scientific and tourist attention since the mid-nineteenth century resulting in a long research tradition. This high mountain landscape is built of crystalline rocks, and the climate is characterised by low temperatures depending on elevation and by a pronounced precipitation gradient from north to south. This chapter is focussed on the Pasterze Glacier, and the glacial history from the Last Glacial Maximum to the glacier's retreat since the Little Ice Age maximum is discussed. Significant findings of fossil wood and peat provide important palaeoclimatic information for the Holocene at Pasterze Glacier. Glacier length changes have been measured since 1879, revealing one of the longest records of glacier retreat worldwide. Data are also presented on the paraglacial gravitational processes that occur adjacent to the retreating glacier. In the foreseeable future, continuing glacier recession will foster mass wasting processes that are significant natural hazards. These developments are also of importance in the context of environmental protection within the Hohe Tauern National Park.

Keywords

Glacier variations • Glacier monitoring • Little Ice Age • Paraglacial concept • Natural hazards

G. K. Lieb (✉) · A. Kellerer-Pirklbauer
Institute of Geography and Regional Science, University of Graz, Heinrichstrasse 36, 8010 Graz, Austria
e-mail: gerhard.lieb@uni-graz.at

25.1 Introduction

Großglockner (3797 m) is Austria's highest summit and overlooks the Pasterze which is Austria's largest glacier (16.6 km^2 in 2012) and amongst the 10 largest glaciers in the Alps. After the Austrian-Hungarian Emperor, Franz Josef I. visited a viewpoint in AD 1856 (since named the Franz-Josefs-Höhe) the scenery of Großglockner and Pasterze has become an icon of Austria's mountains. It attracted much greater tourism following the construction of a high-alpine access road in 1935 and is now one of Austria's most popular tourist destinations. Additionally, the construction of the road was given an almost mythological character because it was carried out during the global economic crisis of the 1930s and thus was exaggerated to serve as proof of Austria's socioeconomic performance. The combination of high mountain vistas, vast glaciated areas, national myths, good accessibility and—since 1981—the Hohe Tauern National Park provides a landscape ensemble comparable with the most famous ones of the Alps such as Mont Blanc (France), Matterhorn (Switzerland) or the Dolomites (Italy).

Even during the last decades of the eighteenth century, the Großglockner-Pasterze area was made famous by scientists who were interested mainly in plants and minerals in a landscape showing a broad diversity resulting from the occurrence of both carbonate and silicate rocks (Sect. 25.2). In the mid-nineteenth century, the Schlagintweit brothers investigated the Pasterze Glacier and in 1850 published a map which is considered to be one of the oldest detailed cartographic representations of a glacier in the Alps (Lieb and Slupetzky 2011). Since then, research on Austria's largest glacier has been continuous, and the geometric variations of the glacier have been monitored annually since 1879 (Sect. 3.3). In this chapter, we focus on the glacial history and the interactions between glaciers and geomorphological processes from different perspectives based on our continuing, long-lasting research.

C. Embleton-Hamann (ed.), *Landscapes and Landforms of Austria*, World Geomorphological Landscapes,
https://doi.org/10.1007/978-3-030-92815-5_25

25.2 Großglockner and Pasterze Glacier as Part of the Hohe Tauern Range

25.2.1 Overview of Geology, Relief and Climate

The Großglockner mountains (in German "Glocknergruppe") are part of the Hohe Tauern range which forms the main divide of the Eastern Alps over a distance of some 120 km. It consists of metamorphic rocks which are part of the so called "Tauern window" in which the Penninic unit, one of the lowest tectonic parts of the Alps, occurs at the surface (see Chap. "Geological and Tectonic Setting of Austria"). Two rock types dominate the geology around the Großglockner and Pasterze (Fig. 25.1b). Both originate from the deep sea floor of the Penninic Ocean (Höck and Pestal 1994) and form contrasting relief (Fig. 25.2):

- Prasinite: A metamorphic volcanic rock with the appearance of a green schist, forms rock faces, sharp crests and steep summits, amongst them Großglockner. It is a very solid rock that weathers to coarse debris.
- Calcareous mica schist: This rock type was formed originally from deep sea sediments accumulated in the Penninic Ocean during the Cretaceous period. It has a brown to greyish colour and forms steep slopes and rounded summits. Weathering products are fine grained and even prone to aeolian transport.

The Hohe Tauern range has a typical high mountain relief with high absolute elevations and elevation ranges, plus significant gravitational, glacial and periglacial processes. According to Frisch et al. (2000), the Hohe Tauern range belongs to the first "principal geomorphological domain" called "high relief, rugged" which characterises about one half of the Eastern Alps. The ongoing uplift of c. 1 mm/a (Frisch et al. 2000) challenges the existence of palaeosurfaces in this region, firmly contradicting the results of older geomorphic research (Büdel and Glaser 1969). However, a plateau-like area extends over more than 10 km^2 between c. 2900–3200 m and forms the accumulation area of the Pasterze Glacier (Fig. 25.3). The morphogenesis of this plateau is today interpreted as resulting from the presence of gently dipping layers of the underlying calcareous mica schist (Höck and Pestal 1994).

Großglockner is situated near the main divide of the Eastern Alps, which, because of its elevation, is a massive barrier for humid air masses coming from the Atlantic Ocean with predominantly north-western winds causing high amounts of precipitation. Exact values of annual precipitation are not known due to measurement difficulties at the wind-exposed high elevations (e.g. the nearby meteorological station at Hoher Sonnblick, 3105 m), but are estimated to be c. 2500–3000 mm at 3000 m asl. There is a striking difference between the northern and the southern sides of the main divide with a mean annual precipitation of 1087 mm (1961–90 data) at Rauris (916 m) in the north and 846 mm in Heiligenblut (1380 m) on the southern slope, which is therefore characterised by a more continental climate with higher altitudes of the standard elevation zones. In the Großglockner-Pasterze Glacier area, the potential timberline is situated at 2100–2150 m; the altitude of 0 °C mean annual air temperature is around 2500 m; the lower limit of discontinuous permafrost is 2500–2800 m, and the current mean equilibrium line of glaciers is at 2900–3100 m (Lieb 2007).

25.2.2 Pasterze Glacier

The Großglockner mountains are characterised by a widespread glacier cover of c. 46 km^2 (Table 25.1). The Pasterze Glacier comprises one third of this glacierised area. The extraordinary size of this glacier results from the high amounts of precipitation at the main divide, the glacier's high altitude and its position on a high-level plateau upon which huge amounts of snow can accumulate (Figs. 25.1a and 25.3). From these plateau-like areas, the ice flows down a steep slope forming an impressive zone of crevasses and seracs (the "Hufeisenbruch", which means horseshoe icefall). Downvalley there is a well-developed glacier tongue which, despite its rapid retreat (Sect. 3.3), was 4 km long in 2017 and terminates at 2070 m asl (Fig. 25.4). The Pasterze is a good example of a compound Alpine valley glacier.

25.3 Glacial History

25.3.1 Last Glacial Maximum (LGM) and Lateglacial Advances

During the LGM, the Großglockner-Pasterze area was part of the accumulation zone of the Eastern Alpine complex of transection glaciers. The ice flowed from there along the Möll and Drau valleys to the south-east and spread over the Klagenfurt basin, with the glacier terminus near its eastern margin (cf. Chap. "Klagenfurt Basin: A Large Basin in the Alps"). A period of rapid ice mass collapse and glacier retreat followed the LGM maximum but was interrupted by readvances of valley glaciers in the Lateglacial period. Morphological evidence of these readvances is sparse in the Möll Valley due to mass movements on the valley flanks. However, based on unpublished results of geological mapping, the Möll Glacier tongue reached a maximum position north of Mörtschach (c. 19 km downvalley of the LIA

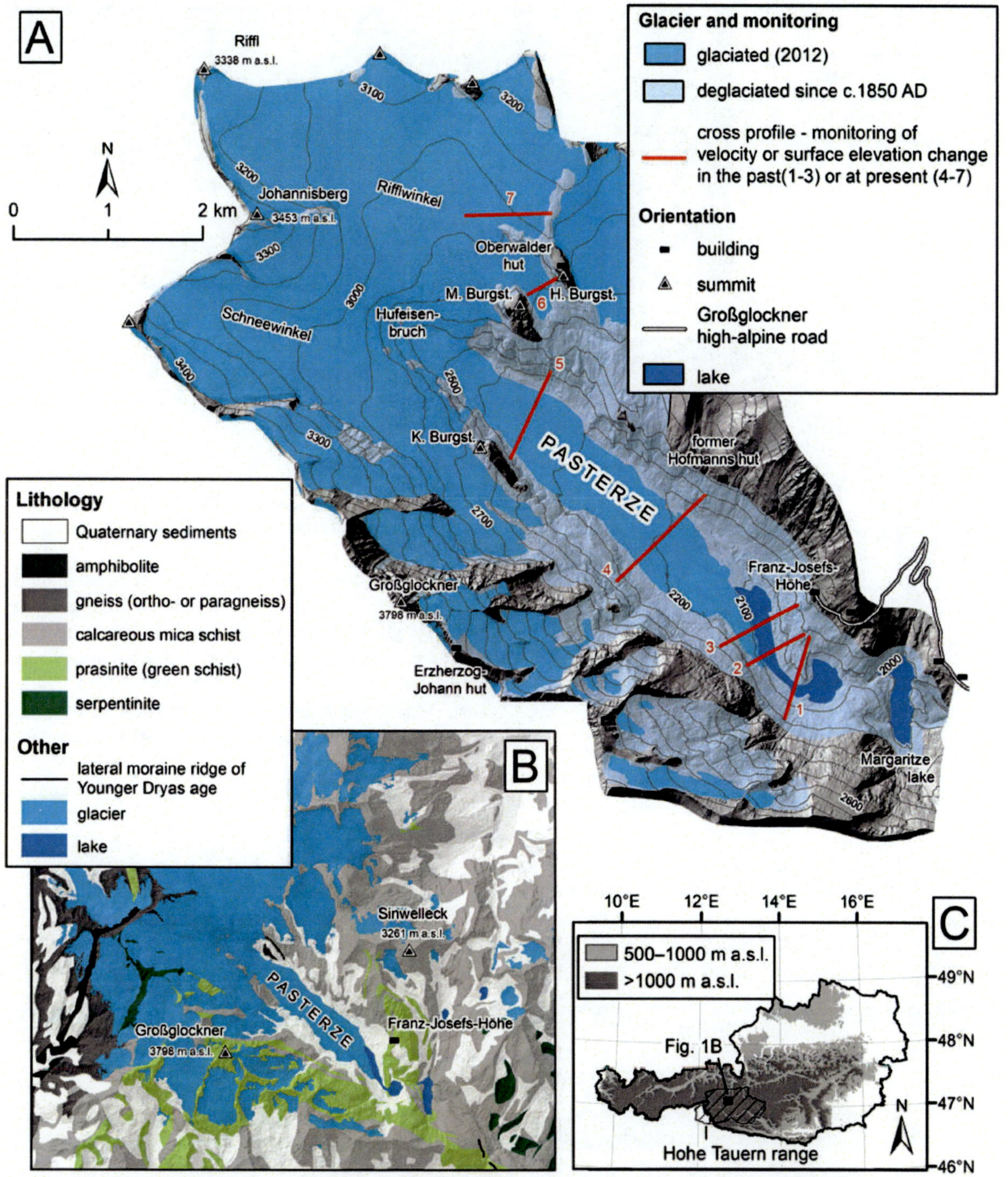

Fig. 25.1 The Großglockner-Pasterze area: **a** the extent of glaciation during the LIA maximum and at present (stage 2012) at the Pasterze Glacier, **b** the main lithological units of the wider area and **c** location in Austria

terminus of Pasterze) in the Gschnitz Stadial (cf. Chap. "The Moraine at Trins and the Alpine Lateglacial") and south-east of Heiligenblut (c. 9 km downvalley of the LIA terminus of Pasterze) in the Egesen Stadial (Younger Dryas) (J. Reitner, personal communication). The glacier extent can be reconstructed from remnants of lateral moraines, e.g. the series of well-developed lateral moraine ridges (Fig. 25.5) on the orographic right side of the Möll Valley, c. 1.5 km

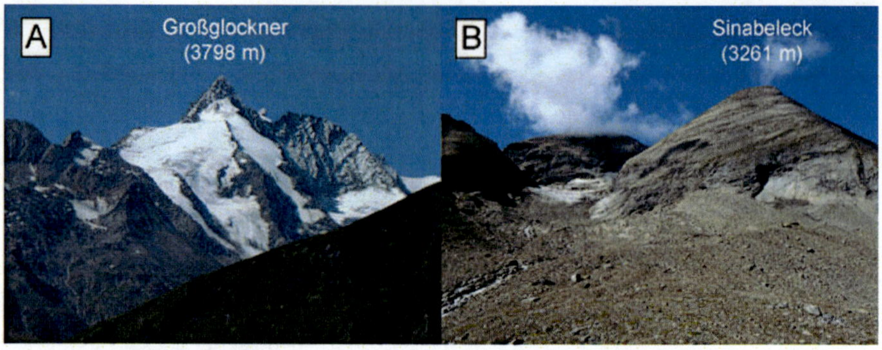

Fig. 25.2 The different shape of the crests and summits in the Großglockner-Pasterze area is due to the different geomorphic responses of prasinite (**a**) and calcareous mica schist (**b**) to weathering and erosion. Photos G. K. Lieb

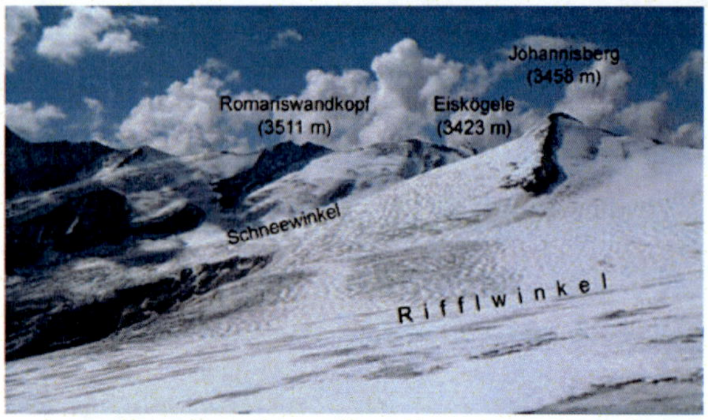

Fig. 25.3 The plateau-like relief in the accumulation area of the Pasterze Glacier is primarily caused by the lithological setting. View from the ascent of Mittl. Bärenkopf to the west. *Photo* G. K. Lieb, 2016

downvalley of the LIA terminus of the Pasterze Glacier. However, these have not yet been absolutely dated, but due to their shape and position they were probably deposited at that time. The stabilisation age of early Younger Dryas moraines is 10Be-dated to 12.8 ka for the Schober Mountains (Reitner et al. 2016) some tens of kilometres to the south of the Pasterze Glacier.

25.3.2 Holocene Glacial History Prior to the Little Ice Age Maximum

Precise knowledge of glacier extent and its variation during the Holocene in the Austrian Alps is far from complete. Ongoing global warming and related glacier recession reveals minerogenic and biogenic sediments in previously glacierised terrain. A number of fragments of *Pinus cembra*, *Larix decidua* and lumps of compressed peat were found in the forefield of the Pasterze Glacier during the 1990s (Slupetzky 1993), and subsequent radiocarbon dating, dendrochronological analyses and palynological research was

Table 25.1 Glacier extent (km^2) in selected areas of the Austrian Alps based on Wakonigg and Lieb (1996), Gross (1987), Fischer et al. (2015), Buckel et al. (2018) and unpublished data gathered by the authors

Region	1850	1920	1969	1998	2009–2012	2015
Pasterze Glacier	26.5	22.6	19.8	18.4	16.6	15.8
Grossglockner mountains	103.6	84.4	68.9	59.8	51.7	45.6
Entire Austrian Alps	941.1	808.0	564.9	470.7	415.4	329.1

Fig. 25.4 "Classical" view of the Pasterze Glacier tongue from Franz-Josefs-Höhe in a westerly direction in 1920 and 2014. The main accumulation area (only partly visible) is connected with the glacier tongue by the icefall, which has become higher and interrupted due to ongoing ice loss. *Photos* Left: Archive G. K. Lieb, right: G. K. Lieb

Fig. 25.5 Lateglacial lateral moraine ridge at the right side of the uppermost Möll Valley near Tröglalm. Besides this ridge, two other ridges developed, indicating that the Egesen Stadial (Younger Dryas) consisted of at least three advance phases or oscillations in this position. *Photo* G. K. Lieb

carried out by Slupetzky et al. (1998) and Nicolussi and Patzelt (2000a, b). Based on their results, it was revealed that substantial periods of smaller glacier extent compared to the year 2000 occurred between 8100 and 6900 cal. BC, around 4800 cal. BC, between 3900 and 3500 cal. BC, around 2700 cal. BC and finally around 1800–1500 cal. BC.

From 2006 onwards, the authors regularly found large peat lumps in the glacier forefield. The temporal break between these and the earlier peat findings suggests that the new material potentially belongs to a different deposit. Pollen analysis and radiocarbon dating by Drescher-Schneider and Kellerer-Pirklbauer (2008) indicates that the studied peat was formed between 3350 and 1430 cal. BC. The pollen flora is dominated by spruce and corresponds with the well-known composition at this elevation (2070 m asl) during the Middle and Late Holocene. The pollen content of one analysed peat sample (dating to 1630–1430 cal. BC) reflects clear human impact on the vegetation of the higher altitudes during the Bronze Age. Their results lead to the conclusion that the Pasterze Glacier was smaller than today, at least between 3370–2200 cal BC and 1940–1430 cal BC. These more recent findings add another period of ice-poor conditions at the Pasterze Glacier to the former record of Nicolussi and Patzelt (2000a, b).

In summer 2014, the so far largest wood fragment, a *Pinus cembra* log, was found in the proglacial sediments of the Pasterze Glacier. It was in two parts, which together measured c. 7 m and weighed 1.7 tons. With the help of the National Park authorities, the cable car operators and a helicopter, it was recovered (Fig. 25.6) and subsequently analysed. The tree was 140 years old before it became buried by ice and sediments c. 6000 years ago (K. Nicolussi, personal communication).

In historic times, an increasing number of text records, paintings, maps and photos have become available and provide information on the youngest Holocene development (see, e.g. Lieb and Slupetzky 2011). The first maximum LIA extent of the Pasterze Glacier documented from historical sources (Paschinger 1948) occurred in the first decades of the seventeenth century. The glacier was almost exactly as large then as in 1851–56, which is also confirmed by fragments of moraine ridges (Patzelt in Büdel and Glaser 1969).

25.3.3 Monitoring of Glacier Retreat Since 1879

The Pasterze reached its last maximum extent during the Little Ice Age (LIA) in 1851–56 (Nicolussi and Patzelt 2000a, b) with a total area of 26.5 km^2 (Table 25.1). Since the termination of the LIA, the glacier has undergone continuous retreat except for short periods of stability and single years of minor advances between 1910 and 1935 (Wakonigg and Lieb 1996) (Fig. 25.7). Even during the cooler and wetter periods of the 1890s, 1920s and 1965–1980, the glacier did not advance substantially, which is attributed to the long response time of the Pasterze Glacier (Wakonigg 1991; Zuo and Oerlemans 1997).

Annual monitoring of the geometrical changes of the Pasterze Glacier started as early as 1879, when F. Seeland setup reference points to measure the length variations of the glacier. Since then, this data series has been interrupted in three years only. Over time, the monitoring of the annual length variations was expanded by additionally recording the variations in surface altitude and surface velocity on five transects across the glacier (Fig. 25.1a) based on simple geodetic measurements (tachymetry, currently DGPS techniques). A selection of recent results is given in Table 25.2 in which all measured parameters demonstrate the strong, and since the 1990s accelerated (see Fig. 25.7) shrinkage of the glacier, which corresponds well with the results of mass balance measurements. Summing up, the Pasterze Glacier behaves similarly to other alpine glaciers and even to most of the glaciers worldwide (Zemp et al 2008).

Current research focuses on the potential of modern techniques for finding out more about the future situation of Pasterze Glacier. Included are geophysical measurements in order to provide data on the underground topography and ice thickness, together with terrestrial laser scanning and the use of unmanned aerial systems (UAS) to monitor glacier retreat, ice collapse and mass transport processes (e.g., Seier et al. 2017).

Fig. 25.6 Recovery of the c. 6000 years old pine tree in the proglacial area of the Pasterze Glacier in June 2015. *Photos* A. Kellerer-Pirklbauer, 2015

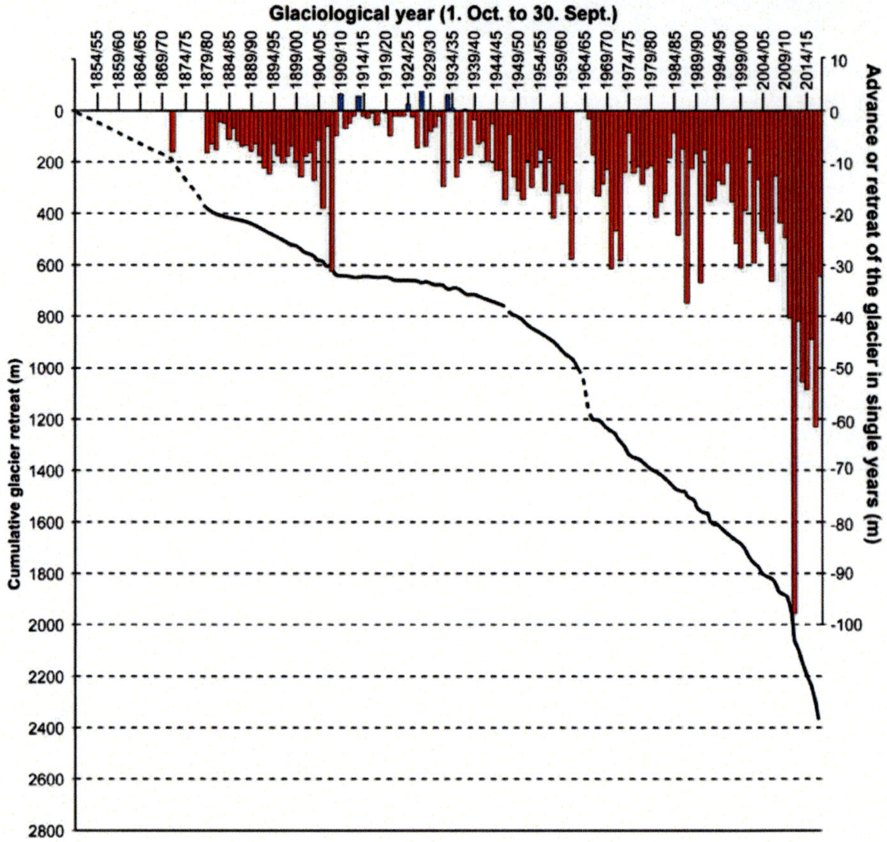

Fig. 25.7 The retreat of the Pasterze Glacier during the period 1852–2018 (single years and cumulative) based on annual field measurements

Table 25.2 Selected results of the standard monitoring programme at the tongue of the Pasterze Glacier (carried out by the Institute of Geography and Regional Science, University of Graz). The data of surface lowering and velocity refer to the transects 3–5 situated on the glacier tongue (see Fig. 25.1a)

Parameter	Measurement technique	Mean annual values 2008/09–2017/18
Length variation	Distance measurement from defined points in defined directions	Retreat of terminus: 47.1 m
Variation of surface elevation	DGPS measurement of points along defined transects[a]	Surface lowering of the glacier tongue: 5.1 m
Variation of surface velocity	Measurement of displacement of stones positioned at the transect points	Surface velocity of glacier tongue: 8.6 m

Annotation [a]DGPS = Differential Global Positioning System; until 2013 these measurements were carried out by analogous tachymetry

25.4 Paraglacial Processes

Paraglaciation is a conceptual model, which was introduced by Church and Ryder (1972) to address those accelerated processes occurring during the transition from previously glaciated landscapes to new fluvially dominated systems. Embleton-Hamann and Slaymaker (2012) give an overview of the history of the concept (including further references) and its relevance for the Austrian Alps. Accordingly, the central concept concerning paraglacial processes is that over a given, though variable period, the intensity of morphodynamics increases significantly following glacier recession and subsequently slowly decreases until it reaches a "normal" level. The geomorphic processes involved are non-glacial and may comprise any erosional, denudational or mass wasting process.

In the surroundings of the shrinking Pasterze Glacier, various measurements (Avian et al. 2018) and field observations show the ongoing paraglacial adaptation of the landscape to post-glacial conditions with obviously highly active erosional and gravitational processes. The proglacial area near the glacier tongue is dominated by glaciofluvial and lacustrine processes that interact with land surface changes caused by slowly melting stagnant ice (Fig. 25.8a). On the valley flanks, which have become ice-free since the middle of the nineteenth century, glacial deposits are eroded by gully incision and mobilised by debris flows (Fig. 25.8b).

Large areas of unconsolidated sediments and rock faces have been exposed since the end of the LIA related to the substantial lowering of the glacier surface. Deglaciation can cause debuttressing of the rock faces, altering the stress field that exists within the rock masses. Changes in the stress field as well as in the thermal regime (e. g. freeze–thaw processes) affect weathering and mass wasting processes. Rockfalls of various scales are typical geomorphic processes acting at such rock faces. A spectacular example happened at the Mittlerer Burgstall (M. Burgst. in Fig. 25.1a) where a large part of a steep rock ridge collapsed (Fig. 25.8b). The main event happened in 2007, but smaller events occurred until 2009. This substantial change in the morphology is schematically illustrated in Fig. 25.9a and visually documented in Fig. 25.9b. Reasons for the triggering of the main event are glacier surface lowering causing debuttressing, but also changes in the permafrost conditions at this site, located at c. 3000 m asl. Furthermore, a warm winter 2006/2007 and joint-rich rocks favoured the destabilisation of rock masses (Kellerer-Pirklbauer et al. 2012). The total amount of the displaced mass is in the order of 500,000 m^3 as revealed by detailed photogrammetric analyses (Kaufmann et al. 2015).

To what extent mass movements in the wider surroundings of the Pasterze Glacier, which already became ice-free in the pre-Holocene, may be considered as paraglacial responses with a long relaxation time remains unclear. An example is the Wasserradkopf rockslide (Fig. 25.10) situated some 5 km south-east of the current glacier terminus of the Pasterze (Weiss 1981). The onset of the sliding process, which stretches over a vertical distance of more than 1000 m, is not known. However, the technical problem of keeping the roadway of the Glocknerstraße stable remains an ongoing challenge.

Fig. 25.8 Intense paraglacial activity in the proglacial area of the Pasterze Glacier. **a** glaciofluvial and lacustrine processes near the glacier terminus, **b** erosional and gravitational processes on the orographic left lateral slope above the glacier tongue. *Photos* G. K. Lieb, 2016

25.5 The Future of the Pasterze Glacier

25.5.1 Increasing Debris Cover of the Glacier Tongue

Debris-cover of glaciers is well-known from many high-relief mountain environments where mass wasting processes deliver large volumes of debris to glacier surfaces. Glacier shrinkage causes an increase of deglaciated slopes, enhancing the rate of (paraglacially mobilised)

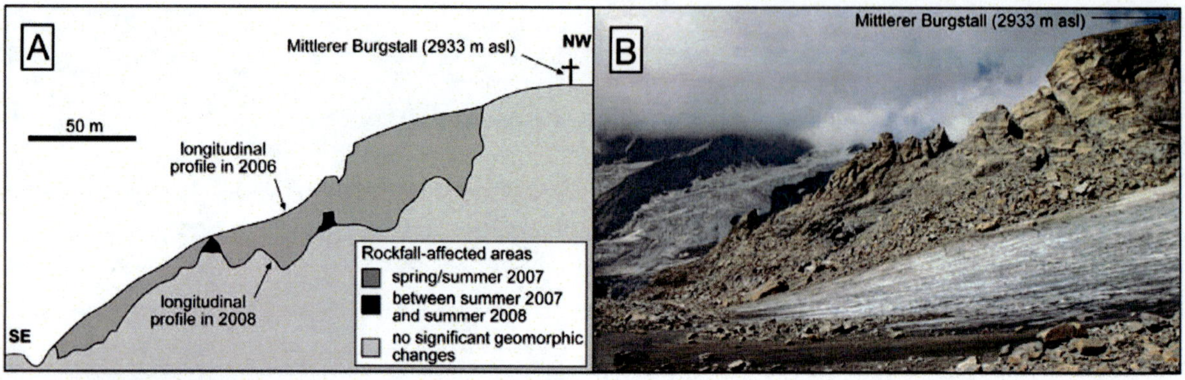

Fig. 25.9 Mittlerer Burgstall, a former nunatak of the Pasterze Glacier during the LIA, changed its shape significantly as a result of a series of rockfalls, which occurred between 2007 and 2009. *Photo* A. Kellerer-Pirklbauer, 2007

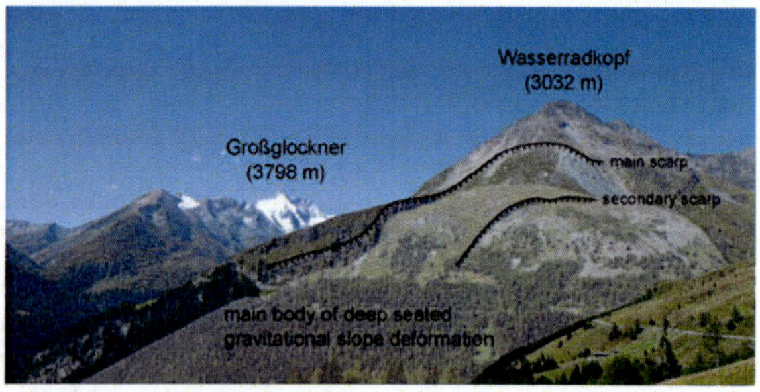

Fig. 25.10 The Wasserradkopf rockslide as seen from Kasereck in a NW direction. Besides the main body of the mass movement the scarp of a secondary slide and areas of increased gravitational activity on the upper slopes are visible (interpretation based on Weiss 1981 and Höck and Pestal 1994). *Photo* G. K. Lieb

debris input either from bedrock or pre-existing sediment stores onto the glacier surface. Consequently, supraglacial debris cover tends to increase over time. Furthermore, at the Pasterze Glacier, the debris input from the two adjacent valley slopes above the glacier tongue differs, reflecting variations in topography (differences in debris entrainment and transport) and lithology (differences in weathering susceptibility). To a certain extent, the increase of debris input is also caused by the release of debris due to permafrost degradation, especially on the right lateral slope of the Pasterze Glacier.

The orographic right part of the glacier tongue is extensively covered by a debris mantle (consisting predominantly of prasinite and calcareous mica schist) with an increasing surface extent shown in Fig. 25.11. At present, more than 70% of the glacier tongue is covered by debris (Kaufmann et al. 2015). The debris layer on the glacier surface influences ablation rates, substantially affecting the glacier topography, the pattern of glacier crevasses and the flow field (Kaufmann et al. 2015; Kellerer-Pirklbauer and Kulmer 2019). Debris thicknesses of up to 1–2 cm increase ablation due to the lowering of the albedo, but a thicker debris cover shields the ice from insolation and atmospheric heat, and ablation rates drop. Therefore, adjacent clean and debris-covered glacier ice reacts quite differently to climate change, resulting in higher surfaces of the debris-covered parts. Kellerer-Pirklbauer et al. (2008) found that in the case of the Pasterze Glacier a continuous debris cover of 15 cm in thickness reduces ablation rates by 30–35%. But, the over-all effects of a debris mantle on glacier surface lowering rates are difficult to establish. There is no doubt that the areal extent as well as the thickness of the debris cover will further increase in future. The implications of the debris cover on the downwasting of Pasterze Glacier will be an interesting aspect of future monitoring.

25.5.2 From a Valley Glacier to Several Cirque Glaciers

The Pasterze Glacier has retreated since the LIA maximum in 1851–56 without any significant interruption (Fig. 25.7). Since the 1990s, retreat rates have accelerated as a consequence of strongly negative mass balances. For example, Kaufmann et al. (2015) calculated the specific mass balance as −1174 mm water equivalent per year for the entire glacier in the period 2003–09 and the surface velocity at two test areas on the glacier showed significant deceleration of c. 31% from the period 2003–2006 to 2006–2009 near the glacier terminus, which corresponds well with the results of the annual monitoring programme. This information leads to the conclusion that the glacier tongue is going to become a large dead ice body in the near future. This prediction is supported by the current collapse (Fig. 25.12) of the glacier terminus (Seier et al. 2017; Kellerer-Pirklbauer and Kulmer 2019) and by an increasing number of ice-free rock areas in the Hufeisenbruch icefall (Fig. 25.4), indicating a decreasing ice-supply from the accumulation area.

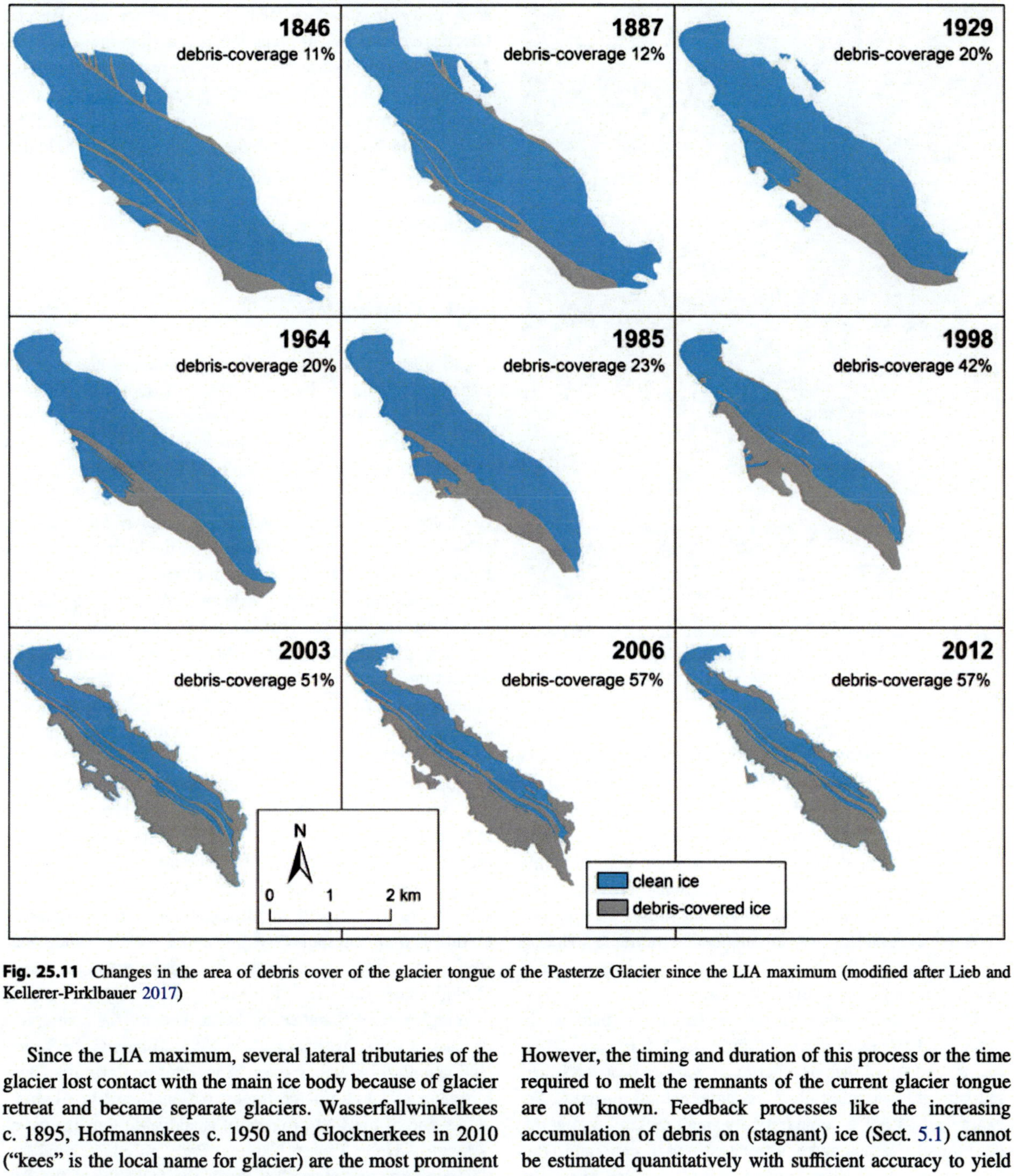

Fig. 25.11 Changes in the area of debris cover of the glacier tongue of the Pasterze Glacier since the LIA maximum (modified after Lieb and Kellerer-Pirklbauer 2017)

Since the LIA maximum, several lateral tributaries of the glacier lost contact with the main ice body because of glacier retreat and became separate glaciers. Wasserfallwinkelkees c. 1895, Hofmannskees c. 1950 and Glocknerkees in 2010 ("kees" is the local name for glacier) are the most prominent examples of a process that will continue in the future. Models developed by Zemp et al. (2006) reveal that an increase of the mean annual air temperature of 2 °C above the 1971–1990 mean will reduce the accumulation area of Pasterze (as depicted in Fig. 70 in Böhm et al. 2007) to half of its current size. The consequence will be the transition from a large, interconnected accumulation area into several single glaciers situated at elevations above c. 3000 m. However, the timing and duration of this process or the time required to melt the remnants of the current glacier tongue are not known. Feedback processes like the increasing accumulation of debris on (stagnant) ice (Sect. 5.1) cannot be estimated quantitatively with sufficient accuracy to yield exact forecasts.

25.5.3 The Importance of Future Changes for Tourism and Hazards

Those responsible for regional tourism are concerned about the future of touristic demand. Glacier retreat increases the

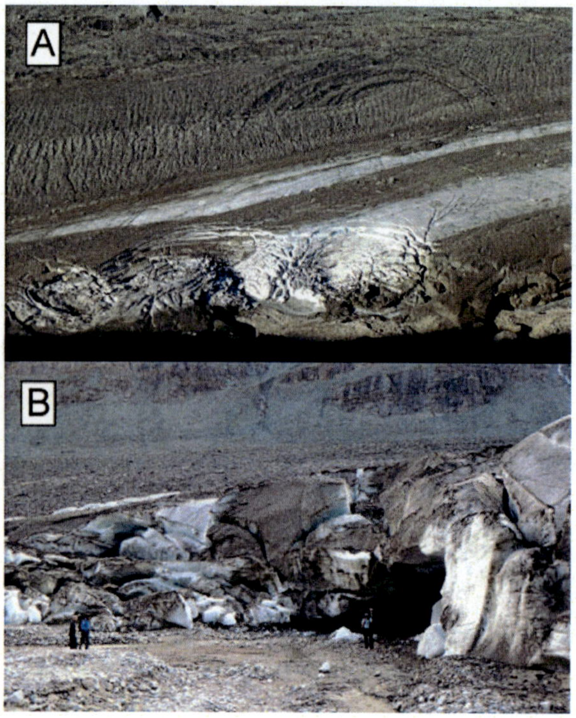

Fig. 25.12 Collapsing tongue of the Pasterze Glacier. **a** Overview of the orographic left part of the glacier tongue with pronounced collapse structures in 2014, **b** Close view of the terminus in 2016. *Photos* G. K. Lieb

distance of the glacier tongue from the Franz-Josefs-Höhe viewpoint and debris cover hides increasing amounts of the glacier surface. However, there are other attractions for tourists in the high mountains, e.g. alpine plant and animal life. Furthermore, as in other glaciated regions of the Alps, the tourist companies have already made provisions for deglaciation by establishing other attractions like exhibitions, educational trails or architectonic landmarks. Thus, it can be concluded that glacier retreat will probably not cause severe economic problems to tourism.

Much more attention, however, has to be paid to the paraglacial processes discussed in Sect. 25.4. In cases where rockfalls, debris flows or other mass wasting processes meet touristic infrastructure they become natural hazards. For example, in 1999, a comparatively small rockfall (a few m^3) onto the Gamsgrubenweg, a frequently used panoramic trail from Franz-Josefs-Höhe towards the glacier, led to its closure by the regional authorities. In order to avoid future hazards, a very expensive solution was found: since 2003, the new trail runs through a series of tunnels.

Kern et al. (2012) used a model approach to detect areas which are prone to gravitational processes and calculated the length of trails and routes possibly affected by these processes. They report that 60.6% of the total length of trails and routes in the area near Großglockner and Pasterze Glacier are situated in areas where gravitational processes are very likely to occur. Additionally, results from a moderate scenario for 2030 (continued glacier retreat and a slight rise of the lower limit of permafrost) indicated 63.8% of the trails would be affected. Obviously, the geomorphological consequences of ongoing climate change will render the maintenance of high-alpine infrastructure more challenging in the future.

25.6 Conclusions

Rapid glacier recession and its consequences are currently transforming the landscape of the high mountain environment significantly and will continue to do so in the near future. In the case of Großglockner and Pasterze, these changes affect a landscape with only limited human impact. With the exception of the touristic road ("Großglockner Hochalpenstraße"), some hotels and alpine huts, a short cable car and a reservoir belonging to the hydropower plant Kaprun (power station on the northern side of the main divide), no infrastructure has been built, ensuring that the largest part of the area retains a natural character.

These restrictions are the result of a long history of nature protection, which dates back to 1918 when A. Wirth, a wood manufacturer from Villach (Carinthia), purchased an area of 40 km^2 comprising the entire Pasterze Glacier and donated it to the Austrian Alpine Association combined with the stipulation "that the Großglockner region shall be preserved for the future as a nature park". The background of this important donation was Wirth's fear of damage to nature caused by the construction of touristic infrastructure that was being considered at that time. Many projects including an elevator to the summit of Grossglockner and summer-ski facilities were subsequently prevented by the landowner. With changing paradigms in environmental policy Großglockner and Pasterze together with a protected area in the neighbouring Schober mountains became the centrepiece of Austria's first national park in 1981. Today, Hohe Tauern National Park stretches over 1856 km^2 in Carinthia, Salzburg and Tyrol and is the largest national park in Austria.

As one of the first projects of sustainable tourism development of Hohe Tauern National Park, an educational trail ("Gletscherweg Pasterze") through the forefield of the Pasterze Glacier was established in 1983. This trail was renewed within the framework of a modern didactical concept in 2004, and a new guide-book was published (Lieb and Slupetzky 2004). Outlining the glacial history, the trail allows visitors to observe a broad variety of geomorphic processes discussed in this chapter, especially the changes occurring in areas which recently became free of ice.

References

Avian M, Kellerer-Pirklbauer A, Lieb GK (2018) Geomorphic consequences of rapid deglaciation at Pasterze Glacier, Hohe Tauern Range, Austria, between 2010 and 2013 based on repeated terrestrial laser scanning data. Geomorphology 310:1–14

Böhm R, Auer I, Schöner W, Hynek B, Kroisleitner C, Weyss G (2007) Gletscher im Klimawandel—Vom Eis der Polargebiete zum Goldbergkees in den Hohen Tauern. Zentralanstalt für Meteorologie und Geodynamik, Vienna

Buckel J, Otto JC, Prasicek G, Keuschnig M (2018) Glacial lakes in Austria—distribution and formation since the Little Ice Age. Glob Planet Change 164:39–51

Büdel J, Glaser U (eds) (1969) Neue Forschungen im Umkreis der Glocknergruppe. Wissenschaftliche Alpenvereinshefte 21, Munich

Church M, Ryder JM (1972) Paraglacial sedimentation: a consideration of fluvial processes conditioned by glaciation. Bull Geol Soc Am 83:3059–3072

Drescher-Schneider R, Kellerer-Pirklbauer A (2008) Gletscherschwund einst und heute: Neue Ergebnisse zur holozänen Vegetations- und Gletschergeschichte der Pasterze (Hohe Tauern, Österreich). In: Abhandlungen der Geologischen Bundesanstalt 62 (=Proceedings of the DEUQUA-Meeting 2008, University of Natural Resources and Applied Life Sciences, Vienna, August–September 2008), pp 45–51

Embleton-Hamann C, Slaymaker O (2012) The Austrian Alps and Paraglaciation. Geogr Ann Ser B 94:7–16. https://doi.org/10.1111/j.1468-0459.2011.00447.x

Fischer A, Seiser B, Stocker-Waldhuber M, Mitterer C, Abermann J (2015) Tracing glacier changes in Austria from the Little Ice Age to the present using a lidar-based high-resolution glacier inventory in Austria. Cryosphere 9:753–766

Frisch W, Székely B, Kuhlemann J, Dunkl I (2000) Geomorphological evolution of the Eastern Alps in response to Miocene tectonics. Zeitschrift für Geomorphologie N.F. 44:103–138

Gross G (1987) Der Flächenverlust der Gletscher in Österreich 1850–1929–1969. Zeitschrift für Gletscherkunde und Glazialgeologie 23 (2):131–141

Höck V, Pestal G (1994) Geological map of Austria 1:50.000, GK sheet 153 Grossglockner. Geological Survey of Austria, Vienna

Kaufmann V, Kellerer-Pirklbauer A, Lieb GK, Slupetzky H, Avian M (2015) Glaciological studies at Pasterze Glacier (Austria) based on aerial photographs 2003–2006–2009. In: Yang X, Li J (eds) Monitoring and modelling of global changes: a geomatics perspective. Springer, pp 173–198

Kellerer-Pirklbauer A, Kulmer B (2019) The evolution of brittle and ductile structures at the surface of a partly debris-covered, rapidly thinning and slowly moving glacier in 1998–2012 (Pasterze Glacier, Austria). Earth Surf Process Landforms 44:1034–1049. https://doi.org/10.1002/esp.4552

Kellerer-Pirklbauer A, Lieb GK, Avian M, Gspurning J (2008) The response of partially debris-covered valley glaciers to climate change: The example of the Pasterze Glacier (Austria) in the period 1964 to 2006. Geogr Ann 90A(4):269–285

Kellerer-Pirklbauer A, Avian M, Lieb GK, Carrivick J (2012) Climate change and rock fall events in high mountain areas: numerous and extensive rock falls in 2007 at Mittlerer Burgstall, Central Austria. Geogr Ann Ser B 94:59–78. https://doi.org/10.1111/j.1468-0459.2011.00449.x

Kern K, Lieb GK, Seier G, Kellerer-Pirklbauer A (2012) Modelling geomorphological hazards to assess the vulnerability of alpine infrastructure: The example of the Großglockner-Pasterze area, Austria. Austrian J Earth Sci 105(2):113–127

Lieb GK, Kellerer-Pirklbauer A (2017) Die Pasterze, Österreichs größter Gletscher, und seine lange Messreihe in einer Ära massiven Gletscherschwundes. In: Fischer A, Patzelt G, Achrainer M, Groß G, Lieb GK, Kellerer-Pirklbauer A, Bendler G, Gletscher im Wandel—125 Jahre Gletschermessdienst des Alpenvereins. Springer Spektrum, Berlin, pp 31–51

Lieb GK, Slupetzky H (2004) Gletscherweg Pasterze. Naturkundlicher Führer zum Nationalpark Hohe Tauern, 2, Innsbruck

Lieb GK, Slupetzky H (2011) Die Pasterze. Der Gletscher am Großglockner. Pustet, Salzburg

Lieb GK (2007) Southeastern and Central Austria. Field guide to the HMRSC-IX (2006) Post-symposium excursion. In: Kaufmann V, Sulzer W (eds) Proceedings of the 9th international symposium on high mountain remote sensing cartography, vol 43. Grazer Schriften der Geographie und Raumforschung, pp 275–292

Nicolussi K, Patzelt G (2000a) Untersuchungen zur holozänen Gletscherentwicklung von Pasterze und Gepatschferner (Ostalpen). Zeitschrift für Gletscherkunde und Glazialgeologie 36:1–87

Nicolussi K, Patzelt G (2000b) Discovery of early-Holocene wood and peat on the forefield of Pasterze Glacier, Eastern Alps, Austria. The Holocene 10:191–199

Paschinger V (1948) Pasterzenstudien. Carinthia II, IX. Sonderheft, Klagenfurt

Reitner JM, Ivy-Ochs S, Drescher-Schneider R, Hajdas I, Linner M (2016) Reconsidering the current stratigraphy of the Alpine Lateglacial: implications of the sedimentary and morphological record of the Lienz area (Tyrol/Austria). E&G Q Sci J 65:113–144

Seier G, Kellerer-Pirklbauer A, Wecht M, Hirschmann S, Kaufmann V, Lieb GK, Sulzer W (2017) UAS-based change detection of the glacial and proglacial transition zone at Pasterze Glacier, Austria. Rem Sens 9:549. https://doi.org/10.3390/rs9060549

Slupetzky H, Krisai R, Lieb GK (1998) Hinweise auf kleinere Gletscherstände der Pasterze (Nationalpark Hohe Tauern, Kärnten) im Postglazial—Ergebnisse von ^{14}C-Datierungen und Pollenanlyse. Wissenschaftliche Mitteilungen aus dem Nationalpark Hohe Tauern 4:225–240

Slupetzky H (1993) Holzfunde aus dem Vorfeld der Pasterze. Erste Ergebnisse von ^{14}C-Datierungen. Zeitschrift für Gletscherkunde und Glazialgeologie 26/2:179–187

Wakonigg H (1991) Die Nachmessungen an der Pasterze von 1879 bis 1990. Arbeiten Aus Dem Institut Für Geographie Der Universität Graz 30:271–307

Wakonigg H, Lieb GK (1996) Die Pasterze und ihre Erforschung im Rahmen der Gletschermessungen. In: Wissenschaft im Nationalpark Hohe Tauern Kärnten. Kärntner Nationalparkschriften 8, Großkirchheim, pp 99–115

Weiss EH (1981) Zur Geologie und Hangmechanik des Raumes zwischen Döllach und Fallbichl (Glocknerstraße-Süd-Rampe) im oberen Mölltal. In: Führer zur Exkursion durch das Glocknergebiet und die Karnischen Alpen in Kärnten 23.-26.9.1981. 3. Sonderheft der Mitteilungen der Österreichischen Bodenkundlichen Gesellschaft, Wien, pp 8–19

Zemp M, Haeberli W, Hoelzle M, Paul F (2006) Alpine glaciers to disappear within decades? Geophys Res Lett 33:L13504

Zemp M, Roer I, Kääb A, Hoelzle M, Paul F, Haeberli W (2008) WGMS (2008): Global glacier changes: facts and figures. World Glacier Monitoring Service, Geneve

Zuo Z, Oerlemans J (1997) Numerical modelling of the historic front variation and the future behaviour of the Pasterze glacier, Austria. Ann Glaciol 24:234–241

Gerhard Karl Lieb is associate professor at the University of Graz, Institute of Geography and Regional Science. He has experience in physical geography of mountain regions with a special focus on permafrost, glaciers and geomorphic processes affected by global change as well as natural hazards. He is one of the two co-leaders of the national glacier monitoring programme of the Austrian Alpine Club. Besides those topics he works on regional geographies (Austria, Alps and Europe) and on didactics of the Austrian school subject geography and economics.

Andreas Kellerer-Pirklbauer is a senior scientist at the University of Graz, Institute of Geography and Regional Science. He has significant experience in the physical geography of mountain regions with a focus on the effects of global change on geomorphic processes and landforms, permafrost, glaciers, mass wasting processes, and related natural hazards. He is the Austrian national representative at the International Association of Permafrost and one of the two co-leaders of the national glacier monitoring programme of the Austrian Alpine Club.

26 Gorges and Slots in Western Carinthia: Their Development and Importance as Geomorphosites

Erich Stocker

Abstract

Nearly all tributaries of the major rivers in western Carinthia emerging from the mountain fronts show knickpoints. They are preconditioned both by glacial overdeepening of the trunk valley and tectonic/isostatic displacements that led to marked base level change and in consequence to sharp incisions of the lowest tributary reaches. Initiation and development of gorges, however, are controlled by very different parameters as demonstrated by the three selected examples. The Ragga Slot (Kreuzeck Mts.) is outstanding in revealing the shaping of a bedrock channel due to a very recent downcutting of the lowest reach of a hanging valley. The Gaisloch Gorge (Gailtal Alps) illustrates an advanced stage in gorge development accomplished by accelerated retreat of its dolomitic sidewalls, creating a spectacular rock scenery. Finally, the Garnitzen Valley (Carnic Alps) represents a complex of gorges, including slots and gaps that are integrated into a glacial through valley. The three presented gorges can be classified as typical geomorphosites. They were opened for visitors at the beginning of the nineteenth century thanks to local initiatives and by the Alpine Club; the gorges/slots of the Ragga River and the Garnitzen River have been declared as natural monuments.

Keywords

Gorges • Slots • Bedrock channels • Scalloped escarpments • Inner gorges • Geomorphosites • Geosites

E. Stocker (✉)
Department of Geography and Geology, University of Salzburg, Hellbrunnerstraße 34, 5020 Salzburg, Austria
e-mail: er.stocker@aon.at

26.1 Introduction

Gorges as unusual features and natural wonders are ranking among the most frequent sites placed under conservation both as natural monuments and landscape protection areas. Considering these features from the scientific point of view, gorges in the past have been regarded often as landforms clearly defined and easy to explain (Machatschek 1973). In fact, they are only briefly described in textbooks, and papers considering them in more detail are few. Additionally, as incised rivers provide a natural insight into geological structures, geologists often focused on this information for explaining regional stratigraphy and tectonics, thereby overlooking to question the development of the gorge itself.

Gorges show generally a high degree of geomorphologic diversity (Speil 1985) and contain a variety of landform successions which suggests different histories of their development. They convey appreciation in understanding both contemporary fluvial processes (including their influence on the shape of bedrock channels) and the historical development of gorges (Wohl 1998).

For studying gorges, which generally have a high density of minor channel landforms and a complex sidewall topography, a high resolution illustration by contour lines on a large-scale map is required. Topographical maps often provide only a generalized overview; in particular dimensions and the shaping of narrow slots cannot be resolved by the usual map surveying methods. Even for the most noted sites, large-scale longitudinal and transverse profiles, or systematic surveys of channel reaches, are not available. In geomorphological research, therefore, analysing these spectacular landforms is a challenge. In this respect the new Atlas of Carinthia (KAGIS) has supplied an essential increase of detail with its high resolution large-scale plans, available with 10 m contour lines.

The three presented gorges display great differences in landforms. While in the Ragga slot sculpting of the bedrock channel is prominent, the highlight of landform development

C. Embleton-Hamann (ed.), *Landscapes and Landforms of Austria*, World Geomorphological Landscapes,
https://doi.org/10.1007/978-3-030-92815-5_26

Fig. 26.1 The Garnitzen Gorge as an example of an inner gorge. It is developed mainly in Devonian limestones and surrounded by the serrated ridges (Ladinian Schlern dolomite) of Mt. Gartnerkofel (2195 m) in the background. *Inset* detail of the lower slot reach with step-pool sequences. *Photos* E. Stocker

in the Gaisloch Gorge lies in the formation of the sidewalls. The entire Garnitzen Valley (Fig. 26.1) constitutes a stepped glacial trough that is combined with a series of gorges and slots, some of which show a development into inner gorges. In Upper Carinthia altogether thirty valley reaches can be classified as distinctively developed gorges which show a broad range of bedrock channel forms and sidewall shapes.

26.2 Geographical and Geological Setting

The examples described belong to different geological units of the Alps (Figs. 26.2 and 26.3). The Kreuzeck Mountains are part of the widespread allochthonous crystalline basement thrust sheet, the "Altkristallin" (Drauzug-Gurktal-thrust sheet system) which includes the Gailtal Alps (Ilickovic and Schuster 2014) and overlies the Subpenninic and Penninic thrusts of the Tauern Window. Hoke (1990) recognized five deformation phases, beginning before the Variscan time. The tectono-stratigraphic complexes are stacked in overturned order, corresponding with a gradual increase of the K/Ar ages from less than 70 Ma in the north to 300–320 Ma in the structurally highest parts of the Altkristallin in the southeast, near the Permo-Mesozoic inconformity line (Hoke 1990).

South of the Kreuzeck Mts the Gailtal Alps stretch in W-E direction. Their mountain chains are mainly built by the "Drauzug", the Permo-Mesozoic cover of the Drauzug-Gurktal-thrust sheet system, situated south of the fault of the Drau valley (Figs. 26.2 and 26.3). The geologic structure is characterized by mostly steeply dipping Triassic strata (Van Bemmelen 1957; Schönlaub et al. 1987, 1989). Adjoining towards the south, the Gail Valley, a broad longitudinal valley, conforms generally to the Periadriatic fault line delimiting the northern border of the Carnic Alps. The mountain chain of the Carnic Alps (Figs. 26.2 and 26.3) is formed by structures of both Variscan and Alpine origin. In the Garnitzen area the sedimentary succession has not been affected by stronger metamorphism and its tectonic structure is generally attributed to thrust and compressive deformations. The stratigraphic sequence of the Paleocarnic Chain

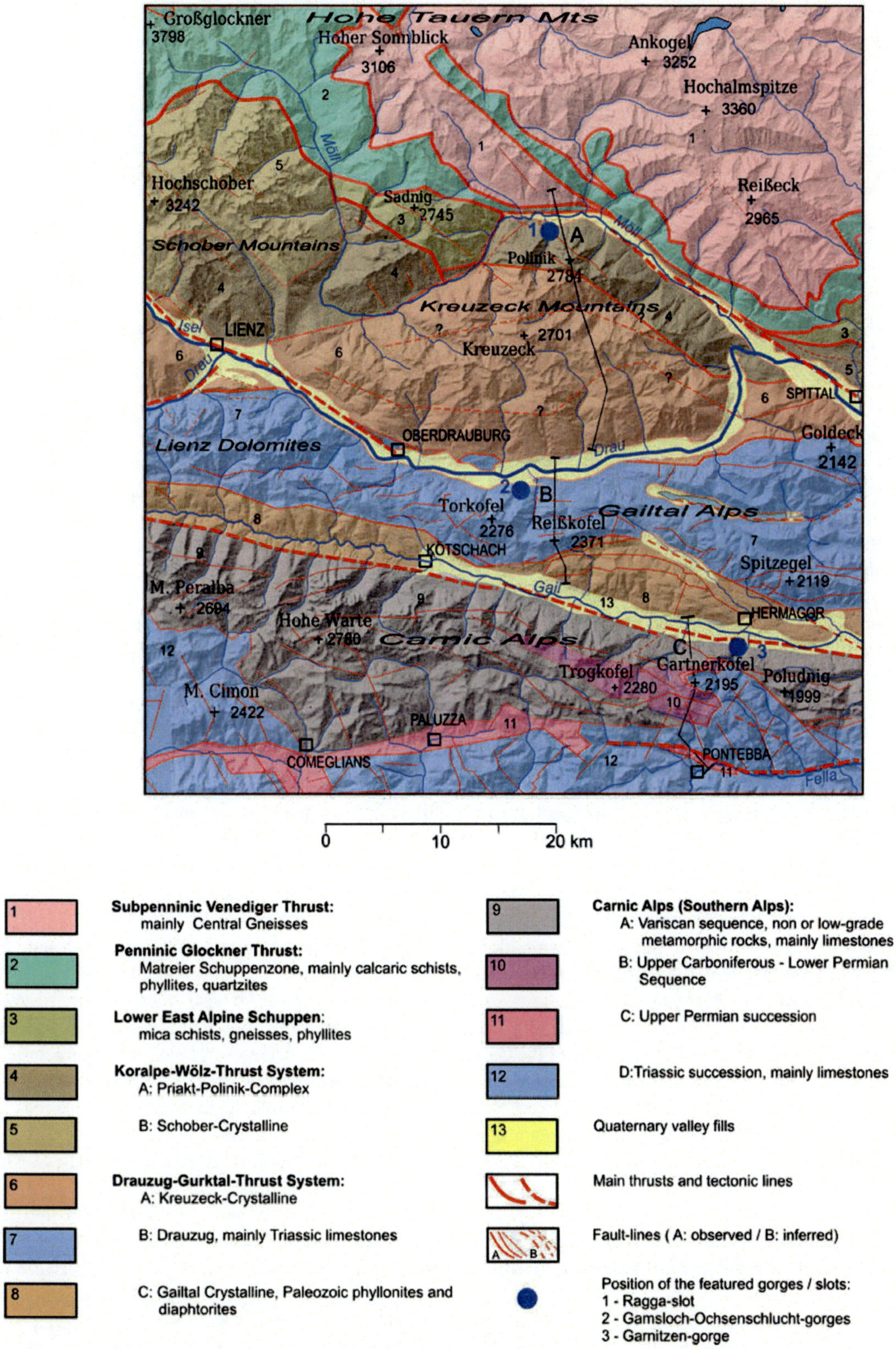

Fig. 26.2 Geology of western Carinthia; position of the gorges and the north–south cross-sections of Kreuzeck Mts (A), Gailtal Alps (B) and Carnic Alps as presented in Fig. 26.3. *Sources* BEV 2005, Braunstingl (2005), Kahler et al. (1959), Kahler and Prey (1963), Linner et al. (2013), Pestal et al. (2006), Schönlaub 1997, Schönlaub et al. (1987, 1989), Schuster (2006), Venturini et al. (2001)

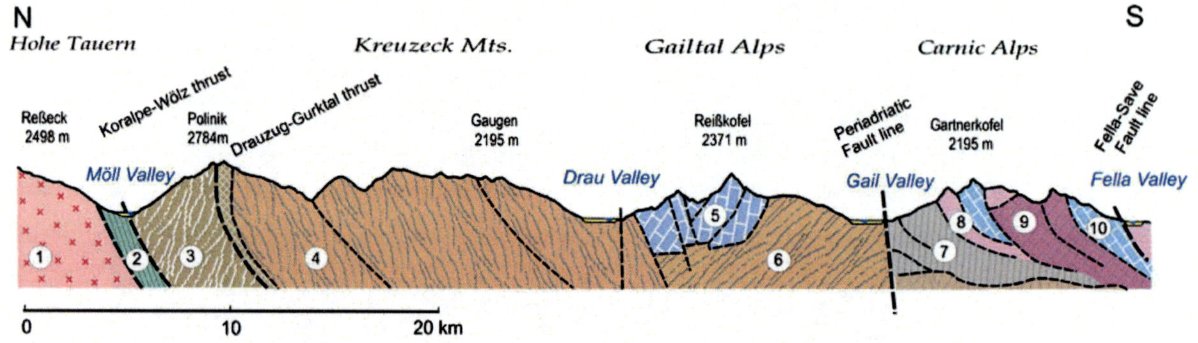

Fig. 26.3 Schematic geological cross-section from north to south. The main valleys of the rivers Möll, Drau, Gail, and Fella confine the geological units as well as the mountain ranges. The indicated numbers are related to those in Fig. 26.2

represents a classical example of the European Paleozoic (Upper Ordovician sandstones and schists, Silurian and Devonian limestones, mostly striped). The stratigraphic series is tectonically structured by thrusts and faults and in detail shows a pile of nappes. The Variscan units are covered by Post-Variscan sequences of Permian and Triassic age (Schönlaub 1987, 1989; Venturini and Pondrelli 2002).

The broad range of rock formations and tectonic structures favoured the development of an equally high degree of geomorphic diversity. The summits of the three mountain groups lie between 2200 m and nearly 2800 m asl. Their central areas are characterized by glacial and periglacial landforms such as cirques, trough valleys, widespread areas with roches mountonnées, rock glaciers, and gelifluction features (Stocker 1989). Many of the trough valleys change in the middle and lower parts into V-shaped valleys with depths between 1000 and 1500 m. Their side slopes underwent gravitational deformations over extensive areas.

One of the most outstanding characteristics of the calcareous mountain chains is the sequence of impressive rock faces consisting of hard Devonian limestone (exemplified in the north wall of Mt. Trogkofel in the Carnic Alps) or serrated ridges developed in Mesozoic dolomites and limestones (Mt. Gartnerkofel in the Carnic Alps (Fig. 26.1) (Schönlaub 1988). Notably the dolomitic rocks of the Gailtal Alps are often affected by accelerated gully incision which can develop into typical alpine badlands (Stocker 2003). These rugged mountains overtop the lower mountain ridges consisting of primarily crystalline rocks.

26.3 Gorges and Slots—Their Characteristics in the Alpine Area of Western Carinthia

Gorges and slots have a potential of high geomorphic diversity (one of the prerequisites of *geomorphosites*), because they usually show a wide range of minor landforms. In the context of gorges and slots, as geomorphosites it is necessary to focus on their singular features in order to finally assess overall geomorphic diversity. Sections 3.2, 3.3 and 3.4 therefore provide a survey of the features of the three selected examples.

26.3.1 Preamble: The Difficulties of a Confounded Terminology

Cross-sections are used to support valley typology; valleys result from interaction of fluvial processes and slope denudation as reported by Fränzle (1968). Therefore, a *slot* is generated by a swift abrasion of the riverbed, creating "perpendicular or even overhanging walls", while a *gorge* is described as an incision which is very narrow, but is widened at the top. They occur primarily in resistant rocks, suggesting either a slower development or a transformation from a slot into a gorge due to reduction in the speed of incision of the riverbed.

The two German terms "Klamm" (slot) and "Schlucht" (gorge) exist both in geomorphological manuals and in naming of locations, if also often confused by colloquial language (the Ragga slot is locally called "gorge", the Gaisloch gorge "slot"). Geomorphological definitions show some contradictions, too: Ahnert (2015) defines "*saw cut gorges*" as a result of rapid fluvial incision into stable rock structures which developed primarily in the alpine area by incision of waterfalls into knickpoints of hanging valleys. Gorges are also explained as initially subglacial incisions (Embleton and King 1971) which, following Kuhle (1991), could be detected by the existence of glacially polished surfaces in the upper parts of the sidewalls. Kuhle (1991) furthermore notes that vertical incision of a slot would only be plausible when followed initial subglacial melt water incision. However, after the retreat of the glaciers, Holocene fluvial incision could also continue the process of slot development.

The causes of rapid down cutting are specified in the gorge definition by Tinkler (2004). Apart from the headward retreat of a knickpoint he mentions the uplifting of a landmass, the outburst of floodwaters, and the superimposition of a channel across resistant rock. On steep mountain sides with high relative relief gorges gorge may easily develop in *transverse* valley reaches that cut through the geological structures.

Whipple et al. (2013) consider gorges/slots (including large steps/falls) as typical features of bedrock channels. Bedrock rivers are characterized by rock outcrops in river bed and banks, even though rock bound reaches are short and intermittent and there is usually a thin and patchy sediment cover which is frequently removed during floods (Whipple et al. 2013). Analyses of channel reaches which comprise a more detailed typology was undertaken by Krzemień (1999) in the Ortler Cevedale massif. The resistance of the channel bed depends on geological structure (mineralogic and lithologic conditions, cleavage, joints ...) and can be analysed at different scales. In particular, investigations of the dimensions of channel reaches (Tinkler and Wohl 1998) can unravel the interactions between geological structures and erosional processes. In alpine areas, most of the studies on bedrock channel features only describe characteristic channel bedforms like *potholes, kolks, pools, undulating walls or grooves* due to shooting water flows. *Kolks* in the German literature (Leser 2003) are generally understood as pools which are eroded into the bedrock, while pools as part of step-pool sequences can occur also in channel reaches consisting of generally coarse boulder accumulations (Knighton 1998). The sidewalls of a slot represent relict lateral margins of bedrock channels, indicating processes of both recent and former erosion. A gorge sidewall, however, can provide evidence of the stages of development, reflecting processes of slope development in the context of structural conditions (Young and Young 1992).

Gorge as a generic term can include *slots, slot canyons* or simply *canyons*; their occurrence in areas which never underwent glaciation suggests that in alpine areas slots may also be created without the immediate agency of glaciers. The initiation of *slot canyons* and slots in semiarid areas by flash floods is intensely discussed. A model for their development based on flow dynamics has been carried out in a laboratory setting (Carter and Anderson 2006). However, the term is also used for slots created under the very different circumstances of subglacial meltwater incision (McEvan et al. 2002).

Since the interior of mountain regions is dominated by bedrock rivers, gorges have frequently developed in the course of the grading of river long profiles in response to tectonic uplift or climatic change. *Inner gorges* are distinctive types of gorges in which continued base level drop led to general downcutting and formation of extended gorges, as analysed by Kelsey (1988) in California. Inner gorges have been studied in alpine environments in the context of stages of rejuvenation of fluvial erosion due to base level fall or changes of environmental conditions. It has been proven that alpine inner gorges have developed over longer periods of the Pleistocene underlying stages of fill in, exposure, and epigenetic incision (Montgomery and Korup 2010; Sanders et al. 2014).

26.3.2 The Ragga Slot, Kreuzeck Mountains

The Ragga Valley (Fig. 26.4) is one of eleven valleys that are descending from the Kreuzeck mountain group to the surrounding main valleys in a radial pattern (Fig. 26.2). Its drainage basin (16.2 km^2) includes the highest peak of Kreuzeck Mts. (Mt. Polinik, 2784 m asl). In contrast to most of the valleys on the south side of the mountain group it is developed as a trough valley in its entire length. It is up to 1600 m deep and its longitudinal profile shows a decrease of about 890 m over a length of 4.9 km, without much variation in slope. The short slot before the river mouth into the Möll Valley has a gradient of 17.5° (Fig. 26.4a). Following the classification of valley slopes (Young 1972) the cross-section of the lowermost Ragga Valley at the position of the Ragga Slot shows three slope units: (i) an upper convex-concave slope that clearly represents the U-shaped glacial trough, (ii) a middle, 50–60° steep slope segment, separated from the upper slope unit by a marked break of slope and indicating an early stage of gorge development (Fig. 26.5a), and (iii) a lower slope unit, characterized by vertical and overhanging walls and marking the youngest stage of the slot development.

The entire gorge incision amounts to 50–60 m in downstream position, the slot depth reaches up to 20–30 m. In the slot section scoured rock reaches frequently alternate with reaches that are covered with gravel, boulders and driftwood. Step-pool sequences on a scale of 1–5 m are centred on two steeper reaches. They are separated by two flatter reaches in which the channel even tends to build riffle-pool sequences (Fig. 26.4a). Pools and laterally developed kolks are the most widely distributed features. They result from corrasion due to concentrated jets of water (Fig. 26.5d).

Frequent changes in flow direction favour the sculpting of kolks, which often form smooth shaped elliptical holes, clearly polished despite their surrounding complex geological structures, such as obliquely dipping joints (Fig. 26.5e). Exceptionally, undulating walls can be observed. However, kolks occur even on rock walls where joints are intersecting and sometimes they show angular edges. Hollows at different levels up to sizes of three metres in diameter can be interpreted as relicts of former lateral kolks. Smaller sculpted

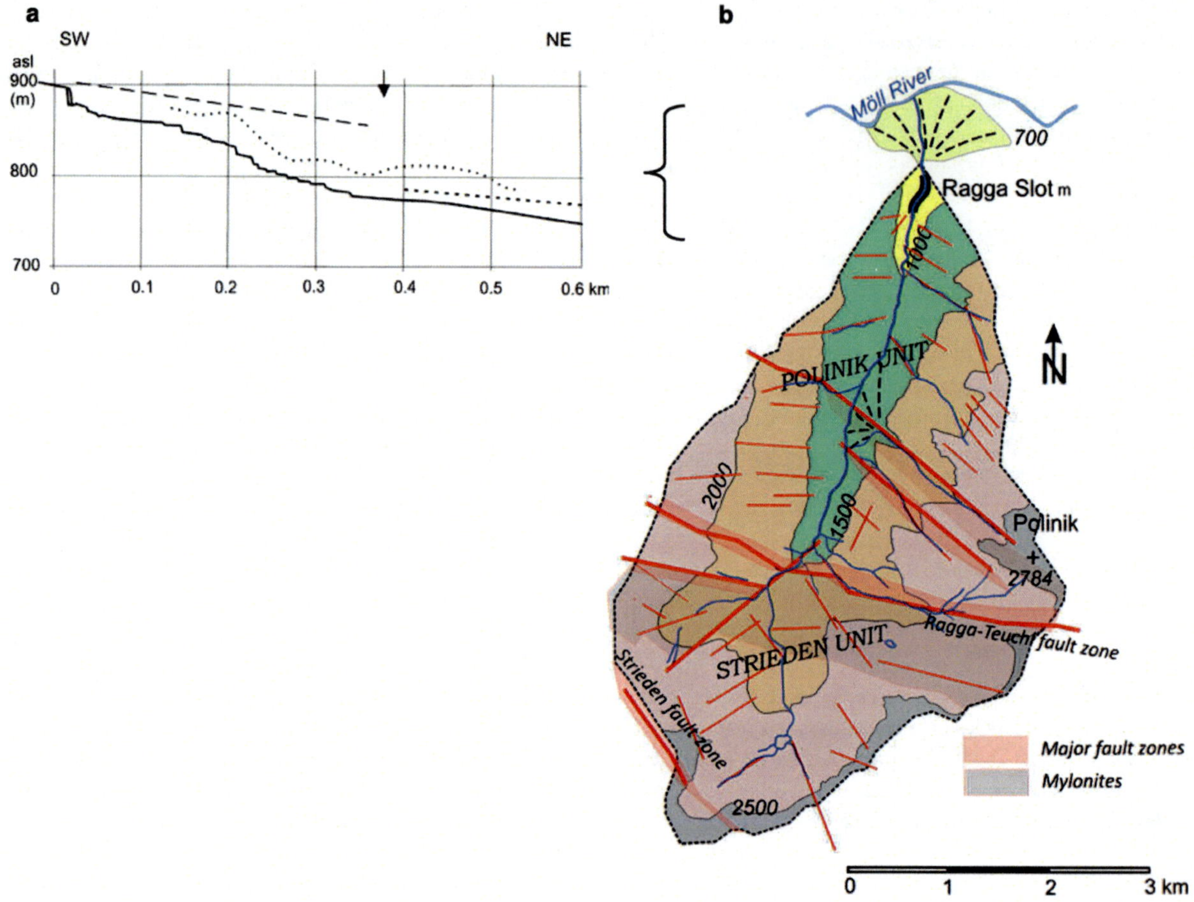

Fig. 26.4 **a** The long profile of the Ragga Slot (0.38 km, mean gradient 17.5°) is characterized by step-pool sequences. The head of the gorge is a knickpoint in resistant pegmatite producing a waterfall. The upper broken line marks the position of the slope break into the gorge, the dotted line shows the upper rim of the slot, the dashed line below shows the surface of the fan head. Its apex lies 70–80 m above the local base level (690 m asl) and is entrenched up to 10 m by the river. **b** The drainage basin of the Ragga River (500 m contour lines, geology following Hoke 1990) shows the great differences in altitude, especially on the side of Mt. Polinik, favouring gully development. The transverse strike of mayor fault lines combined with mylonite zones supports further strong material transfer from the valley sides into the Ragga channel, ensuring a strong coupling of slope and channel processes

forms, such as potholes, seem to be rather weakly developed and are hardly visible on higher parts of the rock sides. Further impressive features are large boulders trapped between narrow rock ledges. In particular a boulder of 2 m in diameter, hanging about 15 m above the channel bed, is one of the most scenic features of the slot (Fig. 26.5c).

The head of the gorge is distinguished by a 17 m high, free waterfall dropping over a thick lens of light reddish pegmatite with a circular plunge pool at the base (Fig. 26.5b). The downstream reach, again incised into the pegmatite, shows a meandering flow path with kolks and very smooth sidewalls. The features, both of the bedrock channel and its sidewalls, indicate that corrasion is the predominant process. Especially along the cascade-like step-pool reaches, concentrated shooting flow implies a high effect of cavitation. Further it may be assumed that the frequent interbedding of gneiss and mica schists and a steeply dipping net of joints and faults favour quarrying and plucking of the loosened rock in between.

26.3.3 The Gaisloch Gorge, Gailtal Alps

The drainage basin of the Jauken Graben (4.9 km^2) is situated on the northern side of the Jauken massif. The source area of the stream lies in a cirque floor at 1300–1400 m asl, surrounded by the 800 m high cirque walls of Mt. Torkofel (2276 m asl) which are built of Triassic Wetterstein Dolomite. After crossing a 250 m high step of Muschelkalk limestone in a steep V-shaped valley (stream gradient of

Fig. 26.5 Above from left to right: **a** a view of the emerging Ragga channel shows that the slot profile stands up clearly from a steep V-shaped gorge; **b** the head of the gorge, a 17 m high waterfall-step into resistant pegmatites; **c** a trapped boulder between rock ledges about 15 m above the actual river bed. Below from left to right: **d** a step-pool sequence; **e** smoothed channel side walls predominantly due to corrasion of concentrated shooting flow, and meandering steered by the direction of joints. *Photos* E. Stocker

32°), it successively forms a deeper incision into hard Carnian limestones and develops a distinctive gorge with some short slots, cut into the Triassic "Hauptdolomite".

The long profile of the gorge has a length of 1300 m and a gradient that decreases from 15° to 7° without pronounced steps. Only the barrage of coarse material in the slot reaches has locally caused downstream steepening of the channel. The graded long profile indicates that a step, originally positioned at the gorge mouth, has retreated upstream. In the last 400 m downstream, cross-sections show a 10–20 m wide valley floor and, 50 m above the floor, a break of slope (Fig. 26.6a, cross-section A); rock faces, however, have mostly retreated up to 100 m. Farther into the gorge, sidewalls become narrower and build alternately some tens of metres long slot reaches, and reaches in which the sidewalls have retreated, enlarging the valley width with gentle basal slopes, generating circularly closed valley sides (Fig. 26.6a, cross-sections B, C and D).

The gorge morphology stands out primarily because of the scenery of the sidewalls of dolomitic breccia. The bedrock channel is covered by an intermittent, but generally mobile fluvial bedload which contains a high portion of boulders partly provided by rock falls (Fig. 26.7a). Bedrock features are visible only over short distances and consist mainly of pools and potholes with generally rough surfaces due to the coarse structure of the dolomites. Reaches with

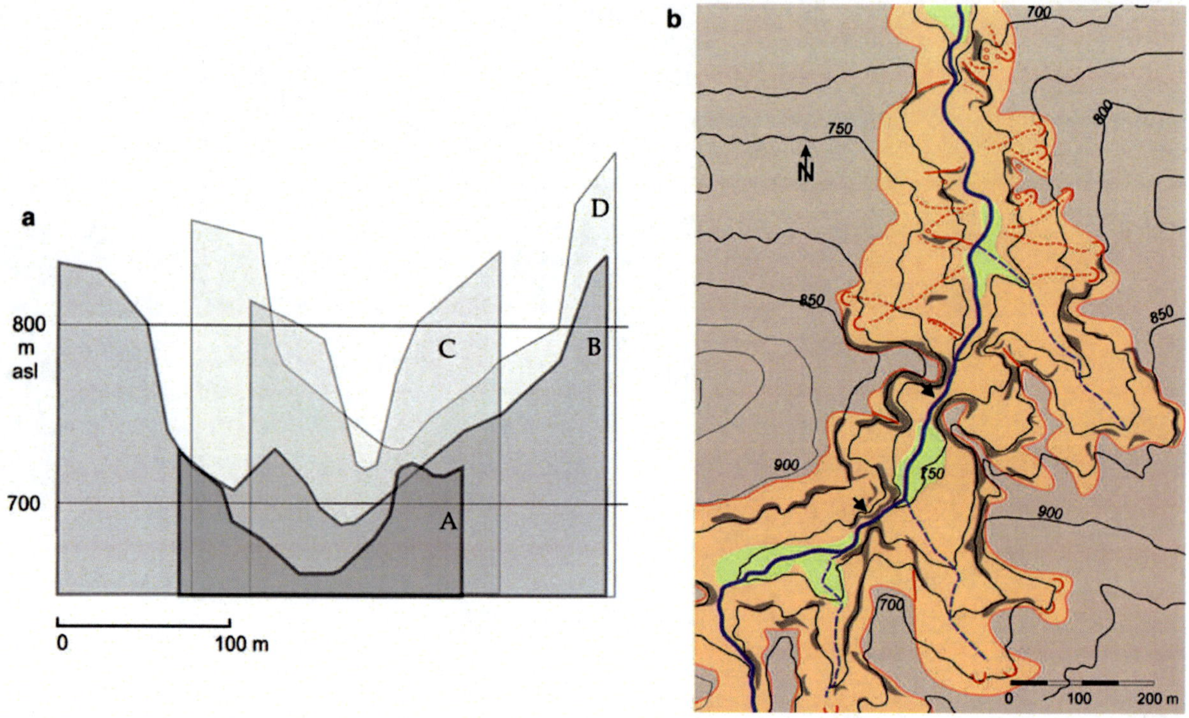

Fig. 26.6 **a** A series of cross-sections through the Gaisloch Gorge shows different stages of retrograding free faces of the gorge walls. Break of slope (A) indicates the original valley mouth, stages of retreated sidewalls (B and D), remnant of the inner slot (C). Contour lines (50 m) and cross-sections are based on KAGIS (2016). **b** The sketch shows the Gaisloch Gorge, which takes up the lower part of the drainage basin. It is developed in brecciated, very steeply dipping Norian Hauptdolomite. The gorge (pale brown) is subdivided into three widened sections which are separated by short slots (arrows). Seepage erosion supported by differential weathering created sidewall retreat, extended into scalloped walls. The most pronounced rock faces are drawn in dark grey, fluvial and slope deposits in light green, gullies in red dotted lines, seepage source areas in red symbols

more abundant sediment show the torrential character of the river bed. The water courses branch and tend to undercut the base of the sidewalls, creating pools or kolks especially in weakened rock. Nevertheless, floods enable transport, causing corrasion of the river bed except in places with strong material supply from sideward gully feeders.

The most spectacular rock features have developed due to differential and cavernous weathering combined with sidewall retreat. Scenic examples of rugged crests, rock towers, pointed peaks, rock castles, pits, tafoni hollows, rock windows and rock figures have been created (Fig. 26.7b, e). Both dense rock fracturing and steep or vertical dips favoured *spring sapping* (Laity 2004), conditioning a differential amount of rock wall retreat and the development of three widened gorge reaches (Figs. 26.6b and 26.7c) in which concentrated seepage created *scalloped escarpments* (Young and Young 1992) (Fig. 26.7d). They are divided by short slot reaches. The first stage of destruction of such slot sidewalls by backward retreat can be observed at the first intermediate slot: an initially generated kolk enlarged extensively by cavernous weathering is undermining the sidewall.

Under similar geologic conditions, the adjacent Ochsenschlucht Gorge shows the same characteristics: widened reaches with extensive retreat of scalloped escarpments alternate with narrow gaps in between.

26.3.4 The Garnitzen Gorge, Carnic Alps

The Garnitzen River, with a drainage basin of c. 8 km length and 21.7 km^2 area, flows through a complex sequence of U-shaped valleys and gorge reaches from the area of Mt. Gartnerkofel (2195 m asl) to the Gail Valley. The drainage basin is developed in west–east striking mountain ridges and shows a trellis drainage pattern (Figs. 26.2 and 26.8). In the highest epigenetic gorge, the river crosses a ridge of resistant Permian Trogkofel limestone, indicating an ancient stage of superposition. Downstream two further impressive gorges follow, crossing the resistant striped limestones of the Devonian Eder Formation. Glacial troughs, on the other hand, characterize the valley head and some of the subsequent reaches. Near the river mouth the trough floor

Fig. 26.7 Gaisloch Gorge. Above from left to right: **a** the river bed near the mouth shows a shallow cover of debris and boulders, its cross-section is moderately widened, and it is characterized by cavernous weathering; **b** rock windows, rugged crests, rock towers, pointed peaks; **c** a slot passage between widened gorge reaches; **d** sidewall retreat developing scalloped escarpments bounded by **e** deeply recessed backwalls. *Photos* E. Stocker

becomes extremely narrow and is flanked by almost 200 m high walls.

The mean river gradient of the 8.8 km long longitudinal profile (Fig. 26.9) is 6.5°. However, it has two steps of 40 m and 100 m, where the gradient rises markedly to 13–15° and the valley cross-profile turns into gorges that contain the most spectacular slot reaches. Their sides are composed of an upper concave slope pointing to a glacial genesis and lower steep gorge/slot walls resulting from more recent fluvial incision.

Channel reaches which follow the strike of the nearly vertically dipping Devonian striped limestones coincide with wider sections of trough valley bottoms, characterized by storage of both glacial and coarse fluvial sediments (Fig. 26.10). It is conspicuous that the channel is developed nearly entirely in the coarse sediment cover with boulders of more than 2 m across, originating from rockfall and Pleistocene moraine deposits. The scenic value of these reaches results from the abundance of morphological detail in the channel and on the valley sides.

In transverse valley reaches bedrock channels are also partly covered with generally coarse material, but their most distinct features are slot reaches that are even developed in stretches with low gradients, as in the 600 m long “Alte Klause” Gorge.

The bedrock channels of the steeper gorge and slot reaches are characterized by an alternation of waterfall steps and large pools. Bedrock incision has created corrasion

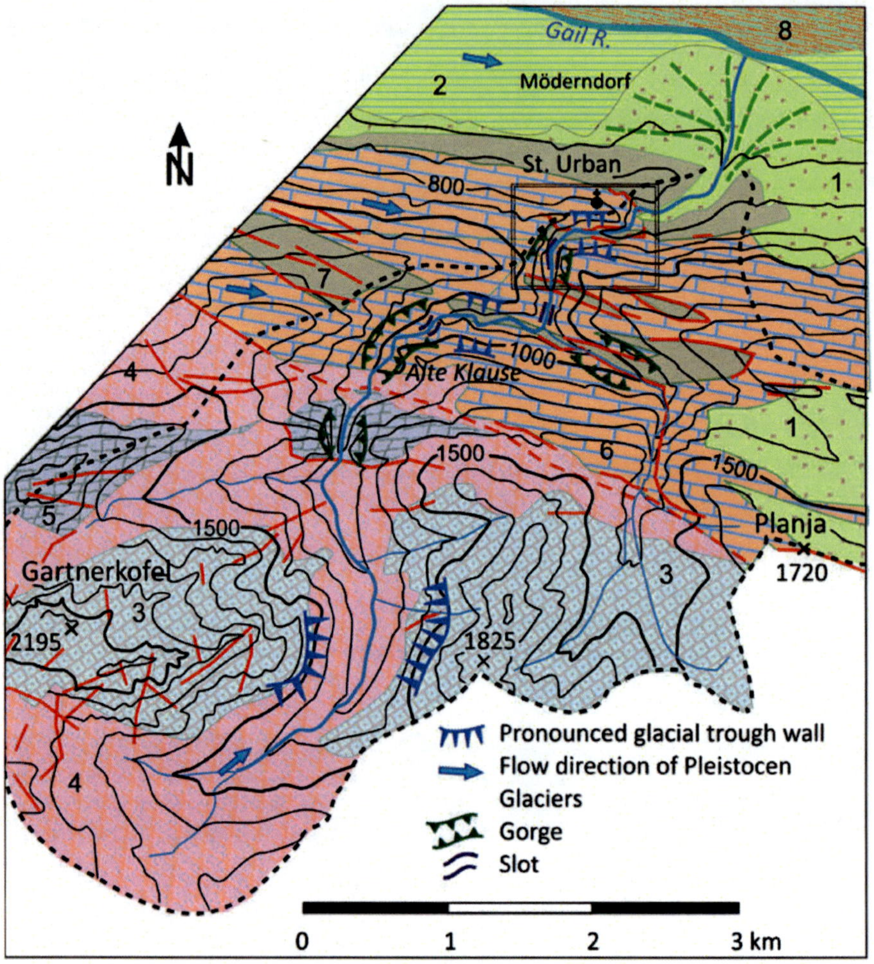

Fig. 26.8 Topography and simplified geological sketch of the Garnitzen Valley. The dashed line shows the catchment watershed. A series of trough valleys, gorges and slots is conditioned by transverse valley reaches through different rock formations (Schönlaub 1987, 1989). Colour code: (1) alluvial fan, colluvial deposits, till; (2) alluvial deposits of the Gail River; (3) Ladinian dolomite; (4) Permomesozoic sandstones and schists; (5) Permian limestones, (6) Devonian striped limestones; (7) Paleozoic schists and phyllites, (8) Gailtal Crystalline, consisting of diaphtoritic mica schists. The square box indicates the boundaries of the more detailed map in Fig. 26.10

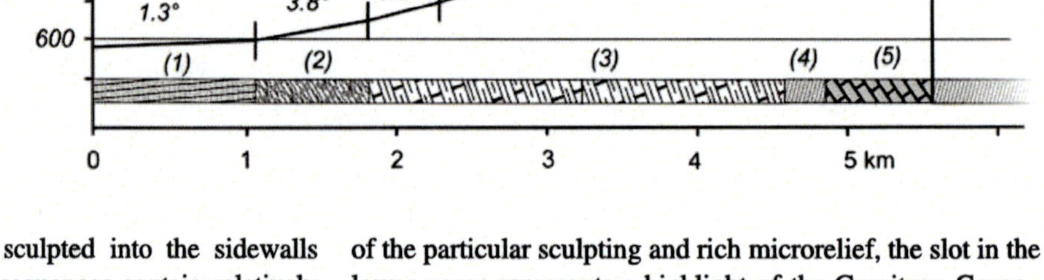

Fig. 26.9 Longitudinal profile and characteristic cross-sections of the Garnitzen Valley. (1) Holocene fluvial and glacial deposits; (2) Paleozoic schists and phyllites; (3) Devonian striped limestones; (4) Banked Permian limestones; (5) Permian limestones

hollows in an arched pattern, sculpted into the sidewalls (Fig. 26.11c, e). The step-pool sequences contain relatively large boulders and driftwood (Fig. 26.11d). Altogether, the depth of recent fluvial incision is only up to 20 m. Because of the particular sculpting and rich microrelief, the slot in the lower gorge represents a highlight of the Garnitzen Gorge.

The lowest reach of the Garnitzen River is incised into schists that are confined by a fault line within the striped

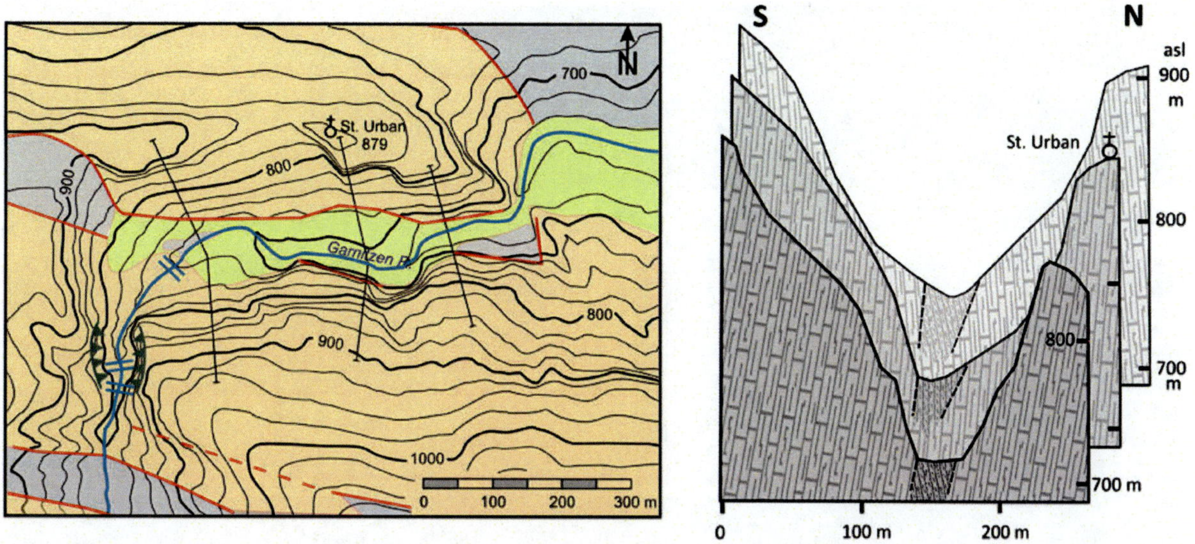

Fig. 26.10 Left: Map of the lower part of the Garnitzen Valley, as delimited in Fig. 26.8, showing the geometry of this U-shaped valley reach. The transverse valley further upstream crosses a ridge of Devonian striped limestones, depicted in light brown (schists in light grey and glacial/alluvial/slope deposits in light green). Right: three cross-sections without exaggeration give an impression of the importance of the geologic structure. Faults and strike within the nearly vertically dipping limestones are W-E directed. Interbedded schists between them favoured glacial downcutting. Sketches based on KAGIS (2016)

limestones. The fault is clearly important for the development of this subsequent outer valley reach. Its peculiarity are the nearly vertical, 190 m high sidewalls (Fig. 26.11a). Considering the width of the valley bottom, the valley cross-profile is that of a typical trough, albeit narrow, rather than a typical gorge. Therefore, systems of steeply dipping joints and the nearly vertical dip create an aspect reminiscent of a canyon (Fig. 26.11b).

26.4 Initiation and Development

Considering the low gradient of the main rivers, it seems paradoxical that most of the gorges occur near the mouths of the tributaries (Fig. 26.2). The 1–3 km wide Möll, Drau, and Gail valleys have gradients between 1 and 6‰, while the slope of their tributaries, even near the mouth, is many times steeper (mainly 5–10° or 88–176‰). This implies that the preexisting valley mouths experienced a base level drop which could have been induced by glacial overdeepening of the main valleys and/or tectonic/isostatic (?) uplift of the mountain margins.

The example of the Ragga slot (Fig. 26.4) shows clearly that its initiation resulted from cutting through the 150 m high step of a hanging valley. During incision the river deposited an alluvial fan (about 600 m long, with a volume of about 11 million m^3). The widening of the valley cross-profile above the vertical slot walls could be evidence for an initial subglacial incision as described by Kuhle (1991). In front of its exit, the long profile diverges strongly from a graded longitudinal profile and proves that even today the Ragga channel is far from being adapted to the local base level and downward erosion is still going on. Recent erosion processes are helped by high peak flow rates that may occur in the Ragga channel. They result from a combination of steep channel slope, drainage basin shape, and the presence of highly active debris flow feeders in the upper valley, which are well coupled to the main channel. Additionally, the narrows of the Ragga Valley favour the build-up of log-jams. Only the high energy of a flood wave following a log-jam burst may explain the presence of boulders trapped more than 10 m above the channel bed.

Similarly, the initiation of the Gaisloch Gorge (Fig. 26.6) started at a former knickpoint of the channel long profile, located at its exit to the main valley. In contrast to the Ragga Valley, the dissection of the former hanging valley step created a relatively well graded concave channel-longitudinal profile due to a more rapid advance of corrasion, quarrying, and plucking. Numerous small feeders of dolomitic and limestone debris have permitted rapid downward erosion, further promoted by the high erodibility of the dolomitic breccia and high probability of intense precipitation events in this area (Stocker 2003). Relict sidewalls about 50 m high near the channel exit indicate the youngest phase of vertical river incision that triggered differential retreat of free faces along the gorge sides. Accelerated headward erosion often generates an upstream extension of gorges that are coupled with systems of densely spaced

Fig. 26.11 Garnitzen Gorge. Above: **a**, **b** views of the narrow trough valley section in the Devonian striped limestones. Below: the most impressive slot reach showing **c** cascades with rock pools, **d** step-pool sequences in a channel reach with large boulders in the gorge crossing the Permian limestones, **e** river side corrasion in folded limestone. *Photos* E. Stocker

ravines and alpine gully systems, locally grown into badland-like features. While the bedrock channel has a veneer of mobile material and lacks spectacular sculpturing, the distinctive feature of the Gaisloch Gorge is the highly diverse microrelief of its sideslopes.

The tributaries draining the northern side of the Carnic Alps are arranged in a typical parallel drainage pattern and show similar knickpoints immediately before entering the main valley of the Gail River. However, the Garnitzen tributary is different and lacks a pronounced gorge before leaving the mountains. The trend of the drainage system diagonal to the strike of the geological units (Fig. 26.8) created a trellis drainage pattern (Morisawa 1985). Valley development was controlled by both superimposition and structural adaption leading to a sequence of strike- and transverse valley reaches. The gorges of the Garnitzen Valley are situated further upstream, mainly along transverse valley sections. The typical alternation of steps and basins in valleys that had been shaped by Pleistocene glaciers is expressed in the Garnitzen Valley through a wide, U-shaped upper reach, followed by wide trough cross-profiles in the subsequent reaches of its middle and lower course.

The long profile in the vicinity of the valley mouth is graded, and it is possible that the Pleistocene Gail Glacier may have reinforced scouring of the valley mouth. The ridge between the Gail- and the Garnitzen valleys reaches an altitude of 900–1200 m asl, whereas the ice surface of the former Gail Glacier lay at 1650 m asl (Van Husen 1987). In addition, the direction of the lowermost Garnitzen trough valley section deviates only slightly from the flow direction of the Pleistocene Gail Glacier (Fig. 26.8). Reinforced glacial downcutting of the mouth of the Garnitzen Valley would have increased the height difference to its upstream following transverse valley reach. In this way the original gorge reach near the mouth had been shifted into the next interior transverse valley reach promoting the development of inner gorges exemplified by three spectacular gorge and slot reaches, which alternate with glacial troughs.

However, in the Gailtal Alps inner gorges are more common, mainly because of active headward extension of gorge reaches and mountain gully systems, as in the 6 km long Ochsenschlucht gorges adjoining the Gaisloch Gorge. Further examples of extended inner gorges are the Kreutzen River gorge (15 km) and even the Gail Valley gorge of 21 km length above Kötschach, incised into crystalline rocks. It has a graded longitudinal profile from 15 to 2‰ and runs parallel to the periadriatic fault line, but c. 1–3 km north of it (Fig. 26.2).

26.5 Gorges as Natural Monuments and Geosites

Like waterfalls, gorges are regarded as extraordinarily attractive landforms. In referring to specific peculiarities Panizza (2009) qualified outstanding landscapes on the basis of type, scale and level, as landscapes with accentuated geomorphic diversity, which can be observed also in the present examples of the alpine heritage. They constitute also classical "geomorphosites" resulting both from their high scientific value and their additional values as aesthetic, ecologic, social/economic, and cultural-historical features (Reynard 2009). The focus on diversity assessment reveals substantial enlargements in geomorphic knowledge.

Thanks to the geological significance of the Carnic region and the effort of numerous researchers since the eighteenth century, one of the first Geoparks in Austria has been founded. Now its area of 830 km^2 contains not only the Austrian part of the Carnic Alps but also parts of the Gailtal Alps and the Lienz Dolomites. Only since 1978 the Garnitzen Gorge has been protected by means of recognition as a natural monument. Like many other gorges it was already opened by a trail in the nineteenth century (the Ragga Slot has been passable since 1882). Just as with other trails in difficult terrain, the Alpine Club and local organizations have played an important role in constructing and maintaining trails, often destroyed by catastrophic floods or by rock falls and snow avalanches. By efforts of the Geologische Bundesanstalt both the Ragga Slot and the Garnitzen Gorge have been equipped with geological teaching trails. The Carnic Alps Geopark, containing 79 geosites, primarily of geological interest and equipped with a modern visitor centre, has been member of the European Geopark Network since 2012 and the UNESCO Global Geoparks since 2015. The cultural and historic values are visualized in panels which show historical mining (Ragga Valley), the extraction of lime in historical furnaces (Gaisloch Gorge), traditional wood drifting, and the church of St. Urban (situated at the edge of the 200 m sidewall of the Garnitzen Gorge).

26.6 Conclusion

With a broad range of features of fluvial erosion, bedrock channels and sidewall development the three presented gorges reveal a high variety of landforms. The common characteristics of nearly all gorges and slots of western Carinthia is their initiation along a step created between the relatively steep tributaries and the main valleys with their low gradients in thick alluvial sediments.

In particular, the confluents from the central high mountains into the bottom of the Möll River clearly reveal the initiation of gorges and slots due to dissection of the steps of hanging valleys. Gorges that extend a few kilometres into the valleys may be explained either by a longer period available for downcutting, or by a more rapid development and retreat of knickpoints conditioned primarily by geologic structures. The tributaries of the main rivers of Drau and Gail show stronger aggradation and greater retreat of former knickpoints, originally situated at the mountain front. In the Gailtal Alps pronounced back- and downcutting of streams triggered the development of distinct gully systems in the upper reaches and led to rapid sidewall retreat resulting in scalloped escarpments.

In the Carnic Alps the Garnitzen Gorge has a singular position. Its general orientation trends diagonally to the striking ridges of Devonian limestones (which additionally are dipping nearly vertically). Therefore it developed partly as a subsequent valley and partly as a transverse valley, triggering a subdivision into basins and steps. Holocene gorge- and slot development is centred on the steps and possibly initiated subglacially in the Pleistocene. Basins, on the other hand, are storage areas both for glacially eroded material and for rockfall deposits containing boulders of extraordinary dimensions. The gorges and slots are integrated in a glacial trough valley which in the lowest reach is especially deep and narrow, thereby creating a spectacular valley entrance.

Acknowledgements I wish to express my gratitude to Prof. Michael Church who kindly took over the reading of my manuscript and provided me with valuable and helpful comments and suggestions.

References

Ahnert F (2015) Einführung in die Geomorphologie. Ulmer, Stuttgart
BEV (2005) Austrian Amap fly. Version 4.0, Bundesamt für Eich- und Vermessungswesen. Wien.

Braunstingl R (2005) Geologische Karte von Salzburg 1:200.000. Geologische Bundesanstalt (ed) Wien
Carter C, Anderson R (2006) Fluvial erosion of physically modeled abrasion-dominated slot canyons. Geomorphology 81:1–2
Dixon J (2011) Geomorphology: deceptively old alpine gorges. Nat Geosci 4:8–9
Embleton C, King C (1971) Glacial and periglacial geomorphology. Arnold, London
Feldman S, Harris A, Fairbridge R (1968) Drainage Patterns. In: Fairbridge RW (ed) Encyclopedia of geomorphology, Dowden, Hutchinson & Ross, Inc. Pennsylvania, Stroudsburg, pp 284–291
Fränzle O (1968) Valley evolution. In: Fairbridge RW (ed) Encyclopedia of geomorphology, Dowden, Hutchinson & Ross Inc. Pennsylvania, Stroudsburg, pp 1183–1189
Hoke L (1990) The Altkristallin of the Kreuzeck Mountains, SE Tauern Window, Eastern Alps—Basement crust in a convergent plate boundary zone. Jb Geol B-A 133:5–87
Husen D van (1987) Die Ostalpen und ihr Vorland in der letzten Eiszeit (Würm) 1:500.000. In: Geol B-A (ed) Wien
Ilickovic T, Schuster R (2014) Bericht 2014 über geologische Aufnahmen auf Blatt 181 Obervellach. Jb Geol B-A 157:310–314
Kahler F, Prey S (1963) Erläuterung zur Geologischen Karte des Naßfeld-Gartnerkofel-Gebietes in den Karnischen Alpen. Geol. B-A (ed) Wien
Kahler F, Prey S, Heritsch H (1959) Geologische Karte des Naßfeld-Gartnerkofel-Gebietes in den Karnischen Alpen 1:25.000. Geologische B-A (ed) Wien
Kelsey H (1988) Formation of inner gorges. CATENA 15:433–458
Knighton, D. (1998): Fluvial forms and processes. A new perspective. Arnold, London, 383 pp
Krzemień K (1999) Structure and dynamics of the high-mountain channel of river Plima in the Ortler-Cevedale Massif (South Tirol). In: Krzemień K (ed) River cannels—pattern, structure and dynamics. Prace Geograficzne 104, Institute of Geography of the Jagiellonian University, Kraków, pp 41–55
Kuhle M (1991) Glazialgeomorphologie. Wissenschaftliche Buchgesellschaft, Darmstadt
Laity J (2004) Groundwater. In: Goudie A (ed) Encyclopedia of geomorphology. Routledge, London, pp 497–500
Land Kärnten (2016) Kärnten Atlas. KAGIS Standartausgabe 1:10.000, Klagenfurt
Leser H (2003) Geomorphologie. Westermann, 8. ed. Braunschweig
Linner M, Reitner J, Pawlik, W (2013) Geologische Karte der Republik Österreich 1:50.000, 179 Lienz. Geologische Bundesanstalt (ed) Wien
Machatschek F (1973) Geomorphologie. Teubner, Stuttgart
McEven L, Matthews J, Shakesby R et al (2002) Holocene gorge excavation linked to boulder fan formation and frost weathering in a Norwegean alpine preglaciofluvial system. Arct Antarct Alp Res 34:346–358
Montgomery D, Korup O (2010) Preservation of inner gorges through repeated Alpine glaciations. Nat Geosci 4:62–67
Morisawa M (1985) Rivers form and process. Longman, London, New York
Panizza M (2009) The geomorphodiversity of the dolomites (Italy): a key of geoheritage assessment. Geoheritage 1:33–42
Pestal G, Rataj W, Reitner J et al (2006) Geologische Karte der Republik Österreich 1:50.000, 182 Spittal an der Drau. Geologische Bundesanstalt (ed) Wien
Putiš M, Bezak V, Kohut M et al (1997) Bericht 1996 über geologische Aufnahmen im Altkristallin auf Blatt 181 Obervellach. Jb Geol B-A140: 345–348
Reynard E (2009) Geomorphosites: definition and characteristics. In: Coratza P, Regolini-Bissig G (eds) Reynard E. Geomorphosites, Pfeil, München, pp 9–20
Sanders D, Wischounig D, Gruber A et al (2014) Inner gorge–slot canyon system produced by repeated stream incision (eastern Alps): Significance for development of bedrock canyons. Geomorphology 214:465–484
Schönlaub H, Bär R, Beckmann U, Boeckelmann K et al (1987) Geologische Karte der Republik Österreich 1:50.000, 198 Weissbriach. Geologische Bundesanstalt (ed) Wien
Schönlaub H, Auferbauer H, Deutsch A et al (1989) Geologische Karte der Republik Österreich 1:50.000, 199 Hermagor. Geologische Bundesanstalt (ed) Wien
Schönlaub H (1988) Vom Urknall zum Gailtal. 500 Millionen Jahre Erdgeschichte in der Karnischen Region. Geologische Bundesanstalt, Wien
Schönlaub H (1997) Geologische Karte der Republik Österreich 1:50.000. 196 Obertilliach, Geologische Bundesanstalt (ed) Wien
Schuster R (2006) Erläuterungen zu Blatt 182 Spittal an der Drau. Geol B-A (ed) Wien
Speil R (1985) An tosenden Wassern. Klammen und Schluchten in Österreich. Styria. Graz Wien Köln
Stocker E (2003) Conditions of alpine gully development as exemplified by the Austrian Alps. Studia Geomorphologica Carpatho-Balcanica 37:5–27
Stocker E (1989) Strukturen de Mikroreliefs in der subalpin-alpinen Höhenstufe in Abhängigkeit von den Standortverhältnissen. In: Hüser K et al (ed) Bayreuther Geowissenschaftliche Arbeiten 14, Naturwissenschaftliche Ges. Bayreuth, Bayreuth, pp 47–57
Tinkler K (2004) Gorge and Ravine. In: Goudie A (ed) Encyclopedia of geomorphology. Routledge, London, pp 486–487
Tinkler K, Wohl E (1998) A primer on bedrock channels. In: Tinkler K, Wohl E (eds) Rivers over rock. Fluvial processes in bedrock channel. Geophysical monograph 107, American Geophysical Union, Washington, pp 1–18
Van Bemmelen R (1957) Beitrag zur Geologie der westlichen Gailtaler Alpen (Kärnten, Österreich) (erster Teil). Jb Geol Reichsanst 100 (2):179–212
Venturini, C Pondrelli M (2002) La catena ercinica non metamorfica. Le Alpi Meridionali centro-orientali nella catena alpina. In: Società Geologica Italiana (ed), Guide Geologiche Regionali. Alpi e Prealpi Carniche e Giulie. Friuli Venezia-Giulia, Roma, pp 71–75
Venturini C et al (2001) Geologic Map of the Carnic Alps 1:25.000. Dipartimento di Scienze della Terra e Geologico-Ambientali, Università di Bologna (ed.). Bologna, Udine
Whipple K, DiBiase R, Crosby B (2013) Bedrock rivers. In: Shroder J, Wohl E (eds) Treatise on geomorphology, fluvial geomorphology, vol 9. Academic Press, San Diego, pp 550–573
Wohl, E (1998) Bedrock channel morphology in relation to erosional processes. In: Tinkler K, Wohl E (eds) Rivers over rock. Fluvial processes in bedrock channel. Geophysical monograph 107, American Geophysical Union, Washington, pp 133–151
Young A (1972) Slopes. Longman, London, New York
Young R, Young A (1992) Sandstone landforms. Springer Series in Physical Environment 11, Springer, Berlin Heidelberg

Erich Stocker is a retired associate professor of physical geography at the University of Salzburg, Department of Geography and Geology. His research in Alpine and Mediterranean areas is focused on slope development, geomorphological mapping and monitoring of environmental change on alpine and subalpine slopes in the Eastern Alps.

27 Rock Glaciers in the Austrian Alps: A General Overview with a Special Focus on Dösen Rock Glacier, Hohe Tauern Range

Andreas Kellerer-Pirklbauer, Gerhard Karl Lieb, and Viktor Kaufmann

Abstract

Rock glaciers are prominent landforms in mountain regions and as such indicate permafrost conditions at present or in former times. So-called active rock glaciers consist of debris and ice and creep slowly downslope thereby forming flow structures with ridges and furrows. This geomorphological expression of permafrost creep is commonly preserved after the complete melt-out of the ice component of the now relict rock glacier body. Rock glaciers are widespread in the Austrian Alps, with c. 2900 relict rock glaciers containing no ice at present and c. 1700 intact rock glaciers which contain ice and indicate the widespread permafrost occurrence. A very recent inventory lists even more than 5700 rock glaciers and related landforms. One of the best studied active rock glaciers in Austria is the Dösen Rock Glacier located in the Central Austrian Alps. It has been investigated during several national and international projects since 1993, dealing with permafrost conditions and distribution, surface kinematics, internal structure and age. Significant ground surface warming of the rock glacier body occurred since 2007, accompanied by general acceleration of the rock glacier surface flow velocity. Relative age dating of the landform indicates a long and variable formation history over a period of several thousand years. Selected topics with a geomorphological focus and the relationships between permafrost and environmental conditions are communicated to the public by an educational trail which has been established at this rock glacier and its vicinity.

A. Kellerer-Pirklbauer (✉) · G. K. Lieb
Institute of Geography and Regional Science, University of Graz, Heinrichstrasse 36, 8010 Graz, Austria
e-mail: andreas.kellerer@uni-graz.at

V. Kaufmann
Institute of Geodesy, Graz University of Technology, Steyrergasse 30, 8010 Graz, Austria

Keywords

Rock glacier evolution • Rock glacier monitoring • Ground temperature • Dösen Valley • Hohe Tauern National Park

27.1 Introduction

Rock glaciers are prominent landforms in alpine regions which typically evolve over a period of several centuries to millennia and indicate present or past permafrost conditions (Barsch 1996; Haeberli et al. 2006; Berthling 2011). Rock glaciers are commonly characterized by a flow structure similar to lava flows with longitudinal and concave-downward bent transversal ridges and furrows at the surface. In case a rock glacier front moves into a steep terrain, the rock glacier body might start to disintegrate (Avian et al. 2009) or even completely collapse (Schoeneich et al. 2015). Rock glaciers can be classified into active, inactive, pseudo-relict and relict, depending on their activity and permafrost-depended ice content (Barsch 1996; Kellerer-Pirklbauer 2008a, 2019) (Fig. 27.1). Active rock glaciers contain widespread ice and move downslope with velocities in the range between a few centimetres to some metres per year. Along with the thaw of permafrost these rock glaciers might transform, first, to inactive rock glaciers (no movement, but still widespread permafrost ice), then to pseudo-relict rock glacier (no movement, sporadic permafrost ice) and, finally, to relict rock glaciers (no movement, no permafrost). Inactive and active rock glaciers are often combined to the class 'intact', particularly when surface movement data are lacking (Barsch 1996). The ice component within a rock glacier may have formed by the transformation of snow via firn to ice ('sedimentary ice' like in glaciers; Haeberli and Vonder Mühll 1996), or by freezing of water ('congelation ice'). The term 'rock glacier' is misleading because rock glaciers in most cases are not in a

C. Embleton-Hamann (ed.), *Landscapes and Landforms of Austria*, World Geomorphological Landscapes,
https://doi.org/10.1007/978-3-030-92815-5_27

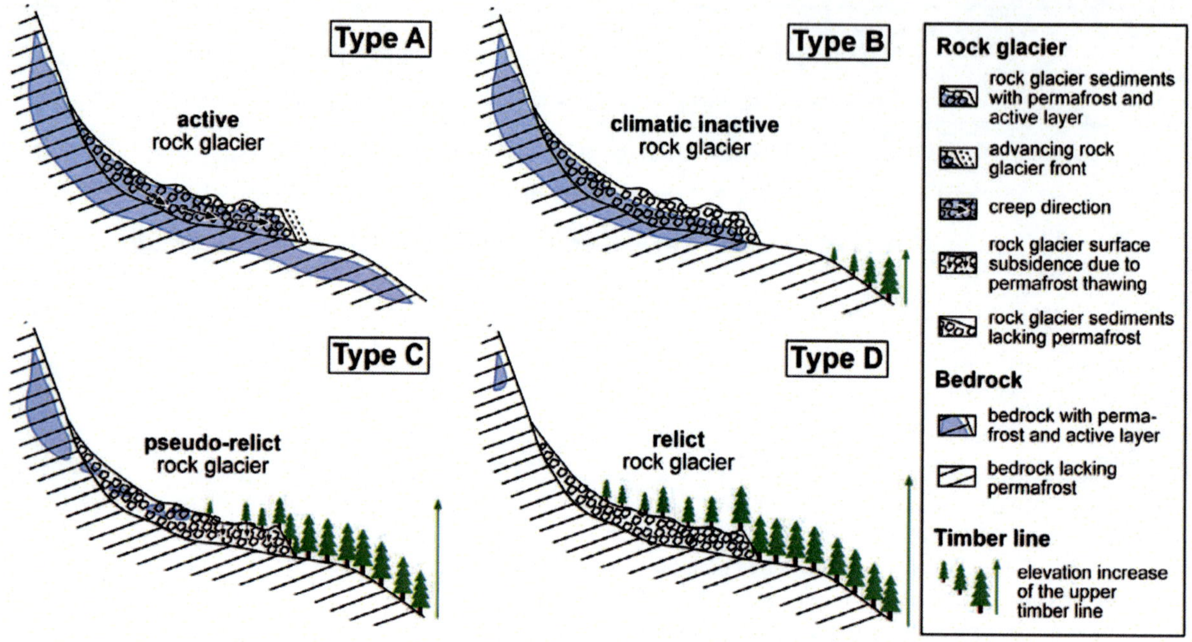

Fig. 27.1 Schematic diagram of rock glacier classification according to activity and permafrost presence: Types A, B, and D are defined as proposed in Barsch (1996). Type C is considered as an intermediate type between B and D, i.e. a rock glacier which looks visually relict (with collapse features due to ground subsidence processes; partly covered by vegetation) but contains sporadic to discontinuous permafrost (modified after Kellerer-Pirklbauer 2008a, Fig. 5.2). The indicated changes of the timber line should be seen as a proxy for changes in vegetation

direct genetic relation with glaciers. Glaciers predominantly consist of sedimentary ice whereas rock glaciers are a mixture of debris and ice of mostly different origin as described above (Barsch 1996). Rock glacier derived from talus sediments (i.e. talus rock glaciers) are genetically different from rock glaciers derived from moraine deposits at the glacial-proglacial transition zone (i.e. debris rock glaciers) (cf. Barsch 1996). However, there is a continuum between these two types.

Permafrost is a thermal phenomenon in periglacial environments and is widespread in the Austrian Alps (Fig. 27.2). Permafrost is defined as ground material (soil and rock unrelated to ice content) that remains at or below 0 °C for at least two years (van Everdingen 1998). The superficial layer which thaws every summer is termed 'active layer'. Its thickness is in the order of metres in the Austrian Alps, depending on elevation, slope aspect, slope angle and especially substrate and seasonal snow cover. In contrast, vegetation only plays a minor role because permafrost in the Alps almost entirely occurs at elevations where only a very sparse vegetation cover exists.

A permafrost model covering the entire Alps in a homogenous approach is the 'Alpine Permafrost Index Map' or APIM (Boeckli et al. 2012). According to this index, permafrost in Austria covers 484 (index $\geq$ 0.9) to 2907 km^2 (index $\geq$ 0.1) (Fig. 27.2a). APIM describes semi-quantitatively the occurrence of permafrost and varies from 'permafrost nearly in all conditions' to 'permafrost only in very favourable conditions'. Most permafrost areas are to be found in the federal province of Tyrol (66%), followed by Salzburg (16%) and Carinthia (10%). Permafrost areas in Austria are substantially larger than glaciated areas which cover only 361.5 km^2 (Paul et al. 2021).

Rock glaciers can be found in many mountain regions of the Austrian Alps. Rock glacier research in Austria was initiated in the Ötztal Alps in the 1920s when the active rock glacier Innere Ölgrube was first described (Finsterwalder 1928). In 1938 Wolfgang Pillewizer began to measure flow velocities at the Hochebenkar rock glacier—also located in the Ötztal Alps—thereby initiating the worldwide longest record of rock glacier flow velocity data (e.g. Schneider and Schneider 2001). Several rock glacier inventories have been elaborated since the 1990s (see Sect. 27.2). Some statistical data are summarized in Fig. 27.2b. By far most of the 4550 inventoried rock glaciers are located in the federal province of Tyrol with 3145 rock glaciers (55% thereof relict), followed by the provinces Carinthia (n = 483; 74% thereof relict), Styria (n = 415; 98% thereof relict) and Salzburg (n = 304, 78% thereof relict). Two hundred two rock glaciers were mapped in Vorarlberg (72% thereof relict) and finally one relict rock glacier was mapped—according to Lieb et al. (2012)—in the province of Upper Austria. A very

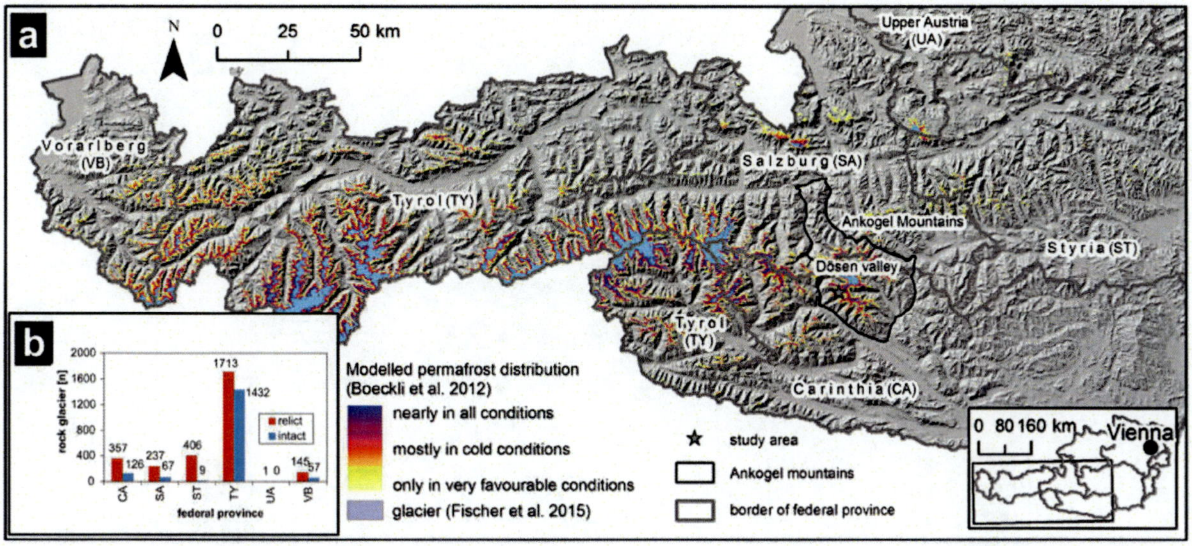

Fig. 27.2 Permafrost, rock glaciers and glaciers in Austria. **a** Spatial extent of permafrost and glaciated areas in Austria. The location of the Dösen Rock Glacier is within the Ankogel Mountains. **b** Number of relict and intact (active and inactive) rock glaciers in the different Austrian federal provinces according to Lieb et al. (2012) [for CA, SA, ST and UA], Krainer and Ribis (2012) [for TY], and Stocker (2012) [for VB]. Abbreviations of federal provinces: CA = Carinthia; SA = Salzburg, ST = Styria, TY = Tyrol, UA = Upper Austria, VB = Vorarlberg

recent rock glacier inventory lists even more than 5700 rock glaciers and related landforms—such as protalus rampars—in Austria (Wagner et al. 2020). In contrast to the high number of rock glaciers only few of them have been studied in more detail regarding to, e.g. movement, internal structure, nourishment processes or landform age (see Krainer et al. 2012 for references and details). One of the best studied rock glaciers in the Austrian Alps is the Dösen Rock Glacier in the Dösen Valley, Ankogel Mountains (Fig. 27.2a), where field-based research was initiated as early as 1993. In the following (1) an overview of the spatial distribution and densities of the rock glaciers in Austria is given and (2) the Dösen Rock Glacier is presented as a key study site for detailed rock glacier research.

27.2 Density and Distribution of Rock Glaciers in the Austrian Alps

Rock glacier density—expressed as the number of rock glaciers per grid cell—for entire Austria (Fig. 27.3) was calculated in ArcGIS10.1. This was accomplished by combining a regular grid with a spatial extent of 10 × 10 km covering entire Austria with the centroid of the rock glacier polygons of three published inventories. The three inventories have been elaborated by (i) Lieb et al. (2012)—analysed in Kellerer-Pirklbauer et al. (2012)—regarding the federal provinces of Salzburg (SA), Upper Austria (UA), Styria (ST), and Carinthia (CA), (ii) by Krainer and Ribis (2012) considering the federal province of Tyrol (TY) and finally (iii) by Stocker (2012) for the federal province of Vorarlberg (VB). The GIS data of the former inventory is accessible via the Pangaea database (Lieb et al. 2012). The data of the latter two inventories have been kindly provided by K. Krainer and K. Stocker, respectively. Pseudo-relict rock glaciers were not accounted separately in the inventories because field data (geophysics or boreholes) would be necessary to distinguish them from relict ones. Furthermore, active and inactive rock glaciers were combined to intact ones in our analysis. A new airborne laser scanning-based rock glacier inventory for central and eastern Austria was recently elaborated and published (Kellerer-Pirklbauer et al. 2016; Wagner et al. 2020). However, the dataset was not available during the time when the density calculation was accomplished.

Figure 27.3 illustrates the spatial distribution of 4550 inventoried rock glaciers, differentiating between all (Fig. 27.3a), relict (Fig. 27.3b), and intact ones (Fig. 27.3c). Maximum density values in terms of the number of rock glaciers per 100 km^2 are 129 for all, 86 for intact, and 75 for the relict ones. The maximum cell value of all three classes in Fig. 27.2 was quantified for the federal province of Tyrol. In contrast, only a few rock glaciers in Styria (n = 9) were classified as intact although the number of relict rock glaciers in this province is high (n = 406). The mean density for all grid cells with at least one rock glacier within a given cell is 19.6 rock glaciers per 100 km^2 for all (232 relevant grid cells), 12.8 for relict (223 grid cells), and 13.3 for intact (127

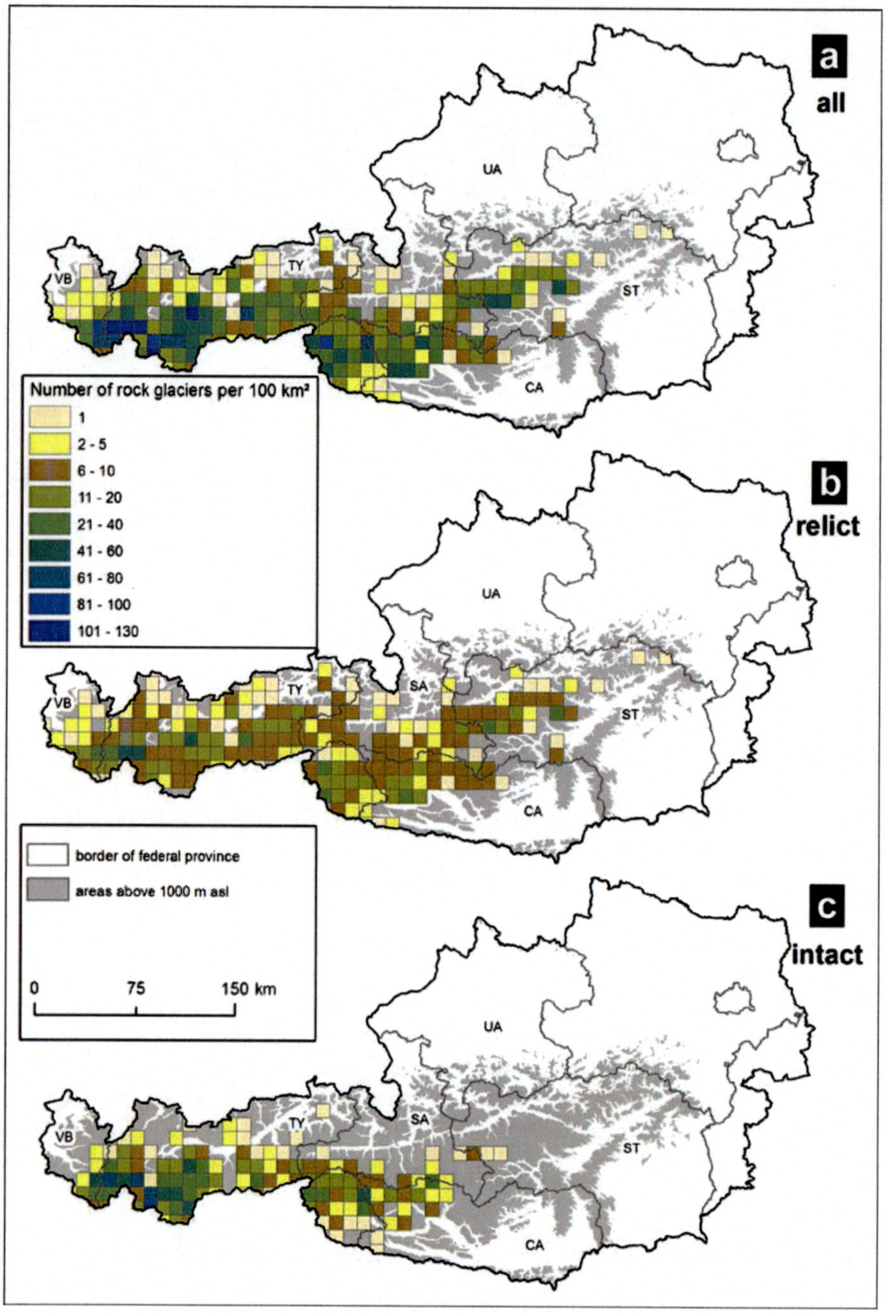

Fig. 27.3 Rock glacier distribution and density in the Austrian Alps based on the three inventories mentioned in the text; abbreviations of federal countries see Fig. 27.2. **a** distribution of all rock glaciers. **b** distribution of relict rock glaciers. **c** distribution of intact rock glaciers. Areas in Austria above 1000 m asl are indicated for topographical orientation

grid cells) ones. For comparison, rock glacier densities in the eastern Swiss Alps exceed 30 active rock glaciers per 210 km^2; i.e. > 14.3 active rock glaciers per 100 km^2 (Barsch 1980). The mean density of a further rock glacier-rich area in the West of Greenland, Disko Island, is 24 per 100 km^2 (Humlum 1998). These comparisons show that the Austrian Alps might be regarded as a region with very abundant rock glaciers. The federal province of Tyrol is by far the one with the highest number of both relict and intact rock glaciers. The ubiquity of relict rock glaciers in the federal provinces of Styria (cf. Figure 29.1 in Chap. 29) and Carinthia can be explained by the long period of existence of favourable conditions for rock glacier evolution during the Alpine Lateglacial period (Kellerer-Pirklbauer et al. 2012).

27.3 Dösen Rock Glacier

27.3.1 Geographical Setting of the Rock Glacier

The Dösen Rock Glacier is one of the 1691 inventoried intact rock glaciers of Austria. This rock glacier is located at the head of Dösen Valley which belongs to the Ankogel Mountains of the Hohe Tauern Range in Central Austria (Fig. 27.2a). The rock glacier occupies the altitude belt from 2340 to 2620 m asl, covers an area of 0.2 km^2, is 950 m long and up to 300 m wide. The main creep direction of the rock glacier is from east to west (Fig. 27.4). Several other smaller rock glaciers, a cirque lake (Dösener Lake) and distinct terminal moraines of presumably Younger Dryas age (Lieb 1996) are located further downstream of the rock glacier (Fig. 27.5). The material building up the rock glacier is metamorphic rock, primarily granitic gneiss forming large very angular to angular blocks at the surface.

27.3.2 Geomorphology of the Rock Glacier

The Dösen Rock Glacier is a typical tongue-shaped rock glacier as can be seen in plain view (Fig. 27.5) or in the terrestrial images (Fig. 27.4). Particularly in the lower and central part of the rock glacier the surface morphology is characterized by very distinct flow structures with typical longitudinal and transverse, downward bent furrows and ridges giving the rock glacier a lava flow appearance. This geomorphic structure is the visual expression of internal deformation and inherent creep.

The flow appearance of the rock glacier surface is very well depicted in the slope angle map (Fig. 27.6). Orange and red colours indicate the steep (30–50°) front of flow structures within the rock glacier body but also at its frontal and lateral areas. Yellowish colours indicate areas within the rock glacier surface which are substantially less inclined. Steeper and flatter areas alternate along a longitudinal profile of the rock glacier. The frontal slope of the rock glacier with the spring beneath it (Fig. 27.4b) has a slope angle commonly exceeding 40°, in a few places even 50°. Such a configuration at the front is typical for active rock glaciers.

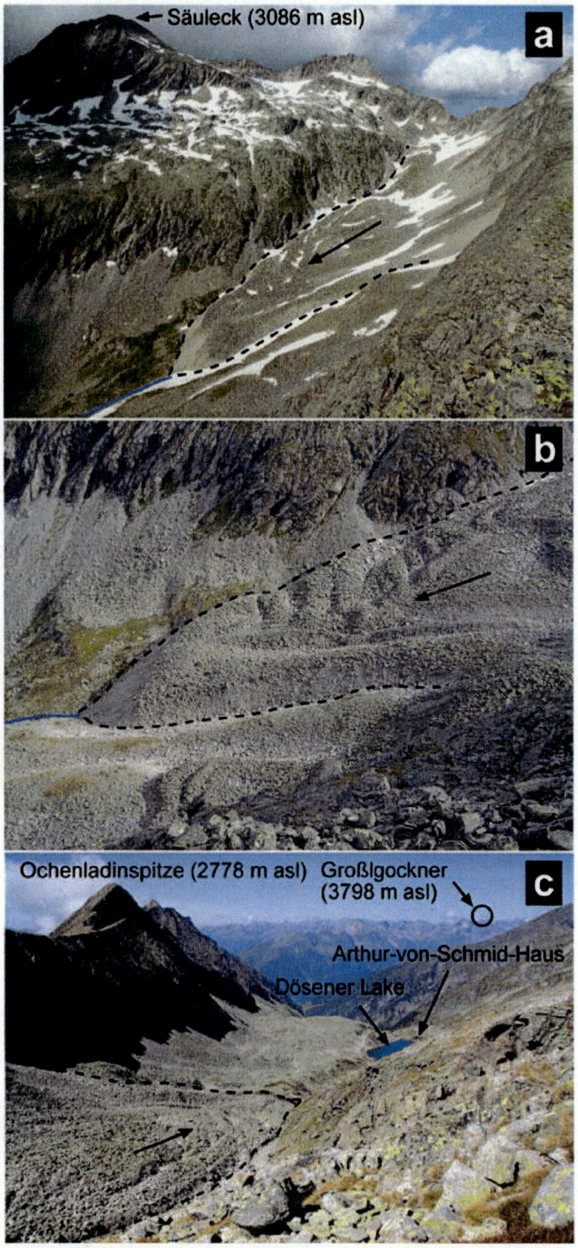

Fig. 27.4 Terrestrial photographs of the active Dösen Rock Glacier and its vicinity. The black arrow indicates the movement direction and the dashed line the margin of the rock glacier. The blue line retraces the creek emanating from the rock glacier terminus. Note the characteristic flow structures of the rock glacier, the snow patches in the depressions (in **a**), and the widespread lichen cover of the blocks (greenish color). **a** view towards NE. **b** View towards N. **c** View towards W with the highest mountain of Austria (Grossglockner) in the far background (47 km away). Photographs by the authors

27.3.3 Climate and Ground Temperature Conditions at the Rock Glacier

The lower limit of permafrost in alpine regions is very much linked to climate and its relationship with the ground temperature. The ground temperature itself is influenced not only by atmospheric conditions, but also by topography, the seasonal snow layer, and the type of substrate (bedrock, coarse- or fine-grained material, soil). In Europe, one might expect discontinuous permafrost (continuity of permafrost

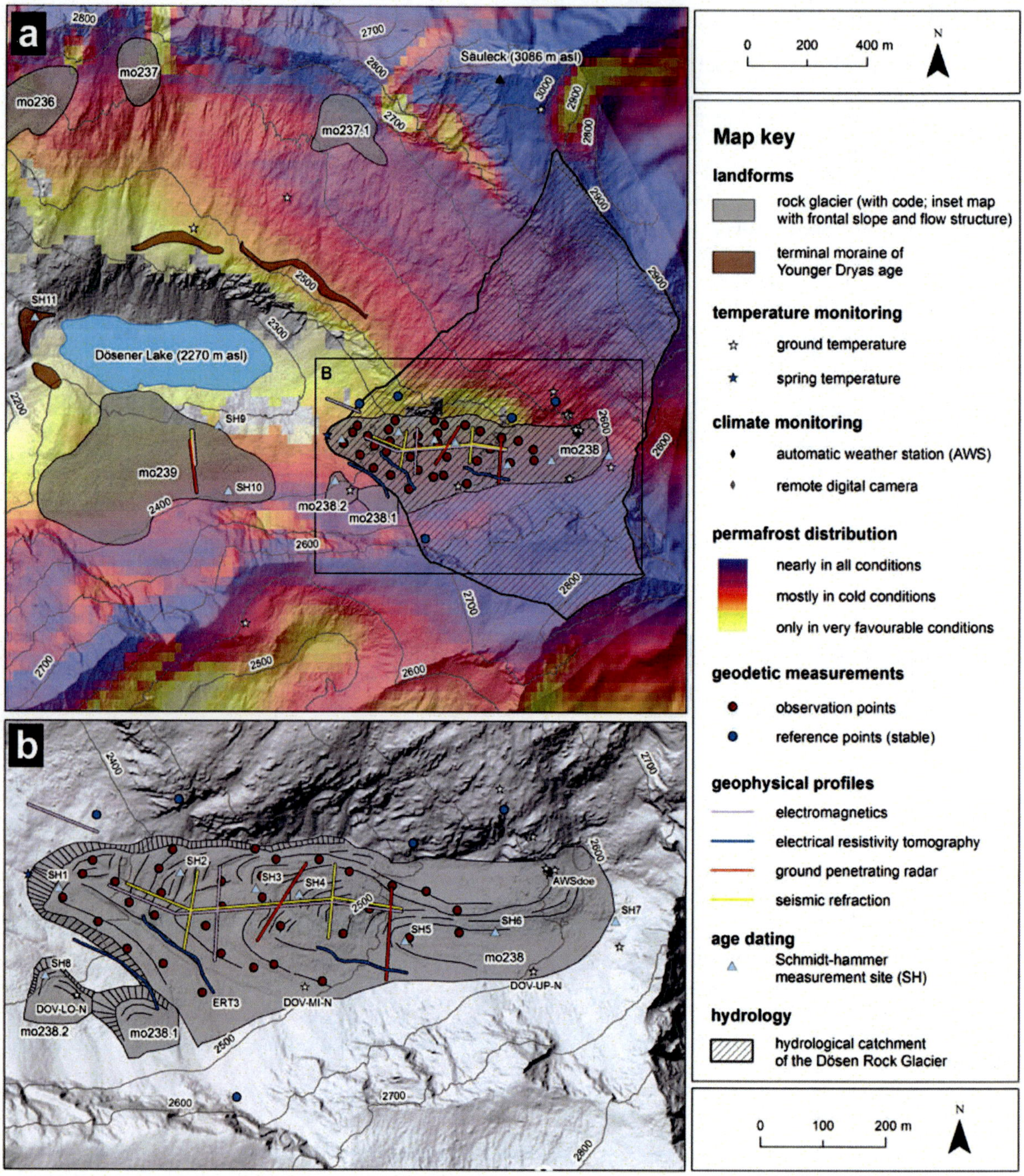

Fig. 27.5 The Dösen Rock Glacier and its vicinity: **a** overview with the valley head of the Dösen Valley including the mountain Säuleck, which is the highest summit in the hydrological catchment of the rock glacier. **b** detailed map of the rock glacier and its close vicinity. Measurement locations, monitoring sites and profiles where geophysical measurements were carried out in the past are indicated and partly labelled. Rock glacier codes according to the inventory by Lieb et al. (2012). Permafrost distribution according to Boeckli et al. (2012)

coverage 50–90% of a given area) at a mean annual air temperature (MAAT) of −1 to −2 °C (Barsch 1978; Haeberli et al. 1993; Humlum 1998). The lower limit of sporadic permafrost (permafrost coverage 10–50%) can be expected at a MAAT of 0 °C (Barsch, 1978; Humlum and Christiansen 1998).

An automatic weather station was installed in 2006 at a large block in the upper part of the rock glacier (Fig. 27.5b)

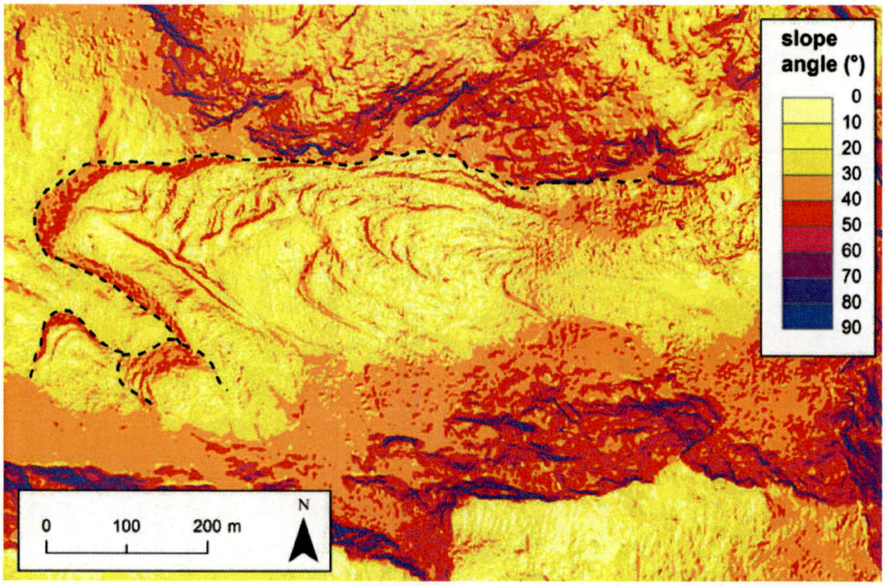

Fig. 27.6 Slope angle map of the Dösen Rock Glacier and its vicinity based on a 1 m digital elevation model derived from airborne laser scanning data. Note the distinct flow structures with steep fronts and slopes at the downvalley side with gradients in the range of 30° to more than 50°. The dashed line marks the margin of the rock glaciers

to continuously record air temperature, air humidity, wind speed, wind direction, and global radiation. In addition, a remote digital camera used for rock fall and snow monitoring in the rooting zone of the rock glacier was installed in 2006 (Fig. 27.5b). The MAAT measured at this weather station during the ten year period 2007–2017 was −1.62 °C (Fig. 27.7a) indicating discontinuous permafrost. According to the Alpine-wide regional permafrost modelling approach APIM (Fig. 27.5a), permafrost is widespread at the site of the rock glacier and the headwalls behind. According to APIM, permafrost does widely exist in the entire head of the Dösen Valley, although some areas exposed to higher radiation are in less permafrost-favourable conditions. This general pattern is in good agreement with earlier permafrost estimates by Lieb (1996), who claimed the existence of permafrost at north-exposed, blocky sites at 2270 m asl, whereas on the south-exposed slopes the lower limit of permafrost was predicted for elevations of 2950 m asl.

The monitoring of ground temperature at the surface and at the near surface (at depths of up to 3 m) at 12 monitoring sites (Fig. 27.5) was also initiated in 2006 and has been operating successfully since then. The 12 monitoring sites are located at different elevations, aspects, and substrates and log automatically ground temperatures at 1 h-intervals using miniature temperature data loggers (Kellerer-Pirklbauer 2016). Deviations of mean annual values (here of the period 1 August to 31 July) from the mean value of the ground surface temperature (MAGST) during the entire period 2007–2017 are shown in Fig. 27.7b–d for three sites at or very close to the rock glacier. The results indicate remarkable interannual variations in both air and ground surface temperatures. Both air temperature and rock glacier monitoring sites lacking a long-lasting seasonal snow cover (DOV-LO-N) show a statistically significant warming trend over the 10 year period. This is in agreement with the warming trend of ground temperatures in central and eastern Austria (Kellerer-Pirklbauer 2016). In contrast, the two sites at the rock glacier surface which are strongly influenced by the long-lasting seasonal snow cover do not show any sign of a clear and statistically significant trend in warming or cooling. Long seasonal snow cover causes an effective de-coupling of ground temperature from air temperature during autumn, winter, and spring masking any trends in atmospheric warming. Furthermore, although the DOV-LO-N site is the lowest one of the three ground temperature monitoring sites depicted in Fig. 27.7, it is by far not the warmest one. Hence, there is no decrease of the MAGST with elevation at the rock glacier sites which we might normally assume. This circumstance is related to the dominance of local topoclimatic conditions which are more important for ground thermal conditions than the common decrease of air temperature with altitude.

Five different types of geophysical methods (seismic refraction, ground penetrating radar, electromagnetics, electrical resistivity tomography, and very low frequency electromagnetic measurements) were applied so far at Dösen Rock Glacier in order to reveal its internal structure and its ground thermal conditions (Schmöller and Fruhwirth 1996; Kellerer-Pirklbauer et al. 2014). Figure 27.5 shows the locations of the geophysical profile measurements at the rock

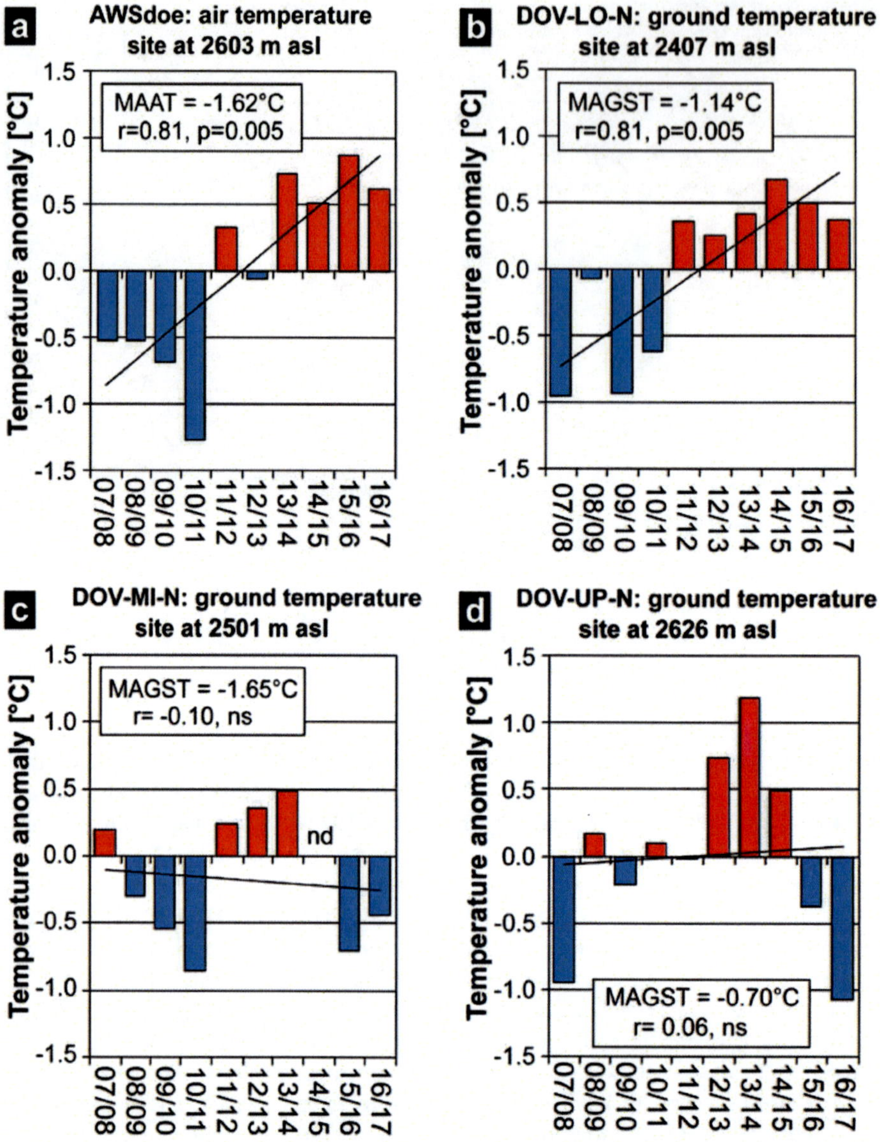

Fig. 27.7 Air (**a**) and ground surface temperature deviations or anomaly (**b–d**) from the mean value during the monitoring period (2007/08–2016/17) based on mean annual values (period August 1 to July 31) at the automatic weather station AWSdoe and at the three monitoring sites at Dösen Rock Glacier (DOV-LO-N, DOV-MI-N, and DOV-UP-N). For locations see Fig. 27.5. Note the statistically significant warming in A and B compared to a much more irregular temperature signal at C and D. MAAT = mean annual air temperature; MAGST = mean annual ground surface temperature; ns = not significant; nd = no data

glacier and its close vicinity. The results indicate that permafrost exists below a several metres thick active layer. The active layer consists of an upper blocky surface layer with open pore space (supportive for ground cooling) and a lower layer with meltwater and finer-grained material covering the permafrost body. The permafrost thickness in the profiles is 10–40 m, whereas the total thickness of the rock glacier is in the order of 30–40 m. The existence of massive sedimentary ice—as known from other rock glaciers in the region (Kellerer-Pirklbauer and Kaufmann 2017)—has not been detected so far. However, field and remote sensing-based evidence—such as massive surface lowering in the rooting zone of the rock glacier– suggests the existence of such massive sedimentary ice bodies at least at its upper part.

27.3.4 Kinematics of the Rock Glacier

Monitoring the kinematics of a rock glacier helps to better understand the general creep pattern of such a peculiar landform. The different measurement methods of rock glacier flow velocity are described by Kaufmann and Kellerer-Pirklbauer (2015). The methods mentioned can be classified into three main groups: photogrammetric/image-based methods, geodetic methods (using a total station or a Global Navigation Satellite System/GNSS-based), and laser scanning. In this section, we briefly describe the successful application of aerial photogrammetry and geodesy at this rock glacier.

Aerial photogrammetry is a powerful method to obtain three-dimensional and area-wide information on horizontal and vertical rock glacier surface displacement. Aerial photographs used at the Dösen Rock Glacier date from 1954, 1969, 1975, 1983, 1993, 2006, and 2010. The computation of horizontal displacement vectors is based on 2D image matching of multi-temporal orthophotos. Figure 27.8a depicts the mean annual horizontal flow velocity for the time period 2006–2010. Maximum flow velocity for this period amounts to 41.4 cm/a. In addition, surface elevation changes can be computed by subtracting two digital elevation models from different years. Limited elevation accuracy and other objections, such as seasonal snow cover and shadows, made successful 'mass balance analyses' (as common for normal ice glaciers) for the entire landform very difficult to perform so far. Such a mass balance approach was, however, successfully applied at a nearby rock glacier (Kellerer-Pirklbauer and Kaufmann 2017). Rock glacier surface lowering at Dösen Rock Glacier due to permafrost degradation (melting of congelation or sedimentary ice) was roughly estimated to be a few centimetres (up to max. 10 cm) per year which is the same order of magnitude as measured at the nearby Hinteres Langtalkar Rock Glacier (Kellerer-Pirklbauer and Kaufmann 2017).

The usage of a geodetic network consisting of a few stable reference points located outside the rock glacier and a set of well-distributed observation points on the rock glacier itself is common for geodetic rock glacier monitoring. At this rock glacier a geodetic network consisting of observation points (107 in total; 34 of them stabilized with brass bolts) on boulders of the rock glacier surface and stable reference points (11) positioned in the surrounding was set up already in 1995 and later expanded (Kaufmann 1998). Since then geodetic field measurements have been carried out on an annual basis (apart from 2003). Figure 27.8b depicts the annual movement of 34 observation points at the rock glacier surface for the period 2016–2017. The maximum flow velocity has been always measured in recent years in the central lower part (at point no. 15) and amounted to

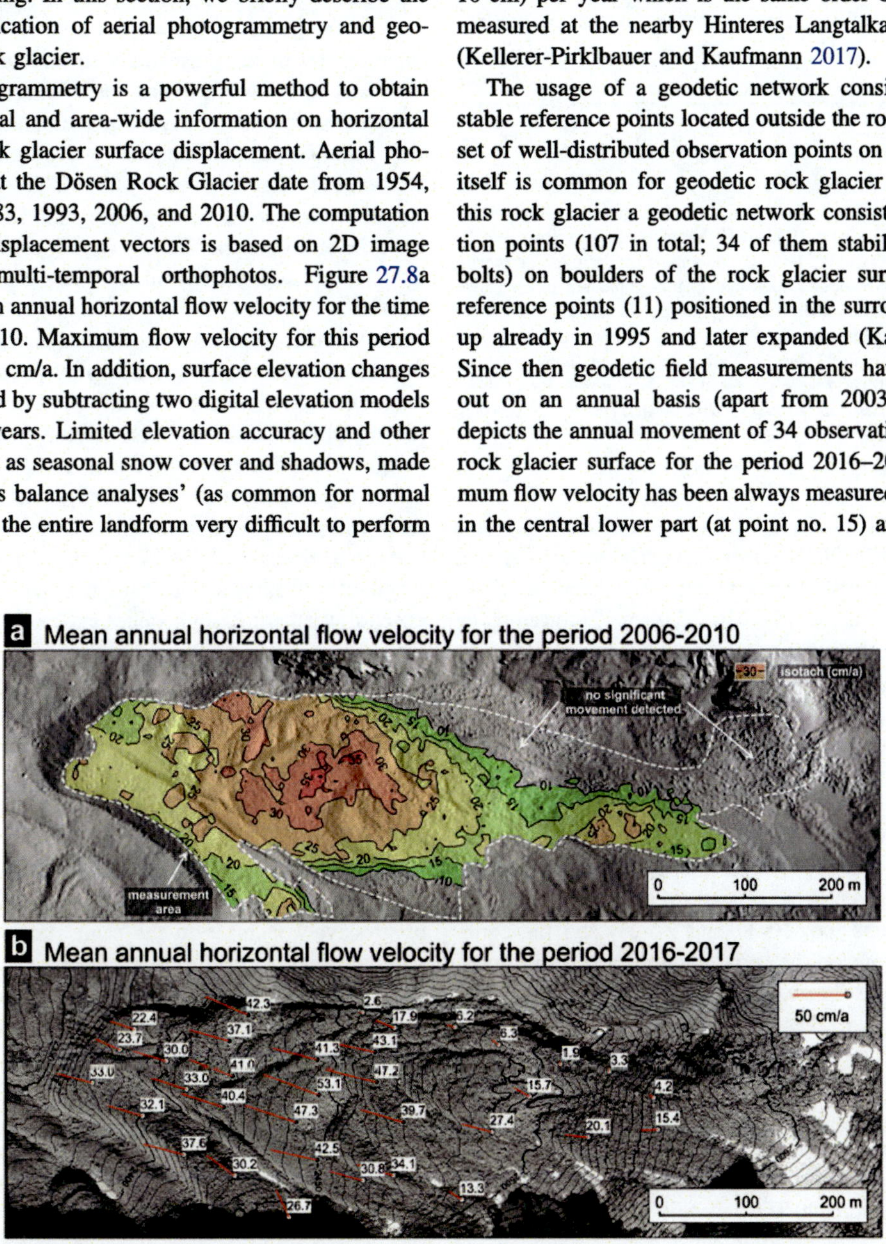

Fig. 27.8 Horizontal flow velocities at Dösen Rock Glacier based on photogrammetric (**a**) and geodetic (**b**) measurements. **a** Mean annual horizontal flow velocity for the period 2006–2010. Equidistance of isotachs is 5 cm/a. Maximum flow velocity amounts to 41.4 cm/a. Computed velocities smaller than 9.6 cm/a are statistically non-significant. The shaded relief is based on an airborne laser scanning-derived digital elevation model (acquisition year 2010) with a grid spacing of 1 m × 1 m and provided by Land Kärnten/KAGIS. **b** Annual movement (2016–2017) of the 34 observation points at the rock glacier. The maximum flow velocity has been measured close to the centre of the rock glacier and amounts to 53.1 cm/a. Observation points indicated in black are used for mean value calculation (see Fig. 27.9) Aerial photograph (21 September 2010), BEV, Vienna

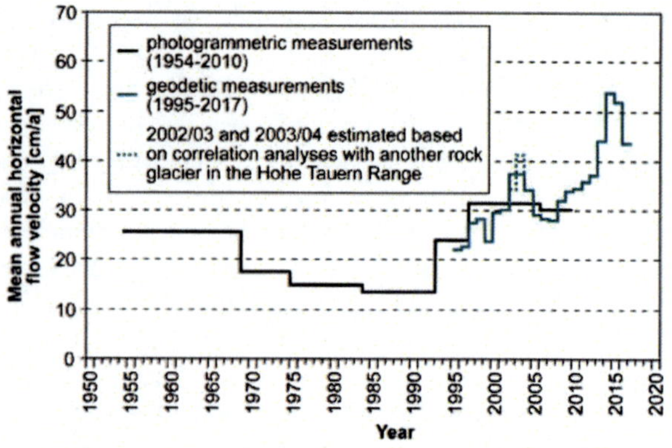

Fig. 27.9 Mean annual horizontal flow velocity of Dösen Rock Glacier during the period 1954–2017 based on photogrammetrically derived flow velocities (1954–2010) and annual geodetic measurements (1995–2017). The velocities indicated in this graph are mean values derived from 11 representative observation points (10–17, 21–23; see black observation points in Fig. 27.8b). Geodetic measurements were not accomplished in 2003 due to funding restrictions. Thus velocities for 2002/03 and 2003/04 were estimated based on correlation analyses with the Hinteres Langtalkar Rock Glacier located c.35 km to the west (cf. Kaufmann 2016)

65.9 cm/a in 2014–2015, 64.7 cm/a in 2015–2016 and 53.1 cm/a in 2016–2017, respectively. The mean velocities (derived from all point measurements) calculated for 2014–2015 (53.6 cm/a) and 2015–2016 (52.0 cm/a) showed by far the highest flow velocities ever measured at this rock glacier. These values are four times higher than respective values measured between the 1970s and the early 1990s. Interestingly, the mean surface flow velocity decreased in 2016–2017 to a value of 43.2 cm/a.

Figure 27.9 summarizes the velocity measurements carried out at Dösen Rock Glacier, depicting mean annual horizontal flow velocity rates for the period 1954–2017. The mean velocity given here is based on 11 selected geodetic observation points (no. 10–17 and 21–23) and the corresponding locations on the aerial photographs, respectively. As shown in this figure, the horizontal flow velocity of Dösen Rock Glacier has changed significantly over time with clear periods of acceleration but also deceleration. An explanation for these changes in flow velocity is given by Delaloye et al. (2008) and Kellerer-Pirklbauer and Kaufmann (2012) who reveal synchronous velocity changes for several rock glaciers in the Alps and attributed this behaviour to climatic forcing. The generally low velocity rates at this rock glacier compared to other rock glaciers in the region (Kaufmann and Kellerer-Pirklbauer 2015) point towards the dominance of internal deformation, low ice content, and the absence of distinct sliding processes (e.g. Hausmann et al. 2012; Avian et al. 2009).

27.3.5 Estimation of Rock Glacier Age

Rock glacier sediments reflect debris accumulations which were formed, deposited, transported, and deformed during a time span of several hundreds to thousands of years (Barsch 1996). Dating rock glaciers is not straightforward but can be best achieved by combining absolute/numeric and relative age-determination methods (Haeberli et al. 2003). Photogrammetry of the present-day surface velocities in combination with the analysis of hardness variations of weathering rinds of blocks at the rock glacier surface using the Schmidt-hammer can be a successful combination (Kellerer-Pirklbauer et al. 2008). Furthermore, Schmidt-hammer rebound values can be calibrated with surfaces of presumably known age as it has been applied at the Dösen Rock Glacier in the past (Kellerer-Pirklbauer, 2008b).

A Schmidt-hammer is a portable and light instrument traditionally used for concrete stability testing by recording a rebound value (*R*-value) of a spring-loaded bolt impacting a given surface. This method has been applied in glacial and periglacial studies for relative rock surface dating since the 1980s (e.g. Matthews and Shakesby 1984; Shakesby et al. 2006). The received *R*-value gives a relative measure of the surface hardness and provides indirect information on the length of time since surface exposure and, hence, defines the onset of weathering. High values indicate a younger age and vice-versa.

A number of sites at the Dösen Rock Glacier and its close vicinity were measured applying this method, focusing particularly on sites along the central flow line of the rock glacier (SH1-SH7 in Fig. 27.5b). This measurement setup was chosen because of the common assumption that blocks at the rock glacier surface are passively transported in a rather stable position from the talus slope at the rooting zone towards the front. Therefore, the age of the blocks should progressively increase in the same direction. In addition, rock material at the front of an advancing rock glacier tumbles down and gets buried by the advancing rock glacier itself, hence the surface of a rock glacier is generally younger compared to the entire landform (Haeberli et al. 2003).

Results at the surface of Dösen Rock Glacier yielded mean *R*-values ranging from 35.4 at the rock glacier front (SH1) to 47.7 on the active talus (SH7) in the rooting zone covering a *R*-value range of 12.3 and indicating a relatively steady decrease in *R*-values between the rooting zone of the rock glacier and its front (Fig. 27.10). Looking into the results in more depth, a step-like *R*-value pattern can be observed. Similar R-values have been measured at two adjacent Schmidt-hammer measurement sites (e.g. SH3 and

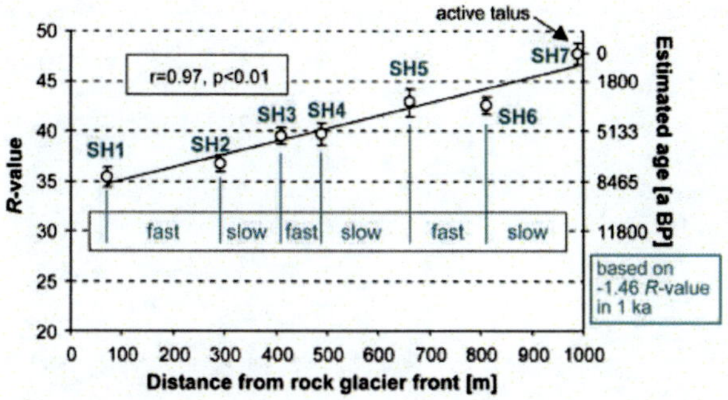

Fig. 27.10 Relative age dating applying the Schmidt-hammer exposure age dating (SHD) method at Dösen Rock Glacier. For each site the arithmetic mean of 50 individual Schmidt-hammer readings and the 95% confidence limits are depicted. The code numbers in the graph refer to the measurement sites depicted in Fig. 27.5b. The left/primary y-axis relates to the measured rebound value (*R*-value). The right/secondary y-axis relates to the estimated age of the measured blocks at the rock glacier surface

SH4 as well as SH5 and SH6) which is in contrast to different values in between these pairs (i.e. between SH4 and SH5). This suggests alternating periods of rapid and slow rock glacier movement. Such an unsteady long-term movement behaviour of rock glaciers is to some extent confirmed by findings of Krainer et al. (2015), who pointed out that active rock glaciers do not move with constant flow velocities but periods of lower and higher activity seem to alternate with periods of relative inactivity (cf. Sect. 3.4).

The conversion of *R*-values to estimated ages was accomplished by using a tentative age-calibration curve of this area (Kellerer-Pirklbauer 2008b). This curve is based on two surfaces of most likely known age (SH7 = an active talus slope; SH11 = a terminal moraine of Younger Dryas age, site SH11 in Fig. 27.5a), yielding a linear *R*-value decrease of −1.46 per 1 ka. The stabilization age of the early Younger Dryas moraines is ^{10}Be-dated to around 12.5 ka for this region (Bichler et al. 2016). The initiation of the Dösen Rock Glacier occurred after the retreat of the (most likely) Younger Dryas glacier which had deposited the moraine ridge some 1.1 km downvalley of the rock glacier front. The regression line indicates that this rock glacier started to form sometime before 8.4 ka BP (estimated age of the lowest Schmidt-hammer measurement point at the rock glacier surface SH1). Recent velocity data from the rock glacier show mean annual velocity rates of 13.4 –53.6 cm/a during the period 1954–2017 (Fig. 27.9). If these velocities are taken as constant (questionable) over time and combined with the length of the rock glacier (950 m), age estimates of 1.8–7.1 ka can be calculated. These estimated ages are between similar to substantially lower if compared to the age estimate based on the Schmidt-hammer approach depicted in Fig. 27.10. The results presented here are well in accordance with absolute dating results of active rock glaciers, indicating long evolutionary histories as for instance shown by Krainer et al. (2015) who dated 10,300-year old ice near the base of a rock glacier in Southern Tyrol (Italy).

27.3.6 Experiencing the Dösen Rock Glacier

The Dösen Rock Glacier is situated in the Hohe Tauern National Park. The task of this national park is, besides nature protection, to promote research and communicate its results to the public. Accordingly, the investigations carried out by the scientific institutions the authors belong to were supported by the national park authorities already from the beginning in the 1990s. As a consequence, the idea came up to present the topic of permafrost and rock glaciers and their interaction with changing climatic conditions to national park visitors. The Institute of Geography and Regional Science of the University of Graz took over this task and developed a concept of a nature trail. This trail was finally established in cooperation with the Austrian Alpine Association and a guide-book was also published (Lieb and Nutz 2009). Thus interested persons can easily learn about this topic and experience high mountain environments by hiking the trail which leads through the entire Dösen Valley up to the rock glacier front. At the high alpine hut Arthur-von-Schmid Haus the hiker will be rewarded with the stunning alpine scenery depicted in Fig. 27.11 and the first glimpses of the Dösen Rock Glacier.

27.4 Conclusions

In this chapter, we have provided a general overview of both the permafrost and rock glacier distribution in Austria and, additionally, have presented the Dösen Rock Glacier as a key research site for in-depth permafrost and rock glacier science in the Austrian Alps. The national-scale rock glacier density analysis showed that rock glaciers are frequently found in the Austrian Alps. Intact rock glaciers dominate in western and south-western Austria whereas relict rock glaciers can be found all over the Austrian Alps apart from the lower-elevated mountain ranges of the eastern and northern Alpine fringe.

The Dösen Rock Glacier has been the object of research and monitoring projects since the 1990s when field-based research was initiated within a project that focused on geomorphological description, internal structure and permafrost content, rock glacier movement, and permafrost distribution. Annual geodetic measurements were initiated during that time, too, and are still ongoing, providing valuable long-term and almost continuous rock glacier velocity data.

Fig. 27.11 Terrestrial photograph of the picturesque head of the Dösen Valley with the alpine hut Arthur-von-Schmid-Haus (2275 m asl), the Dösener Lake (a cirque lake; 2270 m asl), the mountain Säuleck (3086 m asl) and the Dösen Rock Glacier east of the lake with late lying snow patches accentuating the rock glacier morphology. View towards east. Photograph (01.08.2008) by the authors

A network of automatic devices for monitoring the thermal regime of permafrost was installed in 2006 and is still operating. Additional geophysical studies have been performed since then, allowing for an improved characterization of the internal structure and permafrost distribution of the rock glacier. Permafrost thickness has been estimated to be between 10 m (at the margins) and 40 m (central and upper parts). Ground and air temperature monitoring reveals a statistically significant warming since 2007. In particular, the seasonal snow cover strongly influences the ground thermal regime and masks clear trends in warming at some locations.

Horizontal flow velocities at the rock glacier surface varied substantially since the 1950s with the lowest values between the 1970s and the early 1990s. This lower velocity values are related to cooler climatic conditions. The currently high values are related to warmer permafrost conditions and a higher content of liquid water lubricating the rock glacier system.

The application of the Schmidt-hammer exposure-age dating method revealed that rock glacier initiation occurred in the early Holocene. Thus, the evolution of the rock glacier progressed over several thousand years with apparent periods of slower and faster movement. Present movement rates seem to be rather high if compared to the mean velocities during the Holocene.

Based on the research work described a first educational trail focused on permafrost and permafrost-related landforms in the Hohe Tauern National Park was established. Furthermore, the Dösen Rock Glacier has recently become a site for long-term monitoring of ecosystems in the Hohe Tauern National Park. Both aspects highlight the equal importance of both basic research and knowledge transfer to the general public and other scientific communities.

Acknowledgements Research at Dösen Rock Glacier has been funded in the past by the Austrian Science Foundation (FWF) through two projects (project no. P09565 and FWF P18304-N10), by the European

Regional Development Fund, Alpine Space Programme (project no 18-1-3-I), by the Austrian Academy of Sciences (project permafrost), and by the Hohe Tauern National Park authority in Carinthia. Isabelle Gärtner-Roer is very much thanked for her constructive criticism on an earlier version of this paper.

References

Avian M, Kellerer-Pirklbauer A, Bauer A (2009) LiDAR for monitoring mass movements in permafrost environments at the cirque Hinteres Langtal, Austria, between 2000 and 2008. Nat Hazard Earth Syst 9:1087–1094

Barsch D (1978) Active rock glaciers as indicators for discontinuous alpine permafrost. An example from the Swiss Alps. In: Brown J (ed) Proceedings of the 3rd international permafrost conference, vol 1. National Research Council of Canada, Ottawa, pp 349–352

Barsch D (1980) Die Beziehung zwischen der Schneegrenze und der Untergrenze der aktiven Blockgletscher. In: Jentsch C, Liedtke H (eds) Höhengrenzen in Hochgebirgen, vol 29. Arb Geogr Inst Univ Saarlandes, pp 119–133

Barsch D (1996) Rock Glaciers: Indicators for the present and former geoecology in high mountain environments, vol 16. Springer Series in Physical Environment, Springer, Berlin

Berthling I (2011) Beyond confusion: rock glaciers as cryoconditioned landforms. Geomorphology 131:98–106. https://doi.org/10.1016/j.geomorph.2011.05.002

Bichler MG, Reindl M, Reitner JM, Drescher-Schneider R, Wirsig C, Christl M, Hajdas I, Ivy-Ochs S (2016) Landslide deposits as stratigraphic markers for a sequence-based glacial stratigraphy: a case study of a Younger Dryas system in the Eastern Alps. Boreas 45:537–551. https://doi.org/10.1111/bor.12173

Boeckli L, Brenning A, Gruber S, Noetzli J (2012) Permafrost distribution in the European Alps: calculation and evaluation of an index map and summary statistics. Cryosphere 6:807–820. https://doi.org/10.5194/tc-6-807-2012

Delaloye R, Perruchoud E, Avian M, Kaufmann V, Bodin X, Ikeda A, Hausmann H, Kääb A, Kellerer-Pirklbauer A, Krainer K, Lambiel C, Mihajlovic D, Staub B, Roer I, Thibert E (2008) Recent Interannual Variations of Rockglaciers Creep in the European Alps. In: Kane DL, Hinkel KM (eds) Proceedings of the ninth international conference on permafrost, University of Alaska, Fairbanks, USA, June–July 2008, pp 343–348

Finsterwalder S (1928) Begleitworte zur Karte des Gepatschferners. Zeitschrift für Gletscherkunde 16:20–41

Fischer A, Seiser B, Stocker-Waldhuber M, Mitterer C, Abermann J (2015) Tracing glacier changes in Austria from the Little Ice Age to the present using a lidar-based high-resolution glacier inventory in Austria. Cryosphere 9:753–766. https://doi.org/10.5194/tc-9-753-2015

Haeberli W, Vonder Mühll D (1996) On the characteristics and possible origins of ice in rock glacier permafrost. Z Geomorphol Supp 104:43–57

Haeberli W, Guodong C, Gorbunov AP, Harris SA (1993) Mountain permafrost and climate change. Permafrost Periglac 4:165–174

Haeberli W, Hallet B, Arenson L, Elconin R, Humlum O, Kääb A, Kaufmann V, Ladanyi B, Matsuoka N, Springman S, Vonder Mühll D (2006) Permafrost creep and rock glacier dynamics. Permafrost Periglac 17:189–214. https://doi.org/10.1002/ppp.561

Haeberli W, Brandovà D, Burga C, Egli M, Frauenfelder R, Kääb A., Maisch M (2003) Methods for absolute and relative age dating of rock-glacier surface in alpine permafrost. In: Haeberli W, Brandová D (eds) Proceedings of the eight international conference on permafrost, Zurich, Switzerland, pp 343–348

Hausmann H, Krainer K, Brückl E, Ullrich C (2012) Internal structure, ice content and dynamics of Ölgrube and Kaiserberg rock glaciers (Ötztal Alps, Austria) determined from geophysical surveys. Austr J Earth Sci 105(2):12–31

Humlum O (1998) The climatic significance of rock glaciers. Permafrost Periglac 9:375–395

Humlum O, Christiansen HH (1998) Mountain climate and periglacial phenomena in the Faroe Islands, SE North Atlantic Ocean. Permafrost Periglac 9:189–211

Kaufmann V (2016) 20 years of geodetic monitoring of Dösen rock glacier (Ankogel Group, Austria): a short review. Joannea Geol Paläont 12:37–44

Kaufmann V, Kellerer-Pirklbauer A (2015) Active rock glaciers in a changing environment: Geomorphometric quantification and cartographic presentation of rock glacier surface change with examples from the Hohe Tauern Range, Austria: In: Kriz K (ed) 16 years of the Commission on Mountain Cartography (CMC) of the International Association of Cartography (ICA), Wiener Schriften zur Geographie und Kartographie, vol 21, pp 179–190

Kaufmann V (1998) Deformation analysis of the Doesen rock glacier (Austria). In: Lewkowicz AG, Allard M (eds) Proceedings of the seventh international conference on permafrost, Yellowknife, June 1998, pp 551–556, Canada, Collection Nordicana, p 55

Kellerer-Pirklbauer A, Kaufmann V (2012) About the relationship between rock glacier velocity and climate parameters in central Austria. Austr J Earth Sci 105(2):94–112

Kellerer-Pirklbauer A, Kaufmann V (2017) Deglaciation and its impact on permafrost and rock glacier evolution: new insight from two adjacent cirques in Austria. Sci Total Environ. https://doi.org/10.1016/j.scitotenv.2017.10.087

Kellerer-Pirklbauer A, Wangensteen B, Farbrot H, Etzelmüller B (2008) Relative surface age-dating of rock glacier systems near Hólar in Hjaltadalur, northern Iceland. J Quat Sci 23:137–151

Kellerer-Pirklbauer A, Lieb GK, Kleinferchner H (2012) A new rock glacier inventory in the eastern European Alps. Austr J Earth Sci 105(2):78–93

Kellerer-Pirklbauer A, Wagner T, Winkler G (2016) Inventarisierung von blockgletscherverdächtigen Formen und deren Hydrologischen Einzugsgebieten in den Steirischen Niederen Tauern mit Hilfe von hochaufgelösten Geländemodellen. Joannea Geol Paläont 12:53–62

Kellerer-Pirklbauer A, Avian M, Kaufmann V, Niesner E, Kühnast B (2014) Climatic-induced spatio-temporal change of kinematics and ground temperature of rock glaciers and permafrost in the Hohe Tauern Range, Austria. In: Rutzinger M, Heinrich K, Borsdorf A, Stötter J (eds) permafrost—Austrian Permafrost Research Initiative. Final Report, pp 17–39. https://doi.org/10.1553/IGF-1

Kellerer-Pirklbauer A (2008a) Aspects of glacial, paraglacial and periglacial processes and landforms of the Tauern Range, Austria. Unpublished Doctoral Thesis, University of Graz

Kellerer-Pirklbauer A (2008b) The Schmidt-hammer as a relative age dating tool for rock glacier surfaces: examples from Northern and Central Europe. In: Kane DL, Hinkel KM (eds) Proceedings of the ninth international conference on permafrost, University of Alaska, Fairbanks, USA, June–July 2008, pp 913–918

Kellerer-Pirklbauer (2016) A regional signal of significant recent ground surface temperature warming in the periglacial environment of Central Austria. In: Günther F, Morgenstern A (eds) XI. International conference on permafrost—book of abstracts, 20–24 June 2016, Potsdam, Germany, pp 1025–1026

Kellerer-Pirklbauer A (2019) Long-term monitoring of sporadic permafrost at the eastern margin of the European Alps (Hochreichart, Seckauer Tauern range, Austria). Permafrost Periglac 30:260–277. https://doi.org/10.1002/ppp.2021

Krainer K, Ribis M (2012) A rock glacier inventory of the Tyrolean Alps (Austria). Austr J Earth Sci 105(2):32–47

Krainer K, Kellerer-Pirklbauer A, Kaufmann V, Lieb GK, Schrott L, Hausmann H (2012) Permafrost research in Austria: history and recent advances. Austr J Earth Sci 105(2):2–11

Krainer K, Bressan D, Dietre B, Haas JN, Hajdas I, Lang K, Mair V, Nickus U, Reidl D, Thies H, Tonidandel D (2015) A 10,300-year-old permafrost core from the active rock glacier Lazaun, southern Ötztal Alps (South Tyrol, northern Italy). Quat Res 83:324–335. https://doi.org/10.1016/j.yqres.2014.12.005

Lieb GK (1996) Permafrost und Blockgletscher in den östlichen österreichischen Alpen. Arb Inst Geographie Univ Graz 33:9–125

Lieb GK, Nutz M (2009) Blockgletscherweg Dösental. Naturkundlicher Führer zum Nationalpark Hohe Tauern, 21, Innsbruck.

Lieb GK, Kellerer-Pirklbauer A, Kleinferchner H (2012) Second rock glacier inventory (RGI2) of Central and Eastern Austria. https://doi.org/10.1594/PANGAEA.869805

Matthews JA, Shakesby RA (1984) The status of the 'Little Ice Age' in southern Norway: relative-dating of Neoglacial moraines with Schmidt hammer and lichenometry. Boreas 13:333–346

Paul F, Rastner P, Azzoni RS, Diolaiuti G, Fugazza D, Le Bris R, Nemec J, Rabatel A, Ramusovic M, Schwaizer G, Smiraglia C (2021) Glacier shrinkage in the Alps continues unabated as revealed by a new glacier inventory from Sentinel-2, Earth Syst Sci Data 12:1805–1821. https://doi.org/10.5194/essd-12-1805-2020

Schmöller R, Fruhwirth RK (1996) Komplexgeophysikalische Untersuchung auf dem Dösener Blockgletscher (Hohe Tauern, Österreich). Arb Inst Geographie Univ Graz 33:165–190

Schneider B, Schneider H (2001) Zur 60jährigen Messreihe der kurzfristigen Geschwindigkeitsschwankungen am Blockgletscher im Äusseren Hochebenkar, Ötztaler Alpen. Tirol Zeitschrift Für Gletscherkunde und Glazialgeologie 37(1):1–33

Schoeneich P, Bodin X, Echelard T, Kaufmann V, Kellerer-Pirklbauer A, Krysiecki JM, Lieb GK (2015) Velocity change of rock glaciers and induced hazards. In: Lollino G, Giordan D, Thuro K, Carranza-Torres C, Wu F, Marinos P, Delgado C (eds) Engineering geology for society and territory—Volume 1. Springer International Publishing Switzerland 2015, pp 223–227. https://doi.org/10.1007/978-3-319-09300-0_42

Shakesby RA, Matthews JA, Owen G (2006) The Schmidt hammer as a relative-age dating tool and its potential for calibrated-age dating in Holocene glaciated environments. Quat Sci Rev 25:2846–2867

Stocker K (2012) Blockgletscher in Vorarlberg und in der Verwallgruppe. Museumsverein Jahrbuch Vorarlberger Landesmuseum 2012:124–139

Van Everdingen RO (1998) Multi-language glossary of permafrost and related ground-ice terms (revised May 2005). National Snow and Ice Data Center/World Data Center for Glaciology, Boulder

Wagner T, Pleschberger R, Kainz S, Ribis M, Kellerer-Pirklbauer A, Krainer K, Philippitsch R, Winkler G (2020) The first consistent inventory of rock glaciers and their hydrological catchments of the Austrian Alps. Austrian J Earth Sci 113/1:1–23. https://doi.org/10.17738/ajes.2020.0001

Andreas Kellerer-Pirklbauer is a senior scientist at the University of Graz, Institute of Geography and Regional Science. He has significant experience in the physical geography of mountain regions with a focus on the effects of global change on geomorphic processes and landforms, permafrost, glaciers, mass wasting processes, and related natural hazards. He is the Austrian national representative at the International Association of Permafrost and one of the two co-leaders of the national glacier monitoring programme of the Austrian Alpine Club.

Gerhard Karl Lieb is associate professor at the University of Graz, Institute of Geography and Regional Science. He has experience in physical geography of mountain regions with a special focus on permafrost, glaciers and geomorphic processes affected by global change as well as natural hazards. He is one of the two co-leaders of the national glacier monitoring programme of the Austrian Alpine Club. Besides those topics he works on regional geographies (Austria, Alps and Europe) and on didactics of the Austrian school subject geography and economics.

Viktor Kaufmann is an associate professor at the Institute of Geodesy, Graz University of Technology. His current areas of interest are geomorphometric studies applying various remote sensing techniques, close-range photogrammetry for the documentation of cultural heritage, and digital photogrammetry. He has set up a long-term monitoring programme for rock glaciers in the Austrian Alps.

Check for updates

28 The Ore of the Alps UNESCO Global Geopark (Salzburg) Geosites and Geotourism

Horst J. Ibetsberger and Hans Steyrer

Abstract

The Ore of the Alps UNESCO Global Geopark has been a member of the European and Global Geopark Network since 2014. It covers an area of 212 km^2 and includes two major geological units of the Alps, the Greywacke Zone and the Northern Calcareous Alps. Both components are associated with a long history of geological research revealing important phases in the Variscan and Alpine orogenies in this region of the Eastern Alps. The region is also the focus of fundamental research work on Quaternary sediments and landforms. Within the Geopark, numerous geosites document the geology of the basement rocks, the more recent features generated in the landscape by geomorphic processes during the Pleistocene and Holocene epochs, and the influence of human activities including remnants of Neolithic settlements, copper mining since the Bronze Age and modern road- and tunnel building activities. The Geopark and its backbone —the "Copper Trail"—provide a sound basis for geotourism, which is seen as an important branch of soft and sustainable tourism.

Keywords

UNESCO Global Geopark • Greywacke Zone • Northern Calcareous Alps • Show mines • Geotourism • Geosite

H. J. Ibetsberger (✉)
GeoGlobe Häupl & Ibetsberger OG, Statzenbachgasse 5, 5202 Neumarkt am Wallersee, Austria
e-mail: horst.ibetsberger@geoglobe.at

H. Steyrer
Department of Geography and Geology, Geology Working Group, Salzburg University, Hellbrunner Straße 34, 5020 Salzburg, Austria

28.1 Introduction

The Ore of the Alps UNESCO Global Geopark (212 km^2) is situated in the Pongau district of the federal state of Salzburg and includes the four communities Bischofshofen, Mühlbach am Hochkönig, Hüttau and St. Veit im Pongau (Fig. 28.1), with a population of 17,000 inhabitants (Amt der Salzburger Landesregierung 2013). The Geopark area extends from the Northern Calcareous Alps to the Central Alps. Its core region encompasses a mountainous area named "Innergebirg" (meaning "inner mountains") that is developed within the Greywacke Zone. Whereas the high-Alpine Hochkönig mountain range in the Northern Calcareous Alps reaches an altitude of 2941 m asl and includes a permanent glacier, the "Übergossene Alm" ("covered Alp" glacier), the Innergebirg region is characterized by pastures and forests extending up to an altitude of c. 2000 m asl. The Salzach Valley, which crosses the Geopark, is deeply incised into the soft rocks of the Greywacke Zone. All other valleys include extremely steep-sided gorges. The relative relief in the Geopark is very high, with a variation in altitude of almost 2400 m (Geopark Erz der Alpen 2018).

The Geopark area has been continuously populated since approximately 5300 BP. During the Bronze Age, the region was one of the most important sites for copper mining in Europe (e.g. Eibner 2016). In the Middle Ages, mining activities were extended to include gold, iron, lead and zinc. In the 1970s, all mining activities ceased, but the mines still exist today as a number of fascinating show mines. These, and all the spectacular geosites (henceforth referred to as GS), provide the basis of the UNESCO Global Geopark, which has been a member of the European Geopark and Global Geopark Network since 2014 and UNESCO Global Geopark since 2015 (Ibetsberger et al. 2011, 2018; Ibetsberger 2015).

The goals of UNESCO Global Geoparks are defined as follows: they "give international recognition for sites that promote the importance and significance of protecting the

C. Embleton-Hamann (ed.), *Landscapes and Landforms of Austria*, World Geomorphological Landscapes,
https://doi.org/10.1007/978-3-030-92815-5_28

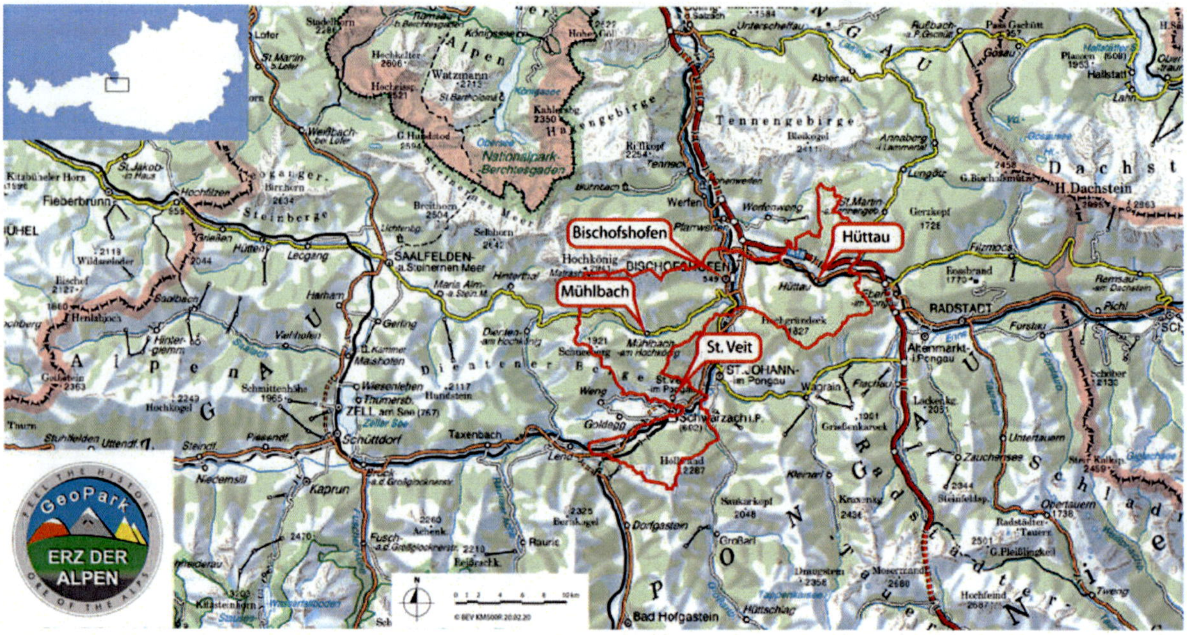

Fig. 28.1 The boundary of the Ore of the Alps UNESCO Global Geopark (thick red line) in the mountainous area of Salzburg, with the four communities Bischofshofen, Mühlbach am Hochkönig, Hüttau and St. Veit im Pongau. *Graphics* GeoGlobe, 2016

Earth's geodiversity through actively engaging with the local communities" (UNESCO Global Geoparks 2017).

28.2 Geographical Setting

28.2.1 Geology

The Geopark area covers two major geological units of the Alps, the Greywacke Zone in the south and the Northern Calcareous Alps in the north, both with a long history of research (Fig. 28.2).

The term *Grauwacke* has been used in Europe for about 200 years and originates from the Harz Mountains (Germany), where it was used to describe dark Paleozoic sandstones. Similar rocks occur within extensive areas of the Alps and these were named *Grauwackenzone,* probably by German miners in the nineteenth century. Until the 1930s, the early Palaeozoic sandstones and phyllites within the Alps were described as *Grauwacken*. The vague definition of the *Grauwackenzone* finally led to the rejection of the term (e.g., Cornelius 1952), but as Dunbar and Rodgers (1958) stated: "… the name has refused to die …", and is still in use.

Within the Geopark area the predominant rocks of this unit are slates, phyllites and quartzites (Fig. 28.2). During the twentieth century, both the tectonic structure and the stratigraphic evolution of the Greywacke Zone were the focus of many investigations. Work on metamorphic processes undertaken e.g., by Hoinkes et al. (1999) showed that regional metamorphism in the Greywacke Zone is of very low to low grade and that it is predominantly Eoalpine (i.e. Cretaceous) in age. Metamorphic conditions, transitional between sub-greenschist and greenschist facies, are recorded at higher structural levels in the Western Greywacke Zone (e.g. around Mitterberg/Mühlbach am Hochkönig; Kralik et al. 1987) and generally increase to a lower greenschist facies grade towards the lower structural units.

More recently, the abandoned copper mines at Larzenbach/Hüttau have sparked renewed interest, because native gold and unusual gold-oxysulphides were found, both associated with tetrahedrite (Kucha and Raith 2009). Tetrahedrite is a copper antimony sulfosalt mineral and is the antimony end-member of the continuous solid solution series with arsenic-bearing tennantite. Pure endmembers of the series are rarely, if ever, seen in nature. Of the two forms, the antimony rich phase tetrahedrite is the more common.

The Northern Calcareous Alps have been the focal site of geological mining for more than 5000 years, starting with salt-mining at Hallein and Hallstatt, flint-mining in the Tyrol and near Vienna, and the exploration of oil shales for medical purposes (Seefelder Ölschiefer).

Modern geological exploration started towards the end of the eighteenth century, with stratigraphic investigations (e.g., Bohadsch 1782) and reached a first zenith in the

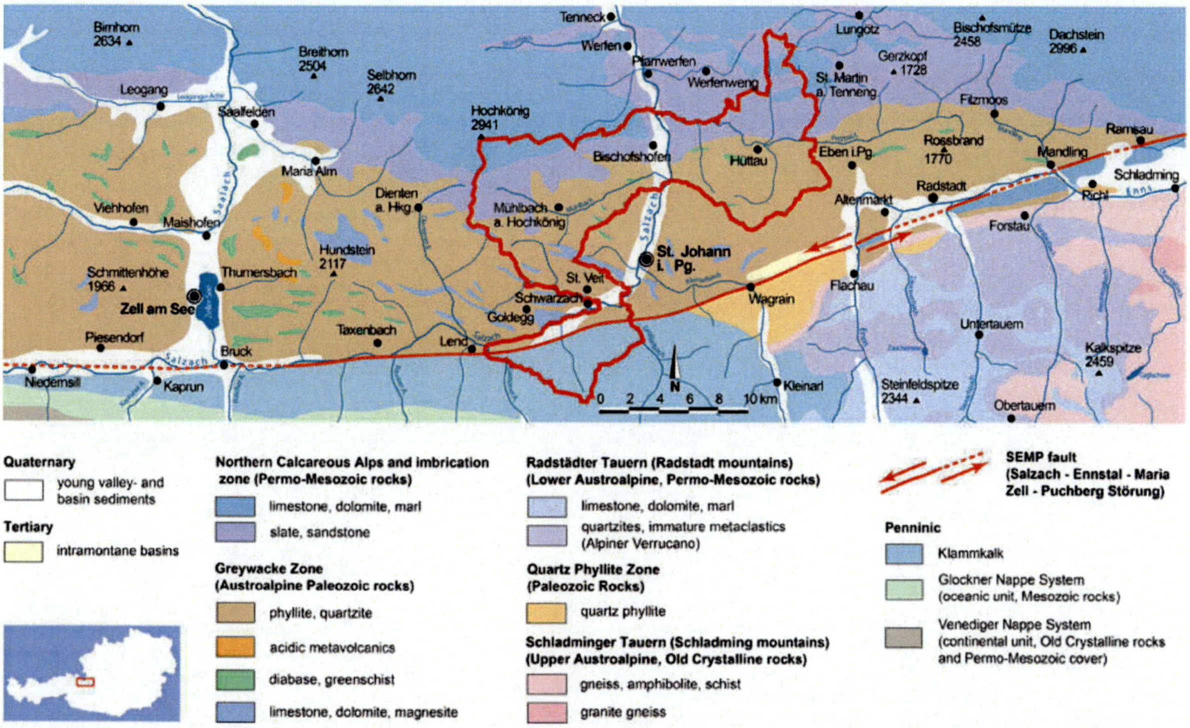

Fig. 28.2 Map of the Greywacke Zone in Salzburg (Weber 1997) with the boundary of the Ore of the Alps UNESCO Global Geopark (thick red line). *Graphics* GeoGlobe, 2017

nineteenth century with the work of von Buch (1802), Sedgwick and Murchinson (1831), Boué (e.g., 1826) and many others. Systematic geological mapping of the Northern Calcareous Alps began after the foundation of the Geologische Reichsanstalt in Vienna (1849) and is still ongoing. The state of knowledge at the end of the twentieth century was summarized by Tollmann (1976, 1985).

New input towards understanding of the structure, deformation history, kinematics and palaeogeography of the Alps resulted from the TRANSALP project ("a Transect through a Young Collisional Orogen"), initiated in 1998. Results for the Northern Calcareous Alps have been summarized by Behrmann and Tanner (2006).

28.2.2 Quaternary History

The Pleistocene was characterized by multiple glaciations, which submerged large parts of the Alps (e.g. Van Husen 1987). The glacial geomorphology of this vast, formerly glaciated landscape, has been the object of investigation for more than a century. Brückner (1886) and Penck and Brückner (1901, 1909) were the first to investigate the landforms of the Salzach Glacier, which is closely linked to the Geopark area. More recent investigations were undertaken by Gamerith and Heuberger (1999) and van Husen (2000). Gamerith and Heuberger (1999) investigated the upper level of the ice cover at the Last Glacial Maximum (LGM) and detected a c. 500 kg granite-gneiss erratic boulder at 1920 m asl. This boulder was transported in ice from the Hohe Tauern to the Geopark area.

The catchment area of the Salzach Glacier encompassed the northern ramp of the Hohe Tauern, with mountains higher than 3000 m asl. The glaciation of the "inner mountain area" of Salzburg can be compared with the present situation in the Coast Range of Alaska. Only peaks and ridges above 2000–2200 m asl protruded above the ice surface as nunataks. Therefore, the whole area of the lower mountain ranges of the Greywacke zone including Hochgründeck (1827 m asl, GS 9 in Fig. 28.5) was ice covered. This is why they have the appearance of rounded hills, similar to mountains in the Alpine foreland. The only mountain range, which rose above the ice surface, was the Hochkönig (GS 30 in Fig. 28.5), with an altitude of 2941 m asl. Measured from the base of the overdeepened Salzach Valley, c. 100 m below its current level, up to the foothill of Hochkönig, the thickness of the ice cover was c. 1500 m (Fig. 28.3).

Richter (1888) described the recent glaciers of the Northern Calcareous Alps within the province of Salzburg. One of the most important subjects of his studies was the

Fig. 28.3 Reconstruction of the ice covering of the Geopark area at the LGM. In the background, the Hochkönig mountain range (2941 m asl), which rose above the ice surface, 22 000 years BP. *Graphics* GeoGlobe, 2013

"covered Alp" glacier at Hochkönig (Fig. 28.4). He wrote that it is a typical example of a plateau glacier of Norwegian style. At the end of the nineteenth century, it extended 4.33 km from east to west and between 1.05 and 1.5 km from north to south and covered an area of 5.5 km^2. Today, the glacier covers an area of 1.5 km^2 (GS 32 in Fig. 28.5).

With the beginning of the Lateglacial (19,000–11,600 BP), the temperature began to increase (Heuberger 1968). This involved the onset of the melting period of Alpine glaciers. Around 16,000 BP, the last remnants of dead ice could be found in the Geopark area. Glaciers existed only at higher altitudes, extending to the foothills of the Hochkönig range. Around 14,500 BP, the climate had almost attained current temperatures. The valley regions of the Geopark were repopulated by vascular plants – pine forests could be found from the valley bottom up to an altitude of the terrace of St. Veit at 750 m asl. The ages of pinewood fragments, which were dated using the 14^C method, were assigned to the warm temperate Bölling Interstadial period (Slupetzky 1975).

With the beginning of the Holocene at 11,600 BP, a new landscape was born. The rivers accumulated material at the bottom of the overdeepened valleys and undercut the over-steepened slopes. Mass movements such as debris flows, slides and rockfalls occurred. An example of ongoing high morphological activity is provided by the slopes of Fritz Valley. The Reit tunnel, part of the Tauern Highway (A 10) near the Geopark village Hüttau, is therefore a permanent building site (GS 97, 98 in Fig. 28.5).

28.2.3 Climatic Conditions

The whole Geopark area is influenced by the alpine climate. This is characterized by relatively short, cool summers and long, cold winters with a lot of snow. The highest amounts of precipitation are measured along the northern rim of the Alps. The entire Geopark area is situated in the transition belt between the border zone and the inner alpine zone. The precipitation is lower in the valleys and increases with the altitude. Bischofshofen (550 m asl) shows an average precipitation of 1009 mm (1997–2010), whereas 1277 mm is recorded for Mühlbach (924 m asl). The temperature decreases as the altitude increases. Bischofshofen shows an average temperature of 8.1 °C and Mühlbach of 6.2 °C (1997–2010). Between the winters 1902/03 and 2015/16, the number of days with a snow cover at Arthurhaus–Mitterberg (1503 m asl), which belongs to Mühlbach, increased from 153 to 172 or 12% in 113 years of record-keeping (Aigner 2018).

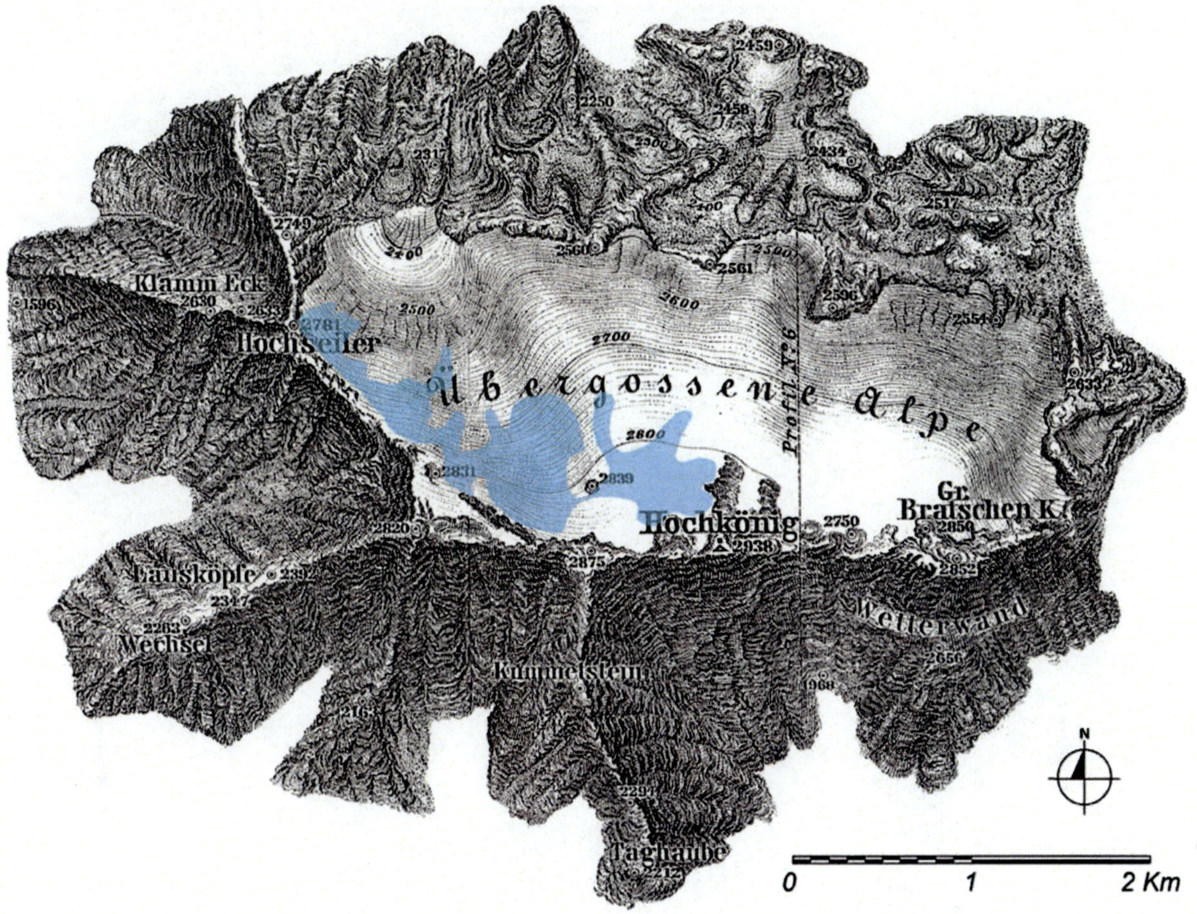

Fig. 28.4 The area of the "covered Alp" glacier at Hochkönig according to Richter (1888) and the current extent of the plateau glacier 2014 (blue-tinted area according to ÖMAP 2014). *Graphics* GeoGlobe, 2016

28.3 Features of the Geopark: Valuable Ores, Mining History and Landscape Experience

The geosites (GS) – which are the geological and geomorphological highlights – in the Ore of the Alps UNESCO Global Geopark are manifold. The number of geosites increases every year, because of new findings emerging from scientific investigations. Some of them are self-explanatory; many, however, are provided with detailed information that is also accessible on the homepage of the Geopark (Erz der Alpen 2018).

The "Copper Trail" forms the backbone of the Geopark. This involves a hike, passing many geosites, of four or five days from Hüttau to Bischofshofen, Arthurhaus, Mühlbach am Hochkönig and continues to St. Veit (Fig. 28.5). The Visitor Centre in Bischofshofen, opened in 2015, provides detailed information about a guided tour along the "Copper Trail". The Visitor Centre also presents the mining history and is home to a special annual exhibition.

28.3.1 Geology of the Ore Mineralization

The copper ore deposit of Mitterberg is situated in the northern part of the Greywacke zone, near the base of the Northern Calcareous Alps. Three lithostratigraphic units can be distinguished in the former mining district:

1. The Grey Sequence of early Palaeozoic age, made up of mainly light and locally dark phyllites, together with subordinate greenschists and diabases.
2. The Violet Sequence of Upper Carboniferous age, consisting of violet-red argillaceous slates, phyllites, quartzites and conglomerates with intercalated graphitic schists and accordant lenses of uranium mineralization.

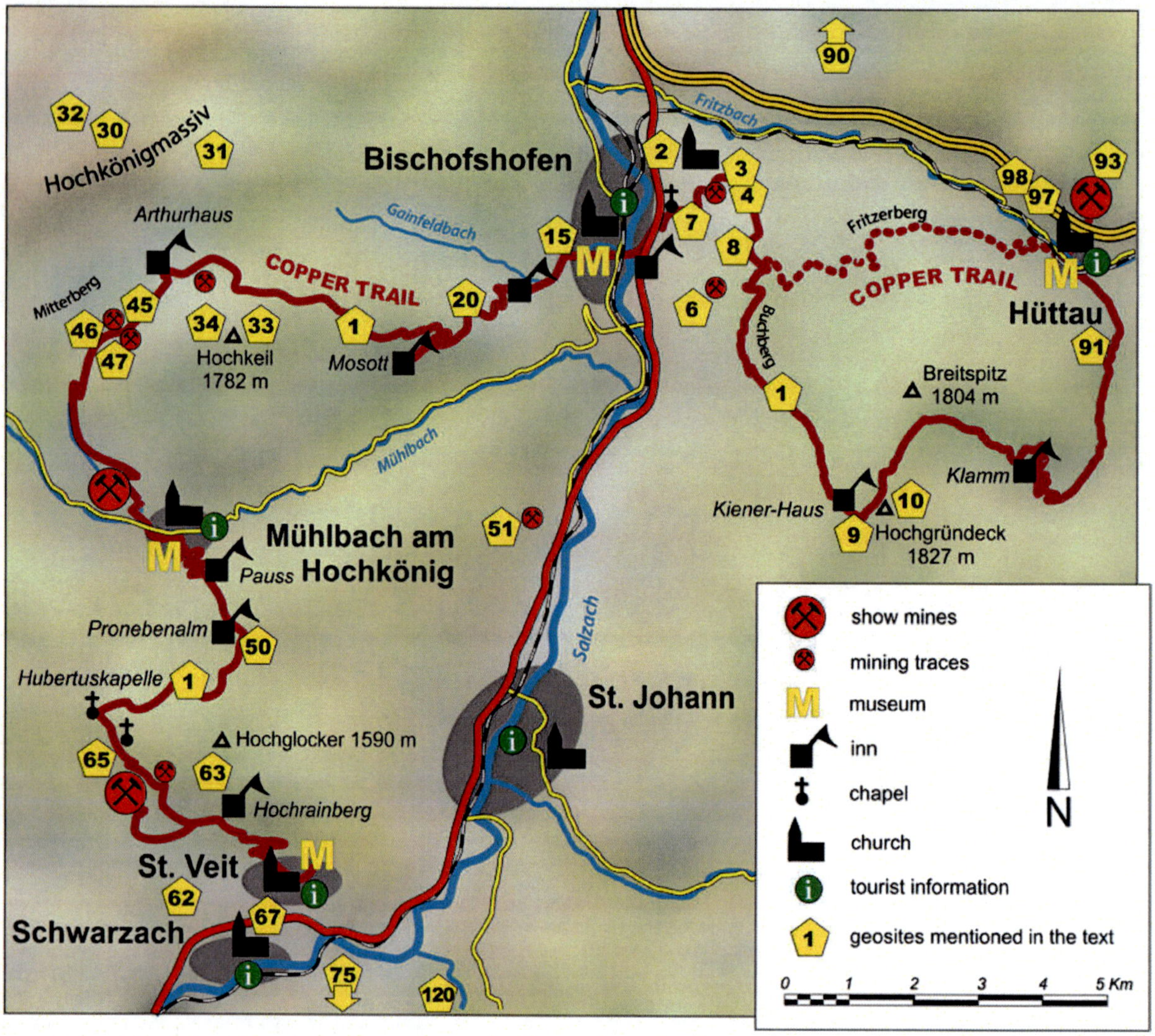

Fig. 28.5 The "Copper Trail", which leads from Hüttau to Bischofshofen, Arthurhaus, Mühlbach am Hochkönig continuing to St. Veit, combines, in a 4 or 5 days' hike, the highlights of the Ore of the Alps UNESCO Global Geopark. *Graphics* GeoGlobe, 2013 and H. Steyrer, 2016

3. The Green Sequence, assigned to the Upper Permian, consisting of greenish-grey argillaceous slates and quartzites, locally containing anhydrite and gypsum.

Transition zones are possible between the Violet Sequence and the Green Sequence. Stratigraphically above the Green Sequence, the Werfen Formation consists of sandstones, quartzites, areno-argillaceous schists and conglomerates. The latter represent the base of the Northern Calcareous Alps (Günther et al. 1993).

The geological and tectonic conditions of the mining area have been intensively investigated in the last few decades.

The Mitterberg main lode (Mitterberger Hauptgang, Fig. 28.6) crosscuts the rocks of the Grey and Violet Sequences as a discordant, 0.2 to 4 m thick ore vein, but does not intersect those of the Green Sequence.

The ore vein strikes west- to northwest and is c. 11 km long. It dips 40° to 80° in a southerly direction. At some sites, overlying and underlying tectonic fragments, as well as short cross veins, occur. In the dip direction, the main lode extends to a mineable depth of 520 m asl. The mining operations were affected mainly by the segmentation of the main lode by six-step faults with downthrows to the west. The fissure veins of the southern district on both sides of the Salzach Valley are mainly concordant with bedding in the phyllites of the Grey Sequence. They strike in a northwesterly direction and extend for up to 3 km.

Fig. 28.6 The steeply south dipping Mitterberg main lode - northern part of "Mariahilfstollen" (GS 47 in Fig. 28.5) – with chalcopyrite (cpy) and pockets of quartzitic (qtz) and carbonate (carb) gangue minerals. The surrounding rocks are phyllites (phyll). *Photo* G. Feitzinger, 2014

The metalliferous mineral assemblages of the Mitterberg ore are extremely diverse. More than 80 different minerals were recorded over the last hundred years (Günther et al. 1993).

The principal copper ore is chalcopyrite, occurring in the western area as irregular ore veins, several metres thick. As mentioned before, tetrahedrite has recently been found at shallower depths (Kucha and Raith 2009). Some parts of the mining area are characterized by a nickel-bearing ore named Gersdorffite. This ore was important for producing nickel products such as cube-nickel and nickel concentrates.

28.3.2 Mining, Use of Regional Resources Since the Bronze Age

A distinctive feature of the Ore of the Alps UNESCO Global Geopark is the mining history. The whole area was famous for mining for more than 5000 years and all local families were involved in mining activities, either as miners, carpenters, wood suppliers or as workers in ore smelting companies. The Geopark region had an outstanding significance for copper extraction and production, not only for Salzburg, but also for the whole of Europe (Feitzinger et al. 2003; SFB HiMAT 2010; Eibner 2016; Pernicka et al. 2016; Ibetsberger 2018; Ibetsberger and Kutil 2019). Based on recent scientific work, the copper used for the Nebra Sky Disc (Himmelsscheibe von Nebra, Germany) originated from the Mitterberg mining district, Mühlbach am Hochkönig (Fig. 28.7a/b, GS 33, 34 in Fig. 28.5).

The oldest copper mine, the Arthur mine (Arthur Stollen: GS 51 in Fig. 28.5) shows traces of prehistoric activities. Together with the Sunnpau mine in St. Veit (Lindenthaler 1991; GS 65 in Fig. 28.5), they were of European-wide significance, until copper (Copper-Age) was replaced by iron (Iron-Age).

In addition to the above-mentioned montane-archaeological attractions, classical archaeological remains were discovered in the Götschenberg–Bischofshofen area. The Götschenberg was settled in the Neolithic, 5500 BP. Archaeological investigations resulted in the unearthing of artefacts, such as needles and pendants, which were made of local copper from the Mitterberg mine near Bischofshofen. The prehistoric "Pongau Castle", above the Gainfeld

a

b

Fig. 28.7 a/b Reconstruction of prehistoric mining activities in the Mitterberg area of Mühlbach am Hochkönig, Ore of the Alps UNESCO Global Geopark. Drawing by G. Feitzinger, 1990

Waterfall, which was built c. 3800 BP, has a history similar to that of the settlement at the Götschenberg site. Both were founded as trader settlements for local copper. The people who lived there were buried at the "black death cemetery" in Bischofshofen, where 556 graves were later excavated (Hörmann 2001).

In addition to the prehistoric mines, the historic mines such as Larzenbach–Hüttau (GS 93 in Fig. 28.5) or the later mines of Mitterberg–Mühlbach am Hochkönig (GS 46, 47 in Fig. 28.5) are also important. Additionally, the oldest traces of mining activities including ancient private pits, open diggings, smelting sites, etc. are visible along the "Copper Trail" which threads through the entire Geopark.

The Geopark has an extraordinarily long tradition of mining, but nowadays all mining activities belong to history. The copper smelting hut Außerfelden (today Mitterberghütten in the community of Bischofshofen), one of the major polluters of the region in the past, closed in 1931. All other mining activities in Mühlbach terminated in 1977 (Gruber and Ludwig 1982).

28.3.3 Geomorphological Key Sites in the Geopark

The Geopark area can be divided into two principal geomorphological domains, which are characterized by distinct elevations and angles of slope:

- The Calcareous Alps are characterized by high and rugged relief due to substantial tectonic uplift since the Miocene and spectacular landscapes mainly shaped by karstification. The karst landscape with sinkholes, karren, karst springs and subsurface drainage as well as the high permeability of karstified rocks greatly influence the distribution of flora and fauna as well as human life in the Geopark area.
- In contrast, the phyllite-dominated Greywacke Zone forms lower mountain ranges and is characterized by rounded landforms, caused mainly by the low resistance of the rocks to Pleistocene glacial erosion.

The large-scale landforms of the Geopark area are mainly structural. As a consequence of Variscan and Alpine mountain building processes, folds, thrusts, normal and transform faults can be observed throughout the area. The main tectonic and geomorphic element is the SEMP zone. This includes the E-W trending Salzach longitudinal fault, which continues as the Ennstal fault in the Enns Valley to the east. North of Bischofshofen, the Salzach Valley changes its direction to S–N, forming a narrow gorge that cuts through the Calcareous Alps. N-S directed faults, directly linked to Alpine nappe stacking, are common along the northern flank of the Hohe Tauern and are responsible for another spectacular gorge in the area, the Liechtenstein Gorge (Fig. 28.8, GS 120 in Fig. 28.5), one of the longest and deepest wild water ravines in the Alps (Geologische Bundesanstalt 2005, 2009; Ibetsberger and Hilberg 2017).

Fig. 28.8 The Großarl River cuts through solid limestone, forming the Liechtenstein Gorge, (GS 120 in Fig. 28.5). *Photo* M. Häupl, 2014

Medium and smaller scale landforms in the area are the result of glacial, glaciofluvial, periglacial, fluvial and gravitational processes.

The Geopark area features numerous traces of the former glaciation. Most typical are the glacial trough valleys (e.g. Mühlbach Valley) with steeply sloping flanks. Additionally, kettle depressions (Fig. 28.9a: Paarseen-Schuhflicker-Heukareck, corresponding to GS 75 in Fig. 28.5), roche moutonnées (Lehen, GS 3 in Fig. 28.5), ice marginal terraces (Buchberg, GS 2 in Fig. 28.5) or rock bars that have been smoothed and striated by the ice are visible. Thick covers of till with erratic rock fragments up to one cubic metre in size characterize the landscape. Most of the erratic boulders were transported by the glacier from the Hohe Tauern. At some sites, the silt and clay of the lodgement till affected the drainage leading to the development of moors and swamps (Ibetsberger et al. 2010).

One of the most interesting raised bog areas is the Troyboden (GS 34 in Fig. 28.5) near the Arthurhaus, a

b

Fig. 28.9 **a** Nature conservation area Paarseen-Schuhflicker-Heukareck, a glacial kettle depression with small lakes (GS 75 in Fig. 28.5). *Photo* TVB Schwarzach – St. Veit, 2013. **b** Double ridge at Hochgründeck (GS 10 in Fig. 28.5). *Photo* GeoGlobe, 2013

community of Mühlbach. The nature conservation area of the province of Salzburg consists of the Langmoos and Sulzbachmoos and shows traces of mining. Pollen stratigraphic analysis has revealed that prehistoric dump material was deposited at the bog. Because of ideal preservation conditions, a lot of Bronze-Age findings were detected. The Troyboden is one of the most important sites of recovery in the Eastern Alps.

Fig. 28.10 The Tennengebirge (left, GS 90 in Fig. 28.5) and the "Bischofsmütze" (middle) present a geomorphological contrast to the Hochgründeck (right, GS 9 in Fig. 28.5). *Photo* GeoGlobe, 2014

Another geomorphological highlight is the double ridge of Hochgründeck (1827 m asl) (GS 10 in Fig. 28.5). The Hochgründeck mountain is built of weak slates, which yield to gravitational mass movements, which leads to a classical mountain splitting. The creeping of the slopes occurs in the range of some millimetres per year, but has been ongoing since the Lateglacial. Small ponds are temporarily visible in between the double ridge (Fig. 28.9b).

At the beginning of the Holocene (11,600 BP), the area of the Geopark was totally free of ice, with temperatures similar to those of the present day. Subsequently, the glacial landforms were transformed by fluvial erosion. The results were deeply incised creeks with waterfalls and cascades, which can be found at many sites in the Geopark area (Fischer Creek: GS 7, Gainfeld Waterfall: GS 15, Putzn Creek: GS 67, all in Fig. 28.5). One of the most beautiful waterfalls, the Gainfeld waterfall, can be visited along the "Waterfall Trail", which has been furnished with numerous information panels.

Spectacular panoramic views are granted to visitors all over the Geopark area. The mountains of the Northern Calcareous Alps (Hochkönig, Tennen- and Hagengebirge, Dachstein) can be seen as well as the peaks of the Hohe Tauern Range. Some panoramic views serve as perfect examples of the geomorphological difference between the rugged limestone rocks and the slates of the gentle "Grass Mountains" of the Greywacke Zone (Fig. 28.10).

28.4 Goals and Challenges of the Geopark

28.4.1 Geotourism in the Alps—A New Quality

Geotourism is an important branch of sustainable and soft tourism. It is based on the principle to observe the landscape holistically, that means to consider geological and geomorphological phenomena in connection with the natural and cultural heritage of the region. This approach includes involvement of the local population. They should "experience the Geopark idea" in which the main goal is to raise awareness and to conserve and promote their natural and cultural treasures. This is the fundamental principle that gives a region its distinctive profile signalling efficiency, sympathy, creativity, transparency and credibility. On the other hand, geotourism brings socioeconomic benefit to the local population. The geological attractions and special landscape features should be integrated in tourist programmes as an important asset of nature-based tourism. The brand name "UNESCO Global Geopark" is a designation which poses special obligations and challenges for the region, which are necessary to conserve an unspoiled area for future generations.

Megerle (2008) described the term "Geotourism" as a part of thematic tourism, which is based on the registration, processing, valorization and commercialization of the wide range of earth and landscape history based topics, including their interaction with botany, zoology, cultural history and actual land use. What is absolutely necessary here is the comprehensive approach to imparting knowledge related to regional features. Geotourism will foster a significant contribution to regional development. At the conference GEOTOP 2011 in Nördlingen/Germany, three facts, namely access, activities and experience, which play a crucial role for the development of a destination, were discussed. It is very important to define interfaces with other touristic offers within a region and to combine them for visitors in an attractive way. This holistic approach is the only possible means for creating nature-based, innovative touristic products.

The potential of Ore of the Alps UNESCO Global Geopark is unique and manifold. Guests can hike along geotrails, investigate the geosites, study the information panels (Fig. 28.11), visit museums and show mines, relax in "oases of silence and tranquillity", eat delicious products from local farms and sit together, talk and laugh with the people living in the region. Apart from these assets, the most important issue is that geotourism protects nature. This is the key message, which should be conveyed (in accordance with fun and event elements) to the younger generation. Geotourism and protecting nature should follow the principles of "less is more", "use all senses", "tell and feel but don't take away" and "act in an interdisciplinary manner". These four guidelines should be conveyed to all visitors. The "kids" are our potential for the future to raise awareness, conserve and promote the geological and other heritage for many further generations.

28.4.2 Recent Landscape Degradation

The main role for the people living and working in the Geopark area involves the tertiary sector in which 68% of the workforce are involved in this field of economy, either in commerce or tourism. Winter and summer tourism is

Fig. 28.11 Info Panel on the glacial terrace of Buchberg with the Tennengebirge (GS 90 in Fig. 28.5). *Photo* N. Guggenberger, 2017

responsible for more than 50% of the added value (Amt der Salzburger Landesregierung 2013). Next to the people who are directly involved in the tourism industry, many other people earn a noticeable indirect income by offering rooms, apartments, Alpine huts, houses, etc. The four Geopark communities recorded 404.410 overnight stays in 2011/12. The main tourism community is Mühlbach, which is a member of the "Sportwelt Amade", one of the largest winter sports areas in the Austrian Alps.

However, as a result of increasing numbers of visitors, the natural environment is suffering from the excessive burden of the "ever" growing economy and tourism industry. The real ecological challenge occurs during the winter months. Winters with less snow in the mountains – a result of climate change – require more and more artificial snowmaking installations. The skiing slopes, thanks to the snow-covering system and their perfect maintenance, should always provide entertaining downhill […] experiences for the guests. However, this system can only be maintained at the cost of severe human impacts on the natural environment, e.g. the construction of water reservoirs on high altitude slopes. Artificial snowmaking is the only way to "whiten" the slopes during a "green" winter season. However, the artificial snow cover is very thin. Therefore, the daily snow grooming activities result in the increasing compaction and degradation of the soil. During summer, rainwater which is no longer absorbed by the soil flows across the surfaces of the ski slopes, resulting in extensive denudation, and channel and gully erosion.

The duty of the Ore of the Alps UNESCO Global Geopark is to warn the stakeholders in the tourism industry of the future consequences of this policy.

Artificial snow production not only depends on water storage in artificial lakes, but also needs a lot of electricity. More powerful electric lines are necessary. At the moment, the people of the Geopark are opposing the construction of a proposed 380 kV power line, which is planned to cut through the Geopark area. Giant electricity pylons should not be built in an area, which is considered unique on account of its landscape.

28.5 Conclusions

The Geopark Ore of the Alps is one of three UNESCO Global Geoparks in Austria – the others are Styrian Eisenwurzen and Karawanken (Hejl et al. 2018). Central to the development of a Geopark is its regional identity, as defined by its geological history, its cultural heritage, its present-day culture and the natural landscape of the region. The story of this regional identity is brought to life by local residents, businesses, organizations and educational authorities working together to give it a concrete shape. And interestingly, when people know more about their local history, they tend to adopt a more sustainable approach to their own environment. The ultimate aim of a Geopark is to use the regional identity as a source of inspiration for the future, by translating it into sustainable economic regional development. Strengthening tourism is an important point for that process. Many regions look for a second pillar in tourism, which will become more and more necessary in the future.

References

Aigner G (2018) Aktuelle Wintertemperatur und Schneemessreihen von der Schmittenhöhe und aus Zell am See. https://www.zukunft-skisport.at/site/assets/files/1019/studie_schmittenhohe_wintertemp_sch_nee_forum-zukunft-skisport_2018.pdf. Accessed 12 Mar 2018

Amt der Salzburger Landesregierung (2013) Bevölkerung Land Salzburg—Stand und Entwicklung 2013, Salzburg

Behrmann JH, Tanner DC (2006) Structural synthesis of the Northern Calcareous Alps. TRANSALP Segment Tectonophysics 414(1–4):225–240

Bohadsch J (1782) Bericht über seine auf allerhöchsten Befehl im Jahre 1763 unternommene Reise nach dem oberösterreichischen Salzkammerbezirk. Abh. Privatges. Böhmen 5:91–227

Boué A (1826) Beschreibung der secundären Gebirge am Nordrande der Alpen. Teutschland Geognost.—Geol. Dargestellt 3(3):438–469

Brückner E (1886) Die Vergletscherung des Salzachgebietes. Geographische Abhandlungen 1(1):1–183

Cornelius HP (1952) Gesteine und Tektonik im Ostabschnitt der nordalpinen Grauwackenzone, vom Alpen-Ostrand bis zum Aflenzer Becken. Mitt. Geol. B.-A. 42–43(1949–50):1–234

Dunbar CO, Rodgers J (1958) Principles of stratigraphy. Wiley & Sons, New York

Eibner C (2016) The copper ore of Mitterberg. Eur Geoparks Mag 13:53

Feitzinger G, Ibetsberger H, Vetters W (2003) Bergbau und Rohstoffe im Land Salzburg. Neumarkt

Gamerith W, Heuberger H (1999) Daten zur Eisstromhöhe des eiszeitlichen Salzachgletschers im Salzachquertal zwischen Schwarzach-St.Veit und Salzburg. Mitt. Ges. f. Salzburger Landeskunde 139:317–341

Geologische Bundesanstalt (2005) Salzburg—Geologische Karte 1:200.000. Salzburg, Wien

Geologische Bundesanstalt (2009) Salzburg—Geologische Karte 1:200.000 Erläuterungen. Salzburg, Wien

Geopark Erz der Alpen (2018) http://geopark-erzderalpen.at. Accessed 14 Jan 2018

Gruber F, Ludwig KH (1982) Salzburger Bergbaugeschichte. Salzburg, München

Günther W, Eibner C, Lippert A, Paar W (1993) 5000 Jahre Kupferbergbau Mühlbach—Bischofshofen, Mühlbach a. Hkg. Bischofshofen

Hejl E, Ibetsberger H, Steyrer H (eds) (2018) UNESCO Geoparks in Austria. Pfeil Verlag, München, pp 1–177

Heuberger H (1968) Die Alpengletscher im Spät- u. Postglazial. Eiszeitalter und Gegenwart 19:270–275

Hoinkes G, Koller F, Rantisch G, Dachs E, Höck V, Neubauer F, Schuster R (1999) Alpine metamorphism of the Eastern Alps. Schweiz Mineral Petrogr Mitt 79:155–181

Hörmann F (2001) Chronik Bischofshofen Bd. 1, 2. Hallwang

Ibetsberger H (2015) Copper so close to the heavens. Eur Geoparks Mag 12:52

Ibetsberger H, Hilberg S (2017) Liechtensteinklamm—deep gorge and mystic springs. Eur Geoparks Mag 14:48
Ibetsberger H, Kutil E (2019) The Nebra Sky disc—a bronze age calendar. Eur Geoparks Mag 16:40
Ibetsberger H, Jäger P, Häupl M (2010) Der Zerfall des Salzachgletschers und die nacheiszeitliche Entwicklung des Salzburger Gewässersystems aus der Sicht der Wiederbesiedlung der Salzburger Gewässer mit Fischen. Gewässerschutz 14:7–54
Ibetsberger H, Pichler H, Häupl M (2011) Pongau-Erz der Alpen—Der erste Geopark im Salzburger Land entsteht. Schriftenreihe der Deutschen Gesellschaft für Geowissenschaften 76:18–19
Ibetsberger H, Feitzinger G, Steyrer H (2018) UNESCO-Geopark Ore of the Alps. In: Hejl E, Ibetsberger H, Steyrer H (eds) UNESCO Geoparks in Austria. Natur- u. Kulturerlebnisführer der Universität Salzburg, Pfeil Verlag, München
Ibetsberger H (2018) "Music was my first love …". Eur Geoparks Mag 15: 28
Kralik M, Krumm H, Schramm JM (1987) Low grade and very low grade metamorphism in the Northern Calcareous Alps and in the Greywacke Zone; illite crystallinity data and isotopic ages. In: Fluegel HW, Faupl P (eds) Geodynamics of the Eastern Alps. Franz Deuticke, Vienna, Austria, pp 164–178
Kucha H, Raith JG (2009) Gold-oxysulphides in copper deposits of the Greywacke Zone, Austria: a mineral chemical and infrared fluid inclusion study. Ore Geol Rev 35(1):87–100
Lindenthaler K (1991) Heimatbuch St. Veit. Salzburg
Megerle H (ed) (2008) Geotourismus: Innovative Ansätze zur touristischen Inwertsetzung und nachhaltigen Regionalentwicklung. Stuttgart
Penck A, Brückner E (1901–1909) Die Alpen im Eiszeitalter. Leipzig
Pernicka E, Lutz J, Stöllner T (2016) Bronze Age copper produced at Mitterberg, Austria, and its distribution. Archeologica Austriaca 100:19–55
Richter E (1888) Gletscher der Ostalpen. Handbücher zur Deutschen Landes- u. Volkskunde 3:1–306
Sedgwick A, Murchinson R (1831) A sketch of the structure of the Eastern Alps. Trans Geol Soc London (2) 3, Tl. 2:301–420
SFB HiMAT (ed) (2010) Abstracts of the 5th milestone meeting des Spezialforschungsbereiches HiMAT, Mühlbach at Hochkönig, 7–10 Oct 2010
Slupetzky H (1975) Erste Radiokarbondatierungen im Pongau im Bereich des inneralpinen Salzachgletschers (Vorbericht). Anzeiger math.-naturw. Klasse der Österreichischen Akademie der Wissenschaften 10:154–160
Tollmann A (1976) Der Bau der Nördlichen Kalkalpen. Monographie der Nördlichen Kalkalpen, Bd. 3. Deuticke, Wien
Tollmann A (1985) Geologie von Österreich, Band 2—Außerzentralalpiner Anteil. Deuticke, Wien
UNESCO Global Geoparks (2017) http://www.unesco.org/new/en/natural-sciences/environment/earth-sciences/unesco-global-geoparks/. Accessed 25 Feb 2018
Van Husen D (2000) Geological processes during the quaternary. Mitt der Österr Geolog Gesellschaft 92:135–156
Van Husen D (1987) Die Ostalpen in den Eiszeiten. Verlag der Geologischen Bundesanstalt (Wien), pp 1–24
von Buch L (1802) Geognostische Beobachtungen auf Reisen durch Deutschland und Italien, Band 1. Haude & Spener (Berlin), pp 1–320
Weber L (ed) (1997) Handbuch der Lagerstätten der Erze, Industrieminerale und Energierohstoffe Österreichs—Erläuterungen zur Metallogenetischen Karte von Österreich 1:500.000 unter Einbeziehung der Industrieminerale und Energierohstoffe. Arch. f. Lagerstättenforschung Geol B.-A. (Wien) 19:1–607

Horst J. Ibetsberger is a self-employed geoscientist and scientific director of the company GeoGlobe, which specialises in the conception and implementation of geo- and ecotourism projects. He is also scientific head of the Ore of the Alps UNESCO Global Geopark. In addition, he is a lecturer of physical geography at the Department of Geography & Geology, Salzburg University.

Hans Steyrer is an associate professor at the Department of Geography & Geology at Salzburg University, Austria. His research interests include structural geology and analogue modelling.

Check for updates

The Variability and Uniqueness of Cirque Landscapes in the Schladminger Tauern

29

Christine Embleton-Hamann and Christian Semmelrock

Abstract

Geomorphologists celebrate the mountain group of the Schladminger Tauern for its superb landforms of glacial erosion. Hikers celebrate it for its scenic quality. This non-glacierized mountain group straddling the main divide of the Eastern Alps in Styria was analysed with the aid of a geomorphological map showing all glacial and periglacial features. The main focus of this chapter is the cirque landscape. A review of cirque development models is followed by analyses of a data base of 76 cirques in the Schladminger Tauern. Three prominent results are noted: (i) cirques of the study area are larger than elsewhere, but most conspicuous is their relatively deep incision; (ii) there has been an above average rate of glacial downcutting; and (iii) optimal Pleistocene conditions for cirque formation were determined from analysis of cirque aspects. The superb cirque development of the study area may be explained by its intermediate altitudinal position on the main divide of the Eastern Alps: during glacial maxima the mountain group was covered by the large Alpine complex of transection glaciers. Much longer durations of the glacial periods, however, were characterized by a local glaciation including intervals of separate cirque glaciers.

Keywords

Cirque • Glacial erosion • Morphometry • Allometry • Scenic quality

C. Embleton-Hamann (✉) · C. Semmelrock
Department of Geography and Regional Research, University of Vienna, Universitätsstr. 7, 1010 Vienna, Austria
e-mail: Christine.embleton-hamann@univie.ac.at

29.1 Geographical Setting

The Schladminger Tauern are situated within the lower section of the main divide of the Eastern Alps, the Niedere (="lower") Tauern range. The Niedere Tauern have no present day glaciers or permanent snow patches like the Hohe (="higher") Tauern immediately to the west. The range is separated by cols into four mountain groups, of which the Schladminger Tauern group is highest, culminating in 2862 m asl (=Hochgolling, for place names see Fig. 29.1). Drainage is directed to the river Enns in the north and to the river Mur in the south.

This mountain group and especially the Klafferkessel region in its centre is famous for its scenic quality and most of the area in Fig. 29.1 is a Styrian nature reserve. In addition to high amenity value, the study area exhibits glacial landforms in textbook-perfection. Its cirques will be the main focus of this chapter. On a larger scale, glacial erosion has steepened ridges and summits into arêtes and pyramidal horns, whereas valleys were turned into classical glacial troughs with U-shaped cross profiles (Figs. 29.2 and 29.3). Tributaries typically occupy hanging valleys: the river Riesach that cascades with a 70 m high waterfall down to the main valley is a fine example (Fig. 29.1). On the smaller scale we find roches moutonnées with smoothed up-ice sides and steepened, cliffed lee sides (e.g. on the lip of the Riesach Valley), and on the smallest scale there are striated and polished rock surfaces. There are also landforms of glacial deposition like moraine ridges, but due to the high alpine position of the study area they play a minor role.

Next to landforms of glacial erosion the study area is rich in rock glaciers as features of the periglacial regime (for a detailed description of rock glaciers in Austria see chapter "Rock Glaciers in the Austrian Alps: A General Overview with a Special Focus on Dösen Rock Glacier, Hohe Tauern Range"). The map in Fig. 29.1 exhibits altogether 34 rock glaciers, originally recorded by Lieb in 1995 (Lieb 1996). Fossil rock glaciers are dominant and account for almost

C. Embleton-Hamann (ed.), *Landscapes and Landforms of Austria*, World Geomorphological Landscapes,
https://doi.org/10.1007/978-3-030-92815-5_29

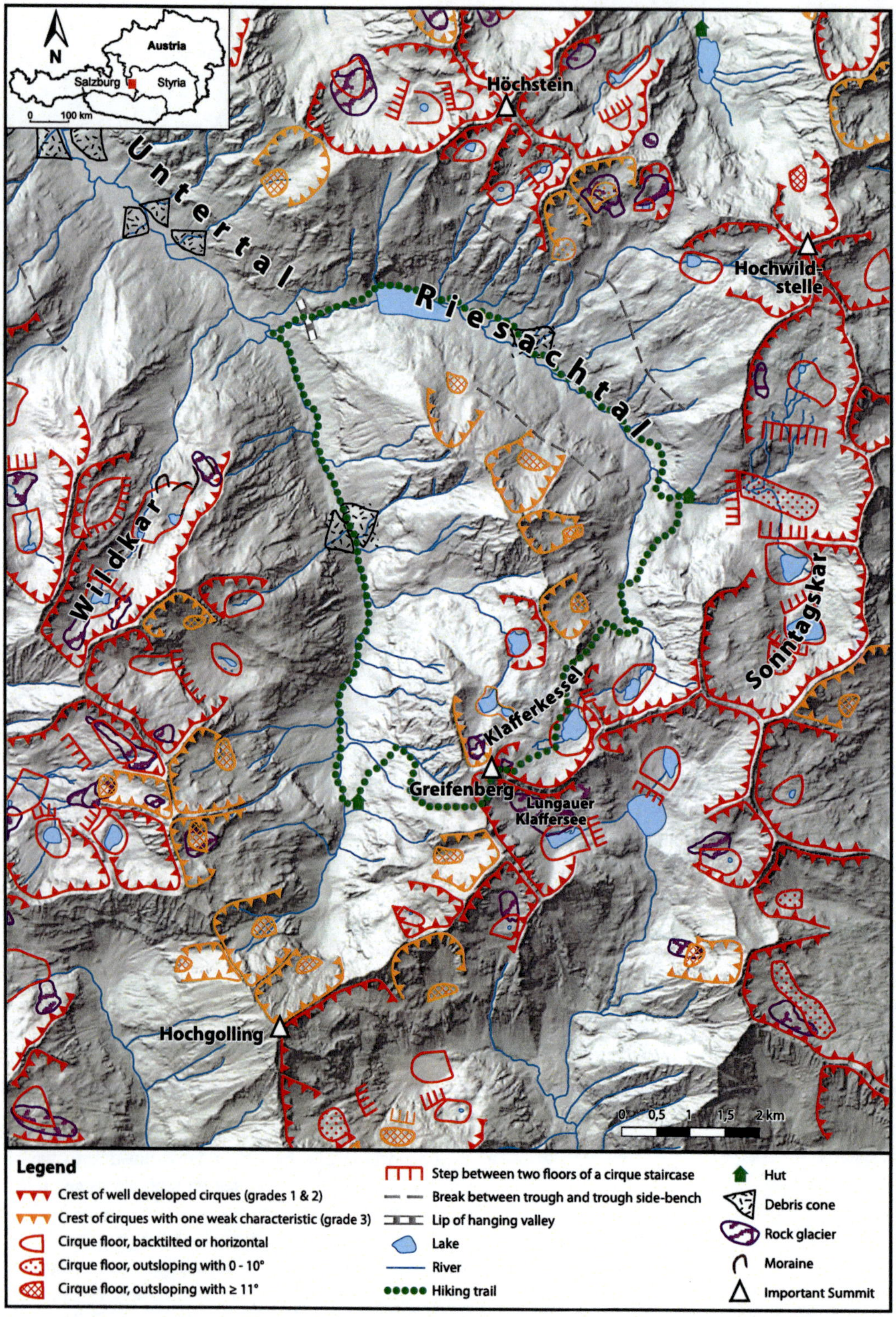

Fig. 29.1 Geomorphological Map. Source of rock glacier distribution: Lieb et al. (2010). Source of background map: https://www.data.gv.at/katalog/dataset/9a6653e0-d5d3-11e3-9c1a-0800200c9a66 Cartography and Layout: C. Semmelrock and W. Lang

Fig. 29.2 Oblique view of the northern part of the study area draining to the Enns Valley. The foreground is dominated by the Klafferkessel with its numerous lakes and glacier polished rock surfaces. In the middle distance the glacial trough of the Untertal stands out, accompanied by a prominent rock shoulder on its left side. Note the valley asymmetry of the Untertal: its right side is steep and short while the crest of its left side lies outside of the photo. Background: Enns Valley and beyond it the Northern Limestone Alps. *Photo* Österreich Werbung/Homberger

85% of them. Ten years later, Kellerer-Pirklbauer assumed that the five remaining intact rock glaciers of 1995 might still contain permafrost, but were no longer able to move because of substantial internal ice losses brought about by climate warming (Kellerer-Pirklbauer 2005).

The study area consists of polymetamorphic ortho- and paragneisses of the so-called Schladming Complex which belongs to the lowermost unit of the Upper Austroalpine nappe pile. The uniform and resistant gneisses of the Schladming Complex favoured preservation of the typically steep sided glacial landforms, whereas in adjacent mountain groups with less resistant bedrock the distinct shape of troughs and cirques became obliterated by mass movements. Lithology also promoted the development of rock glaciers, as the gneisses break down to coarse, blocky debris, which supports permafrost growth.

During the Last Glacial Maximum (LGM) the Schladminger Tauern group was submerged by ice, with only a few nunataks rising above the glacier surface. According to van Husen the ice surface in the Enns Valley lay at 2000 m asl and in the summit area of the Schladminger Tauern at 2500 m asl (Mandl et al. 2014). The ice was connected to the Eastern Alpine transection glacier complex, but situated at its margin. Only about 45 km further east, the abrupt decline of summit elevation coupled with increased climatic dryness resulted in a change to local glaciation.

Downwasting of the LGM ice occurred rapidly and at large scale. Glacier cover receded from the Enns Valley to the mountain valleys of the Schladminger Tauern (Mandl et al. 2014). Around 16 000 years ago the cooling interval of the Gschnitz Stadial initiated a prominent glacier readvance (for a description of the type locality of the Gschnitz advance see chapter "The Moraine at Trins and the Alpine Lateglacial"). Cirque glaciers in valley-head positions merged to feed valley glaciers descending to 1000 m asl in the larger (e.g. Untertal, see Figs. 29.1 and 29.2) and to 1100–1200 m asl in the smaller valleys. Associated with Gschnitz end moraines are overdeepened rock basins, as recognized in all reconnaissance drillings within the Schladminger Tauern (Mandl et al. 2014).

All younger Lateglacial readvances terminated at cirque floor level (cf. Table 7 in Senarclens-Grancy 1962), indicating a period of individual cirque glaciation during the

Fig. 29.3 The foreground shows the lakes of the northern part of the Klafferkessel. Note the arête of the southern side of the Riesach Valley in the middle ground and the pyramidal horns of Höchstein, 2543 m asl (left) and Hochwildstelle, 2747 m asl (right), topping the crest of the northern side of the Riesach Valley in the background. The geomorphological map in Fig. 29.1 depicts how the shape of these two summits results from cirques receding towards each other. View from Greifenberg towards NNE. *Photo* H. Meindl

outgoing Lateglacial. In summary the average conditions of the Schladminger Tauern during glacial periods were characterized by short intervals of thick ice cover during the glacial maxima and long stages of intermediate ice cover similar to the localized valley and cirque glaciers of the Gschnitz Stadial (Mandl et al. 2014).

Glacier response to Holocene climate fluctuation, especially to the cooling period of the Little Ice Age remains unclear. Without doubt the higher parts of the Schladminger Tauern experienced periglacial conditions. Senarclens-Grancy (1962) suggested a Little Ice Age snowline at 2400 m asl, with firn fields expanding in the relatively small mountain area above it. In shaded positions, and where avalanches transferred additional snow masses to the base of the summit slopes, small glaciers may have developed. Within the study area Senarclens-Grancy mapped three fresh looking moraine ridges at elevations of 2370–2400 m asl and attributed them to a possible pocket-sized Little Ice Age readvance.

Present-day climate is dominated by westerly winds, a mean annual air temperature reaching 0 °C at an elevation of 2100–2200 m asl (Kellerer-Pirklbauer 2005) and by a mean annual precipitation rising from 1000 to 1700 mm at the highest areas around the main divide (BMLFUW 2014).

29.2 Review of Ideas on Cirque Formation

The cirque is one of the most characteristic forms of glacial erosion. Evans and Cox (1995) define its overall form as "a hollow, open downstream but bounded upstream by the crest of a steep slope (headwall), which is arcuate in plan around a more gently sloping floor". Textbooks of Glacial Geomorphology usually describe the best developed cirque type, namely the "armchair" cirque. Classic armchair cirques have a steep headwall that curves around an overdeepened basin, often containing a lake or a bog. This basin (the cirque floor) is separated from the steep mountain terrain below by a distinct convex break of slope, the cirque threshold (Fig. 29.4). But the attributes of an armchair cirque are hardly "regular" features. For instance, many typical cirques that fully meet the definition given above have outward-sloping floors. Sometimes cirques are nested, i.e. arranged in cirque staircases with precipitous slopes separating two or more individual cirque floors following above and behind each other (Fig. 29.5).

Cirques are formed by both subglacial and subaerial processes. The latter include periglacial weathering and gravitational mass movements that attack the headwall, aid headwall recession and contribute to plan enlargement of cirques.

Cirque formation starts with glacial occupation of pre-existing mountain side hollows of various origin, like structural benches, landslide scars, gully heads or nivation niches (formed by erosion processes associated with long-lasting snow patches). Once a glacier is established, subglacial erosion will enlarge and reshape the original mountainside concavity.

Subglacial erosion works by two different means, namely abrasion and quarrying. The latter is commonly estimated to be the more powerful process in bedrock erosion (cf. Spotila 2013). Abrasion or scouring is achieved by debris imbedded in the basal ice of a glacier and dragged over the underlying bedrock by glacier flow. Quarrying or plucking addresses the detachment of large blocks from the bedrock floor. This occurs on the downglacier side of protuberances of the bed by gradual widening of joints and fissures until blocks of rock become loose and are pulled away by the moving glacier (Benn and Evans 2010).

Optimal conditions for abrasion and plucking are different for warm based mid-latitude and cold based

Fig. 29.4 Lungauer Klaffersee Cirque with the morphometric variables length = 950 m, width = 1200 m, amplitude = 260 m and height range = 418 m. It is a well enclosed armchair cirque with a steep headwall and a distinct threshold, followed by a 170 m high drop to lake Zwerfenbergsee. A part of the headwall is visible to the right, and lake Zwerfenbergsee is just visible to the left of the cirque lake. There are partly overgrown ridges of a rock glacier visible in front of the lake. View from Greifenberg towards WSW. *Photo* H. Meindl

Fig. 29.5 The cirque stairway of the Sonntagskar. The view focuses on the two central levels of the stairway with their lake-filled, overdeepened basins and distinct thresholds. The highest cirque floor of the stairway is concealed in this picture. The drop to the lowest cirque floor is visible in the foreground (cf. map in Fig. 29.1). *Photo* TVB Schladming—G. Stocker

subpolar/polar glaciers, which are frozen to their floors and where there is no subglacial meltwater. The following account refers to warm based alpine glaciers only and is largely based on the review by Spotila (2013) that presents the current state of research and provides further references.

Basically, subglacial erosion is generated by the movement of the glacier over its bedrock floor. Early cirque development models focused on the idea that glacier movement is maximized by rotational flow (Lewis 1949), initiated by the weight of winter snow accumulations on the upper part of the glacier and summer ablation losses on the lower end of the glacier. In steep and small glaciers equilibrium would be restored by rotational slip. Shortly afterwards this concept was buoyed by the first direct observation of rotational movement in tunnels dug into a steep Norwegian cirque glacier (McCall 1972).

Fuller understanding of glacial erosion processes was gained by studies of the subglacial hydrological system. Surface meltwater reaches the glacier bed via crevasses and the bergschrund (a large crevasse that forms as ice moves away from the cirque headwall). Water at the bed not only accelerates glacier sliding, which in general is important for erosion, but also promotes quarrying by water pressure fluctuations. As the pore pressure in bedrock joints is controlled by subglacial water pressure, joints will be widened by pressure fluctuation, eventually allowing the glacier to pluck away a block of rock. Another important function of subglacial meltwater drainage is the removal of debris, thereby exposing the bedrock to further erosion.

Many empirical studies have shown that cirque widening by headwall recession occurs faster than cirque deepening (cf. Sect 4.2). Important erosional agents at the headwall are subaerial frost-shattering and mass movements. But Hooke (1991) presented evidence that subglacial erosion also plays a significant role in headwall retreat. He could prove that rapid quarrying takes place at the base of the headwall, where the bergschrund permits periodic entry of meltwater. As a consequence the headwall becomes oversteepened and susceptible to mass movements, thus reinforcing subaerial mass wasting. A second positive feedback loop between subaerial and subglacial processes can be found in the frost debris produced at the headwall and reaching the glacier bed via crevasses, where the angular clasts become important tools for abrasion.

Processes of cirque formation and rates of cirque growth have also been the focus of intense study in the context of the glacial "buzzsaw" hypothesis. This hypothesis deals with the question of how high mountains may become and posits that glacial erosion may be compared to a chain saw truncating summits and high lying areas at the same rate as the mountain range is rising by tectonic uplift. Mitchell and Montgomery (2006) describe the main mechanism by which the buzzsaw is supposed to operate: cirque development associated with wearing back and cutting down the high topography will reduce the areal extent of the "roof" of a mountain range and turn it into a zone of predominant steep and unstable slopes. Subsequently weathering and mass wasting, especially in positions of coalescing cirque headwalls, will decrease peak and watershed altitude.

29.3 A Survey of the Schladminger Tauern Cirques: Methods

29.3.1 Geomorphometry and Allometry

A widely used scientific method to analyse cirque development is geomorphometry. This is the quantification of size and shape of relief elements, followed by a statistical analysis of the data. Geomorphometry is divided into "general geomorphometry" considering large-scale data of whole landscapes, e.g. altitudinal range and "specific geomorphometry" dealing with the measurement and analysis of the dimensions of individual surface features like cirques (Evans 2004). Measurement of plan form, length, width, height and orientation of cirques from aerial photos, laser scans, digital terrain models, topographical maps or in the field and subsequent statistical analysis of these data is used to test scientific theories on cirque formation, open questions about the control of cirque form (such as altitude, aspect, lithology, relative relief) and to define the range of variation in cirque size.

Measurements of length, width and depth of a cirque are frequently further used for allometric data analysis. Allometry is a general concept relating the shape and size of landforms. This approach was originally applied to cirques by Olyphant (1981) and systematically tested by Evans (2006, 2010) and Mîndrescu and Evans (2014) in order to find out whether the vertical dimension (height) of a cirque increases at the same rate as its horizontal dimensions (length, width). If all dimensions change at the same rate, development is said to be isometric and cirque shape will stay uniform over time. If cirques change shape as they grow, because one dimension increases faster than the others, the development is allometric. Note, however, that allometric analyses of cirques involve accepting the ergodic principle (the substitution of space for time), as variations of cirque size in a regional cirque population are used instead of actual cirque development ages, which are impossible to establish.

29.3.2 Data Sources and Definition of Morphometric Variables

Data were measured from 1:25 000 topographic maps with contour intervals of 20 m, with ancillary use of air photos. The map source consisted of sheets 3324 West and 3324 Ost of the "ÖK 50 UTM", produced by the Austrian mapping agency. The sampling area covers all cirques depicted in Fig. 29.1 belonging to the watershed of the Untertal, draining north to the river Enns. For the purpose of an unbiased sample, the cirques of an equal proportion of the watershed draining south to the river Mur were also included. Within the delimited area all cirques whether perfectly or imperfectly developed were measured. In the case of nested cirques, all identifiable sub-cirques were also recorded. The resulting data base has a total of 76 cirques.

Each cirque was first delimited from the surrounding slopes by tracing its crest and its threshold. Cirque delimitation was completed by connecting the end points of crest and threshold perpendicular to slope inclination. Afterwards the median axis of the cirque, dividing its area into two equal halves was drawn (Fig. 29.6).

After cirque identification and delimitation on the map the following key morphometric values were measured: cirque length, defined as the length of the median axis, cirque width, defined as the longest line at right angles to the median axis and cirque orientation as the azimuth of the median axis in outward direction. Altitude was recorded for three points: the intersection point of the median axis with the crest, the highest point on the crest and the lowest point on the floor (E_{axis}, E_{max} and E_{min} in Fig. 29.6). Finally the maximum gradient of the headwall and the minimum

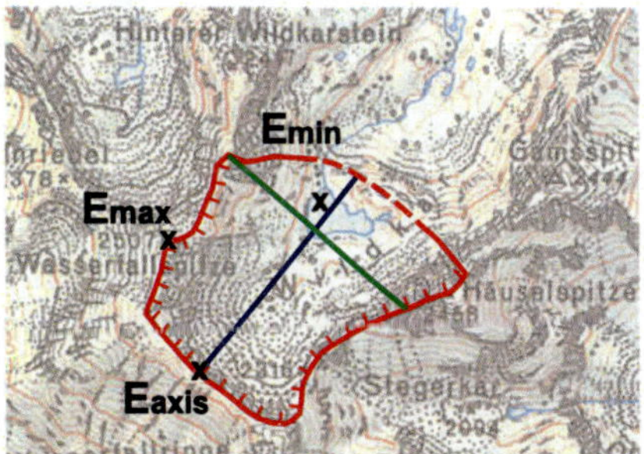

Fig. 29.6 Definitions of measured variables: Cirque crest (solid line with ticks) and threshold (dashed line) are indicated. The dark blue line represents the median axis (at the same time cirque length) and the green line cirque width. E_{axis}, E_{max} and E_{min} mark the three points of altitude reading. Minimum gradient is 0° (horizontal lake level), and maximum was measured to the east of the lake, where a hundred metre drop of the cirque shoulder over a horizontal distance of 67.5 m produces a gradient of 56°. Topographic base map: Alpenvereinskarte 1:50 000, sheet 61

gradient of the floor were calculated from the spacing of the contour lines. For cirques containing a lake or a bog the minimum gradient was recorded as 0°.

From these initial measurements six additional variables were derived: amplitude ($E_{axis} - E_{min}$), height range ($E_{max} - E_{min}$), gradient of the median axis (arctan (amplitude/length)), length/amplitude ratio, length/height range ratio and length/width ratio. The gradient of the median axis approximates the slope of the glacier once filling the cirque, and length/amplitude and length/height ratios are frequently used measures of cirque incision. Last, in order to define overall quality, the existence of a cirque lake and cirque "grade", a subjective assessment of the degree of cirque development, was recorded. For the variable grade the first three grades of the five-fold classification of Evans were adopted: grade 1 cirques are classic armchair cirques with a steep headwall curving around a deeply excavated floor (e.g. the Lungauer Klaffersee cirque in Fig. 29.4), grade 2 cirques are well defined, but are more open or have an outsloping floor and grade 3 cirques are definite features, with no debate over their cirque status, but one or more characteristics may be weak (Evans and Cox 1995; Mîndrescu and Evans 2014).

29.4 A Survey of the Schladminger Tauern Cirques: Results

29.4.1 Comparison with Other Cirque Landscapes

A simple comparison of average morphometric values across the international literature shows that the cirque landscape of the study area is exceptional. Its densely packed cirques separated by narrow arêtes and pyramidal peaks have a regional cirque density of one cirque per 1.8 km^2 which is far higher than the three density measures so far published and summarized by Evans (2006) (Maritime Alps: one cirque per 4 km^2, English Lake District and three mountain groups in Wales: one per 5–7 km^2).

The 76 cirques of the study area are also larger than elsewhere. Mean and median values are 928 m (800) for length, 959 m (863) for width, 420 m (390) for amplitude and 524 m (506) for height range. Figure 29.7 compares these mean sizes with those of similarly defined cirque data sets. Only the cirques of Kamchatka are slightly wider. In all other dimensions the Schladminger Tauern cirques show the highest values and especially their height range exceeds all previous statistics. Consequently they also have the lowest length/height range ratio in Fig. 29.7.

A marked characteristic of the cirques of the study area is therefore their deep incision. In Fig. 29.7 height range was chosen for comparing vertical dimensions, as there are few

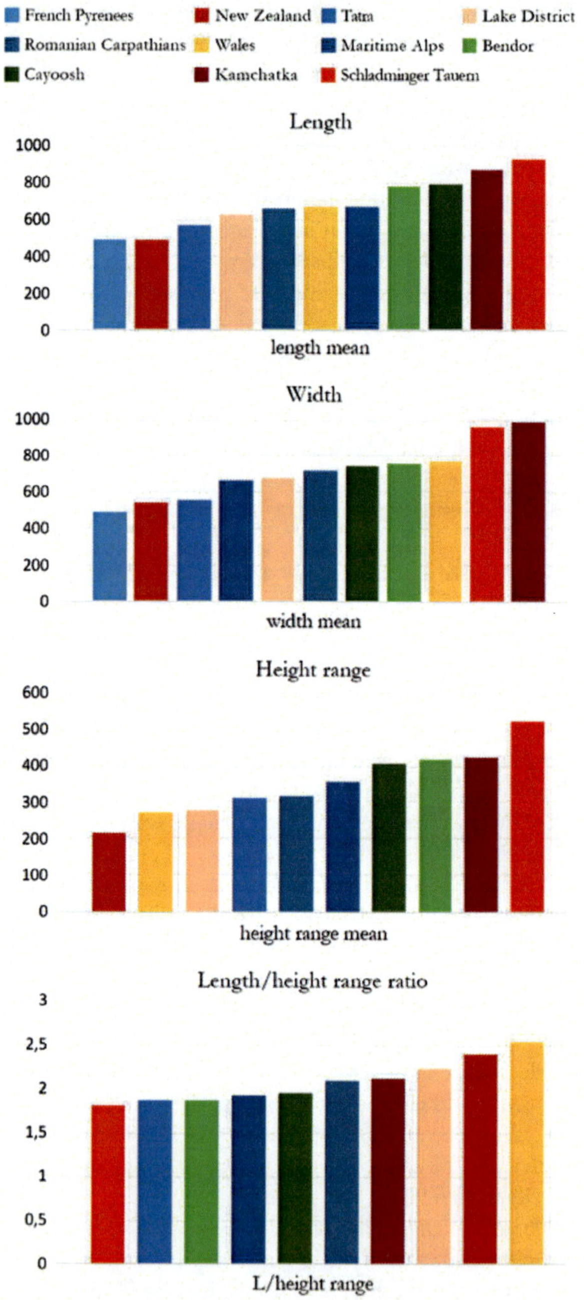

Fig. 29.7 Comparison of cirque size characteristics. Data sources: French Pyrenees: Delmas et al. (2014), New Zealand: after Mîndrescu and Evans (2014), Table 2, Tatra: Křížek and Mida (2013), Lake District: Evans and Cox (1995), Romanian Carpathians: Mîndrescu and Evans (2014), Wales: Evans (2006), Maritime Alps: Federici and Spagnolo (2004), Bendor and Cayoosh Ranges (British Columbia): after Mîndrescu and Evans (2014), Table 2, Kamchatka: Barr and Spagnolo (2013)

data for amplitude as a measure of cirque depth. Only Evans and Cox (1995), Evans (2006), Delmas et al. (2014), and Mîndrescu and Evans (2014) provide mean values for amplitude, which range from 215 to 272 m. The 420 m mean amplitude of the study area is substantially higher, further confirming deep cirque incision. Deep glacial scouring is associated with steep cirque glaciers (see Sect. 29.2). For the Schladminger Tauern cirques axial gradient, a good proxy of former glacier gradient, averages 25°.

Unfortunately further variables indicating high cirque quality, like headwall (maximum) gradient or floor (minimum) gradient are inconclusive. Headwall gradients are strongly dependent on the quality of maps or DTMs from which they were measured. Floor (minimum) gradient on the other hand does not allow calculation of a comparable mean value for the study area, as half of its cirques have a horizontal or back-tilted floor.

For scenic quality two factors are especially important, namely the presence of water and strong relief contrasts. Attractive cirque landscapes are therefore characterized by their many lakes and the relief contrasts arising between near vertical cirque headwalls and near horizontal cirque floors, multiplied in the case of cirque stairways (Embleton-Hamann 1994) (Fig. 29.5).

Amongst hikers the Klafferkessel region is praised for its numerous lakes—every second cirque has one (Figs. 29.1, 29.2, 29.3 and 29.5). Compared with the rock-basin lake frequency in other cirque landscapes this proportion of 50% is followed by a proportion of 42% in the Rezetat group in the Romanian Carpathians (Mîndrescu and Evans 2014), but then values drop sharply to 18% in the Lake District (Evans and Cox 1995), 11% in Kamchatka (Barr and Spagnolo 2013) and 9% in the Maritime Alps (Federici and Spagnolo 2004). Interesting in this context is the strong influence of granitic bedrock on lake frequency that Mîndrescu and Evans (2014) found, as the uniform migmatitic gneisses of the study area are comparable to granites.

Nested cirques are frequent and include cirque stairways with as many as four separate floors in valley-head positions and two to three floors in valley-side positions (see, for instance, the south-north descending "Sonntagskar" and for valley-side cirques the "Wildkar" in Fig. 29.1). There are altogether 29 "inner" (upper) cirques nested within 16 larger "outer" cirques, together 59% of the total, a proportion that is again higher than values reported from other regions. In general, a high concentration of nested cirques developed in ranges that rose well above the Quaternary glacier equilibrium line (Mîndrescu and Evans 2014; Evans 2006). Thus there are higher proportions of nested cirques to be found in active orogenic belts with high relief (e.g. at least 40% in the French Pyrenees (Delmas et al. 2014) and 27% in the Retezat group (Mîndrescu and Evans 2014)) than in old crystalline massifs (e.g. 11% in Wales (Evans 2006)).

29.4.2 Size-Shape Interrelations, or Allometric Versus Isometric Cirque Development

Allometric analysis was performed following the method of Evans (2006): cirque size was obtained by calculating the cube root of (length × width × amplitude). The variables length, width and amplitude, which have positively skewed frequency distributions (mean values are higher than median values—see Sect. 4.1), were logarithmically transformed. Results are shown in Table 29.1 and Fig. 29.8: length has an exponent of 1.14 ± 0.11, the lower 95% confidence limit of 1.03 is above 1.0, thus length increases faster than overall cirque size. Amplitude with an exponent around 1.0 (1.08 ± 0.17) is showing almost isometric behaviour. The width exponent is 0.78 ± 0.18. Its upper confidence limit is below 1.0, so width increases with size, but more slowly than length, amplitude and overall cirque size.

The 95% confidence bands of the allometric exponents are however very broad. Little reliance can be placed on the difference between length and amplitude, as their confidence intervals overlap. In the case of width the greater scatter in the data set produces a rather low R^2 value, measuring goodness of fit of the regression. Therefore, the pattern of allometric growth shown in Fig. 29.8 can only indicate a trend, but this trend is again unusual. So far, almost all morphometric cirque studies produced depth exponents that are significantly below 1.0 (Evans 2010) indicating that as cirques grow in length, width and overall size, their incision advances at a much slower rate. In the study area length is increasing fastest, but followed by an above average rate of down cutting. This picture is consistent with the results of Sect. 4.1: the Schladminger Tauern cirques are remarkably deep. Their many lakes not only enhance landscape beauty, but also attest to considerable glacial overdeepening of cirque floors.

Table 29.1 Schladminger Tauern: exponents for logarithmic regressions of size variables on size, confidence limits, R2 measures of fit and standard deviation for each variable

Variable	Exponent	95% conf	R^2, %	St. dev
Length	1.14	1.03–1.25	85.72%	0.18
Width	0.78	0.61–0.96	52.44%	0.29
Amplitude (cirque depth)	1.08	0.91–1.24	70.26%	0.27

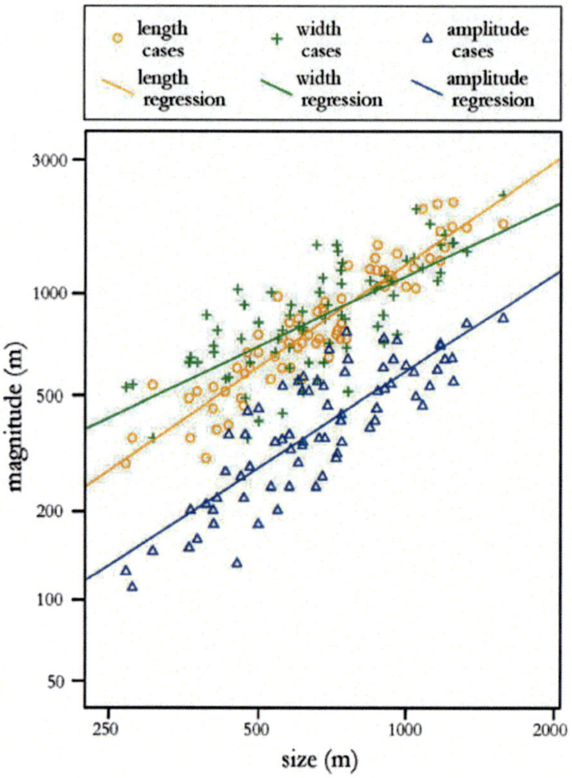

Fig. 29.8 Allometric relationships: logarithmic regressions of length, amplitude and width on overall cirque size

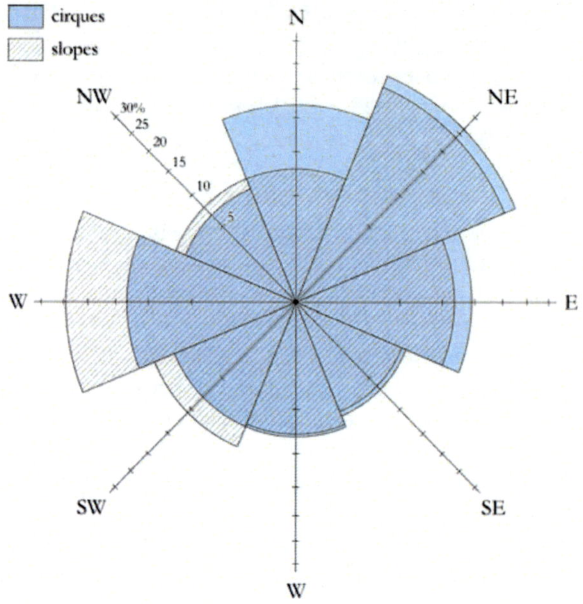

Fig. 29.9 Distribution of slope and cirque aspects in the study area

29.4.3 The Influence of Aspect or Asymmetry Versus Symmetry in Cirque Distribution

Since orientation partly determines how active a glacier is, affecting its total budget and mass balance, and therefore also its erosional competence, one would expect a greater number of cirques together with better cirque development stages in aspects that favour glacier growth. In the Alps this is the northward aspect, where direct solar insolation and therefore ablation is lowest, and secondarily the eastward direction, where under predominant west-winds greater snowdrift accumulations may occur (Evans 2013). At first sight the mean aspect for the cirques of the study area meets these expectations, as it is 30°, or roughly NNE. Also cirque frequency in relation to the distribution of slope aspects (Fig. 29.9) portrays this preference: North facing slopes accommodate far more cirques than expected, followed by northeast to east facing slopes. On the other hand, the west facing slopes and to a lesser extent the southwest facing slopes of the study area fall short. In all other directions cirque frequency more or less corresponds to available slope area.

This picture is however strongly influenced by the valley asymmetry of the Eastern Alps in general and the valley asymmetry of the Schladminger Tauern in particular. Already in an earlier study, comprising all well-developed armchair cirques of the Eastern Alps this topographic control amplified the eastern aspect in cirque orientation (Embleton and Hamann 1988; Hamann and Embleton 1988). In Fig. 29.1 the valley asymmetry of the study area is especially well visible in the western half of the map: steep and short west facing valley sides (forbidding cirque development) are set against gently inclined and long east facing valley sides (with space for cirque development). This topographic asymmetry and associated higher cirque density in eastern and northern aspects was early recognized by Aigner (1930) and triggered a scientific debate of its origins that lasted for three decades until it finally dried up unresolved. As influence of tectonic structure or variable lithology must be excluded, most authors assumed a preglacial valley asymmetry or speculated on various climatic influences. Even high glacial erosion rates, substantially lowering ridges and pushing them further into the mountain mass, comparable to the modern buzzsaw theory (see Sect. 29.2), were expressed (Lucerna 1924).

The mean vector of a data set is however not very meaningful for assessing the climatic influence on cirque distribution. Such analyses are normally based on the data variation around the mean vector aspect. In a circular data set this is provided by the resultant vector length, or "vector strength". Vector strength varies from 0% (=no preferred cirque orientation, cirque distribution is symmetric) to 100% (=cirques are limited to one climatically favoured aspect and cirque distribution is strongly asymmetric). The vector strength of the cirques in the study area is 30%, indicating a weak cirque asymmetry. Vector strength as a measure of local climatic asymmetry is however strongly influenced by topographic constraints as described above. Compared to other mountain ranges a vector strength of 30% is also very low. Evans (1977, Table 2) compiled the vector strengths of 79 cirque data sets. The frequency distribution of this compilation is characterized by a mean of 57%, a lower quartile of 33% and an upper quartile of 67%. Thus the mean vector strength of the cirques in the study area lies in the lower quartile, not far from the commonly agreed starting point of symmetric cirque distribution at 20%.

In addition, climatically favoured positions do not show any better developed cirques. The effects of aspect on cirque development were assessed by analysis of variance. Because of small numbers the climatically favourable aspects N, NE and E, which also show the highest cirque numbers in the study area, were pooled and tested against all other aspect classes. The result is that aspect has no significant influence on cirque form variables.

The cirque distribution of the study area is therefore characterized by a very low asymmetry, which is commonly ascribed to the intensity of glaciation: mountain ranges rising above the Pleistocene snowline have a more symmetric cirque distribution than ranges that were only just high enough to nourish glaciers (Evans 1977; Mîndrescu and Evans 2014).

29.5 Conclusions

Compared with other cirque populations around the world the cirques of the Schladminger Tauern stand out due to their density, nested arrangement, sizeable dimensions, abundant rock-basin lakes, and mostly because of their deep incision. Statistical analysis of cirque aspects shows an almost symmetrical cirque development.

The exceptional quality of the Schladminger Tauern cirques resulting from this survey may be explained by a combination of two factors. Firstly, by the delimitation of the study area embracing the most famous cirque landscape of the Niedere Tauern range. Extension of the survey to adjacent areas with softer rock types would probably lower mean values. Secondly, as cirque enlargement is thought to primarily occur during periods of restricted glaciation, when glaciers are largely constrained to their cirques, the intermediate altitude of the study area on the main divide of the Eastern Alps plays an important role. During the Pleistocene the Hohe Tauern range to the west was submerged by a complex of transection glaciers, while the eastern continuation of the Niedere Tauern range was only marginally glaciated. In the transition zone of the study area, however, separate cirque glaciers existed over long periods before and after each glacial maximum.

The cirque landscape described in this chapter may be visited on a one-day or on a two-day alpine tour. Overnight stays are possible at either the Golling or the Preintaler Hut. Both huts and the trail are indicated in Fig. 29.1.

Acknowledgements We are indebted to Martin Wenk for considerable help with data analysis and visualization and to Olav Slaymaker for corrections made to the early version of this chapter.

References

Aigner A (1930) Das Karproblem und seine Bedeutung für die alpine Geomorphologie. Z Geomorph V:201–223

Barr ID, Spagnolo M (2013) Palaeoglacial and palaeoclimatic conditions in the NW Pacific, as revealed by a morphometric analysis of cirques upon the Kamchatka Peninsula. Geomorphology 192:15–29

Benn DI, Evans DJA (2010) Glaciers and glaciation, 2nd edn. Hodder Education, London

BMLFUW—Bundesministerium für Land—und Forstwirtschaft, Umwelt und Wasserwirtschaft (2014) Mittlerer Jahresniederschlag 1981–2010. http://ehyd.gv.at

Delmas M, Gunell Y, Calvet M (2014) Environmental control on alpine cirque size. Geomorphology 206:318–329

Embleton C, Hamann C (1988) A comparison of cirque forms between the Austrian Alps and the Highlands of Britain. Z Geomorph NF Suppl Bd 70:75–93

Embleton-Hamann C (1994) The role of geomorphological mapping in scenery appraisal. In: Proceedings of the national science council—part c: humanities and social sciences, vol 4/2. Taiwan, pp 231–245

Evans IS (1977) World-wide variations in the direction and concentration of cirque and glacier aspects. Geogr Ann 59A:151–175

Evans IS (2004) Geomorphometry. In: Goudie A (ed) Encyclopedia of geomorphology, vol 1. Routledge London, New York, pp 435–439

Evans IS (2006) Allometric development of glacial cirque form: geological, relief and regional effects on the cirques of Wales. Geomorphology 80:245–266

Evans IS (2010) Allometry, scaling and scale-specificity of cirques, landslides and other landforms. Trans Jpn Geomorphol Union 31 (2):133–153

Evans IS (2013) Glacial landforms, erosional features: major scale forms. In: Elias SA (ed) Encyclopedia of quaternary science, vol 1. Elsevier Amsterdam, pp 847–864

Evans IS, Cox NJ (1995) The form of glacial cirques in the English Lake District. Z Geomorph NF 39(2):175–201

Federici PR, Spagnolo M (2004) Morphometric analysis on the size, shape and areal distribution of glacial cirques in the maritime Alps (western French-Italian Alps). Geogr Ann 86A:235–248

Hamann C, Embleton C (1988) Morphometrische Analyse der Kare in den Salzburger Zentralalpen. In: Salzburger Geographische Arbeiten, vol 17. Salzburger Geographisches Institut, Salzburg, pp 57–74

Hooke RL (1991) Positive feedbacks associated with erosion of glacial cirques and overdeepenings. Geol Soc Am Bull 103:1104–1108

Kellerer-Pirklbauer A (2005) Alpine permafrost occurrence at its spatial limits: first results from the eastern margin of the European Alps. Nor Geogr Tidsskr 59:184–193

Křížek M, Mida P (2013) The influence of aspect and altitude on the size, shape and spatial distribution of glacial cirques in the High Tatras (Slovakia, Poland). Geomorphology 198:57–68

Lewis WV (1949) Glacial movement by rotational slipping. Geogr Ann 31:146–158

Lieb GK (1996) Permafrost und Blockgletscher in den östlichen österreichischen Alpen. In: Arbeiten aus dem Institut für Geographie der Karl-Franzens-Universität Graz, vol 33. Graz, pp 9–125
Lieb GK, Kellerer-Pirklbauer A, Kleinferchner H (2010) Rock glacier inventory of Central and Eastern Austria elaborated within the PermaNET project. Unpublished digital data set of the Department of Geography and Regional Science, Karl-Franzens-Universität Graz
Lucerna R (1924) Der Klafferkessel in den Schladminger Alpen. Zeitschrift Des Deutschen Und Österreichischen Alpenvereins 55:54–78
Mandl GW, Hejl E, van Husen D (2014) Erläuterungen zu Blatt 127 Schladming. Geologische Bundesanstalt, Wien
McCall JG (1972) The flow characteristics of a cirque glacier and their effect on glacial structure and cirque formation. In: Embleton C (ed) Glaciers and glacial erosion. Macmillan, London, pp 205–228
Mîndrescu M, Evans IS (2014) Cirque form and development in Romania: allometry and the buzzsaw hypothesis. Geomorphology 208:117–136
Mitchell SG, Montgomery DR (2006) Influence of a glacial buzzsaw on the height and morphology of the Cascade Range in central Washington State, USA. Quatern Res 65:96–107
Olyphant GA (1981) Allometry and cirque evolution. Geol Soc Am Bull 92(1):679–685
Senarclens-Grancy W (1962) Beiträge zur Eingliederung der Moränen der Schladminger Tauern, der Mittelennstaler Moore und der Ramsau- oder Ennstalterrasse bei Schladming in das alpine Jungquartär. Jahrbuch Geol Bundesanstalt 105:65–128
Spotila JA (2013) Glacially influenced tectonic geomorphology: the impact of the glacial buzzsaw on topography and orogenic systems. In: Shroder J, Owen LA (eds) Treatise on geomorphology, vol 5. Academic Press San Diego, pp 300–317

Christine Embleton-Hamann is a retired professor at the Department of Geography and Regional Research at the University of Vienna. Her main interest is in alpine environments. Within this field she focusses on human-environment interactions with research topics like human impact on geomorphic processes, assessment of the scenic quality of landscapes, and geomorphological hazards. A second set of interest concerns the communication of geomorphological knowledge to a broader audience, in the pursuit of which objective she has written a well-received textbook on geomorphology. She is a former President of the Austrian Research Association on Geomorphology and served on the Executive Boards of the IAG and several IAG and IGU Working Groups.

Christian Semmelrock is currently working as a planner for a geodata office in Freiburg im Breisgau / Germany. He studied Geography in Graz (Austria), Geneve (Switzerland) and Vienna (Austria) with a focus on geomorphology and risk assessment.

30 The Erzberg Area: A Mining Landscape in Styria

Stefan Premm and Christine Embleton-Hamann

Abstract

Iron ore has been mined in the Erzberg area for at least 900 years, possibly as long as 1500 years. This mining history can be divided into three periods of impact on the landscape. Due to the inefficiency of the early smelting techniques and the low production levels, the first period remained free of serious impacts. A huge demand for energy for ore smelting characterized the second period, resulting in supra-regional deforestation for charcoal production. The third period of mechanized surface mining is marked by the creation of large man-made landforms. In the Erzberg area, both historical documentation and modern recording of the mining activities are exceptionally good and provide an opportunity for (i) a qualitative identification and description of the associated landscape changes and (ii) a first attempt to quantify them. Qualitative analysis of the documents provided a detailed description of how distinct anthropogenic impacts and landforms have emerged and identified the procedures that caused mediaeval supra-regional deforestation. A slope exposure opened along a creek allows interpretation of the geomorphic effects of the historical forest depletion. Quantification of the extent of deforestation was carried out for a forest area close to the mining site. The results allow a reconstruction of the condition of the forest at this site over two centuries. For quantification of the earth material moved by man in mechanized surface mining denudation rates for the Erzberg site were computed and compared with the sediment budget of a nearby river catchment. This comparison suggests that on the local scale man is two to three orders of magnitude more effective in transforming the landscape than geomorphic processes.

S. Premm (✉) · C. Embleton-Hamann
Department of Geography and Regional Research, University of Vienna, Universitätsstr. 7, 1010 Vienna, Austria
e-mail: s.premm@gmx.at

Keywords

Impacts of mining • Deforestation • Anthropogenic landforms • Historical forest inspections • Sediment budget

30.1 Introduction

The region addressed in this chapter has a long mining history, suitably portrayed by place names like "Erzberg" (meaning "ore mountain") for a peak towering 730 m over the town of "Eisenerz" (meaning "iron ore").

In mining landscapes, large amounts of ore and waste are moved. Intentional removal of rock is greater than in construction and, together with unintentional soil erosion from agricultural fields, adds up to more sediment movement than would occur naturally (Brown et al. 2017). The amount of material removed is influenced by the mining technique: Underground mining tends to remove less material than surface mining, since much low-grade rock may be left underground by various stope-and-fill methods. In surface mining, earth material extraction and movement is not only greater, but has the additional direct impact of creating excavated man-made landforms, such as sculpted mountain sides, or man-made accumulation landforms, such as waste and slag deposits (Mossa and James 2013).

Of course, mining also has environmental impacts beyond these direct geomorphic effects. A notable example of this impact was associated with mediaeval iron ore extraction, where charcoal was needed for the subsequent smelting of the ore. Charcoal production, in turn, led to severe depletion of the mediaeval forests. Stripping of forests from mountainous land greatly accelerates sediment movement due to soil erosion and mass movements. Thus, historical mining also had an indirect impact on the geomorphic processes of large areas, extending far beyond the actual site of mining.

C. Embleton-Hamann (ed.), *Landscapes and Landforms of Austria*, World Geomorphological Landscapes,
https://doi.org/10.1007/978-3-030-92815-5_30

The long mining history of the Erzberg area can be divided into three periods of varying geomorphic impact, either indirect, by change of geomorphic processes, or direct, by altering the relief and replacing it with man-made landforms (Table 30.1).

First period: the beginning of ore mining at the Erzberg remains unclear. Historical documents from the twelfth century account for its existence. However, a significant volume of evidence points to the start of ore extraction dating back to 400 or even 500 years earlier. During this early period, the ore was mined on the surface of the Erzberg in small open pits, and the waste was thrown back into the pits. The impact on the environment remained slight until the beginning of the thirteenth century, when a new smelting technology boosted both iron production and charcoal demand. The fourteenth century already saw a severe timber shortage (Stepan 1924).

Second period: in this period underground mining commenced, expanding around 1650 with the use of black powder. Only iron ore was sought and removed from the mountain, waste material was deposited either underground or left on adjacent slopes. The first documented man-made accumulation forms consisted of slag tailings along the streets of Eisenerz, but these were subsequently re-worked/removed between c. 1816 and 1836 due to their high ore content (Ferro 1847).

The most important impact of this second period is likely to have been deforestation. This impact will be discussed in detail in Sect. 30.3.

Third period: improved furnaces, based on coke and stone coal for smelting, brought the exploitation of forests to an abrupt end. Charcoal delivery to the Erzberg area was discontinued in 1903 (Premm and Embleton-Hamann 2014). Significantly, the extraction of material increased 20 times by weight (25 times by volume) compared to the end of the second period. The reason for this is that with mechanized mining greater amounts of waste rock were removed (Table 30.1). The associated landform change during this period will be addressed in Sect. 30.4.

30.2 Geographical Setting

The northern part of the study area lies within the Northern Calcareous Alps, represented by the Kaiserschild mountain group in the west, reaching a maximum altitude of 2085 m asl, and the Hochschwab massif further east, rising up to 2277 m asl (for all place names see Fig. 30.1). Also part of the Hochschwab massif is the Polster and the higher ridges immediately east of the Erzberg that appear greyish white in Fig. 30.1. Further south the Greywacke landscape zone with the Erzberg follows. The Erzberg (1465 m asl) belongs to the Eisenerzer Alps, which culminate in the Eisenerzer Reichenstein (2165 m asl). South-east of the Eisenerzer Alps, the summit heights of the Greywacke mountains decrease markedly.

The main river basin of the study area is that of the Erzbach with its tributaries, the Ramsau and the Gsoll creek. It drains towards the Northern Alpine Foreland, whereas the Laming and Vordernberger valleys in the east are orientated towards the south.

Within the geological/tectonic units of the Eastern Alps, the study area is part of the Juvavic and Noric-Tirolic nappes of the Austroalpine nappe pile. The Kaiserschild mountain group

Table 30.1 Mining periods and associated impacts on the landscape of the Erzberg area

Period	Mining technique	Estimate of annually removed material	Direct landform change	Impact on geomorphic processes
1st period until AD 1500, at least 400 years, but probably around 800 years	Manual surface mining in shallow pits	No estimate possible (minimal)	No	Local clear-cutting for charcoal production with timber shortage starting in the Fourteenth century
2nd period AD 1500 until AD 1900, 400 years	Surface and underground mining with increase of the underground sector over time	Sixteenth century: Ore: 27 000 t$^{(1)}$/7 400 m^{3} $^{(2)}$ Nineteenth century: Ore: 381 000 t$^{(3)}$/100 395 m^{3} $^{(4)}$	Minor	Severe supra-regional deforestation
3rd period AD 1900 until present	Mechanized surface mining	1900–2016$^{(5)}$: Ore: 1 918 730 t/504 929 m^{3} $^{(4)}$ Waste: 5 704 184 t/2 097 126 m^{3} $^{(6)}$ $\sum$: 7 622 914 t/2 602 055 m^{3}	Major	No

Sources (1) estimate by Stepan (1924); (2) volume calculated assuming a density of 3.65 for limonite; (3) calculated from data in Redlich (1917); (4) calculated with a density of 3.8 for siderite; (5) data provided by VA Erzberg GmbH; (6) calculated with an average density of 2.72 for the rocks of the Greywacke zone

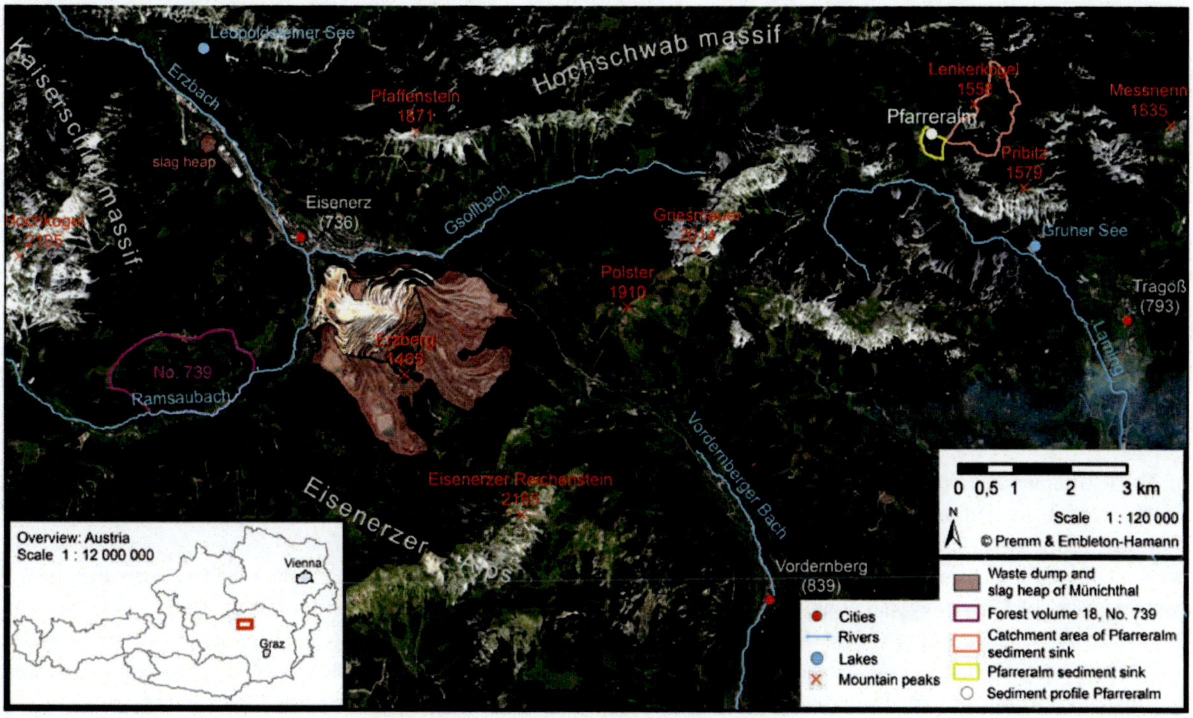

Fig. 30.1 Overview of the Erzberg region based on a SENTINEL-2 RGB-image from 2016

and the Hochschwab massif in the north belong to the Juvavic nappe, regionally consisting of Triassic limestones and dolomites. Light coloured rocks and near vertical rock faces are prominent landscape elements of this zone (Fig. 30.2).

The Eisenerzer Alps and the lower mountain ridges to their south-east are members of the highest tectonic unit of the Noric nappe. They are composed of much older (Palaeozoic) schistose rocks, but again contain significant

Fig. 30.2 Panoramic view of the western part of the study area. The prominent peak behind lake Leopoldstein (Mt. Pfaffenstein, 1871 m asl) belongs to the Hochschwab massif. Note the light-coloured rock walls of Mt. Pfaffenstein and to the left of the lake, which is typical for the limestones of the Juvavic nappe. To the right, the town of Eisenerz and the man-made pyramid of the Erzberg are visible. The Eisenerzer Alps are located in their background. *Photo* Bwag/Commons (2015), publicly available Google Photo

amounts of limestone (Bryda et al. 2013). However, the mountain scenery within the Devonian limestone of the Noric nappe in the south is different from that of the Triassic limestone in the north. Frequently intercalated with thin-bedded slates, the Devonian limestone is not prone to form rock walls and its reddish-brown colour also contributes to a different landscape appearance (Fig. 30.2).

In places, the Devonian limestone is mineralized with siderite. This is the iron carbonate with an average iron content of 33.5% that is mined at the Erzberg. Although the Erzberg is considered the biggest siderite deposit in the world, mining here is demanding. The high-quality ore bodies were impacted by the large-scale deformations of the Variscan and Alpine orogenies that left them rather dispersed.

The Quaternary left distinct glacial landscape features. Glacial erosion formed the trough valleys of lake Leopoldstein and of the upper Laming Valley with their overdeepened rock basins. Borehole information suggests an overdeepening of 180 m at lake Leopoldstein and of 200 m at the Laming Valley north of Grüner See (Fabiani 1984).

During the Last Glacial Maximum (LGM), the ice accumulation on top of the Hochschwab plateau fed two well-nourished valley glaciers in the study area. One followed the Laming Valley and terminated 2.5 km south of Tragöß. The other descended through the catchment of lake Leopoldstein and came to a halt at the present lake outflow. As a result, there were two different situations during the LGM within the study area: while the eastern part was buried under thick ice masses, the central parts of the Erzberg area in the west remained ice-free (Bryda et al. 2013). However, a number of erratics found in the Erzbach Valley indicate that during previous glacials, the Erzberg area was also submerged in ice. One such erratic crowned with a chapel can be visited within Eisenerz at the location "Gradstein".

Modern-day climate in the region is characterized by moderately cold winters and cool summers. In the valleys, the mean July temperature is 14 to 16 °C (Eisenerz: 16.5 °C), whereas the mean January temperature is −3 to −4 °C. At high altitude areas around 2000 m asl seasonal temperatures range between 10 to 12 °C (July) and −4 to −6 °C (January). Total annual precipitation averages 1500 to 1600 mm in the valleys and, with c. 2200 mm, is relatively high in the summit region (Prettenthaler et al. 2010; data averaged for the years 1971 to 2000).

30.3 The Impact of Mining on Forests

From the beginning of the thirteenth century onwards, increasing amounts of charcoal were needed for the smelting process and further processing of the iron (Table 30.1). Between 16 m^3 and 25 m^3 of charcoal were required just for the smelting and refining of 1 000 kg of raw iron. In addition, charcoal was used in the forges of the blacksmiths to produce iron products like horseshoes and tools. In the proximity of settlements withdrawal of construction and firewood as well as clear-cuttings for new farmland and pastures were the major reasons for forest interferences. However, the production of charcoal for the iron industry of the Erzberg also consumed remote forest areas over a period of more than 500 years, which makes the production of charcoal by far the largest interference with forest vegetation in the Eastern Alps during the last millennium (Hafner 1979; Hlubek 1846; Mayrhofer and Hampl 1958).

30.3.1 Forest Exploitation for Charcoal Production

Until the middle of the thirteenth century, forest disturbances adjacent to settlements and a few scattered smelteries were localized but intensive (e.g. clear-cutting of small areas). More remote sites consisted of primeval forests, which were unused or only extensively used for hunting, gathering and grazing (Hafner 1979). The demand for charcoal was largely covered by the local farming population. Small amounts of wood were charred in rectangular charcoal kilns for supplementary income (Fig. 30.3). But with the increased demand for charcoal in the second half of the thirteenth century, triggered by the rise of the iron ore production at the Erzberg, a forest- and charcoal-industry developed and became more and more efficient in terms of charcoal output. Lumberjack, coal burner and the transportation of logs or charcoal became independent professions with the common

Fig. 30.3 Rectangular charcoal kiln set up and operated by the local peasant population as supplementary income opportunity. This charring method, as shown in the image taken around 1940, was practised basically unaltered for centuries (Hafner 1979)

objective to produce large amounts of the essential charcoal (Drescher-Schneider 2003; Hafner 1979).

This transition immediately changed the type of land use from smaller interferences with locally limited intense utilization to large-scale clear-cuttings. Reckless forest management proliferated and everything was subordinated to the maximization of coal production (Hafner 1979; Johann 2007).

30.3.2 Log Transportation Structures Enabling Large-Scale Clear-Cutting

As the demand increased and clear cutting became the norm, the potential utilization of additional forest areas was investigated on a regular basis with decreed forest inspections beginning in the fifteenth century. Experts documented the measures required to utilize unused areas and made recommendation for the construction of log transportation structures. A common log transport chain started on slopes where logs were cut and transported downslope to the rivers where they could be floated. The transport was carried out in gullies or with log slides (Fig. 30.4) (Hafner 1979, Menhardt 1755–1762, Wessely 1853).

The next step was to drift the logs on the river system. In general, the higher water levels of the spring thaw were used to float logs downstream. However, in some cases dams with floodgates were constructed to create artificial high water levels for drifting during periods with low water levels. The log drift ended near large charcoaling sites, where rake-like structures extending across the rivers stopped the logs in the river flow. At these sites, several round charcoal kilns were operated in alternation (Fig. 30.5) (Hafner 1979; Wessely 1853).

Fig. 30.5 Coal production site with several typical round charcoal kilns set up and operated by professional charcoal burners. The illustration shows the different stages of the charring process, from setting up the kiln in the foreground to the coal extraction in the background (Du Hamel du Monceau et al. 1766)

Fig. 30.4 Wooden log slide around 1923 in a tributary canyon of the Enns Valley debouching c. 8 km west of the Erzbach. Log slides were constructed since the late thirteenth century to transport massive amounts of logs downslope to the river systems (Hafner 1979)

Structures built for log transportation had a drastic impact on the vegetation cover in their catchment area. The costs for construction and maintenance were relatively high, and their efficiency was calculated for a maximum wood yield. Regions with transport structures were deforested as fast as possible in order to make the structures economically efficient. Clear-cuttings started months ahead to ensure an immediate start of the transportation of logs upon completion of the structure (Hafner 1979; Wessely 1853).

A prominent example of the radical change of land use caused by a log transportation structure is the log rake in Großreifling. This rake was built in 1570 to utilize the primeval forests of the 870 km^2 large catchment of the Salza Valley. With the construction of the log rake the clear-cuttings started and additionally a series of log slides and dams were built in the Salza Valley and its tributaries (Hafner 1979). The effects of the Großreifling log rake can, for instance, be demonstrated by the changes in the side catchment of the Lassingbach Valley. Between the construction of the rake in 1570 and the forest inspection in 1761 more than 60% of this 110 km^2 measuring catchment was cleared at least once (Menhardt 1755–1762).

30.3.3 Forest Inspections and the Forest Volumes

A lack of wood was always feared, and several forest inspections were initiated over the centuries. The first forest inspection in 1499 was decreed by Emperor Maximilian I. to ensure the supply of charcoal for the operation of the iron industry at the Erzberg. The most comprehensive inspection was decreed by Empress Maria Theresia in 1745 and performed by a commission of mining and forestry experts from 1754 to 1762. The inspection included forest areas in Styria north of Graz and clearly determined the ownership and the permitted utilization for every single forest lot. The results of this inspection were published in 28 volumes (Menhardt 1755– 1762, forest volumes) summarizing 20 636 forest lots in total.

The exceptional feature of this particular forest inspection is an estimation of the condition of every forest lot subdivided into four growth stages. This makes it especially valuable for an assessment of the historic vegetation cover on the local to regional scale. As an example, Fig. 30.6 shows record number 739 from forest volume 18, describing a forest lot close to the Erzberg (see Fig. 30.1). The unit used to describe the four growth stages is "barrels" ("Faß") of charcoal, which could potentially be produced. In this case, at the time of the inspection (AD 1760), 400 barrels could be produced immediately (mature forest, 1% of lot no. 739), 600 barrels could be produced in around 40 years (half-grown trees, 2% of lot no. 739), 3 000 barrels could be produced in around 70 years (bushes and small trees, 10% of lot no. 739) and finally 26 000 barrels of charcoal could be produced in around 120 years (clear-cuttings, 87% of lot no. 739), which would imply that the site had just been cut.

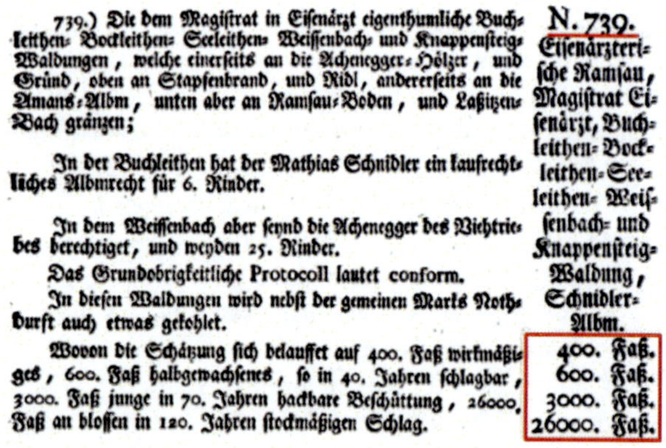

739.) Die dem Magiſtrat in Eiſenärzt eigenthumliche Buchleithen- Bockleithen- Seeleithen- Weiſſenbach- und Knappenſteig-Waldungen, welche einerſeits an die Achenegger-Hölzer, und Gründ, oben an Stapſenbrand, und Ridl, andererſeits an die Amans-Albm, unten aber an Ramſau-Boden, und Laßitzen-Bach gränzen;

In der Buchleithen hat der Mathias Schnidler ein kaufrechtliches Albmrecht für 6. Rinder.

In dem Weiſſenbach aber ſeynd die Achenegger des Viehtriebes berechtiget, und weyden 25. Rinder.

Das Grundobrigkeitliche Protocoll lautet conform.

In dieſen Waldungen wird nebſt der gemeinen Marks Nothdurft auch etwas gekohlet.

Wovon die Schätzung ſich belauffet auf 400. Faß wirkmäßiges, 600. Faß halbgewachſenes, ſo in 40. Jahren ſchlagbar, 3000. Faß junge in 70. Jahren hackbare Beſchüttung, 26000. Faß an bloſſen in 120. Jahren ſtockmäßigen Schlag.

N. 739.
Eiſenärzterische Ramſau, Magiſtrat Eiſenärzt, Buchleithen- Bockleithen- Seeleithen- Weiſſenbach- und Knappenſteig-Waldung, Schnidler-Albm.

400. Faß.
600. Faß.
3000. Faß.
26000. Faß.

Fig. 30.6 Record number 739 of forest volume 18, a forest lot near the Erzberg, reflects a detailed picture of the forest condition in the eighteenth century (Menhardt 1755–1762)

To outline the regional forest condition in the catchment area of the Erzberg iron industry, a summary of the potential coal quantities and the average percentages for the four growth stages is given in Table 30.2. The first line of the table contains the summation of all 28 forest volumes. The second line shows the values and percentages in volume 18, which is the volume with the forest lots directly surrounding the Erzberg. Clear-cuttings (cuttable in 120 years), areas with bushes and small trees (cuttable in 70 years) as well as areas with half-grown trees (cuttable in 40 years) are remarkably higher in volume 18 compared to the average percentages over all 28 volumes. Consequently, the percentage of mature forest areas (cuttable) of volume 18 is less than half of the average over all volumes. These differences clearly express the location of these forest lots next to the Erzberg and the higher population pressure of this supra-regional mining centre.

The forest volumes 1 to 28 are covering an area of c. 10 900 km^2, which is two-thirds of the federal state of Styria. The records of volume 18 cover valleys around the Erzberg with c. 490 km^2.

By carefully considering the four growth stages, a very detailed insight into several centuries of land-use history and vegetation condition is achievable. For example, the growth stage "cuttable in 70 years" implies that this particular area was clear-cut around 50 years before the inspection (assuming 120 years for new forest to be cuttable again). Furthermore, the growth stage "cuttable in 40 years" means that this area was cleared around 80 years before. Finally, the growth stage "cuttable" means either that this area had never been cleared before, or that this area was cleared at least 120 years prior to inspection. In any case, a closer look at the construction dates of log transportation structures to which the particular forest lot gravitates helps to complete the picture of logging history.

30.3.4 Geomorphic Effects of Deforestation

The geomorphic effects of the intense charcoal production and deforestation are still observable today in the Erzberg

Table 30.2 Relative forest areas in the four growth stages of the 28 forest volumes compared to volume 18

	Number of entries	Cuttable, mature forest m^3	%	Cuttable in 40 years m^3	%	Cuttable in 70 years m^3	%	Clear-cuttings, mature in 120 y. m^3	%
Volumes 1–28	20 636	21 752 709	*48*	9 247 391	*20*	7 356 460	*17*	6 755 631	*15*
Volume 18	847	667 698	22	890 723	*29*	802 428	*26*	690 030	*23*

"Barrels" of charcoal were converted into m^3 of charcoar with 1 barrel = 4 Wiener Metzen = 0.246 m^3 (Data from Menhardt (1755–1762): compiled and summarized by S. Premm)

area. Where the debris cones or alluvial fans at the mouths of small side catchments with heavily used forest lots are dissected by the modern creek, the incision quite often exhibits a series of sediment layers. A more comprehensive investigation of these peculiar deposits shows that traces of land use and especially of charcoal production (e.g. layers of charcoal, buried topsoil, layers with coarse gravel) are very common in the Erzberg area (Premm and Embleton-Hamann 2014).

At the Pfarreralm in the Laming Valley near Tragöß (see Fig. 30.1), it was possible to open up an almost 2.5 m deep section (Fig. 30.7). This area was a charcoal source for centuries. Here the coal was produced immediately at the logging site and transported over a mountain pass south of the Griesmauer to the iron processing centre in Eisenerz, as documented in forest volume 4, entry No. 96 "Pfarreralm" (Menhardt 1755–1762).

The coarse debris at the bottom of the sediment profile is overlain by a buried topsoil layer of c. 50 cm thickness (180–235 cm). At 190 cm, a fragment of charred wood was extracted from this ancient soil layer and radiocarbon-dated to 7338–7066 calBC. Above this historic soil layer, the signs of charcoal production become clearly visible. A 25 cm thick layer of debris (155–180 cm), which most likely was deposited after logging activities upstream, is followed by 55 cm of charcoal (100–155 cm). The great thickness of this charcoal layer suggests the existence of a charcoal kiln at or close to this location. This interpretation is supported by two charcoal samples, radiocarbon-dated almost in the same time range (between the late fifteenth and the middle of the seventeenth century). Above the charcoal layer (20–100 cm), a series of debris layers of variable texture is interrupted by a thin layer of organic-rich (soil) material (60–65 cm), which is interpreted as deposited soil material from the upper parts of the catchment. It contained a sample of charred wood with a radiocarbon date of 1491–1659 calAD, comparable to the time range of the two samples in the thick coal layer. The present topsoil (0–20 cm) consists of very coarse debris mixed with organic matter and roots.

Fig. 30.7 Sediment profile at the Pfarreralm with a prominent coal layer and a series of coarse debris deposits above an ancient buried topsoil layer (Photo, recording and dating of the section: S. Premm)

30.4 Man-Made Landforms

During the third period of mining in the Erzberg area, starting at the turn of the nineteenth to the twentieth century, a unique man-made landform was excavated. It is pyramid-shaped with forty sculpted benches at height intervals of 20–24 m, exhibiting a play of colours from red to dark brown (Fig. 30.8c), which forms a rusty red landmark visible from a distance (Fig. 30.2).

A painting from 1649 portrays the Erzberg as a forest-covered peak, perfectly blending into the neighbouring landscape. The painting depicts 28 adit openings, but

Fig. 30.8 Changing appearance of the Erzberg as seen from the north. **a** (top left): photograph taken in 1875 by A. Kurka, © VA Erzberg GmbH, used with permission. **b** (top right): 1:1 000 scale model of the Erzberg, manufactured by the Peter Koch company in 1914, © Technisches Museum Wien, used with permission. **c** (bottom): aerial photo dated 1996, © VA Erzberg GmbH, used with permission

few isolated steps (Angel 1939). Two hundred years later Ferro (1847) describes the mountain as forested to the summit with spruce forest, with some parts interrupted by brown rock walls, where the ore was mined as in a quarry (Ferro 1847). In the eighteenth century, rock walls were excavated during summer and left exposed so that the siderite could weather to limonite, because smelting of siderite only became possible after c. 1750 (Stepan 1924).

The first available photograph of the site dates from 1875 (Fig. 30.8a). In this photograph, the mountain is still

Fig. 30.9 Slagheap of Münichthal. It was emplaced between 1901 and 1946, consists of 1.8 Mio m^3 of slag, with a basal diameter of 300 m and a height of 65 m (Bundesministerium für Nachhaltigkeit und Tourismus 2012). Publicly available Google Photo

partly forested, but that situation would not last much longer. In the 1890s, the mining technique changed from underground mining to surface mining in terrace form, and by 1907 the north and the west face of the mountain had taken on the shape of a geometrical pyramid. It had 60 terraces with an average rise of 12 m each (Fig. 30.8b). In 1925, the summit of the Erzberg, originally at 1537 m asl, was blasted and subsequently lowered to its current elevation of 1465 m asl by the upward extension of mining terraces. In 1928, the height of the terrace risers was doubled (24 m), thereby reducing the number of terraces, and the pyramid-shaped excavation area assumed its present form (Fig. 30.8c).

Associated with the mechanized removal of huge amounts of ore and waste rock was the development of a large man-made accumulation landform along the margins of the excavation area, mainly to the south and west (for the aerial view of the extent of the deposition area see Fig. 30.1). To the east, the waste rock forms a tiered "talus" apron (Fig. 30.8c). To the west, the upper valley (former headwaters) of the Erzbach was deforested and used as a tailings area. The model of the Erzberg from 1913 (Fig. 30.8b) still shows the original valley. Since then up to 250 m of fill has been accumulated (Fig. 30.8c).

A second type of man-made accumulation landform has resulted from the deposition of slag produced in the smelting process. The first slag piles/deposits were emplaced during the first half of the seventeenth century, but later removed (see Introduction). A prominent man-made landform of today is the slag heap of Münichthal, a borough of Eisenerz (Fig. 30.9; for location see Fig. 30.1).

The mining company VA Erzberg GmbH provided valuable data on the annual tonnage of ore and waste rock removed between 1891 and 1916. This database makes it possible to compute a man-made denudation rate for the Erzberg mining site in order to compare the work of humans with natural geomorphic processes. An estimated 7 622 914 t/a was removed from an area of approximately 4 km^2 (cf. Table 30.1, third mining period). The specific sediment yield is therefore 1 905 728.5 t/a/km^2. In order to convert mass to volume, a density of 3.8 was applied for the ore component (25% of the annual tonnage) and of 2.72 for the waste rock (75% of the annual tonnage). The denudation rate in mm/year after adjusting for rock density is 625 mm (see Table 30.3, first "Erzberg" line).

The results of a sediment budget for a nearby unglaciated mountain catchment (Rascher et al. 2018) may serve as comparator. It was carried out on the lower Johnsbach River, which, like the Erzbach (Fig. 30.1), is a tributary of the Enns River, situated c. 20 km west of the study area. The climatic situation is comparable, but it needs to be noted that two features of the lower Johnsbach Valley are different from the Erzbach Valley: it is steeper and solely developed in the brittle Wetterstein dolomite, which is particularly prone to weathering. As the sediment budget model by Rascher et al. (2018) can only provide sediment volumes, denudation rates in lines 2 and 3 of Table 30.3 were estimated by dividing sediment production per year by catchment area.

Table 30.3 Natural and mining related denudations in the Erzberg region

	Catchment area (m^2)	Sediment production (m^3/a)	Denudation rate mm/a	Erzberg/lower Johnsbach Valley
Erzberg(1)			625	× 739
Erzberg(2)	4 000 000	2 602 055	650	× 768
Lower Johnsbach Valley	13 000 000	11 000	0.846	

Erzberg(1): denudation rate computed; Erzberg(2): denudation rate estimated in the same way as for the lower Johnsbach Valley. *Data sources* (Rascher et al. 2018) for the lower Johnsbach Valley; VA Erzberg GmbH for the Erzberg (cf. Table 30.1)

These preliminary estimates suggest that mining activities at the Erzberg area are up to 750 times more effective in transforming the landscape than "natural" fluvial processes. Unfortunately, there are no "natural" rivers left in central Europe, and the Johnsbach River is no exception. It is impacted by gravel mining in two of its side catchments, which was only stopped in 2008. The former mining excavations are still trapping sediments and prohibiting a straightforward delivery of the sediment to the lower Johnsbach River. Once these two side catchments will be fully connected, the sediment budget model by (Rascher et al. 2018) predicts a doubled sediment flux of 21 000 m^3 per year in the lower Johnsbach Valley, with an associated denudation rate of 1.64 mm/a. The respective ratio Erzberg/lower Johnsbach Valley would then be c. 400. In either case, sediment moved within the mining site is two to three orders of magnitude higher than the sediment flux in nearby river catchments, and mining has become an important agent of landscape change in this region.

30.5 Conclusions

With the proposal of the Anthropocene as a new geological epoch in which human influence on Earth dominates over natural processes, many papers on the role of man as a geomorphic agent have appeared. Most have attempted to estimate human alteration of geomorphic processes and landforms on a global basis. There is general agreement that modern human activity moves more sediment than natural processes, mostly by soil erosion in the course of agricultural use, but also due to construction and mining activities. The latter are, however, local procedures with local transport rates much higher than averaged global rates. Although the extent of landscape disturbance in the vicinity of mines may be substantial, it is not yet fully quantified (Brown et al. 2017).

The Erzberg area with its long mining history provides an opportunity for a first quantification. Firstly, historical documents, including forest inspection reports between 1499 and 1762, allow a reconstruction in time and space of the depletion of the forest for ore smelting. The results show that the extent of mediaeval clear-cutting with its severe impact on erosion was substantial. For instance, in the period 1680–1760 78% of the forest area in the valleys surrounding the Erzberg became depleted (see Table 30.2, entries for forest volume 18). As a consequence, strongly disturbed sections with buried soil horizons and charcoal layers between slope waste deposits are commonly observed in the Erzberg area. Secondly, based on the detailed records of the mining company, a "sediment budget" for modern mechanized surface mining can be computed and compared with the sediment budget of a nearby river catchment. This comparison suggests that on the local scale man is two to three orders of magnitude more effective in transforming the landscape than geomorphic processes.

30.6 Recommendations for Visiting the Mining Landscape of the Erzberg Area

Today the Erzberg mining site is open for visitors. Besides the impressive view of the anthropogenically modified Erzberg, the tour includes historical information on the mining activities of the past centuries as well as a ride through the surface mining area with an adapted heavy haulage truck. Blasting in the course of the actual ore mining can also be observed from a safe distance.

As described in Sect. 30.2, the area also offers a number of scenic lake landscapes such as around Lake Leopoldstein. The name originates from the nearby located Schloss Leopoldstein, a castle built in the seventeenth century and renovated by the dynasty of the Wittelsbacher in the late nineteenth century. Geomorphic highlights on a hiking tour around the lake are the views over the steep-sided trough valley and a series of distinct Würm moraine ridges at the lake outflow. The glacially overdeepened rock basin of the lake has a thick sediment fill of basal till, lacustrine sediments and fluvial sediments deposited by the Seebach, flowing in from the east. Therefore, the lake is only 31 m at its deepest today (Bryda et al. 2013).

A third area that is worth a visit is the Laming Valley around Pfarreralm and Grüner See. Natural Exposures along creeks may show similar sequences of sediment as depicted in

Fig. 30.10 View over Grüner See (Green Lake) in northeast direction. The mountain in the background is the Messnerin, with the scar of a Lateglacial rockslide in its middle part. The rockslide debris is visible in the middle ground, where it forms the lakeshore. As the lake is sheltered from wind, its water surface is frequently smooth and reflects the surrounding forest in an emerald to bluish-green colour. *Photo* Orthaber (2008) publicly available Google Photo

Fig. 30.7. Grüner See (Fig. 30.10) is not only a scenic point but also has an interesting geomorphic setting. South of Grüner See, rockslide deposits with their typical rough topography block the Laming Valley, exhibiting blocks of several cubic metres. The material originates from the south-east side of the Messnerin (1835 m) and collapsed at the beginning of the Lateglacial on top of a decaying glacier. The lake depression is situated at the fringe of the rockslide deposits and was created by the melting out of dead ice (Bryda et al. 2013). The water level of Grüner See is directly dependent on the groundwater level of the Laming Valley and is highest during thaw season in spring and early summer.

References

Angel F (1939) Unser Erzberg. In: Mitt Naturw Ver Stmk, vol 75. Graz, pp 227–321

Brown AG, Tooth S, Bullard JE (2017) The geomorphology of the Anthropocene: emergence, status and implications. ESPL 42:71–90

Bryda G, Van Husen D, Kreuss O, Koukal V, Moser M, Pavlik W, Schönlaub HP, Wagreich M (2013) Geologische Karte der Republik Österreich 1:50000. Erläuterungen zu Blatt 101 Eisenerz. Geologische Bundesanstalt Wien, Wien

Bundesministerium für Nachhaltigkeit und Tourismus (2012) Altlast ST27: Schlackenhalde Münichthal. Accessed 25th June 2019. https://altlasten.gv.at/atlas/verzeichnis/Steiermark/Steiermark-ST27.html

Bwag/Commons (2015) Nordwestansicht des Leopoldsteiner Sees. Publicly available Google Photo. Accessed 10th May 2019. https://commons.wikimedia.org/w/index.php?title=Category:Photographs_by_Bwag/Panoramic_views&from=E#/media/File:Eisenerz_-_Leopoldsteinersee.JPG

Drescher-Schneider R (2003) Die Vegetations—und Besiedlungsgeschichte der Region Eisenerz auf der Basis pollenanalytischer Untersuchungen im Leopoldsteiner See und in der Eisenerzer Ramsau. In: Klemm S. Montanarchäologie in den Eisenerzer Alpen, Steiemark. Archäologische und naturwissenschaftliche Untersuchungen zum prähistorischen Kupferbergbau in der Eisenerzer Ramsau. Mitteilungen der Prähistorischen Kommission Band 50. Verlag der Österreichischen Akademie der Wissenschaften, Wien, pp 174–197

Du Hamel du Monceau HL, Oelhafen von Schöllenbach CC, Winterschmidt AW (1766) Von Fällung der Wälder und gehöriger Anwendung des gefällten Holzes. Oder wie mit dem Schlag=Holz, dann halb= und ganz ausgewachsenem Ober=Holz, umzugehen, und alles bekannte Holz richtig zu schätzen und anzuschlagen ist. A. W. Winterschmidt, Nürnberg

Fabiani E (1984) Hydrogeologische und quartärmorphologische Untersuchungen im Hochschwabgebiet. In: Exkursionsführer, 5. Jahrestagung d. Österr. Geol. Ges. in Eisenerz 1984. Graz, Wien, pp 2–7

Ferro F (1847) Die kaiserlich-königliche Innernberger Hauptgewerkschaft und ihr Eisenwerks-Betrieb in Steiermark und Oesterreich bis zum Jahre 1845. In: Ein Jahrbuch für den österreichischen Berg– und Hüttenmann III-V, pp 197–380

Hafner F (1979) Steiermarks Wald in Geschichte und Gegenwart—Eine forstliche Monographie. Österreichischer Agrarverlag, Wien

Hlubek FX (1846) Die Landwirthschaft des Herzogthumes Steiermark. Festgabe für die Mitglieder der X. Versammlung deutscher Land– und Forstwirthe. J. A. Kienreich, Graz

Johann E (2007) Traditional forest management under the influence of science and industry: the story of the alpine cultural landscapes. For Ecol Manage 249:54–62

Mayrhofer RJ, Hampl F (1958) Frühgeschichtliche Bauernrennfeuer im südöstlichen Niederösterreich. Arbeitsbericht über die Grabungen 1950 und 1955 und Naturwissenschaftlich-technologische Untersuchungen der Grabungsfunde. Archiv für ur–und frühgeschichtliche Bergbauforschung Nr II. Archaeologia Austriaca, Beiheft 2. Franz Deuticke, Wien

Menhardt G (1755–1762) General-, Wald-, Bereit-, Berain—und Schätzungs-Commissions-Beschreibung im Erb-Herzogthum Steyer. Waldtomus, 28 Bände. Gregori Menhardt Verlag, Steyr

Mossa J, James LA (2013) Impacts of mining on geomorphic systems. Treatise Geomorphol 13:74–95

Orthaber F (2008) Grüner See. Publicly available Google Photo. Accessed 10th May 2019. https://commons.wikimedia.org/wiki/File:Gr%C3%BCner_See,_spiegelt_seine_Umgebung.jpg

Premm S, Embleton-Hamann C (2014) Historische Waldnutzung und geomorphologische Folgewirkungen im Einzugsgebiet des Steirischen Erzberges. In: Mitteilungen der Österreichischen Geographischen Gesellschaft 156. Wien, pp 171–190

Prettenthaler F, Podesser A, Pilger H (2010) Klimaatlas Steiermark. Periode 1971–2000. Eine anwenderorientierte Klimatographie. Österreichische Akademie der Wissenschaften, Wien

Rascher E, Rindler R, Habersack H, Sass O (2018) Impacts of gravel mining and renaturation measures on the sediment flux and budget in an alpine catchment (Johnsbach Valley, Austria). Geomorphology 318:404–420

Redlich KA (1917) Der steirische Erzberg. In: Mitteilungen der Geologischen Gesellschaft Wien. Wien, pp 1–62

Stepan E (1924) Der Steirische Erzberg und seine Umgebung. Österreichische Druck und Verlagsgesellschaft, Wien, Ein Heimatbuch

Wessely J (1853) Die österreichischen Alpenländer und ihre Forste. Zweiter Theil: Forststatistik der Oesterreichischen Kronlaender Kaernthen, Krain, Salzburg, Steiermark, Tirol und Vorarlberg. Wilhelm Braumüller, k. k. Hofbuchhändler, Wien

Stefan Premm graduated in Theoretical and Applied Geography at the University of Vienna by investigating the potential of historical data and archive material for current issues in geomorphology and hydrology. His present PhD studies aim to reconstruct the effects of historical land use in the Erzberg region by combining centuries old archive data and geomorphological sediment records. He is employed in an engineering office where he is concerned with the statistical evaluation of soil parameters and nutrient distribution.

Christine Embleton-Hamann is a retired professor at the Department of Geography and Regional Research at the University of Vienna. Her main interest is in alpine environments. Within this field she focusses on human-environment interactions with research topics like human impact on geomorphic processes, assessment of the scenic quality of landscapes, and geomorphological hazards. A second set of interest concerns the communication of geomorphological knowledge to a broader audience, in the pursuit of which objective she has written a well-received textbook on geomorphology. She is a former President of the Austrian Research Association on Geomorphology and served on the Executive Boards of the IAG and several IAG and IGU Working Groups.

31 Dobratsch: Landslides and Karst in Austria's Southernmost Nature Park

Gerhard Karl Lieb and Christian Bauer

Abstract

Dobratsch is a karst massif surrounded by tectonic faults near the southern border of Austria. It has been declared a nature park due to its outstanding natural features among which the "landslide landscape" is of special importance. It came into existence by a sequence of landslides which on the one hand shaped the south face of the mountain (summit elevation 2166 m asl) by detachment scarps and on the other hand transformed the valley beneath it into a hummocky deposition area extending over some 20 km^2. One of the landslides occurred in AD 1348 and was the reason for the widespread recognition of the mountain and its natural hazards in legends and historical records until today. The Dobratsch is mainly built of Mesozoic limestones which are intensively karstified. Dolines and especially vertical caves are the most prominent karst features. From the hydrological point of view, the karst aquifer is used for supplying drinking water to nearby city of Villach.

Keywords

Natural hazards • Landslide • Karst • Nature Park

31.1 Introduction

Dobratsch is the name of an isolated mountain massif at the western fringe of the Klagenfurt Basin in the Austrian federal state of Carinthia. The elevation of its summit is 2166 m asl, resulting in a remarkable vertical difference of 1665 m from the city of Villach situated at its eastern foot. The massif has a rather rectangular shape with a length of 17 km in WNW-ESE-direction and a width of 4–5 km. It is predominantly built of Mesozoic limestones which are intensively karstified, giving the mountain an outstanding density of surface as well as underground karst features. However, Dobratsch is above all famous for its southern face landslides, including a huge medieval one.

Because the mountain is so close to densely populated areas, there has been intense interaction between geomorphology and different socioeconomic activities throughout recorded history. For instance, the damages caused by the medieval landslide gave rise to several myths and legends which are regionally well-known until today. Besides landslides and rockfalls, also avalanches have to be considered in dealing with natural hazards. On the other hand, karst vulnerability represents a challenge to regional planning. Above all, Dobratsch is an attractive tourist destination due to the natural variety and the great views it offers over most of southern Austria and beyond.

31.2 Tectonic and Geological Setting

The Dobratsch massif belongs to the Gailtal Alps, a mountain range of the Southern Calcareous Alps which is separated from the Carnic Alps by the broad valley of the Gail River (Fig. 31.1). Its course is predetermined by one of the most important fault lines of the Austrian Alps, the Periadriatic lineament (PAL). As is the case with most large faults, there are also neighbouring fault lines running parallel or subparallel to the main one. An example of such a fault is the Bleiberg fault which is morphologically represented by the deeply incised valley of Bad Bleiberg. This valley runs roughly in a W–E-direction and borders the Dobratsch massif to the north. From a hydrological point of view, the valley is drained by two creeks, the Weißenbach flowing to the east and the Noetschbach to the west. In between, there is a low mountain pass which is barely noticeable when one drives through the valley. Its elevation of slightly above

G. K. Lieb (✉) · C. Bauer
Institute of Geography and Regional Science, University of Graz, Heinrichstrasse 36, 8010 Graz, Austria
e-mail: gerhard.lieb@uni-graz.at

C. Embleton-Hamann (ed.), *Landscapes and Landforms of Austria*, World Geomorphological Landscapes,
https://doi.org/10.1007/978-3-030-92815-5_31

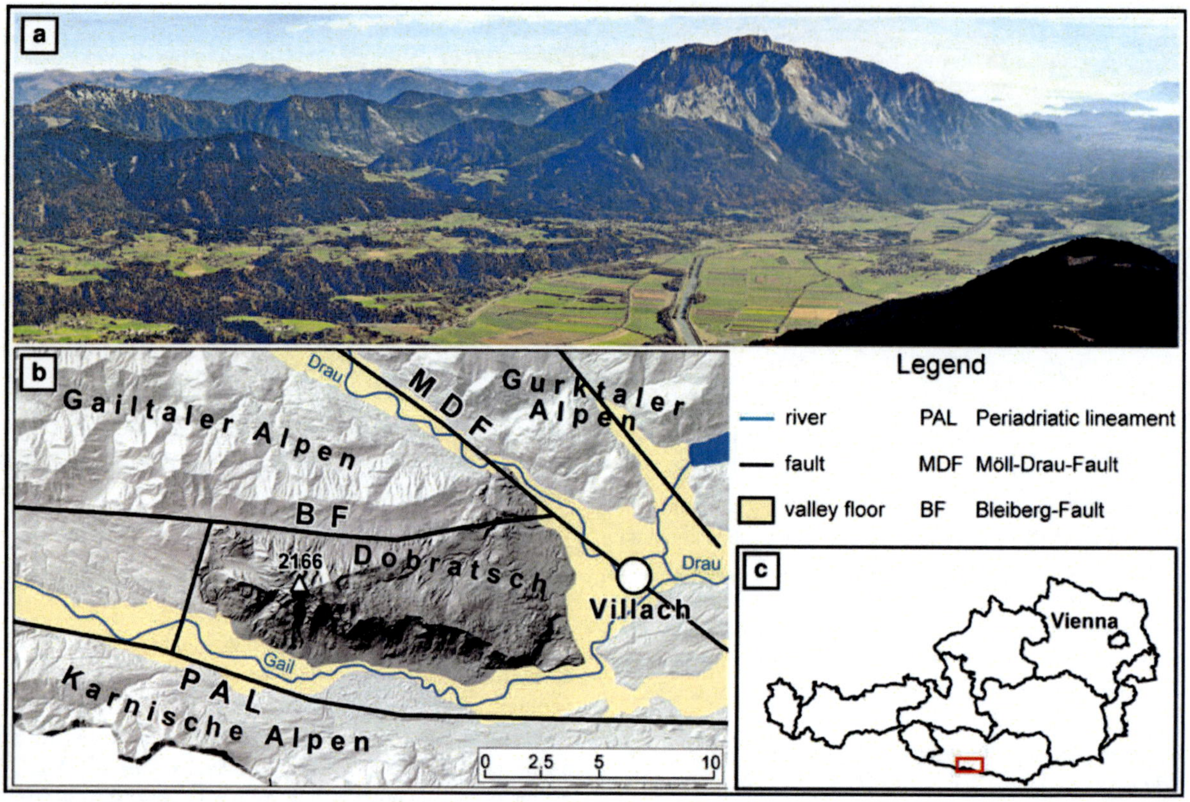

Fig. 31.1 Dobratsch massif. **a** Aerial view of the Dobratsch from SW illustrating the remarkable prominence of the massif; the broad valley bottom in the foreground is due to the fact that the Gail River has been dammed by the landslide deposits which fell down the south face of the massif (Photo C. Fatzi). **b** Tectonic setting featuring the Dobratsch massif as a horst-like structure surrounded by faults. Hillshade is based on an ALS-derived DTM, kindly provided by KAGIS (geographical information service of the federal government of Carinthia). **c** Position of map B within Austria

920 m asl gives the summit of Dobratsch 1250 m relative relief.

The Noetschbach only runs 3.5 km within the valley of Bad Bleiberg to the west, then changes its direction to a southern one and forms a gorge-like valley which separates the Dobratsch massif from the rest of the Gailtal Alps. Also, the course of this valley follows a fault line which runs orthogonally to the aforementioned faults, roughly N–S. Similar directions can also be found within the Dobratsch massif east of its summit in fault lines which run in NNW–SSE direction and have been described by Holler (1976). At each of those faults, the respective eastern block has moved downwards (Schulz 1982) forming a series of step faults. This tectonic setting is responsible for the character of the eastern slope of the Dobratsch massif which resembles a stairway. An early example of a geomorphological description of the plateau and a discussion of its morphogenesis can be found in Czermak (1951).

From a geological point of view, the Dobratsch massif and the Gailtal Alps belong to the so-called Drauzug. This term is used to denote the predominantly sedimentary Austroalpine rock units north of the PAL and south of the Drau Valley. Stratigraphy consists of rocks ranging in age from the Permian to the Triassic, resting on a crystalline base that is not exposed at the surface. The sequence of rocks is similar to that of other Austroalpine units, with the Wetterstein limestone (of Ladinian age) as the most widespread and thickest rock layer. In terms of the geomorphological considerations with focus on landslides and karst, the following aspects of geology and tectonics are of special relevance:

- The stratigraphy consists of metamorphic rocks at the base (weakly-metamorphosed sandstones and schists), which are overlain by Permo-Mesozoic rocks. The latter are dominated by Triassic carbonates: Alpiner Muschelkalk and Wettersteinkalk (partly dolomite).
- Due to overthrusting during the Alpine orogeny, the Mesozoic stratigraphic sequence occurs twice at the southern side of the Dobratsch massif. This leads to the situation that the rocks of the Raibl formation (Carnian) form a ductile base for the thick limestone layers (of Anisian and Ladinian age) towering above them.

Fig. 31.2 Aerial view of the Dobratsch massif and its surroundings from the southeast. **a** = landslide deposition area which stands out due to its forest cover; **b** = Dobratsch plateau which is sharply demarcated by the detachment scarps of the landslides; **c** = valley of Bad Bleiberg. Note the asymmetry of the massif with steeper slopes from **b** to **a** than from **b** to **c**. Photo: C. Fatzi

- The entire rock sequence slightly dips to the north giving the Dobratsch massif an asymmetrical shape, with steep slopes looking to the south and a comparatively gently slope declining to the north (Fig. 31.2).

31.3 The Landslides

31.3.1 History of Research

Besides, e.g. Fernpass, Tschirgant, Köfels (see chap. "Giant "Bergsturz" Landscapes in the Tyrol"), Almtal and Wildalpen the landslides of Dobratsch are among the largest and best known in Austria, not only to the scientific community, but also to the general public. Not only is the landslide landscape spectacular to a lay observer, but the local community is well aware of a large, well-documented landslide event that occurred here in AD 1348 (Fig. 31.3a, b). The term "landslide landscape" will be used in the following text to label the entire landslide-affected terrain, from the detachment areas at the south face of Dobratsch massif to the deposition zone in the Gail Valley below. Furthermore, the awareness of the uniqueness of the landslide landscape was an important aspect in initiating environmental protection, finally leading to the establishment of a nature park (Sect. 31.5).

As already indicated, the landslides from Dobratsch did not have to be detected by scientists as they were regionally well known since the event, which took place on 25 January 1348. There are a lot of historical sources about the event most of them overestimating the effects of the landslides and mixing the information with the contemporaneous earthquake which was the main trigger of the landslide. For instance, a frequently cited chronicle (text in Golob et al. 2013, 15) reports incorrectly the destruction of 17 villages by the 1348 landslide, although the displaced rock masses did not reach beyond older landslide deposits which due to their unfavourable relief had never been settled (and still are not). Till (1907) wrote the first monographic study on the landslide landscape and recognized that there was more than one event.

Abele (1974) compared all large landslides of the Alps and critically evaluated available information, primarily by geomorphometric techniques. His comprehensive dataset allows comparison of the size of Alpine landslides showing that according to the area covered by landslide deposits (24 km^2) Dobratsch ranks third in the Alps, behind the landslides of Flims and Siders in Switzerland (Abele 1974). A further step in developing the knowledge of the landslide was the geological map provided by Anderle (1977), although the author was obviously more interested in the geology of solid rocks than of younger deposits. In the 1980s, the PhD thesis of Brandt (1981) and especially the critical reanalysis of the historical sources carried out by Neumann (1988a; b) (for more details see Sects. 3.3, 3.4) appeared. The different contributions in Golob et al. (2013, including references) provide an overview of current knowledge but do not replace a comprehensive study which is still needed. Smaller rockslide events in the last years (the latest one so far on 16 January 2015) have attracted attention

Fig. 31.3 Dobratsch "landslide landscape" as seen from the observation platform on the upper rim of the detachment scarp called Rote Wand (= red rockface). The vertical difference to the Gail River is almost 1000 m; the valley floor is widely covered by forests indicating the extent of the landslide deposits. **a** View in south-eastern direction with slide planes of older landslides in the foreground; **b** View in south-western direction with the small town Arnoldstein. In the background of both photographs are the Julian Alps in neighbouring Slovenia (**a**) and Italy (**b**). Photos G. K. Lieb

leading to local hazard evaluations based on analysis of preparatory factors or process modelling (e.g. Knafl 2015).

For preparing this chapter the extent of the landslide deposits was mapped using an ALS-derived DTM (1 × 1 m resolution), kindly provided by KAGIS (Fig. 31.4). The area where the hummocky deposits are visible at the surface (dark green in Fig. 31.4) comprises 19.75 km^2. This figure is by far lower than the 24 km^2 communicated by Abele (1974). Obviously, Abele's delineation of the deposition area also included landslide deposits that are nowadays covered by fluvial sediments, together with parts of the slopes. The light green area in Fig. 31.4 (8.23 km^2) shows the slopes which were affected by the sliding process and which are currently covered by talus and debris cones, both indicating the high intensity of recent geomorphodynamics (Sect. 3.3).

31.3.2 Landslide Susceptibility of the Dobratsch Massif

Landslide susceptibility means the propensity of an area to experience this type of physical hazard. As landslides are a manifestation of slope instability, susceptibility combines all destabilizing conditions. These include preparatory factors, such as oversteepening of slopes by erosion or undercutting, and preconditions such as material properties and slope geometry (Crozier 2004).

Addressing landslide susceptibility at the southern side of Dobratsch Krainer (in Golob et al. 2013, 62) emphasized the following three aspects which refer to the information given in Sect. 31.2.

- Steep south face of Dobratsch massif, mainly caused by (i) glacial erosion during the Pleistocene and supported by (ii) calcareous rocks and (iii) the dip of the strata to the north.
- Concerning lithology it is important that ductile rocks (like, e.g., the Raibl Formation) underlie the solid ones, which means that the base of the calcareous rocks of the upper part of Dobratsch is weak. This lithological precondition is often present in rockfall areas of the Calcareous Alps (cf. chap. "The World Heritage Site Hallstatt-Dachstein/Salzkammergut: A Fascinating Geomorphological Field Laboratory").
- The intensive friction of the rocks, both because of the tectonic setting (neighbourhood of PAL and the existence of transversal faults) and the gravitational tension in the oversteepened rock face, make rocks highly susceptible to gravitational processes (Fig. 31.5a, b).

The trigger of the AD 1348 event was an earthquake, which destroyed the city of Villach situated directly east of Dobratsch (Neumann 1988b; Lenhardt 2007). According to Pichorner (in Golob et al. 2013), there is evidence for high amounts of precipitation in the months before the event which might have decreased slope stability by high water pressure in the rock fissures. Because of the location at the seismically active PAL, also earlier landslide events may have been finally triggered by earthquakes.

31.3.3 Landslide Chronology

Till (1907) distinguished one prehistoric and one historic landslide from AD 1348 and assigned the detachment areas

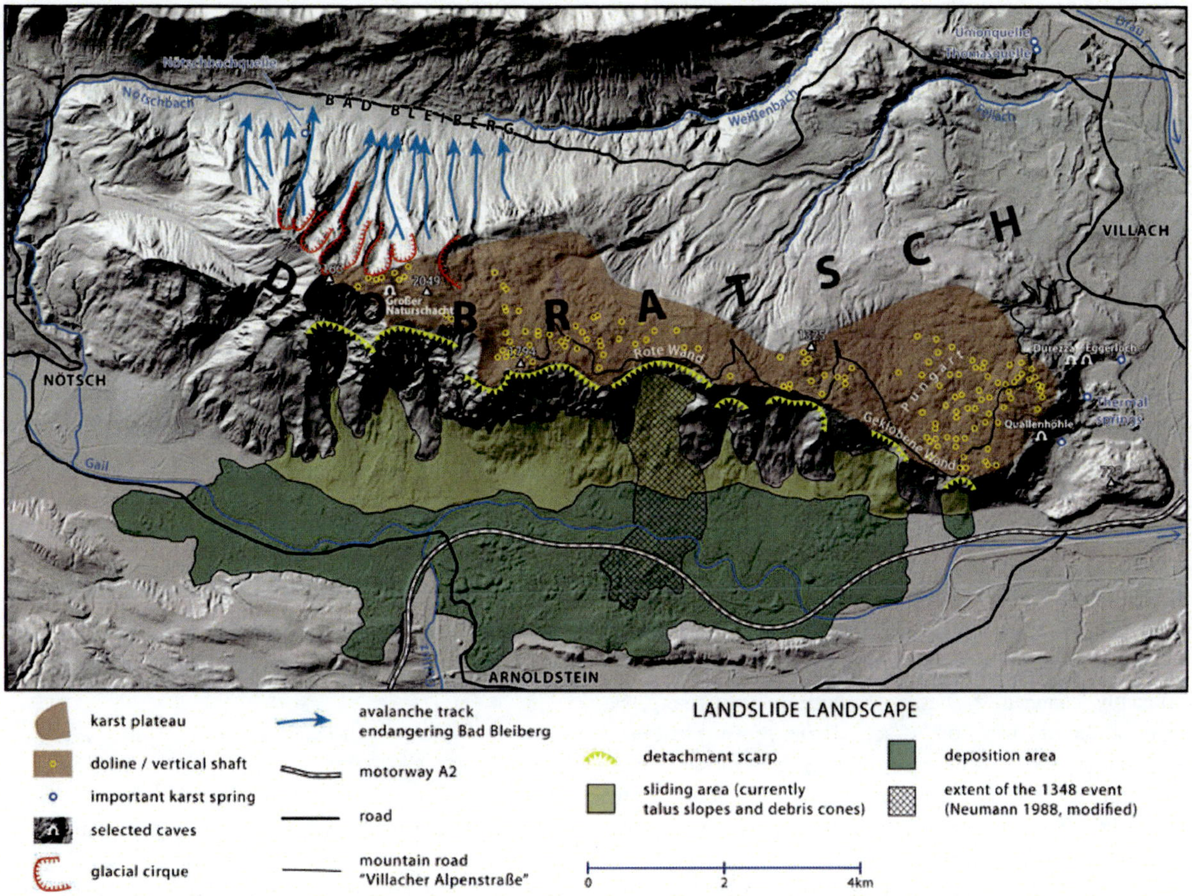

Fig. 31.4 Overview map of Dobratsch massif showing the main elements of the "landslides landscape" and some other features described in this chapter. The hillshade is based on an ALS-derived DTM, kindly provided by KAGIS (geographical information service of the federal government of Carinthia). Drawing: V. Damm & C. Bauer

Fig. 31.5 Example of a tension crack parallel to the upper margin of the already existing detachment cavity of Rote Wand in May 2014 (**a**). At exactly this crack rock failure occurred causing a rockfall with a volume of 6000 m^3 on 16.01.2015 creating a talus apron on the slope below (**b**, Aug. 2016). Photos G. K. Lieb

in the south face of Dobratsch to these two events. This distinction has been accepted by the subsequent authors, among them Anderle (1977) who showed the extent of the two landslide generations in his map. Abele (1974) dealt with the question of age of the prehistoric landslides and proposed an event during the late Würm glaciation with the rock masses falling onto glacier ice which covered at least the western part of today's landslide landscape. As a proof for this assumption, he mentioned the hummocky relief east of the Gailitz River which he interpreted as toma relief (toma = steep, isolated hill) originating from the contact of landslide deposits with stagnant ice of a retreating glacier. However, the origin of similar landforms at other landslides has more recently been explained without the existence of ice (e.g. Abele 1997; van Husen et al. 2007). No current research on this problem has been carried out here.

The main argument for the distinction of older and younger sections of the landslide landscape was the lack of soil and looser vegetation cover of the latter (Fig. 31.6a, b). However, Neumann (1988a) questioned this statement and pointed out that the deposition area of the landslide has been used as forest and pasture over centuries and thus was significantly changed by human activities. Furthermore, he reported the age of an oak (found during power plant construction in 1959 and embedded in landslide deposits) which according to ^{14}C-dating was buried some 7800 years ago by a landslide. Thus, he concluded that there were not only two landslides but obviously more events.

The map shown in Lenhardt (2007, based on Brandt 1981) assigns 6 detachment scarps—among them Rote Wand and Kranzwand which are both visible in Fig. 31.6b—to the AD 1348 event. However, from historic sources the only landslide site which can definitely be assigned to the AD 1348 event is the area shown in Fig. 31.4 (beneath the large detachment scarp of Rote Wand). D. Neumann's mapping was checked and confirmed in the field by the authors in 2016. The deposits cover about 1.6 km^2 but have no clear boundary with older deposits, neither on the surface nor within the deposits. The latter statement can be made because deep excavations for the construction of the motorway (completed in 1984) did not show any stratigraphical distinction in the sediment body (Neumann 1988a).

31.3.4 Effects of the Landslides

According to the reanalysis of historical sources of the AD 1348 landslide by Neumann (1988b) far too high damages and losses of lives were reported in the older literature. We now know that the landslide itself did not affect settlements which never existed there (see above) but instead fell onto older landslide deposits. The deposits dammed the Gail River creating a lake of which remnants were visible at the surface until the eighteenth century, in a location named Seewiese (=lake meadow). However, in the Gail valley bottom two villages located above the landslide landscape had to be abandoned in the following years because of "dramatically deteriorated hydrological conditions" (Neumann in Golob et al 2013). Probably, there were no fatalities at all directly caused by the landslide. In this respect, the earthquake was far worse.

Nevertheless, the deposition area of the landslide has a particularly striking appearance with a chaotic, hummocky mesorelief (in the older as well as in the younger parts) and large boulders (Fig. 31.6a, b). Thus, as already mentioned, settlements and intense types of land utilization were prevented and the entire area contrasts with the deforested bottom of the Gail Valley upstream and downstream of the landslide landscape because of its forest cover. This dissimilarity to the neighbouring areas is also reflected by toponyms like Schütt (=gravel deposit) or Steingröffel (=sterile stone accumulation). However, the increased gradient of the Gail River due to the landslide barrier was the reason for constructing hydropower plants in 1911 and 1959 (Neumann in Golob et al. 2013).

31.4 Karst Features

Looking at the distribution map of karst caves in Austria (Fig. 3 in chap. "Karst Landscapes in Austria"), the Southern Calcareous Alps in general show a much lesser number than the Northern Calcareous Alps—except for one area and this is the Dobratsch massif. By the beginning of 2019, more than 210 caves are documented in the Austrian Cave Register (for details see chap. "Karst Landscapes in Austria"). There is a positive correlation between the high number of caves and a long history of research. Trimmel (1963, 1964) gave a first comprehensive overview of the status of knowledge and pointed to the striking high density of karst features and caves. Spötl (2016) confirmed an outstanding cave abundance together with the dominance of caves with high vertical dimensions. The Große Naturschacht and the Eggerloch, 112 m and 123 m deep, represent the deepest caves in the area so far. In some of the caves, perennial snow and ice exist (e.g. Großer Naturschacht in the summit area of the Dobratsch). There is a high concentration of caves in the eastern part of the Dobratsch plateau called Pungart (location see Fig. 31.4) and its slope towards the Klagenfurt Basin (e.g. Fig. 31.7b). Here, the longest cave (c. 700 m) of the Dobratsch massif, the so-called Eggerloch, is situated. One of the vertical shafts, the so-called Durezza (cf. Fig. 31.4), is also an important archaeologic site. Human remains (bone fragments from at least 36 children and 102 adults) indicate the use of the shaft as a burial site in the La Tène period (Galik 1998).

Fig. 31.6 Surface of the landslide deposits in the location "Steinernes Meer" (=sea of stones) showing a "fresh" appearance with coarse blocks, lack of soil and sparse vegetation (open pine woodland). Although at this specific site the sediments were deposited by the AD 1348 event, older parts of the "landslide landscape" look similar, making it a questionable indicator of age. **a** View in WNW direction, with the south face of the Dobratsch summit area representing older detachment scarps in the background; **b** View in northern direction with the Rote Wand detachment scarp of 1348 and another one of possibly the same age in the background. Photos G. K. Lieb

Spötl (2016) interprets the occurrence of caves in the context of the roughly N-S striking faults mentioned in Sect. 31.2. Furthermore, he discusses the possibility of cave development in the easternmost part of the Dobratsch massif by rising thermal water (referring to hypogene speleogenesis, cf. chap. "Karst Landscapes in Austria").

The faults also favour the development of karst landforms at the surface among which dolines (Fig. 31.7a) occur most frequently. There is no comprehensive analysis of the distribution, shape and size of the dolines so far. Preliminary conclusions can be drawn from the morphometric delineation based on the 10 m ALS-DTM and from observations during geographical excursions:

- Dolines occur all over the area indicated as plateau in Fig. 31.4.
- Fig. 31.4 illustrates dolines (as point elements) with an area larger than 200 m^2 and a depth of more than 2 m. There are slight concentrations at three levels of the plateau: (i) in its uppermost part, east of the summit (c. 2000 m asl); (ii) in the vicinity of the upper end of the touristic road (see Sect. 31.5) near Rosstratte (c. 1750 m asl) where they are best visible because the forest at these elevations has been cleared (probably already in the Middle Ages) in order to provide alpine pastures; (iii) in the densely wooded lowest part of the plateau called Pungart (c. 900 m asl).
- The concentrations of the dolines coincide with the concentration of vertical shafts, which especially applies to the Pungart plateau and points to the existence of local fault systems.

Another special karst feature is the hummocky ground of cultivated pastures. On the Dobratsch plateau, it can for instance be observed around Rosstratte (Fig. 31.7c). The hummocks consist of regolith and calcareous slope debris and have an average diameter of 2 to 4 m and alternate with pits, thereby exhibiting a remarkable regularity in the general distribution pattern. In the German geomorphological literature, this pit and mound microrelief is termed

Fig. 31.7 Surface and subsurface karst features of the Dobratsch massif. **a** Dolines of medium size bound to local faults on the upper part of the Dobratsch plateau (c. 2000 m asl). Photo G. K. Lieb; **b** The cave Quallenhöhle at Pungart. Photo T. Exel; **c** View to the east from the first bend of the hiking trail starting at Rosstratte. The slopes below the forest and the slopes of the large doline in the foreground exhibit a Buckelwiesen microrelief. Its pit-to-mound vertical relief is 15–17 cm. This figure is low when compared to the mode of 50 cm in distinctively developed hummocky pastures elsewhere in the Austrian Calcareous Alps. Albeit observable, the Buckelwiesen on the Dobratsch is thus not very prominent. Photo C. Embleton-Hamann

"Buckelwiese" (literally: "hummocky meadow"). Buckelwiesen are initiated by the uprooting of trees during storms and then further developed by a concentrated limestone solution operating in the rootstock-pits, fostering their gradual subsidence, thereby giving the mounds between them increasing prominence (Embleton-Hamann 2004). As natural forest regeneration is likely to destroy a newly formed, small windthrow relief, Buckelwiesen genesis further affords that the blow-down site remains open until a concentric soil water flow in the pits and a self-enforcing solutional amplification of the microrelief is established. The data for mean solutional denudation rates in the Alps (cf. Plan 2016) suggests that this will take a minimum of 400–500 years. There is obviously a third causal factor involved in their genesis. This is the early presence of people, salvaging the windthrow timber and using the newly formed forest-gaps as pasture for cattle. Thus, the Buckelwiesen pastures may also be regarded as a special cultural heritage of the European Alps, which already had a substantial population during the Middle Ages (Embleton-Hamann 2013).

As most karst areas, the Dobratsch massif represents a karst aquifer. Because of the northerly tilting of the rock layers, it is mainly drained to the north, with the largest karst springs occurring near Bad Bleiberg (Noetschbachquelle) and northwest of the city of Villach (Thomasquelle, Unionquelle, Fig. 31.4). These two springs have a mean discharge of 400 l/s (Poltnig et al. 1996), providing 80% of the drinking water supply (Spötl 2016) of Villach with 61.000 inhabitants. Thus, the city is one of many examples of cities in Austria (among them the capital Vienna) which get their water from karst massifs. Because of the vulnerability of the karst water system, respective measures, especially the establishment of protected areas, have been taken in good accordance with other types of protection (see Sect. 31.5). Special features of karst hydrology are the thermal springs at the western outskirts of Villach, which have been used as spa already since Roman times. They are bound to the karst system with a deep-reaching water circulation along N–S orientated faults. These considerations lead to the interactions of the nature and society at the Dobratsch.

31.5 Dobratsch Massif and Humans

Golob et al. (2013) explored interactions of men and environment at Dobratsch in many ways. In addition to the above-mentioned provision of drinking water, the close relationship between the city of Villach and the mountain situated next to it also includes aspects like the use as destination in leisure time or an icon of the construction of nature contrasting city life. In German, such relations are referred to as "Hausberg" (i.e. something like home mountain). These interactions shall be demonstrated for agriculture, tourism, natural hazards and nature protection:

- The second name of Dobratsch, Villacher Alpe, reminds one of the fact that the upper parts of the mountain have been used as an alpine pasture by the citizens of Villach at least since the sixteenth century. As in other Alpine regions, cattle still grazes on the alpine meadows which have been expanded by lowering the timberline. The presence of people, which probably dates back to medieval times, is also proved by the construction of two churches at the summit already at the end of the seventeenth century.
- Since the turn of the eighteenth to the nineteenth-century Dobratsch has frequently been hiked, inducing the construction of a shelter hut near the summit already in 1810. The main motivation to visit the summit was pilgrimage to the churches and above all the extraordinary panoramic view which the mountain offers because of its geographical position—but not rock climbing due to the brittle rocks favouring gravitational processes. Since around 1900 several ideas to make the mountain accessible for mass tourism came up and finally resulted in the opening of a road in 1965 with its upper end at Rosstratte (1733 m asl); because of its pure touristic purpose, the course of the road was designed to touch as many natural sights and viewpoints as possible (such as Rote Wand, Fig. 31.3). In the period 1965–2002, the plateau up to Zehnernock (1956 m asl) and the north-eastern slopes were additionally used as a skiing resort which was then closed for environmental reasons (see below).
- Dobratsch comprises all aspects of high mountain relief including glacial cirques (Fig. 31.4), ecosystems (at least up to the alpine zone) and gravitational processes. Concerning natural hazards not only rock- and landslides but also debris flows and avalanches have to be mentioned. The latter especially occur on the northern side of the massif, endangering the small town of Bad Bleiberg which has developed there in spite of this hazard due to the occurrence of lead and zinc which have been mined from the Middle Ages until the 1980s (nowadays the main economic activity is tourism). The avalanche danger is very high because of (i) bowl-shaped glacial cirques (ii) in a lee position—the predominant wind direction is SW—collecting a lot of snow, (iii) the amount of which is high because of the position of the mountain in the precipitation-rich Southern Alps, and (iv) the high inclination of the slopes which is not interrupted by any less steep section. Thus, during heavy snowfall events avalanches can reach Bad Bleiberg (Figs. 31.4 and 31.8b). On 24.02.1879, avalanches destroyed 21 houses and killed 39 persons. Therefore, around 1900 first protection measures were taken, and during the period 1969–1985, the avalanche protection service implemented an integrative protection concept including protection structures at the plateau and in the detachment areas of the avalanches (Fig. 31.8a, b) as well as hazard zonation (Letter 2014).
- The first protected area around Dobratsch was established in 1942 within the landslide deposition area called "Schütt". This reflects the high regard for these ecosystems and their biodiversity. More protected areas of different kinds were added over time, finally leading to the

Fig. 31.8 Permanent avalanche barriers built after 1969 prohibit the detachment of avalanches which at the northern side of Dobratsch massif endanger the town of Bad Bleiberg more than 1000 m below. **a** View in western direction towards the summit of Dobratsch, **b** in north-western direction towards Bad Bleiberg. Photos G. K. Lieb

idea of protecting the entire mountain in a comprehensive way. For this purpose, the concept of an Austrian "nature park" (with the main management goals of protection, recreation, education and regional development) was debated since the 1980s and finally established by the regional government of Carinthia in 2002. Today the nature park comprises an area of 72.5 km^2 which equals the entire mountain massif with the exception of the lowest (forested) parts of the northern slope. The landslide landscape only partly belongs to the nature park but is strictly protected as a NATURA 2000 area. When in the 1990s, the decision had to be made whether the skiing resort should be relaunched (by establishing artificial snow-making) or closed, the second option was turned into reality in 2002. Although this decision was supported by financial considerations, it can be considered as sustainable, especially in the context of karst vulnerability (Poltnig et al. 1996, 89).

31.6 Conclusions

Dobratsch massif is a very good example of a karst mountain of the Southern Alps (in a geomorphological sense). It has become famous because of the landslide landscape in which rock falls still occur and will also occur in the future. As has been shown, this landslide landscape still reveals a lot of scientific problems to be solved. In contrast to the incomplete knowledge on the geomorphological processes and their age, biodiversity has been intensively investigated leading to results that helped to establish environmental protection. Thus, Dobratsch has become the southernmost nature park of Austria, a fact which also meets the requirements of drinking water protection in a karst environment. Finally, Dobratsch is excellently accessible and thus can be recommended as a high mountain excursion goal – not only from a geomorphological point of view.

Acknowledgements We would like to thank Christine Embleton-Hamann for her contributions about landslide susceptibility and especially about the Buckelwiesen microrelief.

References

Abele G (1974) Bergstürze in den Alpen, ihre Verbreitung, Morphologie und Folgeerscheinungen. Wissenschaftliche Alpenvereinshefte 25, München, p 230

Abele G (1997) Rock slide movement supported by the mobilization of groundwater-saturated valley floor sediments. Z. Geomorph., N.F. 41:1–20

Anderle N (1977) Geologische Karte der Republik Österreich 1:50.000, sheet 200, Arnoldstein Geol B-A, Wien (with explanatory notes)

Brandt A (1981) Bergstürze an der Villacher Alpe (Dobratsch)/ Kärnten/Österreich. Unpubl PhD thesis, Hamburg

Crozier M (2004) Landslide. In: Goudie A (ed) Encyclopedia of geomorphology, vol 2. Routledge, London and New York, pp 605–608

Czermak E (1951) Die Raxlandschaft in den östlichen Gailtaler Alpen. In: Geographische Gesellschaft in Wien, Geographisches Institut der Universität Wien (ed) Geographische Studien (Festschrift J. Sölch), pp 31–35, Wien

Embleton-Hamann C (2004) Processes responsible for the development of a pit and mound microrelief. CATENA 57(2):175–188

Embleton-Hamann C (2013) An exploration of the role of human activity in the generation and maintenance of hummocky meadows ("Buckelwiesen" landscapes) in the European Alps. In: 8th International Conference on Geomorphology (2013) Abstracts vol 451

Galik A (1998) Gräber tief im Fels des Tscheltschnigkogels? Hallstattzeitliche Menschen und Tierknochen aus der Schachthöhle Durezza bei Villach. Kärnten Geol Paläontol Mitt Innsbruck 23:61–66

Golob B, Jungmeier M, Kreimer E (ed) (2013) Natur und Mensch in der Schütt. Die Bergsturzlandschaft im Naturpark zwischen Dobratsch und Gail. Verlag des Naturwissenschaftlichen Vereins für Kärnten, Klagenfurt, p 296

Holler H (1976) Gedanken zum Bau des Dobratsch in den östlichen Gailtaler Alpen. Carinthia II 166(86):43–55

Knafl K (2015) GIS-gestützte Modellierung von Auswirkungen und Worst-Case-Szenario größerer Sturzprozesse (Fels- und Bergsturz) im Dobratsch-Rupa-Gebiet. Unpubl. Master thesis, University of Graz, p 98

Lenhardt WA (2007) Earthquake-triggered landslides in Austria – Dobratsch revisited. Jahrb Geol Bundesanst 147:193–199

Letter MA (2014) Lawinenschutzvorkehrungen am Dobratsch—Wie gut ist die Gemeinde Bad Bleiberg vor Lawinenkatastrophen gesichert? Unpubl bachelor thesis, University of Graz, p 49

Neumann D (1988) Lage und Ausdehnung des Dobratschbergsturzes von 1348. In: Neues aus Alt-Villach. Museum der Stadt Villach, 25. Jahrbuch 1988:69–77

Neumann W (1988) Zu den Folgen des Erdbebens von 1348, 2.Teil: im Gailtal bei Arnoldstein. In: Neues aus Alt-Villach. Museum der Stadt Villach, 25. Jahrbuch 1988:9–68

Plan L (2016) Oberflächenkarstformen. In: Spötl C, Plan L, Christian E (eds) Höhlen und Karst in Österreich. Oberösterreichisches Landesmuseum, Linz, pp 23–34

Poltnig W, Probst G, Zojer H (1996) Untersuchungen zur Speicherung und zum Schutz von Karstwässern der Villacher Alpe (Kärnten). Mitteilungen der Österreichischen Geologischen Gesellschaft 87:75–90

Schulz O (1982) Kritische Bemerkungen zum tektonischen Bau der Villacher Alpe (Dobratsch). Verhandlungen der Geologischen Bundesanstalt 1982(2):85–94

Spötl C (2016) Südliche Kalkalpen. In: Spötl C, Plan L, Christian E (eds) Höhlen und Karst in Österreich. Oberösterreichisches Landesmuseum, Linz, pp 719–730

Till A (1907) Das große Naturereignis von 1348 und die Bergstürze des Dobratsch. Mitt. der k.k. Geographischen Gesellschaft in Wien 50:534–645

Trimmel H (1963) Die Höhlen der Villacher Alpe (1. Bericht). Carinthia II 73:115–124

Trimmel H (1964) Dobratsch-Alpenstraße und Karst im Gebiet des Dobratsch (Kärnten). Die Höhle 15:35–39

van Husen D, Ivy-Ochs S, Alfimov V (2007) Mechanism and age of late glacial landslides in the Calcareous Alps: The Almtal, Upper Austria. Aust J Earth Sci 100:114–126

Gerhard Karl Lieb is associate professor at the University of Graz, Institute of Geography and Regional Science. He has experience in physical geography of mountain regions with a special focus on permafrost, glaciers and geomorphic processes affected by global change as well as natural hazards. He is one of the two co-leaders of the national glacier monitoring programme of the Austrian Alpine Club. Besides those topics he works on regional geographies (Austria, Alps and Europe) and on didactics of the Austrian school subject geography and economics.

Christian Bauer is senior lecturer for physical geography at the University of Graz, Institute of Geography and Regional Science. His research focuses on geomorphology and geomorphometry in karst terrains, applied geomorphological topics using GIS (natural hazards) as well as human-environment interactions. His interdisciplinary research experience involves different disciplines, including Geosciences, System Science and History.

Check for updates

32 Klagenfurt Basin: A Large Basin in the Alps

Gerhard Karl Lieb

Abstract

The Klagenfurt basin is the second-largest basin of the Alps and hosts the socioeconomically most important subregion of Austria's federal state Carinthia. This chapter focusses on the basin's extraordinary variety of landforms and their evolution. The basin itself is a flexural basin that developed by a downwarping of the crust under the load of the northward moving Karawanken Mountains. During the LGM, almost the entire basin was covered by an ice sheet, which was important for the development of the basin's landscape, characterized by many lakes such as Lake Wörthersee. The last glaciation also explains why rivers such as the Gurk do not follow a N–S course—as would be expected because of the flexure—but rather show a peculiar river network. Two exemplary subregions of the basin that are of special morphological interest are portrayed. The Sattnitz plateau consists of Neogene conglomerates, which are interpreted as sediments from the Karawanken Mountains. However, the areas of erosion and sedimentation are separated from each other by the current valley of the Drau River. This section of the Drau Valley is therefore younger than other sections and only has developed during the Pleistocene. Finally, the Krappfeld, a northern subbasin of the Klagenfurt basin, is discussed with special attention to its glaciofluvial sediment filling and its position to the neighbouring glaciation systems during the Pleistocene.

Keywords

Inner-Alpine basin • Last glacial maximum • Glacial lake • Neogene sediments • Terrace • Drainage pattern development

32.1 Introduction

The Klagenfurt basin (Fig. 32.1) is considered to be the largest inner-Alpine basin of the Alps and the second-largest basin of the Alpine orogen, only surpassed by the Vienna Basin, which stretches across the Austrian border to Slovakia and the Czech Republic, separating the Alps from the Carpathians. Taking the delimitation that is most frequently used in the regional geographical literature (e.g., Paschinger 1976; Seger 2010), it covers an area of some 2000 km^2. In terms of the hydrological network, the entire basin belongs to the catchment of the Drau River, which flows through the basin from west to east, collecting all the tributary rivers from the neighbouring mountains (centripetal drainage system).

From a historical point of view, this specific geo-hydrological setting of valleys directed towards the central basin facilitated to create a political territory that was easily accessible. The first historically ascertainable territory that had its capital in the basin was the Roman Empire's *Provincia Noricum*, a successor of a Keltic kingdom Noricum that encompassed large parts of nowadays Austria and adjacent regions. Today, the basin is part of the Austrian federal state Carinthia with its capital Klagenfurt am Wörthersee. The Klagenfurt basin is the most important subregion of this federal state in terms of population, economy, culture and administration. With about 350,000 inhabitants, almost two-thirds of the entire population of Carinthia live in the basin (Seger 2010). According to the perception of most Austrians, this "heart" of Carinthia is well-known and popular, especially for its lakes, which of course are of geomorphological interest, too (Sect. 3.2).

G. K. Lieb (✉)
Institute of Geography and Regional Science, University of Graz, Heinrichstrasse 36, 8010 Graz, Austria
e-mail: gerhard.lieb@uni-graz.at

C. Embleton-Hamann (ed.), *Landscapes and Landforms of Austria*, World Geomorphological Landscapes,
https://doi.org/10.1007/978-3-030-92815-5_32

Fig. 32.1 Stagnant cold air combined with mist and fog is a typical element of winter weather in the Klagenfurt basin. From elevated positions one can appreciate the dimensions of the basin well—a view from the Saualpe to the south-west towards the Southern Alpine mountains limiting the basin in the south. Photo G. K. Lieb

32.2 Geographical and Geological Setting

32.2.1 Topographical Overview and Subdivision

The Klagenfurt basin is situated in the south of Austria and belongs, as already mentioned, to the Drau catchment. In the north and the east, the Central Alps form the margin of the basin; they are predominantely built of metamorphic rocks and are characterized by a moderate mountainous relief. The Southern Alps border the basin in the west and the south; they are dominated by carbonatic rocks and appear as impressive high mountains (Fig. 32.1). This gives the basin an asymmetrical shape corresponding to its tectonic evolution (Sect. 2.2). The outline of the basin, as shown in Fig. 32.2, was defined morphologically based on steep slopes that surround the entire basin (cf. slope map in Seger 2010) and show a sudden increase in elevation. Only at its south-eastern margin in the low hills at the Austrian–Slovenian border the limitation of the basin is less well pronounced.

Due to its areal extent and manifold geology, the basin has a diversified morphology, which is differentiated in the following way (Fig. 32.2):

- Plains and terraces: Along the rivers, (glacio-)fluvial accumulation has formed plains that have partly been cut into terraces, especially in the eastern (Jauntal) and the northernmost parts of the basin (Krappfeld; Sect. 3.5).
- Hills: Up to relative elevations of 300 m, the relief has been classified as hilly landscape, built of solid rocks overlain by glacial and glaciofluvial deposits in most cases (Sect. 2.3).
- Mountains: In contrast to other basins, the Klagenfurt basin also comprises mountains with a maximum relative elevation of 600 m and maximum absolute heights slightly over 1000 m (highest summit: Lippekogel, 1079 m). Most of these are built of crystalline basement rocks.
- Sattnitz plateau: A special relief feature is the plateau-like ridge of Sattnitz and its equivalents south and east of the river Drau that consist of Neogene conglomerates deposited during the uplift of the Karawanken Mountains (Sect. 3.4).
- Lakes: Glacial troughs that form the lake basins were not specifically indicated in Fig. 32.2 in order to avoid crowded map lettering (Sect. 3.2).

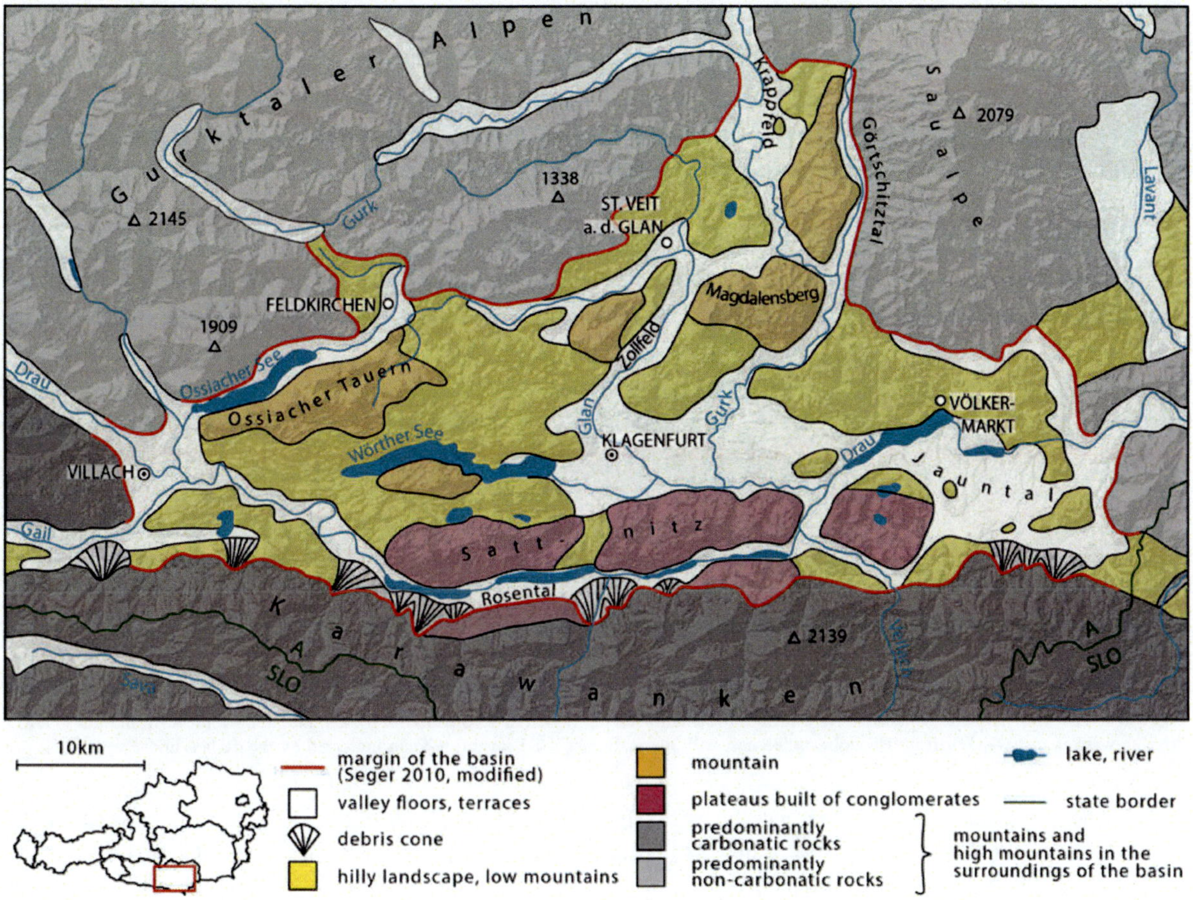

Fig. 32.2 Map of the geomorphological subdivision of the Klagenfurt basin and its surroundings (partly using the map of Seger 2010; drawing by V. Damm)

32.2.2 Tectonic Processes and Geomorphological Development

The development of the Klagenfurt basin began in the Miocene as a consequence of tectonic processes referred to as "lateral extrusion" (see Chapter "Geomorphological Landscape Regions of Austria"). The central part of the Eastern Alps east of the Hohe Tauern range, which is situated between the large strike-slip-faults of the Salzach–Enns–Mariazell–Puchberg line (SEMP) in the north and the Periadriatic Lineament (PAL) in the south (Fig. 6 in chapter "River and Valley Landscapes"), moved eastwards. This movement caused a complex regional pattern of faults, horst and graben structures as well as laterally tilted and rotated crustal blocks. According to Kuhlemann et al. (2003), the counterclockwise rotation of crustal blocks in a later period of the extrusion phase (after 16 Ma) helped to open the depressions in the north of the PAL such as the Klagenfurt basin.

The Klagenfurt basin can be regarded as a W–E trending flexural basin, which is characterized by a deflection of the Austro-Alpine tectonic unit north of the PAL (Nemes et al. 1997). The reason for this flexure was the northward movement of the Karawanken Mountains and the load that consequently warped the crust downwards. This process resulted in a significant sediment input as well as in an overthrust of Miocene basin sediments during the Pliocene and early Quaternary. From a geomorphological point of view, the following striking features can be explained by this tectonic development:

- The already mentioned asymmetrical shape of the basin is a visual expression of the northward overthrusting of the Karawanken block that gives the northern margin of the Karawanken Mountains the appearance of a steep thrust front (Fig. 32.3).
- The morphologically delimitated basin, as shown in Fig. 32.2, is much larger than the depression nowadays filled with Neogene sediments, which comprises a strip not wider than 10–15 km (Nemes et al. 1997) between the line Villach-Klagenfurt in the north and the northern front of the Karawanken Mountains in the south.

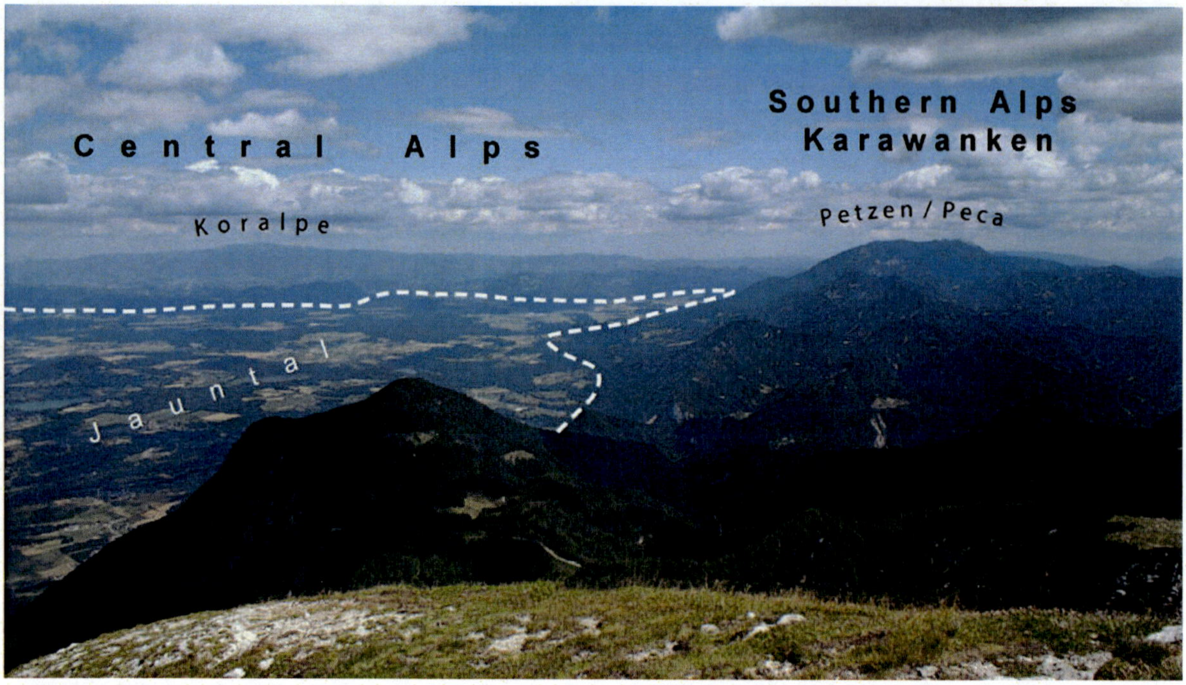

Fig. 32.3 The steep northern front of the Karawanken Mountains has developed in accordance with the northern movement (from right to left) of the Karawanken chain—view from Hochobir to the east with the margin of the Klagenfurt basin indicated by the dashed line. Photo G. K. Lieb

- The Neogene sediments are of great morphological significance because they form the isolated Sattnitz plateau (Sect. 3.4) and its outliers.

32.2.3 Glaciation During the Last Glacial Maximum (LGM)

During the coldest periods of the Pleistocene, the outflows of the Eastern Alpine transection glacier complex almost completely covered the Klagenfurt basin with ice. During the Last Glacial Maximum (LGM, according to Monegato et al. (2007) about 26–19 ka ago) huge Alpine ice masses entered the basin around Villach for the last time, the largest ones following the Drau and Gail valleys. They formed a broad glacier tongue, which had its terminus near the eastern margin of the basin in the vicinity of Bleiburg and Griffen. Based on moraines and ice-marginal complexes, Penck and Brückner already did the first reconstruction of this glacier in 1903/1909. For the modern picture of the LGM ice cover in the Klagenfurt basin see the detailed map of the Eastern Alpine ice cover in Fig. 4, Chapter "The Imprint of Quaternary Processes on the Austrian Landscape". Originally published by van Husen (1987), this map was continuously updated and finally digitized by Ehlers et al. (2011).

At the northern fringe of the basin, the ice masses caused the deviation of rivers flowing down from the partly glaciated mountains east of Villach. Remnants of this former drainage pattern are either valley bottoms in which there is no longer any river (Fig. 32.4) or complicated river courses (see the example of the Gurk River in Sect. 3.3). Furthermore, also within the border of the LGM glaciation, a complicated system of moraines, ice-marginal sediments and landforms, especially meltwater channels (mapped in detail first by Bobek 1959 and Lichtenberger 1959) offers insights into the retreat of the glacier. Apart from short oscillations in the beginning, this retreat took place rapidly and resulted in a widespread collapse of the ice mass. Lateglacial re-advances occurred only in the mountain valleys at higher elevations and thus far outside of the basin (van Husen 1988).

Beyond the terminal moraines, in the Krappfeld (Sect. 3.4) and in particular along the Drau River east of Völkermarkt, huge amounts of glaciofluvial sediments were accumulated, which Eicher (1986) mapped in detail. Since their deposition, these sediments have been dissected to terraces of almost 100 m in height along the Drau River. Today, they are considerable obstacles for the traffic network.

32.3 Diversity of Landforms

32.3.1 Overview

The Klagenfurt basin exhibits a great variety of landforms that developed since the Neogene (Fig. 32.2). In the early

Fig. 32.4 The broad valley bottom near Griffen has to be explained by sedimentation processes of a large river that no longer exists. This so-called palaeovalley of Haimburg-Griffen (cf. Fig. 32.7 and Sect. 3.3 for details) is an example of the drainage system along the margin of the Drau Glacier during the LGM. View from Griffen castle to the south. Photo G. K. Lieb

phase of basin evolution during the Miocene, depositional processes were predominant because of the subsidence of the basin and contemporaneous erosion in the Karawanken Mountains during their northward motion and uplift. In the Pleistocene, the phases of erosion and deposition often alternated. The last period of substantial reshaping of the relief was the LGM with glaciation of almost the entire basin (Sect. 2.3) and the subsequent glacial retreat. The following sections give an exemplary insight into topics concerning the entire basin (Sects. 3.2 and 3.3) and specific geomorphological settings of two subregions (Sects. 3.4 and 3.5).

32.3.2 Evolution and Distribution of Lakes

Lakes are landscape features that are most appreciated by tourists, and they are therefore intensively used for the purpose of touristic marketing, which was, for example, the reason for renaming the Carinthian capital from Klagenfurt to Klagenfurt am Wörthersee in 2007. Adding the largest and best-known Carinthian lake to the town name raised visibility!

As Table 32.1 shows, only five lakes exceed a surface of 1 km^2, so that less than 2% of the area of the entire basin can be attributed to lake surfaces. All natural lakes owe their origin to the Pleistocene glaciation and the subsequent deglaciation. Most of the depressions today filled with water have been carved by selective glacial erosion. Glacial erosion was more effective where the basement rocks were weak, whereas areas with underlying hard rocks resisted deep downcutting. In this context, the fault system in the basin's bedrock plays an important role. For instance, lake Wörthersee has the shape of a zigzag line in plan view. This was explained by Paschinger (1937) with the direction of faults. The same explanation applies to lake Ossiach, which stretches along SW-NE trending faults at the northern margin of the basin (Fig. 32.5). The overdeepening effect of glacial erosion is proved by the relatively high lake depths (Table 32.1).

Small lake depressions were formed as kettle holes. During the short phase of ice-decay after the LGM, the tongue of the Drau Glacier collapsed into stagnant ice masses with shrinking surfaces because of the lack of ice supply and, finally, into isolated dead-ice bodies. Due to the abundance of meltwater and easily erodible sediments (such as moraines and glaciofluvial gravel), which had not yet been stabilized by vegetation, large quantities of sediments were mobilized and frequently deposited at the margins or even on top of the dead-ice bodies. The subsequent melting out of such buried ice masses caused the formation of dead-ice holes (kettles) (Fig. 32.6).

As all other lakes, the lakes in the Klagenfurt basin are ephemeral features on a geologic timescale. Paschinger (1976) pointed out that there was a greater number of lakes immediately after the glacier retreat and that almost all lakes were considerably larger at that time. This is proved by the existence of delta sediments at many locations as well as by the existence of beat bogs in former shallow lake basins.

Table 32.1 Ten largest natural lakes in the Klagenfurt basin (from Sampl 1988)

Name	Area (km^2)	Max. depth (m)	Elevation (m asl)
Lake Wörthersee	19.39	85.2	439
Lake Ossiach	10.79	52.6	501
Lake Faak	2.20	29.5	554
Lake Keutschach	1.33	15.6	506
Lake Klopein	1.11	48.0	446
Längsee	0.75	21.4	548
Turnersee	0.44	13.0	481
Gösselsdorfer See	0.32	3.0	469
Forstsee	0.29	35.0	605
Rauschelesee	0.19	12.0	514

Fig. 32.5 Lake Ossiach is a lake of glacial origin that developed along a fault system. The steep slope to the left rises to mount Gerlitzen (1909 m) and represents part of the northern margin of the Klagenfurt basin. View from the Dobratsch massive to the east. Photo G. K. Lieb

Only those lakes without a larger river flowing through them have survived the 20 millennia of silting up since their formation. Drill holes with depths of more than 200 m (Kahler 1958) prove the existence of former glacially shaped overdeepened depressions, which became infilled, in most cases soon after the LGM. Also, various artificial lakes can be found in the basin—the largest ones are the reservoirs of the hydropower plants along the Drau River.

32.3.3 Drainage Pattern at the Northern Edge of the Basin

Due to the flexure of the Austroalpine crystalline basement at the northern fringe of the basin, an overall N–S flow direction of the rivers can be expected there. However, in most cases the orientation of the valleys does not fulfil this expectation (Fig. 32.2). An important reason for redirection

Fig. 32.6 Kraiger Lake is an example of a small lake (350 × 150 m) that developed as a kettle, probably during a very early stage of glacier retreat after the LGM in the north-eastern part of the Klagenfurt basin. View from the surrounding moraine ridge to the south. Photo G. K. Lieb

of rivers is Pleistocene glacial sediments (Sect. 2.3) hindering the river to follow its course towards the lowest point in the centre of the basin. The Gurk River is an example, as it is blocked at three locations (A, B and C in Fig. 32.7) and can only follow the natural slope quite far in the east of the basin (Eicher 1982):

- North-west of Feldkirchen (A in Fig. 32.7): The most pronounced diversion of the Gurk River starts at the low watershed between the Gurk and the Tiebel rivers (flowing to Lake Ossiach). The watershed is composed of glacial and glaciofluvial sediments that enable groundwater flow from the upper Gurk Valley to the springs of Tiebel, which therefore have an extraordinary discharge compared to their small superficial catchment area (Litscher 1977; Weiss 1977). For more details on the geological, geomorphological and hydrological setting see Eicher (1976, 1982), who also points out that the headward erosion of the Tiebel river system will lead to a stream capture and thus to a redirection of the Gurk River to the south-east in the near geological future. The sediment blockade during several phases of the Pleistocene at site A forced the Gurk River to erode a narrow valley ("Enge Gurk", translating to "narrow Gurk") into the flexural slope of the basin and thus to drain into a north-eastern direction. However, subglacial meltwater incision during a glaciation greater than the LGM (e.g., Mindel) might have contributed to the formation of this narrow valley reach as well (van Husen 2012).
- The other two points of river diversion (B and C in Fig. 32.7) are located in the north-east and east of the town St. Veit an der Glan: Here, the flow direction of the Gurk River to the basin's centre near Klagenfurt is blocked by the moraine ridges of the Drau Glacier (shown in Fig. 32.9), forcing the river to flow to the south-east.
- Finally, the Gurk River succeeds in flowing directly to the Drau River at location D (Fig. 32.7) where it cuts through the Würm-age moraines. During the LGM, however, the moraine ridges had redirected the precursor of the river even further to the east where it formed the broad palaeovalley of Haimburg-Griffen. Its south-eastern part (E in Fig. 32.7) is shown in Fig. 32.4.

In conclusion, the course of the Gurk River with its multiple changes in flow direction is to a large extent the result of Pleistocene glacial sedimentation processes. The blocking of valleys by glaciers did not only occur during the maximum extents of the Drau Glacier—most recently in the LGM—but also during ice-decay at the margin of melting ice masses.

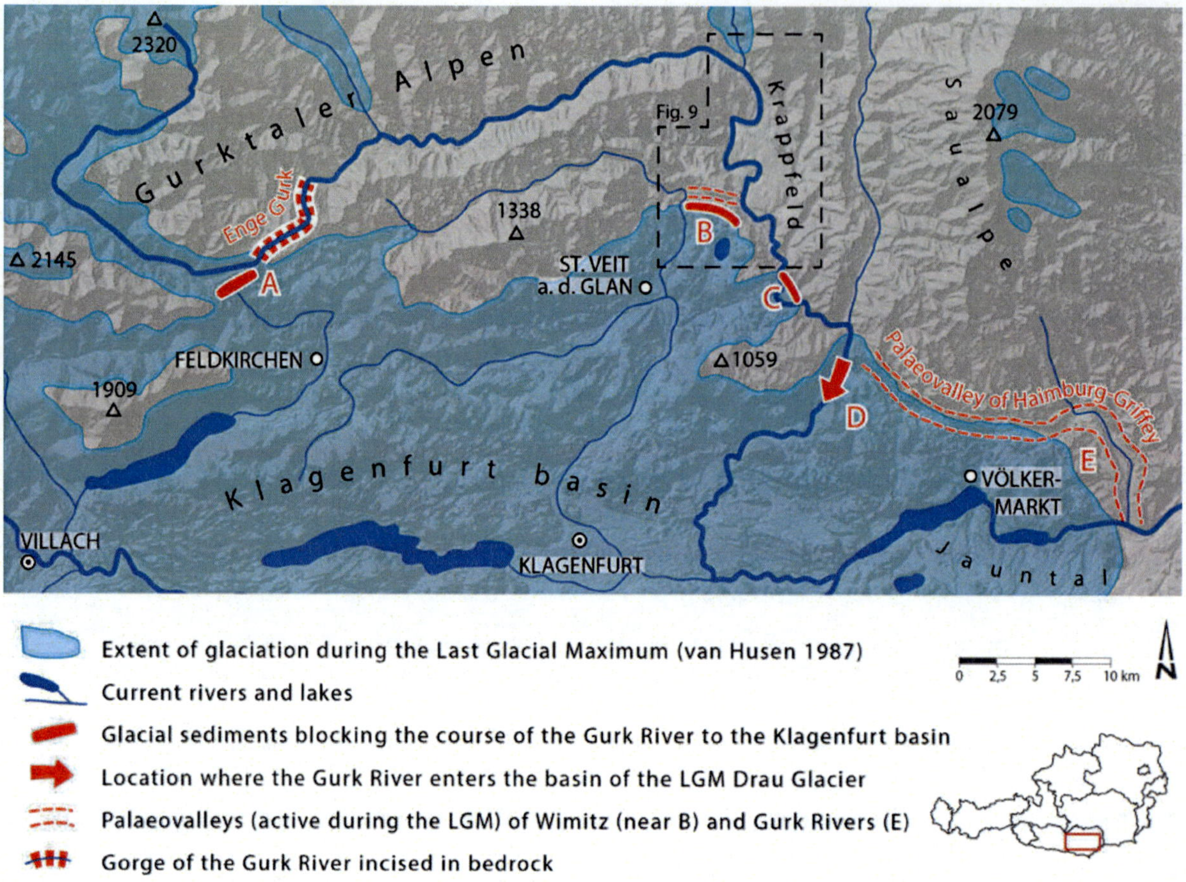

Fig. 32.7 Course of the Gurk River and the locations where glacial deposits prevent it from flowing straight towards the centre of the Klagenfurt basin (drawing by C. Sonnleitner)

32.3.4 The Sattnitz Plateau

Sattnitz is the local name of a 4–5 km wide plateau-like ridge, which stretches in a W–E direction over a distance of approximately 40 km. In the north, it is bounded by the so-called Keutschach Lake Valley ("Keutschacher Seentalung") and the lower Glan Valley; in the south by the Drau Valley, which is called Rosental in this part (Figs. 32.2 and 32.8). The river Drau also borders the Sattnitz ridge in the west and in the south-east. Here, the Sattnitz forms a steep erosional scarp with conspicuous rock faces (Fig. 32.8). The plateau itself can be divided into independent parts separated by depressions running in SE–NW to S–N directions. The largest of these depressions is that of Hollenburg ("Hollenburger Senke") where nowadays a federal highway connects the southern regions of Carinthia as well as Slovenia (via the Loibl pass) with the Carinthian capital Klagenfurt am Wörthersee. Furthermore, this depression was a mediaeval trade route along which Klagenfurt was founded.

The Sattnitz plateau consists of Neogene sediments covering crystalline basement rocks. With the subsequent uplift of the Karawanken Mountains, the amount of sediments as well as the size of the particles increased. The main unit of the Neogene sequence is the so-called Sattnitz conglomerate with a thickness of about 100 m in the west and more than 400 m in the east. The conglomerate also occurs further west and further east of the continuous Sattnitz plateau, e.g., near the south-western margin of the Jauntal plain (Fig. 32.2). Lithologically, the gravel is clearly dominated by calcareous components (in most outcrops >70%). Stratigraphic structures indicate that the sediments were originally deposited in a braided river system draining to the east (Krainer 2006). Large areas of the plateau are covered by Pleistocene till. In the depression of Hollenburg, a multiple sequence of glaciofluvial and glacial sediments has been preserved that was interpreted by Paschinger (1930) as a sedimentological record of almost all major alpine glaciations. Further details on geology can be found in Kahler (1953, 1962) and Krainer (2006, *cum lit.*).

Fig. 32.8 The plateau-like ridge of Sattnitz is composed of Neogene sediments. The upper photograph taken from the summit of Hochobir in NW direction clearly shows the separation of the Sattnitz from the area of origin of the sediments in the Karawanken Mountains (left) by the valley of the Drau River, named Rosental in this section. The lower photo taken from the summit of Magdalensberg in S direction illustrates the plateau-like character of Sattnitz. Each of the arrows indicates one of the transversal depressions mentioned in the text. Photo G. K. Lieb

From a morphogenetic point of view, the Sattnitz conglomerate (and the entire Neogene sequence) originally extended across a much larger area than today and represents some kind of pediment in the foreland of the Karawanken overthrust (Sect. 2.2). In this context, it is worth mentioning that the Karawanken Mountains were also thrust over the Neogene sediments, which were deformed and tilted as a consequence. In the final stage, even the distal parts of the Sattnitz conglomerate were slightly deformed and were subjected to tectonical stress that created faults in the meanwhile consolidated conglomerate mass. These faults guided the development of the mentioned depressions on the Sattnitz plateau and of the erosional channel of the river Drau forming the steep eastern margin of the plateau (Krainer 2006).

An interesting geomorphological question is how the Sattnitz plateau was separated from the Karawanken Mountains (Fig. 32.8), which yielded the sediments. Obviously during the sedimentation of the Sattnitz conglomerate the Drau Valley south of the Sattnitz (Rosental) did not yet exist. It follows a tectonic fault running parallel to the PAL north of the Karawanken that was only activated by the end of the Neogene. In the Pleistocene, this fault facilitated the glacial erosion of a depression, which is used by the Drau River today. Van Husen (1988) suggested that the erosion of the Rosental might have started with an ice-marginal river flowing along the foot of the Karawanken Mountains during an early Pleistocene glaciation. Once the Rosental was formed, large debris cones built of sediments from the Karawanken pushed the bed of the Drau River to the north causing lateral erosion and a steepening of the south side of the Sattnitz.

32.3.5 Example of a Subbasin: Krappfeld

Krappfeld is the name of a plain (c.10 × 5 km) in the northernmost part of the Klagenfurt basin (Fig. 32.2). Figure 32.9 gives an overview of the glacial and glaciofluvial landforms that dominate the area. The Krappfeld represents a bay-like subbasin of the Klagenfurt basin that already remained in low elevations in an early stage of Alpine orogeny. This is proved by Cretaceous (Gosau group) to

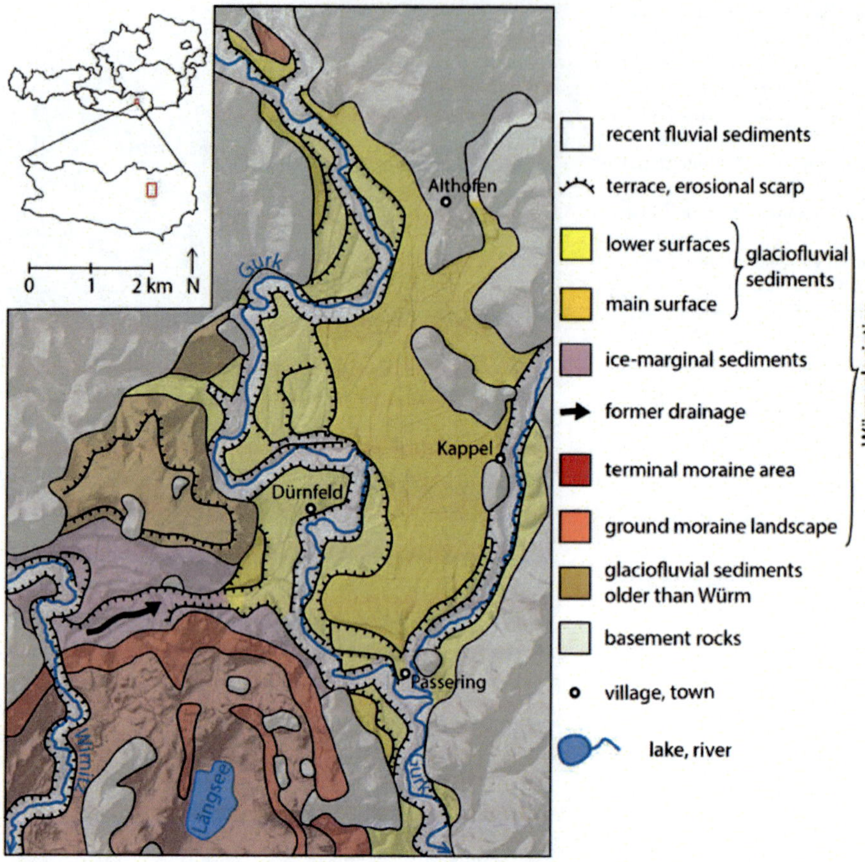

Fig. 32.9 Pleistocene moraines and terraces in the Krappfeld area (simplified from Thiedig et al. 1999; drawing by V. Damm)

Palaeogene sediments covering the Austroalpine crystalline basement in the mountainous area between Krappfeld and the Görtschitz Valley (Fig. 32.2). The subsidence in this part of the Klagenfurt basin was governed by brittle tectonics bound to a system of normal faults running N–S, rather than by the flexural processes described in Sect. 2.2. As a result, the Krappfeld subbasin has the character of a graben (map inset on tectonics in Thiedig et al. 1999). Its different tectonic style was caused by the tilting of the crystalline Saualpe block during the lateral extrusion tectonics of the Alps in the Neogene (Frisch et al. 2000). Thus on both sides of the Saualpe pull-apart-basins developed along fault systems stretching NNW–SSE.

To understand the distribution of Pleistocene sediments and landforms, Krappfeld's proximal position to the LGM glaciation must be taken into account. As Figs. 32.7 and 32.9 show, the central Krappfeld plain remained ice-free during the LGM and probably also during the former Pleistocene cold phases—at least no moraines from those periods have been mapped to date. A lobe of the LGM Drau Glacier (the outline of which was forced by the adjacent mountains) reached the south-western part of the area (Fig. 32.9). Note that the mapping category "terminal moraine area" in Fig. 32.9 does not only exhibit terminal moraines in a strict sense but also comprises other ice-marginal sediments (for details see Thiedig et al. 1999). South–west of the terminal moraines, a typical ground moraine landscape made up of subglacial till can be observed including ice-marginal landforms and moraine ridges that were deposited during short oscillations in the first stages of glacier retreat after the LGM (Fig. 32.10).

Another "terminal moraine area" can be found at the northern edge of the map in Fig. 32.9. Thiedig et al. (1999) recognized a former glacier terminus from the occurrence of erratics, a large kettle hole (van Husen 1989) and the onset of glaciofluvial sediment deposition, forming the gravel body of the Würmian "low terrace" (="Niederterrasse" sensu Penck and Brückner 1903/1909) (Fig. 32.11). It was the terminus of a glacier tongue of the Mur Glacier system. Its ice masses overspilled the relatively low watershed between the catchment areas of the rivers Mur in the north and Drau in the south at six sites (Seger 2010). Thus, the Krappfeld received glaciofluvial sediments from two large LGM Alpine valley glacier systems.

Fig. 32.10 Lake Längsee developed in a depression eroded by one of the lateral glacier tongues of the Drau Glacier. It is surrounded by an amphitheatre of moraines. View from St. Georgen am Längsee in a NW direction. Photo G. K. Lieb

Together with the poor drainage of the subbasin, caused by the sediment blockade of the Gurk River (Sect. 3.3), this resulted in a 20–25 m thick layer of glaciofluvial sediments. Consequently, the Krappfeld area is dominated by a large flat surface that formed as an outwash plain during the Würm (Fig. 32.11). After the LGM, the Gurk River and its tributaries dissected this plain and formed a system of terraces with up to four different levels reflecting phases of erosion and intermitted accumulation (van Husen 1989). The age of the individual terraces is difficult to determine. In the south-west, the sediments of the Würmian terrace turn into ice-marginal sediments in which the former course of the Wimitz River is preserved (black arrow in Fig. 32.9; palaeovalley indicated near location B in Fig. 32.7). West of the village Dürnfeld is a large terrace (with the local name "Auf der Eben", meaning flat surface), which rises 100 m above the level of the Würmian terrace and consists of older gravel, most probably belonging to the Riss and Mindel glaciations (van Husen 1989). This morphologically prominent sediment body suggests that the Krappfeld also had an ice-marginal position during the larger Riss and Mindel glaciations.

32.4 Conclusions

The Klagenfurt basin contains a large variety of landforms from which only a selection are presented in this chapter. The importance of the Pleistocene epoch for shaping the present-day morphology of the basin is obvious. The Pleistocene landforms and sediments had a significant influence for the development of settlements and economic activities from Roman times onwards and even play an important role for some aspects of nowadays' economic use such as tourism around the lakes, extraction of gravel, and the location of agricultural activities (Paschinger 1976).

Nevertheless, geomorphological research has not been intensive in recent years, which is surprising when looking at the diversified morphology and the number of unsolved research problems. For instance, the tectonic evolution of the basin is not as well recognized as it is in other Alpine basins and there is only limited knowledge especially regarding evidence of older glaciations and their extents as well as the exact chronology of the glacier advance towards the LGM maximum extent and the subsequent ice-decay. Thus,

Fig. 32.11 The low terrace ("Niederterrasse") in the northern part of the Krappfeld provides good locations for space-consuming factories as exemplified by the industrial buildings on the left. View from the Althofen old town to the north-west with the Gurktal mountains in the background. Photo G. K. Lieb

geomorphologists will find multiple research opportunities in the future.

References

Bobek H (1959) Der Eisrückzug im östlichen Klagenfurter Becken. Mitteilungen der Geographischen Gesellschaft Wien 101:3–36

Ehlers J, Gibbard PL, Hughes PD (eds) (2011). Quaternary glaciations-extent and chronology: a closer look. Developments in Quaternary Science. Elsevier

Eicher H (1976) Zur Funktion der Würmhochstände im Gebiet der oberen Gurktalung einschließlich der Nahtstelle Gurk- und Draugletscher. Mitteilungen der Österreichischen Geologischen Gesellschaft 69:209–245

Eicher H (1986) Die Drauterrassen-Zwischenfluren und ihre Stellung zum hochglazialen Draugletscher-Rückzug im Jaunfeld. Arbeiten aus dem Institut für Geographie der Karl-Franzens-Universität Graz 27:61–77

Eicher H (1982): Zur Flußnetzentwicklung im Klagenfurter Becken und in der Norischen Senke. In: Carinthia II, 172/92, Klagenfurt, pp 153–172

Frisch W, Székely B, Kuhlemann J, Dunkl I (2000) Geomorphological evolution of the Eastern Alps in response to Miocene tectonics. In: Zeitschrift für Geomorphologie N.F. vol 44, pp 103–138

Kahler F (1953) Der Bau der Karawanken und des Klagenfurter Beckens. Carinthia II 63:5–78

Kahler F (1958) Die Tiefe der Felsoberfläche in den Senken des Klagenfurter Beckens. Carinthia II, 68:5–8, Klagenfurt

Kahler F (1962) Geologische Karte der Umgebung von Klagenfurt 1:50.000. Geologische Bundesanstalt, Wien

Krainer K (2006) Geologie des Sattnitz-Höhnerückens. In: Golob B, Zwander H (eds) Die Sattnitz. Konglomerat der Natur im Süden Kärntens. Verlag des Naturwissenschaftlichen Vereins für Kärnten, Klagenfurt, pp 93–108

Kuhlemann J, Scholz T, Frisch W (2003) Postcollisional stress field changes in Eastern Carinthia (Austria). Mitt Österr Geol Ges 94:55–61

Lichtenberger E (1959) Der Rückzug des Würmgletschers im mittleren Klagenfurter Becken und Krappfeld. Mitteilungen der Geographischen Gesellschaft Wien 101:37–62

Litscher H (1977) Die Ergebnisse der geotechnischen Untersuchungen des Jahres 1976 im Gebiet der Prekowa-Tiebel. In: Carinthia II 167/87, pp 105–112, Klagenfurt

Monegato G, Ravazzi C, Donegana M, Pini R, Calderoni G, Wick L (2007) Evidence of a two-fold glacial advance during the last glacial maximum in the Tagliamento end moraine system (Eastern Alps). Quatern Res 68:284–302

Nemes F, Neubauer F, Cloetingh S, Genser J (1997) The Klagenfurt basin in the Eastern Alps: an intra-orogenic decoupled flexural basin? Tectonophysiscs 282:189–203

Paschinger H (1930) Die glaziale Verbauung der Sattnitzsenke in Kärnten. In: Zeitschrift für Gletscherkunde 18, Leipzig, pp 116–140
Paschinger H (1937) Zur Entstehung des Wörthersees. In: Carinthia II 127/147, pp 37–41
Paschinger H (1976) Kärnten. Eine geographische Landeskunde. Erster Teil: Allgemeine Darstellung. Verlag des Landesmuseums für Kärnten, Klagenfurt, p 322
Penck A, Brückner E (1903/1909) Die Alpen im Eiszeitalter. Tauchnitz-Verlag, Leipzig, vol 1–3, p 1199
Sampl H (1988) Kärnten—ein Land der Seen. In: Mildner P, Zwander H (eds) Kärnten—Natur. Die Vielfalt eines Lande im Süden Österreichs. Verlag des Naturwissenschaftlichen Vereins für Kärnten, Klagenfurt, pp 63–90
Seger M (2010) Kärnten. Landschaftsräume—Lebensräume. Ed by Geschichtsverein für Kärnten and Naturwissenschaftlicher Verein für Kärnten, Klagenfurt am Wörthersee, p 492
Thiedig F, van Husen D, Pistotnik J (1999) Geologische Karte der Republik Österreich 1:50.000, sheet 186, St. Veit an der Glan Geol B-A, Wien
Weiss EH (1977) Zur Hydrogeologie des Grundwasseraustrittes der Gurk unterhalb der Prekowa (Tiebelursprung). In: Carinthia II 167/87, pp 95–104, Klagenfurt.
van Husen D (1987) Die Ostalpen in den Eiszeiten. Geol B-A, Wien, p 24 (with map of LGM glaciation of the Eastern Alps 1:500.000)
van Husen D (1988) Zur quartären Entwicklung in Kärnten. In: Mildner P, Zwander H (eds) Kärnten—Natur. Die Vielfalt eines Landes im Süden Österreichs. Verlag des Naturwissenschaftlichen Vereins für Kärnten, Klagenfurt, pp 259–268
van Husen D (1989) Die Entwicklung des Krappfeldes und seiner weiteren Umgebung im Pliozän und Pleistozän. In: Arbeitstagung Geol B-A, Wien, pp 107–119
van Husen D (2012) Zur glazialen Entwicklung des oberen Gurktales. In: Jb. der Geologischen Bundesanstalt 152/1–4, pp 39–56, Wien

Gerhard Karl Lieb is associate professor at the University of Graz, Institute of Geography and Regional Science. He has experience in physical geography of mountain regions with a special focus on permafrost, glaciers and geomorphic processes affected by global change as well as natural hazards. He is one of the two co-leaders of the national glacier monitoring programme of the Austrian Alpine Club. Besides those topics he works on regional geographies (Austria, Alps and Europe) and on didactics of the Austrian school subject geography and economics.

33 Geomorphological Evidence of Past Volcanic Activity in the Southeast of Austria

Andreas Kellerer-Pirklbauer and Ingomar Fritz

Abstract

Active volcanoes do not exist in Austria today. However, remnants of volcanoes that were active during two main eruption periods between about 23 and 1 million years ago are prominent geomorphic features in the southeast of Austria. In this chapter, we discuss a complex landform evolution, in which sedimentological, tectonic, volcanic, and erosional processes acted together to create the present landscape. The study area is the Styrian basin and its vicinity, where 31 volcanic sites are located. Today, nothing is left of the primary volcanic landscape due to erosional processes lowering the original surface by some hundreds of metres since the end of the eruptions. The two main volcanic landforms present today are (i) maar-diatreme volcanoes, which form prominent and steep mountains consisting of basaltic tuff and maar lake sediments, and (ii) residuals of complex volcanoes consisting of tuff, scoria and massive basalt layers, partly buried by post-eruption sediments. The former are smaller in their spatial extent and monogenetic, whereas the latter are polygenetic, rather large, but partially buried by sediments. Some of the former volcanoes are characterized by well-developed planation surfaces that developed during the last million years. Hypsometric analyses were carried out for six volcano remnants (i) for morphometric characterization and (ii) to compare them with active volcanoes. The results revealed that only well-eroded diatreme volcanoes such as Güssing can be suitably characterized by means of a hypsometric analysis. In all other cases, this approach fails due to the complex formation and erosion history of the volcanoes.

Keywords

Volcano remnant • Maar-diatreme volcano • Stratovolcano • Planation surface • Hypsometric analysis

33.1 Introduction

In this chapter, we discuss volcanic landforms located in the southeast of Austria, i.e., in the federal provinces of Styria and Burgenland (Fig. 33.1). This part of Austria was influenced by volcanic activity over a period of several million years ending around 1.71 ± 0.72 Ma ago (cf. Table 33.1). Since the beginning of volcanic activity in the study area, the region has been influenced by large-scale subsidence followed by a period of uplift, causing formation and destruction of both volcanic and non-volcanic landforms.

Most landforms are the result of both, tectonic processes creating an initial relief and of the sculpturing effects of erosion by wind, water, gravity and ice destroying the previously built-up relief. However, volcanic landforms are special, being the result of enhanced constructive and opposing destructive forces acting on a more local to regional scale compared to tectonic forces (Francis 1993). Constructive processes operate as long as a volcano is active. The active phase might be an extremely short period, forming small landforms such as single lava flows or scoria cones. Volcanic activities might also last intermittently for several thousands of years, forming large volcano edifices (Martí et al. 1994; Rhodes and Lockwood 1995).

Francis (1993) pointed out that erosion of new volcanic material begins on a volcano as soon as the volcano starts growing and even before its new lavas and pyroclastic deposits have cooled (see also Cotton 1944). A large volcano may experience several phases of rapid construction

A. Kellerer-Pirklbauer (✉)
Institute of Geography and Regional Science, University of Graz, Heinrichstrasse 36, 8010 Graz, Austria
e-mail: andreas.kellerer@uni-graz.at

I. Fritz
Centre of Natural History, Geology and Palaeontology, Universalmuseum Joanneum, Weinzöttlstrasse 16, 8045 Graz, Austria

C. Embleton-Hamann (ed.), *Landscapes and Landforms of Austria*, World Geomorphological Landscapes,
https://doi.org/10.1007/978-3-030-92815-5_33

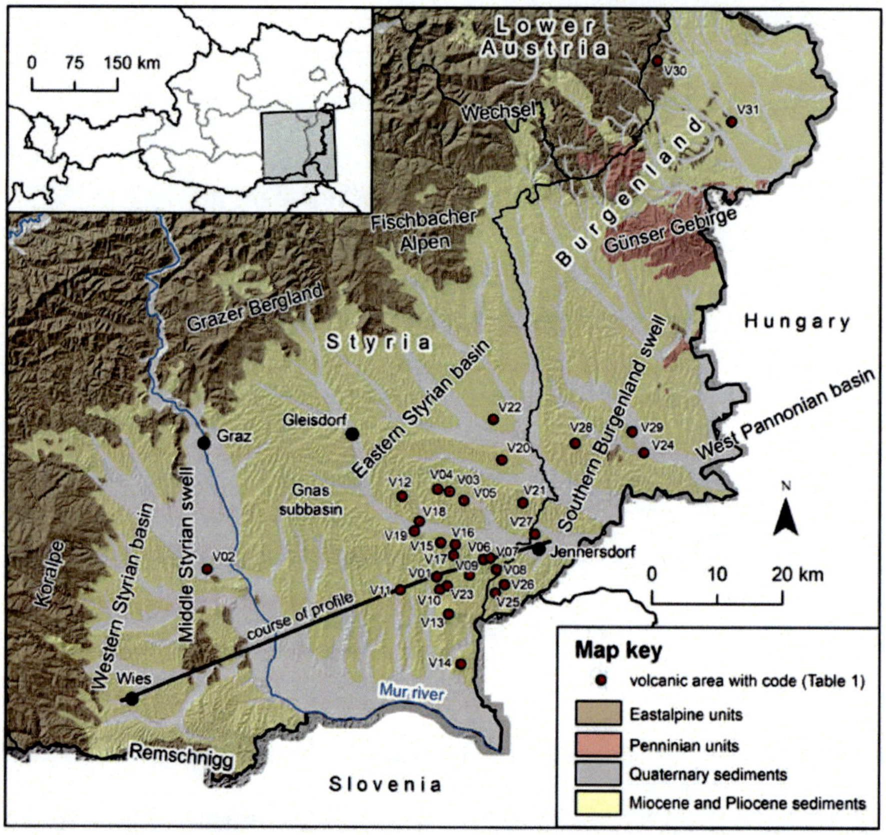

Fig. 33.1 Topographical and geological setting of the south-eastern part of Austria with the main geological basins and swells (see text for details). All volcanic areas listed in Table 33.1 and the course of the geological cross-profile shown in Fig. 33.3 are indicated. Geology based on Weber (1997)

during which the rate of build-up exceeds the rate of erosion. On the Hawaiian Islands, for instance, phases of construction alternated with phases of massive flank collapses, which are traceable in the Pacific Ocean (Moore et al. 1994). On large volcanoes, erosion may be active on one slope, whereas new lava and tephra are being added to another.

Tectonic forces change the relative position and elevation of the sea level. In case of prevailing subsidence and formation of a basin within surrounding mountains, sediments from these mountains might accumulate in the basin, causing complete or partial burial of a volcano.

The aim of this chapter is (i) to give an overview of the tectonic, volcanic and denudational chronology of the Styrian basin since the Miocene and (ii) to describe its geomorphic legacy on the present landscape in qualitative and quantitative analyses.

33.2 Study Region

The study area is located in the western part of the Pannonian basin system (Schönlaub 2000). The Styrian basin is about 100 km long, 60 km wide, up to 4 km deep (Gross et al. 2007) and is surrounded by the Remschnigg mountains to the south-west, the Koralpe and Grazer Bergland to the west and northwest, and the Fischbacher Alpen and Wechsel mountains to the north (Fig. 33.1). The basin is separated by the NNE-SSW striking Southern Burgenland swell from the Western Pannonian basin and internally divided by the Middle Styrian swell into a Western and an Eastern Styrian basin. Because of geological, morphological and volcanological similarities, we also consider the volcano remnants located at the Southern Burgenland swell and those north of the Günser Gebirge in this chapter.

The Southern Burgenland swell was formed due to different subsidence rates of the basins mainly during the Middle Miocene, but continued to form until the Late Miocene (Gross et al. 2007). The swell stretches from the settlement of St. Anna in the southwest to the Günser Gebirge in the northeast and is composed of phyllitic shales, green shales and limestone shales of the Silurian and carbonates of the Devonian (Schönlaub 2000). These Palaeozoic units are widely covered by transgressive Miocene sediments of brackish to freshwater origin (Flügel and Heritsch 1968; Gross et al. 2007).

Table 33.1 List of volcanic areas in the study area with names, location, estimated age ranges and relevant references. For the location of sites see Fig. 33.1

Code	Lat (°N)	Long (°E)	Name	Municipality/-ies	Estimated age (in Ma)	References*
V01	46.8893	15.9092	Gleichenberger Kogeln	Bad Gleichenberg, Feldbach	12.2[4]–17.2[2] 22.97 ± 1.93	1, 2, 4, 6 3
V02	46.8955	15.4456	Weitendorf	Wildon, Dobl-Zwaring	14.0–17.5	1, 2, 4
V03	47.0032	15.9331	Riegersburg	Riegersburg	3.71	7
V04	47.0061	15.9097	Altenmarkt	Riegersburg		
V05	46.9915	15.9621	Stang	Fehring, Riegersburg		
V06	46.9129	16.0023	Fehring (Burgfeld, Heißberg, Waxenegg, Zinsberg)	Fehring	5.15	7
V07	46.9148	16.0161	Beistein	Fehring	5.82	7
V08	46.8997	16.0283	Aschbuch	Fehring, Kapfenstein		
V09	46.8919	15.9743	Kapfensteiner Kogel	Kapfenstein	4.68–4.86	5, 7
V10	46.8720	15.9155	Bad Gleichenberg (Albrechtshöhe, Sulzberg, Wierberg)	Bad Gleichenberg	5.26	7
V11	46.8710	15.8382	Gnas	Gnas	2.88	7
V12	46.9961	15.8402	Edelsbach	Edelsbach bei Feldbach	3.52	7
V13	46.8390	15.9332	Stradner Kogel	Bad Gleichenberg, St. Anna, Straden, Tieschen	1.71 ± 0.72	4
V14	46.7727	15.9587	Königsberg/Klöch	Tieschen, Klöch	2.56 ± 1.2 2.17–2.56	4, 5
V15	46.9350	15.9170	Steinberg (Mühldorf)	Feldbach	2.64–3.05	4
V16	46.9325	15.9457	Pertlstein	Fehring	3.32	7
V17	46.9177	15.9424	Forstkogel	Feldbach	7.51	7
V18	46.9631	15.8744	Auersberg (Gniebing)	Feldbach	2.87	7
V19	46.9499	15.8647	Unterweissenbach/Kalvarienberg (Gniebing)	Feldbach	2.27–2.73	4, 7
V20	47.0461	16.0375	Stadtbergen	Fürstenfeld	6.48	7
V21	46.9888	16.0795	Stein (Großsteinberg/Hiebüchl)	Loipersdorf bei Fürstenfeld		
V22	47.1002	16.0199	Jobst	Bad Blumau		
V23	46.8773	15.9302	Muhrn (Bairisch Kölldorf)	Bad Gleichenberg		
V24	47.0571	16.3227	Güssing	Güssig	5.03–6.3	5, 7
V25	46.8682	16.0271	Neuhaus (several smaller areas around Neuhaus)	Neuhaus am Klausenbach	3.11–3.76	4, 5
V26	46.8794	16.0445	Neuhaus (Wolfsriegel north)	Neuhaus am Klausenbach		
V27	46.9471	16.1041	Grieselstein (two parts)	Jennersdorf		
V28	47.0695	16.1842	Limbach	Kukmirn	5.72	7
V29	47.0853	16.2993	Tobaj	Tobaj	4.98	7
V30	47.5807	16.3472	Pauliberg	Kobersdorf	10.5–12.3	4
V31	47.4996	16.5000	Oberpullendorf	Oberpullendorf	11.1	4

* References: 1—Lippolt et al. (1975), 2—Steininger and Bagdasarjan (1977), 3—Kolmer (1980), 4—Balogh et al (1994), 5—Seghedi et al. (2004), 6—Bojar et al. (2008), 7—Bojar et al (2013)

Today, the Styrian basin and the Southern Burgenland swell form undulating lowlands with elevations between 200 and 600 m asl and comprise Neogene and Pleistocene sediments covering the basement and the volcanic rocks. The highest summit in the area, the Stradner Kogel, is 609 m asl high. The lowest point in the area is located at the border to Slovenia, where the Mur River leaves Austria (200 m asl). In general, the difference in elevation between the two ridges or hills and the valley in between is in the order of 100–200 m. The main valley bottoms are mostly wide and flat (e.g. Mur

Valley in Fig. 33.1), smaller valleys are often V-shaped. Steeper slopes are mostly related to volcanic rocks.

Climatic conditions are moderately continental with snow-poor winters, and the precipitation maximum is in summer. Mean annual precipitation values are between 800 (in the SE) and 1000 (near the mountains) mm. Mean annual air temperatures are influenced by local topographical conditions and are in the order of 8–9 °C (Prettenthaler et al. 2010).

33.3 Geological History

Until the Late Miocene, the palaeogeographical environment of the Styrian basin was marine, belonging to the Paratethys (Gross et al. 2007). The basin formation is connected to continental escape tectonics of alpidic crustal wedges at the final collision stage of the Adriatic plate with the European plate during the Late Oligocene to Miocene (Neubauer and Genser 1990; Ratschbacher et al. 1991). The development of the Carpathian chain and the Pannonian basin system in the Miocene was controlled by retreating subduction in front of the orogen and by back-arc extension associated with the diapiric upraise of the astenosphere (Kovac et al. 2000; Konecny et al. 2002). Lateral escape took place along large, E-W trending strike-slip faults, generating small pull-apart basins. Simultaneously, isostatic uplift of thickened continental crust was associated with gravitational sliding of higher parts of the lithosphere along flat downthrow faults. This led to the exposure of deeper units (Penninic unit). Downthrow faults are regarded as responsible for horizontal block tilting, which caused asymmetric, N-S striking extensional structures like the Styrian basin (Neubauer and Genser 1990; Neubauer et al. 1995). The tectonic evolution of the Pannonian basin system was accompanied by magmatism of variable composition that is documented by two main volcanic phases in the Styrian basin.

33.3.1 Basin Development Since the Miocene

Based on Kollmann (1965), Gross et al. (2007) provided a summary of basin development, which included a stratigraphic chart of the Neogene basin fill of the Styrian basin. The ages published for the different stages of the Miocene vary between the different parts of the Central Paratethys (Kovac et al. 2018). Here we follow mainly the ages presented by Gross et al. (2007). Figure 33.2 gives a summary of the evolution of the Styrian basin and its surrounding basement as judged from Miocene to Quaternary sediments, planation surfaces (here used as a purely descriptive term for a low-relief plain cutting across various rocks and structures) and age constraints. The filling of the basin started in the Early Miocene during the Ottnangian (c.18.3–17.0 Ma) with limnic-fluvial sediments (Figs. 33.2 and 33.3). The alluvial fan and delta sediments of that time are locally coal-bearing (Ebner and Sachsenhofer 1991). Weathered ash-tuff layers (bentonite), intercalated in this sedimentary succession, have also been reported (Gross et al. 2007). Volcanic activity was probably initiated even earlier, around the Oligocene/Miocene transition some 23 Ma ago (Table 33.1), as judged from dates presented by Kolmer (1980). However, later dating results at the same volcano (VO1) revealed younger ages, therefore the onset of volcanism in the study area is not known with certainty (see Table 33.1).

In Karpatian times (c.17.0–16.0 Ma), the increase of tectonic activity caused considerable subsidence, which led to a marine transgression at the basin margin. Besides the formation of the Middle Styrian swell, volcanic activity increased and a compound volcano, consisting of acidic and intermediate material, formed in the Gnas subbasin (Fig. 33.1). The maximum marine extent occurred during the Early Badenian, at 16–15 Ma. Coral reefs and limestone platforms (Leitha Limestone) developed around higher-elevated basement mountains. Some of these former reefs are karstified today (Friebe 1990; Bauer 2015). The westernmost volcanic landform in the Styrian basin was formed 17.5–14.0 Ma ago in Weitendorf (V02 in Fig. 33.1).

The almost complete isolation of the Paratethys from the Mediterranean and the Indopacific (Rögl 2001) during the Sarmatian (c.12.8–11.6 Ma) caused substantial environmental changes. Friebe (1994) inferred a repeatedly oscillating sea level because of mixed siliciclastic-carbonate sedimentation. The Pannonian (c.11.6–7.1 Ma) started with an extensive regression and tectonic fragmentation, causing the origin of horst and graben structures as shown schematically in Fig. 33.2. The coarse sediments of the Pannonian are suitable aquifers and contain large amounts of groundwater. This geological context was of high relevance for the phreatomagmatic eruptions during the Late Miocene to Pleistocene. Ebner and Sachsenhofer (1995) pointed out that during the Middle to Late Miocene a final, rather strong subsidence was followed by the beginning of the uplift of the lowland.

In summary, a massive sedimentary pile developed between ~18 and 11 Ma in the Styrian basin accompanied by a period of active volcanism from 23.0 (or at least 17.5; cf. Table 33.1) to 10.5 Ma ago. According to Ebner and Sachsenhofer (1995), tectonic movement inverted around 5–6 Ma, causing the end of sedimentation and the beginning of the uplift history of the Styrian basin.

33.3.2 Miocene Volcanic Period

The strong crustal extension during the Karpatian was accompanied by volcanic activity, which continued until the

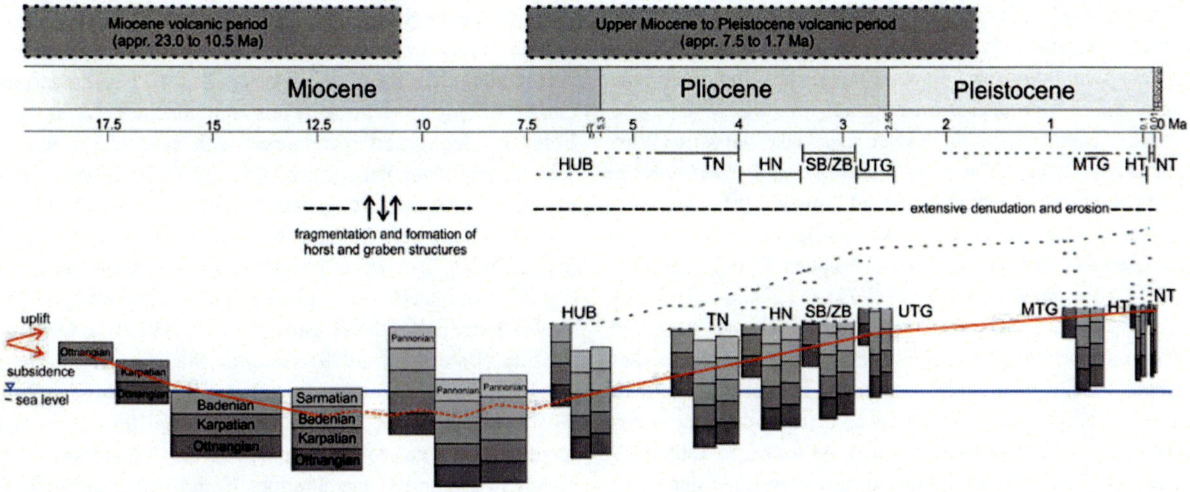

Fig. 33.2 Tectonic and sedimentary evolution of the Styrian basin and its surrounding basement derived from Miocene to Quaternary sediments, planation surfaces and age constraints. The approximate temporal extents of the two main volcanic phases are indicated (based on dates in Table 33.1). Abbreviation of the different planation surfaces and stream terraces (from old/more elevated to young/less elevated): HUB = Hubenhalt niveau, TN = Trahütten niveau, HN = Hochstraden niveau, SB/ZB = Stadelberg/Zahrerberg niveau, UTG = Upper terrace group, MTG = Middle terrace group, HT = high terrace gravels, NT = low terrace gravels (after Wagner et al. 2011; Fig. 33.8, modified)

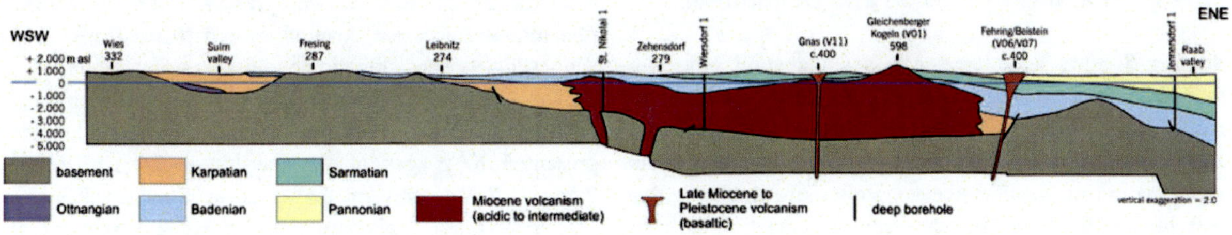

Fig. 33.3 Geological cross-profile of the Styrian basin based on deep boreholes (selected ones are indicated in the profile). For the location of the profile see Fig. 33.1. Profile modified after Ebner and Sachsenhofer (1991, appendix 2)

end of the Early Badenian in the Styrian basin (Handler et al. 2006) and until the Pannonian in Burgenland (Balogh et al 1994). Eruptions of acidic to intermediate composition occurred in the Gnas subbasin. Polygenetic volcanism (i.e., several eruptions form the volcano edifice) prevailed during this first phase of volcanism. Ebner and Sachsenhofer (1991) speculated about a subduction-related origin of the potassium-rich subalkaline-alkaline magma. Huge shield volcanoes developed, which were later nearly completely buried by younger sediments. The knowledge of the sediment-buried volcanic structures stems from drillings (e.g. petroleum exploration; examples on Fig. 33.3) and geophysical surveys (Kröll et al. 1988). Younger alkali-basaltic volcanism forming alkaline effusive rocks of Early Pannonian age (c. 12.3–10.5 Ma) are located north of the Günser Gebirge in the area of Pauliberg and Oberpullendorf (V30 and V31 in Fig. 33.1 and Table 33.1). Figure 33.1 and Table 33.1 give an overview of the 31 relevant volcanic areas. As indicated in this table, only few volcanic areas (V01, V02, V30 and V31) are attributed to this Miocene volcanic phase.

33.3.3 Late Miocene to Pleistocene Volcanic Period

The second phase of volcanic activity, this time primarily of alkali-basaltic composition, occurred between ~7.5 and 1.7 Ma (Table 33.1). In Styria and southern Burgenland, rising magma from the Earth's mantle came into contact with groundwater contained, for instance, in the coarse Pannonian sediments (Pöschl 1991; Fritz 1996; Gross et al. 2007). Hence, phreatomagmatic eruptions occurred, creating funnel-shaped craters in the pre-volcanic surface and producing pyroclastic rocks. These rocks erupted explosively and contain fragmented juvenile and lithic clasts in different compositions. Bomb-sag-structures and low-angle-cross-stratification as well as the existence of accretionary and armoured lapilli indicate phreatomagmatic eruptions and are evident at several volcanic remnants in the Styrian basin.

Presumably, hundreds of rapidly succeeding eruptions produced huge amounts of tephra that were primarily deposited in layers, building up maar-diatreme-volcanoes

(Lorenz 1986). Subsequently, the material solidified and formed volcanic tuff. After the volcanic activity had ceased, craters filled with water, forming maar lakes. Into these maar lakes tephra from the surrounding tuff-rims was re-deposited, forming characteristic fine-clastic maar lake deposits (White and Ross 2011). The once deeper-lying harder volcanic pyroclastic rocks were uncovered through the subsequent erosion of the surrounding loose sedimentary rock. At some localities—Stradner Kogel (V13), Klöch (V14), Steinberg (V15), Altenmarkt (V04), Stein (V21) and Neuhaus (V25 and V26)—basaltic extrusions and/or intrusions are preserved. The volcanoes from this activity period commonly had a short eruptive history and were presumably in most cases monogenetic volcanoes like scoria cones or maar-diatreme volcanoes (Németh and Kereszturi 2015). As shown in Table 33.1, most volcanic areas are attributed to this volcanic phase.

33.4 The Visual Legacy of the Volcanic and Erosional History

33.4.1 The Role of Erosion and Denudation

Starting 5–6 Ma ago, large-scale sedimentation processes diminished and erosion began to prevail. Almost no sediments of the Pliocene age are preserved today (Gross et al. 2007; Wagner et al. 2011). The Styrian basin was never glaciated during the Pleistocene. Relics of landforms that date back to the Late Miocene—such as planation surfaces (cf. Figs. 33.2 and 33.4)—are still preserved today. Geological and geomorphological investigations of the Styrian basin started in the nineteenth century. One of the pioneers working on the reciprocal effects of erosion and build-up (including volcanism) on the Styrian basin was Arthur Winkler-Hermaden. His publication of 1957 reflects the state of knowledge at that date (Winkler-Hermaden 1957).

With the beginning of uplift, several distinct morphological levels were formed in the Styrian basin and the surrounding basement. Wagner et al. (2011) summarized earlier findings for these levels and discussed their ages based on dating and correlations with cave levels, stream terraces and planation surfaces (Fig. 33.3). The two highest planation surfaces (Hubenhalt niveau/HUB and Trahütten niveau/TN) are only found in the mountains surrounding the Styrian basin. Two lower planation surfaces also exist in the Styrian basin itself. These are the Hochstraden niveau (HN) —also termed Kalkleiten-Möstl (Hilber 1912) or Gebirgsrandflur (Untersweg, 1979)—and the Stadelberg/Zahrerberg level (SB/ZB). The elevation above the present-day Mur River (Fig. 33.1) is 325–450 m asl and 180–300 m asl, respectively. The volcanic rocks at the Königsberg near Klöch (V14) and at the Stradner Kogel (V13) are of high relevance in this regard, because planation surfaces developed across them. At about 550 m asl, a well-developed planation surface formed at the Stradner Kogel, giving rise to the name "Hoch(=high)straden" niveau. At a lower level of the Stradner Kogel mountain and in the nearby Klöch area, the SB/ZB level is well-preserved (Fig. 33.4). Both planation surfaces are covered by red loams (Fink 1961), suggesting warmer climatic conditions during formation.

The age of the HN and SB/ZB levels must obviously post-date the volcanic eruptive phase. Balogh et al. (1994) presented K/Ar-ages of the basaltic rocks at Klöch of 2.56 ± 1.2 Ma, and of the ones at Stradner Kogel of 1.71 ± 0.72 Ma. This age range gives a maximum age constraint on the SB/ZB level and suggests that its main formation is younger than the upper limit of the basalt age at both sites, i.e., <2.43 Ma.

As described in Sect. 3.3, the top of the volcanic rock/sediment pile is formed by maar lake sediments. The altitude at which maar sediments are found today (e.g. at Altenmarkt/V04, Burgfeld near Fehring/V06,

Fig. 33.4 Zahrerberg area near Klöch with a well-developed planation surface (Zahrerberg level) consisting of basaltic rocks; **a** the southern part of the planation surface with the view towards west (note the approximate delineation of the slightly inclined and moderately incised planation surface), **b** basalt rock outcrop at the southern margin of the planation level. The Zahrerberg area is intensively used for wine growing due to favourable climate and soil conditions. Photographs by the authors

Gleichenberger Kogeln/V10 and Gnas/V11) can therefore be used to estimate denudation rates since the time of volcanic activity. The result of c. 300 m is consistent with data suggested for comparable settings in the Pannonian basin (Kereszturi et al. 2011).

In summary, erosional processes prevailed during the Pliocene and Pleistocene periods at higher levels of volcanic and non-volcanic landforms. Large-scale sedimentation processes acted only in the main valleys and their tributaries (Wagner et al. 2011). Rates of sedimentation and erosion in the Styrian basin during the Pliocene and the Quaternary are not known.

33.4.2 The Anthropogenic Influence on Volcanic Landforms

Human activities have modified the volcanoes in the region to some extent through surface mining. Figure 33.5 shows two examples of basalt quarries in the Styrian basin. The first example is from a basalt quarry slightly north of the settlement of Klöch. Gross et al. (2007) described well-bedded tuff-layers forming the basis of the volcanic succession and being in discordant contact to the Sarmatian sediments. Intrusive nepheline-basanites and pyroclastic rocks are exposed in the quarry. At the upper part of the open mine, weakly bedded, red tuff breccia with some small lava flows are exposed, forming the top of the volcanic succession. The second example is from the Steinberg area about 2 km SE of Feldbach. The volcanic material at this site covers sediments of Pannonian age (Heritsch 1968). Figure 33.6 depicts one example of the use of volcanic rocks (lapilli tuff) in the study region for construction purposes, in this case to build the church in the municipality of Riegersburg.

On an annual basis, the Austrian Federal Ministry for Sustainability and Tourism (www.bmnt.gv.at) publishes the output of the national mining industry. According to the reports for 2014–2016, basaltic rock was mined at one site in the federal province of Burgenland (V30/Pauliberg) and at five sites in Styria (V14/Klöch, V15/Steinberg/Mühldorf, V13/Stradner Kogel, V02/Weitendorf and V01/Gleichenberger Kogeln) during the reporting periods. As shown in the inset graph in Fig. 33.5, the amount of mined basalt increased from 2012 to 2014, but has decreased slightly since then. On average, 1.57 Mio. t/yr of basalt were mined during the 5-year period 2012–2016.

33.4.3 Visible and Obscured Volcanic Remnants

33.4.3.1 General Conditions

There are many places around and even within the Styrian basin with magnificent views over the morphology of the basin (Fig. 33.7). Prominent features visible from far away are distinct ridges or hills, often with flat summit areas. Such rounded and often forested mountains are called "Kogel". Because of their greater resistance to weathering as compared to the predominantly non-solid clastic sediments, the volcanic rocks stand out against their surroundings. Examples are the Gleichenberger Kogeln (V01; a double summit culminating at 598 and 563 m asl), Kapfensteiner Kogel (V09; 461 m asl), Stradner Kogel (V13; 609 m asl) or the Königsberg-Zarawald-Seindl area (V14; 462 m asl). The most spectacular volcanic forms in the study area are the steep diatremes of Riegersburg (V03; 484 m asl) and Güssing (V24/318). Both of them were used as locations for fortifications in the past and have impressive castles at their tops (Fig. 33.8).

Fig. 33.5 Examples of anthropogenic influence on the volcanoes in the Styrian basin due to surface mining: **a** basalt quarry near Klöch (V14), view towards S; **b** basalt quarry at Steinberg (V15), view towards NW, note the town of Feldbach and the volcano remnants of Unterweissenbach/Kalvarienberg (V19) left (south) of Feldbach; **c** depicts the amount of mined basalt in Austria (all six relevant quarries are located in the study region) during the period 2012–2016 (for data source see text). Photographs by the authors

Fig. 33.6 An example of the use of volcanic rocks in the study region for construction purposes. The church of the settlement of Riegersburg (**a**) is made of basaltic tuff (**b**), which was quarried from the nearby diatreme of Riegersburg (V03). Note the handrail in B for scale. Photographs by the authors

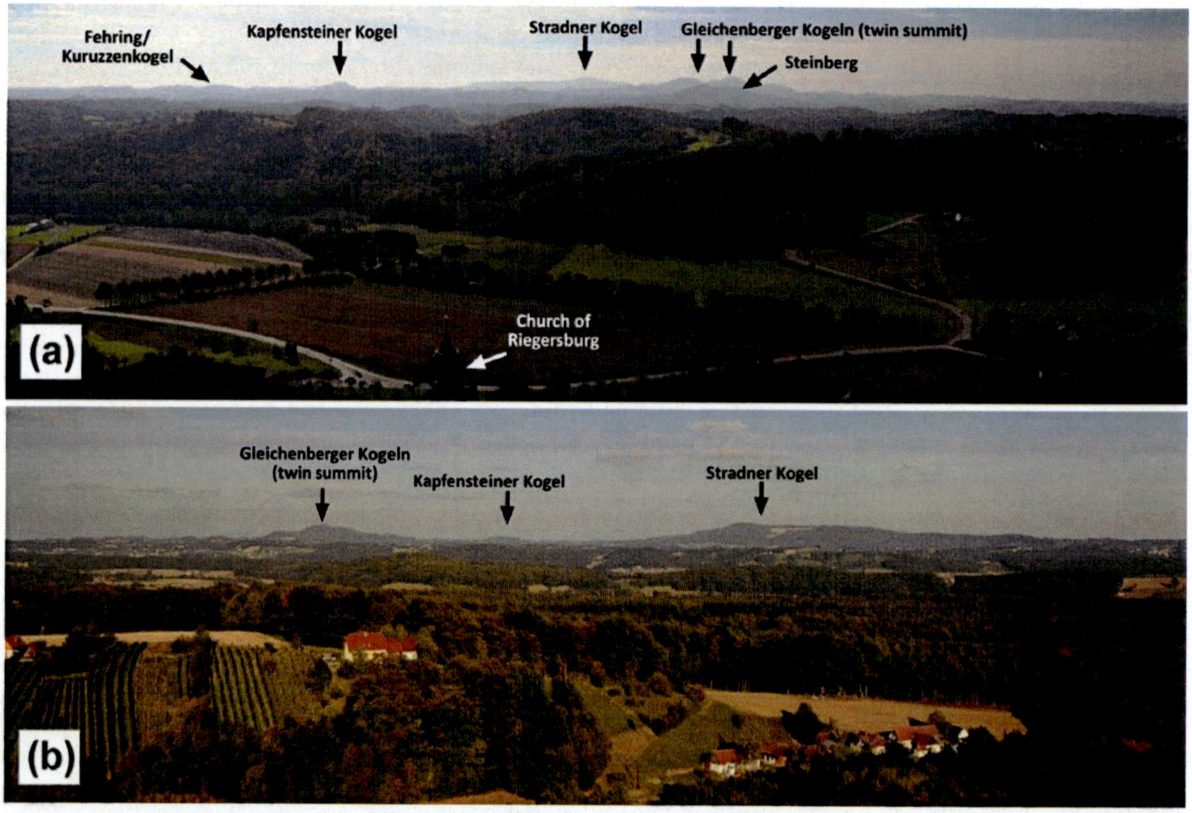

Fig. 33.7 Examples of prominent volcanic landforms in the Styrian basin; **a** view over the basin from the Riegersburg (V03) diatreme towards south over the hilly landscape; **b** view over the basin from a lookout point at St. Peter am Ottersbach towards east. The hills in the foreground of both images consist of Miocene sediments. The pronounced mountains in the background are mainly of volcanic origin. Photographs by the authors

Whereas smaller diatremes, probably derived from smaller and/or older events, clearly rise above the surrounding surface, larger and complex remnants of maar volcanoes are difficult to detect in the landscape using only morphological evidence. The two neighbouring volcanoes of Riegersburg (V03) and Altenmarkt (V04) serve as good

Fig. 33.8 Diatreme of Riegersburg (V03), with the settlement of Riegersburg at its foot (at 320–380 m asl) and a mighty, medieval castle on the top (max. 484 m asl). The church shown in Fig. 33.6 is visible near the right margin of the image. Photograph by the authors. For an aerial view of the Riegersburg from the opposite direction see Fig. 8b in Chap. 1

examples of this difference. V03 is a very well-exposed diatreme (Fig. 33.8), whereas V04 is a complex volcano of a maar-volcano system that is hardly recognizable in the present landscape. Volcanoes, which produced—after an initial phreatomagmatic eruption—scoria cones with partly huge basaltic intrusions and lava lakes (Klöch-Königsberg, Steinberg) or lava flows (Stradner Kogel) are particularly impressive and represent good examples of inverted relief formed during the post-eruption erosional period. Lava flows, which formerly filled the valleys, are now supporting high-elevated hills or ridges.

33.4.3.2 Morphometric Analysis of Selected Volcanic Remnants

Six selected volcanoes are considered in this section in more detail. By means of calculating hypsometric curves for these six volcano remnants, it was aimed to (i) morphometrically characterize them focusing on similarities and differences, and (ii) to compare them with active volcanoes to bring them into a wider geomorphic context. In general, differences in the shape of hypsometric curves between landscapes arise because of different geological conditions and geomorphic processes that shaped the landscape. Table 33.2 gives an overview of the six volcanos, also describing how an area at and around the volcano, subject to morphometric analysis, was delineated. Since the present topography of the volcanic area is often rather similar to that of the surrounding area that consists of non-volcanic rocks and lacks distinct geomorphometric borderlines, the delineation of the six analysed areas is, to some extent, arbitrary. However, in all cases the central part of the studied area (polygon or rectangle) is also the central part of the former volcano. For each of the six volcanoes, a hypsometric analysis was accomplished in ArcGIS and involved the calculation of hypsometric curves using an open-source digital elevation model (www.data.gv.at) with a grid resolution of 10 × 10 m.

Hypsometric curves from one stratovolcano and three shield volcanoes were additionally analysed to bring the six hypsometric curves calculated in our study into a broader volcanological context. Hypsometric data from the shield volcanoes were taken from Bleacher and Greeley (2008). Hypsometric data from the stratovolcano (Mt. Fuji) were calculated using SRTM elevation data (pixel resolution 30 × 30 m). Only the volcano edifice above 1200 m asl (and

Table 33.2 Overview of the six analysed volcanoes within the study area with code, name, and description of the delineation of the morphometrically analysed area

Code	Location	Delineation of morphometrically analysed area (cf. Fig. 33.9)	Area (km^2)	Elevation (m asl)			
				max	min	mean	diff
V03	Riegersburg	Mainly along valleys/gullies	1.57	474	293	341	181
V24	Güssing	Rectangle around the volcano; the surrounding area is flat	1.44	319	211	223	108
V04	Altenmarkt	Rectangle; surrounding topography is similar to the one at the volcano	4.41	446	302	359	144
V14	Klöch-Königsberg	Rectangle; surrounding topography is similar	11.56	462	240	323	222
V13	Stradner Kogel	Mainly along valleys/gullies	48.86	608	232	335	376
V01	Gleichenberger Kogeln	Mainly along valleys/gullies	10.64	598	273	379	325

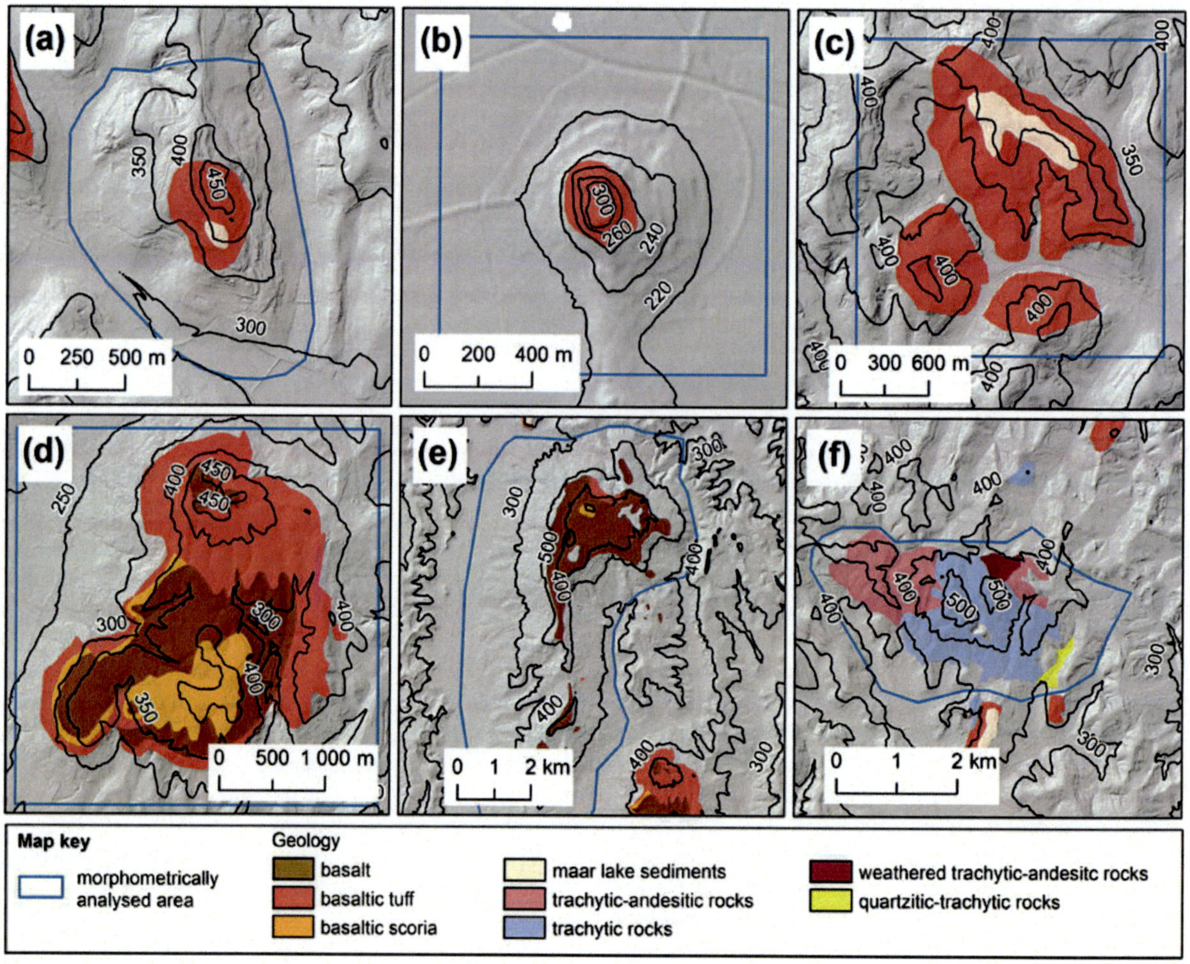

Fig. 33.9 Geological and topographical conditions (depicted as hillshade maps) at six volcanic areas in the southeast of Austria. **a** V03/Riegersburg, **b** V24/Güssing, **c** V04/Altenmarkt, **d** V14/Klöch-Königsberg, **e** V13/Stradner Kogel, **f** V01/Gleichenberger Kogeln. The areas used for calculating the hypsometric curves are indicated. Data sources: geological map of the Geological Survey of Austria, Vienna, provided via GIS-Steiermark, and Fritz (1996). Note that scale differs

up to 3776 m asl) was analysed. Above this elevation, Mt. Fuji is a well-developed stratovolcano. All four volcanoes are "active" following the definition by the Global Volcanism Program; i.e., they erupted at least once during the Holocene.

The first three volcanoes are examples of strongly eroded diatremes. Riegersburg is a very special landmark of Styria and well known as a volcano remnant. The volcano in Güssing is very similar to Riegersburg in its present appearance, whereas Altenmarkt is an example of a diatreme with still preserved lake sediments of the maar. Erosional processes since the end of the volcanic eruptions have resulted in surface lowering in the order of several hundreds of metres at all three sites, causing the complete erosion of the ejecta rings and complete (Riegersburg, Güssing) or at least partial erosion (Altenmarkt) of the maar structures.

The fourth volcano at Klöch-Königsberg is an example of a strongly eroded complex monogenetic volcano, which started with an initial phreatomagmatic phase that was soon followed by scoria ejection phases and the formation of a lava lake. The fifth volcano is the Stradner Kogel, with the highest peak in the study area (608 m asl). This volcano was mainly formed by effusive volcanism, with lava flows covering the Sarmatian sediments and now representing an example of inverted relief.

The Gleichenberger Kogeln is a stratovolcano that was once possibly of similar shape as today's Etna in Sicily. A huge volcanic massif with a base diameter of some 30 km emerged around 23.0—or 17.2 (see Table 33.1 for different efforts at age dating)—to 12.2 Ma ago at this site. We only know the approximate dimensions and expansion of this volcano, with its SiO_2-rich magma forming trachytic to andesitic rocks, from drilling and geophysical investigations. During its period of activity, environmental conditions also changed (Fig. 33.2). The ocean advanced during that time from the southeast, flooding the entire Styrian basin. The peaks of this mighty stratovolcano became active volcanic islands rising more than 2000 m above the sea level. Later on, this volcanic massif was covered by different sediments (limestones, gravel, sand and clay), possibly even entirely. Subsequent erosion uncovered the remnants of the volcano. Today, we only see small peaks of a former huge volcanic system. Local sources of thermal (for spa usage) and mineral waters are related to this volcanic activity and the fault systems, but will not be discussed here.

Figure 33.9 depicts the geological and topographical conditions at the six selected volcanic areas and also shows the delineations for calculating the hypsometric curves. Riegersburg and Güssing solely consist of basaltic tuff. In

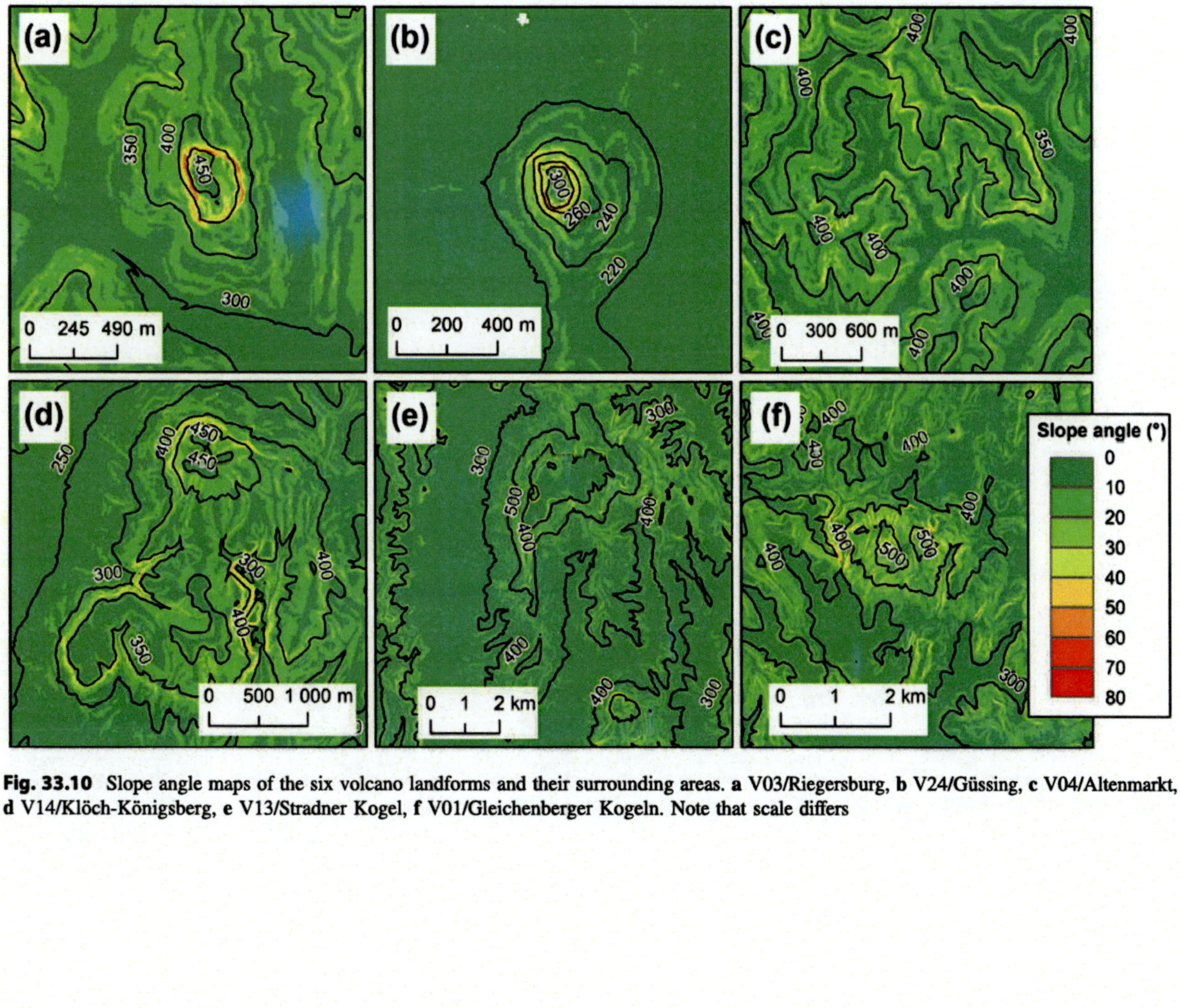

Fig. 33.10 Slope angle maps of the six volcano landforms and their surrounding areas. **a** V03/Riegersburg, **b** V24/Güssing, **c** V04/Altenmarkt, **d** V14/Klöch-Königsberg, **e** V13/Stradner Kogel, **f** V01/Gleichenberger Kogeln. Note that scale differs

contrast, at Altenmarkt, basaltic tuff but also maar sediments occur. The geology at Klöch/Königsberg is more diverse, with tuffs, basalts, tuff in close contact to basalt and scoria basalt. Geological conditions at the Stradner Kogel are comparable to those at Klöch/Königsberg, with basalts, basaltic tuff and scoria basalt. In contrast, geological conditions at the Gleichenberger Kogeln are substantially different, with trachytic to andesitic rocks with breccia, tuff and weathered tuff ("trass") at the Gleichenberger Kogeln site itself and basalts and basaltic tuff of a younger age (i.e., from the second volcanic period) in the vicinity of the volcano. However, the boundaries of the volcanic rocks are not clearly traceable in the topography.

The hypsometric curve of the volcano in Altenmarkt is of particular interest, because major valleys are cut into the volcanic rocks, dividing the former volcano into three

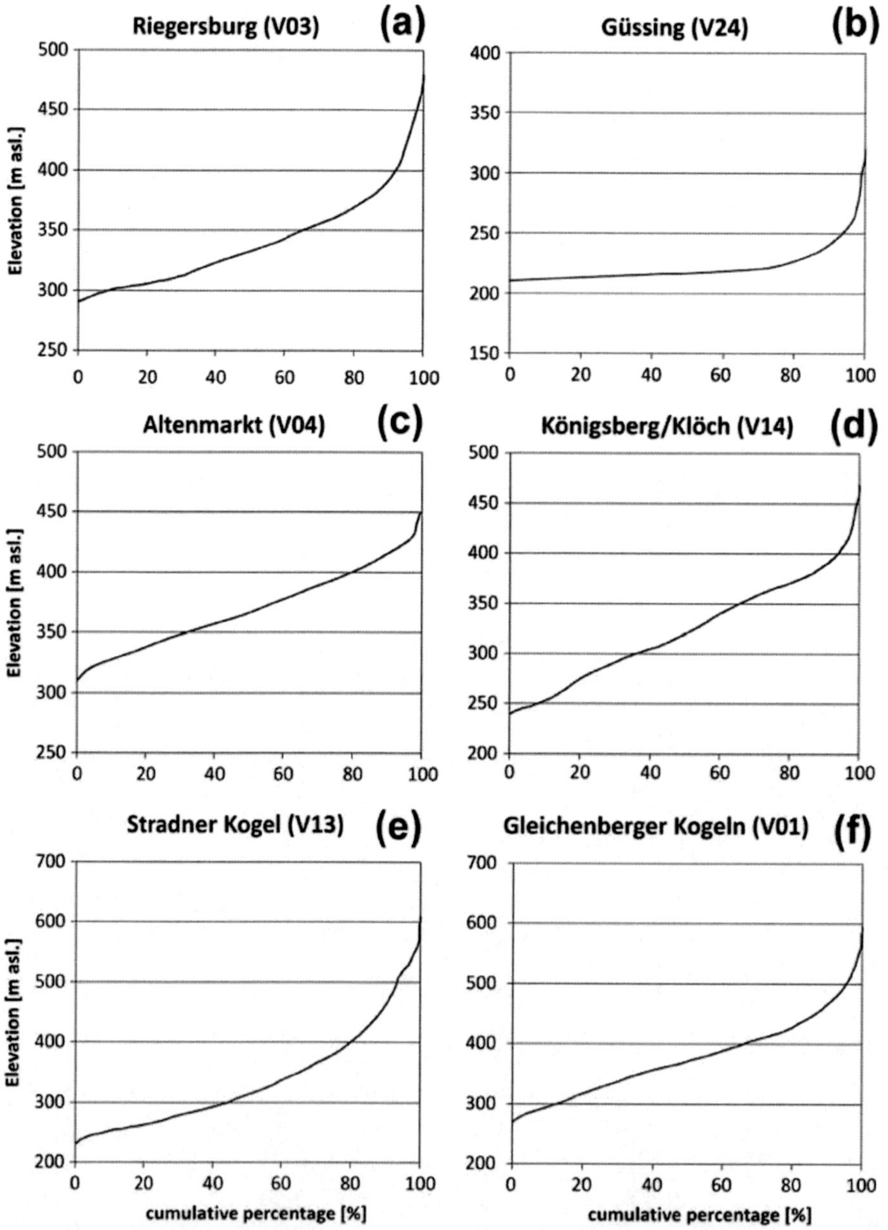

Fig. 33.11 Hypsometric curves of six former volcanoes. **a** V03/Riegersburg, **b** V24/Güssing, **c** V04/Altenmarkt, **d** V14/Klöch-Königsberg, **e** V13/Stradner Kogel, **f** V01/Gleichenberger Kogeln. The areas used for calculating the hypsometric curves are shown in Fig. 33.9

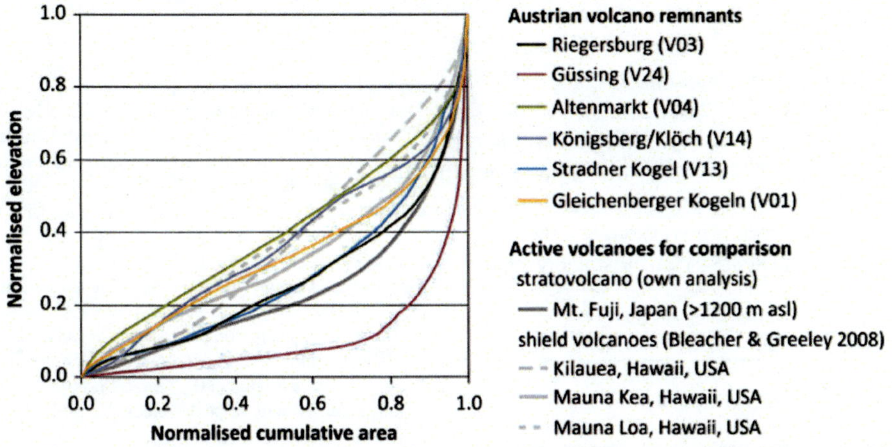

Fig. 33.12 Normalized hypsometric curves of six selected volcanic areas in the study region. For comparison, the normalized hypsometric curves for four active and well-developed volcanoes (one stratovolcano and three shield volcanoes) are additionally plotted

different parts. Figure 33.10 depicts slope angle conditions at the six studied volcanoes. The steepest slopes in five of the six cases (the exception is Altenmarkt) are found at the margins of the volcanic rock terrain, illustrating the generally higher resistance of volcanic rocks against erosion compared to the surrounding clastic sediments. At Altenmarkt, no clear slope-geology-relationship can be determined.

Figure 33.11 depicts the hypsometric curves of the six volcanoes. The largest vertical difference between the lowest and the highest parts of the volcanic areas are at Stradner Kogel and the Gleichenberger Kogeln, exceeding 300 m. In contrast, rather low-relief values have been calculated for the volcanoes in Riegersburg, Altenmarkt and particularly Güssing. The Güssing volcano is special because its surroundings are flat in all directions, apart from the south with some minor hills. Hence, Güssing is a good example of a steep diatreme prominently standing out above its surrounding. The hypsometric curve of the geologically comparable Riegersburg has a very different overall shape. However, the curve at Riegersburg above 400 m asl (6% of the total area) is almost identical to the curve at Güssing above 240 m asl (also 6%), suggesting that the higher parts of the Riegersburg diatreme are morphologically very similar to those at Güssing. The hypsometric curve of Altenmarkt has a rather constant slope. This is also true for the central parts of the hypsometric curves for the Königsberg/Klöch area and the Gleichenberger Kogeln, although in these two cases the final rise of the curve becomes strongly concave, indicating rapidly decreasing areal percentages. Finally, the curve of the Stradner Kogel is concave-shaped along its entire course, although with a lower decrease rate in areal percentages at its upper parts related to the rather high share of planation surfaces ("Hochstraden niveau") at the capping lava flow.

Figure 33.12 shows the same hypsometric data as in Fig. 33.11, but in relative dimension. In addition, the hypsometry of one stratovolcano and three shield volcanos are plotted for comparison purposes. The results clearly show that the normalized hypsometric curves of the six volcanoes differ in shape, partly substantially, from each other. This is, of course, a somehow expected result, because millions of years of sedimentation and erosion have obscured and altered the volcanic landforms in Styria and Burgenland. This also highlights the problem that the delineation of a volcanic area subject to hypsometric analysis is not straightforward in case of a strongly eroded and partly sediment-covered volcano. However, for all studied volcanoes we can conclude that (a) at least 70% of the total area is below 50% of the elevation range or (b) only 10% or less of it is in the upper 25% of the elevation range. The Güssing example is exceptional, because a large, steep and strongly eroded diatreme protrudes from an almost flat area.

A comparison of the volcano remnants studied here with active and potentially active volcanoes around the globe naturally faces substantial problems (e.g. scale differences, delineation differences, forming process differences). Nevertheless, such a comparison is valuable, bringing the own results into a wider volcano-morphological context. Maars formed in the recent past are normally not striking landforms in relationship to their surrounding (e.g. the Ukinrek maars formed in 1977; 57.832 °N, 156.510 °W; White and Ross 2011). Therefore, we decided to compare the six volcanoes analysed in this study with volcanoes with well-known volcanoes on Earth. As shown in Fig. 33.12, the curves of the one stratovolcano and the three shield volcanoes are strongly divergent. Our six curves are placed basically in between (apart from the one of Güssing). Four of our six curves are more similar to those of the three shield volcanoes. One exception is the curve of the very gently sloping Kilauea, with some 35% of its area in the upper half of the volcano edifice. Interestingly, the curve of Riegersburg has strong similarities with the curve of the well-developed stratovolcano Mt. Fuji. In summary, the six studied volcanoes are morphometrically (at least as judged from hypsometric curves) inconspicuous when compared with present active volcanoes. Only strongly eroded diatreme volcanoes such as Güssing can be suitably characterized by this approach.

33.5 Conclusions

Landscape evolution of the Styrian basin was very variable during the last 23 Ma, with a complex tectonic (uplift versus subsidence, fragmentation and formation of horst and graben structures), sedimentological (marine, lacustrine, terrestrial or biogenic sediments), erosional (linear erosion versus denudation) and volcanic (subaerial versus phreatomagmatic eruptions, basaltic versus andesitic rocks, valley fill versus inverted relief) history. These processes acted together, causing on the one hand the burial of large volcano edifices by a pile of sediments several thousand metres thick, and on the other hand, the erosion of diatreme-maar volcanoes, exposing their rock-filled fracture and forming steep, rather small mountains, which were used in historical times as locations for fortifications.

Today, only remnants of the former volcanoes are visible and nothing is left of the original volcanic landscape due to the erosion of several hundreds of metres of rock material of different origin. Two main volcanic landforms prevail in the Styrian basin. These are, on the one hand, maar-diatreme volcanoes, which in some instances form steep mountains. They consist of basaltic tuff, are mostly small in their spatial extent and were presumably produced by a single eruption. On the other hand, residuals of complex and large volcanoes consisting of tuff, scoria and massive basalt, formed during longer, polygenetic eruptions, and visually dominate the landscape due to their greater elevation. The highest mountains in the study area, often characterized by erosional planation surfaces of Miocene to Pleistocene age, are all of volcanic origin and form prominent features in the panoramic views of the area.

Delineation of Austrian volcanoes is not straightforward due to the complex evolution histories in terms of formation, erosion and deposition that make the interpretation of results of hypsometric analyses difficult. Only the curves of strongly eroded diatremes protruding from a flat landscape are substantially different from those typical for volcanoes, which are embedded in non-volcanic sediments.

References

Balogh K, Ebner F, Ravasz C (1994) K/Ar-Alter tertiärer Vulkanite der südöstlichen Steiermark und des südlichen Burgenlandes. In: Lobitzer H, Csaszar G, Dauer A (eds) Jubiläumsschrift 20 Jahre Geologische Zusammenarbeit Österreich-Ungarn 2. Geologische Bundesanstalt, Wien, pp 55–72

Bauer C (2015) Analysis of dolines using multiple methods applied to airborne laser scanning data. Geomorphology 250:78–88. https://doi.org/10.1016/j.geomorph.2015.08.015

Bleacher EJ, Greeley R (2008) Relating volcano morphometry to the developmental progression of Hawaiian shield volcanoes through slope and hypsometric analyses of SRTM data. J Geophys Res 113: B09208. https://doi.org/10.1029/2006JB004661

Bojar AV, Halas S, Bojar HP, Szaran J (2008) Isotopic evidence for the origin of an acid sulphate alteration, Styrian basin, Austria. Terra Nova 20:45–51

Bojar HP, Bojar AV, Halas S, Wojtowicz A (2013) K/Ar geochronology of igneous amphibole phenocrysts in Miocene to Pliocene volcaniclastics, Styrian Basin, Austria. Geological Quarterly 53 (3):505–516. https://doi.org/10.7306/gq.1102

Cotton CA (1944) Volcanoes as landscape forms. Whitcombe and Tombs, Christchurch, New Zealand, 415 p

Ebner F, Sachsenhofer RF (1991) Die Entwicklungsgeschichte des Steirischen Tertiärbeckens. Mitteilungen der Abteilung für Geologie und Paläontologie am Landesmuseum Joanneum 49:1–96

Ebner F, Sachsenhofer RF (1995) Palaeogeography, subsidence and thermal history of the Neogene Styrian Basin (Pannonian basin system, Austria). Tectonophysics 242:133–150

Fink J (1961) Die Südostabdachung der Alpen. Mitteilungen der Österreichischen Bodenkundlichen Gesellschaft 6:123–183

Flügel H, Heritsch H (1968) Das Steirische Tertiär-Becken. Lotze F (ed) Sammlung Geologischer Führer 47:1–196. Gebrüder Borntraeger, Berlin, Stuttgart

Francis P (1993) Volcanoes: a planetary perspective. Clarendon Press, 443 p

Friebe JG (1990) Lithostratigraphische Neugliederung und Sedimentologie der Ablagerungen des Badeniens (Miozän) um die Mittelsteirische Schwelle (Steirisches Becken, Österreich). Jahrb Geol Bundesanst 133(2):223–257

Friebe JG (1994) Gemischt siliziklastisch-karbonatische Abfolge aus dem Oberen Sarmatium (Mittleres Miozän) des Steirischen Beckens. Jahrb Geol Bundesanst 137(2):245–274

Fritz I (1996) Die Entwicklung der jungpliozänen Vulkaniklastika im Raum Altenmarkt Riegersburg und ihre Beziehung zu den umgebenden Sedimenten (Oststeirisches Becken). Unpublished PhD-Thesis, University of Graz, 235 p

Gross M, Fritz I, Piller WE, Solimann A, Harzhauser M, Hubmann B, Moser B, Scholger R, Suttner TJ, Bojar HP (2007) The Neogene of the Styrian Basin: Guide to excursion. Joannea—Geologie und Paläontologie 9:117–193

Handler R, Ebner F, Neubauer F, Bojar AV, Hermann S (2006) 40Ar/39Ar dating of the Miocene tuffs from the Styrian part of the Pannonian Basin: an attempt to refine the basin stratigraphy. Geol Carpath 57(6):483–494

Heritsch H (1968) Vulkanische Gesteine vom Steinberg bei Feldbach, Steiermark. Mitteilungen des Naturwissenschaftlichen Vereins für Steiermark 98:16–26

Hilber V (1912) Die Taltreppe, eine geologisch-geographische Darstellung. Self-Published, Graz, p 60

Kereszturi G, Németh K, Csillag G, Balogh K, Kovács J (2011) The role of external environmental factors in changing eruption styles of

monogenetic volcanoes in a Mio/Pleistocene continental volcanic field in western Hungary. J Volcanol Geoth Res 201(1–4):227–240
Kollmann K (1965) Jungtertiär im Steirischen Becken. Mitteilung Der Geologischen Gesellschaft in Wien 57(2):479–632
Kolmer H (1980) Das Rb/Sr-Alter oststeirischer Vulkanite. Mitteilungen des Naturwissenschaftlichen Vereins für Steiermark 110:23–26
Konecny V, Kovac M, Lexa J, Sefara J (2002) Neogene evolution of the Carpatho-Pannonian region: an interplay of subduction and back-arc diapiric uprise in the mantle. European Geosciences Union, Stephan Mueller Special Publication Series 1, Katlenburg-Lindau 105–123.
Kovac M, Marton E, Sefara J, Konecny V, Lexa J (2000) Miocene development of the Carpathian chain and the Pannonian Basin: movement trajectory of lithospheric fragments, subduction and diapiric uprise of asthenospheric mantle. Slovak Geologica Magazine 6(2–3):77–84
Kovac M, Halasova E, Hudackova N, Holcova K, Hynzy M, Jambrich M, Ruman A (2018) Towards better correlation of the Central Paratethys regional time scale with the standard geological time scale of the Miocene Epoch. Geol Carpath 69(3):283–300. https://doi.org/10.1515/geoca-2018-0017
Kröll A, Flügel HW, Seiberl W, Weber F, Walach G, Zych D (1988) Erläuterungen zu den Karten über den prätertiären Untergrund des Steirischen Beckens und der Südburgenländischen Schwelle. Geologische Bundesanstalt, Wien, p 49
Lorenz V (1986) On the growth of maars and diatremes and its relevance to the formation of tuff rings. Bull Volcanol 48:265–274
Lippolth J, Baranyi I, Todt W (1975) Das Kalium-Argon-Alter des Basaltes vom Lavanttal in Kärnten. Der Aufschluß 26:238–242
Martí J, Mitjavila J, Araña (1994) Stratigraphy, structure and geochronology of the Las Cañadas caldera (Tenerife, Canary Islands). Geol Mag 131(6):715–727
Moore JG, Bryan WB, Beeson MH (1994) Giant Hawaiian landslides. Ann Rev Earth Planet Sci 22:119–144
Németh K, Kereszturi G (2015) Monogenetic volcanism: personal views and discussion. Int J Earth Sci 104(8):2131–2146
Neubauer F, Genser J (1990) Architektur und Kinematik der östlichen Zentralalpen—eine Übersicht. Mitteilungen des Naturwissenschaftlichen Vereins für Steiermark 120:203–219. Graz
Neubauer F, Ebner F, Wallbrecher E (1995) Geological evolution of the internal alps, carpathians and of the pannonian basin—an introduction. Tectonophysics 242:1–4
Pöschl I (1991) A model for the depositional evolution of the volcanoclastic succession of a Pliocene maar volcano in the Styrian Basin (Austria). Jahrb Geol Bundesanst 134(4):809–843
Prettenthaler F, Podesser A, Pilger H (2010) Studien zum Klimawandel in Österreich—Klimaatlas Steiermark., Österreichische Akademie der Wissenschaften, Wien, 361 p
Ratschbacher L, Frisch W, Linzer HG, Merle O (1991) Lateral Extrusion in the Eastern Alps 1. Structural Analysis. Tectonics 10:257–271
Rhodes JM, Lockwood JP (eds) (1995) Mauna Loa revealed: structure, composition, history, and hazards. Geophys Monogr 92:360
Rögl F (2001) Mid-Miocene Circum-Mediterranean paleogeography. Berichte des Instituts für Geologie und Paläontologie der Karl-Franzens-Universität Graz 4:49–59
Schönlaub HP (2000) Geologie der Österreichischen Bundesländer—Burgenland. Geologische Bundesanstalt, Vienna, p 130
Seghedi I, Downes H, Szakacs A, Mason PRD, Thirlwall MF, Rosu E, Pecskay Z, Marton E, Panaiotu (2004) Neogene-Quaternary magmatism and geodynamics in the Carpathian-Pannonian region: a synthesis. Lithos 72:117–146
Steininger FF, Bagdasarjan GP (1977) Neue radiometrische Alter mittelmiozäner Vulkanite der Steiermark (Österreich), ihre biostratigraphische Korrelation und ihre mögliche Stellung innerhalb der paläomagnetischen Zeitskala. Verhandlungen der Geologischen Bundesanstalt 1977:85–99
Untersweg T (1979) Morphologie des Schöcklgebietes. Unpublished PhD-Thesis, University of Graz, 315 p
Wagner T, Fritz H, Stüwe K, Nestroy O, Rodnight H, Hellstrom J, Benischke R (2011) Correlations of cave levels, stream terraces and planation surfaces along the river Mur—Timing of landscape evolution along the eastern margin of the Alps. Geomorphology 134:62–78. https://doi.org/10.1016/j.geomorph.2011.04.024
Weber L (ed.) (1997) Metallogenetische Karte von Österreich 1:500.000: unter Einbeziehung der Industrieminerale und Energierohstoffe, Geological Survey of Austria, Vienna
White JDL, Ross PS (2011) Maar-diatreme volcanoes: a review. J Volcanol Geoth Res 201(1–4):1–29
Winkler-Hermaden A (1957) Geologisches Kräftespiel und Landformung. Springer Vienna, 822 p

Andreas Kellerer-Pirklbauer is a senior scientist at the University of Graz, Institute of Geography and Regional Science. He has significant experience in the physical geography of mountain regions with a focus on the effects of global change on geomorphic processes and landforms, permafrost, glaciers, mass wasting processes, and related natural hazards. He is the Austrian national representative at the International Association of Permafrost and one of the two co-leaders of the national glacier monitoring programme of the Austrian Alpine Club.

Ingomar Fritz is a scientist at the Universalmuseum Joanneum, Natural Department and responsible for the collection of Geology and Palaeontology. He is member of the Austrian National Committee within the International Union for Geodesy and Geophysics (IUGG) and the National Correspondent in the International Association of Volcanology and Chemistry of the Earth's Interior (IAVCEI).

Index
Page numbers followed by "f" and "t" refer to figures and tables, respectively.

C. Embleton-Hamann (ed.), *Landscapes and Landforms of Austria*, World Geomorphological Landscapes,
https://doi.org/10.1007/978-3-030-92815-5

M

N

O

P